Springer Proceedings in Mathematics & Statistics

Volume 373

This book series features volumes composed of selected contributions from workshops and conferences in all areas of current research in mathematics and statistics, including operation research and optimization. In addition to an overall evaluation of the interest, scientific quality, and timeliness of each proposal at the hands of the publisher, individual contributions are all refereed to the high quality standards of leading journals in the field. Thus, this series provides the research community with well-edited, authoritative reports on developments in the most exciting areas of mathematical and statistical research today.

More information about this series at http://www.springer.com/series/10533

Alin Bostan · Kilian Raschel
Editors

Transcendence in Algebra, Combinatorics, Geometry and Number Theory

TRANS19 – Transient Transcendence in Transylvania, Braşov, Romania, May 13–17, 2019
Revised and Extended Contributions

Springer

Editors
Alin Bostan
Inria, Université Paris-Saclay
Palaiseau, France

Kilian Raschel
CNRS, Institut Denis Poisson,
Université de Tours
Tours, France

ISSN 2194-1009 ISSN 2194-1017 (electronic)
Springer Proceedings in Mathematics & Statistics
ISBN 978-3-030-84303-8 ISBN 978-3-030-84304-5 (eBook)
https://doi.org/10.1007/978-3-030-84304-5

Mathematics Subject Classification: 05-XX, 11-XX, 14-XX, 33-XX, 68-XX, 82-XX, 05A15, 05A17, 11B65, 11J82, 11J91, 11R23, 13B40, 14H05, 14D07, 33E50, 68Q70, 68W30, 81T08

This Springer imprint is published by the registered company Springer Nature Switzerland AG
The registered company address is: Gewerbestrasse 11, 6330 Cham, Switzerland

Forewords

The general topic of this volume is the emergence of transcendence in various fields of mathematics, such as algebra, combinatorics, geometry and number theory. The volume is composed of 23 chapters, which we grouped into five main thematic parts:

(i) Transcendence in algebraic geometry
(ii) Transcendence in combinatorics and physics
(iii) Transcendence in commutative algebra
(iv) Transcendence in computer algebra
(v) Transcendence in number theory

This is in the natural continuation of the conference Transcendence in Transylvania (aka Trans'19) that we co-organized in May 2019 in Braşov, Romania. The conference gathered international experts from various fields of mathematics and computer science, with diverse interests and viewpoints on transcendence.

A Detour via Gessel's Lattice Path Conjecture. Being (in part) experimental mathematicians, we wish to present our own experience with transcendence on one example, namely Gessel's lattice path conjecture. It reads as follows: Gessel walks are lattice paths confined to the quarter plane $\mathbb{N}^2$, that start at the origin (0, 0) and move by unit steps in one of the following directions: West, East, South-West and North-East; see Fig. 1. Gessel excursions are those Gessel walks which return to the origin. They have been puzzling the combinatorics community since 2001, when Ira Gessel conjectured that for all $n \geq 0$, the excursions (named after him) of even length $2n$ are counted by the hypergeometric formula

$$e(2n) = 16^n \frac{(5/6)_n (1/2)_n}{(2)_n (5/3)_n}, \tag{1}$$

where $(a)_n = a(a+1)\cdots(a+n-1)$ denotes the Pochhammer symbol. (Obviously, there are no Gessel excursions of odd length, that is $e(2n+1) = 0$ for

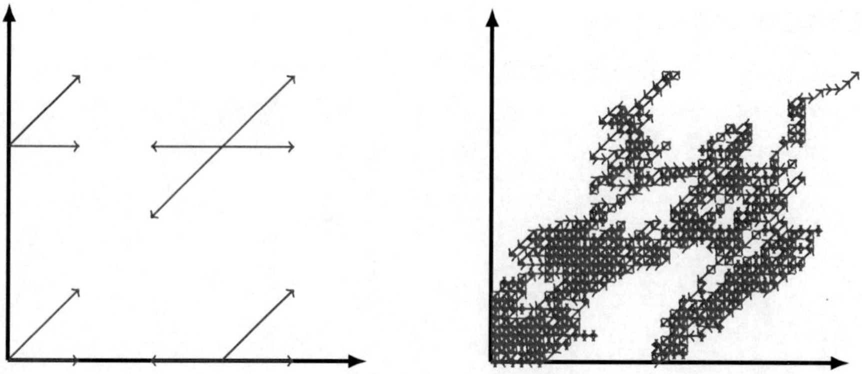

Fig. 1. Gessel walks: the allowed steps (left) and a random Gessel walk (right)

all $n \geq 0$.) In 2008, Kauers, Koutschan and Zeilberger provided a computer-aided proof of this conjecture [KKZ09], which involved computer algebra tools and massive calculations. Apart from its striking simplicity, the sequence $(e(2n))_{n\geq 0}$ in (1) has a nice property: its generating function $\sum_{n\geq 0} e(2n)t^n$ is algebraic. Both facts urge for (direct) combinatorial explanations, which however are still lacking as of 2021.

Now, more generally, for $(i,j) \in \mathbb{N}^2$ and $n \geq 0$, let us call $q(i,j;n)$ the number of Gessel walks of length n ending at the point (i,j); in particular, $e(n) = q(0,0;n)$. A further natural question is the following: Is the complete generating function of Gessel walks

$$Q(x,y;t) = \sum_{i,j,n\geq 0} q(i,j;n)x^i y^j t^n \tag{2}$$

transcendental or algebraic? In particular, what is the nature of the generating function $Q(1,1;t) = \sum_{n\geq 0} q(n)t^n$ for the total number $q(n)$ of Gessel walks with prescribed length n?

Although these combinatorial statements are just examples of transcendence questions, we feel that they played a special role in the elaboration of the present volume, at least for two reasons. First, Gessel's conjecture has received a lot of attention over the past 20 years and has given rise to many interesting methods and results [KKZ09, BK10, KR11, BKR17, BM16, Bud20]. In fact, it is one instance of a more general question consisting in counting lattice walks confined to a given cone. This is a natural and versatile problem, rich in many applications in algebraic combinatorics, queuing theory [CB83], probability theory [FIM17, DW15] and, of course, in enumerative combinatorics [BMM10] via encodings of numerous discrete objects (e.g. permutations, maps, etc.) by lattice walks. Second, from a more personal point of view, both of us (Alin Bostan and Kilian Raschel, editors of the present volume) have worked on these results, initially independently and using totally different ideas, and eventually joined our efforts.

To make a long story short, after several years of unsuccessful attempts, the complete generating function $Q(x, y; t)$ in (2) was proved to be algebraic by the first editor of this volume (AB) together with Manuel Kauers, using computer algebra techniques [BK10]. In many subsequent works, this result was qualified as a true computational *tour de force* and was listed as one of the achievements of modern computer algebra algorithms [Kal21]. At the same time, the second editor (KR) developed techniques coming from complex analysis and probability theory to compute generating functions such as $Q(x, y; t)$, in terms of (intrinsically transcendent) elliptic functions [KR11]. A few years later, both of us, together with Irina Kurkova, were able to offer the first "human proof" of Gessel's lattice path conjecture [BKR17], proving both the closed-form formula (1) and the algebraicity of the series (2). Interestingly, in this proof, the algebraic function Q is expressed as a finite sum of transcendental quantities (related to Weierstrass zeta functions), but the *transcendence is transient*: each of the transcendental summands decomposes as sums/differences of "smaller" algebraic and transcendental parts, and after telescoping summation the transcendental parts magically vanish.

Transcendence at the Crossroads of Many Mathematical Domains. Beyond Gessel's conjecture, transcendence questions have been asked for many other lattice walk models, leading the mathematical community to systematically study small step walk models; see the pioneering work [BMM10]. Tools from diverse horizons have been used: combinatorics [BMM10], functional equations [BM16], complex analysis [FIM17, KR11, BKR17], probability theory [DW15], computer algebra [KKZ09, BK10], Galois theory of difference equations [DHRS18, DHRS20], number theory [BRS14], etc.

This context strongly encouraged us to organize an international conference bringing together experts from these different communities to answer long-standing conjectures, to learn each other's techniques and to plan directions and investigations for the future.

All talks given at the conference, as well as all 23 contributions to this volume, are related to this concept of transcendence, in close relation with other mathematical areas as described at the beginning of our introduction.

Alin Bostan
Kilian Raschel

Acknowledgments

All papers in this volume were refereed very rigorously. We are deeply grateful to all our referees for their time-consuming effort and discipline in evaluating the articles. We thank Ambrose Berkumans, Elizabeth Loew, Hemanth MVN and Robinson dos Santos from Springer for their enthusiasm and for their valuable assistance. We also address our gratitude to the local organizers Nicuşor Minculete, Eugen Pălănea and Diana Savin, as well as to Faculty of Mathematics and Computer Science of Braşov for their support in the organization of the Trans'19 conference. Finally, we thank Herwig Hauser, Bruno Salvy and Sergey Yurkevich for their careful reading of this foreword and for their useful remarks.

Alin Bostan
Kilian Raschel

Lists and Summaries of Papers

We list the articles collected in this volume, and for each of them, we give a brief summary.

First Part: Transcendence in Algebraic Geometry

In Chapter 1, Masanori Asakura describes an algorithm for computing values of Dwork's p-adic hypergeometric function $\mathcal{F}^{\mathrm{Dw}}_{a,b}(t)$ modulo p^n. The algorithm is based on an explicit description of the Frobenius action on the rigid cohomology of the hypergeometric curve $(1-x^N)(1-y^M)=t$, where M and N are the denominators of the rational numbers a and b. The bit complexity of the new algorithm is $O(n^4(\log n)^3)$. A direct application of Dwork's congruences [Dwork69] leads to exponential complexity in n; hence, the new method compares very favourably.

Given a multivariate Laurent polynomial g whose Newton polytope has one interior point, the generating function of the constant term of the powers of g is known to satisfy Dwork's congruences modulo powers of a prime. In Chapter 2, Frits Beukers provides an example of a matrix version of Dwork's congruences in the case when there are more interior points. The proof relies on a geometric approach to these congruences, appearing in two recent papers [BV20, BV21a] written jointly with Masha Vlasenko. While the hypergeometric functions appearing in Dwork's initial congruences are transcendental, the ones occurring in the new matrix version are algebraic.

In Chapter 3, Thomas Dreyfus, Charlotte Hardouin, Julien Roques and Michael F. Singer investigate the main geometric properties (irreducibility, singularities, genus, uniformization) of the so-called *kernel curve*, an object naturally attached to functional equations satisfied by generating functions of walks in the quarter plane. The classification of these generating functions according to their algebraic and differential properties is an active area of research in combinatorics, to which the same authors contributed with two recent important papers [DHRS18, DHRS20].

In Chapter 4, Gwladys Fernandes surveys hypertranscendence results for the solutions of several famous functional equations: Schröder's equation $f(sz) = R(f(z))$, Böttcher's equation $f(z^d) = R(f(z))$ and Abel's equation $f(R(z)) = f(z) + 1$, where R is a rational function. Becker and Bergweiler [BB95] listed all the differentially algebraic solutions of these classes of equations. The proof of this classification result combines various mathematical tools, going from the theory of iteration in complex analysis and dynamical systems, to the algebro-differential notion of coherent families developed by Boshernitzan and Rubel [BR86].

In Chapter 5, Masha Vlasenko gives a short and arithmetically motivated introduction to the definition of the limiting mixed Hodge structures of Deligne and Schmid. In 2016, Golyshev and Zagier [GZ16] proved the first gamma conjecture in mirror symmetry for Fano threefolds of Picard rank 1. Their proof involved computing the so-called *Apéry constants*, related to Apéry's famous proof of irrationality of $\zeta(3)$, for differential operators with maximally unipotent local monodromy. In a recent paper [BV21b], Masha Vlasenko, together with Spencer Bloch, showed that Apéry constants for Picard-Fuchs differential operators are always (explicit) periods, and they linked them to periods of a limiting mixed Hodge structure. The current chapter explains one of the key ingredients used in the proof.

Second Part: Transcendence in Combinatorics and Physics

Chapter 6: *Beck-type identities for Euler pairs of order r*, by Cristina Ballantine and Amanda Welch

Chapter 7: *Quarter-plane lattice paths with interacting boundaries: the Kreweras and reverse Kreweras models*, by Nicholas R. Beaton, Aleksander L. Owczarek and Ruijie Xu

Chapter 8: *Infinite product formulae for generating functions for sequences of squares*, by Christian Krattenthaler, Mircea Merca and Cristian-Silviu Radu

Chapter 9: *A theta identity of Gauss connecting functions from additive and multiplicative number theory*, by Mircea Merca

Chapter 10: *Combinatorial quantum field theory and the Jacobian conjecture*, by Adrian Tanasa

The modern theory of integer partitions has its origins in Euler's discovery in 1748 that the number of partitions of any $n \in \mathbb{N}$ into distinct parts equals the number of partitions of n into odd parts. This is the prototype of all subsequent partition identities, and an elegant proof relies on the identity between transcendental generating functions, $\prod_{n=1}^{\infty}(1+q^n) = \prod_{n=1}^{\infty}\frac{1}{1-q^{2n-1}}$. In 1883, Glaisher found a purely bijective proof of Euler's result [Gla83]. In Chapter 6, Cristina Ballantine and Amanda Welch consider more general partition identities of this type, related to results by Subbarao [Sub71] and Andrews [And17]. They provide both analytic and combinatorial proofs.

In Chapter 7, Nicholas R. Beaton, Aleksander L. Owczarek and Ruijie Xu study lattice walks in the quarter plane with weights associated with visits to the two axes and the origin. They consider two specific models: Kreweras walks (with allowed steps $\{\mathrm{NE}, \mathrm{S}, \mathrm{W}\}$) and reverse Kreweras walks (with allowed steps $\{\mathrm{SW}, \mathrm{N}, \mathrm{E}\}$). Using the so-called *algebraic kernel method*, they prove that for the Kreweras model, the generating function is always D-finite (i.e. solution of a linear differential equation with polynomial coefficients) but possibly transcendental, and for the reverse Kreweras model, the generating function is even algebraic.

Euler's pentagonal number theorem [Eul80] was one of Euler's deepest discoveries. It states the identity between two transcendental power series,

$$\prod_{n\geq 1}(1-q^n) = \sum_{n\geq 0}(-1)^{\lfloor (n+1)/2\rfloor} q^{a_n},$$

where $(a_n)_{n\geq 0} = (0, 1, 2, 5, 7, 12, 15, \ldots)$ is the sequence of integers $m \in \mathbb{N}$ such that $24m+1$ is a square. In Chapter 8, Christian Krattenthaler, Mircea Merca and Cristian-Silviu Radu prove similar product formulas for other transcendental generating functions of the form $\sum_{n\geq 0} \pm q^{a_n}$, for many sequences $(a_n)_{n\geq 0}$ defined by the property that $Pa_n + b^2$ is a perfect square, where P and b are given integers. The proofs rely on the theory of modular functions.

A beautiful identity due to Gauss [Gau66, p. 447, eq. (14)] states that

$$\sum_{n=0}^{\infty} q^{n(n+1)/2} = \prod_{m=1}^{\infty}\frac{1-q^{2m}}{1-q^{2m-1}}.$$

This implies the recurrence relation $\sum_{j=0}^{\infty}(-1)^{j(j+1)/2}\mathrm{pod}(n-j(j+1)/2) = 0$ for all $n > 0$, where $\mathrm{pod}(n)$ is the number of partitions of n into parts not congruent to 2 modulo 4. In Chapter 9, Mircea Merca considers various refinements and generalizations of this identity. They are expressed in terms of the total number $S(k,n)$ of k's in all the partitions of n into parts not congruent to *2* modulo *4*. The proof relies on a truncated theta series identity by Andrews and Merca [AM18].

The Jacobian conjecture states that if a polynomial mapping F from $\mathbb{C}^n$ to itself has Jacobian determinant which is a nonzero constant, then F has a polynomial inverse. In 1980, Wang proved the conjecture when $\deg(F) \leq 2$ [Wan80] and in

1982, Bass, Connell and Wright reduced the general statement to the case $\deg(F) = 3$ [BCW82]. In Chapter 10, Adrian Tanasa gives a short introductory article on the combinatorial quantum field theory (QFT) approach to this notorious open problem, which attempts to fill the gap between the cases of degree 2 and 3. The main result is about reducing the degree of the map involved, but with the caveat that new parameters are introduced and now one has a family of maps to deal with, instead of a single map. This contribution is a summary of the paper [dGST16] by Tanasa with two other collaborators, de Goursac and Sportiello.

Third Part: Transcendence in Commutative Algebra

Chapter 11: *How regular are regular singularities?*, by Herwig Hauser
Chapter 12: *Néron desingularization of extensions of valuation rings*, by Dorin Popescu, with an Appendix by Kęstutis Česnavičius
Chapter 13: *Diagonal representation of algebraic power series: a glimpse behind the scenes*, by Sergey Yurkevich

In Chapter 11, Herwig Hauser offers a conceptual, functional-analytic perspective on regular singular points of linear differential equations with meromorphic coefficients. The general idea is to view an arbitrary regular singular equation as a perturbation of an Euler equation and to construct an isomorphism between their solutions. As the explicit solutions of the Euler equation are well known, one obtains in this way a precise description of the solutions of the original equation. The article sketches how one can recover, using this viewpoint, some classical theorems of Fuchs and Frobenius on the structure of local solutions.

In Chapter 12, Dorin Popescu returns to his celebrated 1986 result [Pop86] on general Néron desingularization for local extensions of valuation rings. In modern language, it says that every regular homomorphism of Noetherian rings is *ind-smooth*. Ind-smoothness is a very useful concept, as it allows, in many concrete problems, to replace a general algebra by a smooth algebra of finite type. The main result of this chapter provides necessary and sufficient conditions for an injective local homomorphism of valuation rings of characteristic zero to be ind-smooth.

In Chapter 13, Sergey Yurkevich gives a detailed account and proof of a celebrated result by Denef and Lipshitz [DL87] that any algebraic power series in n variables can be written as a diagonal of a rational power series in $n + 1$ variables. The proof relies on two important facts: (i) the ring of algebraic power series in n variables is the Henselization of the localization of the ring of polynomials at the maximal ideal generated by the variables; (ii) the Henselization can be described as the direct limit of étale extensions. The article includes a comprehensive proof of these results.

Fourth Part: Transcendence in Computer Algebra

In 1987, David and Gregory Chudnovsky discovered an incredible formula for the number π [CC88]:

$$\pi = \frac{53360\sqrt{640320}}{13591409} \times {}_4F_3\left(\left[\frac{1}{6},\frac{1}{2},\frac{5}{6},\frac{558731543}{545140134}\right];\left[\frac{13591409}{545140134},1,1\right];-\frac{1}{53360^3}\right)^{-1}.$$

This formula is not only spectacular in its appearance, but also the basis of the fastest algorithm used in practice for computing digits of π. In Chapter 14, Jesús Guillera describes a new method, based on elliptic modular functions, which allows to prove this identity automatically, using computer algebra, in a bunch of seconds on a laptop.

The partition sequence $(p(n))_{n\geq 1} = (1,2,3,5,7,\mathbf{11},15,22,30,42,56,\mathbf{77},\ldots)$ counts the number of distinct ways of representing n as a sum of positive integers. A famous result of Ramanujan [Ram00] asserts that the integer $p(11n+6)$ is divisible by 11 for any $n \geq 0$. In Chapter 15, Mark van Hoeij and Cristian-Silviu Radu describe an algorithm that computes a special basis for the space of modular functions for $\Gamma_0(N) := \left\{\begin{pmatrix} a & b \\ c & d \end{pmatrix} \in \mathrm{SL}_2(\mathbb{Z}) : N|c\right\}$ with poles only at infinity. As an application, they obtain a computerized proof of Ramanujan's divisibility result.

A classical result in the theory of modular functions [Sti84, Zag08] asserts that if *h(z)* is a modular function and *g(z)* is a modular form of positive weight, both for a fixed congruence subgroup, and if locally $g(z) = y(h(z))$, then the function *y(t)* is D-finite in *t*. In Chapter 16, Peter Paule and Cristian-Silviu Radu design a guess-and-prove algorithm for finding a linear differential equation satisfied by *y(t)*. This allows for automatic proofs of identities such as $\theta_3(z)^2 = {}_2F_1([1/2,1/2];[1];\lambda(z))$, where $\theta_3(z)$ is Jacobi's theta function $\theta_3(z) = \sum_{n\in\mathbb{Z}} q^{n^2/2} = 1 + 2\sqrt{q} + 2q^2 + 2q^{9/2} + \cdots \quad (q = \exp(2\pi i z), \mathrm{Im}(z) > 0, |q| < 1)$ and $\lambda(z)$ is Legendre's elliptic modular function

$$\lambda(z) = \left(\frac{\sum_{n \in \mathbb{Z}+\frac{1}{2}} q^{n^2/2}}{\theta_3(z)} \right)^4 = 16\sqrt{q} - 128q + 704q^{3/2} - 3072q^2 + \cdots.$$

In Chapter 17, Carsten Schneider and Wadim Zudilin illustrate the general algorithmic process of "creative telescoping" in a Diophantine context. As π is transcendental [Lin82], so is $\zeta(4) = \pi^4/90$; however, quantitative versions of this property remain open, even the quantitative version of the irrationality of $\zeta(4)$. The common method for proving upper bounds on the irrationality measure of zeta values is to construct linear forms in 1 and those values, with rational coefficients. The authors give two such constructions of linear forms in 1 and $\zeta(4)$, and they provide a computer-aided proof that the corresponding linear forms are equal.

Fifth Part: Transcendence in Number Theory

Chapter 18: *Support of an algebraic series as the range of a recursive sequence*, by Jason P. Bell

Chapter 19: *X-coordinates of Pell equations in various sequences*, by Florian Luca

Chapter 20: *A conditional proof of the Leopoldt conjecture for CM fields*, by Preda Mihăilescu

Chapter 21: *Siegel's problem for E-functions of order* 2, by Julien Roques and Tanguy Rivoal

Chapter 22: *Irrationality and transcendence of alternating series via continued fractions*, by Jonathan Sondow

Chapter 23: *On the transcendence of critical Hecke L-values*, by Johannes Sprang

The classical Skolem–Mahler–Lech theorem [Lec53] characterizes the possible supports (i.e. sets of indices with nonzero coefficients) of rational power series over a field of characteristic zero: they are the subsets of $\mathbb{N}$ whose characteristic function is eventually periodic. For algebraic power series in characteristic zero, the analogue of the Skolem–Mahler–Lech theorem is still an open problem. In positive characteristic $p > 0$, the situation is much better understood: the support of an algebraic power series is a p-automatic set [Der07]. It can be an arithmetic sequence, e.g. $A(x) := 1/(1-x) = 1 + x + x^2 + \cdots$, but also a geometric sequence, e.g. $G_p(x) := x + x^p + x^{p^2} + \cdots$. In Chapter 18, Jason Bell addresses the characterization of all cases when the support set is a sequence satisfying a linear recurrence with constant coefficients. He proves that they can be built up from $A(x)$ and $G_p(x)$.

A natural question in the field of Diophantine equations is "intersecting linear sequences", which asks to decide whether two linearly recurrent sequences with constant coefficients share only finitely many common values. Given an integer $d > 1$ which is not a square, it is classical that the *Pell equation* $X^2 - dY^2 = \pm 1$

admits infinitely many solutions $(X_n, Y_n) \in \mathbb{N} \times \mathbb{N}$, where both coordinate sequences $(X_n)_n$ and $(Y_n)_n$ satisfy linear recurrences of order 2 with constant coefficients. In Chapter 19, Florian Luca surveys recent results on special cases of the intersection question; they concern the occurrence of interesting arithmetic sequences, such as Fibonacci numbers $(F_n)_{n\geq 0}$, in X-coordinates of Pell equations. A typical example is the following result by the author and Alain Togbé [LT18] *the equation* $X_n = F_m$ *has at most one positive integer solution* (n, m), *except when* $d = 2$, *for which* $X_1 = F_1 = F_2 = 1$ *and* $X_2 = F_4 = 3$.

In algebraic number theory, the *regulator* of an algebraic number field $\mathbb{K}$ is a positive number which measures the "density" of the units in the ring of algebraic integers of $\mathbb{K}$: if the regulator is small, this means that "there are many units". For an imaginary quadratic field $\mathbb{Q}(\sqrt{D})$ with $D<0$, the regulator is 1. For a real quadratic field $\mathbb{Q}(\sqrt{D})$ with $D > 0$, it is the natural logarithm of the fundamental unit of $\mathbb{K}$; for instance, if $d = 1 \mod 4$, it is $\log((a + b\sqrt{d})/2)$, where (a, b) is the smallest solution of the Pell equation $x^2 - dy^2 = \pm 4$. In arbitrary number fields, the regulator is the determinant of some submatrix of the matrix built by the logarithms of absolute values of all conjugates of a fundamental set of units. Units can be completed p-adically, and – at least for CM fields (these are quadratic imaginary extensions of a field which is real in all of its embeddings, also called "totally real") – Leopoldt's p-adic regulator of $\mathbb{K}$ is defined like above, with respect to the p-adic logarithm. The question is: Will this regulator always be non-trivial? Leopoldt's conjecture [Leo62] is a famous open problem, stating that this is the case: the p-adic regulator of a number field does not vanish. The conjecture was proved in 1967 by Brumer [Bru67], using results of Ax and Baker, for abelian number fields (in particular, for quadratic number fields). Since then, the next step was considered to be a proof of the conjecture for CM fields, but not even the case of solvable extensions is known to hold to this day. In Chapter 20, Preda Mihăilescu provides a (conditional) proof of the statement that for odd primes p, Leopold's conjecture holds for CM fields over $\mathbb{Q}$, and herewith he connects the Leopoldt conjecture to another important conjecture of Iwasawa—building a highly unexpected bridge, which he is prepared to use in subsequent work.

E-functions are D-finite and entire power series subject to some arithmetic conditions; they generalize the exponential function. The class contains most D-finite exponential generating functions (EGFs) in combinatorics and many special functions such as the confluent hypergeometric function ${}_1F_1(\alpha; \beta; z) = \sum_{n=0}^{\infty} \frac{(\alpha)_n}{(\beta)_n} \frac{z^n}{n!}$, with $\alpha \in \mathbb{Q}$ and $\beta \in \mathbb{Q} \setminus \mathbb{Z}_{<0}$. A non-hypergeometric example is the EGF of the Delannoy numbers

$$D(z) := \sum_{n=0}^{\infty} \left(\sum_{k=0}^{n} \binom{n}{k} \binom{n+k}{n} \right) \frac{z^n}{n!},$$

which satisfies $zD''(z) + (1 - 6z)D'(z) + (z - 3)D(z) = 0$. In 1949, Siegel asked [Sie49, Ch. II, Sect. 9] whether any E-function can be expressed in terms of (generalized) confluent hypergeometric series. For instance, $D(z)$ is equal to

$e^{(3-2\sqrt{2})z} \cdot {}_1F_1(1/2\,;1;4\sqrt{2}z)$. In 2004, Gorelov proved that Siegel's question admits a positive answer when the given E-function $f(z)$ satisfies a linear differential equation of order 2 [Gor04]. In Chapter 21, Julien Roques and Tanguy Rivoal provide a new proof of this result. More precisely, they show that $f(z)$ can be written $f(z) = a(z)e^{\mu z}{}_1F_1(\alpha;\beta;\lambda z) + b(z)e^{\mu z}{}_1F_1'(\alpha;\beta;\lambda z)$, where $a(z), b(z) \in \bar{\mathbb{Q}}(z)$, $\lambda \in \bar{\mathbb{Q}}^{\times}$, $\mu \in \bar{\mathbb{Q}}$, and $\alpha \in \mathbb{Q}$, $\beta \in \mathbb{Q}\backslash\mathbb{Z}_{\leq 0}$ are such that $\alpha - \beta \notin \mathbb{Z}$.

In Chapter 22, Jonathan Sondow[1] studies the irrationality and the transcendence of alternating series of two types (I and II); these were introduced by Euler [Eul88, Chap. XVIII], who gave recipes for converting them into equivalent continued fractions. For instance, Euler showed that Leibniz's alternating series $\frac{\pi}{4} = 1 - \frac{1}{3} + \frac{1}{5} - \frac{1}{7} + \cdots$ can be converted into Brouncker's continued fraction $1 + 1^2/(2 + 3^2/(2 + 5^2/(2 + 7^2/(2 + \cdots))))$ for $4/\pi$. The author proves a simple condition for the irrationality of a continued fraction that can be applied to prove the irrationality of constants expressible by alternating series. His main result is that, if a series of type II is equivalent to a *simple* continued fraction, then the sum is transcendental and its irrationality measure exceeds 2. As a consequence, he reproves a result due to Davison and Shallit [DS91]: *Cahen's constant*

$$C = \sum_{n=0}^{\infty} \frac{(-1)^n}{S_n - 1} = 1 - \frac{1}{2} + \frac{1}{6} - \frac{1}{42} + \frac{1}{1806} - \cdots = 0.643410546288338\cdots$$

is transcendental, where $(S_n)_{n\geq 0} = (2, 3, 7, 43, 1807, 3263443, \ldots)$ is Sylvester's sequence, defined by the recursion $S_0 = 2$ and $S_{n+1} = (S_n - 1)S_n + 1$ for $n \geq 0$.

Around 1740, Euler discovered an explicit formula for the values of the zeta function at even integers, $\zeta(2n) = -\frac{(2\pi i)^{2n}}{2(2n)!}B_{2n}$, where B_n are the Bernoulli numbers defined by the generating series $\frac{z}{\exp(z)-1} = \sum_{n=0}^{\infty} B_n \frac{z^n}{n!}$ [Eul40]. Euler's formula shows that the positive critical values of the Riemann zeta function are nonzero rational multiples of powers of the period $2\pi i$. Together with Lindemann's theorem [Lin82], this implies the transcendence of all positive critical values of the Riemann zeta function. Hecke L-functions form a class of functions that are of great importance in number theory. They can be seen as a common generalization of the Riemann zeta function, Dirichlet L-functions and Dedekind zeta functions. In Chapter 23, Johannes Sprang summarizes the results on the algebraicity of critical Hecke L-values up to explicit periods and deduces (conditional) transcendence results for these critical L-values, obtained together with Guido Kings [KS20].

Alin Bostan
Kilian Raschel

[1] This chapter is published posthumously. We thank Wadim Zudilin for his suggestion to include late Jonathan Sondow's work in this volume.

General Description of the Conference

The main topics of the conference Transcendence in Transylvania were algebraic and transcendental aspects of special functions and special numbers arising in combinatorics, number theory, statistical mechanics and probability theory. The meeting was organized by Alin Bostan and Kilian Raschel (scientific committee), Nicusor Minculete, Eugen Păltănea and Diana Savin (local organizers). It was funded by the ERC grant "Elliptic Combinatorics" and by Inria. The Faculty of Mathematics and Computer Science of Transylvania University in Braşov provided useful support in the organization of this event.

About Braşov. Shadowed by mountains and containing a fine Baroque centre, Braşov is one of Transylvania's most appealing cities. Its medieval old town is a marvelous introduction to the Saxon architecture of the region. Thanks to Bram Stoker and Hollywood, Transylvania (from the Latin for "beyond the forest") is famed abroad as the homeland of Dracula, a mountainous place where storms lash medieval hamlets, while wolves—or werewolves—howl from the surrounding woods. But the Dracula image is just one element of Transylvania, whose near 100,000 square kilometres take in alpine meadows and peaks, caves and dense forests sheltering bears and wild boars and lowland valleys where buffalo cool off in the rivers. For the visitor, most striking of all are the stuhls, the former seats of Saxon power, with their medieval streets, defensive towers and fortified churches. One highlight of this corner is the castle at Bran (see the pictures on Fig. 2), which looks just how a vampire count's castle should: a grim facade, perched high on a rock bluff, its turrets and ramparts rising in tiers against a dramatic mountain background. The Carpathian mountains are never far away in Transylvania, and for anyone fond of walking, this is one of the most beautiful while least exploited regions in Europe.

Fig. 2. Posters of the conference (Bran castle)

List of Participants

Seventy-one participants, from several countries and continents, attended the conference physically. David V. Chudnovsky and Gregory V. Chudnovsky gave a lecture by teleconference from New York, USA. The full list of participants is given below.

Fig. 3. Group photo of the conference

- Youssef Abdelaziz, Paris, France
- Axel Bacher, Paris, France
- Cristina Ballantine, Worcester, USA
- Nicholas Beaton, Melbourne, Australia
- Jason Bell, Waterloo, Canada
- Dan Betea, Bonn, Germany

- Frits Beukers, Utrecht, Netherlands
- Alin Bostan, Paris, France
- Kathrin Bringmann, Cologne, Gernmany
- Vasile Brînzănescu, Bucharest, Romania
- Xavier Caruso, Bordeaux, France
- David V. Chudnovsky, New York, USA
- Gregory V. Chudnovsky, New York, USA
- Frédéric Chyzak, Paris, France
- Alexandru Ciolan, Cologne, Germany
- Mihai Ciucu, Bloomington, USA
- Roxana Cojman, Bucharest, Romania
- Andrei Comăneci, Bucharest, Romania
- Éric Delaygue, Lyon, France
- Lucia Di Vizio, Paris, France
- Thomas Dreyfus, Strasbourg, France
- Andrew Elvey Price, Bordeaux, France
- Guy Fayolle, Paris, France
- Ştefan Garoiu, Braşov, Romania
- Tony Guttmann, Melbourne, Australia
- Charlotte Hardouin, Toulouse, France
- Herwig Hauser, Vienna, Austria
- Mark van Hoeij, Tallahassee, USA
- Mioara Joldes, Toulouse, France
- Frédéric Jouhet, Lyon, France
- Christoph Koutschan, Linz, Austria
- Christian Krattenthaler, Vienna, Austria
- Cédric Lecouvey, Tours, France
- Gabriel Lepetit, Grenoble, France
- Florian Luca, Johannesburg, South Africa
- Assia Mahboubi, Nantes, France
- Jean-Marie Maillard, Paris, France
- Marin Marin, Braşov, Romania
- Sergiu Megieşan, Cluj, Romania
- Mircea Merca, Craiova, Romania
- Preda Mihăilescu, Goettingen, Germany
- Nicuşor Minculete, Braşov, Romania
- Marni Mishna, Burnaby, Canad
- Vicenţiu Paşol, Bucharest, Romania
- Cristina Păcurar, Braşov, Romania
- Eugen Păltănea, Braşov, Romania
- Radu Păltănea, Braşov, Romania
- Carine Pivoteau, Marne-la-Vallée, France
- Alexandru Popa, Bucharest, Romania
- Dorin Popescu, Bucharest, Romania
- Cristian-Silviu Radu, Linz, Austria

- Kilian Raschel, Tours, France
- Tanguy Rivoal, Grenoble, France
- Julien Roques, Lyon, France
- Bruno Salvy, Lyon, France
- Diana Savin, Constanţa, Romania
- Michael F. Singer, Raleigh, USA
- Andrea Sportiello, Paris, France
- Johannes Sprang, Regensburg, Germany
- Alexandru Şerban, Braşov, Romania
- Alexandru Ştioboranu, Braşov, Romania
- Adrian Tanasa, Bordeaux, France
- Pierre Tarrago, Paris, France
- Stefan Trandafir, Burnaby, Canada
- George Turcas, Warwick, UK
- Brigitte Vallée, Caen, France
- Daniel Vargas, Lyon, France
- Bianca Vasian, Braşov, Romania
- Masha Vlasenko, Warsaw, Poland
- Michael Wallner, Bordeaux, France
- Jacques-Arthur Weil, Limoges, France
- Sergey Yurkevich, Vienna, Austria
- Alexandru Zaharescu, Urbana, USA

Talks and Abstracts

Semiabelian Varieties and Annihilators of Irreducible Representations

Jason Bell

Let G be a semiabelian variety defined over a field of characteristic zero and let $\Phi : G \to G$ be an endomorphism. We prove that if Y is a subvariety of G that intersects the orbit of every point of G that is periodic under the action of the map Φ, then Y is all of G. As a consequence of this result, we are able to give a topological characterization of the ideals that annihilate the irreducible representations in a class of algebras that "comes from geometry". This is joint work with Dragoş Ghioca.

Dwork's Congruences and the Cartier Operator

Frits Beukers

In his work on zeta functions for families of algebraic varieties, Dwork discovered a number of remarkable p-adic congruences. In this lecture, I discuss some joint work with Masha Vlasenko, in which we give a description of the action of the Cartier operator on the cohomology of the varieties and the resulting Dwork-type congruences. In this work, we only use elementary definitions and arguments.

False Theta Functions and Their Modular Properties

Kathrin Bringmann

In my talk, I will discuss modular properties of false theta functions. Due to a wrong sign factor, these are not directly seen to be modular; however, there are ways to repair this. I will report about this in my talk.

Algebraic Methods for Solutions of Bloch–Iserles Hamiltonian Systems

Vasile Brînzănescu

We shall present some results about the algebraic complete Hamiltonian systems of Bloch–Iserles. By using the spectral curve of such a system, we show that the solutions stay on a commutative algebraic group which is a non-trivial extension of a Prym variety (associated with the spectral curve and to an involution on it) by a multiplicative group $(\mathbb{C}^*)^s$. Finally, we try to give explicit solutions by θ-functions. This is a joint work with Cristina Maria Sandu.

2-adic Differential Equations and Isogenies

Xavier Caruso

Over the last decades, many algorithms were proposed to compute isogenies between elliptic curves. One of them is based on the observation that the rational function giving the isogeny is a solution of an explicit nonlinear differential equation of order 1. In characteristic zero, the latter equation has a unique solution which can be computed easily by a Newton iteration.

In this talk, I will focus on the case where the base field is a finite extension of field of 2-adic numbers $\mathbb{Q}_2$. This situation is particularly interesting because the differential equation we obtain in this case exhibits singularities close to 0, leading to huge numerical instability in the resolution step. In this talk, I will explain how Newton iteration can be modified in order to rub out most of the numerical stability and consequently highly improve the complexity of the isogeny computation.

This is a joint work with Elie Eid and Reynald Lercier.

Calculations of Classical Constants and Special Functions for Fun and Profit (Hardware View)

David V. and Gregory V. Chudnovsky

We describe challenges in contemporary high-performance computing, as exemplified by efforts of large-scale mathematical computations (e.g. π calculations) and of special function computations in applied area of mathematical finance and risk management.

Explicit Generating Series for Small-Step Walks in the Quarter Plane

Frédéric Chyzak

Lattice walks occur frequently in discrete mathematics, statistical physics, probability theory and operational research. The algebraic properties of their enumeration generating series vary greatly according to the family of admissible steps chosen to define them: their generating series are sometimes rational, algebraic, D-finite, differentially algebraic or sometimes they possess no apparent equation. This has recently motivated a large classification effort. Interestingly, the involved equations often have degrees, orders and sizes, making calculations an interesting challenge for computer algebra.

In this talk, we study nearest neighbours walks on the square lattice, that is, models of walks on the square lattice, defined by a fixed step set that is a subset of the 8 nonzero vectors with coordinates 0 or ± 1. We concern ourselves with the counting of walks constrained to remain in the quarter plane, counted by length. In the past, Bousquet–Mélou and Mishna showed that only 19 essentially different models of walks possess a non-algebraic D-finite generating series; the linear differential equations have then been guessed by Bostan and Kauers. In this work, we give the first proof that these equations are satisfied by the corresponding generating functions. This allows to derive nice formulas for the generating functions, expressed in terms of Gauss' hypergeometric series, to decide their algebraicity or transcendence. This also gives hope to extract asymptotic formulas for the number of walks counted by lengths.

(Based on joint work with Alin Bostan, Mark van Hoeij, Manuel Kauers and Lucien Pech.)

Lozenge Tilings of Doubly-Intruded Hexagons

Mihai Ciucu

Motivated in part by Propp's intruded Aztec diamond regions, we consider hexagonal regions out of which two horizontal chains of triangular holes (called ferns) are removed, so that the chains are at the same height and are attached to the boundary. In contrast to the intruded Aztec diamonds (whose number of domino tilings contain some large prime factors in their factorization), the number of lozenge tilings of our doubly intruded hexagons turns out to be given by simple product formulas in which all factors are linear in the parameters. We present in fact q-versions of these formulas, which enumerate the corresponding plane partitions-like structures by their volume. We also pose some natural statistical physics questions suggested by our set-up, which should be possible to tackle using our formulas. This is a joint work with Tri Lai.

Length Derivative of Generating Series for Walks in the Quarter Plane

Charlotte Hardouin

A walk in the quarter plane is a path in the square lattice starting at $(0, 0)$ confined in the first quadrant that goes in each cardinal direction with a certain *probabilistic weight*. Its associated generating series is a trivariate formal power series of the form $Q(x, y, t) = \sum q_{i,j,k} x^i y^j t^k$ where $q_{i,j,k}$ is the probability for a walk to end at the point (i, j) in k steps. While the variables x and y are associated with the ending point of the walk, the variable t is associated with its length. In this talk, we study the algebraic differential equations satisfied by the generating series with respect to the t-derivation. We shall show how one can uniformize the problem in a non-Archimedean setting in order to associate with the generating series an auxiliary function satisfying a simple q-difference equation with meromorphic coefficients over a Tate curve. Then, difference Galois theory gives a dictionary between the differential algebraic relations of the series and the orbit configuration of a set of points on the elliptic curve associated with the Tate curve.

Methods from Commutative Algebra in the Study of Algebraic Power Series

Herwig Hauser

This lecture will be of expository type. We will look at various famous theorems about algebraic power series with the perspective of a commutative algebraist.

Factoring Linear Recurrence Operators

Mark van Hoeij

Several computer algebra systems have implementations for finding hypergeometric solutions of linear recurrence equations. This is equivalent to finding first-order factors of linear recurrence operators. This talk will present several approaches to compute higher-order factors of operators in $\mathbb{Q}(x)[\tau]$ where τ is the shift operator.

Enumeration of Diagonally Symmetric Alternating Sign Matrices

Christoph Koutschan

The number of $n \times n$ alternating sign matrices (ASMs) is given by a nice product formula that was conjectured by Mills, Robbins and Rumsey and that was proven by Zeilberger in 1992 and, shortly after, by Kuperberg. Also several symmetry classes of ASMs are enumerated by similar formulas. However, there are some classes which remain mysterious since their counting sequence appears incompatible with a product formula. We study one of these classes, namely ASMs that are symmetric with respect to the main diagonal and find a Pfaffian formula for their refined enumeration.

Elliptic Hypergeometric Series and Applications

Christian Krattenthaler

Hypergeometric series are ubiquitous in classical analysis and many other fields of mathematics. Their q-analogues, called basic hypergeometric series, underly the study of q-series and are invaluable tools in various problem areas of number theory and combinatorics. The theory of both—ordinary and basic hypergeometric series—has been largely developed in the first half of the twentieth century. A relatively recent development is the theory of elliptic hypergeometric series, which originates from work of Frenkel and Turaev on solutions of the Yang–Baxter equation from the end of the 1990s. These elliptic hypergeometric series contain both the ordinary and basic hypergeometric series as special cases. However, interestingly, work on these new series has (also) led to the discovery of identities that were even new when specialized to the ordinary or the basic case.

In my talk, I shall attempt to give an idea of the theory of these various hypergeometric series and then present several recent applications which concern problems in combinatorics, in special functions theory, respectively, in approximation theory.

X-Coordinates of Pell Equations In Various Sequences

Florian Luca

Let $d > 1$ be a square-free integer and (X_n, Y_n) be the nth solution of the Pell equation $X^2 - dY^2 = \pm 1$. Given your favourite set of positive integers U, one can ask what can we say about those d such that $X_n \in U$ for some n. Formulated in this way, the question has many solutions d, since one can always pick $u \in U$ and write $u^2 \pm 1 = dv^2$ with integers d and v such that d is square-free, obtaining in this way that (u, v) is a solution of the Pell equation corresponding to d. What about if we ask that $X_n \in U$ for at least two different ns? Then the answer is very different. For example, if U is the set of squares, then it is a classical result of Ljunggren that the only such d is 1785 for which both X_1 and X_2 are squares. In my talk, I will survey recent results about this problem when U is the set of Fibonacci numbers, Tribonacci numbers, k-generalized Fibonacci numbers, sums of two Fibonacci numbers, rep digits (in base 10 or any integer base $b \geq 2$) and factorials. The proofs use linear forms in logarithms and computations and in the case of factorials results about primes in arithmetic progressions. These results have been obtained in joint work with various colleagues such as J.J. Bravo, C. A. Gómez, S. Laishram, A. Montejano, L. Szalay and A. Togbé and recent Ph.D. students M. Ddamulira, B. Faye and M. Sias.

Truncated Theta Series, Partitions Inequalities and Rogers–Ramanujan Functions

Mircea Merca

My collaboration with George E. Andrews on the truncated version of Euler's pentagonal number theorem has opened up a new study on truncated theta series. Since then, over twenty papers on this topic have followed, and several partition inequalities are derived in this way. In this talk, we present a very general method for proving the non-trivial linear homogeneous partition inequalities. This method does not involve truncated theta series or q-series and connects the non-trivial linear homogeneous partition inequalities with the Prouhet–Tarry–Escott problem. On the other hand, I present an improvement of a conjecture related to a truncated theta series which I gave with G. E. Andrews in 2012. Combinatorial interpretations of this new conjecture give, for each $S \in \{1,2,3,4\}$, an infinite family of linear homogeneous inequalities for the number of partitions of n into parts congruent to $\pm S \bmod 5$. Twenty new identities involving the Rogers–Ramanujan functions $G(q)$ and $H(q)$ are experimentally discovered in this way.

The Charm of Units—The Conjecture of Kummer and Vandiver

Preda Mihăilescu

A classical conjecture of elementary algebraic number theory—which can be stated in a short phrase involving well-known concepts—but yet unsolved since more than a century, is the Kummer–Vandiver conjecture. It states that the p-part of the class group of the maximal real cyclotomic p-th field—thus the extension of $\mathbb{Q}[\zeta + \bar{\zeta}]$ of $\mathbb{Q}$ by the sum of a p-th root of unity ζ and its complex conjugate—is trivial. I solve this conjecture in two steps: the first, more unexpected one, will be presented in a main session—it was first presented in 2017 in some conferences in India and China, then in France and appears in the proceedings of one of these events. It states that the conjecture is proved, provided that a related, asymptotic conjecture due to Ralf Greenberg, is true. The second step will be presented in an off-schedule seminary, which we offer for those interested to see the result in more detail. It contains, of course, the proof of the Greenberg conjecture, for this p-th cyclotomic field.

A Combinatorial Refinement of the Kronecker–Hurwitz Class Number Relation

Alexandru Popa

I will present a refinement, and a new proof, of the classical Kronecker–Hurwitz class number relation, based on a tessellation of the Euclidean plane into semi-infinite triangles labelled by the modular group. This is a joint work with Don Zagier.

The Bass–Quillen Conjecture and Swan's Question

Dorin Popescu

In 1982, R. Swan noticed that, for an answer to the Bass–Quillen conjecture, it would be useful to prove that any regular local ring (R, m, k) is a filtered inductive limit of regular local rings, essentially of finite type over $\mathbb{Z}$. In 1989, we gave a positive answer to the Swan's question when $p = \text{char(k)}$ is either zero, or $p \notin m^2$, or $0 \neq p \in m^2$ but R is excellent Henselian. In all these questions R is excellent. Last year, Kęstutis Česnavičius wanted to know that in general a regular local ring is excellent because this will allow him to reduce the purity conjecture to the case of regular local rings which are complete. So we had to give a complete positive answer to Swan's question.

Linear Independence of Values of *G*-Functions

Tanguy Rivoal

G-functions are holomorphic functions at 0 solutions of linear differential equations with polynomial coefficients, and whose algebraic Taylor coefficients at 0 satisfy certain growth conditions. They were defined and studied by Siegel in 1929, and they can be viewed as generalizations of $\log(1-z)$: they include for instance Gauss hypergeometric series with rational parameters, polylogarithms, algebraic functions, derivatives and primitives of such functions. I will first present classical Diophantine results concerning the values taken by a *G*-function and its derivatives at algebraic points close to 0 (Siegel, Galochkin, Chudnovsky). I will then present a new Diophantine result concerning the dimension of the vector space generated over $\mathbb{Q}$ by the values taken by a *G*-function and (essentially) its primitives at any algebraic point inside their disc of convergence. This is a joint work with Stéphane Fischler.

Finite Automata, Automatic Sets and Difference Equations

Michael F. Singer

A finite automaton is one of the simplest models of computation. Initially introduced by McCulloch and Pitts to model neural networks, they have been used to aid in software design as well as to characterize certain formal languages and number-theoretic properties of integers. A set of integers is said to be m-automatic if there is a finite automaton that decides if an integer is in this set given its base-m representation. For example, powers of 2 are 2-automatic but not 3-automatic. This latter result follows from a theorem of Cobham describing in which sets of integers are m- and n-automatic for sufficiently distinct m and n. In recent work with Reinhard Schaefke, we gave a new proof of this result based on analytic results concerning normal forms of systems of difference equations. In this talk, I will describe this circle of ideas.

The Tangent Method for the Determination of Arctic Curves

Andrea Sportiello

In the paper [CS16] of Filippo Colomo and myself, we pose the basis for a method aimed at the determination of the "arctic curve" of large random combinatorial structures, i.e. the boundary between regions with zero and nonzero local entropy, in the scaling limit. This "basic" version of the tangent method (TM) is strikingly simple, but unfortunately it is not completely rigorous.

Two other versions of the method exist, let us call them the "entropic" tangent method (E-TM) and the "double-refinement" tangent method (2R-TM). In this talk, we shall first briefly review the "basic" TM, then we will introduce the two other methods and explain how the 2R-TM is completely rigorous, but it involves more complex quantities, while the E-TM has essentially the same technical difficulties of the TM, but it is even more heuristic. Finally, we close the circle, by showing how the Desnanot–Jacobi identity applied to the Izergin determinant implies the equivalence between the E-TM and the 2R-TM in the case of the six-vertex model with domain wall boundary conditions.

The Jacobian Conjecture, a Reduction of the Degree via a Combinatorial Physics Approach

Adrian Tanasa

The Jacobian conjecture is a celebrated conjecture stating (since 1939!) that any locally invertible polynomial system in $\mathbb{C}^n$ is globally invertible with polynomial inverse. C. W. Bass et al. (1982) proved a reduction theorem stating that the conjecture is true for any degree of the polynomial system if it is true in degree three. This degree reduction is obtained with the price of increasing the dimension n. I will show in this talk a theorem concerning partial elimination of variables, which implies a reduction of the generic case to the quadratic one. The price to pay is the introduction of a supplementary parameter $0 < n' < n$, parameter which represents the dimension of a linear subspace where some particular conditions on the system must hold. This result was obtained using the so-called intermediate field method in a quantum field theoretical (QFT) reformulation of the Jacobian conjecture. I will first present the general idea of this QFT method and then show how it applies to obtain our reduction result for the Jacobian conjecture.

Degeneration of Frobenius Structures

Masha Vlasenko

It was observed by Dwork that matrices of p-adic Frobenius operators in families of algebraic varieties satisfy differential equations. This led to the notion of Frobenius structures. We study degeneration of Frobenius structures at a singular point. As numerical evaluation shows, entries of degenerate Frobenius matrices may contain special values of p-adic L-functions. Most of our examples will be families of hypersurfaces with hypergeometric Picard–Fuchs equations. We will discuss a relevant conjecture of Candelas, de la Ossa and van Straten.

Zeros of the Riemann Zeta Function on the Critical Line

Alexandru Zaharescu

We discuss some recent developments that allow one to conclude that more than 5/12 of the non-trivial zeros of the Riemann zeta function lie on the critical line.

References

[AM18] George E. Andrews and Mircea Merca. Truncated theta series and a problem of Guo and Zeng. *J. Combin. Theory Ser. A*, 154:610–619, 2018.

[And17] George E. Andrews. Euler's partition identity and two problems of George Beck. *Math. Student*, 86(1–2):115–119, 2017.

[BB95] Paul-Georg Becker and Walter Bergweiler. Hypertranscendency of conjugacies in complex dynamics. *Math. Ann.*, 301(3):463–468, 1995.

[BCW82] Hyman Bass, Edwin H. Connell, and David Wright. The Jacobian conjecture: reduction of degree and formal expansion of the inverse. *Bull. Amer. Math. Soc. (N. S.)*, 7(2):287–330, 1982.

[BK10] Alin Bostan and Manuel Kauers. The complete generating function for Gessel walks is algebraic. *Proc. Amer. Math. Soc.*, 138(9):3063–3078, 2010. With an appendix by Mark van Hoeij.

[BKR17] A. Bostan, I. Kurkova, and K. Raschel. A human proof of Gessel's lattice path conjecture. *Trans. Amer. Math. Soc.*, 369(2):1365–1393, 2017.

[BM16] Mireille Bousquet-Mélou. An elementary solution of Gessel's walks in the quadrant. *Adv. Math.*, 303:1171–1189, 2016.

[BMM10] Mireille Bousquet-Mélou and Marni Mishna. Walks with small steps in the quarter plane. In *Algorithmic probability and combinatorics*, volume 520 of Contemp. Math., pages 1–39. Amer. Math. Soc., Providence, RI, 2010.

[BR86] Michael Boshernitzan and Lee A. Rubel. Coherent families of polynomials. *Analysis*, 6(4):339–389, 1986.

[BRS14] Alin Bostan, Kilian Raschel, and Bruno Salvy. Non-D-finite excursions in the quarter plane. *J. Combin. Theory Ser. A*, 121:45–63, 2014.

[Bru67] Armand Brumer. On the units of algebraic number fields. *Mathematika*, 14:121–124, 1967.

[Bud20] Timothy Budd. Winding of simple walks on the square lattice. *J. Combin. Theory Ser. A*, 172:105191, 59, 2020.

[BV20] Frits Beukers and Masha Vlasenko. Dwork Crystals I. *Int. Math. Res. Not. IMRN*, pages 1–38, 2020.

[BV21a] Frits Beukers and Masha Vlasenko. Dwork Crystals II. *Int. Math. Res. Not. IMRN*, (6):4427–4444, 2021.

[BV21b] Spencer Bloch and Masha Vlasenko. Gamma functions, monodromy and Frobenius constants. *Commun. Number Theory Phys.*, 15(1):91–147, 2021.

[CB83] Jacob Willem Cohen and O. J. Boxma. *Boundary value problems in queueing system analysis*, volume 79 of *North-Holland Mathematics Studies*. North-Holland Publishing Co., Amsterdam, 1983.

[CC88] D. V. Chudnovsky and G. V. Chudnovsky. Approximations and complex multiplication according to Ramanujan. In *Ramanujan revisited (Urbana-Champaign, Ill., 1987)*, pages 375–472. Academic Press, Boston, MA, 1988.

[CS16] F. Colomo and A. Sportiello. Arctic curves of the six-vertex model on generic domains: the tangent method. *J. Stat. Phys.*, 164(6):1488–1523, 2016.

[Der07] Harm Derksen. A Skolem-Mahler-Lech theorem in positive characteristic and finite automata. *Invent. Math.*, 168(1):175–224, 2007.

[dGST16] Axel de Goursac, Andrea Sportiello, and Adrian Tanasa. The Jacobian conjecture, a reduction of the degree to the quadratic case. Ann. Henri Poincaré, 17(11):3237–3254, 2016.

[DHRS18] Thomas Dreyfus, Charlotte Hardouin, Julien Roques, and Michael F. Singer. On the nature of the generating series of walks in the quarter plane. *Invent. Math.*, 213 (1):139–203, 2018.

[DHRS20] Thomas Dreyfus, Charlotte Hardouin, Julien Roques, and Michael F. Singer. Walks in the quarter plane: genus zero case. *J. Combin. Theory Ser. A*, 174:105251, 25, 2020.

[DL87] J. Denef and L. Lipshitz. Algebraic power series and diagonals. *J. Number Theory*, 26 (1):46–67, 1987.

[DS91] J. L. Davison and J. O. Shallit. Continued fractions for some alternating series. *Monatsh. Math.*, 111(2):119–126, 1991.

[DW15] Denis Denisov and Vitali Wachtel. Random walks in cones. *Ann. Probab.*, 43(3):992–1044, 2015.

[Dwo69] B. Dwork. p-adic cycles. *Inst. Hautes Études Sci. Publ. Math.*, (37):27–115, 1969.

[Eul40] L. Euler. De seriebus quibusdam considerationes. *Commentarii academiae scientiarum Petropolitanae*, 12:53–96, 1740.

[Eul80] L. Euler. Evolutio producti infiniti $(1-x)(1-xx)(1-x^3)(1-x^4)(1-x^5)$ [etc.] in seriem simplicem. *Acta Academiae Scientarum Imperialis Petropolitinae*, pages 47–55, 1780.

[Eul88] Leonhard Euler. *Introduction to analysis of the infinite. Book I.* Springer-Verlag, New York, 1988. Translated from the Latin and with an introduction by John D. Blanton.

[FIM17] Guy Fayolle, Roudolf Iasnogorodski, and Vadim Malyshev. Random walks in the quarter plane}, volume 40 of *Probability Theory and Stochastic Modelling*. Springer, Cham, second edition, 2017. Algebraic methods, boundary value problems, applications to queueing systems and analytic combinatorics.

[Gau66] Carl Friedrich Gauss. *Werke. Band II.* Koniglichen Gesellschaft der Wissenschaften, Gottingen, 1866.

[Gla83] J. W. L. Glaisher. A theorem in partitions. *Messenger of Math.*, 12:158–170, 1883.

[Gor04] V. A. Gorelov. On the Siegel conjecture for the case of second-order linear homogeneous differential equations. *Mat. Zametki*, 75(4):549–565, 2004. English translation in Math. Notes 75 (2004), no. 3-4, 513–529.

[GZ16] V. V. Golyshev and D. Zagier. Proof of the gamma conjecture for Fano 3-folds with a Picard lattice of rank one. *Izv. Ross. Akad. Nauk Ser. Mat.*, 80(1):27–54, 2016. English translation in Izv. Math. 80 (2016), no. 1, 24–49.

[Kal21] Erich L. Kaltofen. Foreword [MICA 2016]. *J. Symbolic Comput.*, 105:1–3, 2021. Held at the University of Waterloo, July 16–18, 2016.

[KKZ09] Manuel Kauers, Christoph Koutschan, and Doron Zeilberger. Proof of Ira Gessel's lattice path conjecture. *Proc. Natl. Acad. Sci. USA*, 106(28):11502–11505, 2009.

[KR11] Irina Kurkova and Kilian Raschel. Explicit expression for the generating function counting Gessel's walks. *Adv. in Appl. Math.*, 47(3):414–433, 2011.

[KS20] G. Kings and J. Sprang. Eisenstein-Kronecker classes, integrality of critical values of Hecke L-functions and p-adic interpolation, 2020. Preprint, 69 pages.

[Lec53] Christer Lech. A note on recurring series. *Ark. Mat.*, 2:417–421, 1953.

[Leo62] Heinrich-Wolfgang Leopoldt. Zur Arithmetik in abelschen Zahlkörpern. *J. Reine Angew. Math.*, 209:54–71, 1962.

[Lin82] F. Lindemann. Ueber die Zahl π. *Math. Ann.*, 20(2):213–225, 1882.

[LT18] Florian Luca and Alain Togbé. On the x-coordinates of Pell equations which are Fibonacci numbers. *Math. Scand.*, 122(1):18–30, 2018.

[Pop86] Dorin Popescu. General Néron desingularization and approximation. *Nagoya Math. J.*, 104:85–115, 1986.

[Ram00] S. Ramanujan. Some properties of $p(n)$, the number of partitions of n [Proc. Cambridge Philos. Soc. **19** (1919), 207–210]. In *Collected papers of Srinivasa Ramanujan*, pages 210–213. AMS Chelsea Publ., Providence, RI, 2000.

[Sie49] Carl Ludwig Siegel. *Transcendental Numbers*. Annals of Mathematics Studies, no. 16. Princeton University Press, Princeton, N. J., 1949.

[Sti84] Peter Stiller. Special values of Dirichlet series, monodromy, and the periods of automorphic forms. *Mem. Amer. Math. Soc.*, 49(299): iv+116, 1984.

[Sub71] M. V. Subbarao. Partition theorems for Euler pairs. *Proc. Amer. Math. Soc.*, 28:330–336, 1971.

[Wan80] Stuart Sui Sheng Wang. A Jacobian criterion for separability. *J. Algebra*, 65(2):453–494, 1980.

[Zag08] Don Zagier. Elliptic modular forms and their applications. In *The 1-2-3 of modular forms*, Universitext, pages 1–103. Springer, Berlin, 2008.

Contents

Frobenius Action on a Hypergeometric Curve and an Algorithm for Computing Values of Dwork's p-adic Hypergeometric Functions

Masanori Asakura

Abstract We provide an algorithm for computing special values modulo p^n of Dwork's p-adic hypergeometric functions whose complexity increases at most $O(n^4(\log n)^3)$. This is based on an explicit description of the Frobenius action on the rigid cohomology of a hypergeometric curve $(1-x^N)(1-y^M)=t$.

1 Introduction

Let p be a prime number. Let $\underline{a}=(a_1,\dots,a_s)\in\mathbb{Z}_p^s$ with $s\geq 2$ an integer. Let

$$F_{\underline{a}}(t) := {}_sF_{s-1}\left(\begin{matrix}a_1,\dots,a_s\\ 1,\dots,1\end{matrix};t\right)=\sum_{n=0}^{\infty}\frac{(a_1)_n}{n!}\cdots\frac{(a_s)_n}{n!}t^n\in\mathbb{Z}_p[[t]]$$

be the hypergeometric series where $(\alpha)_n$ denotes the Pochhammer symbol,

$$(\alpha)_n := \alpha(\alpha+1)\cdots(\alpha+n-1),\quad (\alpha)_0:=1.$$

For $\alpha\in\mathbb{Z}_p$, let α' be the Dwork prime which is defined to be $(\alpha+k)/p$ with $k\in\{0,1,\dots,p-1\}$ such that $\alpha+k\equiv 0 \mod p$. The ratio

$$\mathscr{F}_{\underline{a}}^{\mathrm{Dw}}(t) := \frac{F_{\underline{a}}(t)}{F_{\underline{a}'}(t^p)},\quad \underline{a}':=(a_1',\dots,a_s')$$

is called *Dwork's p-adic hypergeometric function*. In his seminal paper [Dw], Dwork discovered a sequence of rational functions which converges $\mathscr{F}_{\underline{a}}^{\mathrm{Dw}}(t)$. More precisely, for a power series $f(t)=\sum_{i\geq 0}a_it^i$, we denote by $[f(t)]_{<m} := \sum_{i<m}a_it^i$ the truncated polynomial. Then Dwork discovered the following congruence relations, which we call the *Dwork congruence*,

M. Asakura (✉)
Department of Mathematics, Hokkaido University, Sapporo 060-0810, Japan
e-mail: asakura@math.sci.hokudai.ac.jp

A. Bostan and K. Raschel (eds.), *Transcendence in Algebra, Combinatorics, Geometry and Number Theory*, Springer Proceedings in Mathematics & Statistics 373,
https://doi.org/10.1007/978-3-030-84304-5_1

$$\mathscr{F}^{\mathrm{Dw}}_{\underline{a}}(t) \equiv \frac{[F_{\underline{a}}(t)]_{<p^s}}{[F_{\underline{a}'}(t^p)]_{<p^s}} \mod p^s\mathbb{Z}_p[[t]].$$

After the work by Dwork, many people studied the congruences, for example, N. Katz [K] developed the congruences in more general situation, and moreover alternative methods are brought in [BV, SS, MV] etc. Thanks to his congruence, $\mathscr{F}^{\mathrm{Dw}}_{\underline{a}}(t)$ is a p-adic holomorphic function in the sense of Krasner (i.e. an element of a Tate algebra), and hence one can define the special value at $t = \alpha \in \mathbb{C}_p := \widehat{\overline{\mathbb{Q}}}_p$ by

$$\mathscr{F}^{\mathrm{Dw}}_{\underline{a}}(t)|_{t=\alpha} = \mathscr{F}^{\mathrm{Dw}}_{\underline{a}}(\alpha) = \lim_{n\to\infty}\left(\left.\frac{[F_{\underline{a}}(t)]_{<p^n}}{[F_{\underline{a}'}(t^p)]_{<p^n}}\right|_{t=\alpha}\right) \tag{1.1}$$

under the condition

$$\left|[F_{\underline{a}'}(t)]_{<p^n}|_{t=\alpha}\right|_p = 1, \quad \forall n \geq 1 \tag{1.2}$$

where $(-)|_{t=\alpha}$ denote the evaluations of rational functions and $|\cdot|_p$ denotes the p-adic valuation on $\mathbb{C}_p$.

Dwork's functions have the origin in the theory of the Monsky-Washnitzer cohomology or *rigid cohomology*. Let $f : X \to S = \mathbb{P}^1_{\mathbb{Z}_p} \setminus \{0, 1, \infty\}$ be the Legendre family of elliptic curves $y^2 = x(1-x)(1-tx)$. Let

$$\Phi : H^1_{\mathrm{rig}}(X/S) \longrightarrow H^1_{\mathrm{rig}}(X/S)$$

be the p-th Frobenius endomorphism on the rigid cohomology group (we refer the book [LS] for rigid cohomology). Then his function $\mathscr{F}^{\mathrm{Dw}}_{\frac{1}{2},\frac{1}{2}}(t)$ appears in a certain representation matrix of Φ. As a result, he obtained a formula which describes a Frobenius eigenvalue in terms of a value $\mathscr{F}^{\mathrm{Dw}}_{\frac{1}{2},\frac{1}{2}}(t)|_{t=\alpha}$ ([VdP, (7.14)]). His formula is often referred to as the unit root formula.

When we compute a value

$$\mathscr{F}^{\mathrm{Dw}}_{\underline{a}}(t)|_{t=\alpha} \mod p^n$$

according to the Dwork congruence, we find a difficulty especially for large n. Indeed the degrees of polynomials increase by exponential order, and hence the coefficients A_i, A_i' get larger very quickly,

$$(a)_{p^n} \sim (p^n)! \sim e^{p^n(n\log p - 1)} \quad \text{(Stirling)}.$$

We note that the bit complexity for computing $(a)_{p^n}$ is $O(n^2p^{2n})$ (by the naive multiplication algorithm). Recall that Dwork's function $\mathscr{F}^{\mathrm{Dw}}_{\frac{1}{2},\frac{1}{2}}(t)$ appears in the representation matrix of Φ. It is possible to derive a value $\mathscr{F}^{\mathrm{Dw}}_{\frac{1}{2},\frac{1}{2}}(t)|_{t=\alpha}$ from computing the Frobenius action on the rigid cohomology of $y^2 = x(1-x)(1-tx)$. In these decades, there have been lots of developments on the theory of rigid cohomology (cf. [Ke1]). In this paper, we employ the paper [KT] by *Kedlaya-Tuitman*. It is a general fact that each entry of a representation matrix of Φ is an overconvergent function. Then the main theorem of [KT] gives effective bounds on the overconvergence which increase by polynomial order with respect to n. In this way we can obtain an algorithm for computing a value $\mathscr{F}^{\mathrm{Dw}}_{\frac{1}{2},\frac{1}{2}}(t)|_{t=\alpha}$ in polynomial running time. To extend the method to the case of $\mathscr{F}^{\mathrm{Dw}}_{a,b}(t)$ we employ a family

$$(1-x^N)(1-y^M)=t$$

instead of the Legendre family, which we call a *hypergeometric fibration*. We shall prove,

Theorem (=Theorem 4.3) *Let p be a prime number such that $p>\max(N,M)$, and put $A=\mathbb{Z}_p[t,(t-t^2)^{-1}]$. Let $f:X\to \operatorname{Spec}A$ be the projective smooth family of curves defined by an affine equation $(1-x^N)(1-y^M)=t$ (see* Sect. 3.1 *for the precise construction). Write $X_{\mathbb{F}_p}:=X\times_{\mathbb{Z}_p}\mathbb{F}_p$ and $A_{\mathbb{F}_p}:=A\otimes_{\mathbb{Z}_p}\mathbb{F}_p$. Let $A^\dagger$ denote the weak completion (cf.* [LP, p.5]*) and write $A^\dagger_{\mathbb{Q}_p}:=A^\dagger\otimes_{\mathbb{Z}_p}\mathbb{Q}_p$. Then the de Rham cohomology group $H^1_{\mathrm{dR}}(X/A)$ is a free A-module, and under the comparison isomorphism*

$$H^1_{\mathrm{rig}}(X_{\mathbb{F}_p}/A_{\mathbb{F}_p})\cong H^\bullet_{\mathrm{dR}}(X/A)\otimes_A A^\dagger_{\mathbb{Q}_p},$$

the p-th Frobenius Φ is explicitly described by power series (4.8), ..., (4.11) *below.*

Using the explicit description of Φ, we have the algorithm for computing values of Dwork's hypergeometric functions.

Theorem *Let $N,M\geq 2$ be integers. Let $a\in\frac{1}{N}\mathbb{Z}$, $b\in\frac{1}{M}\mathbb{Z}$ with $0<a,b<1$. Suppose that $p>\max(N,M)$ (hence $p\neq 2$). Let $W=W(\overline{\mathbb{F}}_p)$ be the Witt ring of $\overline{\mathbb{F}}_p$. Let $\alpha\in W^\times\setminus(1+pW)$ be an arbitrary element satisfying* (1.2). *Then there is an algorithm for computing the special value*

$$\mathscr{F}^{\mathrm{Dw}}_{a,b}(t)|_{t=\alpha}\mod p^nW$$

such that the bit complexity (for fixed a,b,p,α) is at most $O(n^4(\log n)^3)$ as $n\to\infty$.

The algorithm is displayed in Sect. 5.2, together with some examples of values (Example 5.4).

We notice that our algorithm is not a simple consequence of Kedlaya-Tuitman [KT]. Indeed, in order to work out the above algorithm in a practical sense, we need to solve the issue, *how to estimate the supremum norm of the entries of the Frobenius matrix?* As long as the author sees, this is a non-trivial and delicate question as it concerns a $\mathbb{Z}_p$-lattice structure of p-adic cohomology (cf. the proof of Theorem 5.1 below). The author does not know how to get the p-adic estimate in a general situation. We shall give it only for our hypergeometric fibration $(1-x^N)(1-y^M)=t$.

To get the p-adic estimate, we shall give a free A-basis of the de Rham cohomology $H^1_{\mathrm{dR}}(X/A)$ in terms of Čech cocycles, and also a free basis of de Rham cohomology with log pole along the singular fibers. This is a very careful computation, to which we devote ourselves in Sect. 3. There are other examples of hypergeometric motives. For example, a curve $y^N = x^A(1-x)^B(1-tx)^C$ corresponds to the gaussian hypergeometric function ${}_2F_1$ (e.g. [Ar, AO]). The motive $x_1 + \cdots + x_r = y_1 + \cdots + y_s$, $tx_1^{a_1} \cdots x_r^{a_r} = y_1^{b_1} \cdots y_s^{b_s}$ corresponds to ${}_sF_{s-1}$ (e.g. [BCM]). Besides the MAGMA package provides elliptic curves for some ${}_2F_1$'s. However the author finds considerable technical difficulty in solving the issue of p-adic estimate if we choose these motives. Our hypergeometric fibration $(1-x^N)(1-y^M) = t$ is a simple and nice family so that one can solve all the above delicate problems.

Another key step is computations of power series expansions of the Frobenius matrix where we follow the method of Lauder [L] (=the deformation method). However we notice that the power series are centered at the singular fiber rather than a smooth fiber, and then a new technique appears in the computation (e.g. we use the p-adic digamma functions, [A, §2]).

We hope to obtain a generalization of the algorithm for $\mathscr{F}_{\underline{a}}^{\mathrm{Dw}}(t)$ with $s \geq 3$, by discussing the rigid cohomology of a higher dimensional hypergeometric fibration

$$(1 - x_0^{N_0}) \cdots (1 - x_d^{N_d}) = t,$$

though I have not worked out.[1]

2 Dwork's p-adic Hypergeometric Functions

Let p be a prime number. Let $\mathbb{Z}_p$ be the ring of p-adic integers, and $\mathbb{Q}_p$ the fractional field. Let $\mathbb{C}_p$ be the completion of $\overline{\mathbb{Q}}_p$. Write $O_{\mathbb{C}_p} = \{|x|_p \leq 1\}$ the valuation ring.

2.1 Definition

For an integer $n \geq 0$, we denote by $(\alpha)_n$ the Pochhammer symbol, which is defined by

$$(\alpha)_n := \alpha(\alpha+1)\cdots(\alpha+n-1), \quad (\alpha)_0 := 1.$$

Let $s \geq 2$ be an integer. For $(a_1, \ldots, a_s) \in \mathbb{Q}_p^s$ and $(b_1, \ldots, b_{s-1}) \in (\mathbb{Q}_p \setminus \mathbb{Z}_{\leq 0})^{s-1}$, the *hypergeometric power series* is defined to be

$${}_sF_{s-1}\left(\begin{matrix} a_1, \ldots, a_s \\ b_1, \ldots, b_{s-1} \end{matrix}; t\right) := \sum_{n=0}^{\infty} \frac{(a_1)_n \cdots (a_s)_n}{(b_1)_n \cdots (b_{s-1})_{n-1}} \frac{t^n}{n!} \in \mathbb{Q}_p[[t]].$$

[1] The referee pointed out to the author the recent article [Ke2], which almost solves the question completely (except the p-adic estimates on Frobenius). His approach is different from ours, more concise and maybe desirable.

In this paper, we only consider the series

$$F_{\underline{a}}(t) := {}_sF_{s-1}\left(\begin{matrix} a_1, \dots, a_s \\ 1, \dots, 1 \end{matrix}; t\right) = \sum_{n=0}^{\infty} \frac{(a_1)_n}{n!} \cdots \frac{(a_s)_n}{n!} t^n \in \mathbb{Z}_p[[t]]$$

for $\underline{a} = (a_1, \dots, a_s) \in \mathbb{Z}_p^s$.

For $\alpha \in \mathbb{Z}_p$, let α' denote the Dwork prime, which is defined to be $(\alpha + k)/p$ where $k \in \{0, 1, \dots, p-1\}$ such that $\alpha + k \equiv 0 \bmod p$. Define the i-th Dwork prime by $\alpha^{(i)} = (\alpha^{(i-1)})'$ and $\alpha^{(0)} := \alpha$. Write $\underline{a}' = (a_1', \dots, a_s')$ and $\underline{a}^{(i)} = (a_1^{(i)}, \dots, a_s^{(i)})$. *Dwork's p-adic hypergeometric function* is defined to be a power series

$$\mathscr{F}_{\underline{a}}^{\mathrm{Dw}}(t) := \frac{F_{\underline{a}}(t)}{F_{\underline{a}'}(t^p)} \in \mathbb{Z}_p[[t]].$$

A slight modification is

$$\mathscr{F}_{\underline{a}}^{\mathrm{Dw},\sigma}(t) := \frac{F_{\underline{a}}(t)}{F_{\underline{a}'}(t^\sigma)} \in W[[t]]$$

for a p-th Frobenius σ on $W[[t]]$ given by $\sigma(t) = ct^p$, $c \in 1 + pW$, where $W = W(\overline{\mathbb{F}}_p)$ is the Witt ring of $\overline{\mathbb{F}}_p$.

2.2 Dwork's Congruence Relations

In general, neither of the power series $F_{\underline{a}}(t) \bmod p\mathbb{Z}_p[[t]]$ or $\mathscr{F}_{\underline{a}}^{\mathrm{Dw}}(t) \bmod p\mathbb{Z}_p[[t]]$ terminate. Therefore one cannot substitute $t = \alpha \in W$ in $F_{\underline{a}}(t)$ or $\mathscr{F}_{\underline{a}}^{\mathrm{Dw}}(t)$ directly unless $\alpha \in pW$. In his seminal paper [Dw], Dwork showed that there is a sequence of rational functions which converges to $\mathscr{F}_{\underline{a}}^{\mathrm{Dw}}(t)$, namely it is a *p-adic analytic function* in the sense of Krasner.

Theorem 2.1 (Dwork's congruence relations). *For a power series $f(t) = \sum_{i\ge 0} a_i t^i$, we denote $[f(t)]_{<k} = \sum_{0\le i<k} a_i t^i$ the truncated polynomial. Let $\sigma(t) = ct^p$ with $c \in 1 + pW$. Then*

$$\mathscr{F}_{\underline{a}}^{\mathrm{Dw},\sigma}(t) \equiv \frac{[F_{\underline{a}}(t)]_{<p^n}}{[F_{\underline{a}'}(t^\sigma)]_{<p^n}} \mod p^n W[[t]] \tag{2.1}$$

for any $n \ge 1$. Hence for $\alpha \in \mathscr{O}_{\mathbb{C}_p}$ satisfying

$$[F_{\underline{a}'}(t)]_{<p^n}\big|_{t=\alpha} \not\equiv 0 \mod \mathfrak{m}_{\mathbb{C}_p}, \quad \forall n \ge 1 \tag{2.2}$$

where $\mathfrak{m}_{\mathbb{C}_p} := \{|x|_p < 1\}$ is the maximal ideal, one can define a special value of $\mathscr{F}_{\underline{a}}^{\mathrm{Dw}}(t)$ at $t = \alpha$ by

$$\mathscr{F}_{\underline{a}}^{\mathrm{Dw},\sigma}(t)|_{t=\alpha} = \mathscr{F}_{\underline{a}}^{\mathrm{Dw},\sigma}(\alpha) = \lim_{n\to\infty}\left(\frac{[F_{\underline{a}}(t)]_{<p^n}}{[F_{\underline{a}'}(ct^p)]_{<p^n}}\bigg|_{t=\alpha}\right).$$

Remark 2.2. *One cannot substitute $t = \alpha$ in $F_{\underline{a}}(t)$ since it is not a p-adic analytic function. For example, suppose $\underline{a}' = \underline{a}$ and $p \neq 2$, the following is wrong!*

$$\mathscr{F}_{\underline{a}}^{\mathrm{Dw}}(-1) = \frac{F_{\underline{a}}(-1)}{F_{\underline{a}}((-1)^p)} = \frac{F_{\underline{a}}(-1)}{F_{\underline{a}}(-1)} = 1.$$

Proof. When $c = 1$, this is proven in [Dw, p.37, Thm. 2, p.45]. The general case can be reduced to the case $c = 1$ in the following way. Since $\mathscr{F}_{\underline{a}}^{\mathrm{Dw},\sigma}(t) = \mathscr{F}_{\underline{a}}^{\mathrm{Dw}}(t) \cdot F_{\underline{a}'}(t^p)/F_{\underline{a}'}(ct^p)$, it is enough to show that

$$\frac{F_{\underline{a}}(t)}{F_{\underline{a}}(ct)} \equiv \frac{F_{\underline{a}}(t)_{<p^n}}{F_{\underline{a}}(ct)_{<p^n}} \quad \mod p^{n+1}W[[t]]$$

in general. Let $F_{\underline{a}}(t) = \sum_i A_i t^i$. Then the above is equivalent to that

$$\sum_{i+j=m,i,j\geq 0} A_{i+p^n}(c^j A_j) - A_i(c^{j+p^n} A_{j+p^n}) \equiv 0 \mod p^{n+1}$$

for any $m \geq 0$. However this is obvious as $c^{p^n} \equiv 1 \mod p^{n+1}$. □

Corollary 2.3

$$[F_{\underline{a}}(t)]_{<p^n} \equiv [F_{\underline{a}}(t)]_{<p}([F_{\underline{a}'}(t)]_{<p})^p \cdots ([F_{\underline{a}^{(n-1)}}(t)]_{<p})^{p^{n-1}} \mod p\mathbb{Z}_p[[t]]. \tag{2.3}$$

The condition (2.2) *holds if and only if*

$$[F_{\underline{a}^{(i)}}(t)]_{<p}\Big|_{t=\alpha} \not\equiv 0 \mod \mathfrak{m}_{\mathbb{C}_p}, \quad \forall i \geq 0. \tag{2.4}$$

Moreover we have

$$\mathscr{F}_{\underline{a}}^{\mathrm{Dw},\sigma}(t) \in W[t, h(t)^{-1}]^\wedge := \varprojlim_n \left(W/p^n W[t, h(t)^{-1}]\right), \quad h(t) := \prod_{i=0}^{N} [F_{\underline{a}^{(i)}}(t)]_{<p} \tag{2.5}$$

with some $N \gg 0$. In particular this is a p-adic analytic function in the sense of Krasner.

Proof. It follows from (2.1) that one has

$$\frac{[F_{\underline{a}}(t)]_{<p^n}}{[F_{\underline{a}'}(t^p)]_{<p^n}} \equiv \frac{[F_{\underline{a}}(t)]_{<p^n}}{([F_{\underline{a}'}(t)]_{<p^{n-1}})^p} \equiv [F_{\underline{a}}(t)]_{<p} \mod p\mathbb{Z}_p[[t]].$$

Then one can show (2.3) by induction on n. Notice that a set $\{[F_{\underline{a}^{(i)}}(t)]_{<p} \bmod p\}_{i\geq 0}$ of polynomials with $\mathbb{F}_p$-coefficients has a finite cardinal. Therefore (2.4) is a condition for finitely many i's. (2.5) is now immediate. □

The following congruence is a corollary of the Dwork congruence, proven in the same way as the proof of [Dw, p.45, (3.14)].[2]

Theorem 2.4. *Let $j \geq 0$ be an integer. Then*

$$\frac{\frac{d^j}{dt^j}F_{\underline{a}}(t)}{F_{\underline{a}}(t)} \equiv \frac{\frac{d^j}{dt^j}\left([F_{\underline{a}}(t)]_{<p^n}\right)}{[F_{\underline{a}}(t)]_{<p^n}} \mod p^n\mathbb{Z}_p[[t]] \tag{2.6}$$

for all $n \geq 1$. Hence

$$\frac{\frac{d^j}{dt^j}F_{\underline{a}}(t)}{F_{\underline{a}}(t)} \in W[t, h(t)^{-1}]^\wedge, \quad h(t) := \prod_{i=0}^{N}[F_{\underline{a}^{(i)}}(t)]_{<p}$$

is a p-adic analytic function in the sense of Krasner, and one can define the special values by (2.6).

Proof. Write $F(t) := F_{\underline{a}}(t)$ and $F_n(t) := [F_{\underline{a}}(t)]_{<p^n}$. Let $f^{(j)}(t) := \frac{d^j}{dt^j}f(t)$ denote the j-th derivative. Since

$$\frac{f^{(j+1)}}{f} = \frac{f^{(1)}}{f}\frac{f^{(j)}}{f} + \left(\frac{f^{(j)}}{f}\right)'$$

one can reduce (2.6) to the case $j = 1$ as follows,

$$\begin{aligned}\frac{F^{(j+1)}(t)}{F(t)} &= \frac{F^{(j)}(t)}{F(t)}\frac{F^{(1)}(t)}{F(t)} + \left(\frac{F^{(j)}(t)}{F(t)}\right)' \\ &\equiv \frac{F_n^{(j)}(t)}{F_n(t)}\frac{F_n^{(1)}(t)}{F_n(t)} + \left(\frac{F_n^{(j)}(t)}{F_n(t)}\right)' \mod p^n\mathbb{Z}_p[[t]] \\ &= \frac{F_n^{(j+1)}(t)}{F_n(t)}.\end{aligned}$$

We show

$$\frac{F'(t)}{F(t)} \equiv \frac{F_n'(t)}{F_n(t)} \mod p^n\mathbb{Z}_p[[t]] \tag{2.7}$$

by the induction on n. Let $G(t) = F_{\underline{a}^{(1)}}(t)$ and $G_n(t) := [G(t)]_{<p^n}$. Recall the Dwork congruence

$$\frac{F(t)}{G(t^p)} \equiv \frac{F_n(t)}{G_{n-1}(t^p)} \mod p^n\mathbb{Z}_p[[t]].$$

[2] In case $\underline{a}^{(1)} = \underline{a}$, Theorem 2.4 is immediate from [Dw, Lemma (3.4)].

Taking $f \mapsto f' f^{-1}$ on both sides, we have

$$\frac{F'(t)}{F(t)} - pt^{p-1}\frac{G'(t^p)}{G(t^p)} \equiv \frac{F_n'(t)}{F_n(t)} - pt^{p-1}\frac{G_{n-1}'(t^p)}{G_{n-1}(t^p)} \mod p^n\mathbb{Z}_p[[t]]$$

for arbitrary $n \geq 1$. This immediately implies (2.7) in case $n = 1$. Suppose that (2.7) is true for $n-1$. Then

$$\frac{G'(t)}{G(t)} \equiv \frac{G_{n-1}'(t)}{G_{n-1}(t)} \mod p^{n-1}\mathbb{Z}_p[[t]]$$

and hence

$$pt^{p-1}\frac{G'(t^p)}{G(t^p)} \equiv pt^{p-1}\frac{G_{n-1}'(t^p)}{G_{n-1}(t^p)} \mod p^n.$$

Therefore we have

$$\frac{F'(t)}{F(t)} \equiv \frac{F_n'(t)}{F_n(t)} \mod p^n.$$

This completes the proof of (2.7). □

3 Hypergeometric Fibrations

For a smooth scheme X over a commutative ring A, we denote by $H^*_{\mathrm{dR}}(X/A) := \mathbb{H}^*_{\mathrm{zar}}(X, \Omega^\bullet_{X/A})$ the algebraic de Rham cohomology groups.

3.1 Setting

Let $N, M \geq 2$ be an integer. Let W be a commutative ring such that NM is invertible. Suppose that W contains a primitive $\mathrm{lcm}(N, M)$-th root of unity. Later we shall take W to be the Witt ring of a perfect field of characteristic p with $p \nmid NM$. Let $\mathbb{P} := \mathbb{P}^1 \times \mathbb{P}^1 \times \mathbb{P}^1$ be the product of the projective lines over W with homogeneous coordinates $(X_0, X_1) \times (Y_0, Y_1) \times (T_0, T_1)$. We use inhomogeneous coordinates $x := X_1/X_0$, $y := Y_1/Y_0$, $t := T_1/T_0$ and $z := x^{-1}$, $w := y^{-1}$, $s := t^{-1}$. Let $Y_s \subset \mathbb{P}$ be the closed subscheme defined by a homogeneous equation

$$T_0(X_0^N - X_1^N)(Y_0^M - Y_1^M) = T_1 X_0^N Y_0^M$$

over W. Let

$$f_s : Y_s \longrightarrow \mathbb{P}^1 = \mathrm{Proj}\, W[T_0, T_1]$$

be the projection onto the 3rd line. Put $A := W[t, (t - t^2)^{-1}]$, $U := \mathrm{Spec} A \subset \mathbb{P}^1$ and

$$X := f_s^{-1}(U) = \mathrm{Spec} A[x, y]/((1 - x^N)(1 - y^M) - t).$$

Then $X \to U$ is smooth and projective, and a geometric fiber is a connected smooth projective curve of genus $(N - 1)(M - 1)$ (e.g. the Hurwitz formula). An open set $Y_s \setminus f_s^{-1}(s = 0)$ is smooth over W where "$s = 0$" denotes the closed subscheme $\mathrm{Spec} W[s]/(s) \subset \mathbb{P}^1$. There are singular loci $\{s = 1 - z^N = w = 0\}$ and $\{s = z = 1 - w^M = 0\}$ in the affine open set

$$\mathrm{Spec} W[s, z, w]/(s(1 - z^N)(1 - w^M) - z^N w^M) \subset Y_s.$$

All the singularities are of type "$xy = z^k$" where $k = N$ or $k = M$. One can resolve them according to Propositions 7.1 in Appendix B. The fiber $f_s^{-1}(0)$ at $t = 0$ is a relative simple normal crossing divisor (abbreviated NCD) over W (see Appendix B for the definition), and all components are $\mathbb{P}^1$. The fiber $f_s^{-1}(1)$ at $t = 1$ is an integral divisor which is smooth outside the point $(x, y, t) = (0, 0, 1)$. The normalization of $f_s^{-1}(1)$ is the Fermat curve $z^N + w^M = 1$. In a neighborhood of the point $(x, y, t) = (0, 0, 1)$, the fiber $f_s^{-1}(1) \subset Y_s$ is defined by $x^N + y^M - x^N y^M = 0 \Leftrightarrow (x')^N + y^M = 0, x' := x(1 - y^M)^{\frac{1}{N}}$. One can further resolve it according to Propositions 7.2 in Appendix B.

Summing up the above, we have a smooth projective W-scheme Y with a fibration

$$f : Y \longrightarrow \mathbb{P}^1 = \mathrm{Proj} W[T_0, T_1]$$

which satisfies the following conditions. Let $D_0 := f^{-1}(0)$, $D_1 := f^{-1}(1) = \sum_i n_i D_{1,i}$ and $D_\infty := f^{-1}(\infty) = \sum_j m_j D_{\infty,j}$ denote the fibers at $\mathrm{Spec} W[t]/(t)$, $\mathrm{Spec} W[t]/(t - 1)$ and $\mathrm{Spec} W[s]/(s)$ respectively.

(i) f is smooth over $U = \mathrm{Spec} W[t, (t - t^2)^{-1}] \subset \mathbb{P}^1$, and $X = f_s^{-1}(U) = f^{-1}(U)$.
(ii) D_0 and $\sum_i D_{1,i}$ and $\sum_j D_{\infty,j}$ are simple relative NCD's over W.
(iii) The multiplicities n_i of D_1 are either of $1, iN, jM$ with $i \in \{1, \dots, M\}$, $j \in \{1, \dots, N\}$.
(iv) The multiplicities of m_j of D_∞ are integers $\leq \max(N, M)$.
(v) Any components of D_0 or D_∞ are $\mathbb{P}^1$. There is a unique component of D_1 which is not $\mathbb{P}^1$. It is the Fermat curve $z^N + w^M = 1$.

Let $\mu_n := \{\zeta \in W^\times | \zeta^n = 1\}$ denote the group of n-th roots of 1. For $(\zeta_1, \zeta_2) \in \mu_N \times \mu_M$, the morphism $(x, y, t) \mapsto (\zeta_1 x, \zeta_2 y, t)$ extends to an automorphism on Y or X, which we write by $[\zeta_1, \zeta_2]$.

3.2 $H^1_{\mathrm{dR}}(X/A)$

Let U_0 and U_1 be the affine open sets of X defined by $X_0Y_0 \neq 0$ and $X_1Y_1 \neq 0$ respectively,

$$U_0 = \operatorname{Spec} A[x, y]/((1-x^N)(1-y^M) - t),$$

$$U_1 = \operatorname{Spec} A[z, w]/((1-z^N)(1-w^M) - tz^N w^M).$$

Then $X = U_0 \cup U_1$. For $i \in \{1, \dots, N-1\}$ and $j \in \{1, \dots, M-1\}$ let

$$\omega_{ij} := N\frac{x^{i-1}y^{j-M}}{1-x^N}dx = -M\frac{x^{i-N}y^{j-1}}{1-y^M}dy \tag{3.1}$$

be rational relative 1-forms on X/A.

Lemma 3.1. $\omega_{ij} \in \Gamma(X, \Omega^1_{X/A})$.

Proof. Multiplying $x^i y^j$ on

$$t(1-t)\frac{y^{-M}}{1-x^N}\frac{dx}{x} = t\frac{dx}{x} - MN^{-1}(1-x^N)\frac{dy}{y}$$

one sees $\omega_{ij} \in \Gamma(U_0, \Omega^1_{X/A})$. Similarly, using an equality

$$\frac{1}{1-z^N}\frac{dz}{z} = (1-(1-t^{-1})(1-w^M))\frac{dz}{z} - MN^{-1}\frac{dw}{w}$$

one sees $\omega_{ij} \in \Gamma(U_1, \Omega^1_{X/A})$. □

Lemma 3.2. *Let $H^1(X, \mathscr{O}_X)$ be the Zariski cohomology which is isomorphic to the cokernel of the Cech complex*

$$\delta : \Gamma(U_0, \mathscr{O}_X) \oplus \Gamma(U_1, \mathscr{O}_X) \longrightarrow \Gamma(U_0 \cap U_1, \mathscr{O}_X), \quad (u_0, u_1) \longmapsto u_1 - u_0.$$

Write $[f] := f \bmod \operatorname{Im}\delta \in H^1(\mathscr{O}_X)$. Then $H^1(X, \mathscr{O}_X)$ is generated as A-module by elements

$$[x^i y^{j-M}], \quad i \in \{1, \dots, N-1\},\ j \in \{1, \dots, M-1\}.$$

Moreover for any integers k, l, there is an element $\alpha \in A$ such that $[x^{i+kN}y^{j+lM}] = \alpha[x^i y^{j-M}]$ in $H^1(X, \mathscr{O}_X)$.

Proof. We first note that if $k, l \le 0$ or $k, l \ge 0$ then $[x^k y^l] = 0$ by definition. Let i, j be integers such that $1 \le i \le N-1$ and $1 \le j \le M-1$. Since $1-t = x^N + y^M - x^N y^M$, one has

$$(1-t)^k[x^i y^{j-M}] = [x^{kN} \cdot x^i y^{j-M}] = [x^{i+kN}y^{j-M}], \quad \forall k \ge 0. \tag{3.2}$$

Let $l \geq 1$. Then $(1-t)x^{i-N}y^{j-lM} = x^i y^{j-lM} + x^{i-N}y^{j-(l-1)M} - x^i y^{j-(l-1)M}$, and this implies

$$[x^i y^{j-lM}] = [x^i y^{j-(l-1)M}], \quad \forall l \geq 2, \tag{3.3}$$

and for $l = 1$

$$[x^i y^{j-M}] + [x^{i-N} y^j] = 0. \tag{3.4}$$

We claim

$$[x^{i+kN} y^{j-lM}] \in A[x^i y^{j-M}], \quad \forall k \geq 0,\ l \geq 1. \tag{3.5}$$

If $l = 1$, this is nothing other than (3.2). If $k = 0$, this follows from (3.3). Suppose $k \geq 1$ and $l \geq 2$. Then

$$(1-t)[x^{i+(k-1)N} y^{j-lM}] = [x^{i+kN} y^{j-lM}] + [x^{i+(k-1)N} y^{j-(l-1)M}] - [x^{i+kN} y^{j-(l-1)M}].$$

Hence (3.5) follows by induction on $k+l$. In the same way, one can show $[x^{i-kN} y^{j+lM}] \in A[x^{i-N} y^j]$ for all $k \geq 1$ and $l \geq 0$. Therefore $[x^{i-kN} y^{j+lM}] \in A[x^i y^{j-M}]$ by (3.4). This completes the proof. □

Proposition 3.3. (1) *$\Gamma(X, \Omega^1_{X/A})$ is a free A-module with basis*

$$\omega_{ij}, \quad i \in \{1, \ldots, N-1\},\ j \in \{1, \ldots, M-1\}.$$

(2) *$H^1(X, \mathscr{O}_X)$ is a free A-module with basis*

$$[x^i y^{j-M}], \quad i \in \{1, \ldots, N-1\},\ j \in \{1, \ldots, M-1\}.$$

Proof. For a point $s \in U = \operatorname{Spec} A$, we denote the residue field by $k(s)$, and write $X_s := X \times_A \operatorname{Spec} k(s)$. Let $q = 0, 1$. Since $\dim_{k(s)} H^q(\Omega^{1-q}_{X_s/k(s)}) = (N-1)(M-1)$ is constant with respect to s, one can apply [Ha, III,12.9], so that $H^q(X, \Omega^{1-q}_{X/A})$ is a locally free A-module and the isomorphism $H^q(\Omega^{1-q}_{X/A}) \otimes k(s) \cong H^q(\Omega^{1-q}_{X_s/k(s)})$ follows. Obviously $\omega_{ij}|_{X_s} \neq 0$ and they are linearly independent over $k(s)$ since each ω_{ij} belongs to the distinct simultaneous eigenspace with respect to $\mu_N \times \mu_M$. Noticing that $\dim H^0(\Omega^1_{X_s/k(s)}) = (N-1)(M-1)$, one sees that $\{\omega_{ij}|_{X_s}\}_{i,j}$ forms a $k(s)$-basis of $H^0(\Omega^1_{X_s/k(s)})$, and hence that $\{\omega_{ij}\}_{i,j}$ forms a A-basis of $H^0(A, \Omega^1_{X/A})$ by Nakayama's lemma. This completes the proof of (1). In a similar way, the assertion (2) follows by using Lemma 3.2. □

The algebraic de Rham cohomology $H^1_{\mathrm{dR}}(X/A)$ is described in terms of the Cech complexes. Let

$$\begin{array}{ccc} \Gamma(U_0, \mathscr{O}) \oplus \Gamma(U_1, \mathscr{O}_X) & \xrightarrow{d} & \Gamma(U_0, \Omega^1_{X/A}) \oplus \Gamma(U_1, \Omega^1_{X/A}) \\ \downarrow{\scriptstyle \delta} & & \downarrow{\scriptstyle \delta} \\ \Gamma(U_0 \cap U_1, \mathscr{O}_X) & \xrightarrow{d} & \Gamma(U_0 \cap U_1, \Omega^1_{X/A}) \end{array}$$

be a commutative diagram where d is the differential map and δ is given by $(u_0, u_1) \mapsto u_1 - u_0$. Then the de Rham cohomology $H_{\mathrm{dR}}(X/A)$ is isomorphic to the cohomology of the total complex. In particular, an element of $H^1_{\mathrm{dR}}(X/A)$ is given as the representative of a cocycle

$$(f) \times (\omega_0, \omega_1) \in \Gamma(U_0 \cap U_1, \mathscr{O}_X) \times \Gamma(U_0, \Omega^1_{X/A}) \oplus \Gamma(U_1, \Omega^1_{X/A})$$

which satisfies $df = \omega_1 - \omega_0$. Let $\omega_{ij} \in \Gamma(X, \Omega^1_{X/A})$ be as in Proposition 3.3. We denote by the same notation ω_{ij} the element of $H^1_{\mathrm{dR}}(X/A)$ via the natural map $\Gamma(X, \Omega^1_{X/A}) \to H^1_{\mathrm{dR}}(X/A)$, which is the representative of a cocycle

$$(0) \times (\omega_{ij}|_{U_0}, \omega_{ij}|_{U_1}).$$

We construct a lifting

$$\eta_{ij} := (x^i y^{j-M}) \times (\eta^0_{ij}, \eta^1_{ij}) \in H^1_{\mathrm{dR}}(X/A) \tag{3.6}$$

of $[x^i y^{j-M}] \in H^1(\mathscr{O}_X)$ in the following way. A direct computation yields

$$(j-M)(1-t)x^{i-N}y^{j-M-1}dy - d(x^i y^{j-M}) = -\left(\frac{(j-M)t}{M} + \frac{i}{N}(1-x^N)\right)\omega_{ij}. \tag{3.7}$$

Note $x^{i-N}y^{j-M-1}dy = -z^{N-i}w^{M-j-1}dw \in \Gamma(U_1, \Omega^1_{X/A})$, and the right hand side lies in $\Gamma(U_0, \Omega^1_{X/A})$ by Lemma 3.1. Therefore we put

$$\eta^0_{ij} := -\left(\frac{(j-M)t}{M} + \frac{i}{N}(1-x^N)\right)\omega_{ij}, \quad \eta^1_{ij} := -(j-M)(1-t)z^{N-i}w^{M-j-1}dw,$$

then we get the desired cocycle (3.6). By Proposition 3.3 (2) together with liftings (3.6), the natural map $H^1_{\mathrm{dR}}(X/A) \to H^1(\mathscr{O}_X)$ is surjective, and hence one has an exact sequence

$$0 \longrightarrow \Gamma(X, \Omega^1_{X/A}) \longrightarrow H^1_{\mathrm{dR}}(X/A) \longrightarrow H^1(X, \mathscr{O}_X) \to 0.$$

Thus we get the following theorem.

Theorem 3.4. *$H^1_{\mathrm{dR}}(X/A)$ is a free A-module with basis*

$$\omega_{ij}, \eta_{ij} \quad i \in \{1, \ldots, N-1\}, \ j \in \{1, \ldots, M-1\}.$$

3.3 $H^1(\mathscr{Y}, \Omega^\bullet_{\mathscr{Y}/W[[\lambda]]}(\log D))$

Let λ be an indeterminate. Let $\operatorname{Spec} W[[\lambda]] \to \mathbb{P}^1$ be the morphism induced by $\lambda = t$, $1-t$ or t^{-1}. Let

$$\begin{array}{ccc} \mathscr{Y} & \longrightarrow & Y \\ \downarrow & & \downarrow f \\ \operatorname{Spec} W[[\lambda]] & \longrightarrow & \mathbb{P}^1 \end{array}$$

be the base change. Let $D \subset \mathscr{Y}$ denote the central fiber, namely $D = D_0$, D_1 or D_∞ by the notation in Sect. 3.1. The reduced part D_{red} is a relative simple NCD over W. Put $\mathscr{X} := \mathscr{Y} \setminus D$. Define a $\mathscr{O}_{\mathscr{Y}}$-module

$$\Omega^1_{\mathscr{Y}/W[[\lambda]]}(\log D) := \operatorname{Coker}\left[\mathscr{O}_{\mathscr{Y}} \frac{d\lambda}{\lambda} \to \Omega^1_{\mathscr{Y}/W}(\log D)\right]$$

and consider the cohomology group

$$H^1(\mathscr{Y}, \Omega^\bullet_{\mathscr{Y}/W[[\lambda]]}(\log D)) := H^1_{\mathrm{zar}}(\mathscr{Y}, \mathscr{O}_{\mathscr{Y}} \to \Omega^1_{\mathscr{Y}/W[[\lambda]]}(\log D)).$$

Proposition 3.5. *If $N!M!$ is invertible in W, then $\Omega^1_{\mathscr{Y}/W[[\lambda]]}(\log D)$ is a locally free $\mathscr{O}_{\mathscr{Y}}$-module.*

Proof. If $N!M!$ is invertible in W, then each multiplicity of D is invertible in W (see Sect. 3.1). The assertion can be checked locally on noticing that f is given by $(x_1, x_2) \mapsto \lambda = x_1^{r_1} x_2^{r_2}$ with r_1, r_2 integers which are invertible in W. □

Theorem 3.6. *Suppose that W is an integral domain of characteristic zero, and that $N!M!$ is invertible in W. Put*

$$H_\lambda := \operatorname{Im}[H^1(\mathscr{Y}, \Omega^\bullet_{\mathscr{Y}/W[[\lambda]]}(\log D)) \to H^1_{\mathrm{dR}}(\mathscr{X}/W((\lambda)))],$$

$$\operatorname{Fil}^1 H_\lambda := \operatorname{Im}[\Gamma(\mathscr{Y}, \Omega^1_{\mathscr{Y}/W[[\lambda]]}(\log D)) \to H^1_{\mathrm{dR}}(\mathscr{X}/W((\lambda)))].$$

Then H_λ and $\operatorname{Fil}^1 H_\lambda$ are free $W[[\lambda]]$-modules of rank 2 and 1 respectively. More precisely, the following holds.

(1) *If $\lambda = t$, then $\operatorname{Fil}^1 H_\lambda$ has a $W[[\lambda]]$-basis $\{\omega_{ij}\}$ and H_λ has a $W[[\lambda]]$-basis $\{\omega_{ij}, \eta_{ij}\}$ where (i, j) runs over the pairs of integers such that $1 \le i \le N-1$ and $1 \le j \le M-1$.*
(2) *If $\lambda = s = t^{-1}$, then $\operatorname{Fil}^1 H_\lambda$ has a $W[[\lambda]]$-basis $\{\omega_{ij}\}$ and H_λ has a $W[[\lambda]]$-basis $\{\omega_{ij}, s\eta_{ij}\}$.*

(3) *If* $\lambda = 1 - t$, *set*

$$\omega^*_{ij} := \begin{cases} \omega_{ij} & i/N + j/M \geq 1 \\ (1-t)\omega_{ij} & i/N + j/M < 1, \end{cases}$$

$$\eta^*_{ij} := \begin{cases} \eta_{ij} & i/N + j/M \geq 1 \\ (1 - i/N - j/M)t\omega_{ij} - \eta_{ij} & i/N + j/M < 1. \end{cases}$$

Then $\mathrm{Fil}^1 H_\lambda$ *has a* $W[[\lambda]]$*-basis* $\{\omega^*_{ij}\}$ *and* H_λ *has a* $W[[\lambda]]$*-basis* $\{\omega^*_{ij}, \eta^*_{ij}\}$.

The proof of Theorem 3.6 shall be given in later sections.

3.4 Preliminary on Proof of Theorem 3.6

Let $U_{kl} = \mathscr{Y} \cap \{X_k Y_l \neq 0\}$, $k, l \in \{0, 1\}$ be an affine open set. Then $\mathscr{Y} = \bigcup_{k=0,1} \bigcup_{l=0,1} U_{kl}$. The cohomology group $H^i(\mathscr{Y}, \Omega^\bullet_{\mathscr{Y}/W[[\lambda]]}(\log D))$ is isomorphic to the cohomology of the total complex of the double complex

$$\begin{array}{ccc} \bigoplus \Gamma(U_{ab}, \mathscr{O}_{\mathscr{Y}}) & \xrightarrow{\ d\ } & \bigoplus \Gamma(U_{ab}, \Omega^1_{\mathscr{Y}/W[[\lambda]]}(\log D)) \\ \Big\downarrow{\scriptstyle \delta} & & \Big\downarrow{\scriptstyle \delta} \\ \bigoplus \Gamma(U_{ab} \cap U_{cd}, \mathscr{O}_{\mathscr{Y}}) & \longrightarrow & \bigoplus \Gamma(U_{ab} \cap U_{cd}, \Omega^1_{\mathscr{Y}/W[[\lambda]]}(\log D)) \end{array}$$

An element of $H^1(\mathscr{Y}, \Omega^\bullet_{\mathscr{Y}/W[[\lambda]]}(\log D))$ is represented by a cocycle

$$(f_{ab,cd}) \times (\alpha_{ab}) \in \bigoplus \Gamma(U_{ab} \cap U_{cd}, \mathscr{O}_{\mathscr{Y}}) \times \bigoplus \Gamma(U_{ab}, \Omega^1_{\mathscr{Y}/W[[\lambda]]}(\log D))$$

which satisfies $f_{ab,ef} = f_{ab,cd} + f_{cd,ef}$ and

$$\alpha_{cd}|_{U_{ab} \cap U_{cd}} - \alpha_{ab}|_{U_{ab} \cap U_{cd}} = d(f_{ab,cd}).$$

If we replace $\mathscr{O}_{\mathscr{Y}}$ with $\mathscr{O}_{\mathscr{X}}$ and $\Omega^1_{\mathscr{Y}/W[[\lambda]]}(\log D)$ with $\Omega^1_{\mathscr{X}/W((\lambda))}$ in the above, we obtain the algebraic de Rham cohomology group $H^i_{\mathrm{dR}}(\mathscr{X}/W((\lambda)))$. Let $\omega_{ij}, \eta_{ij} \in H^1_{\mathrm{dR}}(X/A)$ be as in (3.1) and (3.6). Then $\omega_{ij}|_{\mathscr{X}} \in H^1_{\mathrm{dR}}(\mathscr{X}/W((\lambda)))$ is represented by

$$(0) \times (\omega_{ij}|_{U_{ab}}) \in \bigoplus \Gamma(U_{ab} \cap U_{cd}, \mathscr{O}_{\mathscr{X}}) \times \bigoplus \Gamma(U_{ab}, \Omega^1_{\mathscr{X}/W((\lambda))}).$$

A cocycle which represents $\eta_{ij}|_{\mathscr{X}}$ is given as follows. We note that

$$x^{i-N} y^{j-M-1} dy = -z^{N-i} w^{M-j-1} dw \in \Gamma(U_{11}, \Omega^1_{X/A})$$

and

$$\begin{aligned} x^{i-N}y^{j-M-1}dy &= -x^{i-N}w^{M-j-1}dw \\ &= -(x^N+(1-t)w^M-x^Nw^M)x^{i-N}w^{M-j-1}dw \\ &= -(1-w^M)x^iw^{M-j-1}dw-(1-t)t^{-1}\frac{N}{M}(1-w^M)^2x^{i-1}w^{M-j}dx \\ &\in \Gamma(U_{01},\Omega^1_{X/A}) \end{aligned}$$

where the 2nd equality follows from $1 = x^N + (1-t)w^M - x^Nw^M$. On the other hand $x^{i-N}y^{j-M-1}dy \notin \Gamma(U_{00}, \Omega^1_{X/A})$ while we have (3.7). Moreover $x^{i-N}y^{j-M-1}dy \notin \Gamma(U_{10}, \Omega^1_{X/A})$ while we have

$$(j-M)z^{N-i}y^{j-M-1}dy-(1-t)d(z^{2N-i}y^{j-M}) = z^N\left(\frac{(j-M)t}{M}+\frac{(i-2N)}{N}(1-t)(1-z^N)\right)\omega_{ij}.$$

Therefore we put

$$\eta^{11}_{ij} := -(j-M)(1-t)z^{N-i}w^{M-j-1}dw,\quad \eta^{01}_{ij} := -(j-M)(1-t)x^{i-N}w^{M-j-1}dw,$$

$$\eta^{00}_{ij} := -\left(\frac{(j-M)t}{M}+\frac{i}{N}(1-x^N)\right)\omega_{ij},$$

$$\begin{aligned} \eta^{10}_{ij} :&= (1-t)z^N\left(\frac{(j-M)t}{M}+\frac{(i-2N)}{N}(1-t)(1-z^N)\right)\omega_{ij} \\ &= (1-y^M+z^Ny^M)\left(\frac{(j-M)t}{M}+\frac{(i-2N)}{N}(1-t)(1-z^N)\right)\omega_{ij} \end{aligned}$$

and

$$f_{00,11} = f_{00,01} := x^iy^{j-M},\quad f_{10,11} = f_{10,01} := (1-t)^2z^{2N-i}y^{j-M} = (1-t)(1-y^M+z^Ny^M)z^{N-i}y^{j-M},$$

$$f_{01,11} := 0,\quad f_{00,10} := x^iy^{j-M}-(1-t)^2z^{2N-i}y^{j-M} = (1-x^N)(x^Ny^M-2x^N-y^M)z^{2N-i}y^j$$

and $f_{11,00} := -f_{00,11}$ etc. Then we get a cocycle

$$(f_{ab,cd})\times(\eta^{ab}_{ij}) \in \bigoplus \Gamma(U_{ab}\cap U_{cd},\mathscr{O}_{\mathscr{X}})\times\bigoplus\Gamma(U_{ab},\Omega^1_{\mathscr{X}/W((\lambda))}) \tag{3.8}$$

which represents $\eta_{ij}|_{\mathscr{X}} \in H^1_{\mathrm{dR}}(\mathscr{X}/W((\lambda)))$.

3.5 Deligne's Canonical Extension

Let $j : \mathrm{Spec}\mathbb{C}((\lambda)) \to \mathrm{Spec}\mathbb{C}[[\lambda]]$. Let $(\mathscr{H}, \nabla)$ be an integrable connection on $\mathrm{Spec}\mathbb{C}((\lambda))$. There is a unique subsheaf $\mathscr{H}_e \subset \mathscr{H}$ which satisfies the following conditions (cf. [Z] (17)).

(D1) $\mathscr{H}_e$ is a free $\mathbb{C}[[\lambda]]$-module such that $j^{-1}\mathscr{H}_e = \mathscr{H}$,
(D2) the connection extends to have log pole, $\nabla : \mathscr{H}_e \to \frac{d\lambda}{\lambda} \otimes \mathscr{H}_e$,
(D3) each eigenvalue α of $\mathrm{Res}(\nabla)$ satisfies $0 \leq \mathrm{Re}(\alpha) < 1$, where $\mathrm{Res}(\nabla)$ is the map defined by a commutative diagram

$$\begin{array}{ccc} \mathscr{H}_e & \xrightarrow{\nabla} & \frac{d\lambda}{\lambda} \otimes \mathscr{H}_e \\ \downarrow & & \downarrow \scriptstyle{\mathrm{Res}\otimes 1} \\ \mathscr{H}_e/\lambda\mathscr{H}_e & \xrightarrow{\mathrm{Res}(\nabla)} & \mathscr{H}_e/\lambda\mathscr{H}_e . \end{array}$$

The extended bundle $(\mathscr{H}_e, \nabla)$ is called *Deligne's canonical extension.*

Let $g : V \to \mathrm{Spec}\mathbb{C}[[\lambda]]$ be a projective flat morphism which is smooth over $\mathrm{Spec}\mathbb{C}((\lambda))$. Let D be the central fiber. Suppose that D_{red} is a NCD. We define a locally free $\mathscr{O}_V$-module

$$\Omega^1_{V/\mathbb{C}[[\lambda]]}(\log D) := \mathrm{Coker}\left[\mathscr{O}_V \frac{d\lambda}{\lambda} \to \Omega^1_{V/\mathbb{C}}(\log D)\right]$$

and $\Omega^k_{V/\mathbb{C}[[\lambda]]}(\log D) := \wedge^k \Omega^1_{V/\mathbb{C}[[\lambda]]}(\log D)$.

Let $U : V \setminus D$ and let $(\mathscr{H}, \nabla) = (H^i_{\mathrm{dR}}(U/\mathbb{C}((\lambda))), \nabla)$ be the Gauss-Manin connection on $\mathrm{Spec}\mathbb{C}((\lambda))$. Then Deligne's canonical extension of $\mathscr{H}$ is given as follows ([S], (2.18)–(2.20)),

$$\mathscr{H}_e \cong H^i(V, \Omega^\bullet_{V/\mathbb{C}[[\lambda]]}(\log D)).$$

Moreover $\exp(-2\pi i \mathrm{Res}_P(\nabla))$ agrees with the monodromy operator on $H_{\mathbb{C}} = \mathrm{Ker}(\nabla^{an})$ around $\lambda = 0$ (cf. [S], (2.21)).

We turn to our family $\mathscr{Y} \to \mathrm{Spec}W[[\lambda]]$. Let $K := \mathrm{Frac}(W)$ be the fractional field. The characteristic of K is zero by the assumption in Theorem 3.6. Put $\mathscr{Y}_K := \mathscr{Y} \times_{W[[\lambda]]} K[[\lambda]]$, $\mathscr{X}_K := \mathscr{X} \times_{W[[\lambda]]} K[[\lambda]]$ and $D_K := D \times_W K$. Let $(H^1_{\mathrm{dR}}(\mathscr{X}_K/K((\lambda))), \nabla)$ be the Gauss-Manin connection on $\mathrm{Spec}K((\lambda))$.

Proposition 3.7. *Let* $\nabla : H^1_{\mathrm{dR}}(X/A) \to Adt \otimes H^1_{\mathrm{dR}}(X/A)$ *be the Gauss-Manin connection. Then*

$$\begin{pmatrix}\nabla(\omega_{ij}) & \nabla(\eta_{ij})\end{pmatrix} = dt \otimes \begin{pmatrix}\omega_{ij} & \eta_{ij}\end{pmatrix} \begin{pmatrix} 0 & (1-i/N)(1-j/M) \\ (t-t^2)^{-1} & (1-i/N-j/M)(1-t)^{-1} \end{pmatrix}.$$

Proof. [A, Proposition 4.3]. □

Proposition 3.8. *Put Deligne's canonical extension*

$$H_{\lambda,K} := H^1(\mathscr{Y}_K, \Omega^\bullet_{\mathscr{Y}_K/K[[\lambda]]}(\log D_K)) \subset H^1_{\mathrm{dR}}(\mathscr{X}_K/K((\lambda))). \tag{3.9}$$

Then the $K[[\lambda]]$-basis is given as follows.

(1) *If $\lambda = t$, then*

$$H_{\lambda,K} = \bigoplus_{i,j} K[[\lambda]]\omega_{ij} \oplus K[[\lambda]]\eta_{ij}$$

where (i, j) runs over the pairs of integers such that $1 \le i \le N-1$ and $1 \le j \le M-1$.

(2) *If $\lambda = s = t^{-1}$, then*

$$H_{\lambda,K} = \bigoplus_{i,j} K[[\lambda]]\omega_{ij} \oplus K[[\lambda]]s\eta_{ij}.$$

(3) *If $\lambda = 1-t$, then*

$$H_{\lambda,K} = \bigoplus_{i,j} K[[\lambda]]\omega^*_{ij} \oplus K[[\lambda]]\eta^*_{ij}$$

where ω^*_{ij} and η^*_{ij} are as in Theorem 3.6 (3).

Proof. The condition (D1) is obvious by Theorem 3.4. It is straightforward from Proposition 3.7 that (D2) and (D3) are satisfied in each case. □

3.6 Proof of Theorem 3.6 (1), (2)

We prove Theorem 3.6 in case $\lambda = t$ and in case $\lambda = s = t^{-1}$. Write $\eta'_{ij} = \eta_{ij}$ in case $\lambda = t$ and $\eta'_{ij} = \lambda\eta_{ij}$ in case $\lambda = s$. Recall from Theorem 3.4 the fact that

$$H^1_{\mathrm{dR}}(\mathscr{X}/W((\lambda))) \cong W((\lambda)) \otimes_A H^1_{\mathrm{dR}}(X/A)$$

is a free $W((\lambda))$-module with basis $\{\omega_{ij}, \eta_{ij};\, 1 \le i \le N-1,\ 1 \le j \le M-1\}$. It follows from Proposition 3.8 that

$$H_\lambda \subset \bigoplus_{i,j} K[[\lambda]]\omega_{ij} + K[[\lambda]]\eta'_{ij} \subset H^1_{\mathrm{dR}}(\mathscr{X}_K/K((\lambda))).$$

Therefore

$$H_\lambda \subset \bigoplus_{i,j} W[[\lambda]]\omega_{ij} + W[[\lambda]]\eta'_{ij} \tag{3.10}$$

as $K[[\lambda]] \cap W((\lambda)) = W[[\lambda]]$. We show the opposite inclusion, namely

$$\omega_{ij},\ \eta'_{ij} \in H_\lambda. \tag{3.11}$$

We first show $\omega_{ij} \in H_\lambda$. There is an inetger $m \geq 0$ such that $\lambda^m \omega_{ij} \in \Gamma(\mathscr{Y}, \Omega^1_{\mathscr{Y}/W[[\lambda]]}(\log D))$. On the other hand $\omega_{ij} \in \Gamma(\mathscr{Y}_K, \Omega^1_{\mathscr{Y}_K/K[[\lambda]]}(\log D))$. Note that $\Omega^1_{\mathscr{Y}/W[[\lambda]]}(\log D)$ is a locally free $\mathscr{O}_{\mathscr{Y}}$-module (Proposition 3.5). Moreover one can check that the map a in the following diagram is injective.

$$\begin{array}{ccccccccc}
0 & \longrightarrow & \Omega^1_{\mathscr{Y}/W[[\lambda]]}(\log D) & \xrightarrow{\lambda^m} & \Omega^1_{\mathscr{Y}/W[[\lambda]]}(\log D) & \longrightarrow & \Omega^1_{\mathscr{Y}/W[[\lambda]]}(\log D)/\lambda^m & \longrightarrow & 0 \\
 & & \downarrow & & \downarrow & & \downarrow{\scriptstyle a} & & \\
0 & \longrightarrow & \Omega^1_{\mathscr{Y}_K/K[[\lambda]]}(\log D) & \xrightarrow{\lambda^m} & \Omega^1_{\mathscr{Y}_K/K[[\lambda]]}(\log D) & \longrightarrow & \Omega^1_{\mathscr{Y}_K/K[[\lambda]]}(\log D)/\lambda^m & \longrightarrow & 0.
\end{array}$$

Therefore we have $\omega_{ij} \in \Gamma(\mathscr{Y}, \Omega^1_{\mathscr{Y}/W[[\lambda]]}(\log D))$ by diagram chase.

Next we show $\eta'_{ij} \in H_\lambda$. Recall from (3.8) the cocycle which represents η_{ij},

$$(f_{ab,cd}) \times (\eta^{ab}_{ij}) \in \bigoplus \Gamma(U_{ab} \cap U_{cd}, \mathscr{O}_{\mathscr{X}}) \times \bigoplus \Gamma(U_{ab}, \Omega^1_{\mathscr{X}/W((\lambda))}).$$

Therefore it is enough to show

$$f_{ab,cd} \in \Gamma(U_{ab} \cap U_{cd}, \mathscr{O}_{\mathscr{Y}}), \quad \eta^{ab}_{ij} \in \Gamma(U_{ab}, \Omega^1_{\mathscr{Y}/W[[\lambda]]}(\log D)).$$

in case $\lambda = t$, and

$$s f_{ab,cd} \in \Gamma(U_{ab} \cap U_{cd}, \mathscr{O}_{\mathscr{Y}}), \quad s\eta^{ab}_{ij} \in \Gamma(U_{ab}, \Omega^1_{\mathscr{Y}/W[[\lambda]]}(\log D)).$$

in case $\lambda = s$. However we have shown that $\omega_{ij} \in \Gamma(\mathscr{Y}, \Omega^1_{\mathscr{Y}/W[[\lambda]]}(\log D))$. Thus this is immediate from the explicit descriptions in Sect. 3.4. This completes the proof of (3.11) and hence Theorem 3.6 (1), (2).

3.7 Proof of Theorem 3.6 (3)

Let $\lambda = 1 - t$. By the same discussion as in Sect. 3.6, one can show

$$H_\lambda \subset \bigoplus_{i,j} W[[\lambda]]\omega^*_{ij} + W[[\lambda]]\eta^*_{ij}, \tag{3.12}$$

and hence it is enough to show

$$\omega^*_{ij},\ \eta^*_{ij} \in H_\lambda. \tag{3.13}$$

If $i/N + j/M \geq 1$, then the same discussion as the proof of (3.11) works. Suppose that $i/N + j/M < 1$. The same discussion still works for showing $\omega^*_{ij} = \lambda\omega_{ij} \in H_\lambda$. The rest is to show that

$$\eta^*_{ij} = (1 - i/N - j/M)t\omega_{ij} - \eta_{ij} \in H_\lambda. \tag{3.14}$$

Recall from (3.8) the cocycle which represents η_{ij},

$$(f_{ab,cd}) \times (\eta^{ab}_{ij}) \in \bigoplus \Gamma(U_{ab} \cap U_{cd}, \mathscr{O}_{\mathscr{X}}) \times \bigoplus \Gamma(U_{ab}, \Omega^1_{\mathscr{X}/W((\lambda))}).$$

Hence

$$(f_{ab,cd}) \times ((\eta^*_{ij})^{ab}) := (f_{ab,cd}) \times ((1 - i/N - j/M)t\omega_{ij} - \eta^{ab}_{ij})$$

represents $\eta^*_{ij} \in H^1_{\mathrm{dR}}(\mathscr{X}/W((\lambda)))$. Each $f_{ab,cd}$ obviously belongs to $\Gamma(U_{ab,cd}, \mathscr{O}_{\mathscr{Y}})$. Therefore it is enough to show that each $(\eta^*_{ij})^{ab} \in \Gamma(U_{ab}, \Omega^1_{\mathscr{Y}/W[[\lambda]]}(\log D))$.

$$\begin{aligned}
(\eta^*_{ij})^{11} &= \left(1 - \frac{i}{N} - \frac{j}{M}\right) t\omega_{ij} + (j - M)(1 - t)z^{N-i}w^{M-j-1}dw,\\
(\eta^*_{ij})^{01} &= \left(1 - \frac{i}{N} - \frac{j}{M}\right) t\omega_{ij} + (j - M)(1 - t)x^{i-N}w^{M-j-1}dw,\\
(\eta^*_{ij})^{10} &= \left(1 - \frac{i}{N} - \frac{j}{M}\right) t\omega_{ij} - (1 - t)z^N \left(\frac{(j - M)t}{M} + \frac{(i - 2N)}{N}(1 - t)(1 - z^N)\right) \omega_{ij},\\
(\eta^*_{ij})^{00} &= \left(\frac{(j - M)t}{M} + \frac{i}{N}(1 - x^N)\right) \omega_{ij} + \left(1 - \frac{i}{N} - \frac{j}{M}\right) t\omega_{ij} - \frac{i}{N}x^N\omega_{ij}\\
&= \frac{i}{N}(1 - t - 2x^N)\omega_{ij}.
\end{aligned}$$

Multiplying $Nz^{N-i}y^j$ on an equality

$$t\frac{y^{-M}}{1 - z^N}\frac{dz}{z} = \frac{M}{N}(1 - t)(1 - z^N)\frac{dy}{y} + t\frac{dz}{z}$$

one has

$$\omega_{ij} = M(1 - t)(1 - z^N)z^{N-i}y^{j-1}dy + Ntz^{N-i-1}y^j dz.$$

The shows $\omega_{ij} \in \Gamma(U_{10}, \Omega^1_{\mathscr{Y}/W[[\lambda]]}(\log D))$. Similarly, using equalities

$$t\frac{x^{-N}}{w^M - 1}\frac{dw}{w} = -\frac{N}{M}(1 - t)(1 - w^N)\frac{dx}{x} - t\frac{dw}{w}$$

$$\frac{1}{1 - z^N}\frac{dz}{z} = (1 - (1 - t^{-1})(1 - w^M))\frac{dz}{z} - \frac{M}{N}\frac{dw}{w},$$

one has

$$\omega_{ij} \in \Gamma(U_{10} \cup U_{01} \cup U_{11}, \Omega^1_{\mathscr{Y}/W[[\lambda]]}(\log D)).$$

We thus have $(\eta^*_{ij})^{ab} \in \Gamma(U_{ab}, \Omega^1_{\mathscr{Y}/W[[\lambda]]}(\log D))$ for $(a, b) = (0, 1), (1, 0)$ and $(1, 1)$. The rest is the case $(a, b) = (0, 0)$, namely we show

$$\omega^*_{ij} = (1 - t)\omega_{ij},\ x^N \omega_{ij} \in \Gamma(U_{00}, \Omega^1_{\mathscr{Y}/W[[\lambda]]}(\log D)),$$

(note that ω_{ij} no longer belongs to $\Gamma(U_{00}, \Omega^1_{\mathscr{Y}/W[[\lambda]]}(\log D))$). However the former is already shown in (3.13), and the latter follows from an equality

$$x^N \omega_{ij} = -Mx^i y^{j-1} \frac{dy}{1 - y^M} = -Mt^{-1}(1 - x^N)x^i y^{j-1} dy.$$

This completes the proof of (3.14) and hence Theorem 3.6 (3).

4 Rigid Cohomology and Dwork's p-adic Hypergeometric Functions

Let $W = W(k)$ be the Witt ring of a perfect field k of characteristic $p > 0$. Let A be a faithfully flat W-algebra. We mean by a p^n-th Frobenius on A an endomorphism σ such that $\sigma(x) \equiv x^{p^n} \bmod pA$ for all $x \in A$ and that σ is compatible with the p^n-th Frobenius on W. We also write x^σ instead of $\sigma(x)$.

For a W-algebra A of finite type, we denote by $A^\dagger$ the weak completion (cf. [LP, p.5]). Namely if $A = W[T_1, \cdots, T_n]$, then $A^\dagger = W[T_1, \cdots, T_n]^\dagger$ is the ring of power series $\sum a_\alpha T^\alpha$ such that for some $r > 1$, $|a_\alpha| r^{|\alpha|} \to 0$ as $|\alpha| \to \infty$, and if $A = W[T_1, \cdots, T_n]/I$, then $A^\dagger = W[T_1, \cdots, T_n]^\dagger / IW[T_1, \cdots, T_n]^\dagger$.

4.1 Rigid Cohomology

Let $W = W(k)$ be the Witt ring of a perfect field k of characteristic $p > 0$. Put $K :=$ Frac(W) the fractional field. For a flat W-scheme V, we denote $V_K := V \times_W K$ and $V_k := V \times_W k$. For a flat W-ring A, we denote $A_K := A \otimes_W K$ and $A_k := A \otimes_W k$ as well.

Let A be a smooth W-algebra, and X a smooth A-scheme. Thanks to the theory due to Berthelot et al., the *rigid cohomology groups*

$$H^*_{\mathrm{rig}}(X_k/A_k)$$

are defined. We refer the book [LS] for the general theory of rigid cohomology. Here we list the required properties. Let $A^\dagger$ be the weak completion of A, and $A_K^\dagger := A^\dagger \otimes_W K$. We fix a p-th Frobenius σ on $A^\dagger$.

- $H^*_{\mathrm{rig}}(X_k/A_k)$ is a finitely generated $A_K^\dagger$-module.
- (Frobenius) The p-th Frobenius Φ on $H^*_{\mathrm{rig}}(X_k/A_k)$ (depending on σ) is defined in a natural way. This is a σ-linear endomorphism :

$$\Phi(f(t)x) = \sigma(f(t))\Phi(x), \quad \text{for } x \in H^*_{\mathrm{rig}}(X_k/A_k),\ f(t) \in A_K^\dagger.$$

- (Comparison with de Rham cohomology) There is the comparison isomorphism with the algebraic de Rham cohomology,

$$H^*_{\mathrm{rig}}(X_k/A_k) \cong H^*_{\mathrm{dR}}(X_K/A_K) \otimes_{A_K} A_K^\dagger.$$

- (Comparison with crystalline cohomology) Let α be a W-rational point of $\mathrm{Spec}A$ (i.e. a W-homomorphism $\alpha : A \to W$). Let $X_\alpha := X \times_{A,\alpha} W$ denote the fiber at α. There is the comparison isomorphism with the crystalline cohomology,

$$H^*_{\mathrm{rig}}(X_k/A_k) \otimes_{A_K^\dagger,\alpha} K \cong H^*_{\mathrm{crys}}(X_{\alpha,k}/W) \otimes \mathbb{Q}.$$

  If α satisfies $\sigma^{-1}(\mathfrak{m}_\alpha A_K^\dagger) = \mathfrak{m}_\alpha A_K^\dagger$ where $\mathfrak{m}_\alpha \subset A$ denotes the ideal defining α, then $\Phi_\alpha := \Phi \bmod \mathfrak{m}_\alpha A_K^\dagger$ agrees with the p-th Frobenius on the crystalline cohomology.

Let $\mathscr{Y}$ be a proper flat scheme over $W[[t]]$ which is smooth over $W((t))$. Let the central fiber D at $t = 0$. Put $\mathscr{X} := \mathscr{Y} \setminus D$. Suppose that D_{red} is a relative NCD over W and the multiplicities of components of D are prime to p. Then there is the comparison isomorphism with the log crystalline cohomology with log pole D ([Ka, Theorem (6.4)]),

$$H^*_{\mathrm{log-crys}}((\mathscr{Y}_{\overline{\mathbb{F}}_p}, D_{\overline{\mathbb{F}}_p})/(W[[t]], (t))) \cong H^*_{\mathrm{zar}}(\mathscr{Y}, \Omega^\bullet_{\mathscr{Y}/W[[t]]}(\log D)). \tag{4.1}$$

Fix a p-th Frobenius $\widehat{\sigma}$ on $W[[t]]$ given by $\widehat{\sigma}(t) = ct^p$ with some $c \in 1 + pW$. Then the $\widehat{\sigma}$-linear p-th Frobenius Φ_{crys} on the crystalline cohomology group is defined in a natural way. Let $A \to W((t))$ be a W-homomorphism, and $A^\dagger \to W((t))^\wedge$ the induced homomorphism where $W((t))^\wedge$ denotes the p-adic completion. Suppose that there is an isomorphism $\mathscr{X} \cong X \times_A W((t))$ and that σ and $\widehat{\sigma}$ are compatible under the map $A^\dagger \to W((t))^\wedge$. Then the Frobenius Φ agrees with Φ_{crys} under the natural map

$$\begin{array}{ccc} H^*_{\mathrm{log-crys}}((\mathscr{Y}_{\overline{\mathbb{F}}_p}, D_{\overline{\mathbb{F}}_p})/(W[[t]], (t))) & \longrightarrow & H^*_{\mathrm{dR}}(X_K/A_K) \otimes_A W((t))^\wedge \\ & & \big\downarrow \cong \\ & & H^*_{\mathrm{rig}}(X_k/A_k) \otimes_{A^\dagger} W((t))^\wedge. \end{array} \tag{4.2}$$

4.2 Explicit Description of Φ by Overconvergent Functions

Let

$$\begin{array}{ccc} X & \longrightarrow & Y \\ \downarrow & & \downarrow f \\ U = \mathrm{Sp}A & \longrightarrow & \mathbb{P}^1 \end{array}$$

be the fibration in Sect. 3.1. In what follows we work over the Witt ring $W = W(\overline{\mathbb{F}}_p)$ with $p > \max(N, M)$. Put $K := \mathrm{Frac}W$ the fractional field.

Let $c \in 1 + pW$ be fixed, and let $\sigma : A^\dagger \to A^\dagger$ be the p-th Frobenius given by $t^\sigma = ct^p$. Let

$$H^1_{\mathrm{rig}}(X_{\overline{\mathbb{F}}_p}/A_{\overline{\mathbb{F}}_p})$$

be the rigid cohomology group, and Φ the σ-linear p-th Frobenius. We shall give an explicit description of Φ.

Lemma 4.1. *Let* $\mathrm{Spec}W[[t]] \to \mathbb{P}^1$, *and put* $\mathscr{Y} := Y \times_{\mathbb{P}^1} \mathrm{Spec}W[[t]]$ *and* $D := f^{-1}(0) \subset \mathscr{Y}$ *the central fiber. Put* $\mathscr{X} := \mathscr{Y} \setminus D$. *Then the natural map*

$$H^1(\mathscr{Y}, \Omega^\bullet_{\mathscr{Y}/W[[t]]}(\log D)) \longrightarrow H^1(\mathscr{Y}, \Omega^\bullet_{\mathscr{Y}/W[[t]]}(\log D)) \otimes W((t)) \cong H^1_{\mathrm{dR}}(\mathscr{X}/W((t)))$$

is injective.

Proof. It is enough to show that $H^1(\mathscr{Y}, \Omega^\bullet_{\mathscr{Y}/W[[t]]}(\log D))$ is t-torsion free. There is an exact sequence

$$0 \longrightarrow \Gamma(\Omega^1_{\mathscr{Y}/W[[t]]}(\log D)) \longrightarrow H^1(\mathscr{Y}, \Omega^\bullet_{\mathscr{Y}/W[[t]]}(\log D)) \longrightarrow H^1(\mathscr{O}_{\mathscr{Y}}).$$

The 1st term is t-torsion free by Proposition 3.5. We show that $H^1(\mathscr{O}_{\mathscr{Y}})$ is a free $W[[t]]$-module. By [Ha, III,12.9], it is enough to show that $\dim_{\kappa(s)} H^1(Y_s, \mathscr{O}_{Y_s}) = (N-1)(M-1)$ for any point $s \in \mathrm{Spec}W[[t]]$ where $\kappa(s)$ is the residue field, and $Y_s := \mathscr{Y} \times_{W[[t]]} \kappa(s)$. If t is invertible in $\kappa(s)$, then Y_s is a smooth fiber, and then one has $\dim_{\kappa(s)} H^1(Y_s, \mathscr{O}_{Y_s}) = (N-1)(M-1)$ as $g(Y_s) = (N-1)(M-1)$. If $t = 0$ in $\kappa(s)$, then $Y_s = D_s := D \times_W \kappa(s)$ is a simple NCD, and then one can directly show that $\dim_{\kappa(s)} H^1(D_s, \mathscr{O}_{D_s}) = (N-1)(M-1)$. □

The Frobenius σ extends on the Frobenius on $K((t))$ as $\sigma(t) = ct^p$. Let Φ_{crys} be the crystalline Frobenius on

$$H^1_{\log-\mathrm{crys}}((\mathscr{Y}_{\overline{\mathbb{F}}_p}, D_{\overline{\mathbb{F}}_p})/(W[[t]], (t))) \cong H^1(\mathscr{Y}, \Omega^\bullet_{\mathscr{Y}/W[[t]]}(\log D)).$$

Let $1 \le i \le N-1$ and $1 \le j \le M-1$ be integers. Put $a_i := 1 - i/N$ and $b_j := 1 - j/M$. Let

$$F_{ij}(t) = F_{a_i,b_j}(t) = {}_2F_1\left(\begin{matrix} a_i, b_j \\ 1 \end{matrix}; t\right)$$

be the hypergeometric power series. It follows from Theorem 3.6 that the elements

$$\widetilde{\omega}_{i,j} := \frac{1}{F_{ij}(t)}\omega_{i,j}, \quad \widetilde{\eta}_{i,j} := -t(1-t)^{a_i+b_j}F'_{ij}(t)\omega_{ij} + (1-t)^{a_i+b_j-1}F_{ij}(t)\eta_{ij} \tag{4.3}$$

forms a $W[[t]]$-basis of

$$H^1(\mathscr{Y}, \Omega^\bullet_{\mathscr{Y}/W[[t]]}(\log D)) \cong \mathrm{Im}[H^1(\Omega^\bullet_{\mathscr{Y}/W[[t]]}(\log D)) \to H^1_{\mathrm{dR}}(\mathscr{X}/W((t)))]$$

where the isomorphism follows from Lemma 4.1.

Theorem 4.2. *Let $\tau_{ij}(t) \in \mathbb{Q}[[t]]$ be defined by*

$$\frac{d}{dt}\tau_{ij}(t) = \frac{1}{t}\left(1 - \frac{1}{(1-t)^{a_i+b_j}F_{ij}(t)^2}\right), \quad \tau_{ij}(0) = 0. \tag{4.4}$$

Let $\psi_p(z)$ be the p-adic digamma function introduced in [A, §2], and log *the Iwasawa logarithmic function (cf. Appendix A). Put*

$$\tau^{(\sigma)}_{ij}(t) = -2\gamma_p - \psi_p(a_i) - \psi_p(b_j) + p^{-1}\log(c) + \tau_{ij}(t) - p^{-1}\tau_{i'j'}(t^\sigma) \in K[[t]] \tag{4.5}$$

where $i' \in \{1, \ldots, N-1\}$ and $j' \in \{1, \ldots, M-1\}$ are integers such that $i'p \equiv i$ mod N and $j'p \equiv j$ mod M. Then

$$\Phi_{\mathrm{crys}}(\widetilde{\omega}_{i'j'}) = p\widetilde{\omega}_{ij} + p\tau^{(\sigma)}_{ij}(t)\widetilde{\eta}_{ij} \tag{4.6}$$

$$\Phi_{\mathrm{crys}}(\widetilde{\eta}_{i'j'}) = \widetilde{\eta}_{ij}. \tag{4.7}$$

Since Φ_{crys} agrees with Φ under the natural map (4.2), Theorem 4.2 implies the following.

Theorem 4.3. *Write $f'(t) = \frac{d}{dt}f(t)$ for a power series $f(t)$. We define*

$$A_{ij}(t) := \frac{F_{i'j'}(t^\sigma)}{F_{ij}(t)} - t(1-t)^{a_i+b_j}F'_{ij}(t)F_{i'j'}(t^\sigma)\tau^{(\sigma)}_{ij}(t) \tag{4.8}$$

$$C_{ij}(t) := (1-t)^{a_i+b_j-1}F_{ij}(t)F_{i'j'}(t^\sigma)\tau^{(\sigma)}_{ij}(t) \tag{4.9}$$

$$B_{ij}(t) := pt^\sigma(1-t^\sigma)\frac{F'_{i'j'}(t^\sigma)}{F_{i'j'}(t^\sigma)}A_{ij}(t) - t\frac{(1-t)^{a_i+b_j}}{(1-t^\sigma)^{a_{i'}+b_{j'}-1}}\frac{F'_{ij}(t)}{F_{i'j'}(t^\sigma)} \tag{4.10}$$

$$D_{ij}(t) := pt^\sigma(1-t^\sigma)\frac{F'_{i'j'}(t^\sigma)}{F_{i'j'}(t^\sigma)}C_{ij}(t) + \frac{(1-t)^{a_i+b_j-1}}{(1-t^\sigma)^{a_{i'}+b_{j'}-1}}\frac{F_{ij}(t)}{F_{i'j'}(t^\sigma)}. \tag{4.11}$$

Under the comparison isomorphism

$$H^1_{\mathrm{rig}}(X_{\overline{\mathbb{F}}_p}/A_{\overline{\mathbb{F}}_p}) \cong H^\bullet_{\mathrm{dR}}(X/A) \otimes_A A_K^\dagger,$$

the p-th Frobenius Φ *is described as follows,*

$$\left(\Phi(\omega_{i'j'})\ \Phi(\eta_{i'j'})\right) = \left(\omega_{ij}\ \eta_{ij}\right) \begin{pmatrix} pA_{ij} & B_{ij} \\ pC_{ij} & D_{ij} \end{pmatrix}.$$

Corollary 4.4. *All the power series* $\tau_{ij}^{(\sigma)}(t)$, $A_{ij}(t)$, $B_{ij}(t)$, $C_{ij}(t)$ *and* $D_{ij}(t)$ *lie in the ring* $W[[t]]$. *In particular,* $A_{ij}(t)$, $B_{ij}(t)$, $C_{ij}(t)$ *and* $D_{ij}(t)$ *lie in the ring* $A_K^\dagger \cap W[[t]] = A^\dagger \cap W[[t]]$.

Proof. Noticing that (4.3) forms a $W[[t]]$-basis, the fact that $\tau_{ij}^{(\sigma)}(t) \in W[[t]]$ is immediate from Theorem 4.2 (4.6) together with the fact that

$$\Phi_{\mathrm{crys}}(\Gamma(\Omega^1_{\mathscr{Y}/W[[t]]}(\log D)) \subset pH^1(\Omega^\bullet_{\mathscr{Y}/W[[t]]}(\log D)).$$

The others follows from this and the definition. □

Remark 4.5. *I don't know a direct proof of Corollary 4.4 (without p-adic cohomology).*

Remark 4.6. *Note that* $a_{i'} = (a_i)'$ *and* $b_{j'} = (b_j)'$ *(Dwork prime). In particular* $n_i := a_i - pa_{i'}$ *and* $m_j := b_j - pb_{j'}$ *are integers* ≤ 0. *We have*

$$\det \begin{pmatrix} pA_{ij} & B_{ij} \\ pC_{ij} & D_{ij} \end{pmatrix} = p\frac{(1-t)^{a_i+b_j-1}}{(1-t^\sigma)^{a_{i'}+b_{j'}-1}} = p\frac{(1-t)^{n_i+m_j-1}}{1-t^\sigma}\left(\frac{(1-t)^p}{1-t^\sigma}\right)^{a_{i'}+b_{j'}} \tag{4.12}$$

with

$$\left(\frac{(1-t)^p}{1-t^\sigma}\right)^{a_{i'}+b_{j'}} = \sum_{n=0}^{\infty} p^n \binom{a_{i'}+b_{j'}}{n} u(t)^n \in (W[t,(1-t)^{-1}]^\dagger)^\times \tag{4.13}$$

where we put $(1-t)^p/(1-t^\sigma) = 1 + pu(t)$. *In particular*

$$\det \begin{pmatrix} pA_{ij} & B_{ij} \\ pC_{ij} & D_{ij} \end{pmatrix}\Bigg|_{t=\alpha} = p \times (\mathit{unit})$$

for $\alpha \in W^\times \setminus (1+pW)$.

4.3 Proof of Theorem 4.2 (4.7)

For integers k, l with $N \nmid k$ and $M \nmid l$ which do not necessarily satisfy that $1 \leq k \leq N-1$ and $1 \leq l \leq M-1$, ω_{kl} denotes $\omega_{k_0 l_0}$ where $k_0 \in \{1, \ldots, N-1\}$ and

$l_0 \in \{1, \dots, M-1\}$ such that $k \equiv k_0 \bmod N$ and $l \equiv l_0 \bmod M$. We apply the same convention to symbols η_{kl}, $\tau_{kl}(t)$, a_k, b_l etc.

Let

$$\nabla : H^1(\mathscr{Y}, \Omega^\bullet_{\mathscr{Y}/W[[t]]}(\log D)) \longrightarrow \frac{dt}{t} \otimes H^1(\mathscr{Y}, \Omega^\bullet_{\mathscr{Y}/W[[t]]}(\log D))$$

be the Gauss-Manin connection. By Proposition 3.7 (or [A, Prop 4.3]),

$$\left(\nabla(\widetilde{\omega}_{i,j})\ \nabla(\widetilde{\eta}_{i,j})\right) = dt \otimes \left(\widetilde{\omega}_{i,j}\ \widetilde{\eta}_{i,j}\right) \begin{pmatrix} 0 & 0 \\ t^{-1}(1-t)^{-a_i-b_j} F_{ij}(t)^{-2} & 0 \end{pmatrix}. \tag{4.14}$$

Using this, one can show

$$\mathrm{Ker}(\nabla) = \bigoplus_{i,j} W\widetilde{\eta}_{ij}.$$

Since $\nabla\Phi_{\mathrm{crys}} = \Phi_{\mathrm{crys}}\nabla$, one has

$$\Phi_{\mathrm{crys}}(\widetilde{\eta}_{ij}) = \sum_{k,l} \alpha_{kl}\widetilde{\eta}_{kl} \tag{4.15}$$

with some constants $\alpha_{kl} \in W$. Let $i : D \to \mathscr{Y}$ be the embedding. Let h be the composition as follows

$$\begin{array}{ccccc} \mathrm{Ker}(\nabla) & & \overset{h}{\searrow} & & \\ \downarrow & & & & \\ H^1(\Omega^\bullet_{\mathscr{Y}/W[[t]]}(\log D)) & \longrightarrow & H^1(\mathscr{Y}, \mathscr{O}_{\mathscr{Y}}) & \xrightarrow[i^*]{} & H^1(D, \mathscr{O}_D). \end{array}$$

Recall from Sect. 3.1 that $D = f^{-1}(0)$ is a simple relative NCD, and the irreducible components are $\{D_{x=\zeta_1}, D_{y=\zeta_2} \mid \zeta_1 \in \mu_N, \zeta_2 \in \mu_M\}$ where $D_{x=\zeta_1} := \{x = \zeta_1\}$ and $D_{y=\zeta_2} := \{y = \zeta_2\}$ and $\mu_n := \{\zeta \in W \mid \zeta^n = 1\}$. Put $P(\zeta_1, \zeta_2) := D_{x=\zeta_1} \cap D_{y=\zeta_2}$ a single point, and $P := \{P(\zeta_1, \zeta_2)\}_{\zeta_1,\zeta_2} \subset D$. There is an exact sequence

$$\bigoplus_{\zeta_1} H^0(\mathscr{O}_{D_{x=\zeta_1}}) \oplus \bigoplus_{\zeta_2} H^0(\mathscr{O}_{D_{y=\zeta_2}}) \to \bigoplus_{\zeta_1,\zeta_2} H^0(\mathscr{O}_{P(\zeta_1,\zeta_2)}) \overset{\delta}{\to} H^1(\mathscr{O}_D) \to 0$$

arising from an exact sequence

$$0 \longrightarrow \mathscr{O}_D \overset{j}{\longrightarrow} \bigoplus_{\zeta_1} \mathscr{O}_{D_{x=\zeta_1}} \oplus \bigoplus_{\zeta_2} \mathscr{O}_{D_{y=\zeta_2}} \overset{u}{\longrightarrow} \bigoplus_{\zeta_1,\zeta_2} \mathscr{O}_{P(\zeta_1,\zeta_2)} \longrightarrow 0$$

where j is the pull-back and u is the map which sends $(f_{\zeta_1})_{\zeta_1} \times (g_{\zeta_2})_{\zeta_2}$ to $((g_{\zeta_2} - f_{\zeta_1})|_{P(\zeta_1,\zeta_2)})_{\zeta_1,\zeta_2}$.

Lemma 4.7. *Let*

$$e_{ij} := (\zeta_1^i \zeta_2^j)_{\zeta_1,\zeta_2} \in \bigoplus_{\zeta_1,\zeta_2} H^0(\mathscr{O}_{P(\zeta_1,\zeta_2)})$$

be an element for $i, j \in \mathbb{Z}$. *Then* $\delta(e_{ij}) = h(\widetilde{\eta}_{ij})$ *for* $i \in \{1, \dots, N-1\}$ *and* $j \in \{1, \dots, M-1\}$. *In particular* $h \otimes \mathbb{Q}$ *is bijective.*

Proof. Recall from (3.8) the cocycle $(f_{ab,cd}) \times (\eta_{ij}^{ab})$ where

$$f_{00,11} = f_{00,01} := x^i y^{j-M}, \quad f_{10,11} = f_{10,01} := (1-t)^2 z^{2N-i} y^{j-M} = (1-t)(1-y^M + z^N y^M) z^{N-i} y^{j-M},$$

$$f_{01,11} := 0, \quad f_{00,10} := x^i y^{j-M} - (1-t)^2 z^{2N-i} y^{j-M} = (1-x^N)(x^N y^M - 2x^N - y^M) z^{2N-i} y^j.$$

Note that $D \subset U_{00} \cup U_{11}$. We have $h(\widetilde{\eta}_{ij}) = [(x^i y^{j-M}|_D)]$ under the isomorphism

$$H^1(\mathscr{O}_D) \cong \operatorname{Coker}\left[\bigoplus_{(a,b)=(0,0),(1,1)} \Gamma(U_{ab}, \mathscr{O}_D) \xrightarrow{d} \Gamma(U_{00,11}, \mathscr{O}_D)\right]$$

where $U_{00,11} := U_{00} \cap U_{11}$ and $d(f_{00}, f_{11}) := f_{11} - f_{00}$. A diagram chase

$$\begin{array}{ccccc} & & \bigoplus_{\zeta_1} \Gamma(U_{ab}, \mathscr{O}_{D_{x=\zeta_1}}) \times \bigoplus_{\zeta_2} \Gamma(U_{ab}, \mathscr{O}_{D_{y=\zeta_2}}) & \xrightarrow{u} & \bigoplus_{\zeta_1,\zeta_2} \Gamma(\mathscr{O}_{P(\zeta_1,\zeta_2)}) \\ & & \Big\downarrow d & & \\ \Gamma(U_{00,11}, \mathscr{O}_D) & \longrightarrow & \bigoplus_{\zeta_1} \Gamma(U_{00,11}, \mathscr{O}_{D_{x=\zeta_1}}) \times \bigoplus_{\zeta_2} \Gamma(U_{00,11}, \mathscr{O}_{D_{y=\zeta_2}}) & & \end{array}$$

$$\begin{array}{ccccc} & & (0, \zeta_1^i w^{M-j}) \times (-\zeta_2^j x^i, 0) & \xrightarrow{u} & (\zeta_1^i \zeta_2^j) = e_{ij} \\ & & \Big\downarrow d & & \\ h(\widetilde{\eta}_{ij}) = (x^i y^{j-M}|_D) & \longrightarrow & (\zeta_1^i y^{j-M}, \zeta_2^j x^i) & & \end{array}$$

yields $\delta(e_{ij}) = h(\widetilde{\eta}_{ij})$. The last statement is an exercise of linear algebra. □

We turn to the proof of (4.7). Apply Φ_{crys} on the equality $h(\widetilde{\eta}_{ij}) = \delta(e_{ij})$ in Lemma 4.7. Since h and δ are compatible with respect to the action of Φ_{crys}, one has

$$h\Phi(\widetilde{\eta}_{ij}) = \sum_{k,l} \alpha_{kl} h(\widetilde{\eta}_{kl}) = \delta \Phi_{\mathrm{crys}}(e_{ij})$$

by (4.15). On the other hand

$$\Phi_{\mathrm{crys}}(e_{ij}) = (\zeta_1^{ip} \zeta_2^{jp})_{\zeta_1,\zeta_2} = e_{ip,jp} \in \bigoplus_{\zeta_1,\zeta_2} H^0(\mathscr{O}_{P(\zeta_1,\zeta_2)})$$

by definition of Φ_{crys}. Therefore one has

$$\sum_{k,l} \alpha_{kl} h(\widetilde{\eta}_{kl}) = \delta(e_{ip,jp}) = h(\widetilde{\eta}_{ip,jp}),$$

and hence $\alpha_{kl} = 1$ if $(k, l) = (ip, jp)$ in $\mathbb{Z}/N\mathbb{Z} \times \mathbb{Z}/M\mathbb{Z}$ and $= 0$ otherwise. This completes the proof of (4.7).

4.4 *Proof of Theorem 4.2 (4.6)*

For $(\zeta_1, \zeta_2) \in \mu_N \times \mu_M$, we denote by $[\zeta_1, \zeta_2]$ the automorphism of $\mathscr{Y}$ given by $(x, y, t) \mapsto (\zeta_1 x, \zeta_2 y, t)$. Since $[\zeta_1, \zeta_2]\Phi_{\mathrm{crys}} = \Phi_{\mathrm{crys}}[\zeta_1, \zeta_2]$, one has

$$\Phi_{\mathrm{crys}}(\widetilde{\omega}_{ij}) \in W[[t]]\widetilde{\omega}_{ip,jp} + W[[t]]\widetilde{\eta}_{ip,jp}.$$

One can further show that there is $g_{ij}(t) \in W[[t]]$ such that

$$\Phi_{\mathrm{crys}}(\widetilde{\omega}_{ij}) = p\omega_{ip,jp} + g_{ij}(t)\widetilde{\eta}_{ip,jp} \tag{4.16}$$

(this can be proved in the same way as the proof of [A, Proposition 4.7]). Thus our goal is to show $g_{ij}(t) = p\tau^{(\sigma)}_{ip,jp}(t)$. Apply ∇ on (4.16). It follows from (4.14) that we have

$$\begin{aligned}
\mathrm{LHS} &= \nabla\Phi_{\mathrm{crys}}(\widetilde{\omega}_{ij}) \\
&= \Phi_{\mathrm{crys}}\nabla(\widetilde{\omega}_{ij}) \\
&= \Phi_{\mathrm{crys}}\left((1-t)^{-a_i-b_j}F_{ij}(t)^{-2}\frac{dt}{t} \otimes \widetilde{\eta}_{ij}\right) \\
&= p(1-t^\sigma)^{-a_i-b_j}F_{ij}(t^\sigma)^{-2}\frac{dt}{t} \otimes \widetilde{\eta}_{ip,jp} \quad \text{(by Theorem 4.2 (4.7))}
\end{aligned}$$

and

$$\mathrm{RHS} = p(1-t)^{-a_{ip}-b_{jp}}F_{ip,jp}(t)^{-2}\frac{dt}{t} \otimes \widetilde{\eta}_{ip,jp} + g'_{ij}(t)dt \otimes \widetilde{\eta}_{ip,jp}.$$

Hence

$$g'_{ij}(t) = \frac{p}{t}\left[\frac{1}{(1-t^\sigma)^{a_i+b_j}F_{ij}(t^\sigma)^2} - \frac{1}{(1-t)^{a_{ip}+b_{jp}}F_{ip,jp}(t)^2}\right]$$

or equivalently

$$g_{ij}(t) = p(C_{ij} + \tau_{ip,jp}(t) - p^{-1}\tau_{ij}(t^\sigma)) \tag{4.17}$$

with C_{ij} a constant. The rest is to show

$$C_{ij} = -2\gamma_p - \psi_p(a_{ip}) - \psi_p(b_{jp}) + p^{-1}\log(c). \tag{4.18}$$

To do this, we recall from [A, §4.4] the *regulator formula*.

For $(\nu_1, \nu_2) \in \mu_N(K) \times \mu_M(K)$, let

$$\xi = \xi(\nu_1, \nu_2) = \left\{ \frac{x-1}{x-\nu_1}, \frac{y-1}{y-\nu_2} \right\} \in K_2(X) \tag{4.19}$$

be a K_2-symbol. According to [A, (4.26)], we have the 1-extension

$$0 \longrightarrow H^1(X/A)(2) \longrightarrow M_\xi(X/A) \longrightarrow A \longrightarrow 0$$

associated to ξ in the category of Fil-F-MIC(A) (see [AM, §2.6] for the details). Let $e_\xi \in \mathrm{Fil}^0 M_\xi(X/A)_{\mathrm{dR}}$ be the unique lifting of $1 \in A$. Let $E_k^{(ij)}(t) \in W[[t]]$ be defined by

$$e_\xi - \Phi_{\mathrm{crys}}(e_\xi) = -N^{-1}M^{-1}\sum_{i=1}^{N-1}\sum_{j=1}^{M-1}(1-\nu_1^{-i})(1-\nu_2^{-j})[E_1^{(ij)}(t)\widetilde{\omega}_{ij} + E_2^{(ij)}(t)\widetilde{\eta}_{ij}]. \tag{4.20}$$

Then one of the main results in [A] is

$$\frac{E_1^{(ij)}(t)}{F_{ij}(t)} = -\mathscr{F}^{(\sigma)}_{a_i,b_j}(t) \tag{4.21}$$

([A, Theorem 4.9]) where $\mathscr{F}^{(\sigma)}_{\underline{a}}(t)$ is the p-adic hypergeometric function of log type introduced in [A, §3].

We turn to the proof of (4.18). Apply ∇ on (4.20). Noticing that $\Phi_{\mathrm{crys}}\nabla = \nabla\Phi_{\mathrm{crys}}$ and

$$\nabla(e_\xi) = \mathrm{dlog}(\xi) = N^{-1}M^{-1}\sum_{i=1}^{N-1}\sum_{j=1}^{M-1}(1-\nu_1^{-i})(1-\nu_2^{-j})\frac{dt}{t}\omega_{ij},$$

one has a differential equation

$$t\frac{d}{dt}E_2^{(ij)}(t) + (1-t)^{-a_i-b_j}F_{ij}(t)^{-2}E_1^{(ij)}(t) = p^{-1}F_{i'j'}(t^\sigma)g_{i'j'}(t^\sigma)$$

by (4.14) and (4.16) where i', j' are integers such that $i' \in \{1, \ldots, N-1\}$ with $pi' \equiv i \bmod N$ and $j' \in \{1, \ldots, M-1\}$ with $pj' \equiv j \bmod M$. Substitute $t = 0$ in the above. We have

$$E_1^{(ij)}(0) = p^{-1}g_{i'j'}(0) = C_{i'j'}.$$

By (4.21),

$$E_1^{(ij)}(0) = -\mathscr{F}^{(\sigma)}_{a_{i'},b_{j'}}(0) = -2\gamma_p - \psi_p(a_{i'}) - \psi_p(b_{j'}) + p^{-1}\log(c),$$

and hence (4.18) as required. This completes the proof of Theorem 4.2 (4.6).

5 Computing Dwork's p-adic Hypergeometric Functions

In this section, we shall give an algorithm for computing special values of Dwork's p-adic hypergeometric functions whose bit complexity increases at most $O(n^4 \log^3 n)$ as $n \to \infty$.

5.1 p-adic Expansions of $A_{ij}(t)$, $B_{ij}(t)$, $C_{ij}(t)$, $D_{ij}(t)$

We keep the setting in Sect. 4.2. Recall Theorem 4.3,

$$\left(\Phi(\omega_{i'j'})\ \Phi(\eta_{i'j'})\right) = \left(\omega_{ij}\ \eta_{ij}\right)\begin{pmatrix} pA_{ij}(t) & B_{ij}(t) \\ pC_{ij}(t) & D_{ij}(t) \end{pmatrix}$$

with $A_{ij}(t), B_{ij}(t), C_{ij}(t), D_{ij}(t) \in A^\dagger \cap W[[t]]$ (Corollary 4.3). By [KT, Theorem 2.1], the overconvergent functions $A_{ij}(t), \ldots, D_{ij}(t)$ have 'nice' p-adic expansions, and this is the key fact in our algorithm. We here write down the necessary statement.

Theorem 5.1. *For an integer $n \geq 1$, define*

$$e_n := \max\{k \in \mathbb{Z}_{\geq 1} \mid \mathrm{ord}_p(p^k/k!) < n\}.$$

Then

$$pA_{ij}(t) \equiv p\frac{(\textit{polynomial of degree} \leq pe_n + p)}{(1-t^\sigma)(1-t)^{pe_n}} \mod p^n W[[t]],$$

$$B_{ij}(t) \equiv \frac{(\textit{polynomial of degree} \leq pe_n + 2p)}{(1-t^\sigma)(1-t)^{pe_n}} \mod p^n W[[t]],$$

$$pC_{ij}(t) \equiv p\frac{(\textit{polynomial of degree} \leq pe_n + p - 1)}{(1-t^\sigma)(1-t)^{pe_n}} \mod p^n W[[t]],$$

$$D_{ij}(t) \equiv \frac{(\textit{polynomial of degree} \leq pe_n + 2p - 1)}{(1-t^\sigma)(1-t)^{pe_n}} \mod p^n W[[t]].$$

Remark 5.2. *Since $p \neq 2$ by the assumption, $e_n < \infty$ for any $n \geq 1$. More precisely*

$$e_n \sim \frac{p-1}{p-2} n \quad \text{as } n \to \infty.$$

Remark 5.3. *The degrees $pe_n + p$ etc. are not optimal.*

We give a self-contained proof of Theorem 5.1 for the sake of the completeness.

Proof. (cf. [KT, p.11–13]). Let $\lambda = t$, $1 - t$ or t^{-1}. Let σ_λ be the p-th Frobenius on $W[[\lambda]]$ given by $\sigma_\lambda(\lambda) = \lambda^p$. Note that σ_λ induces the p-th Frobenius on $A^\dagger = W[t, (t - t^2)^{-1}]^\dagger$. Let Φ_λ denote the σ_λ-linear Frobenius on

$$H^1_{\mathrm{rig}}(X_{\overline{\mathbb{F}}_p}/A_{\overline{\mathbb{F}}_p}) \cong H^1_{\mathrm{dR}}(X/A) \otimes_A A^\dagger_K.$$

Let σ be the Frobenius given by $\sigma(t) = ct^p$ and Φ the σ-linear Frobenius as in Sect. 4.2. Then the relation with Φ_λ is given as follows ([EK, 6.1], [Ke1, 17.3.1]).

$$\Phi(x) - \Phi_\lambda(x) = \sum_{k=1}^{\infty} \frac{(\lambda^\sigma - \lambda^p)^k}{k!} \Phi_\lambda \partial_\lambda^k x, \quad x \in H^1_{\mathrm{dR}}(X/A) \otimes_A A^\dagger_K \tag{5.1}$$

where $\partial_\lambda := \nabla_{d/d\lambda}$. Let $\lambda = 1 - t$. Since $\lambda^\sigma - \lambda^p = pw(\lambda) \in pW[\lambda]$, (5.1) yields

$$\Phi(x) - \Phi_\lambda(x) = \sum_{k=1}^{\infty} \frac{p^k}{k!} w(t)^k \Phi_\lambda \partial_\lambda^k x.$$

Note that $\Phi_\lambda(H_\lambda) \subset H_\lambda$ while $\Phi(H_\lambda) \not\subset H_\lambda$. Since $\partial_\lambda^k(H_\lambda) \subset \lambda^{-k} H_\lambda$, one has $\Phi_\lambda \partial_\lambda^k(H_\lambda) \subset \lambda^{-kp} H_\lambda$ for all $k \geq 0$, and hence

$$\Phi(x) \in \sum_{k=0}^{\infty} \frac{p^k}{k!} \lambda^{-kp} H_\lambda.$$

We thus have

$$\Phi(x) \in \lambda^{-pe_n} H_\lambda + p^n \widehat{H}_\lambda, \quad \forall n \geq 1, \ \forall x \in (H^1_{\mathrm{dR}}(X/A) \otimes_A A^\dagger_K) \cap H_\lambda \tag{5.2}$$

if $\lambda = 1 - t$ where $\widehat{H}_\lambda$ is the p-adic completion of $H_\lambda \otimes_{W[[\lambda]]} W((\lambda))$. Let $\lambda = t^{-1}$. In this case, since $\sigma_\lambda(t) = t^p$, the Frobenius Φ_λ acts on the $W[[\lambda]]$-lattice H_λ. Hence

$$\Phi_\lambda(H_\lambda) \subset H_\lambda, \quad \lambda = t^{-1}. \tag{5.3}$$

Let us prove Theorem 5.1. Since $A_{ij}, B_{ij}, C_{ij}, D_{ij} \in A^\dagger \cap W[[t]]$, one can write

$$pA_{ij}(t) \bmod p^n W[[t]] = \frac{pF_{ij}^A(t)}{(1-t^\sigma)(1-t)^{d_{ij}^A}},$$

$$B_{ij}(t) \bmod p^n W[[t]] = \frac{F_{ij}^B(t)}{(1-t^\sigma)(1-t)^{d_{ij}^B}},$$

$$pC_{ij}(t) \bmod p^n W[[t]] = \frac{pF_{ij}^C(t)}{(1-t^\sigma)(1-t)^{d_{ij}^C}},$$

$$D_{ij}(t) \bmod p^n W[[t]] = \frac{F_{ij}^D(t)}{(1-t^\sigma)(1-t)^{d_{ij}^D}},$$

in $W/p^nW[[t]]$ with $F_{ij}^A(t),\ F_{ij}^B(t), \ldots \in W/p^nW[t]$ polynomials and $d_{ij}^A,\ d_{ij}^B, \ldots \in \mathbb{Z}_{\geq 0}$. Let $\lambda = t^{-1}$. Then H_λ is a free $W[[t]]$-module with basis $\{\omega_{ij}, \lambda\eta_{ij}\}$ (Theorem 3.6 (2)). Therefore it follows from (5.3) that the entries of the 2×2-matrix in below lie in $W[[\lambda]]$,

$$\left(\Phi(\omega_{i'j'})\ \Phi(\lambda\eta_{i'j'})\right) = (\omega_{ij}\ \lambda\eta_{ij}) \begin{pmatrix} pA_{ij} & \lambda^\sigma B_{ij} \\ p\lambda^{-1}C_{ij} & \lambda^\sigma\lambda^{-1}D_{ij} \end{pmatrix}.$$

This implies

$$\begin{cases} \deg(pF_{ij}^A) \leq d_{ij}^A + p \\ \deg(F_{ij}^B) \leq d_{ij}^B + 2p \\ \deg(pF_{ij}^C) \leq d_{ij}^C + p - 1 \\ \deg(F_{ij}^D) \leq d_{ij}^D + 2p - 1. \end{cases} \tag{5.4}$$

Next we give upper bounds of $d_{ij}^A, d_{ij}^B, d_{ij}^C$ and d_{ij}^D. Let $\lambda = 1 - t$ and let $\omega_{ij}^*, \eta_{ij}^*$ be the basis of H_λ in Theorem 3.6 (3). Let

$$\left(\Phi(\omega_{i'j'}^*)\ \Phi(\eta_{i'j'}^*)\right) = (\omega_{ij}^*\ \eta_{ij}^*) \begin{pmatrix} pA_{ij}^* & B_{ij}^* \\ pC_{ij}^* & D_{ij}^* \end{pmatrix}.$$

It follows from (5.2) that we have

$$(1-t)^{pe_n}pA_{ij}^*,\ (1-t)^{pe_n}B_{ij}^*,\ (1-t)^{pe_n}pC_{ij}^*,\ (1-t)^{pe_n}D_{ij}^* \in W[[\lambda]] + p^nW((\lambda))^\wedge. \tag{5.5}$$

If $i/N + j/M \geq 1$, then $(\omega_{ij}^*, \eta_{ij}^*) = (\omega_{ij}, \eta_{ij})$, and if $i/N + j/M > 1$, then

$$(\omega_{ij}^*\ \eta_{ij}^*) = (\omega_{ij}\ \eta_{ij}) \begin{pmatrix} 1-t & lt \\ 0 & -1 \end{pmatrix}$$

where $l := 1 - i/N - j/M$. Therefore if $i/N + j/M \geq 1$ and $i'/N + j'/M \geq 1$, then

$$\begin{pmatrix} pA_{ij} & B_{ij} \\ pC_{ij} & D_{ij} \end{pmatrix} = \begin{pmatrix} pA^*_{ij} & B^*_{ij} \\ pC^*_{ij} & D^*_{ij} \end{pmatrix}.$$

By (5.5), we have $d^A_{ij}, d^B_{ij}, d^C_{ij}, d^D_{ij} \leq pe_n$. If $i/N + j/M < 1$ and $i'/N + j'/M \geq 1$, then

$$\begin{pmatrix} pA_{ij} & B_{ij} \\ pC_{ij} & D_{ij} \end{pmatrix} = \begin{pmatrix} 1-t & lt \\ 0 & -1 \end{pmatrix} \begin{pmatrix} pA^*_{ij} & B^*_{ij} \\ pC^*_{ij} & D^*_{ij} \end{pmatrix}.$$

By (5.5), we have $d^A_{ij} \leq pe_n - 1$ and $d^B_{ij}, d^C_{ij}, d^D_{ij} \leq pe_n$. If $i/N + j/M \geq 1$ and $i'/N + j'/M < 1$, then

$$\begin{aligned} \begin{pmatrix} pA_{ij} & B_{ij} \\ pC_{ij} & D_{ij} \end{pmatrix} &= \begin{pmatrix} pA^*_{ij} & B^*_{ij} \\ pC^*_{ij} & D^*_{ij} \end{pmatrix} \begin{pmatrix} 1-t^\sigma & lt^\sigma \\ 0 & -1 \end{pmatrix}^{-1} \\ &= \frac{1}{t^\sigma - 1} \begin{pmatrix} -pA^*_{ij} & (1-t^\sigma)B^*_{ij} - lt^\sigma pA^*_{ij} \\ -pC^*_{ij} & (1-t^\sigma)D^*_{ij} - lt^\sigma pC^*_{ij} \end{pmatrix}. \end{aligned}$$

By (5.5), we have $d^A_{ij}, d^B_{ij}, d^C_{ij}, d^D_{ij} \leq pe_n$. If $i/N + j/M < 1$ and $i'/N + j'/M < 1$, then

$$\begin{aligned} \begin{pmatrix} pA_{ij} & B_{ij} \\ pC_{ij} & D_{ij} \end{pmatrix} &= \begin{pmatrix} 1-t & lt \\ 0 & -1 \end{pmatrix} \begin{pmatrix} pA^*_{ij} & B^*_{ij} \\ pC^*_{ij} & D^*_{ij} \end{pmatrix} \begin{pmatrix} 1-t^\sigma & lt^\sigma \\ 0 & -1 \end{pmatrix}^{-1} \\ &= \frac{1}{t^\sigma - 1} \begin{pmatrix} (t-1)pA^*_{ij} - ltpC^*_{ij} & \cdots \\ pC^*_{ij} & (t^\sigma - 1)D^*_{ij} + lt^\sigma pC^*_{ij} \end{pmatrix}. \end{aligned}$$

By (5.5), we have $d^A_{ij}, d^B_{ij}, d^C_{ij}, d^D_{ij} \leq pe_n$. In any case one has

$$d^A_{ij}, d^B_{ij}, d^C_{ij}, d^D_{ij} \leq pe_n. \tag{5.6}$$

Theorem 5.1 follows from (5.4) and (5.6). □

5.2 Algorithm for Computing Dwork's p-adic Hypergeometric Functions

For $a, b \in \mathbb{Q}$, let

$$F_{ab}(t) = {}_2F_1\left({a, b \atop 1} ; t \right)$$

be the hypergeometric power series. We give an algorithm for computing the special values

$$\mathscr{F}_{ab}^{\mathrm{Dw},\sigma}(t) := \frac{F_{ab}(t)}{F_{a'b'}(t^\sigma)}, \quad \frac{F'_{ab}(t)}{F_{ab}(t)} \tag{5.7}$$

at $\alpha \in W^\times \setminus (1 + pW)$ modulo p^n.

Notation Let $N, M \geq 2$ be integers, and $p > \max(N, M)$ a prime. Let $W = W(\overline{\mathbb{F}}_p)$. Let $a, b \in \mathbb{Q}$ satisfy that $a \in \frac{1}{N}\mathbb{Z}$ and $b \in \frac{1}{M}\mathbb{Z}$ and $0 < a, b < 1$. Let a' denote the Dwork prime (see Sect. 2.1). Let $c \in 1 + pW$, and let $\sigma : W[[t]] \to W[[t]]$ be the p-th Frobenius given by $\sigma(t) = ct^p$. Following the notation in (4.5) and Theorem 4.3, we define

$$\frac{d}{dt}\tau_{ab}(t) = \frac{1}{t}\left(1 - \frac{1}{(1-t)^{a+b}F_{ab}(t)^2}\right), \quad \tau_{ab}(0) = 0,$$

$$\tau_{ab}^{(\sigma)}(t) = -2\gamma_p - \psi_p(a) - \psi_p(b) + p^{-1}\log(c) + \tau_{ab}(t) - p^{-1}\tau_{a'b'}(t^\sigma) \in W[[t]],$$

and

$$\begin{aligned}
A_{ab}(t) &:= \frac{F_{a'b'}(t^\sigma)}{F_{ab}(t)} - t(1-t)^{a+b}F'_{ab}(t)F_{a'b'}(t^\sigma)\tau_{ab}^{(\sigma)}(t)\\
C_{ab}(t) &:= (1-t)^{a+b-1}F_{ab}(t)F_{a'b'}(t^\sigma)\tau_{ab}^{(\sigma)}(t)\\
B_{ab}(t) &:= pt^\sigma(1-t^\sigma)\frac{F'_{a'b'}(t^\sigma)}{F_{a'b'}(t^\sigma)}A_{ab}(t) - t\frac{(1-t)^{a+b}}{(1-t^\sigma)^{a'+b'-1}}\frac{F'_{ab}(t)}{F_{a'b'}(t^\sigma)}\\
D_{ab}(t) &:= pt^\sigma(1-t^\sigma)\frac{F'_{a'b'}(t^\sigma)}{F_{a'b'}(t^\sigma)}C_{ab}(t) + \frac{(1-t)^{a+b-1}}{(1-t^\sigma)^{a'+b'-1}}\frac{F_{ab}(t)}{F_{a'b'}(t^\sigma)}.
\end{aligned}$$

Let $a^{(k)}$ denote the k-th Dwork prime. Put

$$F^{(k)}(t) := F_{a^{(k)}b^{(k)}}(t), \quad A_\sigma^{(k)}(t) := A_{a^{(k)}b^{(k)}}(t), \ldots, D_\sigma^{(k)}(t) := D_{a^{(k)}b^{(k)}}(t),$$

$$\mathscr{D}F^{(k)}(t) := \frac{(F^{(k)}(t))'}{F^{(k)}(t)}.$$

for $k \geq 0$. Note that $F^{(k)}(t)$ and $\mathscr{D}F^{(k)}(t)$ do not depend on σ. We put

$$E_\sigma^{(k)}(t) := \frac{(1-t)^{a^{(k)}+b^{(k)}}}{(1-t^\sigma)^{a^{(k+1)}+b^{(k+1)}}} = (1-t)^{m_k}\left(\frac{(1-t)^p}{1-t^\sigma}\right)^{a^{(k+1)}+b^{(k+1)}}$$

where $m_k := a^{(k)} - pa^{(k+1)} + b^{(k)} - pb^{(k+1)} \in \mathbb{Z}_{\leq 0}$. Note $E_\sigma^{(k)}(t) \in (A^\dagger)^\times$. We have

$$D_\sigma^{(k)}(t) = pt^\sigma(1-t^\sigma)C_\sigma^{(k)}(t)\mathscr{D}F^{(k+1)}(t^\sigma) + \frac{1-t^\sigma}{1-t}E_\sigma^{(k)}(t)\mathscr{F}_\sigma^{\mathrm{Dw},(k)}(t) \tag{5.8}$$

and

$$\overbrace{\begin{pmatrix} pA^{(k)}_\sigma(t) & B^{(k)}_\sigma(t) \\ pC^{(k)}_\sigma(t) & D^{(k)}_\sigma(t) \end{pmatrix}}^{H^{(k)}_\sigma(t)} \begin{pmatrix} t^\sigma(1-t^\sigma)\mathscr{D}F^{(k+1)}(t^\sigma) \\ -1 \end{pmatrix}$$
$$= \frac{1-t^\sigma}{1-t} E^{(k)}_\sigma(t)\mathscr{F}^{\mathrm{Dw},\sigma}_{a^{(k)}b^{(k)}}(t) \begin{pmatrix} t(1-t)\mathscr{D}F^{(k)}(t) \\ -1 \end{pmatrix}. \tag{5.9}$$

Algorithm

Let $m \geq 1$ be the smallest integer such that $(a^{(m)}, b^{(m)}) = (a, b)$. Let $\alpha \in W^\times \setminus (1 + pW)$ be an arbitrary element satisfying

$$[F^{(k)}(t)]_{<p}|_{t=\alpha} \not\equiv 0 \mod pW, \quad 0 \leq \forall k \leq m-1.$$

Let $\sigma(t) = ct^p$ with $c \in 1 + pW$ arbitrary. The algorithm for computing (5.7) is the following.

Step 1 Let $\beta \in W^\times \setminus (1 + pW)$ satisfy

$$[F^{(k)}(t)]_{<p}|_{t=\beta} \not\equiv 0 \mod pW, \quad 0 \leq \forall k \leq m-1.$$

In **Step 3**, we shall take $\beta = t^\sigma|_{t=\alpha} = c\alpha^p$. Let $\sigma_\beta(t) = \beta^{1-p}t^p$ so that we have $t^{\sigma_\beta}|_{t=\beta} = \beta$. Then we compute the special values

$$pA^{(k)}_{\sigma_\beta}(\beta),\ pC^{(k)}_{\sigma_\beta}(\beta),\ B^{(k)}_{\sigma_\beta}(\beta),\ D^{(k)}_{\sigma_\beta}(\beta) \mod p^n W$$

for each $k = 0, 1, \ldots, m-1$. One can do it in the following way. Compute the power series

$$(1 - t^{\sigma_\beta})(1-t)^{pe_n} A^{(k)}_{\sigma_\beta}(t)$$

until the degree $pe_n + p$, say $F^A(t)$. Then it follows from Theorem 5.1 that

$$pA^{(k)}_{\sigma_\beta}(t) \equiv \frac{pF^A(t)}{(1-t^{\sigma_\beta})(1-t)^{pe_n}} \mod p^n W[[t]]$$

and hence

$$pA^{(k)}_{\sigma_\beta}(\beta) \equiv \frac{pF^A(\beta)}{(1-\beta)^{pe_n+1}} \mod p^n W.$$

The other values are obtained in the same way.

Step 2 We mean $H^{(l)}_{\sigma_\beta} = H^{(l_0)}_{\sigma_\beta}$ for arbitrary $l \in \mathbb{Z}$ where $l_0 \in \{0, 1, \ldots, m-1\}$ such that $l \equiv l_0 \mod m$.

Compute an eigenvector $\mathbf{u}_\beta$ of a 2×2-matrix

$$H^{(k-m)}_{\sigma_\beta}(\beta) \cdots H^{(k-2)}_{\sigma_\beta}(\beta) H^{(k-1)}_{\sigma_\beta}(\beta)$$

whose eigenvalue is a unit. This is unique up to scalar. Indeed, it follows from (5.9) that the vector

$$\begin{pmatrix} \beta(1-\beta)\mathscr{D}F^{(k)}(\beta) \\ -1 \end{pmatrix} \tag{5.10}$$

is an eigenvector of $H^{(1)}_{\sigma_\beta}(\beta) \cdots H^{(m)}_{\sigma_\beta}(\beta)$ whose eigenvalue is

$$\prod_{k=0}^{m-1} E^{(k)}_{\sigma_\beta}(\beta)\mathscr{F}^{\mathrm{Dw},\sigma_\beta}_{a^{(k)}b^{(k)}}(\beta) \in W^\times. \tag{5.11}$$

The other eigenvalue is not a unit as $\det(H^{(0)}_{\sigma_\beta}(\beta) \cdots H^{(m-1)}_{\sigma_\beta}(\beta)) = p^m \times$(unit) by Remark 4.12 (actually the determinant is equal to p^m). Therefore (5.10) is characterized as the eigenvector with the unique eigenvalue which is a unit. We thus have the special value

$$\mathscr{D}F^{(k)}(\beta) = \left.\frac{F'_{a^{(k)}b^{(k)}}(t)}{F_{a^{(k)}b^{(k)}}(t)}\right|_{t=\beta} \mod p^n W$$

for each k.

Step 3 Let $\sigma(t) = ct^p$ be as in the beginning. Take $\beta = t^\sigma|_{t=\alpha} = c\alpha^p$ in **Step 2**. We have

$$\mathscr{D}F^{(1)}(\beta) = \mathscr{D}F^{(1)}(t^\sigma)|_{t=\alpha} \mod p^n W.$$

Compute the special values

$$pC^{(0)}_\sigma(\alpha),\ D^{(0)}_\sigma(\alpha),\ E^{(0)}_\sigma(\alpha) \mod p^n W$$

according to **Step 1**, and

$$E^{(0)}_\sigma(\alpha) \mod p^n W$$

utilizing the expansion

$$\left(\frac{(1-t)^p}{1-t^\sigma}\right)^{a+b} = \sum_{n=0}^{\infty} p^n \binom{a+b}{n} u(t)^n, \quad \frac{(1-t)^p}{1-t^\sigma} = 1 + pu(t).$$

Substitute $t = \alpha$ in (5.8). Then we have the special value

$$\mathscr{F}^{\mathrm{Dw},\sigma}_{ab}(\alpha) = \left.\frac{F_{ab}(t)}{F_{a'b'}(t^\sigma)}\right|_{t=\alpha} \mod p^n W$$

as $E^{(0)}_\sigma(\alpha) \in W^\times$.

Example 5.4. *Let $p = 7$ and $a = b = \frac{1}{3}$ and $\sigma(t) = t^p$. Then $a' = a$ and $b' = b$, and $[F_{\frac{1}{3},\frac{1}{3}}(t)]_{<p} \equiv t^2 - 3t + 1 \bmod p$, so that the values*

$$\mathscr{F}^{\mathrm{Dw}}_{\frac{1}{3},\frac{1}{3}}(t)|_{t=\alpha}, \quad \mathscr{D}F_{\frac{1}{3},\frac{1}{3}}(t)|_{t=\alpha} = \left.\frac{F'_{\frac{1}{3},\frac{1}{3}}(t)}{F_{\frac{1}{3},\frac{1}{3}}(t)}\right|_{t=\alpha}$$

are defined for any $\alpha \in \mathbb{Z}_p$. If $\alpha^{p-1} = 1$ and $\alpha \neq 1$, then $\mathscr{F}^{\mathrm{Dw}}_{\frac{1}{3},\frac{1}{3}}(t)|_{t=\alpha}$ is an algebraic integer which is a Frobenius eigenvalue of the curve $(1 - x^3)(1 - y^3) = \alpha$ defined over $\mathbb{F}_p$. For example, running our algorithm one has

$$\mathscr{F}^{\mathrm{Dw}}_{\frac{1}{3},\frac{1}{3}}(t)|_{t=-1} \equiv 22143577275619763 \mod 7^{20}$$

in a few tens of seconds (the correct value is $\frac{5+\sqrt{-3}}{2}$). One can compute any other values which are not necessarily algebraic numbers. For example

$$\begin{aligned}
\mathscr{F}^{\mathrm{Dw}}_{\frac{1}{3},\frac{1}{3}}(t)|_{t=2} &\equiv 61938722 & \mathscr{D}F_{\frac{1}{3},\frac{1}{3}}(t)|_{t=2} &\equiv 204502290\\
\mathscr{F}^{\mathrm{Dw}}_{\frac{1}{3},\frac{1}{3}}(t)|_{t=3} &\equiv 47654055 & \mathscr{D}F_{\frac{1}{3},\frac{1}{3}}(t)|_{t=3} &\equiv 98505172\\
\mathscr{F}^{\mathrm{Dw}}_{\frac{1}{3},\frac{1}{3}}(t)|_{t=4} &\equiv 225738014 & \mathscr{D}F_{\frac{1}{3},\frac{1}{3}}(t)|_{t=4} &\equiv 275371181\\
\mathscr{F}^{\mathrm{Dw}}_{\frac{1}{3},\frac{1}{3}}(t)|_{t=5} &\equiv 147819186 & \mathscr{D}F_{\frac{1}{3},\frac{1}{3}}(t)|_{t=5} &\equiv 52924949\\
\mathscr{F}^{\mathrm{Dw}}_{\frac{1}{3},\frac{1}{3}}(t)|_{t=6} &\equiv 43011659 & \mathscr{D}F_{\frac{1}{3},\frac{1}{3}}(t)|_{t=6} &\equiv 172929798
\end{aligned}$$

modulo 7^{10}.

5.3 Bit Complexity

We give an upper estimate of the bit complexity of the algorithm displayed in Sect. 5.2.

We review the notion of the bit complexity. A general reference is the text book [BZ]. The bit of a natural number N is defined to be the number of digits of N in binary notation, so it is at most $\log_2(N + 1)$. The bit of $N!$ is at most $\log_2(N! + 1) \sim (\log 2)^{-1} N \log N$ (Stirling). The bit complexity of an algorithm is defined to be the number of single operations to complete the algorithm. The bit complexity of (1-digit)$\pm$(1-digit) or (1-digit)$\times$(1-digit) is 1 by definition. We denote by $M(n, m)$ the bit complexity of multiplication (n-digits) $\times$ (m-digits). We write $M(n) = M(n, n)$. By the naive multiplication algorithm, $M(n, m)$ is $O(nm)$, which means that there is a constant C such that $M(n, m) \leq Cnm$ when $n, m \to \infty$. We sum up the basic results.

- For integers $i, j \geq 0$, the bit complexity of $i \pm j$ is $O(\max(\log i, \log j))$.
- The bit complexity of $i \cdot j$ is $M(\log i, \log j)$ (which is at most $O(\log i \log j)$).
- The bit complexity for computing the remainder ($i \bmod j$) is $M(\log i, \log j)$.

Let a be a fixed rational number. Then the bit complexity of $(a)_i$ is at most

$$\sum_{n=1}^{i} M(n \log n, \log n) \leq O(i^2 (\log i)^2) \tag{5.12}$$

by computing it in the following way

$$(a)_i = (a + i - 1) \cdot (a)_{i-1}, \quad (a)_{i-1} = (a + i - 2) \cdot (a)_{i-2}, \ldots$$

Let a_i, b_j be rational numbers whose denominators and numerators are less than k. Let $f(t) = \sum_{i=0}^{n} a_i t^i$ and $g(t) = \sum_{j=0}^{n} b_j t^j$. Then the bit complexity of computing $f(t) \pm g(t)$ is $O(n \log k)$ The bit complexity of computing $f(t)g(t)$

$$n^2 M(\log k) + O(n^2 \log(kn)) < O(n^2(\log n + (\log k)^2)) \tag{5.13}$$

on noticing that the coefficients of $f(t)g(t)$ are ratios of integers at most nk.

Let us see the bit complexity of our algorithm in Sect. 5.2. Fix p, a, b, c and α. We need to compute the power series

$$\tau_{ab}^{(\sigma)}(t), \quad A_\sigma^{(k)}(t), \ldots, D_\sigma^{(k)}(t), \quad E_\sigma^{(k)}(t) \tag{5.14}$$

until the degree $pe_n + 2p \sim p(p-1)/(p-2)n$. First of all, the bit complexities of computing the constants

$$\gamma_p + \psi_p(a^{(k)}), \quad \gamma_p + \psi_p(b^{(k)}), \quad \log c$$

modulo p^n are small (cf. Appendix A), so that we can ignore them. Moreover the power series $E_\sigma^{(k)}(t)$ is simple, so we can also ignore the bit complexity of computing it.

We observe the bit complexity of computing $\tau_{ab}^{(\sigma)}(t)$. We work in a ring

$$K[t]/(t^{pe_n+2p+1}), \quad K := \operatorname{Frac} W.$$

We begin with the truncated polynomials

$$F_{ab}(t) \in K[t]/(t^{pe_n+2p+1}).$$

By (5.12), the bit complexities of computing all the coefficients are at most

$$\sum_{i=0}^{pe_n+2p} O(i^2(\log i)^2) < O(n^3(\log n)^2).$$

Next we need compute

$$\frac{1}{F_{ab}(t)} = (1+f)(1+f^2)\cdots(1+f^{2^d}) \in K[t]/(t^{pe_n+2p+1}), \quad f := 1 - F_{ab}(t) \tag{5.15}$$

where $d := \lfloor \log_2(pe_n + 2p) \rfloor + 1 \sim \log_2 n$. The denominators and numerators of the coefficients of f^k for $k \leq pe_n + 2p$ are at most

$$\sum_{i_1+\cdots+i_k=l,\, i_r\geq 1} (i_1!\cdots i_k!)^2 < (l!)^2\binom{l-1}{k-1} < (l!)^2 l^{pe_n+2p} < (n!)^2 n^{cn} \tag{5.16}$$

with $c > 0$ a constant. Hence the bit complexities of computing $f^2, \ldots, f^{2^d}$ are at most

$$O(n^2(\log(n!^2 n^{cn}))^2) = O(n^4(\log n)^2)$$

(5.13), and hence the bit complexity of computing (5.15) is

$$O(dn^4(\log n)^2)) = O(n^4(\log n)^3).$$

Summing up the above, the bit complexity of computing $\tau_{ab}^{(\sigma)}(t)$ is $O(n^4(\log n)^3)$.

The power series of $A_\sigma^{(k)}(t), \ldots, D_\sigma^{(k)}(t)$ are obtained by applying standard arithmetic operations (addition, subtraction and multiplication) on polynomials whose coefficients are ratios of integers at most (5.16). Therefore the bit complexities do not exceed $O(n^4(\log n)^3)$. All the algorithms in **Step 1**,..., **Step 3** are standard arithmetic operations on the coefficients in the polynomials (5.14). One concludes that the total bit complexity of the algorithm in Sect. 5.2 is $O(n^4(\log n)^3)$.

6 Appendix A: *p*-adic Polygamma Functions

We give a brief review of p-adic polygamma functions introduced in [A, §2].

Let $r \in \mathbb{Z}$ be an integer. For $z \in \mathbb{Z}_p$, define

$$\widetilde{\psi}_p^{(r)}(z) := \lim_{n\in\mathbb{Z}_{>0}, n\to z} \sum_{1\leq k<n,\, p\nmid k} \frac{1}{k^{r+1}} \tag{6.1}$$

where "$n \to z$" means the limit with respect to the p-adic metric. The existence of the limit follows from the fact that

$$\sum_{1\le k<p^n,\, p\nmid k} k^m \equiv \begin{cases} -p^{n-1} & p \ge 3 \text{ and } (p-1)|m \\ 2^{n-1} & p = 2 \text{ and } 2|m \\ 1 & p = 2 \text{ and } n = 1 \\ 0 & \text{otherwise} \end{cases} \tag{6.2}$$

modulo p^n. Thus $\widetilde{\psi}_p^{(r)}(z)$ is a p-adic continuous function on $\mathbb{Z}_p$. Let $\log(x)$ be the Iwasawa logarithmic function which is characterized as the unique continuous homomorphism $\log : \mathbb{C}_p^\times \to \mathbb{C}_p$ such that $\log(x) = 0$ if $x = p$ or x is a root of unity, and

$$\log(x) = -\sum_{n=1}^{\infty} \frac{(1-x)^n}{n}, \quad \text{if } |1-x|_p < 1.$$

Define the *p-adic Euler constant* by

$$\gamma_p := -\lim_{n\to\infty} \frac{1}{p^n} \sum_{0\le j<p^n,\, p\nmid j} \log(j).$$

We define the r-th *p-adic polygamma function* to be

$$\psi_p^{(r)}(z) := \begin{cases} -\gamma_p + \widetilde{\psi}_p^{(0)}(z) & r = 0 \\ -L_p(r+1, \omega^{-r}) + \widetilde{\psi}_p^{(r)}(z) & r \ne 0 \end{cases} \tag{6.3}$$

where $L_p(s, \omega^m)$ is the p-adic L-function and ω is the Teichmüller character (see [A, Lem 2.3]). If $r = 0$, we also write $\psi_p(z) = \psi_p^{(0)}(z)$ and call it the *p-adic digamma function*.

Concerning Dwork's p-adic hypergeometric functions, we need to compute the special values of $\widetilde{\psi}_p(z) = \psi_p(z) + \gamma_p$ modulo p^n (cf. (4.5)). To do this, the sum (6.1) is not useful because the number of terms increases with exponential order by (6.2). However we can avoid this difficulty by using the following theorem.

Theorem 6.1 ([A, **Thm. 2.5**]). *Let $0 \le i < N$ be integers and suppose $p \nmid N$. Then*

$$\widetilde{\psi}_p^{(r)}\left(\frac{i}{N}\right) = N^r \sum_{\varepsilon\in\mu_N\setminus\{1\}} (1-\varepsilon^{-i})\, \mathrm{ln}_{r+1}^{(p)}(\varepsilon) \tag{6.4}$$

where $\mathrm{ln}_k^{(p)}(z)$ are the p-adic polylogarithmic functions (cf. [C, IV]).

Let $r = 0$. Then

$$\begin{aligned} \ln_1^{(p)}(z) &= -p^{-1} \log \frac{(1-z)^p}{1-z^p} \\ &= \sum_{n=1}^{\infty} \frac{p^{n-1}}{n} w(z)^n, \quad w(z) := p^{-1}\left(1 - \frac{(1-z)^p}{1-z^p}\right). \end{aligned}$$

Using this expansion, one can compute $\widetilde{\psi}_p(i/N) \bmod p^n$ without (6.1).

7 Appendix B: Resolution of Singularities

Let W be a commutative ring. Let X be a smooth W-scheme of relative dimension $d \geq 2$ or its completion along a closed subscheme. A divisor D is called a *relative normal crossing divisor* (abbreviated relative NCD) over W if it is locally defined by $x_1 \cdots x_s = 0$ where $(x_1, \ldots, x_d)$ is a local coordinates over W. Further D is called *simple* if each component is smooth over W.

Proposition 7.1. *Let $n > 0$ be an integer which is invertible in W. Let*

$$X := \operatorname{Spec} W[[x, y, s]]/(sx - y^n) \supset C := \operatorname{Spec} W[[x, y, s]]/(s, y). \tag{7.1}$$

Then there is a proper morphism $\rho : X' \to X$ satisfying the following. Put $D := \rho^{-1}(C)$.

- *X' is smooth over W, and $X' \setminus D \overset{\cong}{\to} X \setminus C$,*
- *$D = E_1 + 2E_2 + \cdots + (n-1)E_{n-1} + nC'$ where E_i are exceptional curves and C' is the proper transform of C,*
- *$E_1 + E_2 + \cdots + E_{n-1} + C'$ is a simple relative NCD over W. The figure is as follows.*

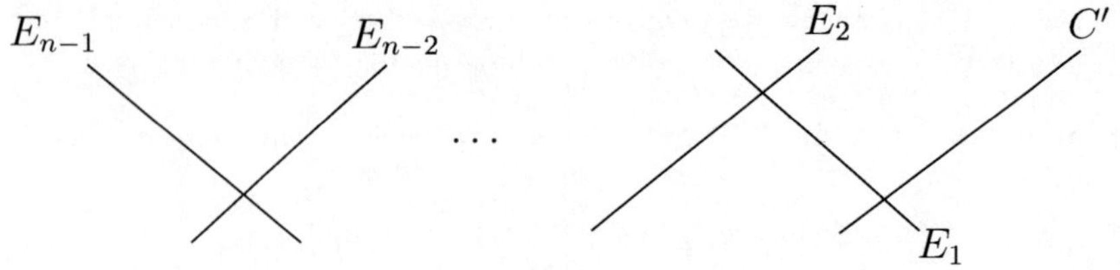

Proof. Let $\rho_1 : X_1 \to X$ be the blow-up with center $(x, y, s) = (0, 0, 0)$. Then X_1 is covered by affine open sets

$$\begin{aligned} U_1 &= \operatorname{Spec} W[[x, y, s]][y_1, s_1]/(s_1 - x^{n-2} y_1^n, xy_1 - y, xs_1 - s) \\ &\cong \operatorname{Spec} W[[x, y, s]][y_1]/(xy_1 - y, x^{n-1} y_1^n - s), \end{aligned}$$

$$U_2 = \operatorname{Spec} W[[x, y, s]][x_2, y_2]/(x_2 - s^{n-2}y_2^n, sy_2 - y, sx_2 - x)$$
$$\cong \operatorname{Spec} W[[x, y, s]][y_2]/(sy_2 - y, s^{n-1}y_2^n - x),$$

$$U_3 = \operatorname{Spec} W[[x, y, s]][x_3, s_3]/(s_3x_3 - y^{n-2}, yx_3 - x, ys_3 - s).$$

U_1 and U_2 are smooth over W. If $n = 2$, there is a unique exceptional curve E such that $E \cap U_1 = \{x = 0\}$, and $\rho_1^{-1}(C) = E + 2C'$ where C' is the proper transform of C. X_1 is smooth over W and $E + C'$ is a simple relative NCD, so we are done. If $n \geq 3$, then the divisor $D_1 := \rho_1^{-1}(C) = E_1 + (n-1)E_2 + nC'$ is as follows.

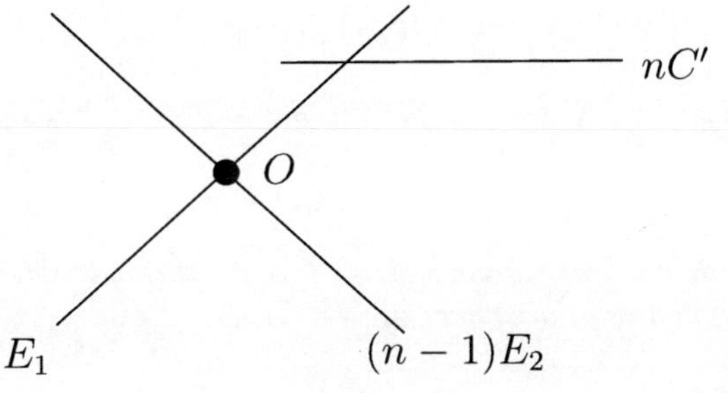

Figure. $n \geq 3$

Here E_1 and E_2 are exceptional curves such that $E_1 \cap U_3 = \{y = x_3 = 0\}$ and $E_2 \cap U_3 = \{y = s_3 = 0\}$, and O is the point $(x_3, y, s_3) = (0, 0, 0)$ in U_3. In a neighborhood of O, X_1 is locally defined by an equation $s_3x_3 = y^{n-2}$. If $n = 3$, then X_1 is smooth, so we are done. If $n \geq 4$, let $\rho_2 : X_2 \to X_1$ be the blow-up at O.

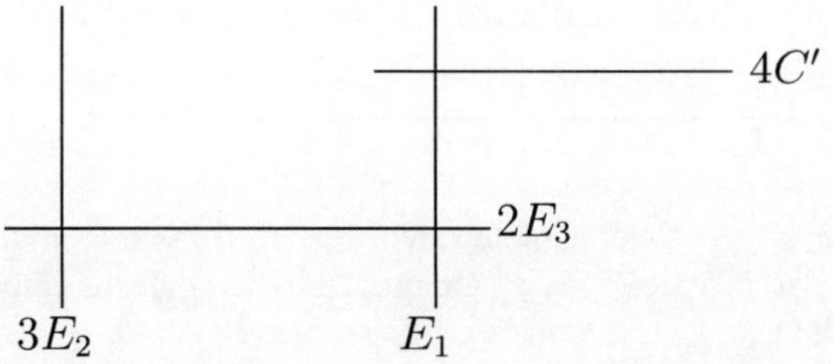

Figure. $n = 4$

If $n = 4$, then X_2 is smooth over W and $\rho_2^{-1}(D_1) = E_1 + 3E_2 + 2E_3 + 4C'$ is as in the figure where E_3 is the unique exceptional curve. So we are done. If $n \geq 5$, then $D_2 = \rho_2^{-1}(D_1) = E_1 + (n-1)E_2 + 2E_3 + (n-2)E_4 + nC'$ is as follows.

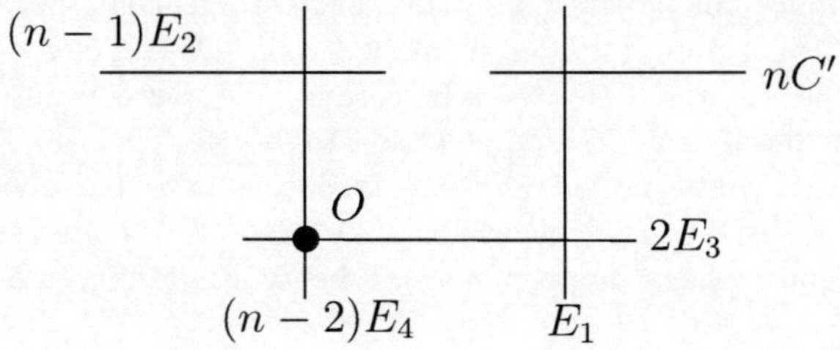

Figure. $n \geq 5$

If $n = 5$ then X_2 is smooth over W, and so we are done. If $n \geq 6$, there is a singular point O. In a neighborhood of O, X_2 is defined by an equation $s_4x_4 = y^{n-4}$, and $E_3 = \{y = x_4 = 0\}$, $E_4 = \{y = s_4 = 0\}$ and $D_2 = \{s_4y^2 = 0\}$. Then we take the blowing-up at O. Continuing this, we finally obtain $\rho : X' = X_n \to X$ with X' a smooth W-scheme such that $\rho^{-1}(C) = E_1 + 2E_2 + \cdots + (n-1)E_{n-1} + nC'$ and $E_1 + \cdots + E_{n-1} + C'$ is a simple relative NCD over W. □

Proposition 7.2. *Let $N, M > 0$ be integers which are invertible in W. Let $d = \gcd(N, M)$. Suppose that W contains a primitive d-th root of unity. Let*

$$X := \operatorname{Spec} W[[x, y]] \supset C := \operatorname{Spec} W[[x, y]]/(x^N + y^M). \tag{7.2}$$

Then there is a proper morphism $\rho : X' \to X$ satisfying the following. Put $D := \rho^{-1}(C)$.

- *X' is smooth over W, and $X' \setminus D \overset{\cong}{\to} X \setminus C$,*
- *$D = \sum n_i D_i$ with D_i smooth over W. Moreover $D = \sum D_i$ is a simple relative NCD over W, and the multiplicities n_i are either of*

$$1, \quad iN, \quad jM, \quad i \in \{1, \ldots, M\}, \; j \in \{1, \ldots, N\}.$$

The figure of $\sum_i D_i$ is as follows, where C' is the proper transform of C which has d-components.

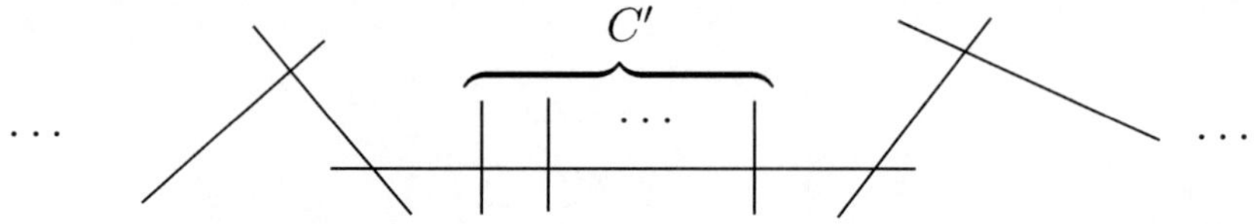

Proof. For integers $a, b, c, d \geq 0$, we denote by $I(a, b; c, d)$ the divisor $\operatorname{Spec} W[[x, y]]/(x^a y^b (x^c + y^d))$ in $\operatorname{Spec} W[[x, y]$. Our goal is to compute the embedded resolution of $I(0, 0; N, M)$.

Let $D = \operatorname{Spec} W[[x, y]]/(x^a y^b (x^c + y^d)) = aD_x + bD_y + D_s \subset X$ where $D_x := \{x = 0\}$, $D_y := \{y = 0\}$ and $D_s := \{x^c + y^d = 0\}$. Let $\rho : X' \to X$ be the blow-up with center $(x, y) = (0, 0)$. Then $D' := \rho^{-1}(D) = (a + b + c)E + aD'_x + bD'_y + D'_s$ where D'_* denotes the proper transform of D_* and E the exceptional curve. In case $c < d$, there is a unique point O which is not normal crossing, and it is locally given by an equation $x^a y^{a+b+c}(x^c + y^{d-c}) = 0$, namely $I(a, a + b + c; c, d - c)$. The multiplicities of D' are $1, a, b, a + b + c$. In case $c > d$, there is also a unique point O such that the divisor D' around O is $I(a + b + d, b; c - d, d)$, and the multiplicities of D_1 are $1, a, b, a + b + d$. In case $c = d$, the divisor $D' = (a + b + c)E + aD'_x + bD'_y + D'_s$ satisfies that $E + D'_x + D'_y + D'_s$ is a simple relative NCD over W, and D'_s has c-components (see the figure). In this case we stop the resolution.

Case $c < d$

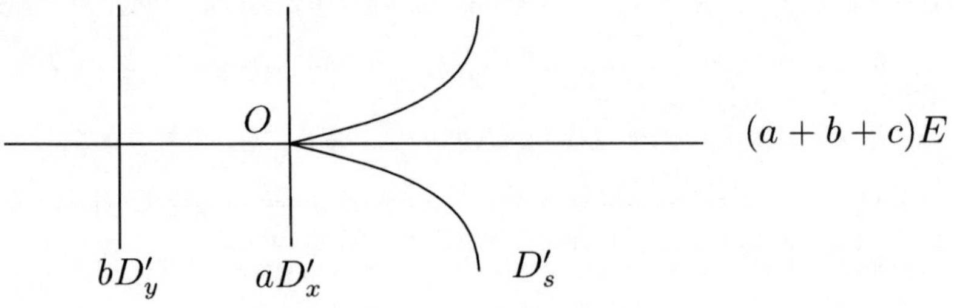

Case $c = d$

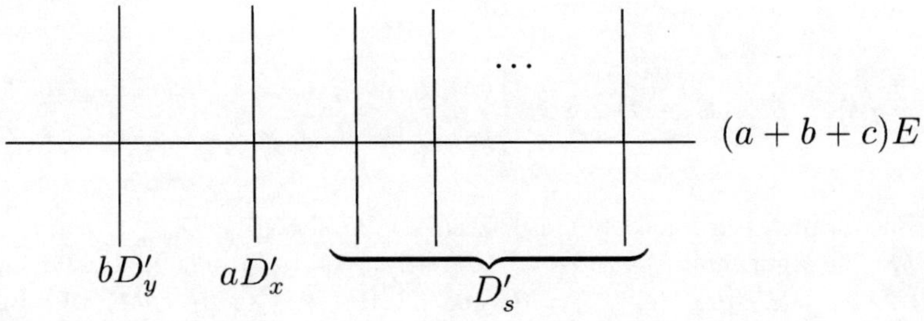

Define

$$(I(a, b; c, d))' := \begin{cases} I(a, a+b+c; c, d-c) & c \le d \\ I(a+b+d, b; c-d, d) & c > d \\ I(a, b; c, d) & cd = 0 \end{cases} \tag{7.3}$$

and $I^{(0)} = I, I^{(i)} = (I^{(i-1)})'$. We begin with $I(0, 0; N, M)$ and consider a sequence $I(a_i, b_i; c_i, d_i) := (I(0, 0; N, M))^{(i)}$

$$I(0, 0; N, M),\ I(a_1, b_1; c_1, d_1),\ \ldots,\ I(a_n, b_n; c_n, d_n)$$

until $c_n d_n = 0$. This corresponds to the sequence of blowing ups at O's as above

$$X_n \longrightarrow X_{n-1} \longrightarrow \cdots \longrightarrow X_0 = X$$

such that the inverse image of C in X_n is supported in a relative simple NCD. Moreover let $D_i \subset X_i$ be the inverse image of C. Then the multiplicities of D_i are either of $1, a_1, \ldots, a_i, b_1, \ldots, b_i$. Therefore if we show Lemma 7.3 below (which is a simple lemma in elementary number theory), then it ends the proof of Proposition 7.2. □

Lemma 7.3. *Let $N, M \geq 1$ be integers and let $I(a_i, b_i; c_i, d_i) := (I(0, 0, ; N, M))^{(i)}$ be defined by* (7.3). *Let n be the minimal integer such that $c_n d_n = 0$.*

(1) *There are integers $A_i, B_i, C_i, D_i \geq 0$ such that $a_i = A_i M$, $b_i = B_i N$, $c_i = C_i N - A_i M$, $d_i = D_i M - B_i N$.*
(2) *A_i, B_i, C_i, D_i are non-decreasing sequences, and $A_n, D_n \leq N$ and $B_n, C_n \leq M$.*

Proof. (1) The assertion is clear for $i = 0$ by putting $(A_0, B_0, C_0, D_0) := (0, 0, 1, 1)$. Suppose that the assertion holds for i. By definition

$$(a_{i+1}, b_{i+1}, c_{i+1}, d_{i+1}) = \begin{cases} (a_i, a_i + b_i + c_i, c_i, d_i - c_i) & c_i \leq d_i \\ (a_i + b_i + d_i, b_i, c_i - d_i, d_i) & c_i > d_i \end{cases}$$
$$= \begin{cases} (A_i M, (B_i + C_i)N, C_i N - A_i M, (A_i + D_i)M - (B_i + C_i)M) \\ ((A_i + D_i)M, B_i N, (B_i + C_i)N - (A_i + D_i)M, D_i M - B_i N). \end{cases}$$

Hence the assertion holds by putting

$$(A_{i+1}, B_{i+1}, C_{i+1}, D_{i+1}) := \begin{cases} (A_i, B_i + C_i, C_i, A_i + D_i) & c_i \leq d_i \\ (A_i + D_i, B_i, B_i + C_i, D_i) & c_i > d_i. \end{cases} \tag{7.4}$$

(2) The former assertion is obvious from (7.4). We show $A_n, D_n \leq N$ and $B_n, C_n \leq M$. The algorithm $(c_0, d_0) \to (c_1, d_1) \to \cdots \to (c_n, d_n)$ is the Euclidean algorithm. Therefore $(c_n, d_n) = (0, \gcd(N, M))$ or $(\gcd(N, M), 0)$. In case $(c_n, d_n) = (0, \gcd(N, M))$, A_n, B_n, C_n, D_n are characterized as the minimal positive integers satisfying $C_n N = A_n M$ and $D_n M - B_n N = \gcd(N, M)$. Hence it turns out that $A_n, D_n \leq N$ and $B_n, C_n \leq M$. The conclusion is the same also in case $(c_n, d_n) = (\gcd(N, M), 0)$. □

Acknowledgements The author would like to express sincere gratitude to Professors Alin Bostan and Kilian Raschel for encouraging the submission to this volume. He is also grateful to Professor Nobuki Takayama for the discussion on the bit complexity of the algorithm.

References

[Ar] N. Archinard, Hypergeometric abelian varieties, Canad. J. Math. **55** (2003), 897–932.
[A] Asakura. M.: *New p-adic hypergeometric functions and syntomic regulators.* arXiv.1811.03770.
[AM] Asakura, M. and Miyatani, K., *Milnor K-theory, F-isocrystals and syntomic regulators.* arXiv:2007.14255.
[AO] Asakura, M. and Otsubo, N.: *CM periods, CM regulators and hypergeometric functions, I.* Canad. J. Math. **70** (2018), 481–514.
[BCM] F. Beukers, H. Cohen, and A. Mellit, *Finite hypergeometric functions,* Pure. Appl. Math. Q. **11** (2015), 559–589.
[BV] Beukers, F., Vlasenko, M.: *Dwork crystals I, II.* arXiv:1903.11155, arXiv:1907.10390.

[BZ] Brent, R., Zimmermann, P.: *Modern Computer Arithmetic*. Cambridge Monographs on Applied and Computational Mathematics, 18. Cambridge University Press, Cambridge, 2011.

[C] Coleman, R.: *Dilogarithms, Regulators and p-adic L-functions*. Invent. Math. **69** (1982), 171–208.

[Dw] Dwork, B.: *p-adic cycles*. Publ. Math. IHES, tome 37 (1969), 27–115.

[EK] Emerton, M., Kisin, M.: *An introduction to the Riemann-Hilbert correspondence for unit F-crystals*. Geometric aspects of Dwork theory. Vol. I, II, 677–700, Walter de Gruyter, Berlin, 2004.

[Ha] Hartshorne, R.: *Algebraic Geometry*. (Grad. Texts in Math. **52**), Springer, 1977.

[Ka] Kato, K.: *Logarithmic structures of Fontaine-Illusie.* Algebraic analysis, geometry, and number theory. Johns Hopkins University Press, Baltimore 1989, 191–224.

[K] Katz, N. *Internal reconstruction of unit-root F-crystals via expansion-coefficients*. With an appendix by Luc Illusie. Ann. Sci. École Norm. Sup. (4) **18** (1985), no. 2, 245–285.

[Ke1] Kedlaya, K.: *p-adic differential equations*. Cambridge Studies in Advanced Mathematics, **125**. Cambridge University Press, Cambridge, 2010.

[Ke2] Kedlaya, K.: *Frobenius structures on hypergeometric equations*, arXiv.1912.13073.

[KT] Kedlaya, K., Tuitman, J.: *Effective convergence bounds for Frobenius structures on connections*. Rend. Semin. Mat. Univ. Padova **128** (2012), 7–16.

[L] Lauder, A.: *Rigid cohomology and p-adic point counting*. J. Théor. Nombres Bordeaux **17** (2005), no. 1, 169–180.

[LP] Lazda, C., Pál, A.: *Rigid cohomology over Laurent series fields*. Algebra and Applications, **21**. Springer, [Cham], 2016. x+267 pp.

[VdP] Van der Put, M., *The cohomology of Monsky and Washnitzer.* Introductions aux cohomologies *p*-adiques (Luminy, 1984). Mem. Soc. Math. France (N.S.) No. 23 (1986), 33–59.

[SS] Samol, K., van Straten, D.: *Dwork congruences and reflexive polytopes*. Ann. Math. Qué. **39** (2015), no. 2, 185–203.

[S] Steenbrink, J.: *Limits of Hodge structures* Invent. Math. **31** (1976), no. 3, 229–257.

[LS] Le Stum, B.: *Rigid cohomology.* Cambridge Tracts in Mathematics, 172. Cambridge University Press, Cambridge, 2007. xvi+319 pp.

[MV] Mellit, A., Vlasenko, M.: *Dwork's congruences for the constant terms of powers of a Laurent polynomial*. Int. J. Number Theory **12** (2016), no. 2, 313–321.

[Z] Zucker, S.: *Degeneration of Hodge bundles (after Steenbrink).* in Topics in transcendental algebraic geometry, 121–141, Ann. of Math. Stud., 106, Princeton Univ. Press, 1984.

A Matrix Version of Dwork's Congruences

Frits Beukers

Abstract In this article we give an example of a matrix version of the famous congruence for hypergeometric functions found by Dwork in 'p-adic cycles'.

1 Introduction

In this paper we shall deal with results of the following type. Let $F(t)$ be an infinite power series with constant term 1 and coefficients in $\mathbb{Z}_p$, the p-adic numbers. Denote by $F_m(t)$ its m-th truncation, i.e. all terms of degree $\geq m$ in $F(t)$ are deleted. We shall be interested whether there are hypergeometric series $F(t)$ for which

$$\frac{F(t)}{F(t^p)} \equiv \frac{F_{p^s}(t)}{F_{p^{s-1}}(t^p)} (\mathrm{mod}\ p^s) \tag{1}$$

for all $s \geq 1$. The first such result was given by Dwork in 'p-adic cycles', [3], for the case of $F(t) = F(1/2, 1/2; 1|t)$. The proof of this result is based on a p-adic study of the coefficients of $F(1/2, 1/2; 1|t)$. Using [3, Cor 1] and [3, Thm 3] one can generalize this approach to other hypergeometric functions whose monodromy around 0 is unipotent (i.e. all β-parameters are 1). The goal of the present paper is to provide a more geometric approach to Dwork's congruences based on the papers [1] and [2] (Dwork crystals I and II), written jointly with Masha Vlasenko. In it we give an elementary approach to the construction of the so-called unit root crystal in Dwork's p-adic theory of zeta-functions of algebraic varieties. As application we present in this paper some hypergeometric examples of (1) in Sect. 3. In Sect. 5 we present main result of this paper,

Work supported by the Netherlands Organisation for Scientific Research (NWO), grant TOP1EW.15.313.

F. Beukers (✉)
Utrecht University, Utrecht, The Netherlands
e-mail: f.beukers@uu.nl

A. Bostan and K. Raschel (eds.), *Transcendence in Algebra, Combinatorics, Geometry and Number Theory*, Springer Proceedings in Mathematics & Statistics 373,
https://doi.org/10.1007/978-3-030-84304-5_2

Theorem 5.1, containing an example of a matrix version of Dwork's congruence. Its proof requires some ideas in addition to [1] and [2].

2 Summary of [1] and [2]

Let R be a characteristic zero domain and p an odd prime such that $\cap_{s\geq 1} p^s R = \{0\}$. Suppose that R is p-adically complete. Let $f \in R[x_1^{\pm 1}, \ldots, x_n^{\pm 1}]$ be a Laurent polynomial and $\Delta \subset \mathbb{R}^n$ its Newton polytope. Let Δ° be its interior. Consider the R-module Ω_f° of differential forms generated over R by

$$\omega_{\mathbf{u}} := (k-1)! \frac{\mathbf{x}^{\mathbf{u}}}{f(\mathbf{x})^k} \frac{dx_1}{x_1} \wedge \cdots \wedge \frac{dx_n}{x_n}, \quad \mathbf{u} \in k\Delta^\circ$$

for all $k \geq 1$. Contrary to [1] and [2] we have now written the elements of Ω_f as differential forms. Let us abbreviate $\frac{dx_1}{x_1} \wedge \cdots \wedge \frac{dx_n}{x_n}$ to $\frac{d\mathbf{x}}{\mathbf{x}}$.

Differential forms in Ω_f° can be expanded as formal Laurent series. To that end we choose a vertex $\mathbf{b}$ of Δ and obtain a Laurent expansion with support in $C(\Delta - \mathbf{b})$, the positive cone generated by the vectors in $\Delta - \mathbf{b}$, and coefficients in R of the form

$$\sum_{\mathbf{k} \in C(\Delta - \mathbf{b})} a_{\mathbf{k}} \mathbf{x}^{\mathbf{k}} \frac{d\mathbf{x}}{\mathbf{x}}, \quad a_{\mathbf{k}} \in R.$$

We denote such forms by Ω_{formal}. The exact forms are denoted by $d\Omega_{\text{formal}}$. We call them formally exact forms and they are characterized by the following lemma of Katz, [4, Lemma 5.1].

Lemma 2.1. *A series $\sum_{\mathbf{k} \in C(\Delta - \mathbf{b})} a_{\mathbf{k}} \mathbf{x}^{\mathbf{k}} \frac{d\mathbf{x}}{\mathbf{x}}$ is a formal derivative if and only if*

$$a_{\mathbf{k}} \equiv 0 (\text{mod } p^{\text{ord}_p(\mathbf{k})}) \quad \textit{for all } \mathbf{k}.$$

Here $\text{ord}_p(k)$ *denotes the p-adic valuation of k and* $\text{ord}_p(\mathbf{k}) = \min(\text{ord}_p(k_1), \ldots, \text{ord}_p(k_n))$.

We define the Cartier operator $\mathscr{C}_p$ on Ω_{formal} by

$$\mathscr{C}_p \left(\sum_{\mathbf{k}} a_{\mathbf{k}} \mathbf{x}^{\mathbf{k}} \frac{d\mathbf{x}}{\mathbf{x}} \right) := \sum_{\mathbf{k}} a_{p\mathbf{k}} \mathbf{x}^{\mathbf{k}} \frac{d\mathbf{x}}{\mathbf{x}}. \tag{2}$$

Using $\mathscr{C}_p$ we have an alternative characterization of formally exact forms which is a direct consequence of Lemma 2.1.

Lemma 2.2. *A series $h \in \Omega_{\text{formal}}$ is a formal derivative if and only if $\mathscr{C}_p^s(h) \equiv 0 (\text{mod } p^s)$ for all integers $s \geq 1$.*

When applied to a rational differential form $\mathscr{C}_p$ acts as

$$\mathscr{C}_p\left(S(\mathbf{x})\frac{d\mathbf{x}}{\mathbf{x}}\right) = \sum_{\mathbf{y}:\mathbf{y}^p=\mathbf{x}} S(\mathbf{y})\frac{d\mathbf{y}}{\mathbf{y}}.$$

The summation extends over all $y_i = \zeta_i x_i^{1/p}, i = 1, \ldots, n$, where each ζ_i runs over all p-th roots of unity. So we see that $\mathscr{C}_p$ sends rational differential forms to rational differential forms. Unfortunately, Ω_f° is not sent to itself. But we have something that comes close. Define the p-adic completion

$$\widehat{\Omega}_f^\circ := \lim_{\leftarrow} \Omega_f^\circ / p^s \Omega_f^\circ.$$

Fix a Frobenius lift σ on R: this is a ring endomorphism $\sigma : R \to R$ such that $\sigma(r) \equiv r^p (\text{mod } p)$ for every $r \in R$. We have

Proposition 2.3. *If $p > 2$ then $\mathscr{C}_p(\Omega_f^\circ) \subset \widehat{\Omega}_{f^\sigma}^\circ$.*

The proof is given in [1, Prop 3.3] and consists of a straightforward computation ending with a p-adic expansion in $\widehat{\Omega}_{f^\sigma}^\circ$.

We shall be interested in $U_f^\circ := \widehat{\Omega}_f^\circ \cap d\Omega_{\text{formal}}$. These are differential forms that are not necessarily exact but become exact when embedded in the formal expansions. Katz refers to them as 'forms that die on formal expansion', [4, Thm 6.2(1.b)]. In [1, Prop 4.2] we find a characterization of the elements of U_f° without any reference to formal expansion.

Proposition 2.4. *With the notations as above we have*

$$U_f^\circ = \{\omega \in \widehat{\Omega}_f^\circ \mid \mathscr{C}_p^s(\omega) \equiv 0 (\text{mod } p^s \widehat{\Omega}_{f^{\sigma^s}}^\circ) \text{ for all } s \geq 1\}.$$

We now come to one of the main results in [1, Thm 4.3]. Let $h = |\Delta^\circ \cap \mathbb{Z}^n|$. Define the Hasse-Witt matrix β_p as the $h \times h$-matrix given by

$$(\beta_p)_{\mathbf{u},\mathbf{v}} = \text{coefficient of } \mathbf{x}^{p\mathbf{u}-\mathbf{v}} \text{ of } f(\mathbf{x})^{p-1}, \quad \mathbf{u}, \mathbf{v} \in \Delta^\circ \cap \mathbb{Z}^n$$

Theorem 2.5. *Suppose* $\det(\beta_p)$ *is invertible in R. Then $\widehat{\Omega}_f^\circ / U_f^\circ$ is a free R-module of rank h with basis $\frac{\mathbf{x}^\mathbf{u}}{f}\frac{d\mathbf{x}}{\mathbf{x}}$, $\mathbf{u} \in \Delta^\circ \cap \mathbb{Z}^n$.*

The remainder of [1] and [2] is then devoted to the construction of p-adic approximations to the $h \times h$-matrix of the Cartier operator. In [2] we give special attention to those approximations that give rise to congruences of the form (1) (in case $h = 1$) and higher.

3 First Examples

In [5] we find a very general theorem providing congruences of the form (1).

Theorem 3.1 (Mellit-Vlasenko). *Let* $g(\mathbf{x}) \in \mathbb{Z}_p[x_1^{\pm 1}, \ldots, x_n^{\pm 1}]$ *be a Laurent polynomial in the variables* $x_1, \ldots, x_n$. *Suppose that the Newton polytope* Δ *of* g *has the origin as unique interior lattice point. For every integer* $r \geq 0$ *denote by* f_r *the constant term of* $g(\mathbf{x})^r$ *and define* $F(t) = \sum_{r \geq 0} f_r t^r$. *Then the congruences (1) hold for all* $s \geq 1$.

In [2, (7)] there is a stronger result with an entirely different proof.

Theorem 3.2 (Beukers-Vlasenko). *With the same notations as in Theorem 3.1 we have*

$$\frac{F(t)}{F(t^p)} \equiv \frac{F_{mp^s}(t)}{F_{mp^{s-1}}(t^p)} (\text{mod } p^s) \tag{3}$$

for all $m, s \geq 1$.

Here is an application.

Corollary 3.3. *Let* $k \geq 2$ *be an integer and* p *an odd prime not dividing* k. *Then (1) holds for the hypergeometric series*

$${}_{k-1}F_{k-2}({}^1\!/\!{}_k, {}^2\!/\!{}_k, \ldots, {}^{(k-1)}\!/\!{}_k; 1, 1, \ldots, 1|t).$$

Proof. Consider

$$g = \frac{1}{k}\left(x_1 + \cdots + x_{k-1} + \frac{1}{x_1 \cdots x_{k-1}}\right).$$

A simple calculation show that f_r is zero if k does not divide r and equal to

$$\frac{1}{k^{kl}} \frac{(kl)!}{(l!)^k} = \frac{({}^1\!/\!{}_k)_l}{l!} \frac{({}^2\!/\!{}_k)_l}{l!} \cdots \frac{({}^{(k-1)}\!/\!{}_k)_l}{l!}$$

if $r = kl$. Hence

$$F(t) = {}_{k-1}F_{k-2}({}^1\!/\!{}_k, {}^2\!/\!{}_k, \ldots, {}^{(k-1)}\!/\!{}_k; 1, 1, \ldots, 1|t^k).$$

Now apply Theorem 3.2 with $m = k$ and replace t^k by t. □

Here is another variation which generalizes Dwork's example

Corollary 3.4. *Let* $k \geq 2$ *be an integer and* p *an odd prime. Then (1) holds for the hypergeometric series*

$${}_{k-1}F_{k-2}({}^1\!/\!{}_2, {}^1\!/\!{}_2, \ldots, {}^1\!/\!{}_2; 1, 1, \ldots, 1|t).$$

Proof. Consider

$$g = 2^{-k}\left(x_1 + \frac{1}{x_1}\right)\cdots\left(x_k + \frac{1}{x_k}\right).$$

A simple calculation shows that f_r is zero if r is odd and equal to

$$\left(\frac{(1/2)_l}{l!}\right)^k$$

if $r = 2l$. Hence

$$F(t) = {}_{k-1}F_{k-2}(1/2, \ldots, 1/2; 1, 1, \ldots, 1|t^2).$$

Now apply Theorem 3.2 with $m = 2$ and replace t^2 by t. □

4 One Variable Polynomials

Let again R be a characteristic zero ring, p an odd prime such that $\cap_{s\geq 1} p^s R = \{0\}$ and suppose R is p-adically complete. Let $\sigma : R \to R$ be a Frobenius lift. It turns out that in the case of one variable polynomials f the theory sketched in Sect. 2 has a very nice simplification that we like to present for general monic $f \in R[x]$ with $f(0) \neq 0$. Let d be the degree of f. We suppose that $d \geq 2$ and that the discriminant of f is invertible in R. The space Ω_f° is given by $\mathcal{O}_f^\circ dx$ where $\mathcal{O}_f^\circ$ is the R-module generated by the forms $l!\frac{x^k}{f^{l+1}}$ with $0 \leq k \leq d(l+1) - 2$. Similarly we define $\mathcal{O}_f$ in the same way but with the inequalities $0 \leq k \leq d(l+1) - 1$. The exact forms in Ω_f° are then given by $d\mathcal{O}_f$. We call them *rational exact forms*.

We define $\mathcal{O}_{\text{formal}} = \frac{1}{x}R[\![1/x]\!]$ and $\Omega_{\text{formal}} = \frac{1}{x}\mathcal{O}_{\text{formal}}dx$. We embed Ω_f° in Ω_{formal} by expansion in powers of $1/x$. The *formally exact forms* are defined by $d\mathcal{O}_{\text{formal}}$.

The interior of the Newton polytope is $\Delta^\circ = (0, d)$ and the cardinality of $\Delta^\circ \cap \mathbb{Z}$ is $d - 1$. So, letting p be an odd prime, the Hasse-Witt matrix $\beta_p(t)$ is a $(d-1) \times (d-1)$-matrix. It turns out that $\det(\beta_p) \equiv \text{disc}(f)^{p-1} (\text{mod } p)$, where $\text{disc}(f)$ is the discriminant of f. By p-adic completeness of R and invertibility of $\text{disc}(f)$ in R we find that $\det(\beta_p)$ is invertible in R. According to Theorem 11 in Dwork crystals I, [1], we know that $\widehat{\Omega}_f^\circ/d\mathcal{O}_{\text{formal}}$ is a free rank $d-1$ module over R with basis $dx/f, xdx/f, \ldots, x^{d-2}dx/f$.

It turns out that in the case $n = 1$ formally exact forms coincide with rational exact forms. More precisely,

Proposition 4.1. *Let $f \in R[x]$ be a monic polynomial and suppose that its discriminant is invertible in R. Then $\Omega_f^\circ \cap d\mathcal{O}_{\text{formal}} = d\mathcal{O}_f$.*

Proof. Clearly $d\mathcal{O}_f \subset d\mathcal{O}_{\text{formal}}$. We first show that every $\omega \in \Omega_f^\circ$ is equivalent modulo $d\mathcal{O}_f$ to a form $Q(x)dx/f$ with $Q(x) \in R[x]$ of degree $\le d-2$. To that end we use the one variable version of the Griffiths reduction procedure. Since p does not divide $\text{disc}(f)$, to every $Q(x) \in R[x]$ of degree $\le N$ there exist polynomials $A, B \in R[x]$ of degrees $\le d-1$ and $\le \max(d-2, N-d)$ respectively, such that $Q = Af' + Bf$.

Let us start with a form $l!Q(x)dx/f^{l+1}$ with $\deg(Q) \le (l+1)d - 2$ and $l > 0$. Write $Q = Af' + Bf$ with $\deg(A) \le d-1$, $\deg(B) \le ld - 2$. Then we obtain

$$
\begin{aligned}
l!\frac{Q(x)}{f^{l+1}}dx &= l!\frac{Af'}{f^{l+1}}dx + l!\frac{B}{f^l}dx \\
&= -d\left((l-1)!\frac{A}{f^l}\right) + (l-1)!\frac{A'}{f^l}dx + l!\frac{B}{f^l}dx \\
&\equiv (l-1)!\frac{lB + A'}{f^l}dx (\text{mod } d\mathcal{O}_f).
\end{aligned}
$$

Note that $\deg(lB + A') \le ld - 2$. By repeating this procedure we see that any $\omega \in \Omega_f^\circ$ is equivalent modulo $d\mathcal{O}_f$ to a form Qdx/f with $Q \in R[x]$ of degree $\le d-2$.

The second part of our proof consists of showing that $Qdx/f \in d\mathcal{O}_{\text{formal}}$ implies that $Q = 0$. Suppose that

$$
\frac{Qdx}{f} = d\left(\sum_{n\ge 0}\frac{a_n}{x^n}\right) = \sum_{n\ge 1} -\frac{na_n}{x^{n+1}}dx.
$$

From this we see that the coefficient of dx/x^{mp^s+1} in the $1/x$-expansion of Qdx/f is divisible by p^s for any $m, s \ge 0$. Let K be the splitting field of f over R and let $\alpha_1, \ldots, \alpha_d \in K$ be the zeros of f. Then there exist $A_1, \ldots, A_d$ in $R[\alpha_1, \ldots, \alpha_d]$ such that

$$
\text{disc}(f)\frac{Qdx}{f} = \sum_{i=1}^{d}\frac{A_i dx}{x - \alpha_i} = \sum_{n\ge 0}(A_1\alpha_1^n + \cdots + A_d\alpha_d^n)\frac{dx}{x^{n+1}}.
$$

We now know that $A_1\alpha_1^{mp^s} + \cdots + A_d\alpha_d^{mp^s}$ is divisible by p^s for all $m \ge 0$. In particular for $m = 0, 1, \ldots, d-1$. Now note that

$$
\begin{aligned}
\det((\alpha_i^{mp^s})_{i=1,\ldots,d;m=0,\ldots,d-1}) &= \prod_{i<j}(\alpha_i^{p^s} - \alpha_j^{p^s}) \\
&\equiv \prod_{i<j}(\alpha_i - \alpha_j)^{p^s} \equiv \text{disc}(f)^{p^s} (\text{mod } p),
\end{aligned}
$$

which is a unit in R. We conclude that $A_i \equiv 0 (\text{mod } p^s)$ for all i and s. Hence $A_i = 0$ for all i and we conclude $Q(x) = 0$, as asserted. □

An immediate corollary is its extension to p-adic completions. Denote $\widehat{\Omega}_f^\circ$ as before and similarly $\widehat{\mathcal{O}}_f$. Then we find,

Proposition 4.2. *Let $f \in R[x]$ be a monic polynomial and suppose that its discriminant is invertible in R. Then $U_f^\circ = \widehat{\Omega}_f^\circ \cap d\mathcal{O}_{\text{formal}} = d\widehat{\mathcal{O}}_f$.*

The operator $\mathscr{C}_p$ is essentially a lift of a Cartier operator which is only well-defined in characteristic p. In [1] and [2] it sufficed to use only the operator $\mathscr{C}_p$ defined above. However, as a new ingredient, we need to consider other lifts. Let $a \in \mathbb{Z}_p$. Define $\mathscr{C}_p^a$ as the operator with the property that $\mathscr{C}_p^a((x-a)^{k-1}dx) = (x-a)^{k/p-1}dx$ if p divides k and 0 if not. In general it acts on rational differential forms as

$$\mathscr{C}_p^a\left(S(x)\frac{dx}{x}\right) = \sum_{y:(y-a)^p=x-a} S(y)\frac{dy}{y}.$$

So we sum over $y = a + \zeta(x-a)^{1/p}$ where ζ runs over the p-th roots of unity. We can compare $\mathscr{C}_p$ and $\mathscr{C}_p^a$ by looking at their action on Ω_{formal}.

Proposition 4.3. *We have $\mathscr{C}_p^a(\Omega_f^\circ) \subset \widehat{\Omega}_{f^\sigma}^\circ$ and*

$$\mathscr{C}_p(\omega) \equiv \mathscr{C}_p^a(\omega) (\text{mod } pd\widehat{\mathcal{O}}_{f^\sigma}) \tag{4}$$

for all $\omega \in \Omega_f^\circ$.

Proof. The fact that the image of $\mathscr{C}_p^a$ lies in $\widehat{\Omega}_{f^\sigma}^\circ$ follows along the same lines as in the proof of [1, Prop 3.3]. Clearly we have $R[\![1/x]\!] \cong R[\![1/(x-a)]\!]$ through the expansion $\frac{1}{x-a} = \sum_{n\geq 0} \frac{a^n}{x^{n+1}}$. Let us prove our second assertion for $\omega_k = (x-a)^{-k-1}dx$ for $k \geq 1$. The full statement then follows by linearity.

Observe that

$$\omega_k = (x-a)^{-k-1}dx = -d\left(\frac{1}{k}(x-a)^{-k}\right).$$

If k is not divisible by p then clearly $\omega_k \in d\mathcal{O}_{\text{formal}}$. Since $\mathscr{C}_p(d\mathcal{O}_{\text{formal}}) \subset pd\mathcal{O}_{\text{formal}}$ we get that $\mathscr{C}_p(\omega_k) \equiv 0 (\text{mod } pd\mathcal{O}_{\text{formal}})$. We have trivially $\mathscr{C}_p^a(\omega_k) = 0$. This proves our statement for k not divisible by p. Suppose now that p divides k. Then

$$\frac{1}{k}(x-a)^{-k} \equiv \frac{1}{k}(x^p - a)^{-k/p} (\text{mod } \mathcal{O}_{\text{formal}})$$

hence, after taking differentials,

$$(x-a)^{-k-1}dx \equiv (x^p-a)^{-k/p-1}x^{p-1}dx (\text{mod } d\mathcal{O}_{\text{formal}}).$$

Application of $\mathscr{C}_p$ gives $\mathscr{C}_p(\omega_k) \equiv \omega_{k/p} (\text{mod } pd\mathcal{O}_{\text{formal}})$. Note that $\omega_{k/p} = \mathscr{C}_p^a(\omega_k)$ when p divides k. Thus we conclude that

$$\mathscr{C}_p(\omega_k) \equiv \mathscr{C}_p^a(\omega_k) (\text{mod } pd\mathscr{O}_{\text{formal}}).$$

By linearity this congruence holds for all $\omega \in \Omega_f^\circ$.

It remains to see that we can replace $pd\mathscr{O}_{\text{formal}}$ by $pd\widehat{\mathscr{O}}_f$. From Proposition 3.6 in [1] it follows that to any $\omega \in \widehat{\Omega}_f^\circ$ there exists $\omega_1 \in \widehat{\Omega}_{f^\sigma}^\circ$ and a polynomial $A(a, \omega)$ such that $\mathscr{C}_p^a(\omega) = \frac{A(a,\omega)}{f^\sigma} + p\omega_1$. Since $\mathscr{C}_p^a(\omega) - \mathscr{C}_p^0(\omega) \in pd\mathscr{O}_{\text{formal}}$ it follows that $A(a, \omega) - A(0, \omega)$ is divisible by p. Hence

$$\frac{1}{p}(\mathscr{C}_p^a(\omega) - \mathscr{C}_p^0(\omega)) \in \widehat{\Omega}_{f^\sigma}^\circ \cap d\mathscr{O}_{\text{formal}} = d\widehat{\mathscr{O}}_{f^\sigma}.$$

The latter equality follows from Proposition 4.2. □

5 A Matrix Example

The examples in the Sect. 3 are all related to the case $h = 1$, one interior lattice point of the Newton polytope Δ. In this section we consider an example of rank $h = 2$.

Theorem 5.1. *Let*

$$\mathscr{Y}(t) = \begin{pmatrix} F(1/3, 2/3, 1/2|t^2) & -\frac{1}{3}tF(7/6, 5/6, 3/2|t^2) \\ -\frac{2}{3}tF(2/3, 4/3, 3/2|t^2) & F(1/6, 5/6, 1/2|t^2) \end{pmatrix}.$$

Denote by $\mathscr{Y}_m(t)$ the m-th truncated version of $\mathscr{Y}(t)$, i.e. we drop all term starting with t^m. Then, for all primes $p > 3$ and all $m, s \geq 1$ we have

$$\mathscr{Y}_{mp^s}(t) \begin{pmatrix} \epsilon_p & 0 \\ 0 & 1 \end{pmatrix} \mathscr{Y}_{mp^{s-1}}(t^p)^{-1} \equiv \mathscr{Y}(t) \begin{pmatrix} \epsilon_p & 0 \\ 0 & 1 \end{pmatrix} \mathscr{Y}(t^p)^{-1} (\text{mod } p^s).$$

Here $\epsilon_p = 1$ if 3 is a square modulo p and -1 if not.

For the proof of this theorem, given at the end of this section, we require the one variable polynomial $f = x^3 - x - t \in R[x]$ with $R = \mathbb{Z}_p[\![t]\!]$, where p is a prime with $p > 3$. As Frobenius lift we take $g(t)^\sigma = g(t^p)$ for all $g(t) \in R$. The discriminant of f equals to $4 - 27t^2$, and hence it is invertible in R.

We define the 2×2-matrix Λ_p with entries in R by

$$\mathscr{C}_p \begin{pmatrix} dx/f \\ xdx/f \end{pmatrix} \equiv \Lambda_p \begin{pmatrix} dx/f^\sigma \\ xdx/f^\sigma \end{pmatrix} (\text{mod } d\widehat{\mathscr{O}}_f). \tag{5}$$

The relation of Λ_p with hypergeometric functions is obtained by period maps. To that end we consider

$$l!\frac{x^{k-1}dx}{f^{l+1}} = l!\frac{x^{k-1}dx}{(x^3-x)^{l+1}}\sum_{r\geq 0}\binom{r+l}{l}\frac{t^r}{(x^3-x)^r},$$

and then take termwise the residue at $x = 0$. We could rephrase this procedure by saying that we expand $x^{k-1}dx/f^{l+1}$ as two-sided Laurent series in $R[\![x, t/x]\!]$ and then take the residue at $x = 0$. Similarly we can take residues at $x = \pm 1$ (i.e. by expanding in Laurent series in $x \mp 1$). The result is again a power series in t. As long as $0 < k < 3(l+1)$ the terms of the series have no residue at ∞ and therefore the sum of the residues at $0, 1, -1$ of the series is 0. We carry out the residue computations for $l = 0, k = 1, 2$. A straightforward calculation shows that

$$\mathrm{res}_{x=0}\frac{dx}{(x^3-x)^{r+1}} = \begin{cases} 0 & \text{if } r \text{ is odd} \\ -\binom{3n}{n} & \text{if } r = 2n \end{cases}$$

$$\mathrm{res}_{x=0}\frac{xdx}{(x^3-x)^{r+1}} = \begin{cases} 0 & \text{if } r \text{ is even} \\ \binom{3n+1}{n} & \text{if } r = 2n+1 \end{cases}.$$

Denote $\mathrm{res}_{\pm}\omega = \mathrm{res}_{x=1}\omega - \mathrm{res}_{x=-1}\omega$. Then we obtain

$$\mathrm{res}_{\pm}\frac{dx}{(x^3-x)^{r+1}} = \begin{cases} 0 & \text{if } r \text{ is even} \\ -\frac{3}{2}\frac{(7/6)_n(5/6)_n}{(3/2)_n n!}\left(\frac{27}{4}\right)^n & \text{if } r = 2n+1 \end{cases}$$

$$\mathrm{res}_{\pm}\frac{xdx}{(x^3-x)^{r+1}} = \begin{cases} 0 & \text{if } r \text{ is odd} \\ \frac{(1/6)_n(5/6)_n}{(1/2)_n n!}\left(\frac{27}{4}\right)^n & \text{if } r = 2n \end{cases}.$$

Let us denote the period map obtained by taking *minus* the residue at 0 by ρ_0 and the one by taking the difference of the residues at ± 1 by $\rho_{\pm}$. We summarize

$$\rho_0(dx/f) = F(1/3, 2/3, 1/2|27t^2/4).$$

$$\rho_0(xdx/f) = -tF(2/3, 4/3, 3/2|27t^2/4).$$

$$\rho_{\pm}(dx/f) = -\frac{3}{2}tF(7/6, 5/6, 3/2|27t^2/4).$$

$$\rho_{\pm}(xdx/f) = F(1/6, 5/6, 1/2|27t^2/4).$$

A crucial property of $\rho_0, \rho_{\pm}$ is that they vanish on exact forms, i.e. $d\widehat{\mathcal{O}}_f$. This is because residues of exact forms are zero, which is a special case of [2, Prop 2.2].

Proposition 5.2. *For every* $\omega \in \widehat{\Omega}^{\circ}_f$ *we have* $\rho_0(\mathscr{C}_p(\omega)) = \rho_0(\omega)$ *and* $\rho_{\pm}(\mathscr{C}_p(\omega)) = \rho_{\pm}(\omega)$.

Proof. Let $\omega \in \widehat{\Omega}_f^\circ$. Expand it in $R[\![x, t/x]\!]dx$. The value of ρ_0 is minus the coefficient of dx/x. By definition of $\mathscr{C}_p$ this value is the same for $\mathscr{C}_p(\omega)$, hence our first assertion follows. Similarly we can see that the residue at 1, which we denote by ρ_1, has the property $\rho_1(\mathscr{C}_p^1(\omega)) = \rho_1(\omega)$. It follows from Proposition 4.3 that $\mathscr{C}_p^1(\omega) \equiv \mathscr{C}_p(\omega) (\text{mod } d\widehat{\mathscr{O}}_f)$. Hence $\rho_1(\mathscr{C}_p(\omega)) = \rho_1(\omega)$. The same result holds of course for $\rho_\pm = \rho_1 - \rho_{-1}$. □

Corollary 5.3. *Let*

$$Y(t) = \begin{pmatrix} F(1/3, 2/3; 1/2|27t^2/4) & -\frac{3}{2}tF(7/6, 5/6; 3/2|27t^2/4) \\ -tF(2/3, 4/3; 3/2|27t^2/4) & F(1/6, 5/6; 1/2|27t^2/4) \end{pmatrix}.$$

Let Λ_p be the 2×2 cartier-matrix in (5). Then

$$\Lambda_p = Y(t)Y(t^p)^{-1}.$$

Proof. We start with the equality (5), apply ρ_0 and use $\rho_0 \circ \mathscr{C}_p = \rho_0$ to obtain

$$\begin{pmatrix} \rho_0(dx/f) \\ \rho_0(xdx/f) \end{pmatrix} = \Lambda_p \begin{pmatrix} \rho_0(dx/f^\sigma) \\ \rho_0(xdx/f^\sigma) \end{pmatrix}.$$

Similarly we obtain

$$\begin{pmatrix} \rho_\pm(dx/f) \\ \rho_\pm(xdx/f) \end{pmatrix} = \Lambda_p \begin{pmatrix} \rho_\pm(dx/f^\sigma) \\ \rho_\pm(xdx/f^\sigma) \end{pmatrix}.$$

Our corollary follows from the above evaluations of the periods. □

In order to get Dwork type congruences we also need to introduce a suitable 'period map mod m'. By that we mean an R-linear map $\rho : \widehat{\Omega}_f \to R$ such that $\rho(\widehat{\Omega}_f \cap d\mathscr{O}_{\text{formal}}) \subset mR$ and $\delta \circ \rho \equiv \rho \circ \delta (\text{mod } mR)$ for any derivation δ on R.

For our purposes we use a slight generalization of the period maps we considered in [2, Section 5]. We define $\rho_{0,m}$ by

$$\rho_{0,m}\omega = \rho_0 \left(1 - \frac{t^m}{(x^3 - x)^m}\right) \omega. \tag{6}$$

Similarly we define $\rho_{1,m}$, $\rho_{-1,m}$ and the difference $\rho_{\pm,m}$. As an illustration we elaborate $\rho_{0,m}(dx/f)$. We get

$$\begin{aligned}\rho_{0,m}(dx/f) &= -\mathrm{res}_{x=0}\left(1-\frac{t^m}{(x^3-x)^m}\right)\frac{dx}{x^3-x-t}\\ &= -\mathrm{res}_{x=0}\frac{1}{(x^3-x)^m}\sum_{r=0}^{m-1}(x^3-x)^{m-1-r}t^r\,dx\\ &= -\mathrm{res}_{x=0}\sum_{r=0}^{m-1}\frac{t^r\,dx}{(x^3-x)^{r+1}}\\ &= \sum_{2n<m}\binom{3n}{n}t^{2n}.\end{aligned}$$

The latter polynomial is the truncation of $F(1/3, 2/3, 1/2|27t^2/4)$ truncated at the degree m term. Denote the truncation at degree m of a power series $g(t)$ by $g(t)_m$. Then we obtain

$$\rho_{0,m}\,(dx/f) = F(1/3, 2/3, 1/2|27t^2/4)_m.$$

$$\rho_{0,m}\,(xdx/f) = -(tF(2/3, 4/3, 3/2|27t^2/4))_m.$$

$$\rho_{\pm,m}\,(dx/f) = -\frac{3}{2}(tF(7/6, 5/6, 3/2|27t^2/4))_m.$$

$$\rho_{\pm,m}\,(xdx/f) = F(1/6, 5/6, 1/2|27t^2/4)_m.$$

Lemma 5.4. *We have $\rho_{0,m}(d\widehat{\mathcal{O}}_f) \equiv 0 \pmod m$ and $\rho_{\pm,m}(d\widehat{\mathcal{O}}_f) \equiv 0 \pmod m$.*

Secondly, for any $m \geq 1$ divisible by p we have $\rho_{0,m} \equiv \rho^\sigma_{0,m/p} \circ \mathscr{C}_p \pmod{p^{\mathrm{ord}_p(m)}}$ and $\rho_{\pm,m} \equiv \rho^\sigma_{\pm,m/p} \circ \mathscr{C}_p \pmod{p^{\mathrm{ord}_p(m)}}$. Here $\rho^\sigma_{0,m}$ is defined as in equation (6) but with t replaced by t^p. Similarly for $\rho^\sigma_{\pm,m}$.

Proof. For any $G \in \widehat{\mathcal{O}}_f$ we have

$$\begin{aligned}\rho_{0,m}dG &= -\text{coefficient of } \tfrac{dx}{x} \text{ in } \left(1-\left(\frac{t}{x^3-x}\right)^m\right)dG\\ &\equiv -\text{coefficient of } \tfrac{dx}{x} \text{ in } d\left(1-\left(\frac{t}{x^3-x}\right)^m\right)G \equiv 0 \pmod m.\end{aligned}$$

The applicability of ρ_0 requires that we consider expansions as doubly infinite Laurent series in $R[\![x, t/x]\!]$. For $\rho_{1,m}$ the proof runs similarly.

For the proof of the second part let $\omega \in \widehat{\Omega}_f$. Then we have

$$
\begin{aligned}
\rho_{0,m}(\omega) &= -\text{coefficient of } \tfrac{dx}{x} \text{ in } \left(1-\left(\frac{t}{x^3-x}\right)^m\right)\omega \\
&\equiv -\text{coefficient of } \tfrac{dx}{x} \text{ in } \left(1-\left(\frac{t^p}{x^{3p}-x^p}\right)^{m/p}\right)\omega \pmod{p^{\mathrm{ord}_p(m)}} \\
&\equiv -\text{coefficient of } \tfrac{dx}{x} \text{ in } \mathscr{C}_p\left(1-\left(\frac{t^p}{x^{3p}-x^p}\right)^{m/p}\right)\omega \pmod{p^{\mathrm{ord}_p(m)}} \\
&\equiv -\text{coefficient of } \tfrac{dx}{x} \text{ in } \left(1-\left(\frac{t^p}{x^3-x}\right)^{m/p}\right)\mathscr{C}_p(\omega) \pmod{p^{\mathrm{ord}_p(m)}} \\
&\equiv \rho^\sigma_{0,m/p}\mathscr{C}_p(\omega) \pmod{p^{\mathrm{ord}_p(m)}}.
\end{aligned}
$$

The second step uses the obvious fact that the Cartier transform does not change the coefficient of $\frac{dx}{x}$.

In a similar manner one can show that

$$\rho_{1,m}(\omega) \equiv \rho^\sigma_{1,m/p}\mathscr{C}^1_p(\omega) \pmod{p^{\mathrm{ord}_p(m)}}.$$

Proposition 4.3 tells us that $\mathscr{C}^1_p(\omega) \equiv \mathscr{C}_p(\omega) \pmod{pd\widehat{\mathscr{O}}_{f^\sigma}}$. Together with the first part of our lemma, which implies that $\rho^\sigma_{1,m/p}(pd\widehat{\mathscr{O}}_{f^\sigma}) \equiv 0 \pmod{p^{\mathrm{ord}_p(m)}}$, we get

$$\rho_{1,m}(\omega) \equiv \rho^\sigma_{1,m/p}\mathscr{C}_p(\omega) \pmod{p^{\mathrm{ord}_p(m)}}.$$

In a similar way the statement for $\rho_{\pm,m}$ follows. □

Corollary 5.5. *Let notations be as in Corollary 5.3 Let $Y_m(t)$ be the matrix $Y(t)$, where the entries have been truncated at t^m. Then, for any $m, s \geq 1$,*

$$Y_{mp^s}(t) \equiv (Y(t)Y(t^p)^{-1})Y_{mp^{s-1}}(t^p) \pmod{p^s}.$$

Proof. We start with the equality (5), which holds true modulo $pd\widehat{\Omega}_f$ according to [1, (14)]. Then apply $\rho^\sigma_{0,mp^{s-1}}$ and use $\rho_{0,mp^s} \equiv \rho^\sigma_{0,mp^{s-1}} \circ \mathscr{C}_p \pmod{p^s}$ to obtain

$$\begin{pmatrix} \rho_{0,mp^s}(dx/f) \\ \rho_{0,mp^s}(xdx/f) \end{pmatrix} \equiv \Lambda_p \begin{pmatrix} \rho^\sigma_{0,mp^{s-1}}(dx/f^\sigma) \\ \rho^\sigma_{0,mp^{s-1}}(xdx/f^\sigma) \end{pmatrix} \pmod{p^s}.$$

Similarly we obtain

$$\begin{pmatrix} \rho_{\pm,mp^s}(dx/f) \\ \rho_{\pm,mp^s}(xdx/f) \end{pmatrix} \equiv \Lambda_p \begin{pmatrix} \rho^\sigma_{\pm,mp^{s-1}}(dx/f^\sigma) \\ \rho^\sigma_{\pm,mp^{s-1}}(xdx/f^\sigma) \end{pmatrix} \pmod{p^s}.$$

Our corollary follows from the above evaluations of the mod m periods and $\Lambda_p = Y(t)Y(t^p)^{-1}$. □

We end with the proof of our main theorem.

Proof of Theorem 5.1. The proof follows the same steps as Corollary 5.3, but with the polynomial $f = x^3 - x - 2t/3\sqrt{3}$. This polynomial is defined over $\mathbb{Z}_p[\sqrt{3}][\![t]\!]$ with Frobenius lift σ such that $\sigma(t) = t^p$ and $\sigma(\sqrt{3}) = \epsilon_p\sqrt{3}$. Hence $f^\sigma = x^3 - x - 2\epsilon_p t^p/\sqrt{3}$. We also use the new basis dx/f, $\sqrt{3}xdx/f$ and replace $\rho_\pm$ by $\frac{1}{\sqrt{3}}\rho_\pm$. The adapted version of Corollary 5.3 would then become

$$\Lambda_p = \mathcal{Y}(t)\begin{pmatrix}\epsilon_p & 0\\ 0 & 1\end{pmatrix}\mathcal{Y}(t^p)^{-1}.$$

The remainder of the proof follows the same lines as above. □

We finally give, without proof, the system of differential equations for $\mathcal{Y}(t)$ and its congruence version. Again the proof follows the same lines as in [2].

Theorem 5.6. *We have*

$$\frac{d}{dt}\mathcal{Y}(t) = \frac{1}{3(1-t^2)}\begin{pmatrix}2t & -1\\ -2 & t\end{pmatrix}\mathcal{Y}(t)$$

and

$$\frac{d}{dt}\mathcal{Y}_{mp^s}(t) \equiv \frac{1}{3(1-t^2)}\begin{pmatrix}2t & -1\\ -2 & t\end{pmatrix}\mathcal{Y}_{mp^s}(t) (\mathrm{mod}\ p^s)$$

For all $m, s \geq 1$.

Acknowledgements I would like to thank Ling Long for our discussions which gave rise to this paper. I also like to thank the referees for their valuable feedback and their corrections.

References

1. F. Beukers, M. Vlasenko, *Dwork crystals I*, Int. Math. Res. Notices 2021, 8807–8844, https://doi.org/10.1093/imrn/rnaa119
2. F. Beukers, M. Vlasenko, *Dwork crystals II*, Int. Math. Res. Notices 2021, 4427–4444, https://doi.org/10.1093/imrn/rnaa120
3. B. Dwork, *p-adic cycles*, Publications Mathématiques de l'I.H.É.S. 37 (1969), 27–115
4. N. Katz, *Internal reconstruction of unit-root F-crystals via expansion coefficients. With an appendix by Luc Illusie* Annales scientifiques de l'É.N.S 18 (1985), 245–285
5. A. Mellit, M. Vlasenko, *Dwork's congruences for the constant terms of powers of a Laurent polynomial*, Int. J. Number Theory 12 (2016), 313–321

On the Kernel Curves Associated with Walks in the Quarter Plane

Thomas Dreyfus, Charlotte Hardouin, Julien Roques, and Michael F. Singer

Abstract The *kernel method* is an essential tool for the study of generating series of *walks* in the quarter plane. This method involves equating to zero a certain polynomial - the *kernel polynomial* - and using properties of the curve - the *kernel curve* - this defines. In the present paper, we investigate the basic properties of the *kernel curve* (irreducibility, singularities, genus, uniformization, etc.).

Keywords Random walks · Uniformization of algebraic curves

2010 Mathematics Subject Classification: 05A15 · 30D05

This project has received funding from the European Research Council (ERC) under the European Union's Horizon 2020 research and innovation programme under the Grant Agreement No 648132. The authors would like to thank ANR-19-CE40-0018, ANR-13-JS01-0002-01, ANR-11-LABX-0040-CIMI within the program ANR-11-IDEX-0002-0, LabEx PERSYVAL-Lab ANR-11-LABX-0025-01, the Simons Foundation (#349357, Michael Singer).

T. Dreyfus (✉)
Institut de Recherche Mathématique Avancée, U.M.R. 7501 Université de Strasbourg et C.N.R.S. 7, rue René Descartes, 67084 Strasbourg, France
e-mail: dreyfus@math.unistra.fr

C. Hardouin
Université Paul Sabatier - Institut de Mathématiques de Toulouse, 118 route de Narbonne, 31062 Toulouse, France
e-mail: hardouin@math.univ-toulouse.fr

J. Roques
Univ Lyon, Université Claude Bernard Lyon 1, CNRS UMR 5208, Institut Camille Jordan, 43 blvd. du 11 novembre 1918, 69622 Villeurbanne cedex, France
e-mail: roques@math.univ-lyon1.fr

M. F. Singer
Department of Mathematics, North Carolina State University, Box 8205, Raleigh, NC 27695-8205, USA
e-mail: singer@ncsu.edu

A. Bostan and K. Raschel (eds.), *Transcendence in Algebra, Combinatorics, Geometry and Number Theory*, Springer Proceedings in Mathematics & Statistics 373,
https://doi.org/10.1007/978-3-030-84304-5_3

1 Introduction

Consider a *walk* with small steps in the positive quadrant $\mathbb{Z}^2_{\geq 0} = \{0, 1, 2, \ldots\}^2$ starting from $P_0 := (0, 0)$, that is a succession of points

$$P_0, P_1, \ldots, P_k,$$

where each P_n lies in the quarter plane, where the moves (or steps) $P_{n+1} - P_n$ belong to $\{0, \pm 1\}^2$, and the probability to move in the direction $P_{n+1} - P_n = (i, j)$ may be interpreted as some *weight-parameter* $d_{i,j} \in [0, 1]$, with $\sum_{(i,j)\in\{0,\pm1\}^2} d_{i,j} = 1$. The step set or the *model* of the *walk* is the set of directions with nonzero *weights*, that is

$$\mathcal{S} = \{(i, j) \in \{0, \pm 1\}^2 \mid d_{i,j} \neq 0\}.$$

The following picture is an example of such path:

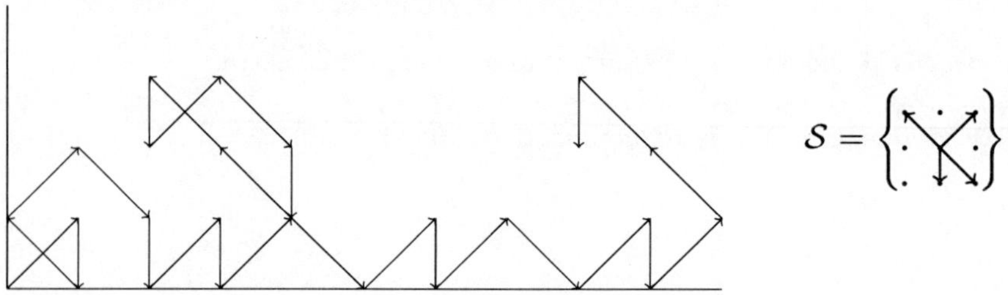

Such objects are very natural both in combinatorics and probability theory: they are interesting for themselves and also because they are strongly related to other discrete structures, see [BMM10, DW15] and references therein.

If $d_{0,0} = 0$ and if the nonzero $d_{i,j}$ all have the same value, we say that the model is *unweighted*.

The *weight of a given walk* is defined to be the product of the *weights* of its component steps. For any $(i, j) \in \mathbb{Z}^2_{\geq 0}$ and any $k \in \mathbb{Z}_{\geq 0}$, we let $q_{i,j,k}$ be the sum of the *weights* of all *walks* reaching the position (i, j) from the initial position $(0, 0)$ after k steps. We introduce the corresponding trivariate generating series[1]

$$Q(x, y, t) := \sum_{i,j,k\geq 0} q_{i,j,k} x^i y^j t^k.$$

The study of the nature of this generating series has attracted the attention of many authors, see for instance [BvHK10, BRS14, BBMR15, BBMR17, BMM10, DHRS18, DHRS20, DR19, DH19, KR12, Mis09, MR09, MM14, Ras12]. The typical questions are: is $Q(x, y, t)$ rational, algebraic, holonomic, etc.? The starting

[1] In several papers it is not assumed that $\sum_{i,j} d_{i,j} = 1$. But after a rescaling of the t variable, we may always reduce to the case $\sum_{i,j} d_{i,j} = 1$.

point of most of these works is the following functional equation, see for instance [DHRS20, Lemma 1.1], and [BMM10] for the *unweighted* case

$$K(x,y,t)Q(x,y,t) = xy + K(x,0,t)Q(x,0,t) + K(0,y,t)Q(0,y,t) + td_{-1,-1}Q(0,0,t)$$

where

$$K(x,y,t) = xy(1 - tS(x,y))$$

with

$$S(x,y) = \sum_{(i,j)\in\{0,\pm1\}^2} d_{i,j}x^i y^j.$$

The polynomial $K(x,y,t)$ is called the *kernel polynomial* and is the main character of the *kernel method.*

Roughly speaking, the first step of the *kernel method* consists in "eliminating" the left hand side of the above functional equation by restricting our attention to the (x,y) such that $K(x,y,t) = 0$. The set E_t made of the (x,y) such that $K(x,y,t) = 0$ is called the *kernel curve*:

$$E_t = \{(x,y) \in \mathbb{C} \times \mathbb{C} \mid K(x,y,t) = 0\}.$$

Thus, for $(x,y) \in E_t$, one has

$$0 = xy + K(x,0,t)Q(x,0,t) + K(0,y,t)Q(0,y,t) + td_{-1,-1}Q(0,0,t), \quad (1)$$

provided that the various series can be evaluated at the given points.

The second step of the *kernel method* is to exploit certain involutive birational transformations ι_1, ι_2 (they are called ζ, η in [FIM17]) of the *kernel curve* E_t of the form

$$\iota_1(x,y) = (x,y') \text{ and } \iota_2(x,y) = (x',y)$$

in order to deduce from (1) some functional equations for $Q(x,0,t)$ and $Q(0,y,t)$. Hence ι_1 and ι_2 switch the roots of the degree two polynomials $y \mapsto K(x,y,t)$ and $x \mapsto K(x,y,t)$ respectively. Concretely, the birational transformations ι_1, ι_2 are induced by restriction to the curve of the involutive birational transformations i_1, i_2 of $\mathbb{C}^2$ given by

$$i_1(x,y) = \left(x, \frac{A_{-1}(x)}{A_1(x)y}\right) \text{ and } i_2(x,y) = \left(\frac{B_{-1}(y)}{B_1(y)x}, y\right)$$

where the $A_i(x) \in x^{-1}\mathbb{Q}[x]$ and the $B_i(y) \in y^{-1}\mathbb{Q}[y]$ are defined by

$$S(x,y) = A_{-1}(x)\frac{1}{y} + A_0(x) + A_1(x)y = B_{-1}(y)\frac{1}{x} + B_0(y) + B_1(y)x,$$

see [BMM10, Section 3], [KY15, Section 3] or [FIM17]. These i_1 and i_2 are the generators of the group of the *walk*; see [BMM10] for details. Note that although i_1 and i_2 do not depend on t, the group generated by the induced involutions ι_1 and ι_2 may depend on t, since the order of $\iota_2 \circ \iota_1$ may depend upon t, see Remark 5.13.

The third step of the *kernel method* is to use the above mentioned functional equations of $Q(x, 0, t)$ and $Q(0, y, t)$ to continue these series as multivalued meromorphic functions. To perform this step, we need an explicit uniformization of the *kernel curve*.

The aim of the present paper is to study the *kernel curve* E_t and the birational transformations ι_1, ι_2. Note that a similar study has been done in the case $t = 1$ in [FIM17] and in the *unweighted* case in [KR12]. The goal of the present paper is to extend these works to the *weighted* case when $t \in]0, 1[$ is transcendental over $\mathbb{Q}(d_{i,j})$. Although many results are similar to [FIM17], the proofs are different. The assumptions we make on t are crucial in many parts of the proof and it is not clear how the proofs of [FIM17] exactly pass to this context.

We could expect to have classification of the geometric properties of $\overline{E_t}$ involving configurations of *weights* independent of t. This paper has been followed by [DR19] where the case for general $t \in]0, 1[$ has been considered. The proofs of the latter paper use continuity arguments with respect the parameter t that permit to deduce many results for algebraic values of t. Such reasoning needs to be very cautious, and it is not trivial to deduce the results for general $t \in]0, 1[$ from the $t = 1$ case. We will mention explicitly every time if the results are correct for arbitrary values of $t \in]0, 1[$.

The paper is organized as follows. In Sect. 2, we describe the *nondegenerate models of walks*. In Sect. 3, we determine the singularities and the genus of the *kernel curve*. In Sect. 4, we establish the basic properties of ι_1 and ι_2. Finally, in Sect. 5, we give an explicit uniformization of the *kernel curve*.

2 Nondegenerate Walks

From now on, we fix $t \in]0, 1[$, that is transcendental over the field $\mathbb{Q}(d_{i,j})$. We start by recalling the notion of *degenerate walks introduced in* [FIM17].

Definition 2.1 A *model of walk* is called *degenerate* if one of the following holds:

- $K(x, y, t)$ is reducible as an element of the polynomial ring $\mathbb{C}[x, y]$,
- $K(x, y, t)$ has x-degree less than or equal to 1,
- $K(x, y, t)$ has y-degree less than or equal to 1.

In what follows we will sometimes represent a *model of walks* with arrows. We will also use dashed arrows for a family of models. For instance, the family of models represented by

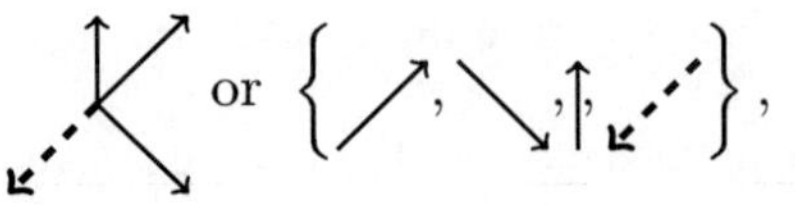

correspond to models with $d_{1,1}, d_{1,-1}, d_{0,1} \neq 0$, $d_{1,0} = d_{0,-1} = d_{-1,1} = d_{-1,0} = 0$, and where nothing is assumed on $d_{-1,-1}$ and $d_{0,0}$. In the following results, the behavior of the *kernel curve* never depends on $d_{0,0}$. This is the reason why, to reduce the amount of notations, we have decided to not associate an arrow to $d_{0,0}$. The following result is the analog of [FIM17, Lemma 2.3.2], that focuses on the case $t = 1$. Our proof differs from the proof of [FIM17, Lemma 2.3.2], which only considered factorization over $\mathbb{R}[x, y]$, while in this paper, we need to prove the absence of factorization over $\mathbb{C}[x, y]$.

Proposition 2.2 *A model of walk is degenerate if and only if at least one of the following holds:*

(1) *There exists* $i \in \{-1, 1\}$ *such that* $d_{i,-1} = d_{i,0} = d_{i,1} = 0$. *This corresponds to the following families of models of walks*

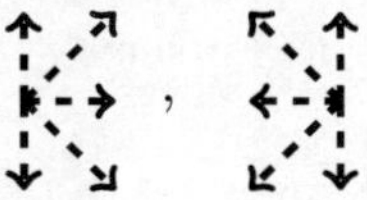

(2) *There exists* $j \in \{-1, 1\}$ *such that* $d_{-1,j} = d_{0,j} = d_{1,j} = 0$. *This corresponds to the following families of models of walks*

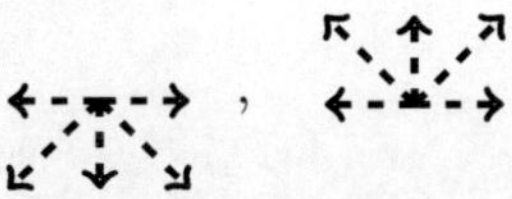

(3) *All the weights are* 0 *except maybe* $\{d_{1,1}, d_{0,0}, d_{-1,-1}\}$ *or* $\{d_{-1,1}, d_{0,0}, d_{1,-1}\}$. *This corresponds to the following families of models of walks*

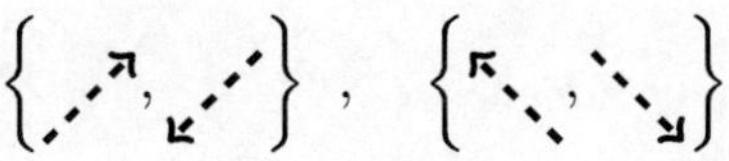

Proof This proof is organized as follows. We begin by showing that (1) (resp. (2)) corresponds to $K(x, y, t)$ having x-degree ≤ 1 or x-valuation ≥ 1 (resp. y-degree ≤ 1 or y-valuation ≥ 1). In these cases, *the model of the walk* is clearly *degenerate*. Assuming (1) and (2) do not hold, we then show that (3) holds if and only if $K(x, y, t)$ is reducible.

Cases (1) and (2). It is clear that $K(x, y, t)$ has x-degree ≤ 1 if and only if $d_{1,-1} = d_{1,0} = d_{1,1} = 0$. Similarly, $K(x, y, t)$ has y-degree ≤ 1 if and only if we have $d_{-1,1} = d_{0,1} = d_{1,1} = 0$. Furthermore, $d_{-1,-1} = d_{-1,0} = d_{-1,1} = 0$ if and only if $K(x, y, t)$ has x-valuation ≥ 1. Similarly, $d_{-1,-1} = d_{0,-1} = d_{1,-1} = 0$ if and only if $K(x, y, t)$ has y-valuation ≥ 1. In these cases, *the model of the walk* is clearly *degenerate*.

Case (3). We now assume that cases (1) and (2) do not hold. This implies that the model belongs to the family of models 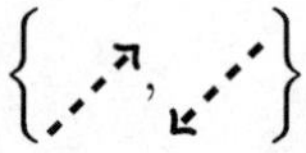 if and only if it belongs to the family of models 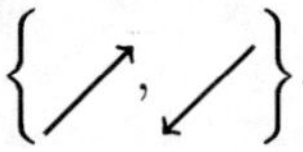. The same holds for the anti-diagonal configuration. If the model belongs to the family of models 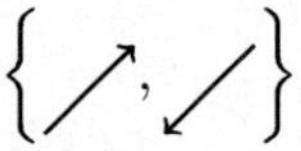, then the *kernel*

$$K(x, y, t) = -d_{-1,-1}t + xy - d_{0,0}txy - d_{1,1}tx^2y^2 \in \mathbb{C}[xy]$$

is a degree two polynomial in xy. Thus it may be factorized in the following form $K(x, y, t) = -d_{1,1}t(xy - \alpha)(xy - \beta)$ for some $\alpha, \beta \in \mathbb{C}$. If the model belongs to the family of models , then

$$K(x, y, t) = -d_{-1,1}ty^2 + xy - d_{0,0}txy - d_{1,-1}tx^2.$$

In this situation, $K(x, y, t)y^{-2} \in \mathbb{C}[x/y]$ may be factorized in the ring $\mathbb{C}[x/y]$, proving that $K(x, y, t)$ may be factorized in $\mathbb{C}[x, y]$ as well.

Conversely, let us assume that *the model of the walk* is *degenerate*. Recall that we have assumed that cases (1) and (2) do not hold, so $K(x, y, t)$ has x- and y-degree two, x- and y-valuation 0, and is reducible. We have to prove that the model belongs to one of the family of models 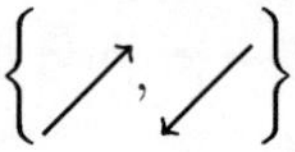 or 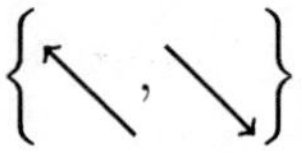. Let us write a factorization

$$K(x, y, t) = -f_1(x, y)f_2(x, y),$$

with $f_1(x, y), f_2(x, y) \in \mathbb{C}[x, y]$ not constant. Let us now prove several lemmas on the polynomials $f_1(x, y), f_2(x, y) \in \mathbb{C}[x, y]$.

Lemma 2.3 *Both* $f_1(x, y)$ *and* $f_2(x, y)$ *have bidegree* (1, 1).

Proof of Lemma 2.3 Suppose to the contrary that $f_1(x, y)$ or $f_2(x, y)$ does not have bidegree $(1, 1)$. Since K is of bidegree at most $(2, 2)$ then at least one of the f_i's has degree 0 in x or y. Up to interchange of x and y and f_1 and f_2, we may assume that $f_1(x, y)$ has y-degree 0 and we denote it by $f_1(x)$. Since we are not in Cases (1) and (2) of Proposition 2.2, the polynomials $d_{-1,-1}t + d_{0,-1}tx + d_{1,-1}tx^2$ and $d_{-1,0}t + (d_{0,0}t - 1)x + d_{1,0}tx^2$ are nonzero. By $K(x, y, t) = -f_1(x)f_2(x, y)$, we find in particular that $f_1(x)$ is a common factor of the nonzero polynomials $d_{-1,-1}t + d_{0,-1}tx + d_{1,-1}tx^2$ and $d_{-1,0}t + (d_{0,0}t - 1)x + d_{1,0}tx^2$. Since t is nonzero, we find that the roots of $d_{-1,-1}t + d_{0,-1}tx + d_{1,-1}tx^2 = 0$ are algebraic over $\mathbb{Q}(d_{i,j})$. On the other hand, since t is transcendental over $\mathbb{Q}(d_{i,j})$, if x is a root of $d_{-1,0}t + (d_{0,0}t - 1)x + d_{1,0}tx^2 = 0$ that is algebraic over $\mathbb{Q}(d_{i,j})$, then the constant term in t has to be zero, proving that $x = 0$. Therefore, they are polynomials with only zero as a potential common roots. So the only potential root of $f_1(x)$ is zero. This means that either $f_1(x)$ has degree 0, i.e. $f_1(x) \in \mathbb{C}$, or x divides $f_1(x)$. In the latter case, x divides $K(x, y, t)$, and we are in Case 1. In both cases, this is a contradiction and proves the lemma.

Lemma 2.4 *The polynomials $f_1(x, y)$ and $f_2(x, y)$ are irreducible in the ring $\mathbb{C}[x, y]$.*

Proof of Lemma 2.4 To the contrary, suppose that we can find a factorization $f_1(x, y) = (ax - b)(cy - d)$ for some $a, b, c, d \in \mathbb{C}$. Since $f_1(x, y)$ has bidegree $(1, 1)$, we have $ac \neq 0$. We then have that

$$0 = K(b/a, y, t) = \frac{b}{a}y - t(\tilde{A}_{-1}(\frac{b}{a}) + \tilde{A}_0(\frac{b}{a})y + \tilde{A}_1(\frac{b}{a})y^2)$$

where $\tilde{A}_i = xA_i \in \mathbb{Q}[x]$. Note that $\tilde{A}_1(x)$ is nonzero because $K(x, y, t)$ has bidegree $(2, 2)$. Equating the y^2-terms we find that $\tilde{A}_1(\frac{b}{a}) = 0$ so $\frac{b}{a}$ is algebraic over $\mathbb{Q}(d_{i,j})$. Equating the y-terms, we obtain that $\frac{b}{a} - t\tilde{A}_0(\frac{b}{a}) = 0$. Using the fact that t is transcendental over $\mathbb{Q}(d_{i,j})$ and $\frac{b}{a}$ is algebraic over $\mathbb{Q}(d_{i,j})$, we deduce $\frac{b}{a} = 0$. Therefore $b = 0$. This contradicts the fact that K has x-valuation 0. A similar argument shows that $f_2(x, y)$ is irreducible.

Lemma 2.5 *Let $\overline{f}_i(x, y)$ denote the polynomial whose coefficients are the complex conjugates of those of $f_i(x, y)$. We may reduce to the case where one of the following cases hold:*

- *there exists $\epsilon \in \{\pm 1\}$ such that $\overline{f_1}(x, y) = \epsilon f_2(x, y)$,*
- *$\overline{f_1}(x, y) = f_1(x, y) \in \mathbb{R}[x, y]$ and $\overline{f_2}(x, y) = f_2(x, y) \in \mathbb{R}[x, y]$.*

Proof of Lemma 2.5 Unique factorization of polynomials implies that since $-K(x, y, t) = f_1(x, y)f_2(x, y) = \overline{f_1}(x, y)\overline{f_2}(x, y)$, there exists $\lambda \in \mathbb{C}^*$ such that

- either $\overline{f_1}(x, y) = \lambda f_2(x, y)$ and $\overline{f_2}(x, y) = \lambda^{-1} f_1(x, y)$;
- or $\overline{f_1}(x, y) = \lambda f_1(x, y)$ and $\overline{f}_2(x, y) = \lambda^{-1} f_2(x, y)$.

In the former case, we have $f_1(x, y) = \overline{\lambda}\,\overline{f_2}(x, y) = \overline{\lambda}\lambda^{-1} f_1(x, y)$ and so $\overline{\lambda}\lambda^{-1} = 1$. This implies that λ is real and replacing $f_1(x, y)$ by $|\lambda|^{-1/2} f_1(x, y)$ and $f_2(x, y)$ by $|\lambda|^{1/2} f_2(x, y)$, we can assume that either $\overline{f_1}(x, y) = f_2(x, y)$ and $\overline{f_2}(x, y) = f_1(x, y)$ or $\overline{f_1}(x, y) = -f_2(x, y)$ and $\overline{f_2}(x, y) = -f_1(x, y)$.

A similar computation in the latter case shows that $|\lambda| = 1$. Letting μ be a square root of λ we have $\mu^{-1} = \overline{\mu}$ so $\lambda = \mu/\overline{\mu}$. Replacing $f_1(x, y)$ by $\mu f_1(x, y)$ and $f_2(x, y)$ by $\overline{\mu} f_2(x, y)$, we can assume that $\overline{f_1}(x, y) = f_1(x, y)$ and $\overline{f_2}(x, y) = f_2(x, y)$.

Let us continue the proof of Proposition 2.2. For $i = 1, 2$, let us write

$$f_i(x, y) = (\alpha_{i,4}x + \alpha_{i,3})y + (\alpha_{i,2}x + \alpha_{i,1}),$$

with $\alpha_{i,j} \in \mathbb{C}$. Equating the terms in $x^i y^j$ with $-1 \leq i, j \leq 1$, in $f_1(x, y) f_2(x, y) = -K(x, y, t)$, we find (recall that $d_{i,j} \in [0, 1], t \in]0, 1[$)

term	coefficient in $f_1(x, y) f_2(x, y)$	coefficient in $-K(x, y, t)$
1	$\alpha_{1,1}\alpha_{2,1}$	$d_{-1,-1}t \geq 0$
x	$\alpha_{1,2}\alpha_{2,1} + \alpha_{1,1}\alpha_{2,2}$	$d_{0,-1}t \geq 0$
x^2	$\alpha_{1,2}\alpha_{2,2}$	$d_{1,-1}t \geq 0$
y	$\alpha_{1,3}\alpha_{2,1} + \alpha_{1,1}\alpha_{2,3}$	$d_{-1,0}t \geq 0$
xy	$\alpha_{1,4}\alpha_{2,1} + \alpha_{1,3}\alpha_{2,2} + \alpha_{1,2}\alpha_{2,3} + \alpha_{1,1}\alpha_{2,4}$	$d_{0,0}t - 1 < 0$
x^2y	$\alpha_{1,4}\alpha_{2,2} + \alpha_{1,2}\alpha_{2,4}$	$d_{1,0}t \geq 0$
y^2	$\alpha_{1,3}\alpha_{2,3}$	$d_{-1,1}t \geq 0$
xy^2	$\alpha_{1,4}\alpha_{2,3} + \alpha_{1,3}\alpha_{2,4}$	$d_{0,1}t \geq 0$
x^2y^2	$\alpha_{1,4}\alpha_{2,4}$	$d_{1,1}t \geq 0$

Let us treat separately two cases.

Case 1: $f_1(x, y), f_2(x, y) \notin \mathbb{R}[x, y]$. So, in this case we have either $\overline{f_1}(x, y) = f_2(x, y)$ or $\overline{f_1}(x, y) = -f_2(x, y)$.

Let us first assume that $\overline{f_1}(x, y) = f_2(x, y)$. Then, evaluating the equality $K(x, y, t) = -f_1(x, y) f_2(x, y)$ at $x = y = 1$, we get the following equality $K(1, 1, t) = -f_1(1, 1) f_2(1, 1) = -|f_1(1, 1)|^2$. But this is impossible because the left-hand term $K(1, 1, t) = 1 - t \sum_{i,j \in \{-1,0,1\}^2} d_{i,j} = 1 - t$ is > 0 whereas the right-hand term $-|f_1(1, 1)|^2$ is ≤ 0.

Let us now assume that $\overline{f_1}(x, y) = -f_2(x, y)$. Equating the constant terms in the equality $f_1(x, y) f_2(x, y) = -K(x, y, t)$, we get $-|\alpha_{1,1}|^2 = d_{-1,-1}t$, so $\alpha_{1,1} = \alpha_{2,1} = d_{-1,-1} = 0$. Equating the coefficients of x^2 in the equality $f_1(x, y) f_2(x, y) = -K(x, y, t)$, we get $-|\alpha_{1,2}|^2 = d_{1,-1}t$, so $\alpha_{1,2} = \alpha_{2,2} = d_{1,-1} = 0$. It follows that the y-valuation of $f_1(x, y) f_2(x, y) = -K(x, y, t)$ is ≥ 2, whence a contradiction.

Case 2: $f_1(x, y), f_2(x, y) \in \mathbb{R}[x, y]$. We first claim that, after possibly replacing $f_1(x, y)$ by $-f_1(x, y)$ and $f_2(x, y)$ by $-f_2(x, y)$, we may assume that $\alpha_{1,4}, \alpha_{2,4}, \alpha_{1,3}, \alpha_{2,3} \geq 0$.

Let us first assume that $\alpha_{1,4}\alpha_{2,4} \neq 0$. Since $\alpha_{1,4}\alpha_{2,4} = d_{1,1}t \geq 0$, we find that $\alpha_{1,4}, \alpha_{2,4}$ belong simultaneously to $\mathbb{R}_{>0}$ or $\mathbb{R}_{<0}$. After possibly replacing $f_1(x, y)$ by $-f_1(x, y)$ and $f_2(x, y)$ by $-f_2(x, y)$, we may assume that $\alpha_{1,4}, \alpha_{2,4} > 0$. Since $\alpha_{1,3}\alpha_{2,3} = d_{-1,1}t \geq 0$, we have that $\alpha_{1,3}, \alpha_{2,3}$ belong simultaneously to $\mathbb{R}_{\geq 0}$ or $\mathbb{R}_{\leq 0}$. Then, the equality $\alpha_{1,4}\alpha_{2,3} + \alpha_{1,3}\alpha_{2,4} = d_{0,1}t \geq 0$ implies that $\alpha_{1,3}, \alpha_{2,3} \geq 0$.

We can argue similarly in the case $\alpha_{1,3}\alpha_{2,3} \neq 0$.

It remains to consider the case $\alpha_{1,4}\alpha_{2,4} = \alpha_{1,3}\alpha_{2,3} = 0$. After possibly replacing $f_1(x, y)$ by $-f_1(x, y)$ and $f_2(x, y)$ by $-f_2(x, y)$, we may assume that $\alpha_{1,4}, \alpha_{2,4} \geq 0$. The case $\alpha_{1,4} = \alpha_{1,3} = 0$ is impossible because, otherwise, we would have $d_{1,1} = d_{-1,1} = d_{0,1} = 0$, which is excluded. Similarly, the case $\alpha_{2,4} = \alpha_{2,3} = 0$ is impossible. So, we are left with the cases $\alpha_{1,4} = \alpha_{2,3} = 0$ or $\alpha_{2,4} = \alpha_{1,3} = 0$. In both cases, the equality $\alpha_{1,4}\alpha_{2,3} + \alpha_{1,3}\alpha_{2,4} = d_{0,1}t \geq 0$ with $\alpha_{1,4}, \alpha_{2,4} \geq 0$, implies that $\alpha_{1,4}, \alpha_{2,4}, \alpha_{1,3}, \alpha_{2,3} \geq 0$.

Arguing as above, we see that $\alpha_{1,2}, \alpha_{2,2}, \alpha_{1,1}, \alpha_{2,1}$ all belong to $\mathbb{R}_{\geq 0}$ or all belong to $\mathbb{R}_{\leq 0}$. Using the equation of the xy-coefficients, we find that $\alpha_{1,2}, \alpha_{2,2}, \alpha_{1,1}, \alpha_{2,1}$ are all in $\mathbb{R}_{\leq 0}$.

Now, equating the coefficients of x^2y in the equality $f_1(x, y)f_2(x, y) = -K(x, y, t)$ we get $\alpha_{1,4}\alpha_{2,2} + \alpha_{1,2}\alpha_{2,4} = d_{1,0}t$. Using the fact that $\alpha_{1,4}\alpha_{2,2}, \alpha_{1,2}\alpha_{2,4} \leq 0$ and that $d_{1,0}t \geq 0$, we get $\alpha_{1,4}\alpha_{2,2} = \alpha_{1,2}\alpha_{2,4} = d_{1,0} = 0$. Similarly, using the coefficients of y, we get $\alpha_{1,3}\alpha_{2,1} = \alpha_{1,1}\alpha_{2,3} = d_{-1,0} = 0$.

So, we have

$$\alpha_{1,4}\alpha_{2,2} = \alpha_{1,2}\alpha_{2,4} = \alpha_{1,3}\alpha_{2,1} = \alpha_{1,1}\alpha_{2,3} = 0.$$

The fact that $K(x, y, t)$ has x- and y-degree 2 and x- and y-valuation 0 implies that, for any $i \in \{1, 2\}$, none of the vectors $(\alpha_{i,4}, \alpha_{i,3})$, $(\alpha_{i,2}, \alpha_{i,1})$, $(\alpha_{i,4}, \alpha_{i,2})$ and $(\alpha_{i,3}, \alpha_{i,1})$ is $(0, 0)$. Since $\alpha_{1,4}\alpha_{2,2} = 0$, we have $\alpha_{1,4} = 0$ or $\alpha_{2,2} = 0$. If $\alpha_{1,4} = 0$, from what precedes, we find

$$\alpha_{1,4} = \alpha_{2,4} = \alpha_{2,1} = \alpha_{1,1} = 0.$$

If $\alpha_{2,2} = 0$ we obtain

$$\alpha_{2,2} = \alpha_{1,2} = \alpha_{1,3} = \alpha_{2,3} = 0.$$

In the first case, the model belongs to the family of models . In the second case, we find that the model belongs to the family of models 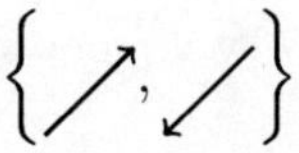.

This completes the proof.

Remark 2.6 The fact $d_{i,j} \in [0, 1]$ are probabilities is crucial in the proof of Proposition 2.2. We do not expect this result to be correct for general $d_{i,j} \in \mathbb{C}$.

Remark 2.7 The "*degenerate models of walks*" are called "*singular*" by certain authors, e.g., in [FIM99, FIM17]. Note also that, in [KR12], "*singular walks*" has a different meaning and refers to *models of walks* such that the associated *kernel* defines a genus zero curve.

Remark 2.8 In [DR19, Proposition 3], the authors show that Proposition 2.2 extends mutatis mutandis to the case when $t \in]0, 1[$ is algebraic over $\mathbb{Q}(d_{i,j})$. Their proof relies on Proposition 2.2 and uses a continuity argument of the parameter t to deduce that Proposition 2.2 stays correct for general values of $t \in]0, 1[$.

From now on, we will only consider *nondegenerate models of walks*. In terms of *models of walks*, this only discards one dimensional problems and *models of walks* in the half-plane restricted to the quarter plane that are easier to study, as explained in [BMM10, Section 2.1].

3 Singularities and Genus of the Kernel Curve

The *kernel curve* E_t is the complex affine algebraic curve defined by

$$E_t = \{(x, y) \in \mathbb{C} \times \mathbb{C} \mid K(x, y, t) = 0\}.$$

We shall now consider a compactification of this curve. We let $\mathbb{P}^1(\mathbb{C})$ be the complex projective line, which is the quotient of $(\mathbb{C} \times \mathbb{C}) \setminus \{(0, 0)\}$ by the equivalence relation $\sim$ defined by

$$(x_0, x_1) \sim (x_0', x_1') \Leftrightarrow \exists \lambda \in \mathbb{C}^*, (x_0', x_1') = \lambda(x_0, x_1).$$

The equivalence class of $(x_0, x_1) \in (\mathbb{C} \times \mathbb{C}) \setminus \{(0, 0)\}$ is denoted by $[x_0 : x_1] \in \mathbb{P}^1(\mathbb{C})$. The map $x \mapsto [x : 1]$ embeds $\mathbb{C}$ inside $\mathbb{P}^1(\mathbb{C})$. The latter map is not surjective: its image is $\mathbb{P}^1(\mathbb{C}) \setminus \{[1 : 0]\}$; the missing point $[1 : 0]$ is usually denoted by ∞. Now, we embed E_t inside $\mathbb{P}^1(\mathbb{C}) \times \mathbb{P}^1(\mathbb{C})$ via $(x, y) \mapsto ([x : 1], [y : 1])$. The *kernel curve* $\overline{E_t}$ is the closure of this embedding of E_t. In other words, the *kernel curve* $\overline{E_t}$ is the algebraic curve defined by

$$\overline{E_t} = \{([x_0 : x_1], [y_0 : y_1]) \in \mathbb{P}^1(\mathbb{C}) \times \mathbb{P}^1(\mathbb{C}) \mid \overline{K}(x_0, x_1, y_0, y_1, t) = 0\}$$

where $\overline{K}(x_0, x_1, y_0, y_1, t)$ is the following bihomogeneous polynomial

$$\overline{K}(x_0, x_1, y_0, y_1, t) = x_1^2 y_1^2 K(\frac{x_0}{x_1}, \frac{y_0}{y_1}, t) = x_0 x_1 y_0 y_1 - t \sum_{i,j=0}^{2} d_{i-1,j-1} x_0^i x_1^{2-i} y_0^j y_1^{2-j}. \quad (3.1)$$

Although it may seem more natural to take the closure of $\overline{E_t}$ in $\mathbb{P}^2(\mathbb{C})$, the above definition allows us to extend the involutions of E_t of Sect. 4 in a natural way as well as allowing us to avoid unnecessary singularities.

We shall now study the singularities and compute the genus of $\overline{E_t}$. Recall that since the *model of walk* under consideration is *nondegenerate*, the polynomial $K(x, y, t)$ is irreducible. We recall that by definition $P = ([a : b], [c : d]) \in \overline{E_t}$ is called a singularity of the irreducible kernel $\overline{E_t}$ if

$$\frac{\partial \overline{K}(a, b, c, d, t)}{\partial x_0} = \frac{\partial \overline{K}(a, b, c, d, t)}{\partial x_1} = \frac{\partial \overline{K}(a, b, c, d, t)}{\partial y_0} = \frac{\partial \overline{K}(a, b, c, d, t)}{\partial y_1} = 0.$$

Note that one can check this condition in any affine neighborhood of a point. For example, if $b, d \neq 0$, the bihomogeneity of $\overline{K}$ implies

$$\begin{aligned} 0 = 2\overline{K}(a/b, 1, c/d, 1, t) &= \frac{a}{b}\frac{\partial \overline{K}(a/b, 1, c, /d, 1, t)}{\partial x_0} + \frac{\partial \overline{K}(a/b, 1, c, /d, 1, t)}{\partial x_1} \\ &= \frac{c}{d}\frac{\partial \overline{K}(a/b, 1, c, /d, 1, t)}{\partial y_0} + \frac{\partial \overline{K}(a/b, 1, c, /d, 1, t)}{\partial y_1}. \end{aligned}$$

Therefore the point $P = ([a : b], [c : d]) \in \overline{E_t}$ is a singular point if and only if

$$\frac{\partial \overline{K}(a/b, 1, c/d, 1, t)}{\partial x_0} = \frac{\partial \overline{K}(a/b, 1, c/d, 1, t)}{\partial y_0} = 0.$$

If $P = ([a : b], [c : d]) \in \overline{E_t}$ is not a singularity of $\overline{E_t}$, then it is called a smooth point of $\overline{E_t}$.

We also recall that $\overline{E_t}$ is called singular if it has at least one singular point. Otherwise, we say that $\overline{E_t}$ is nonsingular or smooth.

The genus of an algebraic curve is a classical notion in algebraic geometry. It is a nonnegative integer that we may attach to a curve, see [Ful84, Section 8.3] for a definition. The study of the genus of $\overline{E_t}$ has been considered in [FIM17]. Proposition 3.1 below shows that the smoothness of $\overline{E_t}$ is intimately related to the value of the genus of $\overline{E_t}$. We define the genus of the *weighted model of walk*, as the genus of its *kernel curve* $\overline{E_t}$.

Recall the following notations from the introduction:

$$\begin{aligned} K(x, y, t) &= xy - txA_{-1}(x) - txA_0(x)y - txA_1(x)y^2, \\ &= xy - tyB_{-1}(y) - tyB_0(y)x - tyB_1(y)x^2, \end{aligned}$$

where $xA_i(x) \in \mathbb{Q}[x]$ and $yB_i(y) \in \mathbb{Q}[y]$. Then we may write

$$\begin{aligned} \overline{K}(x_0, x_1, y_0, y_1, t) &= \overline{C}_1(x_0, x_1, t)y_1^2 + \overline{B}_1(x_0, x_1, t)y_0y_1 + \overline{A}_1(x_0, x_1, t)y_0^2 \\ &= \overline{C}_2(y_0, y_1, t)x_1^2 + \overline{B}_2(y_0, y_1, t)x_0x_1 + \overline{A}_2(y_0, y_1, t)x_0^2. \end{aligned}$$

For any $[x_0 : x_1]$ and $[y_0 : y_1]$ in $\mathbb{P}^1(\mathbb{C})$, we denote by $\Delta_1([x_0 : x_1])$ and $\Delta_2([y_0 : y_1])$ the discriminants of the degree two homogeneous polynomials given by $y \mapsto \overline{K}(x_0, x_1, y, t)$ and $x \mapsto \overline{K}(x, y_0, y_1, t)$ respectively, i.e.

$$\begin{aligned}\Delta_1([x_0 : x_1]) &= \overline{B}_1(x_0, x_1, t)^2 - 4\overline{A}_1(x_0, x_1, t)\overline{C}_1(x_0, x_1, t)\\ &= (x_0x_1 - t^2\Big((d_{-1,0}x_1^2 - \frac{1}{t}x_0x_1 + d_{0,0}x_0x_1 + d_{1,0}x_0^2)^2\\ &\quad -4(d_{-1,1}x_1^2 + d_{0,1}x_0x_1 + d_{1,1}x_0^2)(d_{-1,-1}x_1^2 + d_{0,-1}x_0x_1 + d_{1,-1}x_0^2)\Big)\end{aligned}$$

and

$$\begin{aligned}\Delta_2([y_0 : y_1]) &= \overline{B}_2(y_0, y_1, t)^2 - 4\overline{A}_2(y_0, y_1, t)\overline{C}_2(y_0, y_1, t)\\ &= t^2\Big((d_{0,-1}y_1^2 - \frac{1}{t}y_0y_1 + d_{0,0}y_0y_1 + d_{0,1}y_0^2)^2\\ &\quad -4(d_{1,-1}y_1^2 + d_{1,0}y_0y_1 + d_{1,1}y_0^2)(d_{-1,-1}y_1^2 + d_{-1,0}y_0y_1 + d_{-1,1}y_0^2)\Big).\end{aligned}$$

As we will see in the sequel, see Remark 3.3, $\Delta_1([x_0 : x_1])$ has a double root if and only if $\Delta_2([y_0 : y_1])$ has a double root.

Proposition 3.1 *For nondegenerate models, the following facts are equivalent:*

(1) *the curve $\overline{E_t}$ is a genus zero curve;*
(2) *the curve $\overline{E_t}$ is singular;*
(3) *the curve $\overline{E_t}$ has exactly one singularity $\Omega \in \overline{E_t}$;*
(4) *there exists $([a : b], [c : d]) \in \overline{E_t}$ such that the discriminants $\Delta_1([x_0 : x_1])$ and $\Delta_2([y_0 : y_1])$ have a root $[a : b] \in \mathbb{P}^1(\mathbb{C})$ and $[c : d] \in \mathbb{P}^1(\mathbb{C})$ respectively;*
(5) *there exists $([a : b], [c : d]) \in \overline{E_t}$ such that the discriminants $\Delta_1([x_0 : x_1])$ and $\Delta_2([y_0 : y_1])$ have a double root $[a : b] \in \mathbb{P}^1(\mathbb{C})$ and $[c : d] \in \mathbb{P}^1(\mathbb{C})$ respectively.*

If these properties are satisfied, then the singular point is $\Omega = ([a : b], [c : d])$ where $[a : b] \in \mathbb{P}^1(\mathbb{C})$ is a double root of $\Delta_1([x_0 : x_1])$ and $[c : d] \in \mathbb{P}^1(\mathbb{C})$ is a double root of $\Delta_2([y_0 : y_1])$. If the previous properties are not satisfied, then $\overline{E_t}$ is a smooth curve of genus one.

Proof By [Dui10, Section 3.3.1], the following formula gives the genus of $\overline{E_t}$,

$$g(\overline{E_t}) = 1 - \sum_{P \in \text{Sing}} \frac{m(P)(m(P) - 1)}{2}, \tag{3.2}$$

where $m(P)$ is a positive integer standing for the multiplicity of a point P, that is, some partial derivative of order $m(P)$ does not vanish while for every $\ell < m(P)$, the partial derivatives of order ℓ vanish at P. Since the genus is a nonnegative integer, the above formula shows that $g(\overline{E_t})$ is equal to 0 or 1. This formula shows more precisely that $\overline{E_t}$ is smooth if and only if $g(\overline{E_t}) = 1$. Moreover (3.2) shows that if $\overline{E_t}$ is singular, then there is exactly one singular point that is a double point, and the curve has genus zero. This proves the equivalence between (1), (2) and (3), and the last statement of the proposition.

Let us prove (4) $\Rightarrow$ (3). Assume that the discriminant $\Delta_1([x_0 : x_1])$ (resp. $\Delta_2([y_0 : y_1])$) has a root in $[a : b] \in \mathbb{P}^1(\mathbb{C})$ (resp. $[c : d] \in \mathbb{P}^1(\mathbb{C})$). Let us write

$$\begin{aligned}
&\overline{K}(x_0, x_1, y_0, y_1, t)\\
&= e_{-1,1}(dy_0 - cy_1)^2 + e_{0,1}(bx_0 - ax_1)(dy_0 - cy_1)^2 + e_{1,1}(bx_0 - ax_1)^2(dy_0 - cy_1)^2\\
&+ e_{-1,0}(dy_0 - cy_1) + e_{0,0}(bx_0 - ax_1)(dy_0 - cy_1) + e_{1,0}(bx_0 - ax_1)^2(dy_0 - cy_1)\\
&+ e_{-1,-1} + e_{0,-1}(bx_0 - ax_1) + e_{1,-1}(bx_0 - ax_1)^2.
\end{aligned}$$

Since $([a : b], [c : d]) \in \overline{E_t}$, we have by definition that $\overline{K}(a, b, c, d, t) = 0$, i.e. $e_{-1,-1} = 0$. Since $\Delta_1([x_0 : x_1])$ has a root in $[a : b] \in \mathbb{P}^1(\mathbb{C})$, $K(a, b, y_0, y_1)$ has a double root at $[c, d]$ and so $e_{-1,0} = 0$. Similarly, the fact that $\Delta_2([y_0 : y_1])$ has a root in $[c : d] \in \mathbb{P}^1(\mathbb{C})$ implies $e_{0,-1} = 0$. This shows that

$$\frac{\partial \overline{K}(a, b, c, d, t)}{\partial x_0} = \frac{\partial \overline{K}(a, b, c, d, t)}{\partial x_1} = \frac{\partial \overline{K}(a, b, c, d, t)}{\partial y_0} = \frac{\partial \overline{K}(a, b, c, d, t)}{\partial y_1} = 0,$$

and, hence, $([a : b], [c : d])$ is the singular point of $\overline{E_t}$.

Let us prove (3) $\Rightarrow$ (5). If $\Omega = ([a : b], [c : d])$ is the singular point of $\overline{E_t}$, then $e_{-1,-1} = e_{-1,0} = e_{0,-1} = 0$, and the discriminants $\Delta_1([x_0 : x_1])$ and $\Delta_2([y_0 : y_1])$ have a double root in $[a : b] \in \mathbb{P}^1(\mathbb{C})$ and $[c : d] \in \mathbb{P}^1(\mathbb{C})$ respectively.

The implication (5) $\Rightarrow$ (4) is obvious.

Our next aim is to describe the genus zero *models of walks*.

Lemma 3.2 *The discriminant* $\Delta_2([y_0 : y_1])$ *has a double zero if and only if the model of the walk is included in a closed half plane whose boundary passes through* (0, 0). *In other word, this correspond to models of the walks that belong to one of the following eight families*

Remark 3.3 As the statement of Lemma 3.2 is symmetric with respect to x and y we deduce that the same holds for $\Delta_1([x_0 : x_1])$. We then deduce that $\Delta_1([x_0 : x_1])$ has a double root if and only if $\Delta_2([y_0 : y_1])$ has a double root.

Remark 3.4 In the case $t = 1$, it is proved in [FIM17, Lemma 2.3.10] that, besides the models listed in Lemma 3.2, any *nondegenerate* model such that the drift is zero, i.e.

$$\left(\sum_i i d_{i,j}, \sum_j j d_{i,j}\right) = (0, 0),$$

has a curve $\overline{E_t}$ of genus 0.

Proof The computations seem to be too complicated to be performed by hand, so we used MAPLE.[2] We are going to prove the result with two strategies. This first one is to write the discriminant of the discriminant $\Delta_2([y_0 : y_1])$ and study when the latter is 0. The second strategy consists in decomposing the radical of an ideal into its prime components.

Let us first consider the situation where the double root is at (a, b) where b is not zero. Let us set $y_1 = 1$ and $y_0 = y$ to obtain the specialization $\Delta_2([y : 1])$ of $\Delta_2([y_0 : y_1])$.

The following MAPLE code calculates the discriminant of the discriminant, its degree and order of vanishing in t, and then sets the coefficients of powers of t equal to zero. Solving these equations yields the 8 solutions **S[i]**, i = 1, ...,8 corresponding to the 8 step sets listed in Lemma 3.2.

```
> K := expand(x*y*(1-t*(add(add(d[i, j]*x^i*y^j, i = -1 .. 1), j = -1 .. 1)))):
> DX :=  discrim(K, x):
> DD := discrim(discrim(K,x),y);
> ldegree(DD,t); degree(DD,t);
```

4

12

```
> S := solve({seq(coeff(DD,t,i),i=4..12)},[seq(seq(d[i,j],i=-1..1),j=-1..1)]);
> nops(S);
```

8

```
> S[1];S[2];S[3];S[4];S[5];S[6];S[7];S[8];
```

An alternate approach is to use the *PolynomialIdeals* package

```
> with(PolynomialIdeals):
```

and consider the prime decomposition of the radical of the ideal

```
> J := <seq(coeff(DD,t,i), i=4..12)>:
> PrimeDecomposition(J);
```

$$< d_{-1,-1}, d_{-1,0}, d_{-1,1} >, \quad < d_{-1,-1}, d_{-1,0}, d_{0,-1} >, \quad < d_{-1,-1}, d_{0,-1}, d_{1,-1} >,$$
$$< d_{-1,0}, d_{-1,1}, d_{0,1} >, \quad < d_{-1,1}, d_{0,1}, d_{1,1} >, \quad < d_{0,-1}, d_{1,-1}, d_{1,0} >,$$
$$< d_{0,1}, d_{1,0}, d_{1,1} >, \quad < d_{1,-1}, d_{1,0}, d_{1,1} >.$$

The *PrimeDecomposition* command lists a set of prime ideals whose intersection is the radical of the original ideal. In particular, these ideals have the property that any zero of the original ideal is a zero of one of the listed ideals and vice versa, see [CLO97, Chapter 4, Sect. 6]. These again correspond to the eight step sets listed in

[2] The maple worksheet is available at https://singer.math.ncsu.edu/ms_papers.html.

Lemma 3.2.

We now consider the case where the double root is at (a, b) where $b = 0$, that is, at $(1, 0)$. We will show that this case leads to *models of walks* already mentioned above.

```
> DDX := expand(z^4*subs(y = 1/z, DX)):
```

If $z = 0$ is a double root then the coefficient of 1 and z must be zero

```
> coeff(DDX, z, 0); coeff(DDX, z, 1);
```

$$-4\,t^2 d_{-1,1} d_{1,1} + t^2 d_{0,1}{}^2$$

$$-4\,t^2 d_{-1,0} d_{1,1} - 4\,t^2 d_{-1,1} d_{1,0} + 2\,t^2 d_{0,0} d_{0,1} - 2\,t d_{0,1}$$

Taking into account that t is transcendental over $\mathbb{Q}(d_{i,j})$, we are led to three cases, corresponding to three of the step sets in Lemma 3.2.

$$[d_{0,1} = 0, d_{-1,1} = 0, d_{-1,0} = 0]$$
$$[d_{0,1} = 0, d_{-1,1} = 0, d_{1,1} = 0]$$
$$[d_{0,1} = 0, d_{1,1} = 0, d_{1,0} = 0]$$

Remark 3.5 The proof of Proposition 2.2 proceeds by a direct "hand calculation" while the proof of Lemma 3.2 follows from a simple MAPLE calculation. It would be interesting to have a simple MAPLE based proof of Proposition 2.2 and a hand calculation proof of Lemma 3.2.

Corollary 3.6 *The following holds:*

(1) *The nondegenerate genus zero models of walks are the nondegenerate models whose step set is included in an half space whose boundary passes through* $(0, 0)$. *More precisely, they are nondegenerate models belonging to one of the following families*

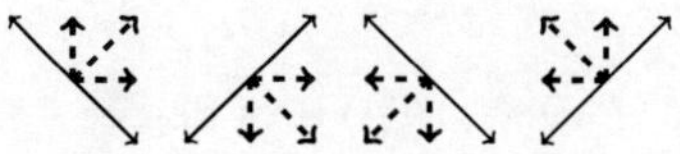

(G0)

(2) *The nondegenerate genus one models of walks are the models whose step set is not included in any half space whose boundary passes through* $(0, 0)$.

Remark 3.7 See also [DR19, Proposition 9] for an extension of Corollary 3.6 to the case when $t \in]0,1[$ is algebraic over $\mathbb{Q}(d_{i,j})$. Their proof relies on the results of the present section and uses a continuity argument with respect to the parameter t to deduce that Corollary 3.6 stays correct for general values of $t \in]0,1[$.

Proof We use Proposition 3.1. We have to determine when there exists $([a:b],[c:d]) \in \overline{E_t}$ such that the discriminants $\Delta_1([x_0:x_1])$ and $\Delta_2([y_0:y_1])$ have a double root $[a:b] \in \mathbb{P}^1(\mathbb{C})$ and $[c:d] \in \mathbb{P}^1(\mathbb{C})$. Lemma 3.2 provides such configurations. By Proposition 2.2, the configurations number $1,3,5,7$ are dismissed since they led to *singular walks*. Then, if we are considering *nondegenerate* genus zero *models of walks*, we are in the four families of models considered in (G0). Furthermore, if the step set is not included in any half space whose boundary passes through $(0,0)$, the configurations of Proposition 2.2 and Lemma 3.2 are excluded and by Proposition 3.1, we are in the genus 1 situation.
Conversely, it remains to prove that if the *models of walks* are in the four families of *models* considered in (G0), the kernel has genus 0. Thanks to Proposition 3.1 it suffices to prove that the discriminants have a common zero in that case. This is the goal of the following Lemma 3.8 and Remark 3.9.

Let us write

$$\Delta_1([x_0:x_1]) = \sum_{i=0}^{4} \alpha_i(t) x_0^i x_1^{4-i}, \quad \text{and } \Delta_2([y_0:y_1]) = \sum_{i=0}^{4} \beta_i(t) y_0^i y_1^{4-i}.$$

where

$$\begin{aligned}
\alpha_0(t) &= t^2 d_{-1,0}^2 - 4t^2 d_{-1,1} d_{-1,-1} \\
\alpha_1(t) &= 2t^2 d_{-1,0} d_{0,0} - 2t d_{-1,0} - 4t^2 d_{-1,1} d_{0,-1} - 4t^2 d_{0,1} d_{-1,-1} \\
\alpha_2(t) &= t^2 d_{0,0}^2 - 2t d_{0,0} + 1 + 2t^2 d_{-1,0} d_{1,0} - 4t^2 d_{-1,1} d_{1,-1} - 4t^2 d_{0,1} d_{0,-1} - 4t^2 d_{1,1} d_{-1,-1} \\
\alpha_3(t) &= -2t d_{1,0} + 2t^2 d_{0,0} d_{1,0} - 4t^2 d_{1,1} d_{0,-1} - 4t^2 d_{0,1} d_{1,-1} \\
\alpha_4(t) &= t^2 d_{1,0}^2 - 4t^2 d_{1,1} d_{1,-1} \\
\beta_0(t) &= t^2 d_{0,-1}^2 - 4t^2 d_{1,-1} d_{-1,-1} \\
\beta_1(t) &= 2t^2 d_{0,-1} d_{0,0} - 2t d_{0,-1} - 4t^2 d_{1,-1} d_{-1,0} - 4t^2 d_{1,0} d_{-1,-1} \\
\beta_2(t) &= t^2 d_{0,0}^2 - 2t d_{0,0} + 1 + 2t^2 d_{0,-1} d_{0,1} - 4t^2 d_{1,-1} d_{-1,1} - 4t^2 d_{1,0} d_{-1,0} - 4t^2 d_{1,1} d_{-1,-1} \\
\beta_3(t) &= -2t d_{0,1} + 2t^2 d_{0,0} d_{0,1} - 4t^2 d_{1,1} d_{-1,0} - 4t^2 d_{1,0} d_{-1,1} \\
\beta_4(t) &= t^2 d_{0,1}^2 - 4t^2 d_{1,1} d_{-1,1}.
\end{aligned}$$

Note that $\Delta_1([x_0:x_1])$ (resp. $\Delta_2([y_0:y_1])$) is of degree 4 and so has four roots counted with multiplicities a_1, a_2, a_3, a_4 (resp. b_1, b_2, b_3, b_4) in $\mathbb{P}^1(\mathbb{C})$. If the discriminant $\Delta_1([x_0:x_1])$ (resp. $\Delta_2([y_0:y_1])$) has a double root; up to renumbering, we can assume that $a_1 = a_2$ (resp. $b_1 = b_2$).

Lemma 3.8 *Assume that the model of the walk is nondegenerate and belongs to the first family of* (G0)

Then, the *walk* has genus zero and the singular point of $\overline{E_t}$ is $\Omega = ([0:1],[0:1])$, that is, $a_1 = a_2 = [0:1]$ (resp. $b_1 = b_2 = [0:1]$) is a double root of $\Delta_1([x_0:x_1])$ (resp. $\Delta_2([y_0:y_1])$). The other roots are distinct from one another and from the double root and are given by

	$a_1 = a_2$	$b_1 = b_2$
	$[0:1]$	$[0:1]$
	a_3	a_4
$\alpha_4(t) \neq 0$	$\left[-\alpha_3(t) - \sqrt{\alpha_3(t)^2 - 4\alpha_2(t)\alpha_4(t)} : 2\alpha_4(t)\right]$	$\left[-\alpha_3(t) + \sqrt{\alpha_3(t)^2 - 4\alpha_2(t)\alpha_4(t)} : 2\alpha_4(t)\right]$
$\alpha_4(t) = 0$	$[1:0]$	$[-\alpha_2(t) : \alpha_3(t)]$
	b_3	b_4
$\beta_4(t) \neq 0$	$\left[-\beta_3(t) - \sqrt{\beta_3(t)^2 - 4\beta_2(t)\beta_4(t)} : 2\beta_4(t)\right]$	$\left[-\beta_3(t) + \sqrt{\beta_3(t)^2 - 4\beta_2(t)\beta_4(t)} : 2\beta_4(t)\right]$
$\beta_4(t) = 0$	$[1:0]$	$[-\beta_2(t) : \beta_3(t)]$

Remark 3.9 We can extend Lemma 3.8 to the other configurations in (G0) by using the following remarks:

(1) Replacing $([x_0:x_1],[y_0:y_1])$ by $([x_0:x_1],[y_1:y_0])$, which corresponds to the variable change $(x,y) \mapsto (x, y^{-1})$, amounts to consider a *weighted model of walk* with *weights* $d'_{i,j} := d_{i,-j}$. This can be used to extend Lemma 3.8 to the second configuration of (G0); for instance, the singular point of $\overline{E_t}$ is $\Omega = ([0:1],[1:0])$ in that case.
(2) Replacing $([x_0:x_1],[y_0:y_1])$ by $([x_1:x_0],[y_1:y_0])$ amounts to consider a *weighted model of walk* with *weights* $d'_{i,j} := d_{-i,-j}$. This can be used to extend Lemma 3.8 to the third configuration of (G0); for instance, the singular point of $\overline{E_t}$ is $\Omega = ([1:0],[1:0])$ in that case.
(3) Replacing $([x_0:x_1],[y_0:y_1])$ by $([x_1:x_0],[y_0:y_1])$ amounts to consider a *weighted model of walk* with *weights* $d'_{i,j} := d_{-i,j}$. This can be used to extend Lemma 3.8 to the fourth configuration of (G0); for instance, the singular point of $\overline{E_t}$ is $\Omega = ([1:0],[0:1])$ in that case.

Remark 3.10 Note that if we consider the x_3, x_4 (resp. y_3, y_4) defined in [FIM17, Chapter 6], we have the equality of sets $\{a_3, a_4\} = \{x_3, x_4\}$ and $\{b_3, b_4\} = \{y_3, y_4\}$, but do not have necessarily $a_i = x_i$, $b_j = y_j$, with $3 \leq i, j \leq 4$.

Proof of Lemma 3.8 We shall prove the lemma for $\Delta_2([y_0:y_1])$, the proof for $\Delta_1([x_0:x_1])$ being similar. By assumption, $d_{-1,-1} = d_{-1,0} = d_{0,-1} = 0$.

Then, $\alpha_0(t) = \alpha_1(t) = 0$. Therefore, the discriminant $\Delta_2([y_0 : y_1])$ has a double root at $[0 : 1]$ and we can write

$$\Delta_2([y : 1]) = \beta_2(t)y^2 + \beta_3(t)y^3 + \beta_4(t)y^4.$$

Since t is transcendental over $\mathbb{Q}(d_{i,j})$, we see that the coefficient of y^2 is nonzero. Therefore $[0 : 1]$ is precisely a double root of $\Delta_2([y_0 : y_1])$. To see that b_3 and b_4 are distinct, we calculate the discriminant of $\Delta_2([y : 1])/y^2$, which is almost the same as the one we considered in the proof of Lemma 3.2. This is a polynomial of degree four in t with the following coefficients:

term	coefficient
t^4	$-16(4d_{-1,1}d_{1,-1}d_{1,1} - d_{1,-1}d_{1,0}^2 - d_{0,0}^2d_{1,1} + d_{0,0}d_{0,1}d_{1,0} - d_{0,1}^2d_{1,-1})d_{-1,1}$
t^3	$-16(2d_{0,0}d_{1,1} - d_{0,1}d_{1,0})d_{-1,1}$
t^2	$16d_{-1,1}d_{1,1}$
t	0
1	0

Let us prove that if $\Delta_2([y_0 : y_1])$ has a double root different from $[0 : 1]$, all the above coefficients must be zero. Recalling that $d_{-1,1}d_{1,-1} \neq 0$, from the coefficient of t^2, we must have $d_{1,1} = 0$. From the coefficient of t^3, we have that $d_{0,1} = 0$ or $d_{1,0} = 0$. From the coefficient of t^4, we get in both cases $d_{0,1} = d_{1,0} = 0$. This implies that the model of the *walk* would be *degenerate*, a contradiction. The formulas for b_3 and b_4 follow from the quadratic formula.

4 Involutive Automorphisms of the Kernel Curve

Following [BMM10, Section 3], [KY15, Section 3] or [FIM17], we consider the involutive rational functions

$$i_1, i_2 : \mathbb{C}^2 \dashrightarrow \mathbb{C}^2$$

given by

$$i_1(x, y) = \left(x, \frac{A_{-1}(x)}{A_1(x)y}\right) \text{ and } i_2(x, y) = \left(\frac{B_{-1}(y)}{B_1(y)x}, y\right).$$

Note that i_1, i_2 are "only" rational functions in the sense that they are a priori not defined when the denominators vanish. The dashed arrow notation used above and in the rest of the paper is a classical notation for rational functions.

The rational functions i_1, i_2 induce involutive rational functions

$$\iota_1, \iota_2 : \overline{E_t} \dashrightarrow \overline{E_t}$$

given by

$$\iota_1([x_0 : x_1], [y_0 : y_1]) = \left([x_0 : x_1], \left[\frac{A_{-1}(\frac{x_0}{x_1})}{A_1(\frac{x_0}{x_1})\frac{y_0}{y_1}} : 1 \right] \right),$$
$$\text{and } \iota_2([x_0 : x_1], [y_0 : y_1]) = \left(\left[\frac{B_{-1}(\frac{y_0}{y_1})}{B_1(\frac{y_0}{y_1})\frac{x_0}{x_1}} : 1 \right], [y_0 : y_1] \right).$$

Again, these functions are a priori not defined where the denominators vanish. However, the following result shows that, actually, this is only an "apparent problem": ι_1 and ι_2 can be extended into endomorphisms of $\overline{E_t}$. We recall that a rational map $f : \overline{E_t} \dashrightarrow \overline{E_t}$ is an endomorphism if it is regular at any $P \in \overline{E_t}$, i.e. if f can be represented in suitable affine charts containing P and $f(P)$ by a rational function with nonvanishing denominator at P. More generally, given X and Y algebraic varieties, we say that $f : X \dashrightarrow Y$ is a morphism if f can be represented by two suitable affine charts containing P and $f(P)$ respectively, by a rational function with nonvanishing denominator at P.

Proposition 4.1 *The rational maps $\iota_1, \iota_2 : \overline{E_t} \dashrightarrow \overline{E_t}$ can be extended into involutive automorphisms of $\overline{E_t}$.*

Proof Note that $\iota_1(x, y)$ is well-defined if the x_i and the y_j are nonzero and if $A_1(\frac{x_0}{x_1})\frac{y_0}{y_1}$ is nonzero. This excludes at most finitely many $(x, y) \in \overline{E_t}$ and, hence, there exists a finite set $\mathcal{S}_0 \subset \overline{E_t}$ such that ι_1 is well defined on $\overline{E_t} \setminus \mathcal{S}_0$. The map ι_1 induces a bijection from $\overline{E_t} \setminus \mathcal{S}_0$ to $\overline{E_t} \setminus \mathcal{S}_1$, where $\mathcal{S}_1$ is a finite set. The same holds for ι_2. We have to prove that $\iota_1, \iota_2 : \overline{E_t} \dashrightarrow \overline{E_t}$ can be extended into endomorphisms of $\overline{E_t}$. According to Proposition 3.1, if the curve $\overline{E_t}$ has genus one, then it is smooth and the result follows from [Har77, Proposition 6.8, p. 43].

It remains to study the case when $\overline{E_t}$ has genus zero. In that case, Proposition 3.1 ensures that $\overline{E_t}$ has a unique singularity Ω. It follows from [Har77, Proposition 6.8, p. 43] that ι_1 and ι_2 can be uniquely extended into morphisms $\overline{E_t} \setminus \{\Omega\} \to \overline{E_t}$ still denoted by ι_1 and ι_2. It remains to study ι_1 and ι_2 at Ω. Let us first assume that the *walk* under consideration belongs to the family of the first configuration of (G0). Lemma 3.8 ensures that $\Omega = ([0 : 1], [0 : 1])$. For $([x : 1], [y : 1]) \in \overline{E_t}$, the equation $K(x, y, t) = 0$ implies that

$$\frac{A_{-1}(x)}{A_1(x)y} = \frac{1}{tA_1(x)} - \frac{A_0(x)}{A_1(x)} - y = \frac{x}{t\widetilde{A}_1(x)} - \frac{\widetilde{A}_0(x)}{\widetilde{A}_1(x)} - y \tag{4.1}$$

where $\widetilde{A}_0(x) = xA_0(x) = d_{-1,0} + d_{0,0}x + d_{1,0}x^2$ and $\widetilde{A}_1(x) = xA_1(x) = d_{-1,1} + d_{0,1}x + d_{1,1}x^2$. Since $d_{-1,1} \neq 0$, $\widetilde{A}_1(x)$ does not vanish at $x = 0$. Since $d_{-1,0} = 0$, $\widetilde{A}_0(x)$ vanishes at $x = 0$. So, (4.1) shows that ι_1 is regular at Ω and that $\iota_1(\Omega) = \Omega$. The argument for ι_2 is similar.

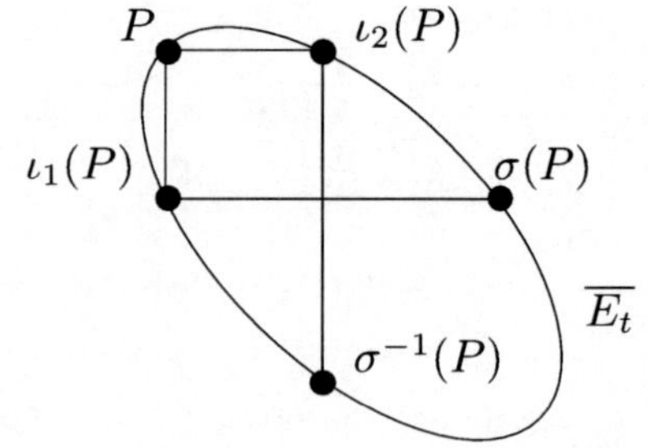

Fig. 1 The maps ι_1, ι_2 restricted to the *kernel curve* $\overline{E_t}$

The other cases listed in (G0) can be treated similarly using a reduction argument as in Remark 3.9.

We also consider the automorphism of $\overline{E_t}$ defined by

$$\sigma = \iota_2 \circ \iota_1.$$

It is easily seen that ι_1 and ι_2 are the vertical and horizontal switches of $\overline{E_t}$ (see Fig. 1), i.e. for any $P = (x, y) \in \overline{E_t}$, we have

$$\{P, \iota_1(P)\} = \overline{E_t} \cap (\{x\} \times \mathbb{P}^1(\mathbb{C})) \text{ and } \{P, \iota_2(P)\} = \overline{E_t} \cap (\mathbb{P}^1(\mathbb{C}) \times \{y\}).$$

We now give a couple of lemmas for later use.

Lemma 4.2 *A point $P = ([x_0 : x_1], [y_0 : y_1]) \in \overline{E_t}$ is fixed by ι_1 if and only if $\Delta_1([x_0 : x_1]) = 0$. A point $P = ([x_0 : x_1], [y_0 : y_1]) \in \overline{E_t}$ is fixed by ι_2 if and only if $\Delta_2([x_0 : x_1]) = 0$.*

Proof Assume that P is fixed by ι_1. Then, the polynomial $[y_0 : y_1] \mapsto \overline{K}(x_0, x_1, y_0, y_1, t)$ has a double root, meaning that the discriminant is zero. This is exactly $\Delta_1([x_0 : x_1]) = 0$. Conversely, $\Delta_1([x_0 : x_1]) = 0$ implies that $[y_0 : y_1] \mapsto \overline{K}(x_0, x_1, y_0, y_1, t)$ has a double root and therefore P is fixed by ι_1. The proof for ι_2 is similar.

The fixed points of ι_1 have y-coordinates that are the double roots of $y \mapsto \overline{K}(x_0, x_1, y, t)$, i.e. they are the roots of the discriminant. By Lemma 3.8 and Remark 3.9, there are 3 points of $\overline{E_t}$ that are fixed by ι_1. A similar statement holds for ι_2. As is shown in the following lemma, one of them plays a particular role.

Lemma 4.3 *Let $P \in \overline{E_t}$. The following statements are equivalent:*

(1) *P is fixed by ι_1 and ι_2;*
(2) *P is a singular point of $\overline{E_t}$;*
(3) *P is fixed by $\sigma = \iota_2 \circ \iota_1$.*

Proof Let $P = ([a : b], [c : d]) \in \overline{E_t}$. From Proposition 3.1, we have that P is a singular point if and only if $\Delta_1([x_0 : x_1])$ and $\Delta_2([y_0 : y_1])$ vanish at $[a : b]$ and $[c : d]$ respectively. We conclude with Lemma 4.2, that (1) is equivalent to (3).

Clearly, (1) implies (3). It remains to prove that (3) implies (1). Assume that $P = (a_1, b_1)$ is fixed by σ. After writing $\iota_1(P) = (a_1, b_1')$ and $\iota_2(\iota_1(P)) = (a_1', b_1')$, it is clear that $\sigma(P) = P$ implies successively $\iota_1(P) = P$ and $\iota_2(P) = P$.

5 Uniformization of the Kernel Curve

We still consider a *weighted model of nondegenerate walk*. The aim of this section is to give an explicit uniformization of $\overline{E_t}$. Thanks to Proposition 3.1, the latter may have genus zero or one. Although there are algorithms to compute such uniformizations, see for instance [vH97, SWPD08], our presentation of explicit uniformizations allows us to understand in detail the pull-backs of σ, ι_1 and ι_2 and therefore their effect on the generating series of the models of walks. Let us start with the genus zero case.

5.1 Genus Zero Case

Let us consider a *nondegenerate weighted model of walks* of genus zero. Thank to Corollary 3.6 combined with Remark 3.9, it suffices to consider the situation where the *nondegenerate model of walk* arises from the following family

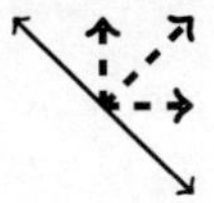

Genus zero curves may be parametrized with maps $\phi : \mathbb{P}^1(\mathbb{C}) \to \overline{E_t}$ which are bijective outside a finite set. The aim of this subsection, achieved with Proposition 5.7, is to find such a parametrization explicitly. Although we could have just written down the formula for this parametrization and verified its properties, we have preferred to explain how the formula arises. This requires a preliminary study of the automorphisms of $\mathbb{P}^1(\mathbb{C})$ obtained by pulling back the maps σ, ι_1 and ι_2 by ϕ, which is done with a series of lemmas preceding Proposition 5.7.

According to Lemma 3.8, $\overline{E_t}$ has a unique singular point $\Omega = (a_1, b_1) = ([0:1], [0:1])$. Moreover $\Delta_1([x_0 : x_1])$ has degree four with a double root at $a_1 = [0:1]$ and the remaining two roots a_3, a_4 are distinct. We let S_3 and S_4 be the points of $\overline{E_t}$ with first coordinate a_3 and a_4 respectively. Similarly, $\Delta_2([y_0 : y_1])$ has degree four with a double root at $b_1 = [0:1]$ and the remaining two roots b_3, b_4 are distinct. We let S_3' and S_4' be the points of $\overline{E_t}$ with second coordinates b_3 and b_4 respectively.

Since $\overline{E_t}$ has genus zero, there is a rational parametrization of $\overline{E_t}$ [Ful89, Page 198, Ex.1], i.e. there exists a birational map

$$\begin{array}{rcl}\phi = (\check{x}, \check{y}) : \mathbb{P}^1(\mathbb{C}) & \dashrightarrow & \overline{E_t} \\ s & \mapsto & (\check{x}(s), \check{y}(s)).\end{array}$$

To simplify the notation, we will abusively denote $(\check{x}, \check{y})$ by (x, y). We will now follow the ideas contained in [FIM17] to produce an explicit uniformization of $\overline{E_t}$ in Proposition 5.7. If we set $t = 1$, we recover the uniformization of [FIM17, Section 6.4.3]. However, it is not clear if their proof can be simply modified to hold in our context, so we preferred to give proofs here with a slightly different strategy.

Lemma 5.1 *The map ϕ is surjective and induces a bijection between $\mathbb{P}^1(\mathbb{C}) \setminus \phi^{-1}(\Omega)$ and $\overline{E_t} \setminus \{\Omega\}$.*

Proof As any nonconstant rational map from $\mathbb{P}^1(\mathbb{C})$ to a projective curve, ϕ is actually a surjective morphism of curves, see [Ful89, Corollary 1, Page 160]. Since Ω is the unique singular point of $\overline{E_t}$, the result follows.

The maps $x, y : \mathbb{P}^1(\mathbb{C}) \to \mathbb{P}^1(\mathbb{C})$ are surjective morphisms of curves as well.

We let $s_3, s_4 \in \mathbb{P}^1(\mathbb{C})$ (resp. $s_3', s_4' \in \mathbb{P}^1(\mathbb{C})$) be such that $S_3 = \phi(s_3)$ and $S_4 = \phi(s_4)$ (resp. $S_3' = \phi(s_3')$ and $S_4' = \phi(s_4')$).

We will need to know the cardinality of $x^{-1}(P)$ (resp. $y^{-1}(P)$) for $P \in \mathbb{P}^1(\mathbb{C})$. This quantity might depend on P but it is a general fact about morphisms of curves that the cardinality of $x^{-1}(P)$ (resp. $y^{-1}(P)$) is constant for P outside a finite subset of $\mathbb{P}^1(\mathbb{C})$. This common value is called the degree of x (resp. y). Inside this finite set, the cardinality can only fall, so is less than the degree.

Lemma 5.2 *The morphisms $x, y : \mathbb{P}^1(\mathbb{C}) \to \mathbb{P}^1(\mathbb{C})$ have degree two.*

Proof We will see that this is a consequence of the fact that $\overline{E_t}$ is a biquadratic curve. Observe that by Lemma 5.1, ϕ induces a bijection between $\mathbb{P}^1(\mathbb{C}) \setminus \phi^{-1}(\Omega)$ and $\overline{E_t} \setminus \{\Omega\}$. Any $(a, b) \in \overline{E_t}$ with $a \neq a_1$ cannot be Ω and therefore has a unique preimage by ϕ. Additionally, let Z be the finite set of zeros of the discriminant Δ_1. Then, for any $a \notin Z$, the cardinality $(\{a\} \times \mathbb{P}^1(\mathbb{C})) \cap \overline{E_t}$ is two. Since $x^{-1}(a) = \phi^{-1}((\{a\} \times \mathbb{P}^1(\mathbb{C})) \cap \overline{E_t})$ and $a_1 \in Z$, it follows that if $a \notin Z$, the cardinality of $x^{-1}(a)$ is two. So, x has degree two. The argument for y is similar.

Since ϕ is a birational map, the involutive automorphisms ι_1, ι_2 of $\overline{E_t}$ induce involutive automorphisms $\tilde{\iota}_1, \tilde{\iota}_2$ of $\mathbb{P}^1(\mathbb{C})$ via ϕ. Similarly, σ induces an automorphism $\tilde{\sigma}$ of $\mathbb{P}^1(\mathbb{C})$. In other words, we have the commutative diagrams

$$\begin{array}{ccc} \overline{E_t} & \xrightarrow{\iota_k} & \overline{E_t} \\ \uparrow \phi & & \uparrow \phi \\ \mathbb{P}^1(\mathbb{C}) & \xrightarrow[\tilde{\iota}_k]{} & \mathbb{P}^1(\mathbb{C}) \end{array} \quad \text{and} \quad \begin{array}{ccc} \overline{E_t} & \xrightarrow{\sigma} & \overline{E_t} \\ \uparrow \phi & & \uparrow \phi \\ \mathbb{P}^1(\mathbb{C}) & \xrightarrow[\tilde{\sigma}]{} & \mathbb{P}^1(\mathbb{C}) \end{array}$$

Note that since by Lemma 5.1 ϕ induces a bijection between $\mathbb{P}^1(\mathbb{C}) \setminus \phi^{-1}(\Omega)$ and $\overline{E_t} \setminus \{\Omega\}$ and Ω is fixed by ι_1, ι_2, see Lemma 4.3, the group generated by ι_1 and ι_2 is isomorphic to the group generated by $\tilde{\iota}_1$ and $\tilde{\iota}_2$. Thus we recover the same group as in [BMM10] for instance. Note that although the cardinal of the group may depends upon t, see Remark 5.13, since the maps ι_1, ι_2 are defined on $\mathbb{Q}(d_{i,j})(t)$, two distinct values of t transcendental over $\mathbb{Q}(d_{i,j})$ lead to isomorphic groups. We summarize some remarks in the following lemmas.

Lemma 5.3 *We have $x = x \circ \tilde{\iota}_1$ and $y = y \circ \tilde{\iota}_2$.*

Proof We obtain $x = x \circ \tilde{\iota}_1$ by equating the first coordinates in the equality $\phi \circ \tilde{\iota}_1 = \iota_1 \circ \phi$ and we obtain $y = y \circ \tilde{\iota}_2$ by equating the second coordinates in the equality $\phi \circ \tilde{\iota}_2 = \iota_2 \circ \phi$.

Lemma 5.4 *Let $P = \phi(s) \in \overline{E_t}$ and let $k \in \{1, 2\}$. We have:*

- *if $\tilde{\iota}_k(s) = s$, then $\iota_k(P) = P$;*
- *if $P \neq \Omega$ and $\iota_k(P) = P$, then $\tilde{\iota}_k(s) = s$.*

Furthermore the map $\tilde{\iota}_1$ (resp. $\tilde{\iota}_2$) has exactly two fixed points, namely s_3 and s_4 (resp. s_3' and s_4').

Proof We have $\iota_k(P) = \iota_k(\phi(s)) = \phi(\tilde{\iota}_k(s))$. The first assertion is now clear, and the second one follows from the fact that ϕ is injective on $\overline{E_t} \setminus \phi^{-1}(\Omega)$. Since $S_3, S_4 \neq \Omega$ are fixed by ι_1, this shows that s_3 and s_4 are fixed by $\tilde{\iota}_1$. Similar proof holds for $\tilde{\iota}_2$.

It remains to prove that there are exactly two points fixed by $\tilde{\iota}_k$. To the contrary, assume that there is a third point fixed by $\tilde{\iota}_k$. Since $\tilde{\iota}_k$ is an automorphism of $\mathbb{P}^1(\mathbb{C})$, i.e. an homography, with three fixed points, it is the identity. This is a contradiction and concludes the proof of the lemma.

Lemma 5.5 *The preimage of Ω by ϕ has two elements.*

Proof We know that $x, y : \mathbb{P}^1(\mathbb{C}) \to \mathbb{P}^1(\mathbb{C})$ have degree two, so $\phi^{-1}(\Omega)$ has one or two elements. Suppose that $\phi^{-1}(\Omega)$ has exactly one element, say s_1. In virtue of $\phi(s_3) = S_3$ and $\phi(s_4) = S_4$, s_1 is different to s_3, s_4. Since $\phi(\tilde{\iota}_1(s_1)) = \iota_1(\phi(s_1)) = \iota_1(\Omega) = \Omega$, we have $\tilde{\iota}_1(s_1) = s_1$. This contradicts Lemma 5.4. Hence, $\phi^{-1}(\Omega)$ has two elements.

From now on, we define $s_1 \neq s_2$ the two preimages of Ω by ϕ, that is

$$\{s_1, s_2\} := \phi^{-1}(\Omega).$$

Lemma 5.6 *The map $\tilde{\iota}_1$ (resp. $\tilde{\iota}_2$) interchanges s_1 and s_2. The map $\tilde{\sigma}$ has exactly two distinct fixed points: s_1 and s_2.*

Proof We have $\phi(s) = \Omega$ if and only if $s = s_1$ or s_2 and the equality $\iota_1(\phi(s)) = \phi(\tilde{\iota}_1(s))$ shows that $\tilde{\iota}_1$ induces a permutation of $\phi^{-1}(\Omega) = \{s_1, s_2\}$. By Lemma 5.4, s_1 is not fixed by $\tilde{\iota}_1$, showing that the permutation is not the identity, i.e. $\tilde{\iota}_1$ interchanges s_1 and s_2.

The proof for $\tilde{\iota}_2$ is similar.

As any homography which is not the identity, $\tilde{\sigma}$ has at most two fixed points in $\mathbb{P}^1(\mathbb{C})$. It only remains to prove that s_1 and s_2 are fixed by $\tilde{\sigma}$, and this is indeed the case because $\tilde{\sigma} = \tilde{\iota}_2 \circ \tilde{\iota}_1$ and $\tilde{\iota}_1, \tilde{\iota}_2$ interchange s_1 and s_2.

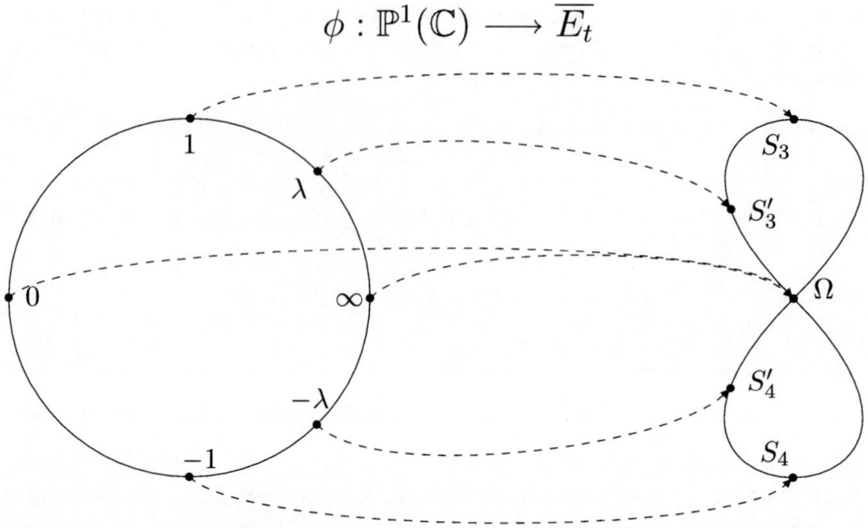

Fig. 2 An idealized representation of the uniformization map used in Proposition 5.7. The left hand side represents the complex Riemann sphere and the right hand side the curve $\overline{E_t}$, seen as an abstract complex algebraic curve

We are now ready to give an explicit expression of ϕ. The coefficients α_i, β_i of the discriminants in this situation are given by the formulas

$$\begin{aligned}
\alpha_0(t) &= \alpha_1(t) = 0\\
\beta_0(t) &= \beta_1(t) = 0\\
\alpha_2(t) &= \beta_2(t) = 1 - 2td_{0,0} + t^2d_{0,0}^2 - 4t^2d_{-1,1}d_{1,-1}\\
\alpha_3(t) &= 2t^2d_{1,0}d_{0,0} - 2td_{1,0} - 4t^2d_{0,1}d_{1,-1}\\
\beta_3(t) &= 2t^2d_{0,1}d_{0,0} - 2td_{0,1} - 4t^2d_{1,0}d_{-1,1}\\
\alpha_4(t) &= t^2(d_{1,0}^2 - 4d_{1,1}d_{1,-1})\\
\beta_4(t) &= t^2(d_{0,1}^2 - 4d_{1,1}d_{-1,1}).
\end{aligned}$$

Note that for $k = 3, 4$, $\beta_k(t)$, may be obtained from $\alpha_k(t)$ by interchanging $d_{1,0}$ with $d_{0,1}$ and $d_{1,-1}$ with $d_{-1,1}$.

Proposition 5.7 *An explicit parametrization $\phi : \mathbb{P}^1(\mathbb{C}) \to \overline{E_t}$ such that*

$$\tilde{\iota}_1(s) = \frac{1}{s},\ \tilde{\iota}_2(s) = \frac{\lambda^2}{s} = \frac{q}{s} \text{ and } \widetilde{\sigma}(s) = qs$$

for a certain $\lambda \in \mathbb{C}^*$ is given by

$$\phi(s) = \left(\frac{4\alpha_2(t)}{\sqrt{\alpha_3(t)^2 - 4\alpha_2(t)\alpha_4(t)}(s + \frac{1}{s}) - 2\alpha_3(t)}, \frac{4\beta_2(t)}{\sqrt{\beta_3(t)^2 - 4\beta_2(t)\beta_4(t)}(\frac{s}{\lambda} + \frac{\lambda}{s}) - 2\beta_3(t)} \right).$$

Moreover, we have, see Fig. 2

$$\begin{aligned} x(0) &= x(\infty) = a_1,\ x(1) = a_3,\ x(-1) = a_4, \\ y(0) &= y(\infty) = b_1,\ y(\lambda) = b_3,\ y(-\lambda) = b_4. \end{aligned}$$

Proof According to Lemma 5.6, $\tilde{\iota}_1$ is an involutive homography with fixed points s_3 and s_4, so there exists an homography h such that $h(s_3) = 1$, $h(s_4) = -1$ and $h \circ \tilde{\iota}_1 \circ h^{-1}(s) = 1/s$. Up to replacing ϕ by $\phi \circ h$, we can assume that $s_3 = 1$, $s_4 = -1$ and $\tilde{\iota}_1(s) = \frac{1}{s}$. Since $s_1 \neq s_2$, we can assume up to renumbering that $s_1 \neq \infty$. Let us consider the homography $k(s) = \frac{s-s_1}{-s_1 s+1}$. Note that k commutes with $s \mapsto 1/s$, so changing ϕ by $\phi \circ k$ does not affect $\tilde{\iota}_1$, and we can also assume that $s_1 = [0:1]$ and $s_2 = [1:0]$. Lemma 5.2 and Lemma 5.3 ensure that the morphism $x : \mathbb{P}^1(\mathbb{C}) \to \mathbb{P}^1(\mathbb{C})$ has degree two and satisfies $x(s) = x(1/s)$ for all $s \in \mathbb{P}^1(\mathbb{C})$. Since the morphism $x : \mathbb{P}^1(\mathbb{C}) \to \mathbb{P}^1(\mathbb{C})$ has degree two, see Lemma 5.2, it follows that

$$x(s) = \frac{a(s+1/s)+b}{c(s+1/s)+d}$$

for some $a, b, c, d \in \mathbb{C}$. We have $x(s_1) = x([0:1]) = a_1 = 0$, $x(s_2) = x([1:0]) = a_1 = 0$, $x(s_3) = x([1:1]) = a_3$ and $x(s_4) = x([-1:1]) = a_4$. The equality $x([1:0]) = 0$ implies $a = 0$. The equalities $x([1:1]) = a_3$ and $x([-1:1]) = a_4$ imply

$$x(s) = \frac{4a_3a_4}{(a_4-a_3)(s+\frac{1}{s}) + 2(a_3+a_4)}.$$

The known expressions for a_3 and a_4 given in Lemma 3.8 lead to the expected expression for $x(s)$.

According to Lemma 5.6, $\tilde{\iota}_2$ is an homography interchanging $[0:1]$ and $[1:0]$, so $\tilde{\iota}_2(s) = \frac{\lambda^2}{s}$ for some $\lambda \in \mathbb{C}^*$. Up to renumbering, we have $s_3' = \lambda$ and $s_4' = -\lambda$. Using the fact that the morphism $y : \mathbb{P}^1(\mathbb{C}) \to \mathbb{P}^1(\mathbb{C})$ has degree two and is invariant by $\tilde{\iota}_2$, and arguing as we did above for x, we see that there exist $\alpha, \beta, \gamma, \eta \in \mathbb{C}$ such that

$$y(s) = \frac{\alpha(\frac{s}{\lambda} + \frac{\lambda}{s}) + \beta}{\gamma(\frac{s}{\lambda} + \frac{\lambda}{s}) + \eta}.$$

The equality $y([1:0]) = 0$ implies $\alpha = 0$. Using the equalities $y(s_3') = y(\lambda) = b_3$ and $y(s_4') = y(-\lambda) = b_4$, and arguing as we did above for x, we obtain the expected expression for $y(s)$.

Remark 5.8 (1) The uniformization is not unique. More precisely, the possible uniformizations are of the form $\phi \circ h$, where h is an homography. However, if one requires that h fixes setwise $\{[0:1], [1:0]\}$ then q is uniquely defined up to inversion.

(2) The real q or q^{-1} specializes for $t = 1$ to the real ρ^2 in [FIM17, Page 178].

The following proposition determines q up to its inverse. We include this for completeness.

Proposition 5.9 [DHRS20], Proposition 1.7, Corollary 1.10). *One of the two complex numbers q or q^{-1} is equal to*

$$\frac{-1+d_{0,0}t-\sqrt{(1-d_{0,0}t)^2-4d_{1,-1}d_{-1,1}t^2}}{-1+d_{0,0}t+\sqrt{(1-d_{0,0}t)^2-4d_{1,-1}d_{-1,1}t^2}}.$$

Furthermore, $q \in \mathbb{R}\setminus\{\pm 1\}$.

Remark 5.10 This implies that σ and $\tilde{\sigma}$ have infinite order (see also [BMM10, FR11]). Because σ is induced on $\overline{E_t}$ by $i_1 \circ i_2$, we find that $i_1 \circ i_2$ has infinite order as well. It follows that the *group of the walk* introduced in [BMM10], which is by definition the group generated by i_1 and i_2, has infinite order. Note that this was proved in [BMM10] using a valuation argument.

Remark 5.11 We stress the fact that since $\phi(s)$, q and $\overline{E_t}$ depend continuously on t and the set of transcendental number over $\mathbb{Q}(d_{i,j})$ in $]0,1[$ is dense in $]0,1[$, we deduce that Proposition 5.7 and Proposition 5.9 stay valid for every $t \in]0,1[$.

5.2 Genus One Case

In this section, we consider the uniformization problem in the genus one context. This problem has been solved in [DR19]. We recall below the main result of [DR19], for the sake of completeness. Let us consider a *nondegenerate model of walk* of genus one. By Proposition 3.1, $\overline{E_t}$ is a smooth curve of genus one and, by Corollary 3.6, this corresponds to the situation where the step set is not included in any half plane whose boundary passes through (0, 0). By [WW96, Chapter XX], it is biholomorphic to $\mathbb{C}/(\mathbb{Z}\omega_1+\mathbb{Z}\omega_2)$ for some lattice $\mathbb{Z}\omega_1+\mathbb{Z}\omega_2$ of $\mathbb{C}$ via some $(\mathbb{Z}\omega_1+\mathbb{Z}\omega_2)$-periodic holomorphic map

$$\begin{aligned}\Lambda : \mathbb{C} &\rightarrow \overline{E_t}\\ \omega &\mapsto (\mathfrak{q}_1(\omega), \mathfrak{q}_2(\omega)),\end{aligned} \tag{5.1}$$

where $\mathfrak{q}_1, \mathfrak{q}_2$ are rational functions of $\wp$ and its derivative $d\wp/d\omega$, and $\wp$ is the Weierstrass function associated with the lattice $\mathbb{Z}\omega_1+\mathbb{Z}\omega_2$:

$$\wp(\omega)=\wp(\omega;\omega_1,\omega_2) := \frac{1}{\omega^2}+\sum_{(\ell_1,\ell_2)\in\mathbb{Z}^2\setminus\{(0,0)\}}\left(\frac{1}{(\omega+\ell_1\omega_1+\ell_2\omega_2)^2}-\frac{1}{(\ell_1\omega_1+\ell_2\omega_2)^2}\right). \tag{5.2}$$

Then, the field of meromorphic functions on $\overline{E_t}$ is isomorphic to the field of meromorphic functions on $\mathbb{C}/(\mathbb{Z}\omega_1+\mathbb{Z}\omega_2)$, which is itself isomorphic to the field of meromorphic functions on $\mathbb{C}$ that are (ω_1,ω_2)-periodic. The latter field is equal to $\mathbb{C}(\wp,\wp')$, see [WW96].

The maps ι_1, ι_2 and σ may be lifted to the ω-plane $\mathbb{C}$. We denote these lifts by $\widetilde{\iota}_1$, $\widetilde{\iota}_2$ and $\widetilde{\sigma}$ respectively. So we have the commutative diagrams

$$\begin{array}{ccc} \overline{E_t} & \xrightarrow{\iota_k} & \overline{E_t} \\ \Lambda\uparrow & & \uparrow\Lambda \\ \mathbb{C} & \xrightarrow[\widetilde{\iota}_k]{} & \mathbb{C} \end{array} \qquad\qquad \begin{array}{ccc} \overline{E_t} & \xrightarrow{\sigma} & \overline{E_t} \\ \Lambda\uparrow & & \uparrow\Lambda \\ \mathbb{C} & \xrightarrow[\widetilde{\sigma}]{} & \mathbb{C} \end{array}$$

The following result has been proved

- in [FIM17, Section 3.3] when $t = 1$,
- in [Ras12] in the *unweighted* case for general $0 < t < 1$, not necessarily transcendental over $\mathbb{Q}(d_{i,j})$,
- in [DR19, Proposition 18] in the *weighted* case, with general $0 < t < 1$, not necessarily transcendental over $\mathbb{Q}(d_{i,j})$.

In what follows, we set $D(\omega) = \Delta_1([\omega : 1])$. Let us introduce $z = 2A(x)y + B(x)$, where $A(x) = t(d_{-1,1} + d_{0,1}x + d_{1,1}x^2)$, and $B(x) = t(d_{-1,0} - \frac{1}{t}x + d_{0,0}x + d_{1,0}x^2)$.

Proposition 5.12 *An explicit uniformization $\Lambda : \mathbb{C} \to \overline{E_t}$ such that*

$$\widetilde{\iota}_1(\omega) = -\omega, \quad \widetilde{\iota}_2(\omega) = -\omega + \omega_3 \quad \textit{and} \quad \widetilde{\sigma}(\omega) = \omega + \omega_3,$$

for a certain $\omega_3 \in \mathbb{C}^*$ is given by

$$\Lambda(\omega) = (x(\omega), y(\omega))$$

where $x(\omega)$ and $y(\omega)$ are given by

	$x(\omega)$	$z(\omega)$
$a_4 \neq [1:0]$	$\left[a_4 + \frac{D'(a_4)}{\wp(\omega) - \frac{1}{6}D''(a_4)} : 1\right]$	$\left[\frac{D'(a_4)\wp'(\omega)}{2(\wp(\omega) - \frac{1}{6}D''(a_4))^2} : 1\right]$
$a_4 = [1:0]$	$[\wp(\omega) - \alpha_2/3 : \alpha_3]$	$[-\wp'(\omega) : 2\alpha_3]$

A suitable choice for the periods (ω_1, ω_2) is given by the elliptic integrals

$$\omega_1 = \mathbf{i}\int_{a_3}^{a_4} \frac{dx}{\sqrt{|D(x)|}} \in \mathbf{i}\mathbb{R}_{>0} \quad \textit{and} \quad \omega_2 = \int_{a_4}^{a_1} \frac{dx}{\sqrt{D(x)}} \in \mathbb{R}_{>0}. \tag{5.3}$$

Note that, according to [DR19, Section 2],

$$\omega_3 = \int_{a_4}^{X_\pm(b_4)} \frac{dx}{\sqrt{D(x)}} \in]0, \omega_2[.$$

Remark 5.13 Contrary to the genus zero situation, the map σ may have finite order.

Acknowledgment The authors want to warmly thank the referees for their detailed and helpful comments.

References

[BBMR15] Olivier Bernardi, Mireille Bousquet-Mélou, and Kilian Raschel, *Counting quadrant walks via Tutte's invariant method*, An extended abstract to appear in *Proceedings of FPSAC 2016*, Discrete Math. Theor. Comput. Sci. Proc., arXiv:1511.04298, 2015.

[BBMR17] Olivier Bernardi, Mireille Bousquet-Mélou, and Kilian Raschel, *Counting quadrant walks via Tutte's invariant method*, Combinatorial Theory (to appear), arXiv:1708.08215, 2017.

[BMM10] Mireille Bousquet-Mélou and Marni Mishna, *Walks with small steps in the quarter plane*, Algorithmic probability and combinatorics, Contemp. Math., vol. 520, Amer. Math. Soc., Providence, RI, 2010, pp. 1–39.

[BRS14] Alin Bostan, Kilian Raschel, and Bruno Salvy, *Non-D-finite excursions in the quarter plane*, J. Combin. Theory Ser. A **121** (2014), 45–63. MR 3115331

[BvHK10] Alin Bostan, Mark van Hoeij, and Manuel Kauers, *The complete generating function for Gessel walks is algebraic*, Proc. Amer. Math. Soc. **138** (2010), no. 9, 3063–3078.

[CLO97] David Cox, John Little, and Donal O'Shea, *Ideals, varieties, and algorithms*, second ed., Undergraduate Texts in Mathematics, Springer-Verlag, New York, 1997, An introduction to computational algebraic geometry and commutative algebra. MR 1417938

[DH19] Thomas Dreyfus and Charlotte Hardouin, *Length derivative of the generating series of walks confined in the quarter plane*, arXiv preprint arXiv:1902.10558 (2019).

[DHRS18] Thomas Dreyfus, Charlotte Hardouin, Julien Roques, and Michael F Singer, *On the nature of the generating series of walks in the quarter plane*, Inventiones mathematicae (2018), 139–203.

[DHRS20] ——, *Walks in the quarter plane, genus zero case*, Journal of Combinatorial Theory, Series A. (2020).

[DR19] Thomas Dreyfus and Kilian Raschel, *Differential transcendence & algebraicity criteria for the series counting weighted quadrant walks*, Publications Mathématiques de Besançon (2019), no. 1, 41–80.

[Dui10] J. Duistermaat, *Discrete integrable systems: QRT maps and elliptic surfaces*, Springer Monographs in Mathematics, vol. 304, Springer-Verlag, New York, 2010.

[DW15] D. Denisov and V. Wachtel, *Random walks in cones*, Ann. Probab. **43** (2015), no. 3, 992–1044. MR 3342657

[FIM99] Guy Fayolle, Roudolf Iasnogorodski, and VA Malyshev, *Random walks in the quarter-plane*, vol. 40, Springer, 1999.

[FIM17] Guy Fayolle, Roudolf Iasnogorodski, and Vadim Malyshev, *Random walks in the quarter plane. Algebraic methods, boundary value problems, applications to queueing systems and analytic combinatorics. 2nd edition, previously published with the subtitle Algebraic methods, boundary value problems and applications.*, vol. 40, Cham: Springer, 2017 (English).

[FR11] G. Fayolle and K. Raschel, *Random walks in the quarter-plane with zero drift: an explicit criterion for the finiteness of the associated group*, Markov Process. Related Fields **17** (2011), no. 4, 619–636.

[Ful84] William Fulton, *Introduction to intersection theory in algebraic geometry*, no. 54, American Mathematical Soc., 1984.

[Ful89] ——, *Algebraic curves*, Advanced Book Classics, Addison-Wesley Publishing Company, Advanced Book Program, Redwood City, CA, 1989, An introduction to algebraic geometry, Notes written with the collaboration of Richard Weiss, Reprint of 1969 original.

[Har77] Robin Hartshorne, *Algebraic geometry*, Springer-Verlag, New York-Heidelberg, 1977, Graduate Texts in Mathematics, No. 52. MR 0463157

[KR12] Irina Kurkova and Kilian Raschel, *On the functions counting walks with small steps in the quarter plane*, Publ. Math. Inst. Hautes Études Sci. **116** (2012), 69–114. MR 3090255

[KY15] Manuel Kauers and Rika Yatchak, *Walks in the quarter plane with multiple steps*, Proceedings of FPSAC 2015, Discrete Math. Theor. Comput. Sci. Proc., Assoc. Discrete Math. Theor. Comput. Sci., Nancy, 2015, pp. 25–36.

[Mis09] Marni Mishna, *Classifying lattice walks restricted to the quarter plane*, J. Combin. Theory Ser. A **116** (2009), no. 2, 460–477. MR 2475028

[MM14] Stephen Melczer and Marni Mishna, *Singularity analysis via the iterated kernel method*, Combin. Probab. Comput. **23** (2014), no. 5, 861–888. MR 3249228

[MR09] Marni Mishna and Andrew Rechnitzer, *Two non-holonomic lattice walks in the quarter plane*, Theoret. Comput. Sci. **410** (2009), no. 38-40, 3616–3630. MR 2553316

[Ras12] Kilian Raschel, *Counting walks in a quadrant: a unified approach via boundary value problems*, J. Eur. Math. Soc. (JEMS) **14** (2012), no. 3, 749–777. MR 2911883

[SWPD08] J Rafael Sendra, Franz Winkler, and Sonia Pérez-Díaz, *Rational algebraic curves*, Algorithms and Computation in Mathematics **22** (2008).

[vH97] Mark van Hoeij, *Rational parametrizations of algebraic curves using a canonical divisor*, Journal of Symbolic Computation **23** (1997), no. 2-3, 209–227.

[WW96] E. Whittaker and G. Watson, *A course of modern analysis*, Cambridge Mathematical Library, Cambridge University Press, Cambridge, 1996, An introduction to the general theory of infinite processes and of analytic functions; with an account of the principal transcendental functions, Reprint of the fourth (1927) edition. MR 1424469

A Survey on the Hypertranscendence of the Solutions of the Schröder's, Böttcher's and Abel's Equations

Gwladys Fernandes

Abstract In 1994, P.-G. Becker and W. Bergweiler [8] listed all the differentially algebraic solutions of three famous functional equations: the Schröder's, Böttcher's and Abel's equations. The proof of this theorem combines various domains of mathematics. This goes from the theory of iteration, which gave birth to these equations, to the algebro-differential notion of coherent families developed by M. Boshernitzan and L. A. Rubel. This survey is an excursion into the history of these equations, in order to enlighten the different pieces of mathematics they bring together and how these parts fit into the result of P.-G. Becker and W. Bergweiler.

1 Introduction

This survey is about three famous functional equations, named after the mathematicians Ernst Schröder, Lucien Böttcher and Niels Henrik Abel. These equations are linked to the local study of the iteration of a rational function $R(z) \in \mathbb{C}(z)$ with complex coefficients around a **fixed point** α. That is a point of $\mathbb{C} \cup \{\infty\}$ such that $R(\alpha) = \alpha$. We can assume, without any loss of generality, that $\alpha = 0$ (see Sect. 2.1 and (8)). Moreover, to avoid the trivial case of Möbius transformations (see (7) for the definition), which is well-understood, we assume that the **degree** of $R(z)$, that is the maximum of the degrees of the coprime polynomials on its numerator and denominator, is at least 2. Then, the equations we are interested in are the following:

1. Let $s = R'(0)$. If $s \neq 0$, then, the Schröder's equation is:

$$f(sz) = R(f(z)), \tag{S}$$

2. If $R(z) = \sum_{n=d}^{+\infty} a_n z^n$, where $d \geq 2$, then the Böttcher's equation is:

G. Fernandes (✉)
Université de Versailles Saint-Quentin-en-Yvelines. 45, avenue des Etats-Unis, 78035 Versailles cedex, France
e-mail: gwladys.fernandes@uvsq.fr

A. Bostan and K. Raschel (eds.), *Transcendence in Algebra, Combinatorics, Geometry and Number Theory*, Springer Proceedings in Mathematics & Statistics 373,
https://doi.org/10.1007/978-3-030-84304-5_4

$$f(z^d) = R(f(z)), \tag{B}$$

3. The Abel's equation is:

$$f(R(z)) = f(z) + 1. \tag{A}$$

In 1994, P.-G. Becker and W. Bergweiler [8] listed all the differentially algebraic solutions of Eqs. (S), (B) and (A). The aim of this survey is to present the proof of this result as a testimony of the richness of the interactions these three equations centralize between various areas of mathematics and how the two authors combine them to get their beautiful result. For more historical details on the theory of iteration, besides the ones given below, we refer to [5] and [2].

Before diving into the history of these equations, let us remind some definitions. First, a formal power series $f(z)$ with coefficients in the complex plane $\mathbb{C}$ is said to be **differentially algebraic** over $\mathbb{C}(z)$ if there exists a non-zero polynomial $P(z, X_0, \ldots, X_n)$ with coefficients in $\mathbb{C}$ such that:

$$P(z, f(z), f'(z), \ldots, f^{(n)}(z)) = 0, \tag{1}$$

where $f^{(n)}(z)$ is the n-th derivative of f with respect to z. We say that f is **differentially transcendental** or **hypertranscendental** over $\mathbb{C}(z)$ if it is not differentially algebraic over $\mathbb{C}(z)$. This notion of differential algebraicity generalises the one of algebraicity. Indeed, a formal power series f with coefficients in $\mathbb{C}$ is said to be **algebraic** over $\mathbb{C}(z)$ if there exists a non-zero polynomial $P(z, X)$ with coefficients in $\mathbb{C}$ such that:

$$P(z, f(z)) = 0, \tag{2}$$

and it is said to be **transcendental** over $\mathbb{C}(z)$ otherwise. Moreover, formal power series $f_1(z), \ldots, f_n(z)$ with coefficients in $\mathbb{C}$ are said to be **algebraically dependent** over $\mathbb{C}(z)$ if there exists a non-zero polynomial $P(z, X_0, \ldots, X_n)$ with coefficients in $\mathbb{C}$ such that:

$$P(z, f_1(z), \ldots, f_n(z)) = 0. \tag{3}$$

If they are not algebraically dependent over $\mathbb{C}(z)$, we say that these functions are **algebraically independent** over $\mathbb{C}(z)$. Thus, a hypertranscendental function is transcendental, and all its derivatives are algebraically independent over $\mathbb{C}(z)$.

Coming back to our topic, the three equations (S), (B) and (A) are introduced for the need of the iteration theory, for which Newton's method is a famous representative, and the research of fixed points of a rational fraction. They reach their zenith with the development of the theory of P. Fatou and G. Julia around 1918, which divides the complex plane into different domains according to the local behaviour of the iterates of a rational fraction, in terms of convergence or divergence to a fixed point. Thus, the rather algebraic problem of iteration of a rational fraction and the determination of its fixed points is linked to the rather analytic theory developed by P. Fatou and G. Julia. The interface between these different domains of mathematics grows in the following years with the work of J. F. Ritt [62] in the 20's, who furnishes the list of all the differentially algebraic solutions of the Schröder's equation near a repelling fixed point (see the definition in Sect. 2.1). This is a first step toward the result of P.-G. Becker and W. Bergweiler [8] we are interested in (Theorem 3.1 of the present paper). After that, the interest for these equations wanes for almost sixty years, with a renewal of enthusiasm in the 80's. Indeed, then, the sets of Fatou and Julia share connexions with dynamical systems and fractals which are very prolific areas at that time. From the algebraic point of view, this surge of interest is visible in 1995 in the result of P.-G. Becker and W. Bergweiler [8], which completes the previous result of J. F. Ritt on the Schröder's equation. The two authors also provide a partial result of Theorem 3.1 one year earlier in [7]. Indeed, they give the list of all the algebraic solutions of the Böttcher's equation (B). In this paper, we present their accomplished result [8], reproduced in Theorem 3.1. This statement provides the list of all the differentially algebraic solutions of the Schröder's, Böttcher's and Abel's equations. The proof of P.-G. Becker and W. Bergweiler relies on the results in [7, 62], a theorem on coherent families developed by M. Boshernitzan and L. A. Rubel [11], and the theory of P. Fatou and G. Julia [25–27, 40].

2 Schröder's, Böttcher's and Abel's Equations: Their Deep Interactions with the Iteration Theory Through History

2.1 The Birth of the Schröder's, Böttcher's and Abel's Equations

The Schröder's equation appears in 1870 in a paper of the same author [68], in link with the strong interest of the author in Newton's method. This consists in finding approximations of real roots of a real-valued function $f(x)$. The idea is the following. Take a real number x_0 and define the following sequence for $n \geq 0$:

$$x_{n+1} = x_n - \frac{f(x_n)}{f'(x_n)}. \tag{4}$$

Graphically speaking, the number x_{n+1} is the abscissa of the intersection point between the tangent of the curve of $f(x)$ at $x = x_n$ and the x-axis. Then, the general principle is that, if one takes x_0 close enough to a root of $f(x)$, then the sequence (x_n) may converge to this root. As mentioned in [2], this problem seems to be one of the oldest processes of iteration in the history of mathematics and we can find tracks of this in ancient Babylone or in the Arab world of the twelfth century.

We shall point out that I. Newton did not present his method with the expression of the sequence (4) but with an equivalent approach based on algebra rather than calculus. An other equivalent formulation is then made by J. Raphson, again without the use of calculus. It is finally T. Simpson who introduces the closest procedure to (4), formally defined in this precise form by J. Fourier.

The contribution of E. Schröder to the further development and generalization of Newton's method results from his idea of turning the discrete problem of the convergence of the sequence (4) into the iteration problem of the function:

$$R(x) = x - \frac{f(x)}{f'(x)}. \tag{5}$$

With this point of view, a zero α of $f(x)$ becomes a fixed point of $R(x)$. This also allows the author to extend Newton's method to the complex plane and to the search of complex fixed points of $R(x)$, or complex zeros of $f(x)$. We will be mainly interested in the case where $R(z)$ is a rational fraction with coefficients in $\mathbb{C}$ of degree at least two. Let $R^n(z)$ denote the n-th iterate of $R(z)$. We remark that if we start with a point z_0 and if $R^n(z_0)$ converges to a point α, then $R(\alpha) = \alpha$. That is, α is a fixed point of $R(z)$. This explains why such points are important in the theory of iteration. Now, by Taylor formula, in a neighbourhood of the fixed point α, the value $|R(z) - \alpha|$ can be approximated by $|R'(\alpha)||z - \alpha|$. Hence, we guess that the behaviour of the sequence $R^n(z)$ is not the same depending whether $|R'(\alpha)| < 1$ or $|R'(\alpha)| > 1$ for example. Indeed, in the first case, the sequence converges to α (it is the fixed point theorem of E. Schröder and G. Koenigs mentioned below), and in the second, the only way for this sequence to converge to α is that $R^N(z) = \alpha$ for some $N \in \mathbb{N}$. That is why we distinguish the following categories of fixed points α of R, according to the value of $|R'(\alpha)|$. We say that a fixed point α is:

1. **attracting** if $0 < |R'(\alpha)| < 1$,
2. **super-attracting** if $|R'(\alpha)| = 0$,
3. **repelling** if $|R'(\alpha)| > 1$,
4. **rationally indifferent** if $R'(\alpha)$ is a root of unity,
5. **irrationally indifferent** if $|R'(\alpha)| = 1$ and if $R'(\alpha)$ is not a root of unity.

E. Schröder establishes (even if his proof is not rigorously explained) the following fixed point theorem in [68]: if $R(z)$ is an analytic function in a neighbourhood of an attracting fixed point α of R, then, there exists a neighbourhood D of α such that:

$$\lim_{n \to +\infty} R^n(z) = \alpha, \forall z \in D.$$

Note that G. Koenigs provides a complete proof of this theorem in [41]. This result can be obtained by applying the Taylor formula to $R(z)$ near $z = \alpha$. As α is attracting, we find the existence of a neighbourhood D of α, and a real number ϵ such that $0 < \epsilon < 1$ and :

$$|R(z) - \alpha| < \epsilon |z - \alpha|, \ \forall z \in D.$$

We deduce that $R(D) \subset D$. By induction on n, we find that $R^n(D) \subset D$, for every integer $n \geq 1$ and the announced convergence in D.

Now, let us assume that the rational fraction $R(z)$ of the Newton's method (5) satisfies the assumption of the fixed point theorem. Then, the aim of E. Schröder is to generalize Newton's method by finding rational fractions $\phi(z)$ distinct from $R(z)$, such that $\phi(\alpha) = \alpha$ and such that $\phi^n(z)$ converges to α when z is close enough to α, in order to improve the rate of convergence of $R^n(z)$. To do so, it is important to understand the iterates of a rational fraction. But these are in general difficult to compute. That is why E. Schröder thinks about finding these $\phi(z)$ such that their iterates are easy to compute, while keeping track of the initial rational fraction $R(z)$. He solves this problem by the use of conjugation. We say that two rational fractions ϕ and R are **conjugated** if there exists an invertible function $F(z)$ such that:

$$R = F^{-1}\phi F, \tag{6}$$

where functional composition is multiplicatively written. Particularly interesting conjugations are those with a **Möbius transformation** F, that is a rational fraction of the following form:

$$F(z) = \frac{az + b}{cz + d}, \tag{7}$$

where $a, b, c, d \in \mathbb{C}$ and $ad - bc \neq 0$. They are of degree one and stable under composition.

Let us stress that the conjugation preserves the notions of fixed points and iteration. Indeed, if α is a fixed point of $R(z)$, of one of the five kinds defined before (attracting, super-attracting, repelling, rationally indifferent or irrationally indifferent) then $F(\alpha)$ is a fixed point of $\phi(z)$ **of the same kind**, and for every $n \in \mathbb{N}$:

$$R^n = F^{-1}\phi^n F.$$

The notion of fixed points is crucial in the study of Eqs. (S), (B) and (A). If a rational fraction R admits a fixed point α, we let $S_{R,\alpha}, B_{R,\alpha}, A_{R,\alpha}$ denote respectively the Schröder's, Böttcher's and Abel's equation (S), (B) and (A) associated to the rational fraction R. The mention of α means that we are interested in the behaviour of the iterations of $R(z)$ in a neighbourhood of the fixed point α.

Furthermore, the conjugation respects the solutions of Eqs. (S), (B) and (A). Indeed, if $R(z)$ is a rational fraction which admits a fixed point α, then we have the following:

1.
$$f \text{ is a solution of } S_{R,\alpha} \Leftrightarrow gf \text{ is a solution of } S_{gRg^{-1},g(\alpha)}.$$

2.
$$f \text{ is a solution of } B_{R,\alpha} \Leftrightarrow gf \text{ is a solution of } B_{gRg^{-1},g(\alpha)}. \tag{8}$$

3.

$$f \text{ is a solution of } A_{R,\alpha} \Leftrightarrow fg^{-1} \text{ is a solution of } A_{gRg^{-1},g(\alpha)}.$$

Let us go back to the approach of E. Schröder to conjugate the rational fraction of Newton's method to find an easier form. The choice of $\phi(z) = sz$ in (6), with a certain $s \in \mathbb{C}^*$ leads to the Schröder's equation:

$$F(R(z)) = sF(z), \qquad (S_0)$$

in which the unknown is the function $F(z)$.

Note that this is slightly different from Eq. (S), with the following link: if f is an invertible solution of Eq. (S), then $F = f^{-1}$ is a solution of Eq. (S_0). E. Schröder also considers the case of $\phi(z) = z + \lambda$, for a fixed $\lambda \in \mathbb{C}$, which leads to the Abel's equation (Eq. (A) is the particular case of $\lambda = 1$):

$$F(R(z)) = F(z) + \lambda. \qquad (A_0)$$

Let us note that the case $\phi(z) = z^d$, where $d \geq 2$ is an integer, gives rise to the following form of the Böttcher's equation:

$$F(R(z)) = F(z)^d. \qquad (B_0)$$

Likewise, if f is an invertible solution of Eq. (B), then $F = f^{-1}$ is a solution of Eq. (B_0).

Moreover, E. Schröder is interested in the so-called *analytic iteration problem*, which he formulates in the following way in [69]. For a given analytic function $\phi(z)$, find a function $\Phi(w, z)$ of two complex arguments, which is continuous (even analytic) in both variables and such that:

$$\begin{aligned} \Phi(w, z) &= \Phi(w - 1, \phi(z)) \\ \Phi(1, z) &= \phi(z). \end{aligned} \qquad (9)$$

Even if E. Schröder seems not to explicitly connect this problem to the resolution of the Schröder's equation, it is worth pointing this link out here, as made in [2]. If there exists an invertible analytic solution F to Eq. (S_0), then, a solution of (9) with $\phi = R$ is given by:

$$\Phi(w, z) = F^{-1}(s^w F(z)). \qquad (10)$$

An other solution can be given based on an invertible analytic solution F of the Abel's equation (A_0) by:

$$\Phi(w, z) = F^{-1}(F(z) + w\lambda). \qquad (11)$$

Finally, a solution can be given based on an invertible analytic solution F of the Böttcher's equation (B_0) by:

$$\Phi(w, z) = F^{-1}\left(F(z)^{d^w}\right). \tag{12}$$

Thus, solving Eqs. (S), (B) or (A) implies solving the analytic iteration problem for the rational fraction $R(z)$. In [41], G. Koenigs proves the existence of an analytic solution F of Eq. (S_0) around an attracting fixed point α, and also (applying his previous reasoning to F^{-1} instead of F) around a repelling fixed point α. Note that in each case, the solution is unique up to a constant multiplier. Let us consider the attracting case. Recall that $s = R'(\alpha)$. The proof of the author consists in showing that the following function is analytic in a neighbourhood D of α.

$$F(z) = \lim_{n\to+\infty} \frac{R^n(z) - \alpha}{s^n}.$$

Indeed, if we write

$$F_n(z) = \frac{R^n(z) - \alpha}{s^n}, \tag{13}$$

we have :

$$\begin{aligned} F_n(R(z)) &= \frac{R^{n+1}(z) - \alpha}{s^n} \\ &= s F_{n+1}(z). \end{aligned}$$

Then, taking the limit when n tends to infinity, we obtain that F is solution of Eq. (S_0).

Now, to prove that F is analytic in a neighbourhood of α, G. Koenigs reduces the problem to the convergence of a series of functions $\sum f_i(z)$. He then uses a result from G. Darboux which states that if the series $\sum f_i(z)$ and $\sum f_i'(z)$ both converge uniformly in a disc, then $\sum f_i(z)$ converges to an analytic function on this disk. Let us reproduce here a shorter proof we can find in [5, page 49] for example, replacing G. Darboux result by a theorem which states that a locally uniform limit of a sequence of analytic functions in an open set D is analytic on D. First, there exist $\sigma \in \mathbb{R}$, such that $s < \sigma < 1$, a real number $r > 0$ little enough, and $A \in \mathbb{R}_+^*$, such that R is analytic in the disc $D(\alpha, r)$ centred at α and of radius r, and such that for all $z \in D(\alpha, r)$:

$$|R(z) - \alpha| < \sigma |z - \alpha|, \tag{14}$$

$$|R(z) - \alpha - s(z - \alpha)| < A|z - \alpha|^2. \tag{15}$$

This arises from the Taylor formula applied to $R(z)$ near $z = \alpha$ and the fact that α is an attracting fixed point of R. We deduce from the first inequality above that $R(D(\alpha, r)) \subset D(\alpha, r)$. Besides, even if this means reducing r we assume that α is the only solution to $R(z) = \alpha$ in $D(\alpha, r)$.

Moreover, we see that for all $z \in D(\alpha, r)$:

$$F_n(z) = (z-\alpha)\prod_{k=0}^{n-1}\frac{R^{k+1}(z)-\alpha}{s(R^k(z)-\alpha)}. \tag{16}$$

The fact that $R(D(\alpha, r)) \subset D(\alpha, r)$ implies by induction that $R^n(D(\alpha, r)) \subset D(\alpha, r)$, for every integer $n \geq 1$. Then, for all $z \in D(\alpha, r)$ we can apply (14) to $R^{k-1}(z)$ and (15) to $R^k(z)$. This gives, for all $z \in D(\alpha, r) \setminus \{\alpha\}$:

$$\left|\frac{R^{k+1}(z)-\alpha}{s(R^k(z)-\alpha)} - 1\right| < \frac{A}{|s|}|R^k(z)-\alpha| =: u_k(z).$$

But it appears that for all $z \in D(\alpha, r) \setminus \{\alpha\}$:

$$\frac{u_{k+1}(z)}{u_k(z)} < \sigma < 1.$$

We conclude that the product of (16) converges uniformly in $D(\alpha, r) \setminus \{\alpha\}$. Moreover, for every $n \in \mathbb{N}$, $F_n(z)$ admits the finite limit 0 when z tends to α. Then (see [67, 7.11 Theorem]), the sequence $(F_n(z))$ of analytic functions over $D(\alpha, r)$ converges uniformly on $D(\alpha, r)$. We deduce that F is analytic in $D(\alpha, r)$. Notice that for every integer $n \geq 1$, $(R^n)'(\alpha) = s^n$. Furthermore, the uniform convergence of $(F_n(z))_n$ near α makes the operations of derivation and limit commute in (13). We deduce that $F'(\alpha) = 1$. Thus F is locally invertible. This result of G. Koenigs strengthen the connection between the Schröder's equation and the theory of iteration.

For his part, P. Fatou points out the connections (10) and (11) between Schröder's and Abel's equations and the analytic iteration problem. Indeed, in [25], P. Fatou remarks that the solution of the Schröder's equation given by G. Koenigs solves this problem. Then, P. Fatou studies the Abel's equation (A_0) in the case of a rationally indifferent fixed point α. Inspired by a previous result from L. Leau [43, 44], P. Fatou proves in [25] the existence of particular domains (that is open connected sets) $L_1, \ldots, L_k$ such that, for all $j \in \{1, \ldots, k\}$, α belongs to the boundary of L_j, $R(L_j) \subset L_j$ and the restriction to L_j of R^n converges uniformly to α on L_j, when n tends to infinity. Such domains are called the **petals** of R and the result is called the *Leau-Fatou Flower Theorem*. Note that a similar result is proved by G. Julia in [40]. In each petal L_j, P. Fatou proves the existence of an analytic solution of the Abel's equation (A_0), and mentions again that this provides a solution to the analytic iteration problem. Let us precise here that Eq. (A) is introduced by N. H. Abel in [1]. The author remarks that it can be turned into a difference equation. Indeed, if f is an invertible solution of Eq. (A), then $F = f^{-1}$ is solution of:

$$F(z+1) = R(F(z)). \tag{$\tilde{A}$}$$

Now, let us consider the Böttcher's equation (B_0), in the case of a super-attracting fixed point α. According to J. F. Ritt [60], the existence of an analytic solution in the neighbourhood of α is due to L. Böttcher in [12–14]. J. F. Ritt provides a short proof of this result in [60]. Let us sketch it here. Even if it means considering, instead of $R(z)$, the conjugation of $R(z)$ with an appropriate Möbius transformation (7), we may assume that $\alpha = 0$. Now, let

$$R(z) = \sum_{n=d}^{+\infty} a_n z^n, \tag{17}$$

where $d \geq 2$. Considering $a_d^{1/(d-1)} R(z/a_d^{1/(d-1)})$, we may assume that $a_d = 1$. We know that there exists a disc D around zero in which the only zero of $R(z)$ in D is the origin. Even if this means reducing the radius of D, we assume that $R(D) \subset D$. This can be deduced from the Taylor formula and the fact that 0 is a super-attracting fixed point of $R(z)$. By induction, we find that $R^n(D) \subset D$, for every integer $n \geq 1$, and that the origin is the only zero of $R^n(z)$ in D. Moreover, the origin is a zero of $R(z)$ of order d. Hence, by induction, the origin is a zero of $R^n(z)$ of order d^n, for every integer $n \geq 1$. Then, for every integer $n \geq 1$:

$$\frac{R^n(z)}{z^{d^n}} \neq 0, \forall n \geq 1, \forall z \in D. \tag{18}$$

Hence, for every integer $n \geq 1$, there exists a d^n-th root of (18) on D, that is an analytic function $g_n(z)$ over D such that

$$g_n(z)^{d^n} = \frac{R^n(z)}{z^{d^n}}, \forall z \in D.$$

Now if we let $h_n(z) = z g_n(z)$, which is analytic over D, we get:

$$h_n(z)^{d^n} = R^n(z), \forall n \geq 1, \forall z \in D.$$

We can then write: $h_n(z) = [R^n(z)]^{1/d^n}$.

Now, we remark that:

$$h_{n+1}(z) = z \prod_{i=0}^{n} \left[g_1\left(R^i(z)\right)\right]^{1/d^i}, \tag{19}$$

Some calculation (see the details in [25, page 188]) prove that the product (19) converges uniformly on D. We deduce that $h_n(z)$ converges uniformly on D to an analytic function $F(z)$ over D. Finally, we have:

$$\begin{aligned}\left[R^n(R(z))\right]^{1/d^n} &= \left[R^{n+1}(z)\right]^{1/d^n}\\ &= \left[\left(R^{n+1}(z)\right)^{1/d^{n+1}}\right]^d.\end{aligned}$$

If we take the limit when n tends to infinity, we obtain that F is solution of (B_0), which concludes the proof.

Let us return to the interest of E. Schröder for the analytic iteration problem. As said before, the author seems not to relate this question to solutions of the Schröder's or Abel's equation in the manner of (10) and (11). But the author has another connection in mind. He links the existence of a solution to the analytic iteration problem (9) to the one of a continuous curve which contains the iterates of $\phi(z)$. This consideration of invariant structures of the complex plane with respect to the iteration is at the heart of the theory developed by P. Fatou and G. Julia. However, the study of E. Schröder and G. Koenigs are limited to a neighbourhood of a fixed point. One of the main innovations of the work of P. Fatou and G. Julia is their idea and tools to investigate the whole complex plane, and in fact the **compactification** $\hat{C} = \mathbb{C} \cup \{\infty\}$ of $\mathbb{C}$, dividing it into zones depending on the behaviour of the sequence of iterates of a rational fraction. Let us note that $\hat{C}$ makes the study of rational fractions a central topic, as they are the only analytic functions over $\hat{C}$. In order to present the main discoveries of P. Fatou and G. Julia, let us first introduce some notations and definitions. As before, and for now on, we let $R(z)$ denote a rational fraction of degree at least two, and we let $R^n(z)$ denote the n-th iterate of $R(z)$.

Quite natural questions arise when considering a point z_0 close to a fixed point α and the sequence of iterates $z_n = R^n(z_0)$. From a local point of view, we can wonder if there exists a neighbourhood of α in which (z_n) converges. From a global point of view, we can study the behaviour of this sequence outside such a neighbourhood, and on its boundary. The work of P. Fatou and G. Julia is about the second point. We may see this as the study of the impact of the point z_0 and its neighbourhood over the convergence of the sequence (z_n). To translate this impact, P. Fatou [25, 26] involves the theory of normal families developed by P. Montel [51]. We say that a family $\mathscr{F}$ of analytic functions defined on a domain D of $\hat{C}$ is **normal** over D if from every infinite subsequence of $\mathscr{F}$, we can extract a sub-sequence of $\mathscr{F}$ which converges uniformly locally on D (that is in every compact set of D). The link with our subject is given by the application of the *Arzela-Ascoli theorem* to $\mathscr{F} = \{R^n\}_n$. Indeed, this states the equivalence for a family $\mathscr{F}$ of continuous functions defined on a domain D of $\hat{C}$ to be normal over D or equicontinuous over D. But, by definition, $\{R^n\}_n$ is **equicontinuous** over D if for every $z \in D$ and every $\epsilon > 0$, there exists $\delta > 0$ such that for every $n \in \mathbb{N}$ and every $z_0 \in D$:

$$|z - z_0| < \delta \Longrightarrow |R^n(z) - R^n(z_0)| < \epsilon.$$

Thus, the notion of equicontinuity exactly transcribes the fact that the behaviour of the sequence (z_n) depends on z_0. In order to understand the boundary of this property, P. Fatou defines and studies the set of all the points of $\hat{C}$ for which the family $\{R^n\}_n$

is not normal (that is there exists no neighbourhood of these points in which the family is normal). He denotes F this set, for *Frontière* (french word for *boundary*). This set is nowadays written as $J(R)$ (or J), for *the Julia set associated to* R, and it is its complement $\hat{C} \setminus J(R)$ that is denoted as $F(R)$ (or F), this time because of the initial letter of Fatou, and called the *Fatou set associated with* R. At the same time, G. Julia defines the set E consisting of all the repelling fixed points of all the iterates R^n, $n \in \mathbb{N}$, and studies the derived set E' composed by all the accumulation points of E, which coincides with J (see Theorem 3.9 below). The study of the set J can provide information on the solutions of functional equations. Let us illustrate this fact with the analytic extension of a solution of the Schröder's equation (S_0) in a neighbourhood of an attracting fixed point α, following a method introduced by P. Fatou [27]. Recall that $s = R'(\alpha)$. According to G. Koenigs, there exists a neighbourhood D_α of α and an analytic function F on D_α such that for all $z \in D_\alpha$:

$$F(R(z)) = sF(z).$$

Even if this means reducing D_α, we assume that $R(D_\alpha) \subset D_\alpha$. This comes from the Taylor formula and the fact that α is attracting. Let us consider $\tilde{D} = R^{-1}(D_\alpha)$. The fact that $R(D_\alpha) \subset D_\alpha$ implies that $D_\alpha \subset \tilde{D}$. Then, let us define for all $\tilde{z} \in \tilde{D}$:

$$\tilde{F}(\tilde{z}) = \frac{1}{s} F(R(\tilde{z})).$$

We see that

$$\tilde{F}(z) = F(z), \forall z \in D_\alpha.$$

Moreover, as $R(\tilde{z}) \in D_\alpha$, for all $z \in \tilde{D}$, we have:

$$\begin{aligned} \tilde{F}(R(\tilde{z})) &= F(R(\tilde{z})), \ \forall z \in \tilde{D} \\ &= s\tilde{F}(\tilde{z}), \ \forall z \in \tilde{D}. \end{aligned}$$

Hence, $\tilde{F}$ extends analytically F on $\tilde{D}$ and remains a solution of the Schröder's equation over $\tilde{D}$. Iterating the process, we can extend F to an analytic function G over the domain of attraction D of α (that is the set of all the elements z such that there exists an integer N such that $R^N(z) \in D_\alpha$), such that G remains a solution of the Schröder's equation over D. Thus, the understanding of properties of $R(z)$, namely the nature of its fixed points, can provide information about a solution $F(z)$ of the Schröder's equation (S_0) .

However, Eq. (S), the other form of the Schröder's equation, also provides a connection with the theory of P. Fatou and G. Julia, maybe even better in some cases. Indeed, when α is a repelling fixed point of R, H. Poincaré [58] proves that the solution f of the Schröder's equation (S) (also called *Poincaré's equation* in this form) can be extended as a meromorphic function over $\mathbb{C}$. Hence, this allows a global study of the structure of the Julia set J. As explained in [6, Theorem 6.3.2], a computation of the coefficients of the Taylor series expansion of a formal solution $f(z)$ of (S) implies that this series has a positive radius of convergence. Hence, there exists $r > 0$ such that $f(z)$ is analytic over the disc $D(\alpha, r)$ centred at α and of radius $r > 0$. Then, one can extend $f(z)$ by induction as follows. The analyticity of $f(z)$ over $D(\alpha, r)$ implies that $R(f(z))$ is meromorphic over $D(\alpha, r)$. Hence, by Eq. (S), so is $f(sz)$. Therefore, $f(z)$ is meromorphic over $D(\alpha, sr)$. By induction, we find that $f(z)$ is meromorphic over $D(\alpha, s^n r)$, for every $n \in \mathbb{N}$. As $|s| > 1$, we obtain that $f(z)$ is meromorphic over $\mathbb{C}$.

The Schröder's equation is also studied from an algebraic point of view. Indeed, J. F. Ritt establishes in [62] the list of all the differentially algebraic solutions of this equation, when $|s| > 1$, that is, in the case of a repelling fixed point of $R(z)$. This is Theorem 3.2 of Sect. 3.2.

2.2 The 80's and the Result of P.-G. Becker and W. Bergweiler

During the 30's, the interest for functional equations (S), (B) and (A) and the theory of iteration is less vigorous. Nonetheless, let us note the work of C. L. Siegel [71] on the existence of a solution to Eq. (S) for some indifferent fixed points, and the one of H. Brolin [15] on the structure of the Julia set. The enthusiasm for this subject rises again sixty years later, during the 80's. This is mainly due to the connections the theory of P. Fatou and G. Julia shares with the active area of dynamical systems and fractals, and the possibility to make computational experiences. Indeed, as said before, the iteration of a rational function $R(z)$ gives rise to a dynamical system, which divides $\hat{C}$ into different areas, depending on the concordance or disparity of the local behaviour of the sequence $\{R^n(z)\}_n$ around a point $z = z_0$. The concordance is formalized by the notion of equicontinuous and normal families. The set of points with this local concordance is the Fatou set $F(R)$ and its complement in $\hat{C}$ is the Julia set $J(R)$. Let us assume that $R(z)$ is of degree at least two. Note that $F(R)$ is open in $\hat{C}$ and $J(R)$ is a closed compact subset of $\hat{C}$. Let us mention that $J(R)$ is always non-empty [25] and perfect, that is, closed and without any isolated point [26, 40]. Moreover, Julia sets provide lots of examples of fractals. These objects are defined by B. Mandelbrot in the 90's [50]. We refer to [18, 21, 57] for more details about what follows. The story of fractals actually goes back to the works of B. Riemann and K. Weierstrass and their discoveries of continuous functions with no derivative, at any point. This created a lot of confusion in the mathematical community which had trouble to apprehend such strange objects (see more details in [2, page 88])! This discomfort was even increased with the work of G. Cantor and his perfect, totally disconnected (that is with all connected components reduced to a point) sets, which questioned the notion of dimension of his time. To deal with the complexity of such objects, the classical topological dimension is replaced by the Hausdorff dimension, which may be a non integral number. Intuitively (see for example [57]), this notion measures the growth of the number of sets of diameter ϵ needed to cover the concerned set, when ϵ tends to zero. The Hausdorff dimension is always greater than or equal to the topological one. B. Mandelbrot defines the fractals as sets for which the Hausdorff dimension is strictly greater than the topological one. For example, the triadic Cantor set has topological dimension 0 and Hausdorff dimension $\log(2)/\log(3)$: it is a fractal. As said before, lots of Julia sets provide examples of fractals. Julia sets still fuel the current research, with different points of view, namely, investigations on their Hausdorff dimension [45, 73], their Lebesgue measure [16], or their computational complexity [24]. Apart from the structure of the Julia set $J(R)$, there is the question of its variation when the coefficients of $R(z)$ depend on a parameter. This question is raised by P. Fatou in [26]. In the case of polynomials of degree two, of the form $R_c(z) = z^2 + c$, where $c \in \mathbb{C}$, there exists a classification of the Julia sets $J(R_c)$. Note that each polynomial of degree two is conjugated to a polynomial of such a form. This classification is encoded by the Mandelbrot set, defined as the set of the complex numbers $c \in \mathbb{C}$ for which $J(R_c)$ is connected. Indeed, the dynamic of $R_c(z)$ changes as c moves from a cardioid to a disc of the Mandelbrot set. For example (see [6, paragraph 1.6]), let us focus on the cardioid C, and the disc D, where $C = u(D(0, 1/2))$, $u(z) = z - z^2$, and $D = D(-1, 1/4)$. The cardioid C and the disc D are part of the Mandelbrot set (see (Fig. 1) below). When $c \in C$, the rational fraction R_c admits a unique attracting fixed point α and a pair (u, v) such that $R_c(u) = v$, $R_c(v) = u$, and u, v are repelling fixed points of R_c^2. Then, when c enters inside D, the point α becomes a repelling fixed point of R_c and the points u, v become attracting fixed point of R_c^2. As an illustration, (Fig. 1) below represents the Mandelbrot set and the form of the Julia sets $J(R_c)$ attached to R_c for some of the points c inside and outside the Mandelbrot set. Note that the Mandelbrot set appears to be connected itself [22].

Besides, the Mandelbrot set M gives rise to the following interesting transcendental result [56]. A certain conformal map $\Phi(z)$, constructed by A. Douady and J. H. Hubbard [22], defined on the complement of M, admits transcendental values $\Phi(\alpha)$ at each algebraic number α of the complement of M. A link with the Böttcher's equation is that $\Phi(c) = f_c(c)$, for every c in the complement of M, where f_c satisfies Eq. (B_0) for $d = 2$ and $R(z) = R_c(z)$. In the same area, the algebraic aspects of the solutions of Eqs. (S), (B) and (A) are also investigated by many authors, along with other kinds of functional equations. The first main result of this type is due to 0. Hölder in 1887. Indeed, the author proves that the Euler's Gamma function defined for every $z \in \mathbb{C}$ such that $\mathrm{Re}(z) > 0$, by:

$$\Gamma(z) = \int_0^{+\infty} t^z e^{-t} \frac{dt}{t},$$

is hypertranscendental over $\mathbb{C}(z)$. As said in [65], the proof of O. Hölder is based on the following functional equation:

$$\Gamma(z+1) = z\Gamma(z).$$

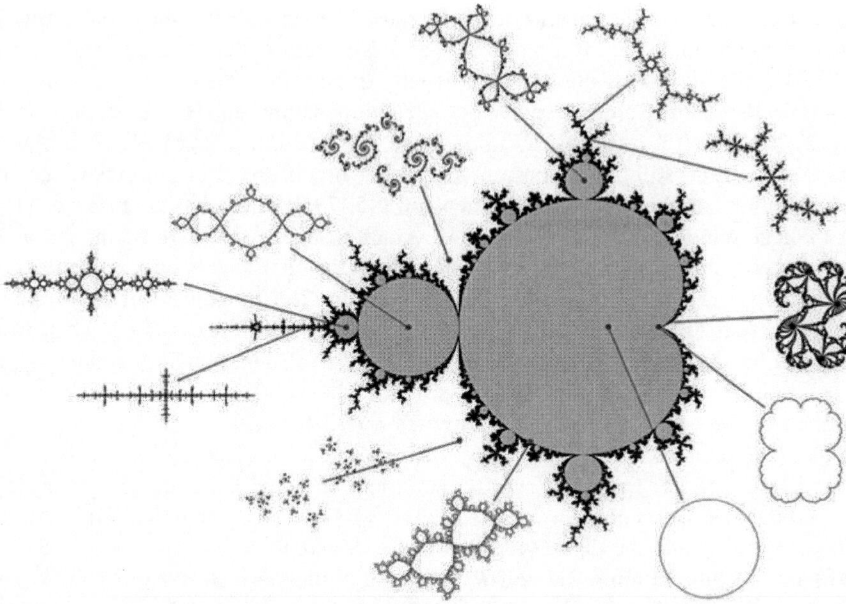

Fig. 1 The Mandelbrot set (in grey) and Julia sets $J(R_c)$ (at the end of the blue lines) for some points $c \in \mathbb{C}$ inside and outside the Mandelbrot set. This image is taken from [19].

This is a non-autonomous version of the Abel's difference equation $(\tilde{A})$, of the form:

$$G(z+1) = R(z, G(z)),$$

where $R(z, X)$ is this time a complex rational fraction of two variables. In [66, Problem 69], L. A. Rubel proposes the study of such a generalised functional equation for the Schröder's equation. This gives rise to further studies. For example, K. Ishizaki [39] considers the case where $R(z, X) = a(z)X + b(z)$, with $a(z), b(z)$ rational fractions. The author proves that every transcendental meromorphic solution of the associated generalised Schöder's equation:

$$G(sz) = a(z)G(z) + b(z),$$

with $|s| \notin \{0, 1\}$, is hypertranscendental.

Concerning the generalised equation of (B), that is:

$$G(z^d) = R(z, G(z)), \tag{20}$$

this is called a d-Mahler equation and was introduced by K. Mahler in 1929 in [47–49]. A solution G of (20) is called a d-Mahler function. Ke. Nishioka [55] proves that a Mahler function is transcendental if and only if it is not rational (see also [56]). The same author provides in [54] a sufficient condition for the hypertranscendence of Mahler functions of order one. Before this time, K. Mahler proved that the d-Mahler function $\sum_{n=0}^{+\infty} z^{d^n}$ is hypertranscendental [48]. This result is generalised in [46] with the study by J. H. Loxton and A. J. Van der Poorten of inhomogeneous linear Mahler systems of order one. Such functional results are motivated by a theorem of K. Mahler [47–49] which establishes, under some assumptions, the equivalence between the transcendence of a Mahler function $f(z)$ and the one of its value $f(\alpha)$ at a non-zero algebraic number α. Thus, results on functional algebraic independence turns into results on algebraic independence of values. More results are obtained using an adapted Galois theory. For short, this theory (see [59] for linear difference equations) associates to a system of linear functional equations an algebraic group, called the *Galois group*, which encodes the algebraic relations between the solutions of the system. If the Galois group is *big enough*, the solutions are algebraically independent. This approach allows K. Nguyen [53] to recover the hypertranscendental result of Ke. Nishioka mentioned above. To illustrate further use of Galois theory for linear Mahler functions, let us mention the result in [23] which provides sufficient conditions for a Mahler function to be hypertranscendental, and the work of J. Roques [63]. Note that these kinds of functional results are still open questions for linear Mahler equations over a function field of positive characteristic (see for example [29–31]). Similarly, the Siegel-Shidlovskii theorem [70], which is a kind of analogue of the Mahler's theorem for certain solutions of linear differential equations over $\mathbb{C}(z)$ called E-functions, motivates the study of the algebraic independence of solutions of such equations. As for Mahler functions, there is a dichotomy between rational and transcendental E-functions. Moreover, the Hrushovski-Feng algorithm [28, 38] computes the Galois group of systems of linear differential equations. More generally, the study of hypertranscendence or algebraic independence of solutions of different types of functional equations is currently a very dynamic area (see for example [20, 36]). This frequently involves the development and use of Galois theories adapted to the different settings. As an illustration, let us mention [37] for linear difference equations, and the work of C. Hardouin for q-difference systems [34], which generalises the work of K. Ishizaki mentioned above, and the development of a general tannakian Galois theory, which in particular apply for τ-difference systems in positive characteristic [35], developed by G. Anderson, W. D. Brownawell and M. Papanikolas [3].

Let us go back to Eqs. (S), (B) and (A). A fruitful link between algebraic properties of solutions of these equations and the iteration theory is given in 1986 by a result of M. Boshernitzan and L. A. Rubel in [11]. This states the equivalence for a solution of Eq. (S), (B) or (A) to be differentially algebraic and the family $\{R^n(z)\}_n$ to be **coherent**. The latter means that all the rational fractions $R^n(z)$ satisfy a same algebraic differential equation (1). Note that L. A. Rubel asked in [66, Problem 27] if there exists a transcendental entire function whose iterates form a coherent family. This is answered negatively by W. Bergweiler in [9].

The theorem of M. Boshernitzan and L. A. Rubel is one of the ingredients of the proof of the result of P.-G. Becker and W. Bergweiler [8] we are interested in. This statement, reproduced here as Theorem 3.1, explicitly lists all the differentially algebraic solutions of Eqs. (S), (B) and (A). Let us note that partial results have already been found. We mentioned earlier the result of J. F. Ritt [62]. But we can also indicate the work of F. W. Carroll [17]. The author considers the case where $R(z)$ is a finite Blaschke product with an attracting fixed point (see for example the survey [32] for the definition and more information about such products). Then, he guarantees that a solution of the Schröder's equation (S_0) is hypertranscendental. Furthermore, P. Borwein examines the case where $d = 2$, and $R(z) = z^2 + c$, with $c > 0$, in the Böttcher's equation (B). Then, the author states that every solution of this equation is hypertranscendental. Moreover, P. Borwein points out that his proof, as well as the one of J. F. Ritt in [62], shares analogies with the proof of the hypertranscendence of the Gamma function by O. Hölder. Finally, P.-G. Becker and W. Bergweiler themselves previously obtained in [7] the list of all the algebraic solutions of the Böttcher's equation (B) when $R(z)$ is conjugated to a polynomial. In fact, this list concerns the solutions of the more general following equation:

$$f(p(z)) = q(f(z)), \tag{21}$$

where $p(z)$, $q(z)$ are polynomials of the same degree $d \geq 2$, and the attracting fixed point is ∞.

In the same paper [7], the authors conjecture that the transcendental solutions of Eq. (21) are in fact hypertranscendental. For the Böttcher's equation (B), Theorem 3.1, their further result, implies that this conjecture is satisfied. To conclude, let us mention some recent works. K. D. Nguyen [52] studies systems of n Böttcher's equations (B) for polynomials $R_1(z), \ldots, R_n(z)$. The author proves a result that links the algebraic independence of some transformations of the solutions f_{R_i} associated to each Böttcher's equation to the conjugacy of some iterates of some polynomials among $R_1(z), \ldots, R_n(z)$. Finally, M. Aschenbrenner and W. Bergweiler prove in [4] the hypertranscendence over $\mathbb{C}(z)$ of the iterative logarithm $\mathrm{itlog}(R)$ of a non-linear rational or entire function R with a rationally indifferent fixed point. The function $\mathrm{itlog}(R)$ is the unique formal power series solution f of the equation:

$$f(R(z)) = R'(z)f(z), \tag{22}$$

The proof of the authors is similar to the one of Theorem 3.1 we present later. Note that the authors also prove that in the case where $R(z)$ is a non-linear entire function, the function $\mathrm{itlog}(R)$ is even hypertranscendental over the ring of entire functions. Note that Eq. (22) is useful to study the iteration of $R(z)$ not only in the petals of $R(z)$, given by the Leau-Fatou Flower theorem mentioned above, but in a neighbourhood of the concerned fixed point.

3 The Complete Classification of Differentially Algebraic Solutions of the Schröder's, Böttcher's and Abel's Equations

3.1 The Statement

Before presenting the statement of P.-G. Becker and W. Bergweiler we are interesting in, let us precisely define its setting. Let $R(z)$ denote a rational fraction with coefficients in $\mathbb{C}$ and of degree at least two. Even if this means replacing $R(z)$ by a conjugate with the appropriate Möbius transformation, we may assume that 0 is a fixed point of $R(z)$. Let us write $s = R'(0)$. The framework considered is the following (see [25, Chapitre II]):

1. If 0 is an attracting, repelling or irrationally indifferent fixed point of $R(z)$, we consider the Schröder's equation (S). The Schröder's equation admits a unique solution of the form

$$f(z) = \sum_{n=1}^{+\infty} a_n z^n, \text{ with } a_1 = 1. \tag{23}$$

   If f converges in a neighbourhood of 0, we say that f is a Schröder function. In the case of an attracting or repelling fixed point, the solution is always convergent, as seen before. But in the case of an irrationally indifferent fixed point, there always exists a formal solution, but not necessarily convergent [71]. The question of the convergence of this formal solution is a dynamic area of research [33, 74].
2. If 0 is a super-attracting fixed point of $R(z)$, there exists an integer $d \geq 2$ such that:

$$R(z) = \sum_{n=d}^{+\infty} b_n z^n, \ b_d \neq 0. \tag{24}$$

   Then, we consider the Böttcher's equation (B). For every $a_1 \in \mathbb{C}$ such that $a_1^{1-d} = b_d$, the Böttcher's equation admits a unique solution of the form

$$f(z) = \sum_{n=1}^{+\infty} a_n z^n. \tag{25}$$

   The function f converges in a neighbourhood of 0, and we say that f is a Böttcher function.
3. If 0 is a rationally indifferent fixed point of $R(z)$, even if this means replacing $R(z)$ by an iterate $R^k(z)$, we may assume that $s = 1$. Indeed, if $s^k = 1$, we have $(R^k)'(0) = s^k = 1$. Let us write:

$$R(z) = z + \sum_{n=d}^{+\infty} b_n z^n, \text{ where } d \geq 2,\ b_d \neq 0. \tag{26}$$

Then, we consider the Abel's equation (A). In each petal given by the Leau-Fatou Flower theorem mentioned earlier, the Abel's equation admits an analytic solution $f(z)$. We say that f is an Abel function.

Let us introduce some vocabulary. Let us remind that we say that two analytic functions S_1 and S_2 are **conjugated** to each other via an invertible analytic function g if $S_1 = g^{-1}S_2g$, where functional composition is denoted multiplicatively. If we say that S_1 and S_2 are **conjugated**, with no precision about g, we mean that g is a Möbius transformation (7).

We are now able to present the result of P.-G. Becker and W. Bergweiler, as it is written in [8]. We give several explanations and comments about Theorem 3.1 directly after its statement.

Theorem 3.1 (P.-G. Becker and W. Bergweiler)
Let $f(z)$ be a Schröder, Böttcher or Abel function. Let us assume that f is differentially algebraic. Then, we have the following.

1. *If f is a Schröder function, then 0 is a repelling fixed point of $R(z)$ and f is a Möbius transformation of a function of one of the following forms:*

   (a) $\exp(\alpha z^r)$. *In this case, $R(z)$ is conjugated to z^d or z^{-d}.*
   (b) $\cos(\alpha z^r + \beta)$. *In this case, $R(z)$ is conjugated to T_d or $-T_d$, where T_d is the d-th Tchebychev polynomial.*
   (c) $\wp(\alpha z^r + \beta)$, $\wp^2(\alpha z^r + \beta)$, $\wp^3(\alpha z^r + \beta)$, $\wp'(\alpha z^r + \beta)$, *where $\wp$ denotes the Weierstrass function,*

   where the constant r is a rational number such that the concerned functions are meromorphic over $\mathbb{C}$, α is a non-zero complex number and β is a fraction of a period of the concerned function.
2. *If f is a Böttcher function, then, f is a Möbius transformation or is a Möbius transformation of a function of one of the following forms:*

   (a) ρz, *where* $\rho^{d-1} = 1$. *In this case, $R(z)$ is conjugated to z^d.*
   (b) $\rho z + \frac{1}{\rho z}$, *where* $\rho^{d-1} = 1$. *In this case, $R(z)$ is conjugated to T_d.*
   (c) $\rho z + \frac{1}{\rho z}$, *where* $\rho^{d-1} = -1$. *In this case, $R(z)$ is conjugated to $-T_d$.*
3. *The function f is not a Abel function.*

   In particular, Abel functions are always hypertranscendental.

The meromorphic condition, for the Schröder functions, is due to the fact that, in the case of a repelling fixed point, every solution of Eq. (S) extends to a meromorphic function on the complex plane, as seen at the end of Sect. 2.1. Remark that, even if the function $z \to z^r$ is not meromorphic on the complex plane when r is not an integer, this case may happen. There is an example below, with the meromorphic function $\cos(\sqrt{2z}) - 1$ on the complex plane, where $r = 1/2$.

We refer to the proof of Theorem 3.8, and references inside, for more details on the rational fraction of the Schröder's equation associated to the Weierstrass function (see also [61]).

We reproduced in Theorem 3.1 the statement as it is written in [8]. The fixed point of the rational fractions $R(z)$ involved in this statement is not necessarily 0. We clarify this point below.

1. If f is a Schröder function
   (a) The rational fractions $R(z) = z^d$ or $R(z) = z^{-d}$ are considered at the repelling fixed point $z = 1$. To move it at the origin, we have to conjugate $R(z)$ with $L(z) = z - 1$. Then we have: $\tilde{R}(z) = (z+1)^d - 1$, or $\tilde{R}(z) = (z+1)^{-d} - 1$, and associated solutions of equation $S_{\tilde{R},0}$ are of the form $\exp(\alpha z^r) - 1$, where α, r are as in the statement of Theorem 3.1. In particular, $\tilde{f}(z) = \exp(z) - 1$ is such that $\tilde{f}(0) = 0$ and $\tilde{f}'(0) = 1$.
   (b) The rational fractions $R(z) = T_d$ or $R(z) = -T_d$ are considered at the repelling fixed point $z = 1$. To move it at the origin, we have to conjugate $R(z)$ with $L(z) = z - 1$. Then we have: $\tilde{R}(z) = T_d(z+1) - 1$, or $\tilde{R}(z) = -T_d(z+1) - 1$, and associated solutions of equation $S_{\tilde{R},0}$ are of the form $\cos(\alpha z^r + \beta) - 1$, where α, β, r are as in the statement of Theorem 3.1. In particular, $\tilde{f}(z) = \cos(\sqrt{2z}) - 1$ is such that $\tilde{f}(0) = 0$ and $\tilde{f}'(0) = 1$.
2. If f is a Böttcher function,
   a. The rational fraction $R(z) = z^d$ is considered at the super-attracting fixed point $z = 0$.
   b. The rational fraction $R(z) = T_d(z)$ is considered at the super-attracting fixed point $z = \infty$. To move it at the origin, we have to conjugate $R(z)$ with $L(z) = 1/z$. Then we have: $\tilde{R}(z) = 1/(T_d(1/z))$, and associated solutions of equation $B_{\tilde{R},0}$ are of the form $1/(\rho z + \frac{1}{\rho z})$, where $\rho^{d-1} = 1$.
   c. The rational fraction $R(z) = -T_d(z)$ is considered at the super-attracting fixed point $z = \infty$. To move it at the origin, we have to conjugate $R(z)$ with $L(z) = 1/z$. Then we have: $\tilde{R}(z) = 1/(-T_d(1/z))$, and associated solutions of equation $B_{\tilde{R},0}$ are of the form $1/(\rho z + \frac{1}{\rho z})$, where $\rho^{d-1} = -1$.

Let us recall the following definitions. The **Weierstrass function** $\wp$ is a meromorphic function over $\mathbb{C}$, attached to a **lattice** $\Lambda \subset \mathbb{C}$, that is a discrete subgroup of $\mathbb{C}$ that contains an $\mathbb{R}$-basis of $\mathbb{C}$, and defined by:

$$\wp(z) := \wp_\Lambda(z) = \frac{1}{z^2} + \sum_{w \in \Lambda, w \neq 0} \left(\frac{1}{(z-w)^2} - \frac{1}{w^2} \right).$$

The function $\wp_\Lambda$ is **periodic** with respect to Λ, that is:

$$\wp_\Lambda(z+\omega) = \wp_\Lambda(z), \ \forall z \in \mathbb{C}, \forall \omega \in \Lambda. \tag{27}$$

Besides, the function $\wp_\Lambda$ satisfies the following algebraic differential equation:

$$\wp_\Lambda'^2 = 4\wp_\Lambda^3 - g_2(\Lambda)\wp_\Lambda - g_3(\Lambda), \tag{28}$$

where $g_2(\Lambda), g_3(\Lambda) \in \mathbb{C}$ depend on Λ and satisfy $g_2(\Lambda)^3 - 27g_3(\Lambda)^2 \neq 0$. For more details, see for example [72].

Finally, **Tchebytchev polynomials** are defined by induction with:

$$T_0(X) = 1, T_1(X) = X, T_{n+2}(X) = 2XT_{n+1}(X) - T_n(X), \forall n \geq 2.$$

Let us note that for every integer $n \geq 0$ and every $x \in [-1, 1]$: $T_n(\cos(x)) = \cos(nx)$.

3.2 *First Ingredient: Preliminary Results Around the (hyper)transcendence of Solutions of Eqs.* (S) *and* (B)

3.2.1 A Result of J. F. Ritt

As mentioned earlier, J. F. Ritt gives in 1926 the list of all the differentially algebraic Schröder functions when 0 is a repelling fixed point of $R(z)$ [62]. These are exactly the functions listed in the first point of Theorem 3.1. Thus, when the Schröder's equation admits a differentially algebraic solution, then, the considered fixed point of the associated rational fraction is always repelling.

Theorem 3.2 (J. F. Ritt)
Let us assume that 0 *is a repelling fixed point of a rational fraction* $R(z)$ *of degree at least two. Let* f *be a solution of the associated Schröder's equation* (S). *If* f *is differentially algebraic, then,* f *is in the list of the first point of Theorem 3.1.*

The proof of this theorem is based on the theories of differentiation and elimination. These techniques allow the author to prove that a solution of Eq. (S) is (after an appropriate change of variables) a solution of a Schwarzian differential equation of the following form:

$$g^{(3)}g' - 3/2(g^{(2)})^2 = A(g)g^{(4)}, \tag{29}$$

where A is a rational fraction.

The author then uses a previous classification of his own [61] for differentially algebraic solutions of (29) to deduce that they are of the desired forms. As noted earlier, P. Borwein in [10] points out that the techniques of this proof share analogies with the proof of the hypertranscendence of the Gamma function by O. Hölder.

3.2.2 A Result of P.-G. Becker and W. Bergweiler

One year before proving Theorem 3.1, P.-G. Becker and W. Bergweiler [7] provided the list of all the algebraic Böttcher functions, when $R(z)$ is conjugated to a polynomial. It is exactly the functions listed in the second point of Theorem 3.1. Thus, the Böttcher functions are transcendental if and only if they are hypertranscendental. This was a part of a conjecture of the two authors in [7]. Let us state this previous result in the case of Böttcher functions. As said before, this applies to the more general equations (21).

Theorem 3.3 (P.-G. Becker and W. Bergweiler)
Let $R(z)$ be a polynomial of degree at least two and let f be a solution of the associated Böttcher's equation (B). *Then, f is algebraic if and only if f is in the list of the second point of Theorem 3.1. In particular, all the algebraic Böttcher functions are rational.*

The proof of Theorem 3.3 is based on the analysis of the finite singularities (algebraic branch points) of the Böttcher function. Indeed, the authors prove that if f is an algebraic solution of (B), different from a Möbius transformation, then its local inverse f^{-1} has exactly two such finite singularities. This result, and [6, Theorem 4.1.2] about exceptional points of a rational fraction (see [6, Definition 4.1.1]), allow them to find the corresponding forms of $R(z)$. Then, the first remarks of the paper [7] provide the Böttcher functions when $R(z) \in \{z^d, T_d, -T_d\}$. Notice that a Möbius transformation of an algebraic function remains algebraic.

3.3 Second Ingredient: The Notion of Coherent Families

Remind that a formal power series $f(z)$ is **differentially algebraic** if there exists a non-zero polynomial $P(z, X_0, \ldots, X_n)$ such that f satisfies (1). Examples of such functions are given by polynomials, algebraic functions, Bessel functions and classical functions as the exponential, logarithm, cosinus or sinus for example. As said before, the first example of hypertranscendental function does not appear before 1887, with the Euler's Gamma function given by O. Hölder. But *most* entire functions, or analytic functions over a domain of $\mathbb{C}$, are hypertranscendental [65, Theorem 5].

In [66, Problem 26"], L. A. Rubel asks the following question: are there any boundary on the growth of entire differentially algebraic functions? More precisely, given an entire differentially algebraic function satisfying an n-order equation (1), do there exist constants A, α such that:

$$|f(z)| \leq A \exp^n(|z|^\alpha)? \tag{30}$$

L. A. Rubel indicates that a strategy to answer the question negatively should be to construct a function f *big enough* (compared to the exponential) such that all its iterates $f^k(z)$ satisfy the same algebraic differential equation. This is the notion of coherent family studied by M. Boshernitzan and L. A. Rubel in [11]. Indeed, recall that a family of functions is said to be **coherent** if all its elements satisfy the same algebraic differential equation (1). As an example, let us consider the family $\{cz^n, c \in \mathbb{C}, n \in \mathbb{N}\}$. This family is coherent because each of its elements satisfies the following algebraic differential equation:

$$zf^{(2)}(z)f(z) + f(z)f'(z) - zf'(z)^2 = 0. \tag{31}$$

However, the family of all polynomials with rational coefficients is proved not to be coherent in [11]. Before stating the main result of M. Boshernitzan and L. A. Rubel, let us mention two important results concerning coherent families and sketch their proof.

Theorem 3.4
Let f be an analytic differentially algebraic function. Then, f satisfies an autonomous algebraic differential equation. This means that there exist a non-zero polynomial $Q(X_0, \ldots, X_n)$, with coefficients in $\mathbb{C}$, independent of z, such that:

$$Q(f(z), f'(z), \ldots, f^{(n)}(z)) = 0. \tag{32}$$

Proof. We can find a proof of this well-known result in [11], and a more detailed reasoning in [64]. Let us gather the explanations here. Let $P(z, X_0, \ldots, X_n)$ be a non-zero polynomial such that $f(z)$ satisfies Eq. (1). Without any loss of generality, we can assume that P is irreducible in $\mathbb{C}[z, X_0, \ldots, X_n]$. The goal is then to remove the variable z from the equation. The notion of the resultant of two polynomials will solve the problem. Let us define the following operator on polynomials $S(z, X_0, \ldots, X_n) \in \mathbb{C}[z, X_0, \ldots, X_n]$, over $\mathbb{C}[z, X_0, \ldots, X_n, X_{n+1}]$:

$$D : S(z, X_0, \ldots, X_n) \longmapsto DS(z, X_0, \ldots, X_{n+1}) = \frac{dS}{dz}(z, X_0, \ldots, X_n) + \sum_{k=0}^{n} \frac{dS}{dX_k}(z, X_0, \ldots, X_n)X_{k+1}.$$

The advantage of this definition is that for all $S \in \mathbb{C}[z, X_0, \ldots, X_n]$ we have:

$$DS(z, f(z), \ldots, f^{(n+1)}(z)) = \frac{d}{dz}[S(z, f(z), \ldots, f^{(n)}(z))]$$

Hence, $DP(z, f(z), \ldots, f^{(n+1)}(z)) = 0$. Hence, if R is the resultant of the polynomials P and DP with respect to the variable z, properties of the resultant guarantee that $R \in \mathbb{C}[X_0, \ldots, X_{n+1}]$ and that there exist $A, B \in \mathbb{C}[z, X_0, \ldots, X_{n+1}]$ such that:

$$R = AP + B(DP). \tag{33}$$

Now, if we specialize (33) at $(z, f(z), \ldots, f^{(n+1)}(z))$, we obtain:

$$R(f(z), \ldots, f^{(n+1)}(z)) = 0. \tag{34}$$

This provides an autonomous algebraic differential equation for f if we prove that R is a non-zero polynomial. To do so, let us assume by contradiction that $R = 0$. Then, P and DP admit a non-constant common factor. As P is irreducible, $P \mid DP$ in $\mathbb{C}[z, X_0, \dots, X_{n+1}]$. Let $T \in \mathbb{C}[z, X_0, \dots, X_{n+1}]$ such that:

$$(DP)(z, X_0, \dots, X_{n+1}) = P(z, X_0, \dots, X_n)T(z, X_0, \dots, X_{n+1}).$$

Then, for all $U(z) \in \mathbb{C}[z]$, we have:

$$(DP)(z, U(z), \dots, U^{(n+1)}(z)) = P(z, U(z), \dots, U^{(n)}(z))T(z, U(z), \dots, U^{(n+1)}(z)). \tag{35}$$

Let $U(z) \in \mathbb{C}[z]$. Let us write $Q(z, U(z), \dots, U^{(n+1)}(z)) = \tilde{Q}(z)$, for all polynomial $Q \in \mathbb{C}[z, X_0, \dots, X_{n+1}]$. Then, by (35) we have:

$$\tilde{P}'(z) = (\tilde{DP})(z) = \tilde{P}(z)\tilde{T}(z). \tag{36}$$

This provides:

$$\tilde{P}(z) \in \mathbb{C}. \tag{37}$$

Now, an argument from linear algebra allows us to conclude that $P \in \mathbb{C}$, which is a contradiction. More precisely, if $x_0, x_1 \in \mathbb{C}$, there is a surjective morphism of $\mathbb{C}$-vector spaces:

$$\begin{aligned} \phi_{x_0,x_1} : \mathbb{C}_{2n+2}[z] &\longrightarrow \mathbb{C}^{2n+2} \\ U(z) &\longmapsto \left(\begin{pmatrix} U(x_0) \\ \vdots \\ U^{(n)}(x_0) \end{pmatrix}, \begin{pmatrix} U(x_1) \\ \vdots \\ U^{(n)}(x_1) \end{pmatrix} \right), \end{aligned} \tag{38}$$

where $\mathbb{C}_{2n+2}[z]$ denotes the $\mathbb{C}$-vector space of complex polynomials of degree less that or equal to $2n+2$. Then, for all $x_0 \in \mathbb{C}$ and for all $(z_0, \dots, z_n) \in \mathbb{C}^{n+1}$, there exists $U(z) \in \mathbb{C}_{2n+2}[z]$ such that $(U(x_0), \dots, U^{(n)}(x_0)) = (z_0, \dots, z_n)$ and $(U(0), \dots, U^{(n)}(0)) = (0, \dots, 0)$. Thus, Eq. (37) successively applied at $(x_0, U(x_0), \dots, U^{(n)}(x_0))$ and at $(0, U(0), \dots, U^{(n)}(0))$ implies that

$$P(x_0, z_0, \dots, z_n) = P(0, \dots, 0),\ \forall (x_0, z_0, \dots, z_n) \in \mathbb{C}^{n+2}.$$

This implies that $P = P(0, \dots, 0) \in \mathbb{C}$ and yields a contradiction. Theorem 3.4 is proved.

□

Another useful result about coherent families is that they are stable with respect to many operations. This is the following statement, proved by M. Boshernitzan and L. A. Rubel in [11].

Theorem 3.5 (M. Boshernitzan and L. A. Rubel)
Let f, g be two analytic differentially algebraic functions. Let P, Q be non-zero differential algebraic polynomials providing an autonomous differential algebraic

relation for f and g respectively. Let us consider the functions :

$$f + g, f - g, f \times g, f/g, fg, fG, fg', \tag{39}$$

where G is a primitive of g, and the composition of applications is denoted multiplicatively. Then, for every function h in this list, there exists a complex autonomous polynomial $T(X_0, \dots, X_n)$, ***which depends only on*** *P and Q* ***(and not on*** *f nor g) such that:*

$$T(h(z), h'(z), \dots, h^{(n)}(z)) = 0. \tag{40}$$

In particular*, coherent families are stable under the operations in (39). In other words, if $\mathscr{F}$ and $\mathscr{G}$ are two coherent families, the family $\{f * g \mid f \in \mathscr{F}, g \in \mathscr{G}\}$ is a coherent family, where $*$ is a fixed operation of the set $\{+, -, \times, \div\}$, or the operation of composition. And the two families $\{fG \mid f \in \mathscr{F}, G' \in \mathscr{G}\}$, $\{fg' \mid f \in \mathscr{F}, g \in \mathscr{G}\}$ are coherent families.*

Proof. In order to explain the reasoning, we only sketch the proof for $h = f + g$ (it is similar in the other cases) and the case where P, Q are of order 1 (to reduce the notations). In other words, we have the equations: $P(f(z), f'(z)) = 0$ and $Q(g(z), g'(z)) = 0$, with $P, Q \in \mathbb{C}[X_0, X_1]$ autonomous. As P is autonomous, if we derive the first equation with respect to the variable z, we obtain:

$$f''(z)S(P)(f) + \tilde{P}(f(z), f'(z)) = 0, \tag{41}$$

where $\tilde{P} \in \mathbb{C}[X_0, X_1]$, and $S(P)$ is the *separant* of the polynomial P. For a polynomial $U(X_0, \dots, X_n)$ in $\mathbb{C}[X_0, \dots, X_n]$, $S(U)$ is defined by:

$$S(U)(X_0, \dots, X_n) = \frac{dU}{dX_n}(X_0, \dots, X_n),$$

where the derivation is made with respect to the biggest variable appearing in U. For every analytic function ϕ, we let $S(U)(\phi) = S(U)(\phi(z), \phi'(z), \dots, \phi^{(n)}(z))$. Similarly, we have:

$$g''(z)S(Q)(g) + \tilde{Q}(g(z), g'(z)) = 0, \tag{42}$$

where $\tilde{Q} \in \mathbb{C}[X_0, X_1]$.

Let us first assume that $S(P)(f) \neq 0$ and $S(Q)(g) \neq 0$. Then, by (41) and (42), there exists a non-zero rational fraction $R \in \mathbb{C}(Z_0, \dots, Z_3)$ such that:

$$h^{(2)} = R(f, f', g, g').$$

Note that R only depends on P and Q. But, $g = h - f$ and $g' = h' - f'$. Thus, there exists a non-zero rational fraction $H_2 \in \mathbb{C}[X_0, \dots, X_3]$. Such that:

$$h^{(2)}(z) = H_2(f(z), f'(z), h(z), h'(z)).$$

Note that H_2 only depends on P and Q. We deduce the existence of non-zero rational fractions $H_3, H_4 \in \mathbb{C}(X_0, \dots, X_3)$ such that:

$$h^{(3)}(z) = H_3(f(z), f'(z), h(z), h'(z)).$$

$$h^{(4)}(z) = H_4(f(z), f'(z), h(z), h'(z)).$$

Note that H_3, H_4 only depend on P and Q.

Now, let us consider for every $i \in \{2, 3, 4\}$, $H_i(X_0, \dots, X_3)$ as

$$H_i(X_0, \dots, X_3) \in K := [\mathbb{C}(X_2, X_3)](X_0, X_1).$$

The transcendence degree of K over $L := C(X_2, X_3)$ is equal to 2. Hence, there exists a non-zero monic polynomial $S(X_2, X_3)(Y_0, Y_1, Y_2) \in L[Y_0, Y_1, Y_2]$, where the Y_i's are formal variables for $i \in \{0, 1, 2\}$, such that:

$$S(X_2, X_3)(H_2(X_0, \dots, X_3), H_3(X_0, \dots, X_3), H_4(X_0, \dots, X_3)) = 0. \qquad (43)$$

Note that the coefficients of S are in L. Hence, S only depends on the $H_i(X_0, \dots, X_3)$, that is, only on P and Q.

Now, if we let $X_0 = f(z)$, $X_1 = f'(z)$, $X_2 = h(z)$, $X_3 = h'(z)$ in (43), we get:

$$S(h(z), h'(z))(h^{(2)}(z), h^{(3)}(z), h^{(4)}(z)) = 0.$$

Now, if we formally replace $h^{(j)}(z)$ by a variable X_j, for every $j \in \{0, \dots, 4\}$, we find a polynomial $T \in \mathbb{C}[X_0, \dots, X_4]$ such that

$$T(h(z), h'(z), h^{(2)}(z), h^{(3)}(z), h^{(4)}(z)) = 0.$$

As, $S \in L[Y_0, Y_1, Y_2]$ is monic with respect to the variables Y_0, Y_1, Y_2, T is a non-zero polynomial. Finally, as S only depend on P and Q, the same is true for T. Hence, T provides a non-zero autonomous differential algebraic relation for $h(z)$, which only depends on P and Q.

Then, it remains to treat the case where $S(P)(f) = 0$, or $S(Q)(g) = 0$.

First, let us notice that there exist integers $k, l \geq 1$ such that the iterates $S^k(P)$ and $S^l(Q)$ are non-zero constants. Indeed, the separant of a polynomial strictly reduces its total degree. Hence, as polynomials, $S^k(P), S^l(Q) \neq 0$. Thus, there exist $k_1 \leq k - 1$ and $l_1 \leq l - 1$ such that:

$$S^{k_1}(P)(f) = 0 \text{ and } S^{k_1+1}(P)(f) \neq 0 \qquad (44)$$
$$S^{l_1}(Q)(g) = 0 \text{ and } S^{l_1+1}(Q)(g) \neq 0.$$

Note that for all $k_1 \leq k$ and $l_1 \leq l$, $S^{k_1}(P)$, $S^{l_1}(Q) \neq 0$, as polynomials. Besides, we notice that k, l only depend on P and Q, but k_1, l_1 depend on f and g. The idea is then to apply the first part of the proof to $S^{k_1}(P)$, $S^{l_1}(Q)$, instead of P, Q respectively, but we need to get rid of the dependence on f and g.

To do so, let us consider all the integers $k_1 \leq k-1$ and $l_1 \leq l-1$. As $S^{k_1}(P)$ and $S^{l_1}(Q)$ are non-zero polynomials, we can consider two formal variables $\tilde{f}_{k_1}$, $\tilde{g}_{k_1}$ and assume that they formally satisfy (44), for k_1, l_1. We deduce from the first part of the proof that there exists a differential algebraic relation satisfied by the formal variable $h_{k_1,l_1} = \tilde{f}_{k_1} + \tilde{g}_{l_1}$, which only depends on $S^{k_1}(P)$, $S^{l_1}(Q)$. Let us note T_{k_1,l_1} the associated non-zero autonomous differential algebraic polynomial, which only depends on P, Q and k_1, l_1. Moreover, for every functions u, v which satisfy (44) for some k_1, l_1, we have:

$$T_{k_1,l_1}(u+v) = 0.$$

Then, let us note:

$$T = \prod_{k_1 \leq k-1; l_1 \leq l-1} T_{k_1,l_1}.$$

Then, T is a non-zero autonomous differential algebraic polynomial which only depends on P and Q. Moreover, $T(f+g) = 0$ and this concludes the proof. □

Let us notice that if f is an invertible analytic function which is differentially algebraic, then f^{-1} is also differentially algebraic. Indeed, for all $n \in \mathbb{N}$, $\left(f^{-1}\right)^{(n)}(z)$ is a rational fraction of $f(w), f'(w), \ldots, f^{(n)}(w)$, where $w = f^{-1}(z)$. But f is differentially algebraic. Therefore, the family $\{f^{(n)}(w)\}_n$ has a finite transcendence degree over $\mathbb{C}(z)$, and so has the family $\left\{\left(f^{-1}\right)^{(n)}(z)\right\}_n$.

We are finally able to state the main result of M. Boshernitzan and L.A. Rubel [11], which links coherent families to the algebraic properties of Schröder, Böttcher and Abel functions.

Theorem 3.6. (M. Boshernitzan and L.A. Rubel)
Let f be a Schröder, Böttcher or Abel function. Let $R(z)$ be the associated rational fraction of degree at least two. Then, f is differentially algebraic if and only if the family $\{R^n(z)\}_{n\in\mathbb{N}}$ is coherent.

The proof of Theorem 3.1 only uses the direct implication of Theorem 3.6, that is why we will only reproduce this part of the proof here.

Proof of the direct implication of Theorem 3.6. Let us assume that f is differentially algebraic. The goal is to prove that the family $\{R^n(z)\}_n$ is coherent. First, let us assume that f is a Schröder function. Then, $g = f^{-1}$ is differentially algebraic and satisfy Equation (S_0) . Thus, we have:

$$R = g^{-1}(sz)g. \tag{45}$$

Then, for all $n \in \mathbb{N}$:

$$R^n = g^{-1}(s^n z)g.$$

We have seen that the family $\{s^n z\}_n$ is coherent (see (31)). Hence, by Theorem 3.5, $\{R^n\}_n$ is coherent.

If f is a Böttcher function, then, $g = f^{-1}$ is differentially algebraic and satisfy Eq. (B_0). Similarly, we obtain

$$R^n = g^{-1}(z^{d^n})g.$$

We have seen that the family $\{z^{d^n}\}_n$ is coherent (see (31)). Hence, by Theorem 3.5, $\{R^n\}_n$ is coherent.

Finally, if f is an Abel function, we find:

$$R^n = f^{-1}(z+n)f. \tag{46}$$

But the family $\{z+n\}_n$ is coherent because $\{n\}_n$ is (see (31)) and the coherence is stable under addition by Theorem 3.5. Hence, again by Theorem 3.5, $\{R^n\}_n$ is coherent. □

3.4 Third Ingredient: The Theory of P. Fatou and G. Julia

The ingredients from the theory of P. Fatou and G. Julia needed in the proof of Theorem 3.1 are stated as Theorems 3.8 and 3.9 below. Let $R(z)$ be a rational fraction of degree at least two. Let $F(R)$ denote the Fatou set of R. Recall that it is the open set of all the elements $z_0 \in \hat{C}$ for which the family $\{R^n(z_0)\}_n$ is normal in a neighbourhood of z_0. A fundamental result about normal families is the following normality criterion from P. Montel [51].

Theorem 3.7 (P. Montel)
Let D be a domain in $\hat{C}$. Let $\Omega = \hat{C} \setminus \{0, 1, \infty\}$. Then, the family

$$\mathscr{F} = \{f : D \longrightarrow \Omega \mid f \text{ is analytic over } D\}$$

is normal over D.

Among other things, this allows the author to prove various theorems from É. Picard. One of them states that an analytic function on a punctured neighbourhood of the origin, which admits an essential singularity at 0, takes all the values of $\hat{C}$, except at most two.

Now, let $J(R) = \hat{C} \setminus F(R)$ denote the Julia set of R. Remind that $J(R)$ is a closed compact subset of $\hat{C}$, which is non empty and perfect. The Fatou set, however, can be empty, as we will see in Theorem 3.8. Let us note that the Fatou and Julia sets are compatible with the notions of iteration and conjugation via a Möbius transformation g. Indeed, we can prove that:

$$F(R^n) = F(R),\ J(R^n) = J(R),\ \forall n \in \mathbb{N}. \tag{47}$$

Moreover, if g is a Möbius transformation and $S = gRg^{-1}$, we have:

$$F(S) = g(F(R)),\ J(S) = g(J(R)). \tag{48}$$

Now, let us discuss the two results used in the proof of Theorem 3.1. The first one deals with the cases where the Fatou set is empty or not.

Theorem 3.8
Let $R(z)$ be a rational fraction of degree at least two.

1. *Assume that $R(z)$ admits a non repelling fixed point α. If α is irrationally indifferent, assume further that the associated Schröder equation* (S) *admits a convergent solution in a neighbourhood of α. Then $F(R) \neq \emptyset$.*
2. *Assume that $R(z)$ is the rational fraction of the Schröder's equation associated with one of the functions $\wp(\alpha z^r + \beta)$, $\wp^2(\alpha z^r + \beta)$, $\wp^3(\alpha z^r + \beta)$, or $\wp'(\alpha z^r + \beta)$ which appears in the first point of Theorem 3.1. Then $F(R) = \emptyset$.*

The second result used by P.-G. Becker and W. Bergweiler to prove Theorem 3.1 is the following.

Theorem 3.9
Let $R(z)$ be a rational fraction of degree d at least two. The set of all the repelling fixed points of all the iterates $R^n(z)$, $n \in \mathbb{N}$, is dense in $J(R)$.

Let us sketch the proof of Theorem 3.8 (see details in [6]).

Proof of Theorem 3.8. First, we can prove that an attracting or super-attracting fixed point of $R(z)$ belongs to $F(R)$. Indeed, based on the Taylor development of R, there exists $\sigma < 1$ and a neighbourhood D of the fixed point α such that:

$$|R(z) - R(\alpha)| \leq \sigma |z - \alpha|,\ \forall z \in D$$

if α is attracting; and even:

$$|R(z) - R(\alpha)| \leq \sigma |z - \alpha|^2,\ \forall z \in D,$$

if α is super-attracting. Secondly, if α is rationally indifferent, $\alpha \in J(R)$. This is proved by P. Fatou [26] and G. Julia [40] (see also [6, Theorem 6.5.1]). But the Leau-Fatou Flower theorem implies the existence of domains $L_1, \ldots, L_k$ called the petals of R such that, for all $j \in \{1, \ldots, k\}$, α belongs to the boundary of L_j, $R(L_j) \subset L_j$ and the restriction to L_j of R^n converges uniformly to α on L_j, when n tends to infinity. This latter property shows that each petal is included in $F(R)$. Hence, $F(R)$ is not empty.

Finally, if α is an irrationally indifferent fixed point of $R(z)$, we can use a theorem stated in [6] which establishes the following equivalence, valid for an indifferent fixed point of $R(z)$:

$$R(z) \text{ is linearizable in a neighbourhood of } \alpha \Leftrightarrow \alpha \in F(R). \tag{49}$$

We say that $R(z)$ is **linearizable** in a neighbourhood D of α if there exists an invertible analytic function g over D such that R is locally conjugated via g to a function of the form:

$$h(z) = \alpha + (z - \alpha)h'(\alpha).$$

But this is precisely the case for R, which is conjugated, via the solution of the associated Schröder's equation (S), to $h(z) = sz$. Then $\alpha \in F(R)$.

Besides, for the second part of the theorem, we only sketch the proof for the Weierstrass function $\wp(z)$, following [6, p. 74]. Recall that this function is **periodic**, that is, satisfies (27). Moreover, this function satisfies the following Schröder's equation:

$$\wp(2z) = R(\wp(z)), \tag{50}$$

for a certain rational fraction $R(z)$. This is the *duplication formula* for elliptic curves (see for example [72, page 54, page 170] and [42, Chapter 1]). Now, let D be a disc in $\mathbb{C}$ and let $U = \wp^{-1}(D)$. Let $\phi(z) = 2z$. Then, $\phi^n(U) = 2^n U$. Hence, for N big enough, $2^N U$ will contain a period parallelogram of the lattice Λ associated with $\wp$. This means that the values that $\wp$ will take in $2^N U$ are exactly the one it takes on $\mathbb{C}$. But it is known, as a consequence of the open mapping theorem, that $\wp(\mathbb{C}) = \hat{C}$. Hence, using (50), we obtain:

$$R^N(D) = R^N(\wp(U)) = \wp(2^N U) = \hat{C}.$$

This implies that $\{R^n\}_n$ is not equicontinuous over D. Indeed, the local behaviour of the iterates does not respect the proximity of antecedent points, because D is sent onto the whole Riemann sphere by R^N. As D is arbitrary, we conclude that $\{R^n\}_n$ is not equicontinuous over any open subset of $\mathbb{C}$. This implies that $J(R) = \hat{C}$. □

Note that, as quickly mentioned in Sect. 2.2, the case of an irrationally indifferent fixed point α was proved by C. L. Siegel in 1942 for particular cases of α, called diophantine fixed points.

Now, let us reproduce the proof of Theorem 3.9, as detailed in [6, Theorem 6.9.2]. First let us recall the following facts. Let $R(z)$ be a rational fraction of degree $d \geq 2$. Then, $R(z)$ is a d-fold map of $\hat{C}$ onto itself. That is, for all $w \in \hat{C}$, the equation

$R(z) = w$ admits exactly d solutions in $\hat{C}$, when counting multiplicities. Now, we say that w is a critical value of $R(z)$, if there exists $z_0 \in \hat{C}$ such that $R(z_0) = w$ and if there is no neighbourhood of z_0 in which $R(z)$ is injective. But there are only finitely many critical values of $R(z)$ in $\hat{C}$. Indeed, for an element $x \in \hat{C}$, there exists a neighbourhood of x in which $R(z)$ is injective if $R'(z)$ has neither a zero nor a pole at x. When w is not a critical value of $R(z)$, there exists exactly d pairwise distinct elements $z_i \in \hat{C}$, $i = 1, \ldots, d$ such that $R(z_i) = w$, for every $i \in \{1, \ldots, d\}$.

Proof of Theorem 3.9. The first part of the proof consists in showing that $J(R)$ is contained in the topological closure of the set of all the fixed points of all the $R^n(z)$. Then, by [6, Theorem 9.6.1], the set of all the non repelling fixed points of all the R^n is finite. We deduce that $J(R)$ is contained in the topological closure of the set $\mathscr{P}$ of all the repelling fixed points of all the $R^n(z)$. Finally, as, for every integer n, each repelling fixed point of $R^n(z)$ is contained in the closed set $J(R^n) = J(R)$, we have that $J(R)$ is the topological closure of $\mathscr{P}$. This gives the result of Theorem 3.9.

It thus remains to show that $J(R)$ is contained in the topological closure of the set of all the fixed points of all the $R^n(z)$. It suffices to consider an open set $\mathscr{N}$ of $\hat{C}$ such that $\mathscr{N} \cap J \neq \emptyset$, and prove that $\mathscr{N}$ contains a fixed point of one of the $R^n(z)$. Let $w \in \mathscr{N} \cap J \neq \emptyset$. We may assume that w is not a critical value of R^2. Indeed, the number of such values is finite and we may thus find a non-critical value of R^2 in a neighbourhood of w in $\mathscr{N} \cap J$. Then, $R^{-2}(w)$ contains at least four distinct points w_j, $j = 1, \ldots, 4$. Indeed, the fact that $d \geq 2$ implies that the degree of R^2 is more than or equal to four. At least three of these points, say w_1, w_2, w_3 are distinct from w. Then, we construct three neighbourhoods $\mathscr{N}_i$ of w_i, $i = 1, 2, 3$, whose topological closures are pairwise disjoint, and a neighbourhood $\mathscr{N}_0 \subset \mathscr{N}$ of w, disjoint from every $\mathscr{N}_i$ and such that $R^2 : \mathscr{N}_i \longrightarrow \mathscr{N}_0$ is a homeomorphism, with reciprocal S_i, for every $i \in \{1, \ldots, 3\}$. But $\{R^n(z)\}_n$ is not a normal family on $\mathscr{N}_0$, because $w \in \mathscr{N}_0 \cap J(R)$. Then [6, Theorem 3.3.6] (which is a corollary of Theorem 3.7) gives the existence of $z_0 \in \mathscr{N}_0$, $n \geq 1$ and $i \in \{1, \ldots, 3\}$ such that:

$$R^n(z_0) = S_i(z_0).$$

We deduce that $R^{2+n}(z_0) = R^2(S_i(z_0)) = z_0$. Hence z_0 is a fixed point of an iterate of $R(z)$, contained in $\mathscr{N}$. □

Note that P. Fatou [26] and G. Julia [40] proved the finiteness of the sets of attracting and rationally indifferent fixed points of all the $R^n(z)$.

3.5 The Proof of Theorem 3.1

In this section, we reproduce the proof of Theorem 3.1 established in [8] by P.-G. Becker and W. Bergweiler . We give detailed explanations about the way that the three ingredients exposed earlier merge. This is a beautiful illustration of the power of the interactions between distinct domains of mathematics. As mentioned before, the statement of Theorem 3.1 belongs to the theory of hypertranscendence of solutions of functional equations. The proof of Theorem 3.1 uses previous results of hypertranscendence and coherent families, along with the theory of iteration of a rational fraction and analytic properties of the sets of Fatou and Julia.

Proof of Theorem 3.1. We will use the notations of (8) and the framework of Sect. 3.1. Let $R(z)$ be a rational fraction of degree at least two, which admits 0 as a fixed point. According to the nature of this fixed point, we let f be either a Schröder solution of $S_{R,0}$ (recall that we assume that this equation admits a convergent solution), a Böttcher solution of $B_{R,0}$, or a Abel solution of $A_{R,0}$. Let us assume that f is differentially algebraic. Our goal is to prove that f cannot be a Abel function and that f is in the list 1 or 2 of the statement of Theorem 3.1, depending on whether f is a Schröder or Böttcher function.

Let us immediately remark that Theorem 3.6 guarantees that the family $\{R^n(z)\}_n$ is coherent. To begin with, if 0 is repelling (that is, f is a Schröder function), we can apply Theorem 3.2 and get that f is in the list of the first point of Theorem 3.1.

Now, let us deal with the case where 0 is not repelling. The goal is to reduce this case to the repelling one. Let us first notice that Theorem 3.8 implies that $F(R) \neq \emptyset$. Moreover, by Theorem 3.9, the set of all the repelling fixed points of all the iterates $R^n(z)$ are dense in $J(R)$. As $J(R)$ is always a non-empty set [25], we deduce that there exist $w \in \hat{C}$ and $k \in \mathbb{N}$ such that w is a repelling fixed point of $R^k(z)$. We may then consider a Schröder solution ϕ of $S_{R^k,w}$. But $\{(R^k)^n\}_n$ is coherent, as a subfamily of the coherent family $\{R^n(z)\}_n$. Hence, by Theorem 3.6, ϕ is differentially algebraic.

Then, Theorem 3.2, implies that there exists a Möbius transformation $g(z)$ such that $g^{-1}R^k g$ is one of the rational fractions which appear in the first point of Theorem 3.1 (with d^k instead of d). Let us note $S(z) = g^{-1}R^k g$. By (47) and (48), the fact that $F(R) \neq \emptyset$ implies that $F(S) \neq \emptyset$.

Hence, by Theorem 3.8, $S(z) \in \{z^{d^k}, z^{-d^k}, T_{d^k}, -T_{d^k}\}$. Besides, as 0 is a non repelling fixed point of $R(z)$, as noticed in Sect. 3.4, $g^{-1}(0)$ is a non repelling fixed point of $S(z)$. But z^{-d^k} does not admit any such fixed points, and all the non repelling fixed points of z^{d^k}, T_{d^k} or $-T_{d^k}$ are super-attracting (a computation shows that a fixed point of a Tchebychev polynomial distinct from ∞ is repelling and that ∞ is a super-attracting fixed point). Hence, $S(z) \in \{z^{d^k}, T_{d^k}, -T_{d^k}\}$, and $g^{-1}(0)$ is a super-attracting fixed point of one of these three polynomials. Note that, by Sect. 3.4 again, this implies that 0 is a super-attracting fixed point of $R^k(z)$ and $R(z)$.

We deduce that f is a solution of $B_{R,0}$. Iterating this equation, we see that f is a solution of $B_{R^k,0}$. Now, by (8), $g^{-1}f$ is a solution ψ of $B_{S,g^{-1}(0)}$. But the fact that $S(z) \in \{z^{d^k}, T_{d^k}, -T_{d^k}\}$ guarantees, via Theorem 3.3 that ψ is algebraic and is a Möbius transformation or a Möbius transformation of one of the functions of the list of the second point of Theorem 3.1. As $f = g\psi$, we deduce that f is a Möbius transformation or a Möbius transformation (composition with the Möbius transformation g) of a function of the list of the second point of Theorem 3.1.

Hence, $f(z)$ is solution of $B_{M,0}$, where $M(z)$ is of the form of the rational fractions appearing in the second point of Theorem 3.1. As $f(z)$ is solution of $B_{R,0}$ and $B_{M,0}$, we have $R = M$ and $R(z)$ is of the form of the rational fractions appearing in the second point of Theorem 3.1.

To conclude, we have proved that f is either a Schröder or Böttcher function (thus f is not a Abel function). Moreover, we have proved that, when f is a Schröder function, f is of the form described in the first point of Theorem 3.1, and when f is a Böttcher function, f is of the form described in the second point of Theorem 3.1. Theorem 3.1 is thus proved. □

Acknowledgement The author would like to warmly thank Alin Bostan, Lucia Di Vizio, and Kilian Raschel for their enthusiastic proposal to write this paper. She would also like to thank the detailed reports of the reviewers which allowed her to improve the fluidity, clarity and precision of this survey. Finally, she would like to thank Lucia Di Vizio for her support and her reviews and corrections of this paper.

References

1. N. H. ABEL – *œuvres complètes de Niels Henrik Abel. Tome II*, Imprimerie de Grøndahl & Son, Christiania; distributed by the Norwegian Mathematical Society, Oslo, 1981, Contenant les mémoirs posthumes d'Abel. [Containing the posthumous memoirs of Abel], Edited and with notes by L. Sylow and S. Lie.
2. D. S. ALEXANDER – *A history of complex dynamics-From Schröder to Fatou and Julia*, Aspects of Mathematics, E24, Friedr. Vieweg & Sohn, Braunschweig, 1994.
3. G. W. ANDERSON, W. D. BROWNAWELL & M. A. PAPANIKOLAS – "Determination of the algebraic relations among special Γ-values in positive characteristic", *Ann. of Math. (2)* **160** (2004), p. 237–313.
4. M. ASCHENBRENNER & W. BERGWEILER – "Julia's equation and differential transcendence", *Illinois J. Math.* **59** (2015), no. 2, p. 277–294.
5. M. AUDIN – *Fatou, Julia, Montel, le grand prix des sciences mathématiques de 1918, et après*, Springer-Verlag, Berlin, 2009.
6. A. F. BEARDON – *Iteration of rational functions-Complex analytic dynamical systems*, Graduate Texts in Mathematics, vol. 132, Springer-Verlag, New York, 1991.
7. P.- G. BECKER & W. BERGWEILER – "Transcendency of local conjugacies in complex dynamics and transcendency of their values", *Manuscripta Math.* **81** (1993), no. 3-4, p. 329–337.
8. —"Hypertranscendency of conjugacies in complex dynamics", *Math. Ann.* **301** (1995), no. 3, p. 463–468.
9. W. BERGWEILER – "Solution of a problem of Rubel concerning iteration and algebraic differential equations", *Indiana Univ. Math. J.* **44** (1995), no. 1, p. 257–267.
10. P. BORWEIN – "Hypertranscendence of the functional equation $g(x^2) = [g(x)]^2 + cx$", *Proc. Amer. Math. Soc.* **107** (1989), no. 1, p. 215–221.
11. M. BOSHERNITZAN & L. A. RUBEL – "Coherent families of polynomials", *Analysis* **6** (1986), no. 4, p. 339–389.
12. L. BÖTTCHER – "Principaux résultats de convergence des itérées et applications à l'analyse (en russe)", *Izv. Kazan* **13** (1904), p. 1–37.
13. — , "Principaux résultats de convergence des itérées et applications à l'analyse (en russe)", *Izv. Kazan* **14** (1904), p. 155–200.
14. — , "Principaux résultats de convergence des itérées et applications à l'analyse (en russe)", *Izv. Kazan* **14** (1904), p. 201–234.
15. H. BROLIN – "Invariant sets under iteration of rational functions", *Ark. Mat.* **6** (1965), p. 103–144 (1965).

16. X. BUFF & A. CHÉRITAT – "Ensembles de Julia quadratiques de mesure de Lebesgue strictement positive", *C. R. Math. Acad. Sci. Paris* **341** (2005), no. 11, p. 669–674.
17. F. W. CARROLL – "Transcendental transcendence of solutions of Schröder's equation associated with finite Blaschke products", *Michigan Math. J.* **32** (1985), no. 1, p. 47–57.
18. J.- L. CHABERT – "Un demi-siècle de fractales: 1870–1920", *Historia Math.* **17** (1990), no. 4, p. 339–365.
19. A. CHÉRITAT – "L'ensemble de mandelbrot", *Images des mathématiques, CNRS* (Novembre 2010).
20. L. DI VIZIO – "Approche galoisienne de la transcendance différentielle", in *Transcendance et irrationalité*, SMF Journ. Annu., vol. 2012, Soc. Math. France, Paris, 2012, p. 1–20.
21. A. DOUADY – "Systèmes dynamiques holomorphes", in *Bourbaki seminar, Vol. 1982/83*, Astérisque, vol. 105, Soc. Math. France, Paris, 1983, p. 39–63.
22. A. DOUADY & J. H. HUBBARD – "Itération des polynômes quadratiques complexes", *C. R. Acad. Sci. Paris Sér. I Math.* **294** (1982), no. 3, p. 123–126.
23. T. DREYFUS, C. HARDOUIN & J. ROQUES – "Hypertranscendence of solutions of Mahler equations", *J. Eur. Math. Soc.* **20** (2018), p. 2209–2238.
24. A. DUDKO & M. YAMPOLSKYY – "On computational complexity of cremer Julia sets.", (2019), arxiv:1907.11047v3 [math.DS]
25. P. FATOU – "Sur les équations fonctionnelles", *Bull. Soc. Math. France* **47** (1919), p. 161–271.
26. —, "Sur les équations fonctionnelles", *Bull. Soc. Math. France* **48** (1920), p. 33–94.
27. —, "Sur les équations fonctionnelles", *Bull. Soc. Math. France* **48** (1920), p. 208–314.
28. R. FENG – "Hrushovski's algorithm for computing the Galois group of a linear differential equation", *Adv. in Appl. Math.* **65** (2015), p. 1–37.
29. G. FERNANDES – "Méthode de Mahler en caractéristique non nulle (thèse)", *tel-02386667*, https://tel.archives-ouvertes.fr/tel-02386667/document.
30. —, "Méthode de Mahler en caractéristique non nulle: un analogue du théorème de Ku. Nishioka", *Ann. Inst. Fourier (Grenoble)* **68** (2018), no. 6, p. 2553–2580.
31. —, "Regular extensions and algebraic relations between values of Mahler functions in positive characteristic", *Trans. Amer. Math. Soc.* **372** (2019), no. 10, p. 7111–7140.
32. S. R. GARCIA, J. MASHREGHI & W. T. ROSS – "Finite Blaschke products: a survey", *Math and Computer Science Faculty Publications* **181** (2018).
33. L. GEYER – "Linearizability of saturated polynomials", *Indiana Univ. Math. J.* **68** (2019), no. 5, p. 1551–1578.
34. C. HARDOUIN – "Hypertranscendance des systèmes aux différences diagonaux", *Compos. Math.* **144** (2008), no. 3, p. 565–581.
35. —, "Unipotent radicals of Tannakian Galois groups in positive characteristic", in *Arithmetic and Galois theories of differential equations*, Sémin. Congr., vol. 23, Soc. Math. France, Paris, 2011, p. 223–239.
36. —, "Galoisian approach to differential transcendence", in *Galois theories of linear difference equations: an introduction*, Math. Surveys Monogr., vol. 211, Amer. Math. Soc., Providence, RI, 2016, p. 43–102.
37. C. HARDOUIN & M. F. SINGER – "Differential Galois theory of linear difference equations", *Math. Ann.* **342** (2008), no. 2, p. 333–377.
38. E. HRUSHOVSKI – "Computing the Galois group of a linear differential equation", in *Differential Galois theory*, Banach Center Publ., vol. 58, Polish Acad. Sci. Inst. Math., Warsaw, 2002, p. 97–138.
39. K. ISHIZAKI – "Hypertranscendency of meromorphic solutions of a linear functional equation", *Aequationes Math.* **56** (1998), no. 3, p. 271–283.
40. G. JULIA – "Mémoire sur l'itération des fonctions rationnelles (8)", *J. Math. Pures Appl.* **1** (1918), p. 47–46.
41. G. KOENIGS – "Recherches sur les intégrales de certaines équations fonctionnelles", *Ann. Sci. École Norm. Sup. (3)* **1** (1884), p. 3–41.
42. S. LANG – *Elliptic curves: Diophantine analysis*, Grundlehren der Mathematischen Wissenschaften [Fundamental Principles of Mathematical Sciences], vol. 231, Springer-Verlag, Berlin-New York, 1978.

43. L. LEAU – "Étude sur les équations fonctionnelles à une ou à plusieurs variables", *Ann. Fac. Sci. Toulouse Sci. Math. Sci. Phys.* **11** (1897), no. 2, p. E1–E24.
44. —, "Étude sur les équations fonctionnelles à une ou à plusieurs variables", *Ann. Fac. Sci. Toulouse Sci. Math. Sci. Phys.* **11** (1897), no. 3, p. E25–E110.
45. G. LEVIN & M. ZINSMEISTER – "On the Hausdorff dimension of Julia sets of some real polynomials", *Proc. Amer. Math. Soc.* **141** (2013), no. 10, p. 3565–3572.
46. J. H. LOXTON & A. J. VAN DER POORTEN – "A class of hypertranscendental functions", *Aequationes Math.* **16** (1977), no. 1-2, p. 93–106.
47. K. MAHLER – "Arithmetische Eigenschaften der Lösungen einer Klasse von Funktionalgleichungen", *Math. Ann.* **101** (1929), no. 1, p. 342–366.
48. —, "Arithmetische Eigenschaften einer Klasse transzendental-transzendenter Funktionen", *Math. Z.* **32** (1930), no. 1, p. 545–585.
49. —, "Uber das Verschwinden von Potenzreihen mehrerer Veränderlichen in speziellen Punktfolgen", *Math. Ann.* **103** (1930), no. 1, p. 573–587.
50. B. MANDELBROT – *Les objets fractals*, Flammarion, Editeur, Paris, 1975, Forme, hasard et dimension, Nouvelle Bibliothèque Scientifique.
51. P. MONTEL – "Sur les familles de fonctions analytiques qui admettent des valeurs exceptionnelles dans un domaine", *Ann. Sci. École Norm. Sup. (3)* **29** (1912), p. 487–535.
52. K. D. NGUYEN – "Algebraic independence of local conjugacies and related questions in polynomial dynamics", *Proc. Amer. Math. Soc.* **143** (2015), no. 4, p. 1491–1499.
53. P. NGUYEN – "Hypertranscedance de fonctions de Mahler du premier ordre", *C. R. Math. Acad. Sci. Paris* **349** (2011), no. 17-18, p. 943–946.
54. K. NISHIOKA – "A note on differentially algebraic solutions of first order linear difference equations", *Aequationes Math.* **27** (1984), no. 1-2, p. 32–48.
55. —, "Algebraic function solutions of a certain class of functional equations", *Arch. Math. (Basel)* **44** (1985), no. 4, p. 330–335.
56. K. NISHIOKA – *Mahler functions and transcendence*, Lecture Notes in Mathematics, vol. 1631, Springer-Verlag, Berlin, 1996.
57. H.- O. PEITGEN & P. H. RICHTER – *The beauty of fractals-Images of complex dynamical systems*, Springer-Verlag, Berlin, 1986.
58. H. POINCARÉ – "Sur une nouvelle classe de transcendantes uniformes", *Journ. de Math. (4)* **6** (1890), p. 313–365.
59. M. VAN DER PUT & M. F. SINGER – *Galois theory of difference equations*, Lecture Notes in Mathematics, vol. 1666, 1997.
60. J. F. RITT – "On the iteration of rational functions", *Trans. Amer. Math. Soc.* **21** (1920), no. 3, p. 348–356.
61. —, "Periodic functions with a multiplication theorem", *Trans. Amer. Math. Soc.* **23** (1922), no. 1, p. 16–25.
62. —, "Transcendental transcendency of certain functions of Poincaré", *Math. Ann.* **95** (1926), no. 1, p. 671–682.
63. J. ROQUES – "On the algebraic relations between Mahler functions", *Trans. Amer. Math. Soc.* **370** (2018), p. 321–355.
64. L. A. RUBEL – "Generalized solutions of autonomous algebraic differential equations", *Canad. Math. Bull.* **29** (1986), no. 3, p. 372–374.
65. —, "A survey of transcendentally transcendental functions", *Amer. Math. Monthly* **96** (1989), no. 9, p. 777–788.
66. —, "Some research problems about algebraic differential equations. II", *Illinois J. Math.* **36** (1992), no. 4, p. 659–680.
67. W. RUDIN – *Principles of mathematical analysis*, third éd., McGraw-Hill Book Co., New York-Auckland-Düsseldorf, 1976, International Series in Pure and Applied Mathematics.
68. E. SCHRÖDER – "Ueber unendlich viele Algorithmen zur Auflösung der Gleichungen", *Math. Ann.* **2** (1870), no. 2, p. 317–365.
69. E. SCHRÖDER – "Ueber iterirte Functionen", *Math. Ann.* **3** (1870), no. 2, p. 296–322.

70. A. B. SHIDLOVSKII – *Transcendental numbers*, De Gruyter Studies in Mathematics, vol. 12, Walter de Gruyter & Co., Berlin, 1989, Translated from the Russian by Neal Koblitz, With a foreword by W. Dale Brownawell.
71. C. L. SIEGEL – "Iteration of analytic functions", *Ann. of Math. (2)* **43** (1942), p. 607–612.
72. J. H. SILVERMAN – *The arithmetic of elliptic curves*, Graduate Texts in Mathematics, vol. 106, Springer, Dordrecht, 2009.
73. F. YANG – "Cantor Julia sets with hausdorff dimension two", (2018), arxiv:1802.01063v1 [math.DS].
74. J.- C. YOCCOZ – "Théorème de Siegel, nombres de Bruno et polynômes quadratiques", no. 231, 1995, Petits diviseurs en dimension 1, p. 3–88.

Hodge Structures and Differential Operators

Masha Vlasenko

These are extended notes of my talk at the IMPANGA seminar in Warsaw on October 11, 2019. The goal was to give a non-technical and arithmetically motivated introduction to the definition of the limiting mixed Hodge structure by Schmid and Deligne. We state several assertions in terms natural to the classical theory of ordinary differential operators and prove them using elementary arguments. References to the geometric context are only made in a few remarks and examples, which can be easily skipped by a reader not familiar with algebro-geometric techniques.

In Sect. 1 we review a classical method of solving linear differential equations near a regular singular point using Laurent series and the logarithm function. This method produces what we call a standard basis in the space of solutions of the differential equation. The key (and well known) observation is that, if the differential operator itself is a polynomial with coefficients in a field $K \subset \mathbb{C}$, then the power series involved in the standard basis also have coefficients in K.

In Sect. 2 we mention geometric (Picard–Fuchs) differential equations and express some period integrals in the standard basis. The upshot of what is done later in Sect. 4 is that the coefficients of such expressions are periods of the limiting Hodge structure. One may wish to skip Sect. 2 on the first reading.

In Sect. 3 we review the definition and basic examples of Hodge structures. In Sect. 4 we consider algebraic families of linear functionals on the space of solutions of a polynomial differential operator and define their limits at a singular point. Our main observation (Lemma 3 and Corollary 4) is that such limits span the K-structure dual to the one determined by the standard basis. We then describe a mixed Hodge structure on the space of solutions, which for geometric differential operators coincides with the limiting mixed Hodge structure constructed by Schmid in [1].

M. Vlasenko (✉)
Institute of Mathematics of the Polish Academy of Sciences, Warsaw, Poland
e-mail: mvlasenko@impan.pl

A. Bostan and K. Raschel (eds.), *Transcendence in Algebra, Combinatorics, Geometry and Number Theory*, Springer Proceedings in Mathematics & Statistics 373,
https://doi.org/10.1007/978-3-030-84304-5_5

1 Solutions Near a Regular Singular Point

Let $K \subset \mathbb{C}$ be a field. A K-*structure* on a $\mathbb{C}$-vector space V is a K-vector space $V_K \subset V$ such that $V = V_K \otimes_K \mathbb{C}$.

Let $K(t)$ be the field of rational functions with coefficients in K. Denote $\theta = t\frac{d}{dt}$ and consider a differential operator

$$L = \theta^r + q_1(t)\theta^{r-1} + \ldots + q_{r-1}(t)\theta + q_r(t) \in K(t)[\theta] \tag{1}$$

whose coefficients satisfy the condition

$$q_j(0) = 0, \qquad 1 \le j \le r. \tag{2}$$

Using classical methods one can construct a set of r linearly independent solutions of (1) in the ring $K[\![t]\!][\log(t)]$, see Lemma 1 below. Later we are going to discuss the K-structure that this construction yields on the space of solutions of L near $t = 0$.

One starts with finding a Laurent series solution $\phi(t) = \sum a_n t^n$. Expanding the coefficients $q_j(t)$ into power series, let us rewrite $L = \sum_{j \ge 0} t^j p_j(\theta)$ with some polynomials $p_j \in K[\theta]$. Condition (2) is equivalent to $p_0(\theta) = \theta^r$, and hence the differential equation $L\phi = 0$ is equivalent to the recurrence relation

$$n^r a_n + p_1(n-1)a_{n-1} + p_2(n-2)a_{n-2} + \ldots = 0.$$

Here we see that the smallest n such that $a_n \neq 0$ can only be $n = 0$, and with the normalization $\phi(0) = 1$ ($a_0 = 1$) there is a unique Laurent series solution which we will denote $\phi_0(t) = \sum_{n \ge 0} a_n t^n \in 1 + tK[\![t]\!]$. Next, one looks for solutions of the shape $\phi(t) = \log(t)\phi_0(t) + \sum_n b_n t^n$, and so on:

Lemma 1 *There exist unique power series* $f_0 \in 1 + t\mathbb{C}[\![t]\!], f_1 \ldots, f_{r-1} \in t\mathbb{C}[\![t]\!]$ *such that*

$$\phi_k(t) = \sum_{j=0}^{k} \frac{\log(t)^j}{j!} f_{k-j}(t), \qquad k = 0, \ldots, r-1, \tag{3}$$

are solutions of the differential operator (1). *Moreover, one has* $f_j \in K[\![t]\!]$ *for all* $0 \le j \le r-1$.

Proof For $0 \le j \le r-1$ consider the differential operator of order $r - j$ given by $L^{(j)} := \frac{\partial^j L}{\partial \theta^j}$, the formal jth derivative of L in θ. One can easily check that expressions (3) are solutions of L if and only if for each k we have

$$L f_k + L^{(1)} f_{k-1} + \frac{1}{2!} L^{(2)} f_{k-2} + \ldots \frac{1}{k!} L^{(k)} f_0 = 0. \tag{4}$$

If we write $L = \sum_{j \ge 0} t^j p_j(\theta)$ with $p_j \in K[\theta]$ of degree at most r and $p_0(\theta) = \theta^r$, then equation $L(\sum a_n t^n) = \sum b_n t^n$ is equivalent to the recurrence relation

$$n^r a_n + p_1(n-1)a_{n-1} + p_2(n-2)a_{n-2} + \ldots = b_n. \tag{5}$$

One can easily see that there is a unique (up to multiplication by a constant) non-zero Laurent series solution $f_0 = \sum_n a_n t^n$ to $Lf_0 = 0$, and this solution is a power series with $a_0 \neq 0$. Moreover, if $a_0 \in K$ then $a_n \in K$ for all $n \geq 0$. Secondly, we observe that $L(K[\![t]\!]) \subset tK[\![t]\!]$ and, more generally, we have

$$L^{(j)}(K[\![t]\!]) \subset tK[\![t]\!], \qquad 0 \leq j \leq r-1.$$

The third observation we can make from formula (5) is that the map

$$L : t\mathbb{C}[\![t]\!] \to t\mathbb{C}[\![t]\!]$$

is invertible and maps $tK[\![t]\!]$ to itself. With these three observations we can now solve (4) by induction on $k = 0, 1, \ldots$ as follows. We start with $k = 0$ and normalize the unique power series solution so that $\phi_0 = f_0 \in 1 + tK[\![t]\!]$. For $k \geq 1$ equation (4) has shape $Lf_k = b$ with $b = -\sum_{j=0}^{k-1}(j!)^{-1}L^{(j)}f_{k-j} \in tK[\![t]\!]$. This latter equation has a unique solution $f_k \in t\mathbb{C}[\![t]\!]$. Moreover, this f_k has coefficients in K, which proves the second assertion of the lemma. □

Example *For the differential operator $L = \theta^r$ the solutions constructed in Lemma 1 are $\phi_k(t) = (k!)^{-1}\log(t)^k$, $0 \leq k \leq r-1$.*

Example *The hypergeometric differential operator $L = \theta^2 - \frac{t}{1-t}\theta - \frac{1}{4}\frac{t}{1-t}$ has solutions*

$$\phi_0(t) = {}_2F_1(\tfrac{1}{2}, \tfrac{1}{2}, 1; t) = \sum_{n=0}^{\infty} \frac{(\frac{1}{2})_n^2}{n!^2} t^n,$$

$$\phi_1(t) = \log(t)\phi_0(t) + \sum_{n=1}^{\infty} \frac{(\frac{1}{2})_n^2}{n!^2} \left(\sum_{k=1}^{n} \frac{1}{k(k-\frac{1}{2})} \right) t^n.$$

In fact the series $f_j(t)$ in Lemma 1 converge in some neighbourhood of $t = 0$. This follows from the classical theorem of Fuchs (see [2, Theorem 2.6]), but one could also estimate the grows of coefficients of $f_j(t)$ directly. Let V be the vector space of solutions to L in a neighbourhood of some regular point $t = t_0$ located close to $t = 0$. By Cauchy's theorem $\dim_{\mathbb{C}} V = r$ and hence functions (3) form a basis in the space of solutions, which we will refer to as *the standard basis*. Note that this basis and the respective K-structure

$$V_K \;=\; K\text{-span of } \phi_0, \ldots, \phi_{r-1} \tag{6}$$

depend on the choice of branch of $\log(t)$ near the base point $t = t_0$.

One can write

$$(\phi_0, \ldots, \phi_{r-1}) = (f_0, \ldots, f_{r-1}) t^{\begin{pmatrix} 0\,1\,0\ldots \\ 0\,0\,1\ldots \\ \ldots \end{pmatrix}},$$

where for a square matrix N we use the matrix-valued function $t^N = \exp(N \log(t)) = \sum_{h \geq 0} (h!)^{-1} N^h \log(t)^h$. Since $\log(t)$ changes to $\log(t) + 2\pi i$ after a counterclockwise turn around the origin, the respective *local monodromy transformation* $\gamma : V \to V$ in the standard basis is given by the matrix

$$\gamma = \exp\left(2\pi i \begin{pmatrix} 0\,1\,0\ldots \\ 0\,0\,1\ldots \\ \ldots \end{pmatrix}\right) = \begin{pmatrix} 1 & 2\pi i & \frac{(2\pi i)^2}{2!} & \ldots \\ 0 & 1 & 2\pi i & \ldots \\ & & & \ldots \end{pmatrix}.$$

Observe that this matrix is *maximally unipotent*, that is $(\gamma - I)^r = 0$ but $(\gamma - I)^k$ is nonzero when $k < r$.

Remark *Condition* (2) *means that the differential operator* (1) *has a regular singularity at $t = 0$ and the local monodromy around this point is maximally unipotent. More generally, operator* (1) *has at most a regular singularity at $t = 0$ if and only if none of the coefficients $q_j(t)$ has a pole at $t = 0$. If this is the case, the solutions $\rho \in \mathbb{C}$ to the algebraic equation $\rho^r + \sum_{j=1}^r q_j(0)\rho^{r-j} = 0$ are called local exponents. (The reader could refer to* [2, §2] *for the local analysis of singularities of linear differential operators.) By Fuchs' theorem, for each local exponent ρ there is a holomorphic function $f(t)$ with $f(0) \neq 0$ such that $t^\rho f(t)$ is a solution of* (1)*. Moreover, one can construct a basis in the space of solutions of* (1) *near $t = 0$ using only functions* $\log(t)$, *t^ρ for local exponents ρ and holomorphic series. Similarly to Lemma 1, the power series involved in this* standard basis *will have coefficients in the field $K(\rho_0, \ldots, \rho_{r-1})$. However in this note we will restrict ourselves to the case of maximally unipotent local monodromy.*

2 Picard–Fuchs Differential Operators

Period integrals are integrals of algebraically defined differential forms over domains described by algebraic equations and inequalities. Period functions are period integrals that depend on a parameter. For example, the elliptic integral

$$t \mapsto \psi(t) := \int_1^\infty \frac{dx}{\sqrt{x(x-1)(x-t)}} \tag{7}$$

is a period function associated to the Legendre family of elliptic curves

$$y^2 = x(x-1)(x-t).$$

Such functions typically satisfy linear differential equations with algebraic coefficients. For example, expanding the integrand in (7) in a power series in t and integrating term by term one finds that

$$\int_1^\infty \frac{dx}{\sqrt{x(x-1)(x-t)}} = \pi\ {}_2F_1(\tfrac{1}{2}, \tfrac{1}{2}, 1; t), \tag{8}$$

and hence this period integral is annihilated by the respective hypergeometric differential operator (see the example given in Sect. 1). Vaguely speaking, Picard–Fuchs differential operators are those that annihilate period functions. An excellent introduction to this topic is given in [3, Chapter II].

Solution spaces of Picard–Fuchs differential operators have a natural $\mathbb{Q}$-structure, which is preserved by the monodromy transformations. This $\mathbb{Q}$-structure consists of period functions. To be slightly more precise, assume we are given a smooth projective family $f : X \to U$ over an algebraic curve $U = \mathbb{A}^1 \setminus \{\text{roots of} Q\}$ for some polynomial $Q \in K[t]$. Then the relative de Rham cohomology $H^m_{dR}(X/U)$ is a module over the ring of regular functions $\mathcal{O}_U = K[t, Q(t)^{-1}]$ equipped with a K-linear transformation $\frac{d}{dt} : H^m_{dR}(X/U) \to H^m_{dR}(X/U)$ (Gauss–Manin connection). Consider the ring $\mathcal{D}_U = \mathcal{O}_U[\frac{d}{dt}]$ of differential operators on U. We further assume there is a class of differential forms $\omega \in H^m_{dR}(X/U)$ and a differential operator $L \in \mathcal{D}_U$ such that $L\omega = 0$. Moreover, we assume that the differential submodule $M = \mathcal{D}_U\omega \subset H^m_{dR}(X/U)$ is isomorphic to $\mathcal{D}_U/\mathcal{D}_U L$. Then the $\mathbb{Q}$-structure on the space V of solutions of L near a regular point is given by period integrals $V_{\mathbb{Q}} = \left\{\int_\gamma \omega\right\}$ over families of topological cycles $\gamma_t \in H_m(X_t(\mathbb{C}), \mathbb{Q})$ where $X_t = f^{-1}(t)$ is the fibre at t.

One passes from this $\mathbb{Q}$-structure to the K-structure discussed in Sect. 1 by expressing period integrals in the standard basis (3):

$$\int_\gamma \omega = \sum_{k=0}^{r-1} \lambda_k\, \phi_k(t).$$

We will return to this setting at the end of Sect. 4, to mention that (under certain assumptions) the coefficients $\lambda_k \in \mathbb{C}$ are periods of the limiting Hodge structure at $t = 0$.

For example, the following period integrals for the Legendre family and $\omega = \frac{dx}{y}$ can be expressed in the standard basis as

$$\begin{aligned} \int_1^\infty \frac{dx}{\sqrt{x(x-1)(x-t)}} &= \pi\, \phi_0(t), \\ \int_t^1 \frac{dx}{\sqrt{x(x-1)(x-t)}} &= -4i\log(2)\,\phi_0(t) + i\,\phi_1(t). \end{aligned} \tag{9}$$

(Here we assume that $0 < t < 1$ and the real branch $\log(t) \in \mathbb{R}$ is chosen in $\phi_1(t)$ in the right-hand side.) Since the space of solutions of this Picard–Fuchs differential operator is 2-dimensional, it should be possible to write each of the other two elliptic integrals $\int_{-\infty}^{0} \omega$ and $\int_0^t \omega$ as a $\mathbb{Q}$-linear combination of the two integrals given above. One can quickly check that in fact $\int_{-\infty}^{0} \omega = \int_t^1 \omega$ and $\int_0^t \omega = \int_1^{\infty} \omega$. We already mentioned how to obtain the first expression in (9). The second expression is trickier. Firstly, we notice that the change of variable $x \mapsto 1 - \frac{1-t}{x}$ transforms the second elliptic integral in (9) into $-i\,\psi(1-t)$ where $\psi(t)$ is the first integral (we introduced this notation in (7)). It remains to show that

$$\pi\,\phi_0(1-t) = 4\log(2)\phi_0(t) - \phi_1(t). \tag{10}$$

In order to prove (10) we shall use the classical relations among hypergeometric series $F(a,b,c;t) := {}_2F_1(a,b,c;t)$ due to Kummer, and also their analytic dependence on the parameters a, b, c.

The following formula is valid for $Re(c) > 0$, $Re(c-a-b) > 0$:

$$\begin{aligned} F(a,b,c;1-t) =& \frac{\Gamma(c)\Gamma(c-a-b)}{\Gamma(c-a)\Gamma(c-b)} F(a,b,a+b-c+1;t) \\ &+ \frac{\Gamma(c)\Gamma(a+b-c)}{\Gamma(a)\Gamma(b)} z^{c-a-b} F(c-a,c-b,1+c-a-b;t). \end{aligned} \tag{11}$$

We take a small $\varepsilon > 0$ and substitute $a = b = \frac{1}{2} - \varepsilon$, $c = 1 - \varepsilon$. Observe that the following linear combination yields the log solution in the limit:

$$\begin{aligned} &\lim_{\varepsilon\to 0}\left(\frac{t^{c-a-b}F(c-a,c-b,1+c-a-b;t) - F(a,b,a+b-c+1;t)}{c-a-b}\right) \\ &= \lim_{\varepsilon\to 0}\frac{1}{\varepsilon}\left(t^{\varepsilon}\sum_{n\ge 0}\frac{(\frac{1}{2})_n^2}{(1+\varepsilon)_n n!}t^n - \sum_{n\ge 0}\frac{(\frac{1}{2}-\varepsilon)_n^2}{(1-\varepsilon)_n n!}t^n\right) \\ &= \sum_{n\ge 0}\frac{t^n}{n!}\lim_{\varepsilon\to 0}\frac{1}{\varepsilon}\left(\frac{t^{\varepsilon}(\frac{1}{2})_n^2}{(1+\varepsilon)_n} - \frac{(\frac{1}{2}-\varepsilon)_n^2}{(1-\varepsilon)_n}\right) \\ &= \sum_{n\ge 0}\frac{t^n}{n!}\lim_{\varepsilon\to 0}\left(\frac{t^{\varepsilon}(\frac{1}{2})_n^2}{(1+\varepsilon)_n}\left(\log(t) - \sum_{k=0}^{n-1}\frac{1}{1+\varepsilon+k}\right) - \frac{(\frac{1}{2}-\varepsilon)_n^2}{(1-\varepsilon)_n}\sum_{k=0}^{n-1}\left(-\frac{2}{\frac{1}{2}-\varepsilon+k} + \frac{1}{1-\varepsilon+k}\right)\right) \\ &= \sum_{n\ge 0}\frac{(1/2)_n t^n}{n!^2}\left(\log(t) - \sum_{k=1}^{n}\frac{1}{k(k-\frac{1}{2})}\right) = \phi_1(t). \end{aligned}$$

(The above computation is a known trick, see the part about Kummer relations in [2, §1].) Therefore the limit of relation (11) when $\varepsilon \to 0$ gives

$$\begin{aligned} \phi_0(1-t) &= \lim_{\varepsilon\to 0}\left(\frac{\Gamma(1-\varepsilon)\Gamma(\varepsilon)}{\Gamma(\frac{1}{2})^2} + \frac{\Gamma(1-\varepsilon)\Gamma(-\varepsilon)}{\Gamma(\frac{1}{2}-\varepsilon)^2}\right)\phi_0(t) + \lim_{\varepsilon\to 0}\left(\frac{\varepsilon\Gamma(1-\varepsilon)\Gamma(-\varepsilon)}{\Gamma(\frac{1}{2}-\varepsilon)^2}\right)\phi_1(t) \\ &= \frac{4\log(2)}{\pi}\phi_0(t) - \frac{1}{\pi}\phi_1(t), \end{aligned}$$

which completes our proof of (10).

Note that taking the limit in (10) as $t \to 1$ one gets the expression

$$\pi = \sum_{n=0}^{\infty} \frac{(\frac{1}{2})_n^2}{n!^2} \sum_{k=n+1}^{\infty} \frac{1}{k(k-\frac{1}{2})}.$$

3 Hodge Structures

Let $m \in \mathbb{Z}$ be an integer. A *pure Hodge structure of weight m* is a $\mathbb{C}$-vector space V with a $\mathbb{Q}$-structure $V_{\mathbb{Q}}$ and a decreasing filtration $\mathcal{F}^\bullet V$ satisfying the conditions that for any integers p, q such that $p + q = m + 1$ there is a direct sum decomposition $V = \mathcal{F}^p \oplus \overline{\mathcal{F}^q}$.

Example *For a compact complex manifold* $\mathfrak{X}$ *and* $0 \le m \le 2 \dim \mathfrak{X}$, *the complexification of the singular cohomology space* $V_{\mathbb{Q}} = H^m(\mathfrak{X}, \mathbb{Q})$ *possesses a pure Hodge structure of weight m. This Hodge structure is a basic construction in Hodge theory.*

The following special case is interesting from the arithmetic perspective. When $\mathfrak{X} = X(\mathbb{C})$ *is given by complex points of a smooth projective algebraic variety* X *defined over a field* $K \subset \mathbb{C}$, *then there is also a natural* K*-structure given by the de Rham cohomology* $V_K = H^m_{dR}(X)$. *If* $V_{\mathbb{Q}}$ *is the rational structure given by singular cohomology then, in general,* $V_K \ne V_{\mathbb{Q}} \otimes_{\mathbb{Q}} K$. *Comparison of these two structures yields* periods *of* X. *The Hodge filtration* $\mathcal{F}^\bullet$ *can be defined algebraically, as a filtration of* V_K.

The reader could check the following simple fact as an exercise:

Example *There is no one dimensional pure Hodge structure of odd weight. For each* $k \in \mathbb{Z}$ *there is a unique up to isomorphism one-dimensional pure Hodge structure of weight* $2k$; *it is denoted by* $\mathbb{Q}(-k)$.

A *mixed Hodge structure* is a $\mathbb{C}$-vector space V with a $\mathbb{Q}$-structure $V_{\mathbb{Q}}$ and two filtrations: an increasing filtration $\mathcal{W}_\bullet V$ defined over $\mathbb{Q}$ (that is, it comes from a filtration on $V_{\mathbb{Q}}$) and a decreasing filtration $\mathcal{F}^\bullet V$ satisfying the condition that for every m the graded piece $gr_m^{\mathcal{W}} V = \mathcal{W}_m / \mathcal{W}_{m-1}$ is a pure Hodge structure of weight m. Here $\mathcal{W}_\bullet$ is called the *weight filtration* and $\mathcal{F}^\bullet$ is called the *Hodge filtration*.

In 1970's Pierre Deligne showed that the cohomology of a complex variety (possibly singular or non-compact) is a mixed Hodge structure. The following example, in which a mixed Hodge structure arises in a limiting process, will serve as a motivation for our computations in Sect. 4.

Example *For a smooth projective family of algebraic varieties* $f : X \to U$ *over a curve* $U \subset \mathbb{A}^1$ *one can consider the family of pure Hodge structures of weight m given by* $V_t = H^m(X_t)$ *where* $X_t = f^{-1}(t)$ *is the fibre at* $t \in U(\mathbb{C})$. *Suppose* $0 \in \mathbb{A}^1 \setminus U$ *is a singular point and we would like to define the limiting Hodge structure* V *at* $t = 0$.

We fix a base point $t = t_0$ *close to* $t = 0$ *and define* $V_{\mathbb{Q}} := (V_{t_0})_{\mathbb{Q}} = H^m(X_{t_0}, \mathbb{Q})$. *Since the local monodromy transformation* $\gamma : V_{t_0} \to V_{t_0}$ *preserves the* $\mathbb{Q}$*-structure* $(V_{t_0})_{\mathbb{Q}}$, *one can think that the* $\mathbb{Q}$*-structure in the family is not varying. However it happens that* $\gamma(\mathcal{F}^p V_{t_0}) \neq \mathcal{F}^p V_{t_0}$ *for some* p, *which is an obstacle for defining the limiting filtration. In* [1] *Schmid introduced a way to remove the monodromy of the Hodge filtration, so that the resulting filtration passes to the limit as* $t \to 0$. *Let us sketch the approach in* [1]. *Take a small punctured disk* $\Delta^* = \Delta \setminus \{0\}$ *and consider its universal covering by an upper half of the complex plane* $e : \mathcal{H} \to \Delta^*$, $e(z) = \exp(2\pi i z)$. *After one chooses a preimage of the base point* $t_0 \in \Delta^*$ *in* $\mathcal{H}$ *(note that this is equivalent to choosing a branch of* $\log(t)$ *near* t_0*), for each* $z \in \mathcal{H}$ *there is now a preferred linear isomorphism* $V_{e(z)} \cong V_{t_0}$. *This yields a family of filtrations* $\mathcal{F}_z^\bullet V$ *on* $V = V_{t_0}$ *indexed by* $z \in \mathcal{H}$. *Let* $\gamma = \gamma_s \gamma_u$ *be the Jordan decomposition of the monodromy transformation into its semisimple and unipotent parts. Consider the nilpotent transformation* $N = \log(\gamma_u) : V \to V$. *Schmid shows (*[1, Theorem 6.16]*) there is a limiting filtration*

$$\mathcal{F}_\infty^p V := \lim_{\mathrm{Im}(z) \to \infty} \exp(-zN) \mathcal{F}_z^p V, \tag{12}$$

and V *equipped with* $\mathcal{F}_\infty^\bullet$ *and the* monodromy weight filtration $\mathcal{W}_\bullet$ *is a mixed Hodge structure. The monodromy weight filtration was defined by Deligne as follows. For a nilpotent transformation of a vector space there is an associated filtration called the Jacobson filtration. If* $\mathcal{L}_\bullet$ *is the Jacobson filtration associated to* $N : V \to V$, *then* $\mathcal{W}_\bullet = \mathcal{L}_{\bullet - m}$.

For later use, we would like to mention that if a nilpotent transformation $N : V \to V$ has one Jordan block and $d = \dim V$ then the respective Jacobson filtration $\mathcal{L}_\bullet V$ is given by

$$\begin{array}{ccccccc} 0 \subset & N^{d-1}(V) & \subset \ldots \subset & N(V) & \subset & V \\ \| & \| & & \| & & \| \\ \mathcal{L}_{-d} \subset \mathcal{L}_{-d+1} & = \mathcal{L}_{-d+2} & \subset \ldots \subset & \mathcal{L}_{d-3} & = \mathcal{L}_{d-2} \subset & \mathcal{L}_{d-1}. \end{array} \tag{13}$$

4 The Limiting Hodge Structure on the Space of Solutions

Let us return to the setting of Sect. 1. We are given a differential operator

$$L = \theta^r + q_1(t)\theta^{r-1} + \ldots + q_{r-1}(t)\theta + q_r(t) \in K(t)[\theta]$$

with rational coefficients satisfying the condition

$$q_j(0) = 0, \qquad 1 \le j \le r.$$

As we explained in Sect. 1, this condition means that $t = 0$ is a regular singularity and the local monodromy of solutions of L around this point is maximally unipotent. We shall now give an alternative construction of the K-structure on the space of solutions of L, which was defined in (6) using the standard basis of solutions.

We fix a punctured disc $\Delta^* = \Delta \setminus \{0\} = \{t \in \mathbb{C} \mid 0 < |t| < \varepsilon\}$ which is sufficiently small to contain no singularities of L. Consider the universal covering

$$e : \mathcal{H} \to \Delta^*, \quad e(z) = \exp(2\pi i z)$$

by the respective upper halfplane $\mathcal{H} = \{z \in \mathbb{C} \mid \mathrm{Im}(z) > -\frac{1}{2\pi}\log(\varepsilon)\}$. It will be convenient to view multivalued solutions of L in Δ^* as functions on $\mathcal{H}$. For an open subset $\mathcal{U} \subset \mathbb{C}$ we denote by $\mathcal{O}^{an}_{\mathcal{U}}$ the ring of complex analytic (holomorphic) functions in $\mathcal{U}$. Consider the solution space

$$V := \{u(z) \in \mathcal{O}^{an}_{\mathcal{H}} \mid (e^*L)u = 0\},$$

where $e^*L = (\frac{1}{2\pi i}\frac{d}{dz})^r + \sum_{j=0}^{r} q_j(e(z))(\frac{1}{2\pi i}\frac{d}{dz})^{r-j}$ is the pullback of L to $\mathcal{H}$. This differential operator is obtained via the substitution $t = e(z)$ in L. By Cauchy's theorem $\dim_{\mathbb{C}} V = r$. The basis in V given by

$$\phi_k(z) = \sum_{j=0}^{k} \frac{(2\pi i z)^j}{j!} f_{k-j}(e(z)), \qquad 0 \le k \le r-1 \tag{14}$$

with $f_j \in \mathcal{O}^{an}_{\Delta}$ defined in Lemma 1 will be called the *standard basis*. The monodromy transformation $\gamma : V \to V$ acts by

$$(\gamma u)(z) = u(z+1). \tag{15}$$

The space of linear functionals on the solution space will be denoted by $V^\vee := Hom_{\mathbb{C}}(V, \mathbb{C})$.

Definition 2 *A family of linear functionals $\pi_z \in V^\vee$ indexed by $z \in \mathcal{H}$ is an* algebraic family *if it is given by*

$$\pi_z : u \mapsto \sum_{j=0}^{r-1} \frac{v_j(e(z))}{(2\pi i)^j}\left(\frac{d^j u}{dz^j}\right)(z) \tag{16}$$

for some rational functions $v_0, \ldots, v_{r-1} \in K(t)$ having no poles in Δ^*. To such a family we associate its symbol $m = \sum_{j=0}^{r-1} v_j(t)\theta^j$, *and we will write $\pi_z = \pi_z(m)$ to denote the respective linear functionals* (16).

An algebraic family of linear functionals with symbol $m = \sum_{j=0}^{r-1} v_j(t)\theta^j$ is said to be analytic at $t = 0$ *if none of the coefficients $v_j \in K(t)$ has a pole at $t = 0$.*

We would like to consider the limits of $\pi_z \in V^\vee$ as $\mathrm{Im}(z) \to +\infty$ in algebraic families. However one can easily check in formula (16) that $\pi_{z+1} \ne \pi_z$, which is an obvious obstruction for taking the above mentioned limit. We will now remove this monodromy using Schmid's formula (12) (see [1, (6.15)]):

Lemma 3 *Let $N = \log(\gamma) = \sum_{h\geq 1}(-1)^{h-1}(\gamma - I)^h/h$ be the logarithm of the local monodromy transformation γ. For an algebraic family of linear functionals $\pi_z = \pi_z(m)$ corresponding to $m = \sum_{j=0}^{r-1} v_j(t)\theta^j$ the family of linear functionals defined by*

$$\pi'_z := \exp(-zN) \circ \pi_z \in V^\vee$$

satisfies $\pi'_{z+1} = \pi'_z$. The limit $\lim_{\mathrm{Im}(z)\to+\infty} \pi'_z$ exists whenever (π_z) is analytic at $t = 0$, in which case one has

$$\lim_{\mathrm{Im}(z_0)\to+\infty} \pi'_z = \sum_{j=0}^{r-1} v_j(0)\phi_j^\vee. \tag{17}$$

Here $\phi_0^\vee, \phi_1^\vee, \ldots$ is the basis in $V^\vee$ dual to the standard basis (14).

Proof Note that every solution $u \in V$ can be uniquely written as

$$u(z) = \sum_{k=0}^{r-1} u_k(e(z))z^k$$

with $u_0(t), \ldots, u_{r-1}(t) \in \mathcal{O}_\Delta^{an}$. The action of N on V is given by

$$N : \sum_k u_k(e(z))z^k \mapsto \sum_k u_k(e(z))k\, z^{k-1}.$$

For $m = \sum_j v_j\theta^j$ we evaluate $\pi'_z = \exp(-zN) \circ \pi_z(m)$ on $u = \sum_k u_k(e(z))z^k \in V$ as follows:

$$\begin{aligned}
\pi'_z(u) &= \sum_{h\geq 0} \frac{(-z)^h}{h!}\pi_z(m)(N^h u) \\
&= \sum_{h\geq 0}\sum_{j,k=0}^{r-1} \frac{(-z)^h}{h!}\frac{v_j(e(z))}{(2\pi i)^j}\left(\frac{d}{dz}\right)^j\left(u_k(e(z))\left(\frac{d}{dz}\right)^h z^k\right) \\
&= \sum_{h\geq 0}\sum_{j,k=0}^{r-1} \frac{(-z)^h}{h!}v_j(e(z))\sum_{s=0}^{j}(\theta^{j-s}u_k)(e(z))\frac{k(k-1)\ldots(k-h-s+1)}{(2\pi i)^s}z^{k-h-s} \\
&= \sum_{j,k=0}^{r-1}\sum_{s=0}^{\min(j,k)} v_j(e(z))(\theta^{j-s}u_k)(e(z))(2\pi i)^{-s}\frac{k!}{(k-s)!}z^{k-s}\sum_{h\geq 0}(-1)^h\binom{k-s}{h} \\
&= \sum_{j\geq k} v_j(e(z))(\theta^{j-k}u_k)(e(z))\frac{k!}{(2\pi i)^k}.
\end{aligned}$$

Note that the function in the last row is periodic in z, so we conclude that $\pi'_{z+1} = \pi'_z$. When all $v_j(t)$ are analytic at $t = 0$ the above expression passes to the limit as $\mathrm{Im}(z)$ grows infinitely:

$$\lim_{\mathrm{Im}(z)\to+\infty} \pi'_z(m)(u) \;=\; \sum_{j=0}^{r-1} v_j(0)u_j(0)\frac{j!}{(2\pi i)^j}.$$

It remains to notice that the linear functional mapping $u = \sum_k u_k(e(z))z^k$ into $u_j(0)(2\pi i)^{-j}\, j!$ coincides with $\phi_j^\vee$. This fact follows from formula (14) because $f_0(0) = 1$ and $f_j(0) = 0$ when $j > 0$. □

The following observation is a key fact relating the K-structure described in Sect. 1 with the limiting process defined in Lemma 3:

Corollary 4 *The vector space*

$$V_K^\vee := K\text{ -span of } \phi_0^\vee, \ldots, \phi_{r-1}^\vee$$

consists of the limits at $t = 0$ of algebraic families of linear functionals on the solution space V.

Proof Since in formula (17) all $v_j(0) \in K$, we obtain $\lim_{\mathrm{Im}(z)\to\infty} \pi'_z \in V_K^\vee$. Conversely, any functional $\sum \lambda_j \phi_j^\vee \in V_K^\vee$ is the limit of the family $\pi'_z(m)$ corresponding to the symbol $m = \sum_j \lambda_j \theta^j$. □

Our next goal is to describe a mixed Hodge structure on $V^\vee$. Recall that $N = \log(\gamma)$ is a nilpotent transformation, and hence images of its powers define a finite filtration on $V^\vee$. We define the weight filtration $\mathcal{W}_\bullet V^\vee$ as follows:

$$\begin{array}{ccccccc} 0 & \subset N^{r-1}(V^\vee) \subset \ldots \subset & N(V^\vee) & \subset & V^\vee \\ \| & \| & \| & & \| \\ \mathcal{W}_{-1} \subset \mathcal{W}_0 & = \mathcal{W}_1 \subset \ldots \subset & \mathcal{W}_{2r-4} & = \mathcal{W}_{2r-3} \subset & \mathcal{W}_{2r-2}. \end{array} \tag{18}$$

This filtration is the shift $\mathcal{W}_\bullet = \mathcal{L}_{\bullet-r+1}$ of the Jacobson filtration associated to N (see (13)). Note that it is is naturally defined on any $\mathbb{Q}$-structure preserved by the local monodromy transformation γ.

To define the Hodge filtration we will identify elements of $V^\vee$ with the limits of families of linear functionals at $t = 0$ and filter families by the order of their symbol in θ. Consider the ring of differential operators $\mathcal{D} = K(t)[\theta]$ and note that symbols $m = \sum_{j=0}^{r-1} v_j(t)\theta^j$ of algebraic families of linear functionals in Definition 2 can be thought as elements of the differential module $M = \mathcal{D}/\mathcal{D}L \cong \sum_{j=0}^{r-1} K(t)\theta^j$. More generally, we would like to consider families with analytic coefficients $v_j \in \mathcal{O}_\Delta^{an}$. Their symbols are elements of

$$\widetilde{M} \;=\; \widetilde{\mathcal{D}}/\widetilde{\mathcal{D}}L \;\cong\; \sum_{j=0}^{r-1} \mathcal{O}_\Delta^{an}\, \theta^j,$$

where $\widetilde{\mathcal{D}} = \mathcal{O}_\Delta^{an}[\theta]$ is the ring of differential operators whose coefficients are holomorphic functions in Δ. For each $z \in \mathcal{H}$ formula (16) yields a map

$$\pi_z : \widetilde{M} \to V^\vee.$$

We now consider a decreasing filtration of $\widetilde{M}$ by $\mathcal{O}_\Delta^{an}$-modules given by

$$\mathcal{F}^p \widetilde{M} := \sum_{j=0}^{r-1-p} \mathcal{O}_\Delta^{an} \theta^j, \quad 0 \le p \le r-1$$

and define a family of filtrations on $V^\vee$ indexed by $z \in \mathcal{H}$ via

$$\mathcal{F}_z^p V^\vee := \pi_z\left(\mathcal{F}^p \widetilde{M}\right), \quad 0 \le p \le r-1.$$

The computation in Lemma 3 applies also to analytic families. An immediate consequence is the following

Corollary 5 *The limiting filtration*

$$\mathcal{F}_\infty^p V^\vee := \lim_{\mathrm{Im}(z)\to\infty} \exp(-zN)\mathcal{F}_z^p V^\vee, \quad 0 \le p \le r-1$$

exists and is given by

$$\mathcal{F}_\infty^p V^\vee = \mathbb{C}\text{-span of } \phi_0^\vee, \dots, \phi_{r-1-p}^\vee.$$

Proposition 6 *Let $\mathcal{W}_\bullet$ be the Jacobson filtration associated to the nilpotent transformation $N = \log(\gamma)$ of $V^\vee$ and shifted by $r-1$. Let $\mathcal{F}_\infty^\bullet$ be the limiting filtration from Corollary 5. Then for any $\mathbb{Q}$-structure on the solution space $V_\mathbb{Q}$ preserved by the monodromy transformation γ the triple $(V_\mathbb{Q}^\vee, \mathcal{W}_\bullet, \mathcal{F}_\infty^\bullet)$ is a mixed Hodge structure.*

Proof Filtration $\mathcal{W}_\bullet$ is given explicitly by (18). Note that $N^j(V^\vee) = Span_\mathbb{C}(\phi_j^\vee, \dots, \phi_{r-1}^\vee)$, and hence the two filtrations are opposite in the sense that $V^\vee = \mathcal{W}_{2k} \oplus \mathcal{F}_\infty^{k+1}$ for each $0 \le k \le r-1$. It follows that the filtration induced by $\mathcal{F}_\infty^\bullet$ on $gr_{2k}^{\mathcal{W}} V^\vee$ is zero in degrees $> k$ and everything in degree k. □

Note that the pure graded pieces of the mixed Hodge structure in Proposition 6 are one dimensional. With the notation explained in Sect. 3 we have

$$gr^{\mathcal{W}} V^\vee \cong \oplus_{j=0}^{r-1} \mathbb{Q}(-k).$$

Let us also mention that one can always find a $\mathbb{Q}$-structure preserved by γ. For example, the reader may check that γ preserves the $\mathbb{Q}$-span of $(2\pi i)^{-k}\phi_k$, $0 \le k \le r-1$. For a deeper example one can turn to Picard–Fuchs differential operators. As we mentioned in Sect. 2, their spaces of solutions possess a natural $\mathbb{Q}$-structure consisting of period functions.

Remark *The construction of this section applies to Picard–Fuchs differential operators with maximally unipotent local monodromy. Namely, suppose* $X \to U$ *is a smooth projective family and in the setting described in Sect. 2 we assume in addition that* ω *belongs to the smallest* $\mathcal{O}_U$*-submodule in the Hodge filtration* $\mathcal{F}^m H^m_{dR}(X/U)$*, and that the operator* $L \in \mathcal{D}_U$ *annihilating* ω *is of order* $r = m + 1$*. Then using the Griffiths' transversality property of a variation of Hodge structure and the fact that the monodromy of* L *is maximally unipotent, one can show that the Hodge filtration on* $M = \mathcal{D}_U/\mathcal{D}_U L \cong \mathcal{D}_U\omega \subset H^{r-1}_{dR}(X/U)$ *is given by* $\mathcal{F}^\bullet M = \sum_{j=0}^{r-1-\bullet} \mathcal{O}_U \theta^j \omega$*. Solutions of* L *can be identified with horizontal sections of the analytification* $M^\vee_{an} = M^\vee \otimes_{\mathcal{O}} \mathcal{O}^{an}$ *of the dual connection* $M^\vee = \mathrm{Hom}_{\mathcal{O}}(\mathrm{M}, \mathcal{O})$*. Over the punctured disk* Δ^**, this identification yields the pairing of elements* $m = \sum_j v_j \theta^j \in M$ *with solutions* $u \in V$*. The result of this pairing is the analytic function* $z \mapsto \pi_z(u)$ *given by the right-hand side in our formula (2). Since the limiting process of Lemma 3 coincides with the one in* [1]*, the mixed Hodge structure in Proposition 6 is the limiting mixed Hodge structure of Deligne and Schmid.*

In this geometric situation, there is a fine notion of the de Rham structure of the limiting mixed Hodge structure. The arguments given in [4, Remark 42] *indicate that it should coincide with our* K*-structure* V_K *spanned by the standard basis in the space of solutions.*

Afterword. The content of this note is essentially covered by Lemma 41 in [4], where we use the ideas explained here to compute periods of limiting Hodge structures. I am grateful to Spencer Bloch for introducing me to this interesting subject. Special thanks to Wadim Zudilin, whose expertise in hypergeometric functions helped to work out the example given in Sect. 2.

Our account of Hodge structures in Sect. 3 is limited to basic definitions and examples. We refer the reader to the book [5] for a systematic exposition of this subject.

This work supported by the National Science Centre of Poland (NCN), grant UMO-2016/21/B/ST1/03084.

References

1. W. Schmid, *Variation of Hodge structure: the singularities of the period mapping*, Inventiones math. 22 (1973), 211–319
2. F. Beukers, *Gauss' hypergeometric function*, Progress in Mathematics 260 (2007), 23–42
3. M. Kontsevich, D. Zagier, *Periods*, Mathematics unlimited – 2001 and beyond (2001), 771–808
4. S. Bloch, M. Vlasenko, *Motivic gamma functions, monodromy and Frobenius constants*, Communications in Number Theory and Physics, Volume 15, Number 1, 91–147, 2021
5. C. A. M. Peters, J.H.M. Steenbrink, *Mixed Hodge structures*, Ergebnisse der Mathematik und ihrer Grenzgebiete. 3. Folge. A Series of Modern Surveys in Mathematics, Springer-Verlag, Berlin, 2008, 470 pp.

Beck-Type Identities for Euler Pairs of Order r

Cristina Ballantine and Amanda Welch

Abstract Partition identities are often statements asserting that the set $\mathcal{P}_X$ of partitions of n subject to condition X is equinumerous to the set $\mathcal{P}_Y$ of partitions of n subject to condition Y. A Beck-type identity is a companion identity to $|\mathcal{P}_X| = |\mathcal{P}_Y|$ asserting that the difference $b(n)$ between the number of parts in all partitions in $\mathcal{P}_X$ and the number of parts in all partitions in $\mathcal{P}_Y$ equals $c|\mathcal{P}_{X'}|$ and also $c|\mathcal{P}_{Y'}|$, where c is some constant related to the original identity, and X', respectively Y', is a condition on partitions that is a very slight relaxation of condition X, respectively Y. A second Beck-type identity involves the difference $b'(n)$ between the total number of different parts in all partitions in $\mathcal{P}_Y$ and the total number of different parts in all partitions in $\mathcal{P}_X$. We extend these results to Beck-type identities accompanying all identities given by Euler pairs of order r (for any $r \geq 2$). As a consequence, we obtain many families of new Beck-type identities. We give analytic and bijective proofs of our results.

Keywords Partitions · Euler pairs · Beck-type identities · Words

MSC 2010: 05A17 · 11P81 · 11P83

C. Ballantine (✉)
Department of Mathematics and Computer Science, College of the Holy Cross,
Worcester, MA 01610, USA
e-mail: cballant@holycross.edu

A. Welch
Department of Mathematics and Computer Science, Eastern Illinois University,
Charleston, IL 61920, USA
e-mail: arwelch@eiu.edu

A. Bostan and K. Raschel (eds.), *Transcendence in Algebra, Combinatorics, Geometry and Number Theory*, Springer Proceedings in Mathematics & Statistics 373,
https://doi.org/10.1007/978-3-030-84304-5_6

1 Introduction

The origin of this article is rooted in two conjectures by Beck which appeared in The On-Line Encyclopedia of Integer Sequences [1] on the pages for sequences A090867 and A265251. The conjectures, as formulated by Beck, were proved by Andrews in [4] using generating functions. Certain generalizations and combinatorial proofs appeared in [7] and [12]. Combinatorial proofs of the original conjectures were also given in [6]. Several additional similar identities were proved in the last two years. See for example [5, 8–10]. In order to define Beck-type identities, we first introduce the necessary terminology and notation.

In this article $\mathbb{N}$ denotes the set of positive integers. Given a non-negative integer n, a *partition* λ of n is a non-increasing sequence of positive integers $\lambda = (\lambda_1, \lambda_2, \ldots, \lambda_k)$ that add up to n, i.e., $\sum_{i=1}^{k} \lambda_i = n$. Thus, if $\lambda = (\lambda_1, \lambda_2, \ldots, \lambda_k)$ is a partition, we have $\lambda_1 \geq \lambda_2 \geq \ldots \geq \lambda_k$. The numbers λ_i are called the *parts* of λ and n is called the *size* of λ. The number of parts of the partition is called the *length* of λ and is denoted by $\ell(\lambda)$.

If λ, μ are two arbitrary partitions, we denote by $\lambda \cup \mu$ the partition obtained by taking all parts of λ and all parts of μ and rearranging them to form a partition. For example, if $\lambda = (5, 5, 3, 2, 2, 1)$ and $\mu = (7, 5, 3, 3)$, then $\lambda \cup \mu = (7, 5, 5, 5, 3, 3, 3, 2, 2, 1)$.

When convenient, we use the exponential notation for parts in a partition. The exponent of a part is the multiplicity of the part in the partition. For example, $(7, 5^2, 4, 3^3, 1^2)$ denotes the partition $(7, 5, 5, 4, 3, 3, 3, 1, 1)$. It will be clear from the context when exponents refer to multiplicities and when they are exponents in the usual sense.

Let S_1 and S_2 be subsets of the positive integers. We define $\mathcal{O}_r(n)$ to be the set of partitions of n with all parts from the set S_2 and $\mathcal{D}_r(n)$ to be the set of partitions of n with parts in S_1 repeated at most $r - 1$ times. Subbarao [11] proved the following theorem.

Theorem 1.1. *$|\mathcal{O}_r(n)| = |\mathcal{D}_r(n)|$ for all non-negative integers n if and only if $rS_1 \subseteq S_1$ and $S_2 = S_1 \backslash rS_1$.*

Andrews [2] first discovered this result for $r = 2$ and called a pair (S_1, S_2) such that $|\mathcal{O}_2(n)| = |\mathcal{D}_2(n)|$ an *Euler pair* since the pair $S_1 = \mathbb{N}$ and $S_2 = 2\mathbb{N} - 1$ gives Euler's identity. By analogy, Subbarao called a pair (S_1, S_2) such that $|\mathcal{O}_r(n)| = |\mathcal{D}_r(n)|$ an *Euler pair of order r*.

Example 1 (Subbarao [11]). Let

$$S_1 = \{m \in \mathbb{N} : m \equiv 1 \ (mod\ 2)\};$$
$$S_2 = \{m \in \mathbb{N} : m \equiv \pm 1 \ (mod\ 6)\}.$$

Then (S_1, S_2) is an Euler pair of order 3.

Note that Glaisher's bijection used to prove $|\mathcal{O}_r(n)| = |\mathcal{D}_r(n)|$ when $S_1 = \mathbb{N}$ and $S_2 = 2\mathbb{N} - 1$ can be generalized to any Euler pair of order r. If (S_1, S_2) is an Euler pair of order r, let φ_r be the map from $\mathcal{O}_r(n)$ to $\mathcal{D}_r(n)$ which repeatedly merges r equal parts into a single part until there are no parts repeated more than $r - 1$ times. The map φ_r is a bijection and we refer to it as Glaisher's bijection.

Given (S_1, S_2), an Euler pair of order r, we refer to the elements in $S_2 = S_1 \backslash r S_1$ as *primitive* elements and to the elements of $r S_1 = S_1 \backslash S_2$ as *non-primitive* elements. We usually denote primitive parts by bold lower case letters, for example $\mathbf{a}$. Non-primitive parts are denoted by (non-bold) lower case letters. If a is a non-primitive part of a partition and we want to emphasize the largest power k of r such that $a/r^k \in S_1$, we write $a = r^k \mathbf{a}$ with $\mathbf{a} \in S_2$ and $k \geq 1$.

Let $\mathcal{O}_{1,r}(n)$ be the set of partitions of n with parts in S_1 such that the set of parts in $r S_1$ has exactly one element. Thus, a partition in $\mathcal{O}_{1,r}(n)$ has exactly one non-primitive part (possibly repeated). Let $\mathcal{D}_{1,r}(n)$ be the set of partitions of n with parts in S_1 in which exactly one part is repeated at least r times.

Let $a_r(n) = |\mathcal{O}_{1,r}(n)|$ and $c_r(n) = |\mathcal{D}_{1,r}(n)|$. Let $b_r(n)$ be the difference between the number of parts in all partitions in $\mathcal{O}_r(n)$ and the number of parts in all partitions in $\mathcal{D}_r(n)$. Thus,

$$b_r(n) = \sum_{\lambda \in \mathcal{O}_r(n)} \ell(\lambda) - \sum_{\lambda \in \mathcal{D}_r(n)} \ell(\lambda).$$

Let $\mathcal{T}_r(n)$ be the subset of $\mathcal{D}_{1,r}(n)$ consisting of partitions of n in which one part is repeated more than r times but less than $2r$ times. Let $c'_r(n) = |\mathcal{T}_r(n)|$. Let $b'_r(n)$ be the difference between the total number of different parts in all partitions in $\mathcal{D}_r(n)$ and the total number of different parts in all partitions in $\mathcal{O}_r(n)$ (i.e., in each partition, parts are counted without multiplicity). If we denote by $\bar{\ell}(\lambda)$ the number of different parts in λ, then

$$b'_r(n) = \sum_{\lambda \in \mathcal{D}_r(n)} \bar{\ell}(\lambda) - \sum_{\lambda \in \mathcal{O}_r(n)} \bar{\ell}(\lambda).$$

In [1], Beck conjectured that, if $S_1 = \mathbb{N}$ and $S_2 = 2\mathbb{N} - 1$, then

$$a_2(n) = b_2(n) = c_2(n)$$

and

$$c'_2(n) = b'_2(n).$$

Andrews proved these identities in [4] using generating functions. Combinatorial proofs were given in [6]. For the case $r \geq 2$, $S_1 = \mathbb{N}$, and $S_2 = \{k \in \mathbb{N} : k \not\equiv 0 \pmod{r}\}$, Fu and Tang [7] gave generating function proofs for

$$a_r(n) = \frac{1}{r-1} b_r(n) = c_r(n) \tag{1}$$

and

$$c'_r(n) = b'_r(n). \tag{2}$$

They also proved combinatorially that $a_r(n) = c_r(n)$. In [12], Yang gave combinatorial proofs of (1) and (2) in the case $r \geq 2$, $S_1 = \mathbb{N}$, and $S_2 = \{k \in \mathbb{N} : k \not\equiv 0 \pmod{r}\}$.

Our main theorems establish the analogous result for all Euler pairs. We will prove the theorems both analytically and combinatorially. We refer to the results in Theorem 1.2 as first Beck-type identities and to the result in Theorem 1.3 as the second Beck-type identity.

Theorem 1.2. *If n, r are integers such that $n \geq 0$ and $r \geq 2$, and (S_1, S_2) is an Euler pair of order r, then*

(i) $a_r(n) = \dfrac{1}{r-1} b_r(n)$

(ii) $c_r(n) = \dfrac{1}{r-1} b_r(n)$.

Theorem 1.3. *If n, r are integers such that $n \geq 0$ and $r \geq 2$, and (S_1, S_2) is an Euler pair of order r, then $c'_r(n) = b'_r(n)$.*

Example 2. We continue with the Euler pair of order 3 from Example 1. We have

$$\mathcal{O}_3(7) = \{(7), (5, 1^2), (1^7)\}; \ \mathcal{D}_3(7) = \{(7), (5, 1^2), (3^2, 1)\};$$

and

$$\mathcal{O}_{1,3}(7) = \{(3^2, 1), (3, 1^4)\}; \ \mathcal{D}_{1,3}(7) = \{(1^7), (3, 1^4)\}.$$

Glaisher's bijection gives us

$$(7) \xrightarrow{\varphi_3} (7)$$

$$(5, 1, 1) \longrightarrow (5, 1, 1)$$

$$(\underbrace{1, 1, 1}, \underbrace{1, 1, 1}, 1) \longrightarrow (3, 3, 1).$$

We note that

$$a_3(7) = |O_{1,3}(7)| = 2, \, c_3(7) = |D_{1,3}(7)| = 2, \text{ and } b_3(7) = 11 - 7 = 4.$$

Thus,

$$\frac{1}{3-1} b_3(7) = a_3(7) = c_3(7).$$

If we restrict to counting different parts in partitions, we see that there are a total of 4 different parts in the partitions of $\mathcal{O}_3(7)$ and a total of 5 different parts in the partitions of $\mathcal{D}_3(7)$. Since $\mathcal{T}_3(7) = \{(3, 1^4)\}$, we have

$$b'_3(7) = 5 - 4 = 1 = |\mathcal{T}_3(7)|.$$

The analytic proofs of Theorems 1.2 and 1.3 are similar to the proofs in [4] and [7], while the combinatorial proofs follow the ideas of [6]. However, the generalizations of the proofs in the aforementioned articles to Euler pairs of order $r \geq 2$ are important as establishing the theorems in such generality leads to a multitude of new Beck-type identities. We reproduce several Euler pairs listed in [11]. For each identity $|\mathcal{O}_r(n)| = |\mathcal{D}_r(n)|$ holding for the pairs below, there are companion Beck-type identities as in Theorems 1.2 and 1.3.

The following pairs (S_1, S_2) are Euler pairs (of order 2).

(i) $S_1 = \{m \in N : m \not\equiv 0 \pmod 3\}$;
$S_2 = \{m \in N : m \equiv 1, 5 \pmod 6\}$.
In this case, the identity $|\mathcal{O}_2(n)| = |\mathcal{D}_2(n)|$ is known as Schur's identity.

(ii) $S_1 = \{m \in N : m \equiv 2, 4, 5 \pmod 6\}$;
$S_2 = \{m \in N : m \equiv 2, 5, 11 \pmod{12}\}$.
In this case, the identity $|\mathcal{O}_2(n)| = |\mathcal{D}_2(n)|$ is known as Göllnitz's identity.

(iii) $S_1 = \{m \in N : m = x^2 + 2y^2 \text{ for some } x, y \in \mathbb{Z}\}$;
$S_2 = \{m \in N : m \equiv 1 \pmod 2 \text{ and } m = x^2 + 2y^2 \text{ for some } x, y \in \mathbb{Z}\}$.

The following is an Euler pair of order 3.

(iv) $S_1 = \{m \in N : m = x^2 + xy + y^2 \text{ for some } x, y \in \mathbb{Z}\}$;
$S_2 = \{m \in N : \gcd(m, 3) = 1 \text{ and } m = x^2 + xy + y^2 \text{ for some } x, y \in \mathbb{Z}\}$.

The following pairs (S_1, S_2) are Euler pairs of order r.

(v) $S_1 = \{m \in N : m \equiv \pm r \pmod{r(r+1)}\}$;
$S_2 = \{m \in N : m \equiv \pm r \pmod{r(r+1)} \text{ and } m \not\equiv \pm r^2 \pmod{r^2(r+1)}\}$.

(vi) $S_1 = \{m \in N : m \equiv \pm r, -1 \pmod{r(r+1)}\}$;
$S_2 = \{m \in N : m \equiv \pm r, -1 \pmod{r(r+1)}$ and
$m \not\equiv \pm r^2, -r \pmod{r^2(r+1)}\}$.
If $r = 2$, this Euler pair becomes Göllnitz's pair in (ii) above.

(vii) Let $r + 1$ be a prime.
$S_1 = \{m \in N : m \not\equiv 0 \pmod{r+1}\}$;
$S_2 = \{m \in N : m \not\equiv tr, t(r+1) \pmod{r^2 + r} \text{ for } 1 \leq t \leq r\}$.
If $r = 2$, this Euler pair becomes Schur's pair in (i) above.

(viii) Let p be a prime and r a quadratic residue $\pmod p$.
$S_1 = \{m \in \mathbb{N} : m \text{ quadratic residue } \pmod p\}$;
$S_2 = \{m \in \mathbb{N} : m \not\equiv 0 \pmod r \text{ and } m \text{ quadratic residue } \pmod p\}$.

Note that each case (v)–(viii) gives infinitely many Euler pairs and therefore leads to infinitely many new Beck-type identities. We also note that in (vii) we corrected a slight error in (3.4) of [11].

Example 3. Consider the Euler pair in (vii) above with $r = 4$. We have
$S_1 = \{m \in N : m \not\equiv 0 \pmod 5\}$;
$S_2 = \{m \in N : m \not\equiv 4t, 5t \pmod{20} \text{ for } 1 \le t \le 4\}$.

Then (S_1, S_2) is an Euler pair of order 4 and we have

$\mathcal{O}_4(7) = \{(7), (6, 1), (3^2, 1), (3, 2^2), (3, 1^4), (3, 2, 1^2), (2^3, 1), (2^2, 1^3), (2, 1^5), (1^7)\}$;
$\mathcal{D}_4(7) = \{(7), (6, 1), (3^2, 1), (3, 2^2), (4, 3), (3, 2, 1^2), (2^3, 1), (2^2, 1^3), (4, 2, 1), (4, 1^3)\}$.

We have $\mathcal{O}_{1,4}(7) = \{(4, 1^3), (4, 2, 1), (4, 3)\}$; $\mathcal{D}_{1,4}(7) = \{(1^7), (2, 1^5), (3, 1^4)\}$.
We note that $a_4(7) = |O_{1,4}(7)| = 3$, $c_4(7) = |D_{1,4}(7)| = 3$, and $b_4(7) = 40 - 31 = 9$, so $\frac{1}{3}b_4(7) = a_4(7) = c_4(7)$.

If we restrict to counting different parts, we see that there are 19 different parts in the partitions of $\mathcal{O}_4(7)$ and 21 different parts in the partitions of $\mathcal{D}_4(7)$. So $b'_4(7) = 21 - 19 = 2 = |\mathcal{T}_4(7)|$ since $\mathcal{T}_4(7) = \{(1^7), (2, 1^5)\}$.

2 Proofs of Theorem 1.2

2.1 Analytic Proof

In this article, whenever we work with q-series, we assume that $|q| < 1$. When working with two-variable generating functions, we assume both variables are complex numbers less than 1 in absolute value. Then all series converge absolutely. The generating functions for $|\mathcal{D}_r(n)|$ and $|\mathcal{O}_r(n)|$ are given by

$$\sum_{n=0}^{\infty} |\mathcal{D}_r(n)| q^n = \prod_{a \in S_1} (1 + q^a + q^{2a} + \cdots + q^{(r-1)a}) = \prod_{a \in S_1} \frac{1 - q^{ra}}{1 - q^a};$$

and

$$\sum_{n=0}^{\infty} |\mathcal{O}_r(n)| q^n = \prod_{\mathbf{b} \in S_2} \frac{1}{1 - q^{\mathbf{b}}}.$$

To keep track of the number of parts used, we introduce a second variable z, where $|z| < 1$. Let

$$\mathcal{D}_r(n; m) = \{\lambda \in \mathcal{D}_r(n) \mid \lambda \text{ has exactly } m \text{ parts}\}$$

and

$$\mathcal{O}_r(n; m) = \{\lambda \in \mathcal{O}_r(n) \mid \lambda \text{ has exactly } m \text{ parts}\}.$$

Then, the generating functions for $|\mathcal{D}_r(n; m)|$ and $|\mathcal{O}_r(n; m)|$ are

$$f_{\mathcal{D}_r}(z,q) := \sum_{n=0}^{\infty}\sum_{m=0}^{\infty} |\mathcal{D}_r(n;m)| z^m q^n = \prod_{a\in S_1} (1 + zq^a + z^2q^{2a} + \cdots + z^{(r-1)}q^{(r-1)a})$$
$$= \prod_{a\in S_1} \frac{1 - z^r q^{ra}}{1 - zq^a};$$

and

$$f_{\mathcal{O}_r}(z,q) := \sum_{n=0}^{\infty}\sum_{m=0}^{\infty} |\mathcal{O}_r(n;m)| z^m q^n = \prod_{\mathbf{b}\in S_2} \frac{1}{1 - zq^{\mathbf{b}}}.$$

To obtain the generating function for the total number of parts in all partition in $\mathcal{D}_r(n)$ (respectively $\mathcal{O}_r(n)$), we take the derivative with respect to z of $f_{\mathcal{D}_r}(z,q)$ (respectively $f_{\mathcal{O}_r}(z,q)$), and set $z = 1$. We obtain

$$\left.\frac{\partial}{\partial z}\right|_{z=1} f_{\mathcal{D}_r}(z,q)$$
$$= \prod_{a\in S_1} \frac{1-q^{ra}}{1-q^a} \sum_{a\in S_1} \frac{-rq^{ra}(1-q^a) + q^a(1-q^{ra})}{(1-q^a)(1-q^{ra})}$$
$$= \prod_{a\in S_1} \frac{1-q^{ra}}{1-q^a} \sum_{a\in S_1} \left(\frac{q^a}{1-q^a} - \frac{q^{ra}}{1-q^{ra}} - (r-1)\frac{q^{ra}}{1-q^{ra}}\right)$$
$$= \prod_{a\in S_1} \frac{1-q^{ra}}{1-q^a} \left(\sum_{a\in S_1} \sum_{\substack{k=1 \\ r\nmid k}}^{\infty} q^{ka} - \sum_{a\in S_1} (r-1)\frac{q^{ra}}{1-q^{ra}}\right);$$

and

$$\left.\frac{\partial}{\partial z}\right|_{z=1} f_{\mathcal{O}_r}(z,q) = \prod_{\mathbf{b}\in S_2} \frac{1}{1-q^{\mathbf{b}}} \sum_{\mathbf{b}\in S_2} \frac{q^{\mathbf{b}}}{1-q^{\mathbf{b}}}.$$

Since $|\mathcal{D}_r(n)| = |\mathcal{O}_r(n)|$, we have

$$\sum_{n=0}^{\infty} b_r(n) q^n = \prod_{\mathbf{b}\in S_2} \frac{1}{1-q^{\mathbf{b}}} \left(\sum_{\mathbf{b}\in S_2} \frac{q^{\mathbf{b}}}{1-q^{\mathbf{b}}} - \sum_{\substack{a\in S_1 \\ k\in\mathbb{N} \\ r\nmid k}} q^{ka} + \sum_{a\in S_1} (r-1)\frac{q^{ra}}{1-q^{ra}}\right).$$

Next we see that

$$\sum_{\mathbf{b}\in S_2}\frac{q^{\mathbf{b}}}{1-q^{\mathbf{b}}}=\sum_{\substack{a\in S_1\\ k\in\mathbb{N}\\ r\nmid k}}q^{ka}. \tag{3}$$

We have

$$\sum_{\substack{a\in S_1\\ k\in\mathbb{N}\\ r\nmid k}}q^{ka}=\sum_{\substack{a\in S_1\\ k\in\mathbb{N}}}q^{ka}-\sum_{\substack{a\in S_1\\ k\in\mathbb{N}}}q^{rka}=\sum_{a\in S_1}\frac{q^a}{1-q^a}-\sum_{a\in S_1}\frac{q^{ra}}{1-q^{ra}}=\sum_{\mathbf{b}\in S_2}\frac{q^{\mathbf{b}}}{1-q^{\mathbf{b}}}.$$

The last equality holds because $S_2=S_1\backslash rS_1$. Therefore, (3) holds.

Then, the generating function for $b_r(n)$ becomes

$$\begin{aligned}\sum_{n=0}^{\infty}b_r(n)q^n&=\prod_{\mathbf{b}\in S_2}\frac{1}{1-q^{\mathbf{b}}}\left((r-1)\sum_{a\in S_1}\frac{q^{ra}}{1-q^{ra}}\right)\\&=\prod_{a\in S_1}(1+q^a+q^{2a}+\cdots+q^{(r-1)a})\left((r-1)\sum_{a\in S_1}\frac{q^{ra}}{1-q^{ra}}\right).\end{aligned}$$

Therefore

$$\sum_{n=0}^{\infty}b_r(n)q^n=\sum_{n=0}^{\infty}(r-1)|\mathcal{O}_{1,r}(n)|q^n=\sum_{n=0}^{\infty}(r-1)|\mathcal{D}_{1,r}(n)|q^n.$$

Equating coefficients results in $a_r(n)=\dfrac{1}{r-1}b_r(n)$ and $c_r(n)=\dfrac{1}{r-1}b_r(n)$.

2.2 *Combinatorial Proof*

2.2.1 $b_r(n)$ as the Cardinality of a Set of Marked Partitions

We start with another example of Glaisher's bijection.

Example 4. We continue with the Euler pair of order 3 from Example 1, but this time use $n=11$.

$\mathcal{O}_3(11)=\{(11),(7,1^4),(5^2,1),(5,1^6),(1^{11})\}$;
$\mathcal{D}_3(11)=\{(11),(9,1^2),(7,3,1),(5^2,1),(5,3^2)\}$.

Thus, $b_3(11)=27-13=14$.
Glaisher's bijection gives

$$
\begin{array}{rcl}
(11) & \xrightarrow{\varphi_3} & (11)\\[1em]
(7, \underbrace{1, 1, 1}, 1) & \longrightarrow & (7, 3, 1)\\[1em]
(5, 5, 1) & \longrightarrow & (5, 5, 1)\\[1em]
(5, \underbrace{1, 1, 1}, \underbrace{1, 1, 1}) & \longrightarrow & (5, 3, 3)\\[1em]
(\underbrace{\underbrace{1, 1, 1}, \underbrace{1, 1, 1}, \underbrace{1, 1, 1}}, 1, 1) & \longrightarrow & (9, 1, 1).
\end{array}
$$

From Glaisher's bijection, it is clear that each partition $\lambda \in \mathcal{O}_r(n)$ has at least as many parts as its image $\varphi_r(\lambda) \in \mathcal{D}_r(n)$.

When calculating $b_r(n)$, we sum up the differences in the number of parts in each pair $(\lambda, \varphi_r(\lambda))$. Write each part μ_j of $\mu = \varphi_r(\lambda)$ as $\mu_j = r^{k_j}\mathbf{a}$. Then, μ_j was obtained by merging r^{k_j} parts equal to $\mathbf{a}$ in λ and thus contributes an excess of $r^{k_j} - 1$ parts to $b_r(n)$. Therefore, the difference between the number of parts of λ and the number of parts of $\varphi_r(\lambda)$ is $\sum_{j=1}^{\ell(\varphi_r(\lambda))} (r^{k_j} - 1)$.

Example 5. In the setting of Example 4, we see that $(7, 3, 1)$ contributes 2 to $b_3(11)$, $(5, 3, 3)$ contributes $2 + 2$ to $b_3(11)$, and $(9, 1, 1)$ contributes 8 to $b_3(11)$. Thus, $b_3(11) = 2 + 4 + 8 = 14$.

Definition 1. Given (S_1, S_2), an Euler pair of order r, we define the set $\mathcal{MD}_{1,r}(n)$ of *marked partitions* of n as the set of partitions in $\mathcal{D}_r(n)$ such that exactly one part of the form $r^k\mathbf{a}$ with $k \geq 1$ has as index an integer t satisfying $1 \leq t \leq r^k - 1$. If $\mu \in \mathcal{D}_r(n)$ has parts $\mu_i = \mu_j = r^k\mathbf{a}$, $k \geq 1$, with $i \neq j$, the marked partition in which μ_i has index t is considered different from the marked partition in which μ_j has index t.

Note that marked partitions, by definition, have a non-primitive part. Then, from the discussion above we have the following interpretation for $b_r(n)$.

Proposition 1. *Let n, r be integers such that $n \geq 1$ and $r \geq 2$. Then,*

$$b_r(n) = |\mathcal{MD}_{1,r}(n)|.$$

Definition 2. An *r-word* w is a sequence of letters from the alphabet $\{0, 1, \ldots r - 1\}$. The *length* of an r-word w, denoted $\ell(w)$, is the number of letters in w. We refer to *position* i in w as the ith entry from the right, where the rightmost entry is counted as position 0.

Note that leading zeros are allowed and are recorded. For example, if $r = 5$, the 5-words 032 and 32 are different even though in base 5 they both represent 17.

We have $\ell(032) = 3$ and $\ell(32) = 2$. The empty bit string has length 0 and is denoted by $\emptyset$.

Definition 3. Given (S_1, S_2), an Euler pair of order r, we define the set $\mathcal{DD}_r(n)$ of *r-decorated partitions* as the set of partitions in $\mathcal{D}_r(n)$ with at least one non-primitive part such that exactly one non-primitive part $r^k\mathbf{a}$ is decorated with an index w, where w is an r-word satisfying $0 \le \ell(w) \le k-1$. As in Definition 1, if $\mu \in \mathcal{D}_r(n)$ has non-primitive parts $\mu_i = \mu_j = r^k\mathbf{a}$ with $i \neq j$, the decorated partition in which μ_i has index w is considered different from the decorated partition in which μ_j has index w.

Thus, for each part $\mu_i = r^{k_i}\mathbf{a}$ of $\mu \in \mathcal{DD}_r(n)$ there are $\dfrac{r^{k_i}-1}{r-1}$ possible indices and for each partition $\mu \in \mathcal{DD}_r(n)$ there are precisely $\dfrac{1}{r-1}\displaystyle\sum_{j=1}^{\ell(\mu)}(r^{k_j}-1)$ possible decorated partitions with the same parts as μ.

The discussion above proves the following interpretation for $\dfrac{1}{r-1}b_r(n)$.

Proposition 2. *Let n, r be integers such that $n \ge 1$ and $r \ge 2$. Then,*

$$\frac{1}{r-1}b_r(n) = |\mathcal{DD}_r(n)|.$$

While it is obvious that $|\mathcal{MD}_{1,r}(n)| = (r-1)|\mathcal{DD}_r(n)|$, to see this combinatorially, consider the map $\psi_r : \mathcal{MD}_{1,r}(n) \to \mathcal{DD}_r(n)$ defined as follows. If $\lambda \in \mathcal{MD}_{1,r}(n)$, then $\psi_r(\lambda)$ is the partition in $\mathcal{DD}_r(n)$ in which the r-decorated part is the same as the marked part in λ. The index of the part of $\psi_r(\lambda)$ is obtained from the index of the part of λ by writing it in base r and removing the leading digit. Clearly, this is an $r-1$ to 1 mapping.

2.2.2 A Combinatorial Proof for $a_r(n) = \frac{1}{r-1}b_r(n)$

To prove combinatorially that $a_r(n) = \dfrac{1}{r-1}b_r(n)$, we establish a one-to-one correspondence between $\mathcal{O}_{1,r}(n)$ and $\mathcal{DD}_r(n)$.

From $\mathcal{DD}_r(n)$ to $\mathcal{O}_{1,r}(n)$:

Start with an r-decorated partition $\mu \in \mathcal{DD}_r(n)$. Suppose the non-primitive part $\mu_i = r^k\mathbf{a}$ is decorated with an r-word w of length $\ell(w)$. Then, $0 \le \ell(w) \le k-1$. Let d_w be the decimal value of w. We set $d_\emptyset = 0$. We transform μ into a partition $\lambda \in \mathcal{O}_{1,r}(n)$ as follows.

Define $\bar{\mu}$ to be the partition whose parts are all non-primitive parts of μ of the form $\mu_j = r^t\mathbf{a}$ with $j \le i$, i.e., all parts $r^t\mathbf{a}$ with $t > k$ and, if μ_i is the pth part of size $r^k\mathbf{a}$ in μ, then $\bar{\mu}$ also has p parts equal to $r^k\mathbf{a}$.

Define $\tilde{\mu}$ to be the partition whose parts are all parts of μ that are not in $\bar{\mu}$.

(1) In $\bar{\mu}$, split one part of size $r^k\mathbf{a}$ into $d_w + 1$ parts of size $r^{k-\ell(w)}\mathbf{a}$ and $r^k - (d_w + 1)r^{k-\ell(w)}$ primitive parts of size $\mathbf{a}$. Every other part in $\bar{\mu}$ splits completely into parts of size $r^{k-\ell(w)}\mathbf{a}$. Denote the resulting partition by $\bar{\lambda}$.
(2) Let $\tilde{\lambda} = \varphi_r^{-1}(\tilde{\mu})$. Thus, $\tilde{\lambda}$ is obtained by splitting all parts in $\tilde{\mu}$ into primitive parts.

Let $\lambda = \bar{\lambda} \cup \tilde{\lambda}$. Then $\lambda \in \mathcal{O}_{1,r}(n)$ and its set of non-primitive parts is $\{r^{k-\ell(w)}\mathbf{a}\}$.

Remark 1. Since $d_w + 1 \leq r^{\ell(w)}$, in step 1, the resulting number of primitive parts equal to $\mathbf{a}$ is non-negative. Moreover, there is at least one non-primitive part in $\bar{\lambda}$.

Example 6. We continue with the Euler pair of order 3 from Example 1. Consider the decorated partition

$$\begin{aligned}\mu &= (1215, 135_{02}, 135, 51, 35, 15, 15, 3)\\ &= (3^5 \cdot 5, (3^3 \cdot 5)_{02}, 3^3 \cdot 5, 3 \cdot 17, 35, 3 \cdot 5, 3 \cdot 5, 3 \cdot 1) \in \mathcal{DD}_3(1604).\end{aligned}$$

We have $k = 3$, $\ell(w) = 2$, $d_w = 2$, and
$\bar{\mu} = (3^5 \cdot 5, 3^3 \cdot 5)$;
$\tilde{\mu} = (3^3 \cdot 5, 3 \cdot 17, 35, 3 \cdot 5, 3 \cdot 5, 3 \cdot 1)$.

To create $\bar{\lambda}$ from $\bar{\mu}$:

(1) Part $135 = 3^3 \cdot 5$ splits into three parts of size 15 and eighteen parts of size 5.
(2) Part $1215 = 3^5 \cdot 5$ splits into eighty one parts of size 15.

This results in $\bar{\lambda} = (15^{84}, 5^{18})$.

To create $\tilde{\lambda}$ from $\tilde{\mu}$:

All parts in $\tilde{\mu}$ are split into primitive parts. Thus, part $3^3 \cdot 5$ splits into twenty seven parts of size 5, part $3 \cdot 17$ splits into three parts of size 17, both parts of $3 \cdot 5$ split into three parts of size 5 each, and part $3 \cdot 1$ splits into three parts of size 1. Part 35 is already primitive so remains unchanged.
This results in $\tilde{\lambda} = (35, 17^3, 5^{33}, 1^3)$. Then, setting $\lambda = \bar{\lambda} \cup \tilde{\lambda}$ results in $\lambda = (35, 17^3, 15^{84}, 5^{51}, 1^3) \in \mathcal{O}_{1,3}(1604)$. The non-primitive part is $15 = 3 \cdot 5$.

From $\mathcal{O}_{1,r}(n)$ to $\mathcal{DD}_r(n)$:
Start with a partition $\lambda \in \mathcal{O}_{1,r}(n)$. In λ there is one and only one non-primitive part $r^k\mathbf{a}$. Let f be the multiplicity of the non-primitive part of λ. We transform λ into an r-decorated partition in $\mathcal{DD}_r(n)$ as follows.
Apply Glaisher's bijection to λ to obtain $\mu = \varphi_r(\lambda) \in \mathcal{D}_r(n)$. Since λ has a non-primitive part, μ will have at least one non-primitive part.
Next, we determine the r-decoration of μ. Consider the non-primitive parts μ_{j_i} of μ of the form $r^{t_i}\mathbf{a}$ (same $\mathbf{a}$ as in the non-primitive part of λ) and $t_i \geq k$. Assume

$j_1 < j_2 < \cdots$. For notational convenience, set $\mu_{j_0} = 0$. Let h be the positive integer such that

$$\sum_{i=0}^{h-1} \mu_{j_i} < f \cdot r^k \mathbf{a} \leq \sum_{i=0}^{h} \mu_{j_i}. \tag{4}$$

Then, we will decorate part $\mu_{j_h} = r^{t_h}\mathbf{a}$. To determine the decoration, let

$$N = \frac{\sum_{i=0}^{h-1} \mu_{j_i}}{r^k \mathbf{a}}.$$

Then, (4) becomes

$$r^k \mathbf{a} N < f \cdot r^k \mathbf{a} \leq r^k \mathbf{a} N + r^{t_h} \mathbf{a},$$

which implies $0 < f - N \leq r^{t_h - k}$.

Let $d = f - N - 1$ and $\ell = t_h - k$. We have $0 \leq \ell \leq t_h - 1$. Consider the representation of d in base r and insert leading zeros to form an r-word w of length ℓ. Decorate μ_{j_h} with w. The resulting decorated partition is in $\mathcal{DD}_r(n)$.

Example 7. We continue with the Euler pair of order 3 from Example 1. Consider the partition $\lambda = (35, 17^3, 15^{84}, 5^{51}, 1^3) \in \mathcal{O}_{1,3}(1604)$. The non-primitive part is 15. We have $k = 1$, $f = 84$.

Glaisher's bijection produces the partition $\mu = (1215, 135^2, 51, 35, 15^2, 3) = (3^5 \cdot 5, 3^3 \cdot 5, 3^3 \cdot 5, 3 \cdot 17, 35, 3 \cdot 5, 3 \cdot 5, 3 \cdot 1) \in \mathcal{DD}_3(1604)$. The parts of the form $3^{r_i} \cdot 5$ with $r_i \geq 1$ are 1215, 135, 135, 15, 15. Since $1215 < 84(3^1 \cdot 5) \leq 1215 + 135$, the decorated part will be the first part equal to $135 = 3^3 \cdot 5$. We have $N = 1215/15 = 81$.

To determine the decoration, let $d = 84 - 81 - 1 = 2$ and $\ell = 3 - 1 = 2$. The base 3 representation of d is 2. To form a 3-word of length 2, we introduce one leading 0. Thus, the decoration is $w = 02$ and the resulting decorated partition is $(1215, 135_{02}, 135, 51, 35, 15, 15, 3) = (3^5 \cdot 5, (3^3 \cdot 5)_{02}, 3^3 \cdot 5, 3 \cdot 17, 35, 3 \cdot 5, 3 \cdot 5, 3 \cdot 1) \in \mathcal{DD}_{1,3}(1604)$.

2.2.3 A Combinatorial Proof for $c_r(n) = \dfrac{1}{r-1} b_r(n)$

We note that one can compose the bijection of Sect. 2.2.2 with the bijection of [7] to obtain a combinatorial proof of part (ii) of Theorem 1.2. However, we give an alternative proof that $c_r(n) = \dfrac{1}{r-1} b_r(n)$ by establishing a one-to-one correspondence between $\mathcal{D}_{1,r}(n)$ and $\mathcal{DD}_r(n)$. This proof does not involve the bijection of [7] and it mirrors the proof of Sect. 2.2.2.

From $\mathcal{DD}_r(n)$ to $\mathcal{D}_{1,r}(n)$:

Start with an r-decorated partition $\mu \in \mathcal{DD}_r(n)$. Suppose the non-primitive part $\mu_i = r^k\mathbf{a}$ is decorated with an r-word w of length $\ell(w)$ and decimal value d_w. Then, $0 \leq \ell(w) \leq k-1$. We transform μ into a partition $\lambda \in \mathcal{D}_{1,r}(n)$ as follows.

Let $\bar{\bar{\mu}}$ be the partition whose parts are all non-primitive parts of μ of the form $\mu_j = r^t\mathbf{a}$ with $j \geq i$, and $k - \ell(w) - 1 < t \leq k$, i.e., all parts $r^t\mathbf{a}$ with $k - \ell(w) - 1 < t < k$ and, if there are $p-1$ parts of size $r^k\mathbf{a}$ in μ after the decorated part, then $\bar{\bar{\mu}}$ has p parts equal to $r^k\mathbf{a}$.

Let $\tilde{\tilde{\mu}}$ be the partition whose parts are all parts of μ that are not in $\bar{\bar{\mu}}$.

In $\bar{\bar{\mu}}$, perform the following steps.

(1) Split one part equal to $r^k\mathbf{a}$ into $r(d_w+1)$ parts of size $r^{k-\ell(w)-1}\mathbf{a}$ and m primitive parts of size $\mathbf{a}$, where $m = r^k - r(d_w+1)r^{k-\ell(w)-1}$. Apply Glaisher's bijection φ_r to the partition consisting of m parts equal to $\mathbf{a}$.
(2) Split all remaining parts of $\bar{\bar{\mu}}$ completely into parts of size $r^{k-\ell(w)-1}\mathbf{a}$.

Denote by $\bar{\bar{\lambda}}$ the partition with parts resulting from steps 1 and 2 above.

Let $\lambda = \bar{\bar{\lambda}} \cup \tilde{\tilde{\mu}}$. Since $r(d_w+1) \geq r$, it follows that $\lambda \in \mathcal{D}_{1,r}(n)$. The part repeated at least r times is $r^{k-\ell(w)-1}\mathbf{a}$.

Remark 2. (i) Since $d_w + 1 \leq r^{\ell(w)}$, the splitting in step 1 can be performed.

(ii) Note that $r \mid m = r^{k-\ell(w)}(r^{\ell(w)} - (d_w+1))$. Thus, if $w \neq \emptyset$, after applying Glaisher's bijection φ_r to the partition consisting of m parts equal to $\mathbf{a}$, we obtain parts $r^j\mathbf{a}$ with $k - \ell(w) \leq j < k$. Since in $\tilde{\tilde{\mu}}$, all parts of the form $r^i\mathbf{a}$ have $i \geq k$ or $i \leq k - \ell(w) - 1$, Glaisher's bijection in step (1) does not create parts equal to any parts in $\tilde{\tilde{\mu}}$.

(iii) If $w = \emptyset$, then, in step (1), we have $m = 0$ and $r^k\mathbf{a}$ splits into r parts equal to $r^{k-1}\mathbf{a}$.

Example 8. We continue with the Euler pair of order 3 from Example 1. Consider the partition $\mu = (32805, (10935)_{0120}, 10935, 1215, 45, 45, 25, 9, 3) = (3^8 \cdot 5, (3^7 \cdot 5)_{0120}, 3^7 \cdot 5, 3^5 \cdot 5, 3^2 \cdot 5, 3^2 \cdot 5, 25, 3^2 \cdot 1, 3 \cdot 1) \in \mathcal{DD}_3(56017)$. Then the decorated part is $\mu_2 = 3^7 \cdot 5$ and the decoration is $w = 0120$. We have $k = 7$, $\ell(w) = 4$, $d_w = 15$. So

$$\bar{\bar{\mu}} = (3^7 \cdot 5, 3^7 \cdot 5, 3^5 \cdot 5);$$
$$\tilde{\tilde{\mu}} = (3^8 \cdot 5, 3^2 \cdot 5, 3^2 \cdot 5, 25, 3^2 \cdot 1, 3 \cdot 1).$$

(1) $3^7 \cdot 5$ splits into

- $r(d_w+1) = 48$ parts of $3^2 \cdot 5$ and
- $m = r^k - r(d_w+1)r^{k-\ell(w)-1} = 3^7 - 48(3^2) = 1755$ parts of 5.

The 1755 parts of 5 merge into two parts of 3645, one part of 1215, and two parts of 135.

(2) $3^7 \cdot 5$ splits into two hundred and forty three parts of $3^2 \cdot 5$ and $3^5 \cdot 5$ splits into twenty seven parts of $3^2 \cdot 5$.

This results in
$\bar{\bar{\lambda}} = (3645^2, 1215, 135^2, 45^{318})$;
$\lambda = \bar{\bar{\lambda}} \cup \tilde{\tilde{\mu}} = (32805, 3645^2, 1215, 135^2, 45^{320}, 25, 9, 3) \in \mathcal{D}_{1,3}(56017)$. The part repeated at least three times is $45 = 3^2 \cdot 5$.

From $\mathcal{D}_{1,r}(n)$ to $\mathcal{DD}_r(n)$:

Start with a partition $\lambda \in \mathcal{D}_{1,r}(n)$. Then, among the parts of λ, there is one and only one part that is repeated at least r times. Suppose this part is $r^k\mathbf{a}$, $k \geq 0$, and denote by $f \geq r$ its multiplicity in λ. As in Glaisher's bijection, we merge repeatedly parts of λ that are repeated at least r times to obtain $\mu \in \mathcal{D}_r(n)$. Since λ has a part repeated at least r times, μ will have at least one non-primitive part.

Next, we determine the decoration of μ. In this case, we want to work with the parts of μ from the right to the left (i.e., from smallest to largest part). Let $\tilde{\mu}_q = \mu_{\ell(\mu)-q+1}$. Consider the parts $\tilde{\mu}_{j_i}$ of the form $r^{t_i}\mathbf{a}$ with $t_i \geq k$. If $t_1 \leq t_2 \leq \cdots$, we have $j_1 < j_2 < \cdots$.

As before, we set $\tilde{\mu}_{j_0} = 0$. Let h be the positive integer such that

$$\sum_{i=0}^{h-1} \tilde{\mu}_{j_i} < f \cdot r^k\mathbf{a} \leq \sum_{i=0}^{h} \tilde{\mu}_{j_i}. \tag{5}$$

Then, we will decorate part $\tilde{\mu}_{j_h} = r^{t_h}\mathbf{a}$. To determine the decoration, let

$$N = \frac{\sum_{i=0}^{h-1} \tilde{\mu}_{j_i}}{r^k\mathbf{a}}. \tag{6}$$

Then, (5) becomes

$$r^k\mathbf{a}N < f \cdot r^k\mathbf{a} \leq r^k\mathbf{a}N + r^{t_h}\mathbf{a},$$

which implies $0 < f - N \leq r^{t_h - k}$.

Let $d = \dfrac{f-N}{r} - 1$ and $\ell = t_h - k - 1$. Note that, by construction, $t_h > k$, and therefore $0 \leq \ell \leq t_h - 1$. Consider the representation of d in base r and insert leading zeros to form an r-word w of length ℓ. Decorate $\tilde{\mu}_{j_h}$ with w. The resulting decorated partition (with parts written in non-increasing order) is in $\mathcal{DD}_r(n)$.

Remark 3. To see that $f - N$ above is always divisible by r, note that if $f = qr + t$ with $q, t \in \mathbb{Z}$ and $0 \leq t < r$, then there are t terms equal to $r^k\mathbf{a}$ in the numerator of N. All other terms, if any, are divisible by $r^{k+1}\mathbf{a}$. Therefore, the remainder of N upon division by r is t.

Example 9. We continue with the Euler pair of order 3 from Example 1. Consider the partition $\lambda = (32805, 3645^2, 1215, 135^2, 45^{320}, 25, 9, 3) \in \mathcal{D}_{1,3}(56017)$. The part repeated at least three times is $45 = 3^2 \cdot 5$. We have $k = 2$ and $f = 320$.

Applying Glaisher's bijection to λ results in

$$\mu = \varphi_3(\lambda) = (3^8 \cdot 5, 3^7 \cdot 5, 3^7 \cdot 5, 3^5 \cdot 5, 3^2 \cdot 5, 3^2 \cdot 5, 25, 3^2, 3 \cdot 1) \in \mathcal{D}_3(56017).$$

The parts of the form $3^{t_i} \cdot 5$ with $t_i \geq 2$ are $3^2 \cdot 5, 3^2 \cdot 5, 3^5 \cdot 5, 3^7 \cdot 5, 3^7 \cdot 5, 3^8 \cdot 5$. Since $3^2 \cdot 5 + 3^2 \cdot 5 + 3^5 \cdot 5 + 3^7 \cdot 5 < 320 \cdot 3^2 \cdot 5 \leq 3^2 \cdot 5 + 3^2 \cdot 5 + 3^5 \cdot 5 + 3^7 \cdot 5 + 3^7 \cdot 5$, the decorated part will be the second part (counting from the right) equal to $3^7 \cdot 5 = 10935$. We have $N = \dfrac{3^2 \cdot 5 + 3^2 \cdot 5 + 3^5 \cdot 5 + 3^7 \cdot 5}{3^2 \cdot 5} = 272$. Thus $d = \dfrac{320 - 272}{3} - 1 = 15$ and $\ell = 7 - 2 - 1 = 4$. The base 3 representation of d is 120. To form a 3-word of length 4, we introduce one leading 0. Thus, the decoration is $w = 0120$ and the resulting decorated partition is

$$\begin{aligned}\mu &= (32805, (10935)_{0120}, 10935, 1215, 45, 45, 25, 9, 3)\\ &= (3^8 \cdot 5, (3^7 \cdot 5)_{0120}, 3^7 \cdot 5, 3^5 \cdot 5, 3^2 \cdot 5, 3^2 \cdot 5, 25, 3^2 \cdot 1, 3 \cdot 1) \in \mathcal{DD}_3(56017).\end{aligned}$$

3 Proofs of Theorem 1.3

3.1 Analytic Proof

We create a bivariate generating function to keep track of the number of different parts in partitions in $\mathcal{O}_r(n)$, respectively $\mathcal{D}_r(n)$.

We denote by $\mathcal{O}'_r(n; m)$ the set of partitions of n with parts from S_2 using m different parts. We denote by $\mathcal{D}'_r(n; m)$ the set of partitions of n with parts from S_1 using m different parts and allowing parts to repeat no more than $r - 1$ times. Then,

$$\begin{aligned}f_{\mathcal{O}'_r}(z, q) := \sum_{n=0}^{\infty}\sum_{m=0}^{\infty} |\mathcal{O}'_r(n; m)| z^m q^n &= \prod_{\mathrm{b} \in S_2} (1 + zq^{\mathrm{b}} + zq^{2\mathrm{b}} + \cdots)\\ &= \prod_{\mathrm{b} \in S_2} \left(1 + \frac{zq^{\mathrm{b}}}{1 - q^{\mathrm{b}}}\right);\end{aligned}$$

and

$$\begin{aligned}f_{\mathcal{D}'_r}(z, q) := \sum_{n=0}^{\infty}\sum_{m=0}^{\infty} |\mathcal{D}'_r(n; m)| z^m q^n &= \prod_{a \in S_1} (1 + zq^a + \cdots + zq^{(r-1)a})\\ &= \prod_{a \in S_1} \left(1 + \frac{zq^a - zq^{ra}}{1 - q^a}\right).\end{aligned}$$

To obtain the generating function for the total number of different parts in all partitions in $\mathcal{O}_r(n)$ (respectively $\mathcal{D}_r(n)$), we take the derivative with respect to z of $f_{\mathcal{O}'_r}(z,q)$ (respectively $f_{\mathcal{D}'_r}(z,q)$), and set $z=1$. We obtain

$$\begin{aligned}\frac{\partial}{\partial z}\bigg|_{z=1} f_{\mathcal{O}'_r}(z,q) &= \sum_{\mathbf{b}\in S_2} \frac{q^{\mathbf{b}}}{1-q^{\mathbf{b}}} \prod_{\mathbf{c}\in S_2,\mathbf{c}\neq\mathbf{b}} \left(1+\frac{q^{\mathbf{c}}}{1-q^{\mathbf{c}}}\right)\\ &= \sum_{\mathbf{b}\in S_2} \frac{q^{\mathbf{b}}}{1-q^{\mathbf{b}}} \prod_{\mathbf{c}\in S_2,\mathbf{c}\neq\mathbf{b}} \left(\frac{1}{1-q^{\mathbf{c}}}\right)\\ &= \prod_{\mathbf{b}\in S_2} \frac{1}{1-q^{\mathbf{b}}} \sum_{\mathbf{b}\in S_2} q^{\mathbf{b}};\end{aligned}$$

and

$$\begin{aligned}\frac{\partial}{\partial z}\bigg|_{z=1} f_{\mathcal{D}'_r}(z,q) &= \sum_{a\in S_1} \frac{q^a-q^{ra}}{1-q^a} \prod_{d\in S_1,d\neq a} \left(1+\frac{q^d-q^{rd}}{1-q^d}\right)\\ &= \sum_{a\in S_1} \frac{q^a-q^{ra}}{1-q^a} \prod_{d\in S_1,d\neq a} \frac{1-q^{rd}}{1-q^d}\\ &= \prod_{a\in S_1} \frac{1-q^{ra}}{1-q^a} \sum_{a\in S_1} \frac{q^a-q^{ra}}{1-q^{ra}}.\end{aligned}$$

Since $|\mathcal{D}_r(n)| = |\mathcal{O}_r(n)|$, we have

$$\sum_{n=0}^{\infty} b'_r(n)q^n = \prod_{a\in S_1} \frac{1-q^{ra}}{1-q^a}\left(\sum_{a\in S_1}\frac{q^a}{1-q^{ra}} - \sum_{a\in S_1}\frac{q^{ra}}{1-q^{ra}} - \sum_{\mathbf{b}\in S_2} q^{\mathbf{b}}\right).$$

Moreover,

$$\begin{aligned}\sum_{a\in S_1}\frac{q^a}{1-q^{ra}} - \sum_{a\in S_1}\frac{q^{ra}}{1-q^{ra}} &= \left(\sum_{a\in S_1} q^a + \sum_{a\in S_1}\frac{q^{(r+1)a}}{1-q^{ra}}\right) - \left(\sum_{a\in rS_1} q^a + \sum_{a\in S_1}\frac{q^{2ra}}{1-q^{ra}}\right)\\ &= \left(\sum_{a\in S_1} q^a - \sum_{a\in rS_1} q^a\right) + \left(\sum_{a\in S_1}\frac{q^{(r+1)a}}{1-q^{ra}} - \sum_{a\in S_1}\frac{q^{2ra}}{1-q^{ra}}\right)\\ &= \sum_{\mathbf{b}\in S_2} q^{\mathbf{b}} + \sum_{a\in S_1}\frac{q^{(r+1)a}-q^{2ra}}{1-q^{ra}},\end{aligned}$$

the last equality occurring because $S_1 = S_2 \sqcup rS_1$.

Therefore,

$$\begin{aligned}
\sum_{n=0}^{\infty} b'_r(n)q^n &= \prod_{a\in S_1} \frac{1-q^{ra}}{1-q^a} \sum_{a\in S_1} \frac{q^{(r+1)a}-q^{2ra}}{1-q^{ra}} \\
&= \sum_{a\in S_1} \frac{q^{(r+1)a}+q^{(r+2)a}+\cdots+q^{(2r-1)a}}{1+q^a+\cdots+q^{(r-1)a}} \prod_{a\in S_1}(1+q^a+\cdots+q^{(r-1)a}) \\
&= \sum_{a\in S_1}(q^{(r+1)a}+q^{(r+2)a}+\cdots+q^{(2r-1)a}) \prod_{d\in S_1, c\neq a} (1+q^d+\cdots+q^{(r-1)d}) \\
&= \sum_{n=0}^{\infty} c'_r(n)q^n.
\end{aligned}$$

3.2 Combinatorial Proof

3.2.1 $b'_r(n)$ as the Cardinality of a Set of Overpartitions

As in Sect. 2.2.1, we use Glaisher's bijection and calculate $b'_r(n)$ by summing up the difference between the number of different parts of $\varphi_r(\lambda)$ and the number of different parts of λ for each partition $\lambda \in \mathcal{O}_r(n)$. For a given $\mathbf{a} \in S_2$, each part in $\varphi_r(\lambda)$ of the form $r^k\mathbf{a}$, $k \geq 0$, is obtained from λ by merging r^k parts equal to $\mathbf{a}$. Therefore, the contribution to $b'_r(n)$ of each $\mu \in \mathcal{D}_r(n)$ equals

$$\sum_{\substack{\mathbf{a}\in S_2 \\ \mathbf{a} \text{ part of } \varphi_r^{-1}(\mu)}} (m_\mu(\mathbf{a})-1),$$

where

$$m_\mu(\mathbf{a}) = |\{t \geq 0 \mid r^t\mathbf{a} \text{ is a part of } \mu\}|.$$

Next, we define a set of overpartitions. An overpartition is a partition in which the last appearance of a part may be overlined. For example, $(5, \bar{5}, 3, 3, \bar{2}, 1, 1, \bar{1})$ is an overpartition of 21. We denote by $\overline{\mathcal{D}}_r(n)$ the set of overpartitions of n with parts in S_1 repeated at most $r-1$ times in which *exactly one* part is overlined and such that part $r^s\mathbf{a}$ with $s \geq 0$ may be overlined only if there is a part $r^t\mathbf{a}$ with $t < s$. In particular, no primitive part can be overlined. Note that when we count parts in an overpartition, the overlined part contributes to the multiplicity. The discussion above proves the following interpretation of $b'_r(n)$.

Proposition 3. *Let $n \geq 1$. Then, $b'_r(n) = |\overline{\mathcal{D}}_r(n)|$.*

3.2.2 A Combinatorial Proof for $c'_r(n) = b'_r(n)$

We establish a one-to-one correspondence between $\overline{\mathcal{D}}_r(n)$ and $\mathcal{T}_r(n)$.

From $\overline{\mathcal{D}}_r(n)$ to $\mathcal{T}_r(n)$:

Start with an overpartition $\mu \in \overline{\mathcal{D}}_r(n)$. Suppose the overlined part is $\mu_i = r^s\mathbf{a}$. Then there is a part $\mu_p = r^t\mathbf{a}$ of μ with $t < s$. Let k be the largest non-negative integer such that $r^k\mathbf{a}$ is a part of μ and $k < s$. To obtain $\lambda \in \mathcal{T}_r(n)$ from μ, split μ_i into r parts equal to $r^k\mathbf{a}$ and $r-1$ parts equal to $r^j\mathbf{a}$ for each $j = k+1, k+2, \ldots, s-1$.

Example 10. We continue with the Euler pair of order 3 from Example 1. Let

$$\mu = (3^8 \cdot 5, 3^7 \cdot 5, \overline{3^7 \cdot 5}, 3^5 \cdot 5, 3^2 \cdot 5, 3^2 \cdot 5, 25, 3^2 \cdot 1, 3 \cdot 1) \in \overline{\mathcal{D}}_3(56017).$$

Then $k = 5$ and $3^7 \cdot 5$ splits into three parts equal to $3^5 \cdot 5$ and two parts equal to $3^6 \cdot 5$. Thus, we obtain the partition

$$\lambda = (3^8 \cdot 5, 3^7 \cdot 5, 3^6 \cdot 5, 3^6 \cdot 5, 3^5 \cdot 5, 3^5 \cdot 5, 3^5 \cdot 5, 3^5 \cdot 5, 3^2 \cdot 5, 3^2 \cdot 5, 25, 3^2 \cdot 1, 3 \cdot 1)$$
$$\in \mathcal{T}_3(56017).$$

The part repeated more than three times but less than six times is $3^5 \cdot 5$.

From $\mathcal{T}_r(n)$ to $\overline{\mathcal{D}}_r(n)$:

Start with a partition $\lambda \in \mathcal{T}_r(n)$. Suppose $r^k\mathbf{a}$ is the part repeated more than r times but less than $2r$ times. Let $\mu = \varphi_r(\lambda) \in \mathcal{D}_r(n)$. Overline the smallest part of μ of form $r^t\mathbf{a}$ with $t > k$. The resulting overpartition is in $\overline{\mathcal{D}}_r(n)$.

Example 11. We continue with the Euler pair of order 3 from Example 1. Let

$$\lambda = (3^8 \cdot 5, 3^7 \cdot 5, 3^6 \cdot 5, 3^6 \cdot 5, 3^5 \cdot 5, 3^5 \cdot 5, 3^5 \cdot 5, 3^5 \cdot 5, 3^2 \cdot 5, 3^2 \cdot 5, 25, 3^2 \cdot 1, 3 \cdot 1)$$
$$\in \mathcal{T}_3(56017).$$

The part repeated more than three times but less than six times is $3^5 \cdot 5$. We have $k = 5$. Merging by Glaisher's bijection, we obtain

$$\mu = (3^8 \cdot 5, 3^7 \cdot 5, 3^7 \cdot 5, 3^5 \cdot 5, 3^2 \cdot 5, 3^2 \cdot 5, 25, 3^2 \cdot 1, 3 \cdot 1) \in \mathcal{D}_3(56017).$$

The smallest part of μ of the form $r^t\mathbf{a}$ with $t > k = 5$ is $3^7 \cdot 5$. Thus we obtain the overpartition

$$\mu = (3^8 \cdot 5, 3^7 \cdot 5, \overline{3^7 \cdot 5}, 3^5 \cdot 5, 3^2 \cdot 5, 3^2 \cdot 5, 25, 3^2 \cdot 1, 3 \cdot 1) \in \overline{\mathcal{D}}_3(56017).$$

Remark 4. We could have obtained the transformation above from the combinatorial proof of part (ii) of Theorem 1.2. In the transformation from $\mathcal{D}_{1,r}(n)$ to $\mathcal{DD}_r(n)$, if part $r^k\mathbf{a}$ is the part repeated more than r times but less than $2r$ times, we have $f = r + s$ for some $1 \le s \le r-1$, $h = s+1$, and $N = s$. Thus $d = 0$ and the decorated part is the last occurrence of the smallest part in the transformed partition μ

that is of the form $r^t\mathbf{a}$ with $t > k$. Thus, in μ, the decorated part $r^t\mathbf{a}$ is decorated with an r-word consisting of all zeros and of length $t-k-1$, one less than the difference in exponents of r of the decorated part and the next smallest part with the same $\mathbf{a}$ factor. Since in this case the decoration of a partition in $\mathcal{DD}_r(n)$ is completely determined by the part being decorated, we can simply just overline the part.

4 Concluding Remarks

In this article, we proved first and second Beck-type identities for all Euler pairs (S_1, S_2) of order $r \geq 2$. Euler pairs of order r satisfy $rS_1 \subseteq S_1$ and $S_2 = S_1 \setminus rS_1$. Thus, we established Beck-type identities accompanying all partition identities of the type given in Theorem 1.1.

At the end of [11], Subbarao mentions that the characterization of Euler pairs of order r given by Theorem 1.1 can be extended to vector partitions. The corrected statement for partitions of multipartite numbers is given in [3, Theorem 12.2] and indeed it has Beck-type companion identities as we explain below.

A multipartite (or s-partite) number $\mathfrak{n} = (n_1, n_2, \ldots, n_s)$ is an s-tuple of non-negative integers, not all 0. We view multipartite numbers as vectors and refer to $n_1, n_2, \ldots, n_s$ as the entries of $\mathfrak{n}$.

A multipartition (or vector partition) $\xi = (\xi^{(1)}, \xi^{(2)}, \ldots, \xi^{(t)})$ of $\mathfrak{n}$ is a sequence of multipartite numbers in non-increasing lexicographic order satisfying

$$\mathfrak{n} = \xi^{(1)} + \xi^{(2)} + \cdots + \xi^{(t)}.$$

We refer to $\xi^{(i)}$, $1 \leq i \leq t$ as the multiparts (or vector parts) of the multipartition ξ and to the number of multiparts t of ξ as the length of ξ, which we denote by $\ell(\xi)$.

Let S_1 and S_2 be sets of positive integers. Given a multipartition ξ of $\mathfrak{n}$ with all entries of all multiparts in S_1, we say that a multipart $\xi^{(i)}$ of ξ is *primitive* if at least one entry of $\xi^{(i)}$ is in S_2. Otherwise, the multipart is called *non-primitive*. We denote by $\mathcal{VD}_r(\mathfrak{n})$ the set of multipartitions $\xi = (\xi^{(1)}, \xi^{(2)}, \ldots, \xi^{(t)})$ of $\mathfrak{n} = (n_1, n_2, \ldots, n_s)$ with all entries of all multiparts in S_1 and such that all multiparts are repeated at most $r-1$ times. We denote by $\mathcal{VO}_r(\mathfrak{n})$ the set of multipartitions $\eta = (\eta^{(1)}, \eta^{(2)}, \ldots, \eta^{(u)})$ of $\mathfrak{n} = (n_1, n_2, \ldots, n_s)$ such that each multipart $\eta^{(i)}$ of η is primitive.

Then, Andrews [3] gives the following theorem mentioning that its proof can be constructed similar to the proof using ideals of order 1 for the analogous result for regular partitions.

Theorem 4.1. *Let S_1 and S_2 be sets of positive integers. Then*

$$|\mathcal{VD}_r(\mathfrak{n})| = |\mathcal{VO}_r(\mathfrak{n})|$$

if and only if (S_1, S_2) is an Euler pair of order r, i.e., $rS_1 \subseteq S_1$ and $S_2 = S_1 \setminus rS_1$.

We note that Glaisher's bijection can be extended to prove Theorem 4.1 combinatorially. The Glaisher type transformation, $v\varphi_r$, from $\mathcal{VO}_r(\mathfrak{n})$ to $\mathcal{VD}_r(\mathfrak{n})$ repeatedly merges r equal multiparts (as addition of vectors) until there are no multiparts repeated more than $r-1$ times. The transformation from $\mathcal{VD}_r(\mathfrak{n})$ to $\mathcal{VO}_r(\mathfrak{n})$ takes each non-primitive multipart (all its entries are from rS_1) and splits it into r equal multiparts, repeating the process until the obtained multiparts are primitive. The remark below is the key to adapting the combinatorial proofs of Theorems 1.2 and 1.3 to proofs of Beck-type identities for multipartitions.

Remark 5. Let $\eta \in \mathcal{VO}_r(\mathfrak{n})$ and $\xi = v\varphi_r(\eta) \in \mathcal{VD}_r(\mathfrak{n})$. Then multipart $\xi^{(i)}$ of ξ was obtained by merging r^k multiparts of η if and only if, when writing all entries of $\xi^{(i)}$ in the form $r^j\mathbf{a}$, the smallest exponent of r in all entries of $\xi^{(i)}$ is k.

To formulate Beck-type identities for multipartitions, let $vb_r(\mathfrak{n})$ be the difference between the number of multiparts in all multipartitions in $\mathcal{VO}_r(\mathfrak{n})$ and the number of multiparts in all multipartitions in $\mathcal{VD}_r(\mathfrak{n})$. Similarly, let $vb'_r(\mathfrak{n})$ be the difference in the total number of different multiparts in all multipartitions in $\mathcal{VD}_r(\mathfrak{n})$ and the total number of different multiparts in all multipartitions in $\mathcal{VO}_r(\mathfrak{n})$. Then, we have the following Beck-type identities for multipartions.

Theorem 4.2. *Suppose (S_1, S_2) is an Euler pair of order $r \geq 2$ and let $\mathfrak{n}$ be a multipartite number. Then*

(i) $\dfrac{1}{r-1}vb_r(\mathfrak{n})$ *equals the number of multipartitions* $\xi = (\xi^{(1)}, \xi^{(2)}, \ldots, \xi^{(t)})$ *of* $\mathfrak{n}$ *with all entries of all multiparts in* S_1 *and such that exactly one multipart is repeated at least r times. Moreover,* $\dfrac{1}{r-1}vb_r(\mathfrak{n})$ *equals the number of multipartitions* $\eta = (\eta^{(1)}, \eta^{(2)}, \ldots, \eta^{(u)})$ *of* $\mathfrak{n}$ *such that each multipart* $\eta^{(i)}$ *of* η *has all entries in* S_1 *and exactly one multipart, possibly repeated, is non-primitive.*

(ii) $vb'_r(\mathfrak{n})$ *equals the number of multipartitions* $\xi = (\xi^{(1)}, \xi^{(2)}, \ldots, \xi^{(t)})$ *of* $\mathfrak{n}$ *with all entries of all multiparts in* S_1 *and such that exactly one multipart is repeated more than r times but less than 2r times.*

The combinatorial proofs of these statements follow the combinatorial proofs of Theorems 1.2 and 1.3 with all references to expression of the form "part $r^k\mathbf{a}$ in the partition μ" changed to "multipart of multipartition μ in which $r^k\mathbf{a}$ is the entry with the smallest exponent of r (among the entries of the multipart)."

To our knowledge, there is no analytic proof of Theorem 4.2.

Acknowledgements We are grateful to the anonymous referees for suggestions that improved the exposition of the article. In particular, one referee suggested the short proof of (3), and another referee alerted us to the correct statement of Subbarao's theorem for vector partitions.

References

1. The On-Line Encyclopedia of Integer Sequences, oeis: A090867 and A265251.
2. G. E. Andrews, *Two theorems of Euler and a general partition theorem*, Proc. Amer. Math. Soc. 20 (1969), 499–502.
3. G. E. Andrews, *The Theory of Partitions*, Cambridge Mathematical Library, Cambridge University Press, Cambridge, 1998. Reprint of the 1976 original.
4. G. E. Andrews, *Euler's partition identity and two problems of George Beck*, Math. Student 86 (2017), no. 1-2, 115–119.
5. G. E. Andrews and C. Ballantine, *Almost partition identities*, Proc. Natl. Acad. Sci. USA 116 (2019), no. 12, 5428–5436
6. C. Ballantine and R. Bielak, *Combinatorial proofs of two Euler-type identities due to Andrews*, Ann. Comb. 23 (2019), no. 3-4, 511–525.
7. S. Fu and D. Tang, *Generalizing a partition theorem of Andrews*, Math. Student 86 (2017), no. 3-4, 91–96.
8. R. Li and A. Y. Z. Wang, *Composition analogues of Beck's conjectures on partitions* European J. Combin. 81 (2019), 210–220.
9. R. Li and A. Y. Z. Wang, *Generalization of two problems of George Beck*, Discrete Math. 343 (2020), no. 5, 111805, 12 pp.
10. R. Li and A. Y. Z. Wang, *Partitions associated with two fifth-order mock theta functions and Beck-type identities*, Int. J. Number Theory 16 (2020), no. 4, 841–855
11. M. V. Subbarao, *Partition theorems for Euler pairs*. Proc. Amer. Math. Soc. 28 (1971), 330–336.
12. Jane Y. X. Yang, *Combinatorial proofs and generalizations of conjectures related to Euler's partition theorem*, European J. Combin. 76 (2019), 62–72.

Quarter-Plane Lattice Paths with Interacting Boundaries: The Kreweras and Reverse Kreweras Models

Nicholas R. Beaton, Aleksander L. Owczarek, and Ruijie Xu

Abstract Lattice paths in the quarter plane have led to a large and varied set of results in recent years. One major project has been the classification of step sets according to the properties of the corresponding generating functions, and this has involved a variety of techniques, some highly intricate and specialised. The famous Kreweras and reverse Kreweras walk models are two particularly interesting models, as they are among the only four cases which have algebraic generating functions.

Here we investigate how the properties of the Kreweras and reverse Kreweras models change when boundary interactions are introduced. That is, we associate three real-valued weights a, b, c with visits by the walks to the x-axis, the y-axis and the origin $(0, 0)$ respectively. These models were partially solved in a recent paper by Beaton, Owczarek and Rechnitzer (2019). We apply the algebraic kernel method to completely solve these two models. We find that reverse Kreweras walks have an algebraic generating function for all a, b, c, regardless of whether the walks are restricted to end at the origin or on one of the axes, or may end anywhere at all. For Kreweras walks, the generating function for walks returning to the origin is algebraic, but the other cases are only D-finite. To our knowledge this is the first example of a quarter-plane model with this property.

Keywords Random walks · Enumeration · Generating functions · Kernel method · Algebraic functions · D-finite functions

N. R. Beaton (✉) · A. L. Owczarek · R. Xu
School of Mathematics and Statistics, The University of Melbourne,
Melbourne 3010, VIC, Australia
e-mail: nrbeaton@unimelb.edu.au

A. L. Owczarek
e-mail: owczarek@unimelb.edu.au

R. Xu
e-mail: ruijiex1@student.unimelb.edu.au

A. Bostan and K. Raschel (eds.), *Transcendence in Algebra, Combinatorics, Geometry and Number Theory*, Springer Proceedings in Mathematics & Statistics 373,
https://doi.org/10.1007/978-3-030-84304-5_7

1 Introduction

Lattice path models with boundary conditions have been studied widely. As a combinatorial problem, they are closely related to probability theory, algebra, complex analysis and statistical physics [3, 7, 9, 13, 15]. The typical goal is to study the properties of random walks with steps in a fixed set S. These properties include the number of paths of a certain length, the generating function, the asymptotic behaviour and bijections with other combinatorial objects.

The generating function of simple walks in the bulk can be written down directly with simple calculations. Once the boundary conditions are introduced, the problem becomes more complicated and some interesting results appear. In [1], Banderier and Flajolet prove that for walks in a half plane (walks with one boundary constraint), the generating function is always algebraic. For quarter-plane walks with small steps, Bousquet-Mélou and Mishna [7] associated each step set S with a group, and proved that among all 79 non-isomorphic (and non-trivial) quarter-plane models, exactly 23 have a finite group and the remaining 56 have an infinite group. A walk model with a finite group can be solved by the kernel method [17] and all of them have D-finite generating functions. Four specific models: Kreweras, reverse Kreweras, double Kreweras, and Gessel, have algebraic generating functions. Models whose associated group is infinite are far more difficult to solve (find an explicit expression for the generating function), but many properties can still be discussed [8, 16]. Recently, walks avoiding a quarter-plane have also been studied [5, 18].

The connection with statistical physics can be seen in a recent publication [19]. The authors studied a two-dimensional model of interacting directed polymers above an impenetrable surface. Two walks starting at the origin can walk north-east or south-east with step length 1. Weights (Boltzmann weights) are assigned when the two walks touch or when either touches the surface.

This interacting directed walk model has a bijection to a certain quarter-plane path problem (see Fig. 1). Let s be the generating variable for half the distance between the two walks, and let r be the generating variable for the distance between the lower walk and the surface. If we have the following bijection:

$$\begin{pmatrix}\nearrow\\ \nearrow\end{pmatrix} \mapsto (\rightarrow) \quad \begin{pmatrix}\nearrow\\ \searrow\end{pmatrix} \mapsto (\nwarrow) \quad \begin{pmatrix}\searrow\\ \searrow\end{pmatrix} \mapsto (\leftarrow) \quad \begin{pmatrix}\searrow\\ \nearrow\end{pmatrix} \mapsto (\searrow)$$

Then, the directed walk with generating variable (s, r) corresponds to a quarter-plane walk whose allowed steps are east, west, northwest and southeast. This quarter-plane walk model has interactions when the walk touches the x axis and the y axis. The weights are the same as in the corresponding directed walk model.

We may expand this idea. As there is a rich body of literature on non-interacting quarter-plane walks, what happens if we introduce the interaction into other quarter-plane walk problems? In a recent work [2], Beaton, Owczarek and Rechnitzer introduced interacting boundary weights to quarter-plane walk models and studied the 23 models which are associated with finite groups. In particular, they assigned weights

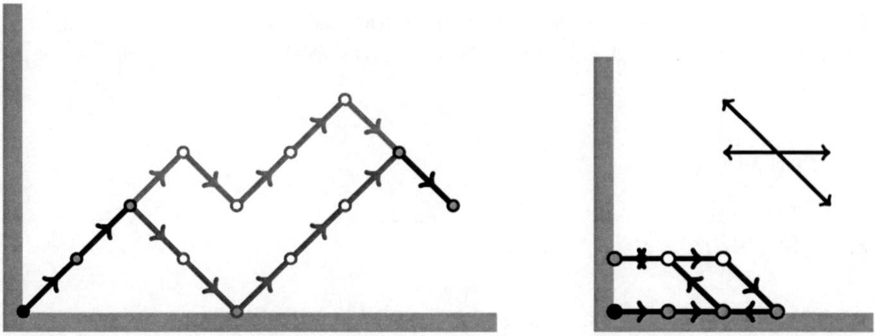

Fig. 1 Left: A pair of interacting directed paths with weights associated with shared vertices and visits to the bottom boundary. Right: The corresponding quarter-plane lattice path with weights associated with visits to the two boundaries.

a and b to the two axes and ab to the origin. Except for a couple of very special cases which can be solved by other means, they essentially applied the same method—a *half-orbit sum* followed by coefficient extraction—to all cases, and explored which could be solved in this way. For some of the models, the generating function stays D-finite for all a, b. For some models, the generating function may be D-finite, may be D-algebraic or may be unsolvable depending on the values of a and b.

In this paper, we focus on two further cases which require more elaborate techniques: the reverse Kreweras and Kreweras models (Models 20 and 19 in [2]). It has been proved that the generating functions of these two models are algebraic without interactions, or with equal interactions on two boundaries (that is, $a = b$) [2, 7]. For arbitrary a, b, it is still unclear, and it is this general case that we address here. (In fact we generalise further by associating weight c with visits to the origin, instead of ab.) For these two models we obtain solutions by delicately combining two commonly-used tools—a *full-orbit sum* in addition to a *half-orbit sum*.

The two other models which are algebraic in the absence of boundary weights—double Kreweras and Gessel walks—are much more technically challenging and still remain unsolved.

The layout of this paper is as follows. In Sect. 2, we will define some notation used in this paper and recall some general definitions. We will present the final result in Sect. 3. In Sects. 4 and 5 we will walk through the whole process of the algebraic kernel method and solve the two problems. We will prove that the generating function of reverse Kreweras walks is still algebraic for all boundary weights, while the generating function of Kreweras walk is shown to be D-finite (and probably not algebraic). In Sect. 6 we will discuss some remaining open problems.

Many calculations in this paper involve complicated rational, algebraic or D-finite expressions, some of which would take multiple pages to be written down explicitly.

For this reason we have produced MATHEMATICA notebooks which go through all calculations and verify all equations. These are available at the first author's website.[1]

2 The Model

2.1 Definitions and Notation

We first define some notation used in this paper. We write $[x^i]f(x)$ for the coefficient of $[x^i]$ in the Laurent series expansion of $f(x)$. Respectively, $[x^>]f(x)$, $[x^<]f(x)$ and $[x^\geq]f(x)$ are those terms with positive, negative and non-negative powers of x. We use the notation $\overline{x} = x^{-1}$ and $\overline{y} = y^{-1}$.

For a ring $\mathbb{K}$, we denote

1. $\mathbb{K}[t]$ as the set of polynomials in t with coefficients in $\mathbb{K}$;
2. $\mathbb{K}[t, t^{-1}]$ as the set of polynomials in t and $1/t$ with coefficients in $\mathbb{K}$;
3. $\mathbb{K}[[t]]$ as the set of formal power series in t with coefficients in $\mathbb{K}$;
4. $\mathbb{K}((t))$ as the set of Laurent series in t with coefficients in $\mathbb{K}$;
5. $\mathbb{K}(t)$ as the set of rational functions in t with coefficients in $\mathbb{K}$.

The definition may extend to multiple variables. For example $\mathbb{R}(x)((t))$ refers to the set of Laurent series in t with coefficients in the ring of rational polynomials of x with real coefficients.

Next, we define functions for counting lattice paths with site interactions on the boundaries.

We denote $q_{n,k,l,h,v,u}$ as the number of walks of length n that start at $(0, 0)$ and end at (k, l) which visit the horizontal boundary (except the origin) h times, the vertical boundary (except the origin) v times and the origin u times. The associated generating function is

$$Q(t; x, y; a, b, c) \equiv Q(x, y) = \sum_n t^n \sum_{k,l,h,v,u} q_{n,k,l,h,v,u} x^k y^l a^h b^v c^u \equiv \sum_n t^n Q_n(x, y). \tag{2.1}$$

We also define line-boundary terms:

$$[y^i]Q(t; x, y; a, b, c) \equiv Q_{-,i}(x) = \sum_n t^n \sum_{k,h,v,u} q_{n,k,i,h,v,u} x^k a^h b^v c^u. \tag{2.2}$$

This is the generating function of walks ending on the line $y = i$. $Q_{i,-}(y)$ is defined similarly.

Furthermore, we define the generating functions of walks ending on a diagonal line:

[1] http://www.nicholasbeaton.com/papers.

$$Q_j^d(x) = \sum_n t^n \sum_{i,h,v,u} q_{n,i,i+j,h,v,u} x^i a^h b^v c^u. \tag{2.3}$$

Note that the variable x in $Q_j^d(x)$ marks the x-coordinate of the endpoint of walks.

Finally we define point boundary terms:

$$[x^i y^j]Q(t; x, y; a, b, c) \equiv Q_{i,j} = \sum_n t^n \sum_{h,v,u} q_{n,i,j,h,v,u} a^h b^v c^u. \tag{2.4}$$

This is the generating function of walks ending at point (i, j).

2.2 The General Functional Equation

Consider a walk starting from the origin with allowed steps $\mathcal{S} \subseteq \{-1, 0, 1\}^2$. This set of steps is usually denoted {N, S, W, E, NE, NW, SE, SW}.

The *step generator* S is

$$S(x, y) = \sum_{(i,j)\in\mathcal{S}} x^i y^j. \tag{2.5}$$

This can be written as

$$S(x, y) = A_{-1}(x)\overline{y} + A_0(x) + A_1(x)y = B_{-1}(y)\overline{x} + B_0(y) + B_1(y)x \tag{2.6}$$

where for example $A_{-1}(x)$ is $[y^{-1}]S(x, y)$, which refers to the steps going southwards, including {S, SE, SW} steps.

Since the quarter-plane walk is restricted in the first quadrant, it is allowed to touch the axes but not cross them. We denote $A(x, y) = A_{-1}(x)\overline{y}$ as the illegal steps crossing the x-axis, $B(x, y) = B_{-1}(y)\overline{x}$ as the illegal steps crossing the y-axis and $G(x, y) = [x^{-1}y^{-1}]S(x, y)\overline{xy}$ as the illegal steps crossing the origin diagonally.

By the geometric properties of quarter-plane walks and the boundary conditions, we can derive a functional equation satisfied by the generating function.

Theorem 1 (Slightly modified version of [2, Theorem 6]). *For a lattice walk restricted to the quarter-plane, starting from the origin, with weight a associated with vertices on the x-axis (except the origin), weight b associated with vertices on y-axis (except the origin) and weight c associated with the origin, the generating function $Q(x, y)$ satisfies the following functional equation:*

$$\begin{aligned} K(x, y)Q(x, y) = \frac{1}{c} + \frac{1}{a}(a - 1 - taA(x, y))Q(x, 0) + \frac{1}{b}(b - 1 - tbB(x, y))Q(0, y) \\ + \left(\frac{1}{abc}(ac + bc - ab - abc) + tG(x, y)\right) Q(0, 0) \end{aligned} \tag{2.7}$$

where $K(x, y) = 1 - tS(x, y)$. *It is called the* kernel *of the generating function.*

Note that the empty walk gets weight 1 instead of c, and more generally the initial vertex of any walk is not considered a visit to the corner. This convention is used so that the weight of the concatenation of a sequence of paths, each starting and ending at the origin, is the product of their weights.

Proof. The proof is essentially the same as that of [2, Theorem 6], except the weight at the origin is c here instead of ab. We present it here in full for completeness.

Define

$$Q^{\dagger}(x, y) = Q(x, y) + \frac{1}{a}Q(x, 0) + \frac{1}{b}Q(0, y) + \left(\frac{1}{c} - \frac{1}{a} - \frac{1}{b}\right)Q(0, 0). \tag{2.8}$$

This generating function counts walks which *do not* end on a boundary once, with the correct weight (in $Q(x, y)$). Walks which *do* end on a boundary are counted twice: once with the correct weight (in $Q(x, y)$), and once with an incorrect weight (underweighted by a factor of a, b or c depending on whether they end on the x-axis, the y-axis or the corner).

Let us now compute $Q^{\dagger}(x, y)$ in a different way. Walks ending on the boundary are counted with the correct weights by

$$Q(x, 0) + Q(0, y) - Q(0, 0). \tag{2.9}$$

Every non-empty walk can be constructed by appending a step to a shorter walk, making sure not to cross the boundaries. If we disregard the possibility of accruing an a, b or c weight with this step, then this gives

$$tS(x, y)Q(x, y) - tA(x, y)Q(x, 0) - tB(x, y)Q(0, y) + tG(x, y)Q(0, 0). \tag{2.10}$$

Walks which do not end on a boundary are counted correctly in (2.10); walks which do end on a boundary are underweighted in the same way as (2.8).

Adding (2.9) and (2.10) thus almost gives $Q^{\dagger}(x, y)$—the only thing missing is the underweighted term for the empty walk. This is $\frac{1}{c}$. Putting this all together, we find

$$\begin{aligned} &Q(x, y) + \frac{1}{a}Q(x, 0) + \frac{1}{b}Q(0, y) + \left(\frac{1}{c} - \frac{1}{a} - \frac{1}{b}\right)Q(0, 0) \\ &= \frac{1}{c} + Q(x, 0) + Q(0, y) - Q(0, 0) + tS(x, y)Q(x, y) - tA(x, y)Q(x, 0) - tB(x, y)Q(0, y) \\ &\qquad + tG(x, y)Q(0, 0). \end{aligned} \tag{2.11}$$

Rearranging gives the result. □

If we let $a = b = c = 1$, we get the functional equation for quarter-plane walks without interactions.

3 Main Results

For reverse Kreweras walks, the allowed steps are {SW, N, E}. So we have

$$S(x, y) = x + y + \overline{xy}, \tag{3.1}$$

and hence

$$A(x, y) = B(x, y) = G(x, y) = \overline{xy}. \tag{3.2}$$

The functional equation (2.7) becomes

$$\begin{aligned}(1 - t(x + y + \overline{xy}))\, Q(x, y) = \frac{1}{c} + \frac{1}{a}(a - 1 - ta\overline{xy})Q(x, 0) + \frac{1}{b}(b - 1 - tb\overline{xy})Q(0, y)\\ + \left(\frac{1}{abc}(ac + bc - ab - abc) + t\overline{xy}\right) Q(0, 0).\end{aligned} \tag{3.3}$$

Our aim is to solve this equation. From previous work [3, 7, 15], we know that when $a = b = c = 1$, the generating function of reverse Kreweras walks is algebraic. We can use both the algebraic kernel method and the obstinate kernel method to prove this. Here, we extend the result to a more general case:

Theorem 2. *For arbitrary a, b, c, the generation function $Q(x, y)$ of reverse Kreweras walks is algebraic.*

We will solve the generating function for reverse Kreweras walks using the algebraic kernel method in the next section. The final solution is an algebraic expression.

Corollary 3. *For Kreweras walks, the generating function $Q_{0,0} \equiv Q(0, 0)$ is algebraic for all (a, b, c).*

Proof. For any particular model, evaluating $Q(x, y)$ at $x = y = 0$ gives the generating function for walks which start and end at the origin. If we reverse a Kreweras walk starting and ending at the origin, we get a reverse Kreweras walk (see Fig. 2), and vice versa. Moreover, the number of visits to the x-axis, y-axis and origin is the same for the original walk and its reversal. So $Q(0, 0)$ is the same for these two models. By Theorem 2 this function is algebraic. □

This property does not hold for general $Q_{i,j}$. However, it will play an important role when solving Kreweras walks.

For Kreweras walks, the allowed steps are {NE, S, W}. So we have

$$S(x, y) = xy + \overline{x} + \overline{y} \tag{3.4}$$

and hence

$$A(x, y) = \overline{y}, \qquad B(x, y) = \overline{x}, \qquad G(x, y) = 0. \tag{3.5}$$

The functional equation reads

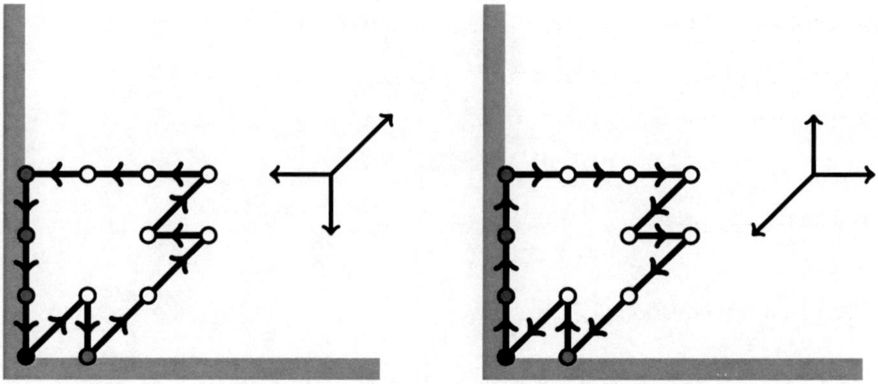

Fig. 2 Examples of reverse Kreweras and Kreweras walks ending at the origin.

$$(1 - t(xy + \overline{x} + \overline{y}))\, Q(x, y) = \frac{1}{c} + \frac{1}{a}(a - 1 - ta\overline{y})Q(x, 0) + \frac{1}{b}(b - 1 - tb\overline{x})Q(0, y) + \frac{1}{abc}(ac + bc - ab - abc)Q(0, 0). \quad (3.6)$$

Our main result regarding Kreweras walks is the following.

Theorem 4. *For arbitrary a, b, c, the generating function $Q(x, y)$ of Kreweras walks is D-finite.*

We will prove this theorem in Sect. 5. The general idea is the same as for reverse Kreweras walks. We will make use of Corollary 3 in this proof.

4 Reverse Kreweras Walks

In our situation, the basic idea of the algebraic kernel method will be to find a linear equation in $Q(x, 0)$, $Q_0^d(\overline{x})$ and $Q_{i,j}$ where the positive and negative powers of x can be separated. One can review the walk-through of algebraic kernel method by solving the Kreweras or reverse Kreweras walks with $a = b = c = 1$ [3, 7, 15]. For arbitrary a, b, c, we still follow the same ideas but the process is more involved.

The whole process of solving reverse Kreweras walks consists of three steps.

- First, we recall the symmetry group of the kernel and use it to obtain the full-orbit sum. We extract the $[x^> y^0]$ and $[x^< y^0]$ parts of the full-orbit sum to obtain two equations (Lemma 5).
- Second, we compute a half-orbit sum and again extract the $[y^0]$ terms. We use the equations obtained from the full-orbit sum to eliminate certain boundary terms, and then once again take the positive and negative parts with respect to x. This yields two equations which have kernel-like form (Lemma 8).

- Finally, we cancel the kernels of these two equations (see Lemma 9) and find a set of linear equations. These equations will provide us the final solution.

This process should be contrasted with the solution to reverse Kreweras walks without boundary weights [7]. There, the full-orbit sum yields no useful information, and instead the half-orbit sum alone is sufficient to solve the model.

4.1 The Symmetry Group

The *symmetry group* of a lattice path model is the set of birational transformations of (x, y) which leave the kernel $K(x, y)$ unchanged [7]. This group is generated by the pair

$$\phi : (x, y) \mapsto \left(\frac{B_{-1}(y)}{x B_1(y)}, y\right) \text{ and } \psi : (x, y) \mapsto \left(x, \frac{A_{-1}(x)}{y A_1(x)}\right). \tag{4.1}$$

For reverse Kreweras walks, these are

$$\phi : (x, y) \mapsto (\overline{xy}, y) \text{ and } \psi : (x, y) \mapsto (x, \overline{xy}). \tag{4.2}$$

The resulting group is isomorphic to D_3:

$$(x, y),\ (\overline{xy}, y),\ (y, \overline{xy}),\ (y, x),\ (\overline{xy}, x),\ (x, \overline{xy}). \tag{4.3}$$

4.2 Full-Orbit Sum

Applying all these symmetries to the functional equation (3.3) yields six equations. For simplicity, we write these equations in a matrix form:

$$K(x, y)\mathbf{Q} = \mathbf{MV} + \mathbf{C}. \tag{4.4}$$

Here $\mathbf{Q}$ is the column vector of all transformed $Q(x, y)$ and $\mathbf{V}$ is the transformed line boundary terms (note that the order of the terms in $\mathbf{V}$ is arbitrary)

$$\mathbf{Q} = \begin{pmatrix} Q(x, y) \\ Q(\overline{xy}, y) \\ Q(y, \overline{xy}) \\ Q(y, x) \\ Q(\overline{xy}, x) \\ Q(x, \overline{xy}) \end{pmatrix}, \qquad \mathbf{V} = \begin{pmatrix} Q(x, 0) \\ Q(0, y) \\ Q(\overline{xy}, 0) \\ Q(0, \overline{xy}) \\ Q(0, x) \\ Q(y, 0) \end{pmatrix}, \tag{4.5}$$

and **M** is the coefficient matrix

$$\mathbf{M} = \begin{pmatrix} A'(x,y) & B'(x,y) & 0 & 0 & 0 & 0 \\ 0 & B'(\overline{xy},y) & A'(\overline{xy},y) & 0 & 0 & 0 \\ 0 & 0 & 0 & B'(y,\overline{xy}) & 0 & A'(y,\overline{xy}) \\ 0 & 0 & 0 & 0 & B'(y,x) & A'(y,x) \\ 0 & 0 & A'(\overline{xy},x) & 0 & B'(\overline{xy},x) & 0 \\ A'(x,\overline{xy}) & 0 & 0 & B'(x,\overline{xy}) & 0 & 0 \end{pmatrix}. \tag{4.6}$$

Here $A'(x, y) = \frac{1}{a}(a - 1 - ta\overline{xy})$ and $B'(x, y) = \frac{1}{b}(b - 1 - tb\overline{xy})$. The column vector **C** contains the point boundary terms $Q(0, 0)$ and some known terms in $\mathbb{R}[x, \overline{x}, y, \overline{y}, t, t^{-1}]$:

$$\mathbf{C} = \frac{1}{abc}\begin{pmatrix} [-ab + ac + bc - abc + tabc\overline{xy}]\, Q(0,0) + ab \\ [-ab + ac + bc - abc + tabcx]\, Q(0,0) + ab \\ [-ab + ac + bc - abc + tabcx]\, Q(0,0) + ab \\ [-ab + ac + bc - abc + tabc\overline{xy}]\, Q(0,0) + ab \\ [-ab + ac + bc - abc + tabcy]\, Q(0,0) + ab \\ [-ab + ac + bc - abc + tabcy]\, Q(0,0) + ab \end{pmatrix}. \tag{4.7}$$

It is straightforward to show that the determinant of **M** equals zero. Thus, there exists a linear combination of these six equations that cancels all variables in **V**. We choose a vector **N** that spans the nullspace:

$$\mathbf{N} = \begin{pmatrix} -y(1 - b + tbx)(1 - a + tay) \\ \overline{x}(1 - a + tay)(tb + xy - bxy) \\ -\overline{x}(1 - b + tby)(ta + xy - axy) \\ y(1 - a + tax)(1 - b + tby) \\ -\overline{x}(1 - a + tax)(tb + xy - bxy) \\ -\overline{x}(1 - b + tbx)(ta + xy - axy) \end{pmatrix}^{\mathsf{T}} \tag{4.8}$$

Multiplying both sides of (4.4) by **N**, we have

$$K(x, y)\mathbf{NQ} = \mathbf{NMV} + \mathbf{NC} = \mathbf{NC} \tag{4.9}$$

since $\mathbf{NM} = \mathbf{0}$. The Eq. (4.9) is called the *full-orbit sum*.

We also have $\mathbf{NC} = \mathbf{0}$ in (4.9). This happens in all non-interacting algebraic models [7], including Kreweras and reverse Kreweras walks. In fact it only happens for algebraic models since for all walks with transcendental generating functions, $\mathbf{NC} \neq \mathbf{0}$ [15]. Note that the full-orbit sum is redundant in the unweighted case because $Q(x, y) = Q(y, x)$; the lack of such symmetry here means that we can extract useful information from (4.9).

Next, divide both sides of (4.9) by $K(x, y)$ and then extract the $[y^0]$ term. Some new boundary terms will appear when performing the extraction.

By geometric properties, we can eliminate some of them and find an equation of the form

$$[y^0]\mathbf{NQ} = 0 = \alpha + \alpha_{0,0}Q(0,0) + \alpha_{0,x}Q(0,x) + \alpha_{x,0}Q(x,0) + \alpha_0^{\mathrm{d}}Q_0^{\mathrm{d}}(\overline{x}) + \alpha_1^{\mathrm{d}}Q_1^{\mathrm{d}}(\overline{x}) \tag{4.10}$$

where the α coefficients are Laurent polynomials in x and t.

For an equation with this form, the terms with positive and negative powers of x can be naturally separated. We then take the $[x^>]$ and $[x^<]$ parts. After some simplifications by boundary conditions, we obtain the following.

Lemma 5. *The* $[x^>]$ *and* $[x^<]$ *parts of* (4.10) *can be written as*

$$[x^> y^0]\mathbf{NQ} = 0 = \beta + \beta_{0,0}Q(0,0) + \beta_{0,1}Q_{0,1} + \beta_{1,0}Q_{1,0} + \beta_{x,0}Q(x,0) + \beta_{0,x}Q(0,x) \tag{4.11}$$

and

$$[x^< y^0]\mathbf{NQ} = 0 = \gamma + \gamma_{0,0}Q(0,0) + \gamma_{0,1}Q_{0,1} + \gamma_{1,0}Q_{1,0} + \gamma_0^{\mathrm{d}}Q_0^{\mathrm{d}}(\overline{x}) + \gamma_1^{\mathrm{d}}Q_1^{\mathrm{d}}(\overline{x}), \tag{4.12}$$

where the β *and* γ *coefficients are Laurent polynomials in* x *and* t.

When $a = b$, (4.11) and (4.12) give $Q(x,0) = Q(0,x)$ and $0 = 0$. This step is omitted when solving the non-interacting or equally-interacting cases [2, 3] since the geometric symmetry is obvious.

4.3 Half-Orbit Sum

We have now obtained some geometric symmetries of reverse Kreweras walks from the full-orbit sum. However, we lost the information contained in the kernel, since the RHS of the full-orbit sum was 0. We will now "regain" this information by taking a half-orbit sum.

Recall the six equations obtained by applying the symmetry group. We rewrite the matrix equation (4.4) slightly differently:

$$K(x,y)\mathbf{Q} = \mathbf{M}_2\mathbf{V}_2 + \mathbf{C}_2 \tag{4.13}$$

Unlike the full-orbit sum, $Q(x,0)$ and $Q(0,x)$ are not included in the vector $\mathbf{V}_2$. Instead, $\mathbf{V}_2$ is

$$V_2 = \begin{pmatrix} Q(0,y) \\ Q(\overline{xy},0) \\ Q(0,\overline{xy}) \\ Q(y,0) \end{pmatrix} \tag{4.14}$$

The terms $Q(x,0)$ and $Q(0,x)$ are treated as constant and we put them in $\mathbf{C}_2$. $\mathbf{M}_2$ contains the corresponding coefficients of $\mathbf{V}_2$ in $\mathbf{M}$. So $\mathbf{M}_2$ is a 6×4 rectangular matrix. The dimension of the null-space is 2. Two orthogonal basis vectors span $\mathbf{M_2}^\mathsf{T}$. We may choose either of them in the following derivation. The vector we choose is

$$\mathbf{N}_2 = \begin{pmatrix} -x(1-b+tbx)(1-a+tay) \\ \overline{y}(1-a+tay)(tb+xy-bxy) \\ 0 \\ 0 \\ -\overline{y}(1-a+tax)(tb+xy-bxy) \\ 0 \end{pmatrix}^\mathsf{T}. \tag{4.15}$$

Notice that $\mathbf{N}_2$ has three zeros and three non-zeros, corresponding to a half-orbit sum.

Then, multiply $\mathbf{N}_2$ on (4.13) and divide by $K(x, y)$. We find

$$\mathbf{N}_2\mathbf{Q} = \frac{\mathbf{N}_2\mathbf{C}_2}{K(x, y)}, \tag{4.16}$$

since $\mathbf{N}_2\mathbf{M}_2 = 0$. The equation (4.16) is called the *half-orbit sum.*

As we did for the full-orbit sum case, we extract the $[y^0]$ of (4.16) to get an equation without the variable y. We can again use boundary conditions to simplify this equation, eventually finding an equation of the form

$$[y^0]\mathbf{N_2Q} = \delta + \delta_{0,0}Q(0,0) + \delta_{x,0}Q(x,0) + \delta_{0,x}Q(0,x) + \delta_0^\mathrm{d}Q_0^\mathrm{d}(\overline{x}) + \delta_1^\mathrm{d}Q_1^\mathrm{d}(\overline{x}) \tag{4.17}$$

where the δ coefficients are polynomials in x and t.

Now, consider the RHS. The RHS contains two parts, $\mathbf{N}_2\mathbf{C}_2$ and $1/K(x, y)$. First consider $\mathbf{N}_2\mathbf{C}_2$. It is a linear equation of $Q(x, 0)$, $Q(0, x)$ and $Q(0, 0)$ with coefficients in $\mathbb{R}(x, t)[y, \overline{y}]$. It can be treated as a function of y whose coefficients are linear terms of $Q(x, 0)$, $Q(0, x)$, $Q(0, 0)$ and $\mathbb{R}(x, t)$, and by examination we find $\mathbf{N}_2\mathbf{C}_2$ only contains y^{-1}, y^0, y^1 terms. So it can be written as

$$\mathbf{N}_2\mathbf{C}_2 = u_{-1}y^{-1} + u_0y^0 + u_1y^1. \tag{4.18}$$

Then consider $\frac{1}{K(x,y)}$. It can be expanded as a power series in t with coefficients that are polynomials in $x, \overline{x}, y, \overline{y}$. By collecting all terms with a given power of y, we can define $K_i = [y^i]\frac{1}{K(x,y)}$.

It follows that the $[y^0]$ term of (4.16) is

$$[y^0]\left(\frac{\mathbf{N}_2\mathbf{C}_2}{K(x, y)}\right) = u_{-1}K_1 + u_0K_0 + u_1K_{-1} \tag{4.19}$$

$$= \epsilon + \epsilon_{0,0}Q(0,0) + \epsilon_{x,0}Q(x,0) + \epsilon_{0,x}Q(0,x) \tag{4.20}$$

where the ϵ coefficients are Laurent polynomials in x, t and K_1, K_0, K_{-1}.

In the next section, we demonstrate how to compute K_i.

4.4 The Roots of the Kernel

The following lemma will be useful.

Lemma 6 (Lemma 7 in [7]). *For a quadratic equation* $K(x, y) = 1 - tS(x, y)$, *where* $S(x, y)$ *is defined as per* (2.6), *we have:* $1/K(x, y)$ *is a formal power series of* t *with polynomial coefficients in* $x, \overline{x}, y, \overline{y}$, *and*

$$\frac{1}{K(x,y)} = \frac{1}{\sqrt{\Delta(x)}}\left(\frac{1}{1-\overline{y}Y_0(x)} + \frac{1}{1-y/Y_1(x)} - 1\right) \tag{4.21}$$

where

$$\Delta(x) = (1 - tA_0(x))^2 - 4t^2 A_{-1}(x)A_1(x) \tag{4.22}$$

and

$$Y_0 = \frac{1 - tA_0(x) - \sqrt{\Delta(x)}}{2tA_1(x)}, \quad Y_1 = \frac{1 - tA_0(x) + \sqrt{\Delta(x)}}{2tA_1(x)}. \tag{4.23}$$

Thus

$$[y^j]\frac{1}{K(x,y)} = K_j = \begin{cases} \frac{Y_0(x)^{-j}}{\sqrt{\Delta(x)}} & \text{if } j \leq 0 \\ \frac{Y_1(x)^{-j}}{\sqrt{\Delta(x)}} & \text{if } j \geq 0. \end{cases} \tag{4.24}$$

Proof. Factoring $K(x, y)$, we have

$$K(x, y) = t\overline{y}A_1(x)(y - Y_0)(y - Y_1) = -tA_1(x)Y_1(1 - \overline{y}Y_0)(1 - y/Y_1) \tag{4.25}$$

Rearranging yields (4.21). Since $Y_0(x)$ has valuation 1 in t and $Y_1(x)$ has valuation -1 in t, the RHS of (4.24) is a formal power series in t. □

Now equating (4.17) and (4.20), using (4.11) and (4.12) to eliminate $Q(0, x)$ and $Q_1^{\mathrm{d}}(\overline{x})$, and substituting the explicit expressions for K_i, gives

$$\begin{aligned}\mu_{x,0}Q(x,0) + \nu_0^{\mathrm{d}}\sqrt{\Delta}Q_0^{\mathrm{d}}(\overline{x}) = (\mu + \nu\sqrt{\Delta}) + (\mu_{0,0} + \nu_{0,0}\sqrt{\Delta})Q(0,0)\\ + (\mu_{0,1} + \nu_{0,1}\sqrt{\Delta})Q_{0,1} + (\mu_{1,0} + \nu_{1,0}\sqrt{\Delta})Q_{1,0},\end{aligned} \tag{4.26}$$

where the μ and ν are all *polynomials*, and

$$\sqrt{\Delta} = \sqrt{t^2\overline{x}\left(x^3 - 4\right) - 2tx + 1}. \tag{4.27}$$

Unfortunately (4.26) still cannot be separated directly, as $\sqrt{\Delta}$ is a Laurent series of x and $Q_0^d(\overline{x})$ is a series of $\overline{x}$ with unknown coefficients. Then the $[x^>]$ of $\sqrt{\Delta(x)}Q_0^d(\overline{x})$ involves an infinite number of unknowns. We need to introduce the canonical factorisation of Δ, which provides a way to deal with $\sqrt{\Delta}$.

4.5 The Canonical Factorisation

Since the kernel $K(x, y)$ and its discriminant Δ do not depend on a, b, c, the following result is identical to the unweighted case. We quote it from [7].

Lemma 7 ([7, Section 4.4]). *We have*

$$\Delta(x) = (1 - tA_0(x))^2 - 4t^2 A_{-1}(x)A_1(x) \tag{4.28}$$

If $\Delta(x)$ has valuation $-\delta$ and degree d in x, then $\Delta(x) = 0$ has $\delta + d$ roots. Exactly δ of them (say $X_1, \dots, X_\delta$) are finite (actually vanish) when $t = 0$ and the remaining d roots ($X_{\delta+1}, \dots, X_{\delta+d}$) have negative valuation in t and thus diverge when $t = 0$.

Then Δ can be factored as:

$$\Delta(x) = \Delta_0 \Delta_-(\overline{x})\Delta_+(x) \tag{4.29}$$

with

$$\Delta_- \equiv \Delta_-(\overline{x}, t) = \Pi_{i=1}^{\delta}(1 - X_i/x) \tag{4.30}$$

$$\Delta_+ \equiv \Delta_+(\overline{x}, t) = \Pi_{i=1+\delta}^{\delta+d}(1 - x/X_i) \tag{4.31}$$

$$\Delta_0 \equiv \Delta_0(t) = (-1)^{\delta} \frac{[\overline{x}^{\delta}]\Delta(x)}{\Pi_{i=1}^{\delta} X_i} = (-1)^d [x^d]\Delta(x)\Pi_{i=\delta+1}^{\delta+d} X_i. \tag{4.32}$$

It can be seen that Δ_0, Δ_+, Δ_- are formal power series in t with constant term 1. We will use this property in following derivation. This factorisation is known as the *canonical factorisation*—see also Gessel's work [10] for more information.

In our case, Δ has one root which is a power series in t and two which are not:

$$X_1 = 4t^2 + 32t^5 + 448t^8 + O(t^{11}) \tag{4.33}$$

$$X_2 = t^{-1} + 2t^{1/2} - 2t^2 + 5t^{7/2} - 16t^5 + \tfrac{231}{4}t^{13/2} - 224t^8 + \tfrac{7293}{8}t^{19/2} + O(t^{11}) \tag{4.34}$$

$$X_3 = t^{-1} - 2t^{1/2} - 2t^2 - 5t^{7/2} - 16t^5 - \tfrac{231}{4}t^{13/2} - 224t^8 - \tfrac{7293}{8}t^{19/2} + O(t^{11}). \tag{4.35}$$

The canonical factorisation is then

$$\Delta_0 = t^2 X_2 X_3 \tag{4.36}$$

$$\Delta_+ = (1 - x/X_2)(1 - x/X_3) \tag{4.37}$$

$$\Delta_- = 1 - X_1\overline{x} \tag{4.38}$$

so that

$$\frac{1}{\sqrt{\Delta_+}} = 1 + tx + t^2x^2 + t^3x^3 + t^4(6x + x^4) + \mathrm{O}(t^5) \tag{4.39}$$

$$\sqrt{\Delta_0\Delta_-} = 1 - 2t^2\overline{x} - 4t^3 - 2t^4\overline{x}^2 + \mathrm{O}(t^5) \tag{4.40}$$

4.6 The Algebraic Solution of Reverse Kreweras Walks

We now take (4.26) and divide by $\sqrt{\Delta_+}$. Each μ term, divided by $\sqrt{\Delta_+}$, produces only non-negative powers of x. Each ν term, now multiplied by $\sqrt{\Delta_0\Delta_-}$ only, produces only powers of x greater than $-i$ for some small i. It is thus possible to compute the positive and negative parts with respect to x. Doing so, rearranging a bit and sending $x \mapsto \overline{x}$ in the second equation gives the following.

Lemma 8.

$$\sigma_{x,0}P_{x,0}Q(x,0) = \sigma + \sigma_{0,0}Q(0,0) + \sigma_{0,1}Q_{0,1} + \sigma_{1,0}Q_{1,0} \tag{4.41}$$

$$\tau_0^{\mathrm{d}}P_0^{\mathrm{d}}Q_0^{\mathrm{d}}(x) = \tau + \tau_{0,0}Q(0,0) + \tau_{0,1}Q_{0,1} + \tau_{1,0}Q_{1,0}, \tag{4.42}$$

where

$$P_{x,0} = (a^2t^2 + x - ax - atx^2 + a^2tx^2)(2abt^2 + 2x - ax - bx - atx^2 - btx^2 + 2abtx^2) \tag{4.43}$$

$$P_0^{\mathrm{d}} = (-at + a^2t + x - ax + a^2t^2x^2)(-bt + b^2t + x - bx + b^2t^2x^2) \tag{4.44}$$

and the τ and σ coefficients are algebraic functions of t and x (and (a, b, c)).

Now observe that (4.41) and (4.42) are in the form of kernel equations. The unknown functions $Q(x, 0)$ and $Q_0^{\mathrm{d}}(x)$ are formal power series of t with polynomial coefficients in x.

Lemma 9. *The kernel $P_{x,0}$ has two quadratic factors. The roots are*

$$x_{1,2} = \frac{\mp\sqrt{-4a^4t^3 + 4a^3t^3 + a^2 - 2a + 1} + a - 1}{2at(a-1)} \tag{4.45}$$

$$x_{3,4} = \frac{\mp\sqrt{(a+b-2)^2 - 8abt^3(2ab - a - b)} + a + b - 2}{2t(2ab - a - b)} \tag{4.46}$$

If $a > 1$ (resp. $0 < a < 1$) then x_1 (resp. x_2) is a formal power series of t and converges as $t \to 0$. Similarly, if $a + b > 2$ (resp. $0 < a + b < 2$) then x_3 (resp. x_4) is a formal power series of t and converges as $t \to 0$.

The kernel P_0^{d} also has two quadratic factors, leading to two pairs of roots

$$x_{5,6} = \frac{\mp\sqrt{-4a^4t^3 + 4a^3t^3 + a^2 - 2a + 1} + a - 1}{2a^2t^2} \tag{4.47}$$

$$x_{7,8} = \frac{\mp\sqrt{-4b^4t^3 + 4b^3t^3 + b^2 - 2b + 1} + b - 1}{2b^2t^2}. \tag{4.48}$$

If $a > 1$ (resp. $0 < a < 1$) then x_5 (resp. x_6) is a formal power series of t and converges as $t \to 0$. Similarly, if $b > 1$ (resp. $0 < b < 1$) then x_7 (resp. x_8) is a formal power series of t and converges as $t \to 0$.

Proof of Theorem 2. For any $a, b > 0$ with $a \neq 1, b \neq 1$ and $a + b \neq 2$, by substituting appropriate roots into (4.41) and (4.42), we get a set of four equations of the form

$$0 = \zeta^{(i)} + \zeta_{0,0}^{(i)} Q(0,0) + \zeta_{0,1}^{(i)} Q_{0,1} + \zeta_{1,0}^{(i)} Q_{1,0} \equiv H_i, \tag{4.49}$$

where i indicates which x_i root has been used in (4.41) or (4.42).

For simplicity assume $a, b > 1$, so that $i \in \{1, 3, 5, 7\}$. The final question, then, is which combination(s) of these equations, if any, give a solution. That is, for which i, j, k is the determinant

$$D_{i,j,k} = \zeta_{0,0}^{(i)}\zeta_{0,1}^{(j)}\zeta_{1,0}^{(k)} - \zeta_{0,0}^{(i)}\zeta_{1,0}^{(j)}\zeta_{0,1}^{(k)} - \zeta_{0,1}^{(i)}\zeta_{0,0}^{(j)}\zeta_{1,0}^{(k)} + \zeta_{0,1}^{(i)}\zeta_{1,0}^{(j)}\zeta_{0,0}^{(k)} + \zeta_{1,0}^{(i)}\zeta_{0,0}^{(j)}\zeta_{0,1}^{(k)} - \zeta_{1,0}^{(i)}\zeta_{0,1}^{(j)}\zeta_{0,0}^{(k)} \tag{4.50}$$

non-zero? Curiously, it appears that $D_{1,3,5} = 0$ while

$$D_{1,3,7} = \frac{16a^6b^4c^2(a-1)(a-2)(a-b)^2(ab-1)t^{10}}{a+b-2} + \mathrm{O}(t^{11}) \tag{4.51}$$

$$D_{1,5,7} = -8a^8b^3c^2(a-1)^2(b-1)^2(a-b)^2(ab-a+1)t^{10} + \mathrm{O}(t^{11}) \tag{4.52}$$

$$D_{3,5,7} = -\frac{16a^7b^4c^2(a-1)^2(b-1)^2(a-b)^2(ab-1)t^{10}}{a+b-2} + \mathrm{O}(t^{11}) \tag{4.53}$$

(These determinants depend on the choice of $\sigma_{x,0}$ and τ_0^{d} from (4.41) and (4.42). What is important is whether they are 0 or not.)

Choosing one of the valid combinations gives algebraic solutions to $Q(0,0)$, $Q_{0,1}$ and $Q_{1,0}$. By back-substitution into (4.41) we then get the solution to $Q(x,0)$, and then by symmetry $Q(0,x)$ (and hence $Q(0,y)$). Finally, the original equation (3.3) yields $Q(x,y)$. (To save space we will not explicitly write down the expression.) Because all involved terms are algebraic, $Q(x,y)$ is too, specifically over $\mathbb{C}[t, x, y, a, b, c]$.

If $a < 1$ or $b < 1$ then the roots which are not power series in t can be swapped out as required. In all cases the determinants which were non-zero remain so. □

4.7 Some Special Cases

There are some special cases where things appear to break down, namely $a = 1$, $b = 1$ and $a + b = 2$. When $a + b = 2$ the second factor in $P_{x,0}$ loses its x term. As a result, neither x_3 nor x_4 are valid power series roots of $P_{x,0}$, and we lose one of our equations in $Q(0, 0)$, $Q_{0,1}$ and $Q_{1,0}$. However, if $(a, b) \neq (1, 1)$, the remaining three equations are still valid, and the solution still emerges.

When $b = 1$, the system simplifies dramatically. One finds that the coefficient $\sigma_{1,0}$ of $Q_{1,0}$ vanishes. The two equations then obtained (either H_1 and H_3, or H_2 and H_4, depending on a) are linearly independent, and the solution follows. Naturally the $a = 1$ case is then just a reflection.

In addition to using the $[x^>]$ and $[x^<]$ parts of (4.26), the $[x^0]$ part also provides an equation (say, H_0) with unknowns $Q(0, 0)$, $Q_{1,0}$, $Q_{0,1}$. This is not in a kernel form (it has no dependence on x or y), but it can be combined with the other equations discussed above. It turns out that the equation sets formed by $\{H_0, H_1, H_7\}$ or $\{H_0, H_5, H_7\}$ have non-zero determinants, and can thus also be used to generate the solutions.

5 Kreweras Walks

We now turn our attention to Kreweras walks, and attempt to obtain the solution using a similar method. We still use the algebraic kernel method, but the process is different because some symmetries have changed.

- First, we recall the symmetry group of the kernel and use it to take the full-orbit sum. We then extract the $[x^> y^0]$ part (Lemma 10).
- Secondly, we take a half-orbit sum and again take the $[y^0]$ terms. We use the equations obtained from the full-orbit sum to eliminate certain boundary terms, and then once again take the positive and negative parts with respect to x. This yields two equations which have a kernel-like form (Lemma 11).
- Thirdly, we cancel the kernels of these two equations (see Lemma 12) and find a set of linear equations.
- Finally, we substitute the result of $Q(0, 0)$ into the equations obtained in the previous step and solve the linear equation set.

The first two steps are the same as for reverse Kreweras walks. The difference is that we are unable to directly solve the problem by the third step. So we need some extra work. We will show how to do this in the following section. Note that, as with reverse Kreweras walks, the unweighted case can be solved with the half-orbit sum alone [7].

For simplicity some notation from the previous sections will be reused—no definitions carry over unless otherwise indicated.

5.1 The Functional Equation and the Symmetry Group

For Kreweras walks, the allowed steps are {NE, S, W}. Recall the functional equation (3.6):

$$(1 - t(xy + \overline{x} + \overline{y}))\, Q(x, y) = \frac{1}{c} + \frac{1}{a}(a - 1 - ta\overline{y})Q(x, 0) + \frac{1}{b}(b - 1 - tb\overline{x})Q(0, y) + \frac{1}{abc}(ac + bc - ab - abc)Q(0, 0). \tag{5.1}$$

The symmetry group for Kreweras walks is the same as for reverse Kreweras walks.

5.2 Full-Orbit Sum

We again apply the symmetries to the functional equation and write them in a matrix form

$$K(x, y)\mathbf{Q} = \mathbf{MV} + \mathbf{C}. \tag{5.2}$$

Here $\mathbf{Q}$ and $\mathbf{V}$ are as per (4.5), while $\mathbf{C}$ is just the constant vector of

$$\frac{(-ab + ac + bc - abc)Q(0, 0)}{abc} + \frac{1}{c}, \tag{5.3}$$

and $\mathbf{M}$ is as per (4.6) but now with $A'(x, y) = \frac{1}{a}(a - 1 - ta\overline{y})$ and $B'(x, y) = \frac{1}{b}(b - 1 - tb\overline{x})$.

For Kreweras walks, we still have $\det(\mathbf{M}) = 0$, which means we can find a linear combination of these equations that cancels all variables in $\mathbf{V}$. The nullspace is spanned by the vector

$$\mathbf{N} = \begin{pmatrix} (at + x - ax)(bt + y - by)(1 - a + atxy)(1 - b + btxy) \\ -\overline{x}(at + x - ax)(bt + x - bx)(bt + y - by)(1 - a + atxy) \\ \overline{x}(at + x - ax)(bt + x - bx)(at + y - ay)(1 - b + btxy) \\ -(bt + x - bx)(at + y - ay)(1 - a + atxy)(1 - b + btxy) \\ \overline{y}(bt + x - bx)(at + y - ay)(bt + y - by)(1 - a + atxy) \\ -\overline{y}(at + x - ax)(at + y - ay)(bt + y - by)(1 - b + btxy) \end{pmatrix}^{\mathsf{T}} \tag{5.4}$$

As with reverse Kreweras, left-multiplying by $\mathbf{N}$ gives

$$K(x, y)\mathbf{NQ} = \mathbf{NC}. \tag{5.5}$$

But now the difference becomes apparent. For reverse Kreweras, we had $\mathbf{NC} = 0$ (also true for unweighted Kreweras walks), but here we have

$$\mathbf{NC} = \frac{t^3\overline{xy}(a-b)(x-y)\left(x^2y-1\right)\left(xy^2-1\right)\left[ab-(ab-ac-bc+abc)Q(0,0)\right]}{c}. \tag{5.6}$$

This is more complicated than reverse Kreweras (except, of course, when $a = b$), but it is still in a form that we can deal with. Analogously to the half-orbit sum in the reverse Kreweras case, we divide (5.5) by the kernel and take the $[y^0]$ part. After simplification by boundary relations, the left hand side is a linear equation of the form

$$[y^0]\mathbf{NQ} = \alpha + \alpha_{0,0}Q(0,0) + \alpha_{0,x}Q(0,x) + \alpha_{x,0}Q(x,0) + \alpha_0^{\mathrm{d}}Q_0^{\mathrm{d}}(\overline{x}) + \alpha_1^{\mathrm{d}}Q_1^{\mathrm{d}}(\overline{x}) \tag{5.7}$$

where the α coefficients are Laurent polynomials in x and t. For the RHS, since $\mathbf{NC}$ does not contain $Q(0, y)$, it can be regarded as a Laurent polynomial of y, namely

$$\mathbf{NC} = v_{-1}y^{-1} + v_0y^0 + v_1y^1 + v_2y^2 + v_3y^3. \tag{5.8}$$

As with reverse Kreweras walks we can also define $K_i = [y^i]\frac{1}{K(x,y)}$. It follows that

$$[y^0]\left(\frac{\mathbf{NC}}{K(x,y)}\right) = v_{-1}K_1 + v_0K_0 + v_1K_{-1} + v_2K_{-2} + v_3K_{-3} \tag{5.9}$$

$$= \eta + \eta_{0,0}Q(0,0) \tag{5.10}$$

where the η coefficients are Laurent polynomials in x, t and the K_i.

The main result obtained from the full-orbit sum is the following.

Lemma 10. *The* $[x^>]$ *part of* (5.7) *can be written as*

$$[x^>y^0]\mathbf{NQ} = \beta + \beta_{x,0}Q(x,0) + \beta_{0,x}Q(0,x) + \beta_{0,0}Q(0,0) + \beta_{1,0}Q_{1,0} + \beta_{2,0}Q_{2,0} + \beta_{3,0}Q_{3,0} \tag{5.11}$$

where the β *coefficients are Laurent polynomials in* t, x. *The* $[x^>]$ *part of* (5.10) *can be written as*

$$[x^>y^0]\left(\frac{\mathbf{NC}}{K(x,y)}\right) = \theta + \theta_{0,0}Q(0,0). \tag{5.12}$$

where θ *and* $\theta_{0,0}$ *are D-finite. That is, the vector space over* $\mathbb{C}[t,x,a,b,c]$ *spanned by all partial derivatives of* θ *(with respect to* t *and* x *only) has finite dimension, and likewise for* $\theta_{0,0}$.

Proof. The former follows immediately from the form of (5.7). For the latter, note that the η coefficients in (5.10) are algebraic functions of t and x, being rational functions of x, t and $\sqrt{\Delta}$, where

$$\Delta = (1 - t\overline{x})^2 - 4t^2x. \tag{5.13}$$

As the positive parts of algebraic functions, we can write θ and $\theta_{0,0}$ into series form and do know that they are D-finite (see, for example, [14, Theorem 3.7]). We have no reason to believe that they are algebraic. □

Equating (5.11) and (5.12), we obtain an equation relating $Q(x, 0)$, $Q(0, x)$ and several x-independent unknowns—the equivalent of (4.11). Unlike reverse Kreweras walks, it will not be necessary to also take the $[x^< y^0]$ part of the full-orbit sum.

5.3 Half-Orbit Sum

Following the same process as for reverse Kreweras walks, we now take the half-orbit sum. The process is nearly the same. We will also have equations similar to (4.13), and (4.14) in this case. The half-orbit sum is

$$K(x, y)\mathbf{Q} = \mathbf{M}_2\mathbf{V}_2 + \mathbf{C}_2 \tag{5.14}$$

where

$$\mathbf{V}_2 = \begin{pmatrix} Q(0, y) \\ Q(\overline{xy}, 0) \\ Q(0, \overline{xy}) \\ Q(y, 0) \end{pmatrix} \tag{5.15}$$

We have two vectors spanning the nullspace. The one we choose is

$$\mathbf{N}_2 = \begin{pmatrix} (at + x - ax)(1 - b + btxy) \\ -\overline{x}(at + x - ax)(bt + x - bx) \\ 0 \\ 0 \\ \overline{y}(bt + x - bx)(at + y - ay) \\ 0 \end{pmatrix} \tag{5.16}$$

and then

$$\mathbf{N}_2\mathbf{Q}_2 = \frac{\mathbf{N}_2\mathbf{C}_2}{K(x, y)}. \tag{5.17}$$

We take the $[y^0]$ term of (5.17) analogously to the reverse Kreweras case, and end up with the equivalents of (4.17) and (4.20):

$$[y^0]\mathbf{N_2Q} = \delta + \delta_{0,0}Q(0,0) + \delta_{x,0}Q(x,0) + \delta_{0,x}Q(0,x) + \delta_0^{\mathrm{d}}Q_0^{\mathrm{d}}(\overline{x}) \tag{5.18}$$

$$[y^0]\left(\frac{\mathbf{N_2C_2}}{K(x,y)}\right) = \epsilon + \epsilon_{0,0}Q(0,0) + \epsilon_{x,0}Q(x,0) + \epsilon_{0,x}Q(0,x). \tag{5.19}$$

As with reverse Kreweras walks, the δ coefficients are Laurent polynomials in t, x while the ϵ coefficients are Laurent polynomials in t, x and the K_i. (The absence of $Q_1^{\mathrm{d}}(\overline{x})$ in (5.18) is why we did not need to take the $[x^< y^0]$ part of the full-orbit sum.)

We then equate (5.18) and (5.19) and use (5.11) and (5.12) to eliminate $Q(0, x)$. This leads to an equation of the form

$$\begin{aligned}\mu_{x,0}Q(x,0) + \nu_0^{\mathrm{d}}\sqrt{\Delta}Q_0^{\mathrm{d}}(\overline{x}) &= (\hat{\mu} + \hat{\nu}\sqrt{\Delta}) + (\hat{\mu}_{0,0} + \hat{\nu}_{0,0}\sqrt{\Delta})Q(0,0)\\ &+(\mu_{1,0} + \nu_{1,0}\sqrt{\Delta})Q_{1,0} + (\mu_{2,0} + \nu_{2,0}\sqrt{\Delta})Q_{2,0} + (\mu_{3,0} + \nu_{3,0}\sqrt{\Delta})Q_{3,0},\end{aligned} \tag{5.20}$$

where the μ and ν coefficients are polynomials in t, x and the $\hat{\mu}$ and $\hat{\nu}$ coefficients are D-finite functions, being polynomials in t, x, and θ or $\theta_{0,0}$ respectively.

We next apply the canonical factorisation to Δ. There are three roots, of which two are Puiseux series which converge at 0 and one is not:

$$X_1 = t + 2t^{5/2} + 6t^4 + 21t^{11/2} + 80t^7 + \tfrac{1287}{4}t^{17/2} + 1344t^{10} + \mathrm{O}(t^{21/2}) \tag{5.21}$$

$$X_2 = t - 2t^{5/2} + 6t^4 - 21t^{11/2} + 80t^7 - \tfrac{1287}{4}t^{17/2} + 1344t^{10} + \mathrm{O}(t^{21/2}) \tag{5.22}$$

$$X_3 = \tfrac{1}{4}t^{-2} - 2t - 12t^4 - 160t^7 - 2688t^{10} + \mathrm{O}(t^{11}) \tag{5.23}$$

Following Lemma 7 we then define

$$\Delta_0 = 4t^2 X_3 \tag{5.24}$$

$$\Delta_+ = 1 - x/X_3 \tag{5.25}$$

$$\Delta_- = (1 - X_1\overline{x})(1 - X_2\overline{x}) \tag{5.26}$$

so that

$$\frac{1}{\sqrt{\Delta_+}} = 1 + 2t^2x + 6t^4x^2 + 16t^5x + \mathrm{O}(t^6) \tag{5.27}$$

$$\sqrt{\Delta_0\Delta_-} = 1 - t\overline{x} - 4t^3 - 2t^4\overline{x} - 2t^5\overline{x}^2 + \mathrm{O}(t^6). \tag{5.28}$$

As with reverse Kreweras walks, we next divide (5.20) by $\sqrt{\Delta_+}$ and extract the $[x^>]$ and $[x^<]$ parts. Unfortunately this has the side effect of introducing $Q_{4,0}$ as another unknown.

Lemma 11

$$\sigma_{x,0} P_{x,0} Q(x,0) = \hat{\sigma} + \hat{\sigma}_{0,0} Q(0,0) + \sigma_{1,0} Q_{1,0} + \sigma_{2,0} Q_{2,0} + \sigma_{3,0} Q_{3,0} + \sigma_{4,0} Q_{4,0} \tag{5.29}$$

$$\tau_0^{\mathrm{d}} P_0^{\mathrm{d}} Q_0^{\mathrm{d}}(x) = \hat{\tau} + \hat{\tau}_{0,0} Q(0,0) + \tau_{1,0} Q_{1,0} + \tau_{2,0} Q_{2,0} + \tau_{3,0} Q_{3,0} + \tau_{4,0} Q_{4,0} \tag{5.30}$$

where

$$P_{x,0} = (ta - ta^2 - x + ax - t^2a^2x^2)(tb - tb^2 - x + bx - t^2b^2x^2) \times (ta + tb - 2tab - 2x + ax + bx - 2t^2abx^2) \tag{5.31}$$

$$P_0^{\mathrm{d}} = (t^2a^2 + x - ax - tax^2 + ta^2x^2)(t^2b^2 + x - bx - tbx^2 + tb^2x^2), \tag{5.32}$$

the σ and τ coefficients are algebraic functions of t, x (in fact $\sigma_{4,0}$ and $\tau_{4,0}$ are polynomials), and the $\hat{\sigma}$ and $\hat{\tau}$ coefficients are D-finite functions of t with non-negative powers of x.

We then have the analogue of Lemma 9.

Lemma 12. *The roots of $P_{x,0}$ are*

$$x_{1,2} = \frac{\mp\sqrt{-4a^4t^3 + 4a^3t^3 + a^2 - 2a + 1} + a - 1}{2a^2t^2} \tag{5.33}$$

$$x_{3,4} = \frac{\mp\sqrt{-4b^4t^3 + 4b^3t^3 + b^2 - 2b + 1} + b - 1}{2b^2t^2} \tag{5.34}$$

$$x_{5,6} = \frac{\mp\sqrt{(a+b-2)^2 - 8abt^2(2abt - at - bt)} + a + b - 2}{4abt^2} \tag{5.35}$$

and the roots of P_0^{d} are

$$x_{7,8} = \frac{\mp\sqrt{-4a^4t^3 + 4a^3t^3 + a^2 - 2a + 1} + a - 1}{2at\,(a-1)} \tag{5.36}$$

$$x_{9,10} = \frac{\mp\sqrt{-4b^4t^3 + 4b^3t^3 + b^2 - 2b + 1} + b - 1}{2bt\,(b-1)}. \tag{5.37}$$

If $a > 1$ (resp. $0 < a < 1$) then x_1 and x_7 (resp. x_2 and x_8) are power series in t; if $b > 1$ (resp. $0 < b < 1$) then x_3 and x_9 (resp. x_4 and x_{10}) are power series in t; and if $a + b > 2$ (resp. $0 < a + b < 2$) then x_5 (resp. x_6) is a power series in t.

For any $a, b > 0$ with $a \neq 1, b \neq 1$ and $a + b \neq 2$, we thus get a set of five equations of the form

$$0 = \zeta^{(i)} + \zeta^{(i)}_{0,0} Q(0,0) + \zeta^{(i)}_{1,0} Q_{1,0} + \zeta^{(i)}_{2,0} Q_{2,0} + \zeta^{(i)}_{3,0} Q_{3,0} + \zeta^{(i)}_{4,0} Q_{4,0} \equiv H_i, \quad (5.38)$$

where i indicates which x_i root has been used in (5.29) or (5.30).

Unfortunately, the determinant of the corresponding 5×5 matrix of coefficients appears to be 0. We can even include a sixth equation (say, H_0), by taking the $[x^0]$ part of (5.20), but there is still no set of five linearly independent equations.

5.4 *Incorporating the Solution to Reverse Kreweras Walks*

By Corollary 3, the generating function $Q(0, 0)$ is the same for Kreweras and reverse Kreweras walks. Since we have solved reverse Kreweras walks, $Q(0, 0)$ is actually known. We substitute it into (5.38) and then we have six equations with four unknowns, and hence 15 different equation sets.

Proof of Theorem 4. Substituting and expanding reveals that (at least) 10 of the equations sets yield a non-zero determinant. For example,

$$D_{1,3,5,7} = \frac{16a^{12}b^5c^4(a-1)^3(a-2)(b-1)^5(a-b)^4(ab-a-b)(2ab-a-b)^5t^{26}}{(a+b-2)^4} + O(t^{27}) \quad (5.39)$$

(As with reverse Kreweras walks, this determinant depends on the the choice of $\sigma_{x,0}$ and τ_0^{d}.)

Any set of equations with a non-zero determinant then leads to a solution for $Q_{1,0}$, $Q_{2,0}$, $Q_{3,0}$ and $Q_{4,0}$. Back-substitution yields $Q(x, 0)$, and by reflective symmetry, $Q(0, y)$. The original functional equation then gives $Q(x, y)$. If $a < 1$ or $b < 1$ then the roots which are not power series in t can be replaced as required, and the resulting determinants remain non-zero.

Since $\zeta^{(i)}_{1,0}$, $\zeta^{(i)}_{2,0}$, $\zeta^{(i)}_{3,0}$ and $\zeta^{(i)}_{4,0}$ are algebraic while $\zeta^{(i)} + \zeta^{(i)}_{0,0} Q(0, 0)$ is D-finite, the resulting solutions to $Q_{1,0}$, $Q_{2,0}$, $Q_{3,0}$ and $Q_{4,0}$ are D-finite. By back-substituting we find that the same is true for $Q(x, 0)$, $Q(0, y)$ and $Q(x, y)$. More specifically, the vector space over $\mathbb{C}[t, x, y, a, b, c]$ spanned by all partial derivatives of $Q(x, y)$ (with respect to t, x, y only) is finite-dimensional. We do not (yet) have a proof that any of these functions are *not* algebraic. □

5.5 *Special Cases*

The special values of a, b work in much the same way as for reverse Kreweras. When $a + b = 2$ the third factor in $P_{x,0}$ loses its x term, and as a result $x_{5,6}$ are not longer valid kernel roots. However, there remain three combinations which do not use these roots, and the solution can be obtained from any of those.

If $b = 1$ then $\sigma_{3,0} = \sigma_{4,0} = \tau_{3,0} = \tau_{4,0} = 0$, and we thus only need two independent equations to solve for the two unknowns $Q_{1,0}$ and $Q_{2,0}$. Using H_1 and H_7 (or H_2 and H_8) suffices. Naturally $a = 1$ is just a reflection of the $b = 1$ case.

When $a = b$, note that (5.6) vanishes, and hence $\eta = \eta_{0,0} = \theta = \theta_{0,0} = 0$. As was the case for reverse Kreweras walks, all coefficients in (5.20) are then algebraic, and it follows that the resulting solution is too. Note that this has already been established for the $c = a^2 = b^2$ case in [2].

6 Discussion

6.1 A Generic Equation

Observe that in solving both Kreweras and reverse Kreweras walks, an equation of the form

$$\Lambda_F \Lambda_{FG} F(x) = \Lambda_{FG} \Lambda_G \sqrt{\Delta} G(\overline{x}) + \sqrt{\Delta} A(x, U_1, \ldots, U_k) + B(x, U_1, \ldots, U_k) \tag{6.1}$$

arose (namely, (4.26) and (5.20)). Here,

- $F(x)$ and $G(\overline{x})$ are series in t whose coefficients are polynomials in x and $\overline{x}$ respectively;
- Λ_F, Λ_{FG} and Λ_G are products of distinct irreducible quadratic polynomials in x;
- Δ is an irreducible cubic polynomial in x (possibly divided by some power of x);
- the U_i are boundary terms of F and G, i.e. $[x^j]F(x)$ or $[x^j]G(\overline{x})$ for some j; and
- $A(\cdots)$ and $B(\cdots)$ are linear combinations of unknowns $U_1, \ldots, U_k$ with coefficients that are polynomial in x, t in (4.26) and D-finite in (5.20).

For both models we then obtained two kernel-like equations by taking either the positive or negative part of (6.1) with respect to x. For reverse Kreweras this led to exactly three linearly independent equations in the U_i, and since $k = 3$, this gave the solution. For Kreweras we obtained four independent equations, but while we initially had $k = 4$, a fifth boundary term emerged when taking the positive and negative parts. It was only because we already knew the (algebraic) solution to one of the U_i that we were able to solve the system.

A generic equation of this form can be contrasted with the results of [6]. Suppose we have a single polynomial equation

$$P(Q(x,t), Q_1, \ldots, Q_k, t, x) = 0, \tag{6.2}$$

and seek to determine all unknowns $Q(x, t)$ and Q_k. It has been proved that under some mild assumptions regarding the form of the equation, if one can find k distinct formal series $X_i(t)$ $(i = 1, \ldots, k)$ that make

$$\frac{\partial P}{\partial x_0} P(x_0, Q_1, \ldots, Q_k, t, X_i)\bigg|_{x_0=Q(X_i,t)} = 0, \tag{6.3}$$

then these $k+1$ series are algebraic. (For a kernel-like equation obtained from a lattice walk problem, Banderier and Flajolet [1] point out that the determinant arising from the kernel method forms a Vandermonde determinant.)

The conditions under which (6.1) is solvable seem to be more complicated. The positive and negative parts can be written as

$$\Lambda_F \Lambda_{FG} \frac{F(x)}{\sqrt{\Delta_+}} = G_+(x, U_1, \ldots, U_k, \ldots, U_\ell) + A_+(x, U_1, \ldots, U_k) + \frac{B(x, U_1, \ldots, U_k)}{\sqrt{\Delta_+}} \tag{6.4}$$

$$\begin{aligned} 0 = \Lambda_{FG} \Lambda_G \sqrt{\Delta_0 \Delta_-} G(\overline{x}) - G_+(x, U_1, \ldots, U_k, \ldots, U_\ell) \\ + \sqrt{\Delta_0 \Delta_-} A(x, U_1, \ldots, U_k) - A_+(x, U_1, \ldots, U_k) \end{aligned} \tag{6.5}$$

where

- $\Delta = \Delta_0 \Delta_+ \Delta_-$ is the canonical factorisation (see Lemma 7);
- $G_+(\cdots)$ is the non-negative part of $\Lambda_{FG} \Lambda_G \sqrt{\Delta_0 \Delta_-} G(\overline{x})$—this can be expressed as a linear combination of boundary terms of $G(\overline{x})$ with polynomial coefficients; and
- $A_+(\cdots)$ is the non-negative part of $\sqrt{\Delta_0 \Delta_-} A(x)$—this is a linear combination of the unknowns U_i with coefficients of the same type as in $A(\cdots)$ (i.e. polynomials or D-finite functions).

Generically, this system is solvable if exactly ℓ linearly independent equations can be obtained by substituting values of x which cancel Λ_F, Λ_{FG} or Λ_G, while leaving $F(x)$ or $G(\overline{x})$ as power series in t. We have yet to understand exactly what constitutes a set of sufficient conditions to guarantee this.

6.2 *Comparison with Other Models*

The models with boundary weights which have been previously solved can be summarised as follows. We use the numbering scheme of [2, 3]. Recall that in [2, 19] the convention $c = ab$ is used, but generalising c does not affect the following. All except the last were solved in [2]; Model 22 was solved in [19].

- Models 1–4 (horizontally and vertically symmetric, e.g. {N, S, E, W});
- Models 5–16 (horizontally symmetric, e.g. {NE, NW, S}) for $b = 1$ only;
- Model 17 {N, W, SE};
- Model 18 {N, S, E, W, NW, SE} for $a = b$ only;
- Models 19 {NE, W, S} and 20 {N, E, SW} for $a = b$ only;
- Model 22 {E, W, NW, SE}.

We may now ask for which other models an equation of the form (6.1) arises. In particular, since the separation of (6.1) into positive and negative parts depends on the canonical factorisation of $\sqrt{\Delta}$, the coefficient of $G(\overline{x})$ has to be a polynomial or a polynomial multiplying $\sqrt{\Delta}$. A factor like $C(x,t) + D(x,t)\sqrt{\Delta}$ (with $C, D \in \mathbb{R}[x,t]$) will prevent us from finding a solution. This is because in an equation of the form

$$\frac{\Lambda_F \Lambda_{FG}}{\Delta_+} F(x) = \left(\frac{C}{\sqrt{\Delta_+}} + D\sqrt{\Delta_0\Delta_-}\right) G(\overline{x}) + \sqrt{\Delta_0\Delta_-} A(x, U_1, \ldots, U_k) + \frac{B(x, U_1, \ldots, U_k)}{\sqrt{\Delta_+}}, \tag{6.6}$$

the first term in the RHS is still a product of a formal series in x and a series in $\overline{x}$.

For all the cases fully solved in [2] (Models 1–4; the $b = 1$ cases of Models 5–16; Model 17; and the $a = b$ cases of Models 18–20), it is straightforward to show that $\det(\mathbf{M})$ (recall (4.4)) is zero.[2] This is also true for Model 22, which was solved in [19]. For all other cases except Kreweras and reverse Kreweras (Models 19 and 20, which we have solved here) and Gessel paths (Model 23), this determinant is nonzero. That is, it is not possible to eliminate all terms of the form $Q(\bullet, 0)$ and $Q(0, \bullet)$ in a full-orbit sum. Without loss of generality, assume that $Q(x, 0)$ is the term left on the RHS of the full-orbit sum. The full-orbit sum can be written as

$$\mathbf{NQ} = \frac{P_1(Q(x,0), x, y, t)}{K(x,y)}, \tag{6.7}$$

where P_1 is a linear function of $Q(x, 0)$ with polynomial coefficients in x, y, t. When taking the $[y^0]$ part of this equation, the LHS has a similar form to (4.17):

$$[y^0]\mathbf{NQ} = \phi + \phi_{0,0} Q(0,0) + \phi_{x,0} Q(x,0) + F(x) + G(\overline{x}). \tag{6.8}$$

Here, $F(x)$ is some (possibly empty) linear combination of unknown series in x (except $Q(x, 0)$), and $G(\overline{x})$ is likewise a linear combination of unknown series in $\overline{x}$. All the coefficients on the LHS are polynomials and do not contain $\sqrt{\Delta}$. But on the RHS, we must apply Lemma 6 when finding the $[y^0]$ part, and therefore we have

$$[y^0]\frac{P_1(Q(x,0), x, y, t)}{K(x,y)} = \frac{1}{\sqrt{\Delta}} P_2(Q(x,0), x, t). \tag{6.9}$$

The coefficient of $Q(x, 0)$ contains $\sqrt{\Delta}$. Combining the LHS and RHS together, we will get an equation with polynomial coefficients on $F(x)$ and $G(\overline{x})$, but a coefficient of the form $C + D\sqrt{\Delta}$ on $Q(x, 0)$.

We now turn to the cases which have been solved using the kernel method, both in this paper and in [2, 19]. For the algebraic cases that we know currently (reverse Kreweras; the $a = b$ case of Kreweras; and the unweighted versions of double Kreweras and Gessel), the full-orbit sum is 0. The half-orbit sum leads to an equation in the form of (6.1) (except Gessel walks, which are solved by different ideas in [4]

[2] Note that in [2] the authors were primarily concerned with solving $Q(0, 0)$.

and [12]) where the coefficient of $Q_0^{\mathrm{d}}(\overline{x})$ is $\sqrt{\Delta}$ times a polynomial. We can walk through the algebraic kernel method and solve the problem.

For the solved D-finite and D-algebraic cases (i.e. all the other cases solved in [2, 19], as well as Kreweras solved here), the RHS of the full-orbit sum does not vanish. But it does not contain $Q(x, 0)$. The orbit sum can be written as

$$\mathbf{NQ} = \frac{P_1(x, y)}{K(x, y)}. \tag{6.10}$$

After taking the $[y^0]$ part of it, we have

$$[y^0]\frac{P_1(x, y)}{K(x, y)} = \frac{1}{\sqrt{\Delta}}P_2(x, t). \tag{6.11}$$

The LHS still has the form of (6.8) (in fact the $Q(x, 0)$ can now be included in $F(x)$). We do have $\sqrt{\Delta}$ in the RHS but the coefficients of unknown series in x or $\overline{x}$ all come from the LHS. So the coefficient of those unknowns are polynomials. We cannot divide the $\sqrt{\Delta_+}$ on both sides since the LHS contains unknown series in $\overline{x}$. However we can directly take the $[x^>]$ or $[x^<]$ part of this equation. This leads to the D-finite terms on the RHS.

After the separation one attempts to cancel the kernels and (hopefully) solve a set of linear equations. Let us temporarily ignore the previous discussion in Sect. 6.1, and assume that we can find a solution. This means if A, B and the Λ terms are algebraic in (6.1), the final solution is algebraic. If they are D-finite, the final solution is D-algebraic. In order to distinguish the D-finite and D-algebraic cases, we need check the coefficients of the unknowns U_i in this equation. By the process of the algebraic kernel method, we find all coefficients of U_i are polynomials of x, t except possibly $Q(0, 0)$. This is because $Q(0, 0)$ is the only one of the U_i which can appear in both sides of the full-orbit sum. So, we conclude, if the coefficient of $Q(0, 0)$ on the RHS of the full-orbit sum is 0, the solution is D-finite. Otherwise, the solution is D-algebraic.

This implies that the values of a, b, c can affect the nature of the generating function. For example, consider simple random walks with step set {N, E, S, W}. The functional equation reads

$$\begin{aligned} K(x, y)Q(x, y) = \frac{1}{c} + \frac{1}{a}(a - 1 - ta\overline{y})Q(x, 0) + \frac{1}{b}(b - 1 - tb\overline{x})Q(0, y) \\ + \frac{1}{abc}(ac + bc - ab - abc)Q(0, 0) \end{aligned} \tag{6.12}$$

Using the algebraic kernel method, we find the RHS of the full-orbit sum to be

$$-\frac{\left(x^2 - 1\right)\left(y^2 - 1\right)z^2[(ab - ac - bc + abc)Q(0, 0) - ab]}{cxyK(x, y)}. \tag{6.13}$$

Then, taking the $[y^0]$ part of both sides, the coefficient of $Q(0, 0)$ in the RHS is

$$z^2(ab - ac - bc + abc)[y^0]\left(\frac{(x^2-1)(y^2-1)}{cxyK(x, y)}\right). \tag{6.14}$$

Notice that $[y^0]\left(\frac{(x^2-1)(y^2-1)}{cxyK(x,y)}\right)$ is D-finite (since we take the positive part in x of a rational function). If $ab - ac - bc + abc = 0$, the solution is D-finite. Otherwise, as discussed above, it is D-algebraic. This is an improvement on the result obtained with the obstinate kernel method in [2, Theorem 1].

We have just discussed the general properties of walks with and without boundary interactions. These simple inferences hold for most models we have studied, except Kreweras walks. Kreweras walks appear to be rather special. Kreweras walks without boundary interactions are algebraic. When interactions are introduced, the full-orbit sum (see (5.11)) behaves more like a D-algebraic model, as the RHS is not 0. Unlike the D-algebraic cases, however, since $Q(x, 0)$ and $Q(0, x)$ both appear in the LHS of full-orbit sum, we still need the half-orbit sum to cancel $Q(0, x)$. The half-orbit sum then behaves more like an algebraic model. For example, observe that (5.20) is in the same form as (4.26), and the coefficient of $Q_0^{\mathrm{d}}(\overline{x})$ is a polynomial times $\sqrt{\Delta}$. However, since some terms in (5.20) are D-finite and not algebraic, the final result ends up being D-finite.

However, in another sense, Kreweras walks are different to the other D-algebraic models. The coefficient of $Q(0, 0)$ in the full-orbit sum (5.6) does not affect the properties of the solution—it will always be D-finite.

We summarise this discussion in Table 1. We know, for non-interacting models, the algebraic and D-algebraic cases can be distinguished just by examining the full-orbit sum. However, for interacting models, this rule is broken by Kreweras walks. Is it still the case that when the full-orbit sum equals 0, the solution must be algebraic? And is there an easy way to determine when the solution is D-finite or D-algebraic?

It is of great interest to know whether other recently discovered techniques which do not use the kernel method—for example, the elliptic functions of Kurkova and Raschel [11]—may be applied to the problems of quarter-plane lattice paths with interacting boundaries.

6.3 Singularities and Asymptotic Behaviour

We have so far only been concerned with the solutions to the generating functions for Kreweras and reverse Kreweras walks, without considering any physical or probabilistic meaning for the weights a, b, c. However, as mentioned in Sect. 1, there are close connections between lattice paths and models in statistical physics and probability theory. It is thus also worth investigating the asymptotic behaviour of the coefficients of the generating functions. This is ongoing work.

Table 1 A summary of the various model types. Here C and D are in $\mathbb{R}[x, t]$.

Model	Full-orbit sum	Coefficient of $Q_0^d(\overline{x})$ in $[y^0]$ of full-orbit sum	Coefficient of $Q_0^d(\overline{x})$ in $[y^0]$ of half-orbit sum	Generating function $Q(x, y, t)$
D-finite model without interactions	Linear equation of $Q(0, 0)$ with coefficients in $\mathbb{R}[x, y, t]$	Polynomial	Not needed	D-finite
Algebraic model without interactions	0	0	$D\sqrt{\Delta}$	Algebraic
D-algebraic model with interactions	Linear equation of $Q(0, 0)$ with coefficients in $\mathbb{R}[x, y, t]$	Polynomial	Not needed	D-algebraic (may become D-finite for certain a, b, c)
Reverse Kreweras walks with interactions	0	0	$D\sqrt{\Delta}$	Algebraic
Unsolved models with interactions	Linear equation of $Q(x, 0)$, $Q(0, 0)$ with coefficients in $\mathbb{R}[x, y, t]$	$C + D\sqrt{\Delta}$	Not needed	Unclear
Kreweras walks with interactions	Linear equation of $Q(0, 0)$ with coefficients in $\mathbb{R}[x, y, t]$	Polynomial	$D\sqrt{\Delta}$	D-finite for all a, b, c

7 Conclusion

We have introduced a model of quarter-plane lattice paths with weights a, b, c associated with visits to the two boundaries and the origin. We have then used the algebraic kernel method to solve the particular cases of reverse Kreweras walks and Kreweras walks.

The final solution of reverse Kreweras walks is algebraic and we have directly solved it for all (a, b, c) (without explicitly writing out the expression, due to its complexity). However, we were unable to solve Kreweras walks directly, and instead needed to make use of the fact that when only considering walks which start and end at the origin, Kreweras and reverse Kreweras generating functions are identical. The overall solution for Kreweras walks is D-finite.

There remain many other quarter-plane lattice path models which have not been solved, and it is unclear which methods may be useful when boundary weights are included. How the values of a, b, c affect the asymptotic behaviour of these models is also an open question.

Acknowledgements NRB and ALO gratefully acknowledge support from the Australian Research Council.

References

1. C. Banderier and P. Flajolet. "Basic analytic combinatorics of directed lattice paths". Theoret. Comput. Sci. 281.1-2 (2002), pp. 37–80. https://doi.org/10.1016/S0304-3975(02)00007-5.
2. N. R. Beaton, A. L. Owczarek, and A. Rechnitzer. "Exact solution of some quarter plane walks with interacting boundaries". Electr. J. Combin. 26.3 (2019), P3.53. https://doi.org/10.37236/8024.
3. M. Bousquet-Mélou. "Walks in the quarter plane: Kreweras' algebraic model". Ann. Appl. Probab. 15.2 (2005), pp. 1451–1491. https://doi.org/10.1214/105051605000000052.
4. M. Bousquet-Mélou. "An elementary solution of Gessel's walks in the quadrant". Adv. Math. 303 (2016), pp. 1171–1189. https://doi.org/10.1016/j.aim.2016.08.038.
5. M. Bousquet-Mélou. "Square lattice walks avoiding a quadrant". J. Combin. Theory Ser. A 144 (2016), pp. 37–79. https://doi.org/10.1016/j.jcta.2016.06.010.
6. M. Bousquet-Mélou and A. Jehanne. "Polynomial equations with one catalytic vari- able, algebraic series and map enumeration". J. Combin. Theory Ser. B 96.5 (2006), pp. 623–672. https://doi.org/10.1016/j.jctb.2005.12.003.
7. M. Bousquet-Mélou and M. Mishna. "Walks with small steps in the quarter plane". Algorithmic Probability and Combinatorics. Contemp. Math. 520. Amer. Math. Soc., Providence, RI, 2010, pp. 1–39. https://doi.org/10.1090/conm/520/10252.
8. T. Dreyfus, C. Hardouin, J. Roques, and M. F. Singer. "On the nature of the generating series of walks in the quarter plane". Invent. math. 213.1 (2018), pp. 139–203. https://doi.org/10.1007/s00222-018-0787-z.
9. G. Fayolle, R. Iasnogorodski, and V. Malyshev. Random Walks in the Quarter-Plane. Applications of Mathematics 40. Springer-Verlag, Berlin, 1999. https://doi.org/10.1007/978-3-642-60001-2.
10. I. M. Gessel. "A factorization for formal Laurent series and lattice path enumeration". J. Combin. Theory Ser. A 28.3 (1980), pp. 321–337. https://doi.org/10.1016/0097-3165(80)90074-6.
11. I. Kurkova and K. Raschel. "New Steps in Walks with Small Steps in the Quarter Plane: Series Expressions for the Generating Functions". Ann. Comb. 19.3 (2015), pp. 461–511. https://doi.org/10.1007/s00026-015-0279-4.
12. I. Kurkova and K. Raschel. "Explicit expression for the generating function counting Gessel's walks". Adv. in Appl. Math. 47.3 (2011), pp. 414–433. https://doi.org/10.1016/j.aam.2010.11.004.
13. I. Kurkova and K. Raschel. "On the functions counting walks with small steps in the quarter plane". Publ. Math. Inst. Hautes Études Sci. 116 (2012), pp. 69–114. https://doi.org/10.1007/s10240-012-0045-7.
14. L. Lipshitz. "D-finite power series". J. Algebra 122.2 (1989), pp. 353–373. https://doi.org/10.1016/0021-8693(89)90222-6.
15. M. Mishna. "Classifying lattice walks restricted to the quarter plane". J. Combin. Theory Ser. A 116.2 (2009), pp. 460–477. https://doi.org/10.1016/j.jcta.2008.06.011.
16. M. Mishna and A. Rechnitzer. "Two non-holonomic lattice walks in the quarter plane". Theoret. Comput. Sci. 410.38-40 (2009), pp. 3616–3630. https://doi.org/10.1016/j.tcs.2009.04.008.
17. H. Prodinger. "The kernel method: a collection of examples". Sém. Lothar. Combin. 50 (2003/04), Art. B50f, 19 pp. https://www.mat.univie.ac.at/~slc/wpapers/s50proding.html.
18. K. Raschel and A. Trotignon. "On Walks Avoiding a Quadrant". Electr. J. Combin. 26 (2019), P3.31. https://doi.org/10.37236/8019.
19. R. Tabbara, A. L. Owczarek, and A. Rechnitzer. "An exact solution of two friendly interacting directed walks near a sticky wall". J. Phys. A 47.1 (2014), pp. 015202, 34. https://doi.org/10.1088/1751-8113/47/1/015202.

Infinite Product Formulae for Generating Functions for Sequences of Squares

Christian Krattenthaler, Mircea Merca, and Cristian-Silviu Radu

Abstract We state and prove product formulae for several generating functions for sequences $(a_n)_{n\geq 0}$ that are defined by the property that $Pa_n + b^2$ is a square, where P and b are given integers. In particular, we prove corresponding conjectures of the second author. We show that, by means of the Jacobi triple product identity, all these generating functions can be reduced to a linear combination of theta function products. The proof of our formulae then consists in simplifying these linear combinations of theta products into single products. We do this in two ways: (1) by using modular function theory, and (2) by applying the Weierstraß addition formula for theta products.

Keywords Jacobi triple product identity · Jacobi theta functions · Weierstraß Relation for theta products · Modular functions

2010 Mathematics Subject Classification Primary 11B65 · Secondary 05A30 · 11F27 · 33D05

C. Krattenthaler and C.-S. Radu—Supported by the Austrian Science Foundation FWF (grant S50-N15) in the framework of the Special Research Program "Algorithmic and Enumerative Combinatorics".

C. Krattenthaler (✉)
Universität Wien, Fakultät für Mathematik, Oskar-Morgenstern-Platz 1, 1090 Vienna, Austria
e-mail: christian.krattenthaler@univie.ac.at
URL: https://www.mat.univie.ac.at/~kratt

M. Merca
Department of Mathematics, University of Craiova, Craiova, Romania;
Academy of Romanian Scientists, Bucharest, Romania

C.-S. Radu
Research Institute for Symbolic Computation, J. Kepler University Linz, 4040 Linz, Austria
URL: https://risc.jku.at/m/cristian-silviu-radu/

A. Bostan and K. Raschel (eds.), *Transcendence in Algebra, Combinatorics, Geometry and Number Theory*, Springer Proceedings in Mathematics & Statistics 373,
https://doi.org/10.1007/978-3-030-84304-5_8

1 Introduction

Relations between infinite q-series and infinite q-products have an honorable history starting with Euler and Gauß and first studied systematically by Jacobi. Many identities of the form

$$\text{infinite } q\text{-series} = \text{infinite } q\text{-product}$$

arise in number theory, analysis, combinatorics, the theory of integer partitions, representation theory of Lie algebras, vertex operator algebras, knot theory, and statistical mechanics. Playing with series and products, Euler discovered the pentagonal number theorem (cf. e.g. [2, Ex. 2.18]),

$$\sum_{n=-\infty}^{\infty} (-1)^n q^{n(3n-1)/2} = (q;q)_\infty, \tag{1.1}$$

which is the very first theorem that is of this type. Here, and in the following, q is a complex number with[1] $|q| < 1$, and the symbol $(a;q)_\infty$ denotes the infinite product

$$(a;q)_\infty := \prod_{i=0}^{\infty}(1 - aq^i).$$

Moreover, we use the short notation

$$(a_1, a_2, \ldots, a_m; q) := \prod_{j=1}^{m}(a_j;q)_\infty.$$

An alternative way to state Euler's identity is as follows:

Let $(a_n)_{n\geq 0} = (0, 1, 2, 5, 7, 12, 15, 22, 26, 35, 40, \ldots)$ *be the sequence of non-negative integers* m *such that* $24m + 1$ *is a square. Then*

$$\sum_{n=0}^{\infty}(-1)^{\lfloor (n+1)/2 \rfloor} q^{a_n} = (q;q)_\infty.$$

In [3], the second author studied—among others—series-product identities of this type and listed several empirically found such formulae, see [3, Ids. 5.1–5.6, Ids. 6.1–6.14]. A typical example (cf. [3, Id. 6.1] and Theorem 7 below) is the following statement:

Let $(a_n)_{n\geq 0} = (0, 4, 7, 10, 21, 26, 33, 59, 61, 95, 108, \ldots)$ *be the sequence of non-negative integers* m *such that* $240m + 1$ *is a square. Then*

---

[1] Alternatively, all definitions and identities may be understood in the sense of formal power series in q.

$$\sum_{n=0}^{\infty}(-1)^{\lfloor (n+2)/4\rfloor}q^{a_n}=\frac{(q,q^7,q^8;q^8)_\infty\,(q^6,q^{10};q^{16})_\infty}{(q,q^4;q^5)_\infty}. \tag{1.2}$$

This paper started by the observation that, by the use of Jacobi's triple product identity, the proofs of all these conjectured formulae can be reduced to the verification of certain identities between (specialised) Jacobi theta functions. Now, it is a folklore fact that, since these theta functions are modular functions (for certain subgroups of $SL_2(\mathbb{Z})$), such identities are routinely verifiable. This is what we did, first. Subsequently, however, we wanted to have a conceptual understanding of the established formulae. Moreover, we were interested in whether there is more than just these, first empirically found, formulae from [3]. Indeed, further computer experiments led us to discover many more such formulae, but still of sporadic nature. Nevertheless, altogether, they helped us to come up with two parametric theorems that subsume several of the earlier empirically found formulae under one roof.

Obviously, parametric theorems cannot be routinely verified anymore. Rather we found that Weierstraß' three-term relation between theta products is the "magic" identity that is behind *all* of our formulae. More precisely, (aside from four formulae that are direct consequences of the Jacobi triple product identity) for one class of formulae the verification consists in a single application of Weierstraß' relation which reduces the sum of two theta products to a single theta product, while for a second class of formulae the verification requires a double application of Weierstraß' relation—a first application to reduce the sum of four theta products to the sum of two theta products, and then another application to reduce the latter to a single theta product.

Our parametric theorems belong both to the first class. At this point in time, we are not able to offer a conceptual understanding for our proofs for the second class of formulae in the sense that we were not able to embed these into parametric families of formulae. Rather, we performed some computer assisted searches for more formulae in the second class that remained unsuccessful, suggesting that there may not be parametric families in the second class but only these sporadic formulae.

Our paper is structured as follows. In the next section we collect the empirically found identities from [3]. (We have slightly altered the order in which they are presented in order to provide a more systematic listing.) They become theorems by the proofs in Sects. 7 and 9. Section 3 lists the additional identities that we found by our computer experiments subsequent to the publication of [3].

The first step in all our proofs is to apply Jacobi's triple product identity that converts the left-hand sides of our identities into a linear combination of products of theta functions. We recall Jacobi's identity in Sect. 4, where we also derive two (well-known) corollaries that we need in some of the proofs.

There follows another section of preparatory character, namely Sect. 5, in which facts from the theory of modular functions are collected that are relevant in our context. Based on them, we explain in Sect. 6 how to routinely verify identities between theta products. In the subsequent section, Sect. 7, we apply this methodology to prove all the theorems from Sects. 2 and 3 (with the exception of the theorems that are special cases of parametric families). We provide full details for Theorem 1, while for all other theorems proofs are given in a stenographic fashion since the pattern is always the same.

Section 8 is again preparatory. There, we recall the earlier mentioned Weierstraß' addition formula, and two of its corollaries that we shall use particularly frequently. Then, in Sect. 9, we present our proofs of all theorems from Sects. 2 and 3 (again with the exception of the theorems that are special cases of parametric families) by the use of the Weierstraß relation. Again, we give full details for the proof of Theorem 1, while for all other theorems proofs are presented only in an abridged fashion.

The contents of Sect. 10 are two parametric families of formulae of the type as in (1.2); one of them consists in Theorem 34 and Corollary 35 (although these contain different statements, their proofs are the same), and the other in Theorem 36 and Corollary 37 (again, these contain different statements, but their proofs are the same). Finally, in Theorem 38 we unify the statements of Theorems 31 and 32, and we provide a uniform proof.

We close the article by mentioning some consequences and open problems in Sect. 11.

2 Generating Functions for Sequences of Squares

Here, we list the empirically found formulae for generating functions for sequences of squares from [3]. They become theorems by the proofs in Sects. 7 and 9, respectively.

Theorem 1 (conjectured in [3, Id. 5.1]) *Let $(a_n)_{n\ge0}$ be the sequence of non-negative integers m such that $840m+361$ is a square. Then*

$$\sum_{n=0}^{\infty}(-1)^{t(n)}q^{a_n}=\frac{(q,q^6,q^7;q^7)_\infty}{(q,q^4;q^5)_\infty}, \tag{2.1}$$

where

$$t(n)=\begin{cases}0, & \text{if } n\equiv 0,1,3,5,10,12,14,15 \pmod{16},\\ 1, & \text{otherwise.}\end{cases}$$

Theorem 2 (conjectured in [3, Id. 5.2]) *Let $(a_n)_{n\ge0}$ be the sequence of non-negative integers m such that $840m+529$ is a square. Then*

$$\sum_{n=0}^{\infty}(-1)^{t(n)}q^{a_n}=\frac{(q,q^6,q^7;q^7)_\infty}{(q^2,q^3;q^5)_\infty}, \tag{2.2}$$

where

$$t(n)=\begin{cases}0, & \text{if } n\equiv 0,2,3,6,9,12,13,15 \pmod{16},\\ 1, & \text{otherwise.}\end{cases}$$

Theorem 3 (conjectured in [3, Id. 5.3, corrected]) *Let* $(a_n)_{n\geq 0}$ *be the sequence of non-negative integers m such that* $840m + 121$ *is a square. Then*

$$\sum_{n=0}^{\infty}(-1)^{\lfloor (n+4)/8\rfloor} q^{a_n} = \frac{(q^2, q^5, q^7; q^7)_\infty}{(q, q^4; q^5)_\infty}. \tag{2.3}$$

Theorem 4 (conjectured in [3, Id. 5.4]) *Let* $(a_n)_{n\geq 0}$ *be the sequence of non-negative integers m such that* $840m + 289$ *is a square. Then*

$$\sum_{n=0}^{\infty}(-1)^{t(n)} q^{a_n} = \frac{(q^2, q^5, q^7; q^7)_\infty}{(q^2, q^3; q^5)_\infty}, \tag{2.4}$$

where

$$t(n) = \begin{cases} 0, & \text{if } n \equiv 0, 1, 3, 5, 10, 12, 14, 15 \pmod{16}, \\ 1, & \text{otherwise.} \end{cases}$$

Theorem 5 (conjectured in [3, Id. 5.5, corrected]) *Let* $(a_n)_{n\geq 0}$ *be the sequence of non-negative integers m such that* $840m + 1$ *is a square. Then*

$$\sum_{n=0}^{\infty}(-1)^{\lfloor (n+4)/8\rfloor} q^{a_n} = \frac{(q^3, q^4, q^7; q^7)_\infty}{(q, q^4; q^5)_\infty}. \tag{2.5}$$

Theorem 6 (conjectured in [3, Id. 5.6]) *Let* $(a_n)_{n\geq 0}$ *be the sequence of non-negative integers m such that* $840m + 169$ *is a square. Then*

$$\sum_{n=0}^{\infty}(-1)^{t(n)} q^{a_n} = \frac{(q^3, q^4, q^7; q^7)_\infty}{(q^2, q^3; q^5)_\infty}, \tag{2.6}$$

where

$$t(n) = \begin{cases} 0, & \text{if } n \equiv 0, 1, 2, 4, 11, 13, 14, 15 \pmod{16}, \\ 1, & \text{otherwise.} \end{cases}$$

Theorem 7 (conjectured in [3, Id. 6.1, corrected]) *Let* $(a_n)_{n\geq 0}$ *be the sequence of non-negative integers m such that* $240m + 1$ *is a square. Then*

$$\sum_{n=0}^{\infty}(-1)^{\lfloor (n+2)/4\rfloor} q^{a_n} = \frac{(q, q^7, q^8; q^8)_\infty \, (q^6, q^{10}; q^{16})_\infty}{(q, q^4; q^5)_\infty}. \tag{2.7}$$

Theorem 8 (conjectured in [3, Id. 6.2]) *Let* $(a_n)_{n\geq 0}$ *be the sequence of non-negative integers m such that* $240m + 49$ *is a square. Then*

$$\sum_{n=0}^{\infty}(-1)^{\lfloor 5n/4\rfloor}q^{a_n}=\frac{(q,q^7,q^8;q^8)_\infty\,(q^6,q^{10};q^{16})_\infty}{(q^2,q^3;q^5)_\infty}. \tag{2.8}$$

Theorem 9 (conjectured in [3, Id. 6.5, corrected]) *Let $(a_n)_{n\geq 0}$ be the sequence of non-negative integers m such that $240m+121$ is a square. Then*

$$\sum_{n=0}^{\infty}(-1)^{\lfloor (n+2)/4\rfloor}q^{a_n}=\frac{(q^3,q^5,q^8;q^8)_\infty\,(q^2,q^{14};q^{16})_\infty}{(q,q^4;q^5)_\infty}. \tag{2.9}$$

Theorem 10 (conjectured in [3, Id. 6.6]) *Let $(a_n)_{n\geq 0}$ be the sequence of non-negative integers m such that $240m+169$ is a square. Then*

$$\sum_{n=0}^{\infty}(-1)^{\lfloor 5n/4\rfloor}q^{a_n}=\frac{(q^3,q^5,q^8;q^8)_\infty\,(q^2,q^{14};q^{16})_\infty}{(q^2,q^3;q^5)_\infty}. \tag{2.10}$$

Theorem 11 (conjectured in [3, Id. 6.3, corrected]) *Let $(a_n)_{n\geq 0}$ be the sequence of non-negative integers m such that $15m+1$ is a square. Then*

$$\sum_{n=0}^{\infty}(-1)^{\lfloor (n+2)/4\rfloor}q^{a_n}=\frac{(q^2,q^6,q^8;q^8)_\infty\,(q^4,q^{12};q^{16})_\infty}{(q,q^4;q^5)_\infty}. \tag{2.11}$$

Theorem 12 (conjectured in [3, Id. 6.4, corrected]) *Let $(a_n)_{n\geq 0}$ be the sequence of non-negative integers m such that $15m+4$ is a square. Then*

$$\sum_{n=0}^{\infty}(-1)^{\lfloor (n+2)/4\rfloor}q^{a_n}=\frac{(q^2,q^6,q^8;q^8)_\infty\,(q^4,q^{12};q^{16})_\infty}{(q^2,q^3;q^5)_\infty}. \tag{2.12}$$

Theorem 13 (conjectured in [3, Id. 6.7]) *We have*

$$\sum_{n=-\infty}^{\infty}q^{n(5n+1)}=\frac{(q,q^9,q^{10};q^{10})_\infty\,(q^8,q^{12};q^{20})_\infty}{(q,q^4;q^5)_\infty}. \tag{2.13}$$

Theorem 14 (conjectured in [3, Id. 6.8]) *We have*

$$\sum_{n=0}^{\infty}\left(q^{n(n+1)}-q^{5n(n+1)+1}\right)=\frac{(q,q^9,q^{10};q^{10})_\infty\,(q^8,q^{12};q^{20})_\infty}{(q^2,q^3;q^5)_\infty}. \tag{2.14}$$

Theorem 15 (conjectured in [3, Id. 6.9]) *We have*

$$1+\sum_{n=1}^{\infty}\left(q^{n^2}+q^{5n^2}\right)=\frac{(q^2,q^8,q^{10};q^{10})_\infty\,(q^6,q^{14};q^{20})_\infty}{(q,q^4;q^5)_\infty}. \tag{2.15}$$

Theorem 16 (conjectured in [3, Id. 6.10]) *We have*

$$\sum_{n=-\infty}^{\infty} q^{n(5n+2)} = \frac{(q^2, q^8, q^{10}; q^{10})_\infty \, (q^6, q^{14}; q^{20})_\infty}{(q^2, q^3; q^5)_\infty}. \tag{2.16}$$

Theorem 17 (conjectured in [3, Id. 6.11]) *We have*

$$\sum_{n=0}^{\infty} \left(q^{n(n+1)} + q^{5n(n+1)+1}\right) = \frac{(q^3, q^7, q^{10}; q^{10})_\infty \, (q^4, q^{16}; q^{20})_\infty}{(q, q^4; q^5)_\infty}. \tag{2.17}$$

Theorem 18 (conjectured in [3, Id. 6.12]) *We have*

$$\sum_{n=-\infty}^{\infty} q^{n(5n+3)} = \frac{(q^3, q^7, q^{10}; q^{10})_\infty \, (q^4, q^{16}; q^{20})_\infty}{(q^2, q^3; q^5)_\infty}. \tag{2.18}$$

Theorem 19 (conjectured in [3, Id. 6.13]) *We have*

$$\sum_{n=-\infty}^{\infty} q^{n(5n+4)} = \frac{(q^4, q^6, q^{10}; q^{10})_\infty \, (q^2, q^{18}; q^{20})_\infty}{(q, q^4; q^5)_\infty}. \tag{2.19}$$

Theorem 20 (conjectured in [3, Id. 6.14]) *We have*

$$\sum_{n=1}^{\infty} \left(q^{n^2-1} - q^{5n^2-1}\right) = \frac{(q^4, q^6, q^{10}; q^{10})_\infty \, (q^2, q^{18}; q^{20})_\infty}{(q^2, q^3; q^5)_\infty}. \tag{2.20}$$

3 More Generating Functions for Sequences of Squares

In this section, we collect the additional formulae for generating functions for sequences of squares that we found when we started our work on this kind of identities. Also these become theorems by the proofs in Sects. 7 and 9, respectively.

Theorem 21 *Let* $(a_n)_{n\ge 0}$ *be the sequence of non-negative integers* m *such that* $120m + 49$ *is a square. Then*

$$\sum_{n=0}^{\infty} (-1)^{\lfloor 5n/4 \rfloor} q^{a_n} = \frac{(q, q^4, q^5; q^5)_\infty}{(q^2, q^3; q^5)_\infty}. \tag{3.1}$$

Theorem 22 *Let* $(a_n)_{n\ge 0}$ *be the sequence of non-negative integers* m *such that* $120m + 1$ *is a square. Then*

$$\sum_{n=0}^{\infty}(-1)^{\lfloor (n+2)/4\rfloor}q^{a_n}=\frac{(q^2,q^3,q^5;q^5)_\infty}{(q,q^4;q^5)_\infty}. \tag{3.2}$$

Theorem 23 *Let $(a_n)_{n\geq 0}$ be the sequence of non-negative integers m such that $168m+121$ is a square. Then*

$$\sum_{n=0}^{\infty}(-1)^{\lfloor 5n/4\rfloor}q^{a_n}=\frac{(q,q^6,q^7;q^7)_\infty}{(q^3,q^4;q^7)_\infty}. \tag{3.3}$$

Theorem 24 *Let $(a_n)_{n\geq 0}$ be the sequence of non-negative integers m such that $168m+1$ is a square. Then*

$$\sum_{n=0}^{\infty}(-1)^{\lfloor (n+2)/4\rfloor}q^{a_n}=\frac{(q^2,q^5,q^7;q^7)_\infty}{(q,q^6;q^7)_\infty}. \tag{3.4}$$

Theorem 25 *Let $(a_n)_{n\geq 0}$ be the sequence of non-negative integers m such that $168m+25$ is a square. Then*

$$\sum_{n=0}^{\infty}(-1)^{\lfloor (n+2)/4\rfloor}q^{a_n}=\frac{(q^3,q^4,q^7;q^7)_\infty}{(q^2,q^5;q^7)_\infty}. \tag{3.5}$$

Theorem 26 *Let $(a_n)_{n\geq 0}$ be the sequence of non-negative integers m such that $48m+1$ is a square. Then*

$$\sum_{n=0}^{\infty}(-1)^{\lfloor (n+2)/4\rfloor}q^{a_n}=\frac{(q^2,q^6,q^8;q^8)_\infty}{(q,q^7;q^8)_\infty}. \tag{3.6}$$

Theorem 27 *Let $(a_n)_{n\geq 0}$ be the sequence of non-negative integers m such that $48m+25$ is a square. Then*

$$\sum_{n=0}^{\infty}(-1)^{\lfloor 5n/4\rfloor}q^{a_n}=\frac{(q^2,q^6,q^8;q^8)_\infty}{(q^3,q^5;q^8)_\infty}. \tag{3.7}$$

Theorem 28 *Let $(a_n)_{n\geq 0}$ be the sequence of non-negative integers m such that $21m+1$ is a square. Then*

$$\sum_{n=0}^{\infty}(-1)^{\lfloor (n+2)/4\rfloor}q^{a_n}=\frac{(q,q^6,q^7;q^7)_\infty(q^5,q^9;q^{14})_\infty}{(q,q^3;q^4)_\infty}. \tag{3.8}$$

Theorem 29 *Let $(a_n)_{n\geq 0}$ be the sequence of non-negative integers m such that $21m+4$ is a square. Then*

$$\sum_{n=0}^{\infty}(-1)^{\lfloor (n+2)/4\rfloor} q^{a_n} = \frac{(q^2, q^5, q^7; q^7)_\infty (q^3, q^{11}; q^{14})_\infty}{(q, q^3; q^4)_\infty}. \tag{3.9}$$

Theorem 30 *Let* $(a_n)_{n\ge 0}$ *be the sequence of non-negative integers* m *such that* $21m + 16$ *is a square. Then*

$$\sum_{n=0}^{\infty}(-1)^{\lfloor 5n/4\rfloor} q^{a_n} = \frac{(q^3, q^4, q^7; q^7)_\infty (q, q^{13}; q^{14})_\infty}{(q, q^3; q^4)_\infty}. \tag{3.10}$$

Theorem 31 *Let* $(a_n)_{n\ge 0}$ *be the sequence of non-negative integers* m *such that* $16m + 1$ *is a square. Then*

$$\sum_{n=0}^{\infty} q^{a_n} = \frac{(q, q^7, q^8; q^8)_\infty (q^6, q^{10}; q^{16})_\infty}{(q, q^3; q^4)_\infty}. \tag{3.11}$$

Theorem 32 *Let* $(a_n)_{n\ge 0}$ *be the sequence of non-negative integers* m *such that* $16m + 9$ *is a square. Then*

$$\sum_{n=0}^{\infty} q^{a_n} = \frac{(q^3, q^5, q^8; q^8)_\infty (q^2, q^{14}; q^{16})_\infty}{(q, q^3; q^4)_\infty}. \tag{3.12}$$

4 The Jacobi Triple Product Identity and Two of Its Consequences

The purpose of this section is, first of all, to state the Jacobi triple product identity which is ubiquitously used in the proofs of our theorems, and, second, to make two corollaries explicit that we need in the proofs of four of our theorems.

The Jacobi triple product identity says that (cf. [2, Eq. (1.6.1)])

$$\sum_{n=-\infty}^{\infty} (-1)^n q^{\binom{n}{2}} z^n = (q, z, q/z; q)_\infty. \tag{4.1}$$

Letting $q \to q^2$ and setting $z = -q$ in (4.1), we obtain

$$\sum_{n=-\infty}^{\infty} q^{n^2} = (q^2, -q, -q; q^2)_\infty.$$

Since we have

$$\sum_{n=1}^{\infty} q^{n^2} = \sum_{n=-\infty}^{-1} q^{n^2},$$

the previous identity implies

$$\sum_{n=1}^{\infty} q^{n^2} = \frac{1}{2}\big((q^2, -q, -q; q^2)_\infty - 1\big). \tag{4.2}$$

We shall need this identity in the proofs of Theorems 15 and 20.

On the other hand, letting $q \to q^2$ and setting $z = -q^2$ in (4.1), we obtain

$$\sum_{n=-\infty}^{\infty} q^{n^2+n} = 2\,(q^2, -q^2, -q^2; q^2)_\infty.$$

Since we have

$$\sum_{n=0}^{\infty} q^{n^2+n} = \sum_{n=-\infty}^{-1} q^{n^2+n},$$

the previous identity implies

$$\sum_{n=0}^{\infty} q^{n^2+n} = (q^2, -q^2, -q^2; q^2)_\infty. \tag{4.3}$$

This identity will be used in the proofs of Theorems 14 and 17.

5 Background on Modular Functions

In this section, we give a brief introduction to modular functions, tailored to our purposes. Let $\mathbb{H} := \{x \in \mathbb{C} : \mathrm{Im}(x) > 0\}$ denote the upper half plane. Roughly speaking, modular functions are (certain) meromorphic functions on $\mathbb{H}$ that are invariant under the action of a subgroup Γ of $\mathrm{SL}_2(\mathbb{Z})$. In our setting, $\Gamma = \Gamma_1(N)$ and $N > 3$, where

$$\Gamma_1(N) := \left\{ \begin{pmatrix} a & b \\ c & d \end{pmatrix} \in \mathrm{SL}_2(\mathbb{Z}) : a, d \equiv 1 \ (\mathrm{mod}\ N) \text{ and } c \equiv 0 \ (\mathrm{mod}\ N) \right\}.$$

The crucial fact on which we base the "methodology" explained in the next section, is the following proposition (cf. [4, Prop. 4.12]).

Proposition 33 *Let f be a non-constant meromorphic function on a compact Riemann surface X. Then*

$$\sum_{p\in X} \operatorname{ord}_p(f) = 0,$$

where $\operatorname{ord}_p(f)$ *is the order of the Laurent series expansion of* f *about the point* p.

For $\gamma = \left(\begin{smallmatrix} a & b \\ c & d \end{smallmatrix}\right)$ and $\tau \in \mathbb{H}$ we define $\gamma\tau := \frac{a\tau+b}{c\tau+d}$. For a subgroup Γ of $\mathrm{SL}_2(\mathbb{Z})$, we let

$$A(\Gamma) := \{f \text{ meromorphic on } \mathbb{H} \ : \ f(\gamma\tau) = f(\tau)\text{for all } \gamma \in \Gamma \text{ and } \tau \in \mathbb{H}\}.$$

A function $f \in A(\Gamma_1(N))$ can be viewed as a meromorphic function $\tilde{f}$ on the Riemann surface $\mathbb{H}/\Gamma_1(N) := \{[\tau] : \tau \in \mathbb{H}\}$. Here, $[\tau] := \{\gamma\tau : \gamma \in \Gamma_1(N)\}$ denotes the orbit of $\tau \in \mathbb{H}$ under the action of the group $\Gamma_1(N)$, and $\tilde{f}([\tau]) := f(\tau)$. One equips $\mathbb{H}/\Gamma_1(N)$ with the quotient topology so that it is a topological space. To make $\mathbb{H}/\Gamma_1(N)$ into a Riemann surface, we follow the recipe in [5, Sec. 1.8] (cf. also [6]). Then $\tilde{f}$ becomes a meromorphic function on $\mathbb{H}/\Gamma_1(N)$. However, as we shall explain in Sect. 6, we want to use Proposition 33, which is an assertion on meromorphic functions on a *compact* Riemann surface. In other words, we would need $\mathbb{H}/\Gamma_1(N)$ to be compact, which it is not. So we need to add some points in order to make it compact. In order to achieve this, we define $\mathbb{H}^* := \mathbb{H} \cup \mathbb{Q} \cup \{i\infty\}$. We extend the action of $\gamma = \left(\begin{smallmatrix} a & b \\ c & d \end{smallmatrix}\right)$ to $\mathbb{H}^*$ as follows. For $\frac{s}{t} \in \mathbb{Q}$, we let $\gamma\frac{s}{t} := \frac{a(s/t)+b}{c(s/t)+d}$ if $c(s/t) + d \neq 0$ and $\gamma\frac{s}{t} := i\infty$ otherwise. Moreover, we let $\gamma(i\infty) = \frac{a}{c}$ if $c \neq 0$ and $\gamma(i\infty) := i\infty$ otherwise. Then $\mathbb{H}^*/\Gamma_1(N) := \{[\tau] : \tau \in \mathbb{H}^*\}$ is a compact topological space when equipped with the quotient topology. More precisely, the topology of $\mathbb{H}^*$ is generated by the topology of $\mathbb{H}$ (which inherits the standard topology in $\mathbb{C}$) and the sets $U_M := \{x \in \mathbb{H} : \mathrm{Im}(x) > M\} \cup \{i\infty\}$, and the sets γU_M for all $\gamma \in \mathrm{SL}_2(\mathbb{Z})$, where M is a positive real number.

In particular $X_1(N) := \mathbb{H}^*/\Gamma_1(N)$ is made into a Riemann surface following the recipe given in [5, Sec. 1.8] (cf. also [6]). Again, this creates a problem: for a function $f \in A(\Gamma_1(N)$, the corresponding function $\tilde{f}$ on the quotient space is not necessarily meromorphic on $X_1(N)$. The problematic points are the cusps $\{[s/t] : s/t \in \mathbb{Q} \cup \{i\infty\}\}$. Given a function $f \in A(\Gamma_1(N))$, in order for $\tilde{f}$ to be meromorphic on $X_1(N)$, we need that, for each reduced fraction $\frac{a}{c} \in \mathbb{Q}$, f can be expressed as a Laurent series with finite principal part in powers of $e^{\frac{2\pi i\gamma^{-1}\tau}{h_c}}$. Here, $h_c := N/\gcd(c, N)$ is called the width of the cusps a/c. When these additional conditions are satisfied we say that f is a modular function for the group $\Gamma_1(N)$, and we write $f \in M(\Gamma_1(N))$. In particular, $f \in M(\Gamma_1(N))$ implies that $\tilde{f}$ is meromorphic on $X_1(N)$. Furthermore, $\operatorname{ord}_{[a/c]}\tilde{f}$ equals the order of the Laurent series of f in powers of $e^{\frac{2\pi i\gamma^{-1}\tau}{h_c}}$. For an arbitrary function $f \in M(\Gamma_1(N))$, the order of $\tilde{f}$ at a point $[\tau_0]$ for $\tau_0 \in \mathbb{H}$ is t, where t is the order of f when expanded in powers of $\tau - \tau_0$.

6 How to "Mechanically" Prove Theta Function Identities

Here we outline the proof strategy for identities equating sums of theta products that we are going to apply in the proofs of our theorems from Sects. 2 and 3 given in the next section.

Let us fix a positive modulus $N > 3$. For $g \in \{0, \dots, N-1\}$ define

$$E_g = E_g(q; N) := q^{NB_2(g/N)/2}(q^g, q^{N-g}; q^N)_\infty, \tag{6.1}$$

where $B_2(x) = x^2 - x + \frac{1}{6}$ is the second Bernoulli polynomial. The function E_g is essentially a (specialised) Jacobi theta function, "essentially" referring to the missing factor $(q^N; q^N)_\infty$. In the following we explain how identities of the form

$$\sum_{j=1}^{r} c_j \prod_g E_g^{a_g^{(j)}} = 0, \tag{6.2}$$

where the c_j's are complex numbers and the $a_g^{(j)}$'s are integers, can be routinely verified if each summand in the sum above is a modular function for the group $\Gamma_1(N)$. It should be noted that the left-hand side of (6.2) is a linear combination of theta products.

Now, to verify the identity (6.2), the following steps have to be performed. For convenience, in the following we write LHS for the left-hand side of (6.2).

STEP 1. According to [8, Prop. 3], $\prod_g E_g^{a_g}(q; N)$ is a modular function in τ for the group $\Gamma_1(N)$, where $q = e^{2\pi i\tau}$, that is, $\prod_g E_g^{a_g}$ is an element of $M(\Gamma_1(N))$ if

$$\sum_g a_g \equiv 0 \pmod{12} \quad \text{and} \quad \sum_g g^2 a_g \equiv 0 \pmod{y(N)}, \tag{6.3}$$

where $y(N) = 2N$ if N is even, and $y(N) = N$ if N is odd. We use this criterion in LHS for each summand in order to check that each summand is a modular function for $\Gamma_1(N)$.

STEP 2. Representatives of the cusps for the group $\Gamma_1(N)$ are computed. (There are only finitely many.) The computer algebra programme *Magma* provides an implementation in form of the function `Cusps(Gamma1(N))`.

STEP 3. For each (representative of a) cusp—except $i\infty$, c say, and $j = 1, 2, \dots, r$, the order of the function $\prod_g E_g^{a_g^{(j)}}$ at c has to be computed. According to [8, Prop. 4], the order $\operatorname{ord}(E_g; c, N)$ of the function E_g at the cusp c of the group $\Gamma_1(N)$ is given by

$$\operatorname{ord}(E_g; c, N) = \frac{1}{2}\gcd(D_c, N) B_2(\{N_c g / \gcd(D_c, N)\}), \tag{6.4}$$

where D_c is the denominator of c and N_c is the numerator of c, while $\{\alpha\}$ denotes the fractional part of the rational number α. The order of $\prod_g E_g^{a_g^{(j)}}$ at c then is

$$\sum_g a_g^{(j)} \text{ord}(E_g; c, N).$$

STEP 4. We obtain a lower bound on the order of LHS at c by taking the minimum of the orders of the individual summands of the sum in LHS.

STEP 5. By Proposition 33, if LHS is not identically zero, then the sum of all the orders equals zero. Hence, (again assuming that LHS is not identically zero) for the function $\widetilde{LHS}$ on the compact Riemann surface $X_1(N)$, we have

$$0 = \text{ord}_{[i\infty]}(\widetilde{LHS}) + \sum_{[c],\, c \text{ a cusp},\, [c] \neq [i\infty]} \text{ord}_{[c]}(\widetilde{LHS}) + \sum_{[p],\, p \text{ not a cusp}} \text{ord}_{[p]}(\widetilde{LHS}).$$

If we sum all the lower bounds that we found in Step 4 over all the cusps different from $i\infty$, then we obtain a lower bound, $-U$ say, on the first sum in the above expression. Furthermore, from the definition of the function E_g it is obvious that it cannot have a singularity at a point $p \in \mathbb{H}$, and thus the order of E_g at p is non-negative. This implies directly that the orders $\text{ord}_{[p]}(\widetilde{LHS})$ are non-negative, yielding the lower bound 0 on the second sum. Everything combined, we see that $\text{ord}_{[i\infty]}(\widetilde{LHS}) \leq U$.

STEP 6. We now verify by direct computation that LHS has the power series expansion $0 + 0q + \cdots + 0q^U + \cdots$. This says that $\text{ord}_{[i\infty]}(\widetilde{LHS}) > U$ (the reader should recall that, under the relation $q = e^{2\pi i\tau}$, the point $\tau = i\infty$ corresponds to $q = 0$), a contradiction to our finding in Step 5 under the assumption that LHS is not identically zero. Consequently, LHS must be the zero function.

7 "Mechanical" Proofs

This section is devoted to the presentation of the proofs of the theorems in Sects. 2 and 3 that are based on the procedure outlined in the previous section. We provide full details for the proof of Theorem 1, while we remain brief for the proofs of the other theorems, all of them being completely analogous. For the theorems which are specialisations of the parametric theorems in Sect. 10, we refer to the proofs given there.

Proof of Theorem 1 STEP 0. We write the sum on the left-hand side of (2.1) explicitly, and then apply the Jacobi triple product identity to obtain an expression that is a linear combination of products of theta functions.

In order to accomplish this, we first observe that squares that are congruent to 361 modulo 840 are of the form S^2, where $S \equiv 19, 61, 79, 89, 121, 131, 149, 191 \pmod{210}$. Consequently, taking the definition of $t(n)$ into account, we have

$$\begin{aligned}\sum_{n=0}^{\infty}(-1)^{t(n)}q^{a_n} &= \sum_{k=0}^{\infty}(-1)^k q^{\frac{1}{840}((210k+19)^2-361)} + \sum_{k=0}^{\infty}(-1)^k q^{\frac{1}{840}((210k+61)^2-361)} \\ &\quad - \sum_{k=0}^{\infty}(-1)^k q^{\frac{1}{840}((210k+79)^2-361)} + \sum_{k=0}^{\infty}(-1)^k q^{\frac{1}{840}((210k+89)^2-361)} \\ &\quad - \sum_{k=0}^{\infty}(-1)^k q^{\frac{1}{840}((210k+121)^2-361)} + \sum_{k=0}^{\infty}(-1)^k q^{\frac{1}{840}((210k+131)^2-361)} \\ &\quad - \sum_{k=0}^{\infty}(-1)^k q^{\frac{1}{840}((210k+149)^2-361)} - \sum_{k=0}^{\infty}(-1)^k q^{\frac{1}{840}((210k+191)^2-361)} \\ &= \sum_{k=0}^{\infty}(-1)^k q^{\frac{105k^2}{2}+\frac{19k}{2}} + \sum_{k=0}^{\infty}(-1)^k q^{\frac{105k^2}{2}+\frac{61k}{2}+4} - \sum_{k=0}^{\infty}(-1)^k q^{\frac{105k^2}{2}+\frac{79k}{2}+7} \\ &\quad + \sum_{k=0}^{\infty}(-1)^k q^{\frac{105k^2}{2}+\frac{89k}{2}+9} - \sum_{k=0}^{\infty}(-1)^k q^{\frac{105k^2}{2}+\frac{121k}{2}+17} + \sum_{k=0}^{\infty}(-1)^k q^{\frac{105k^2}{2}+\frac{131k}{2}+20} \\ &\quad - \sum_{k=0}^{\infty}(-1)^k q^{\frac{105k^2}{2}+\frac{149k}{2}+26} - \sum_{k=0}^{\infty}(-1)^k q^{\frac{105k^2}{2}+\frac{191k}{2}+43}.\end{aligned} \tag{7.1}$$

By performing the replacement $k \to -k-1$, we see that the last sum on the right-hand side of (7.1) can be rewritten as

$$\sum_{k=0}^{\infty}(-1)^k q^{\frac{105k^2}{2}+\frac{191k}{2}+43} = \sum_{k=-\infty}^{-1}(-1)^{k+1} q^{\frac{105k^2}{2}+\frac{19k}{2}}.$$

Thus, it can be combined with the first sum on the right-hand side of (7.1). This is similar for the other sums. As a result, they can be paired so that one obtains four sums over *all* integers k:

$$\begin{aligned}\sum_{n=0}^{\infty}(-1)^{t(n)}q^{a_n} &= \sum_{k=-\infty}^{\infty}(-1)^k q^{\frac{105k^2}{2}+\frac{19k}{2}} + \sum_{k=-\infty}^{\infty}(-1)^k q^{\frac{105k^2}{2}+\frac{61k}{2}+4} \\ &\quad - \sum_{k=-\infty}^{\infty}(-1)^k q^{\frac{105k^2}{2}+\frac{79k}{2}+7} + \sum_{k=-\infty}^{\infty}(-1)^k q^{\frac{105k^2}{2}+\frac{89k}{2}+9}.\end{aligned}$$

Now, as announced, to each of these sums we apply the Jacobi triple product identity (4.1) to get

$$\sum_{n=0}^{\infty}(-1)^{t(n)}q^{a_n} = (q^{105},q^{62},q^{43};q^{105})_\infty + q^4\,(q^{105},q^{83},q^{22};q^{105})_\infty \\ - q^7\,(q^{105},q^{92},q^{13};q^{105})_\infty + q^9\,(q^{105},q^{97},q^{8};q^{105})_\infty. \tag{7.2}$$

Thus, we have to prove the identity

$$0 = (q^{105},q^{62},q^{43};q^{105})_\infty + q^4\,(q^{105},q^{83},q^{22};q^{105})_\infty \\ - q^7\,(q^{105},q^{92},q^{13};q^{105})_\infty + q^9\,(q^{105},q^{97},q^{8};q^{105})_\infty - \frac{(q,q^6,q^7;q^7)_\infty}{(q,q^4;q^5)_\infty}.$$

We divide both sides of the identity by the first term on the right-hand side and obtain

$$0 = 1 + q^4\frac{(q^{105},q^{83},q^{22};q^{105})_\infty}{(q^{105},q^{62},q^{43};q^{105})_\infty} - q^7\frac{(q^{105},q^{92},q^{13};q^{105})_\infty}{(q^{105},q^{62},q^{43};q^{105})_\infty} \\ + q^9\frac{(q^{105},q^{97},q^{8};q^{105})_\infty}{(q^{105},q^{62},q^{43};q^{105})_\infty} - \frac{(q,q^6,q^7;q^7)_\infty}{(q,q^4;q^5)_\infty\,(q^{105},q^{62},q^{43};q^{105})_\infty}.$$

Now we fix $N := 105$. With the notation (6.1), our identity can be written as

$$0 = 1 + \frac{E_{22}}{E_{43}} - \frac{E_{13}}{E_{43}} + \frac{E_8}{E_{43}} - \frac{E_7E_8E_{13}E_{15}E_{20}E_{22}E_{27}E_{28}E_{35}E_{42}E_{48}E_{50}}{E_4E_9E_{11}E_{16}E_{19}E_{24}E_{26}E_{31}E_{39}E_{44}E_{46}E_{51}}. \tag{7.3}$$

In order to rewrite the last term we used that

$$(q^7;q^7) = (q^7,q^{14},q^{21},\dots,q^{105};q^{105}),$$

and similar "blow-ups" for other terms.

STEP 1. We use the criterion (6.3) to see that all summands on the right-hand side of (7.3) are modular functions for $\Gamma_1(105)$. Indeed, for $N = 105$ and $\frac{E_{22}}{E_{43}}$, we have $1 - 1 \equiv 0 \pmod{12}$ and $22^2 - 43^2 \equiv 0 \pmod{105}$. This shows that $\frac{E_{22}}{E_{43}}$ is a modular function for the group $\Gamma_1(105)$. Similarly, we observe that $\frac{E_{13}}{E_{43}}$, $\frac{E_8}{E_{43}}$ and $\frac{E_7E_8E_{13}E_{15}E_{20}E_{22}E_{27}E_{28}E_{35}E_{42}E_{48}E_{50}}{E_4E_9E_{11}E_{16}E_{19}E_{24}E_{26}E_{31}E_{39}E_{44}E_{46}E_{51}}$ are modular functions for the group $\Gamma_1(105)$. Consequently,

$$f := 1 + \frac{E_{22}}{E_{43}} - \frac{E_{13}}{E_{43}} + \frac{E_8}{E_{43}} - \frac{E_7E_8E_{13}E_{15}E_{20}E_{22}E_{27}E_{28}E_{35}E_{42}E_{48}E_{50}}{E_4E_9E_{11}E_{16}E_{19}E_{24}E_{26}E_{31}E_{39}E_{44}E_{46}E_{51}}$$

is a modular function for the group $\Gamma_1(105)$.

STEP 2. We use the computer algebra *Magma* to compute representatives of the cusps for the group $\Gamma_1(105)$. This is done by using the command `Cusps (Gamma1(105))`. The output is

```
[oo,0,1/13,1/12,2/23,1/11,3/32,2/21,1/10,3/29,5/48,2/19,3/28,4/37,1/9,\
5/44,4/35,3/26,8/69,5/43,2/17,3/25,4/33,1/8,6/47,5/39,9/70,4/31,11/84,\
13/99,5/38,7/53,2/15,13/96,8/59,3/22,7/51,1/7,5/34,4/27,29/195,18/121,\
3/20,48/319,79/525,5/33,16/105,7/45,18/115,8/51,4/25,17/105,1/6,6/35,\
11/63,18/103,7/40,8/45,23/129,5/28,7/39,9/50,11/60,9/49,12/65,5/27,\
19/102,30/161,13/69,17/90,4/21,1/5,109/525,27/130,19/91,23/110,22/105,\
47/222,18/85,33/155,3/14,14/65,68/315,13/60,41/189,64/295,29/133,\
19/87,26/119,23/105,9/41,20/91,11/50,11/49,9/40,71/315,23/102,30/133,\
17/75,27/119,8/35,13/56,44/189,7/30,13/55,5/21,6/25,9/35,11/42,59/225,\
37/140,23/87,53/200,13/49,4/15,62/231,51/190,47/175,32/119,368/1365,\
17/63,13/48,29/105,31/112,46/165,41/147,59/210,69/245,2/7,13/45,\
17/63,71/245,7/24,92/315,45/154,43/147,31/105,34/115,29/98,52/175,\
25/84,94/315,19/63,32/105,13/42,24/77,11/35,16/45,113/315,48/133,\
38/105,23/63,11/30,31/84,13/35,37/98,8/21,67/175,523/1365,29/75,\
64/165,41/105,124/315,2/5,43/105,41/100,26/63,31/75,44/105,103/245,\
47/105,16/35,7/15,8/15,19/35,39/70,137/245,47/84,17/30,4/7,97/168,\
41/70,37/63,13/21,152/245,87/140,22/35,19/30,24/35,46/63,11/15,\
23/30,27/35]}.
```

STEP 3. We now compute the order of each summand of LHS at each cusp except $i\infty$. To do this, we have to use (6.4). Assume for example that we want to compute the order at the cusp $27/35$ of the function $\frac{E_{22}}{E_{43}}$. It is convenient to implement the following two functions in *Maple*:

```
B:=proc(x)
x^2-x+1/6
end proc:

orderCusp1:=proc(x,N,g)
local a,c;
a:=numer(x);
c:=denom(x);
igcd(c,N)*B(frac(a*g/igcd(c,N)))/2
end proc:
```

Now, in order to achieve our task we enter in *Maple*:

```
orderCusp1(27/35,105,22)-orderCusp1(27/35,105,43);
2
```

The output means that $\frac{E_{22}}{E_{43}}$ has a zero of order 2 at the cusp $\frac{27}{35}$.

STEP 4. We want to compute a lower bound on the order of f at the cusp $27/35$. In order to do this, we do the above computation for each term in f and take the minimum of all orders. This computation can be simplified by the function

```
orderCuspGroup:=proc(x,N,g)
local ii,order;
order:=0;
for ii from 1 to nops(g) do
order:=order+orderCusp1(x,N,g[ii][1])*g[ii][2];
od;
order
end proc:
```

This function takes as input the cusp representative x, the N—which in our case is 105 —, and the index g from E_g. The product $\frac{E_{22}}{E_{43}}$ is expressed as

```
f1:=[[22,1],[43,-1]];
```

Then we can compute the order of $\frac{E_{22}}{E_{43}}$ at the cusp $27/35$ by typing

```
orderCuspGroup(27/35,105,f1);
2
```

We define the other terms by

```
f2:=[[8,1],[43,-1]];
f3:=[[13,1],[43,-1]];
f4:=[[7,1],[8,1],[13,1],[15,1],[20,1],[22,1],[27,1],[28,1],
[35,1],[42,1],[48,1],[50,1],[4,-1],[9,-1],[11,-1],[16,-1],
[19,-1],[24,-1],[26,-1],[31,-1],[39,-1],[44,-1],[46,-1],[51,-1]];
```

Now we can get a lower bound on the order of f at the cusp $27/35$ by writing

```
min(orderCuspGroup(27/35,105,f1),orderCuspGroup(27/35,105,f2),
orderCuspGroup(27/35,105,f3),orderCuspGroup(27/35,105,f4));
0
```

Hence, our lower bound is 0.

STEP 5. We sum up the lower bounds on the orders at all cusps except $i\infty$. Thus, we obtain an upper bound on the order of $\tilde{f}$ at $[i\infty]$. This is done as follows:

```
cusps:=[0,1/13,1/12,2/23,1/11,3/32,2/21,1/10,3/29,5/48,...];
cnt:=0;
for ii in cusps do
 mn:=min(orderCuspGroup(ii,105,f1),orderCuspGroup(ii,105,f2),
 orderCuspGroup(ii,105,f3),orderCuspGroup(ii,105,f4)); cnt:=cnt+mn; od;
print(cnt);
-148
```

This means that the order of $\tilde{f}$ at $i\infty$ is at most 148, under the assumption that f is not identically zero.

STEP 6. It is routine to verify that $f = 0 + 0q + \cdots + 0q^{148} + \cdots$. This implies that $\mathrm{ord}_{[i\infty]}(\tilde{f}) > 148$. Hence, f must be the zero function. □

Proof of Theorem 2 It is elementary to see that squares that are congruent to 529 modulo 840 are of the form S^2, where $S \equiv 23, 37, 47, 103, 107, 163, 173, 187 \pmod{210}$. If we now do a computation analogous to the one in the proof of Theorem 1, we obtain

$$\sum_{n=0}^{\infty}(-1)^{t(n)}q^{a_n} = (q^{105}, q^{64}, q^{41}; q^{105})_\infty - q\,(q^{105}, q^{71}, q^{34}; q^{105})_\infty \\ + q^2\,(q^{105}, q^{76}, q^{29}; q^{105})_\infty + q^{12}\,(q^{105}, q^{104}, q; q^{105})_\infty. \tag{7.4}$$

Continuing as in the previous proof, we observe that the assertion of the theorem is equivalent to

$$0 = 1 - \frac{E_{34}}{E_{41}} + \frac{E_{29}}{E_{41}} + \frac{E_1}{E_{41}} - \frac{E_1 E_6 E_{15} E_{20} E_{29} E_{34} E_{36} E_{50} E_{14} E_{21} E_{35} E_{49}}{E_2 E_3 E_{12} E_{17} E_{18} E_{23} E_{32} E_{33} E_{37} E_{38} E_{47} E_{52}},$$

where we used the notation (6.1) with $N = 105$. Using (6.3), we see that the right-hand side is a modular function for the group $\Gamma_1(105)$. We estimate the sum of the orders of the right-hand side at the cusps using the same programmes as in the previous proof with modifications of the parameters. The upper bound on the order of the right-hand side at $i\infty$ is also 148, so it suffices to check that the first 148 coefficients of the right-hand side are zero in order to prove the identity, which we checked using a computer. □

Proof of Theorem 3 It is elementary to see that squares that are congruent to 121 modulo 840 are of the form S^2, where $S \equiv 11, 31, 59, 101, 109, 151, 179, 199 \pmod{210}$. If we now do a computation analogous to the one in the proof of Theorem 1, we obtain

$$\sum_{n=0}^{\infty} (-1)^{\lfloor (n+4)/8 \rfloor} q^{a_n} = (q^{105}, q^{47}, q^{58}; q^{105})_\infty + q\,(q^{105}, q^{37}, q^{68}; q^{105})_\infty \qquad (7.5)$$
$$+ q^4\,(q^{105}, q^{23}, q^{82}; q^{105})_\infty + q^{12}\,(q^{105}, q^{2}, q^{103}; q^{105})_\infty.$$

Next, we observe that, using the notation (6.1) with $N = 105$, the assertion of the theorem is equivalent to

$$0 = 1 + \frac{E_{37}}{E_{47}} + \frac{E_{23}}{E_{47}} + \frac{E_2}{E_{47}} - \frac{E_2 E_5 E_7 E_{12} E_{23} E_{28} E_{30} E_{33} E_{35} E_{37} E_{40} E_{42}}{E_1 E_4 E_6 E_{11} E_{24} E_{29} E_{31} E_{34} E_{36} E_{39} E_{41} E_{46}}.$$

Using (6.3), we see that the right-hand side is a modular function for the group $\Gamma_1(105)$. We estimate the sum of the orders of the right-hand side at the cusps using the same programmes as before with modifications of the parameters. The upper bound on the order of the right-hand side at $i\infty$ is also 148, so it suffices to check that the first 148 coefficients of the right-hand side are zero in order to prove the identity, which we checked using a computer. □

Proof of Theorem 4 It is elementary to see that squares that are congruent to 289 modulo 840 are of the form S^2, where $S \equiv 17, 53, 67, 73, 137, 143, 157, 193 \pmod{210}$. If we now do a computation analogous to the one in the proof of Theorem 1, we obtain

$$\sum_{n=0}^{\infty} (-1)^{t(n)} q^{a_n} = (q^{105}, q^{44}, q^{61}; q^{105})_\infty + q^3\,(q^{105}, q^{26}, q^{79}; q^{105})_\infty \qquad (7.6)$$
$$- q^5\,(q^{105}, q^{19}, q^{86}; q^{105})_\infty + q^6\,(q^{105}, q^{16}, q^{89}; q^{105})_\infty.$$

Next, we observe that, using the notation (6.1) with $N = 105$, the assertion of the theorem is equivalent to

$$0 = 1 + \frac{E_{26}}{E_{44}} - \frac{E_{19}}{E_{44}} + \frac{E_{16}}{E_{44}} - \frac{E_5 E_9 E_{14} E_{16} E_{19} E_{21} E_{26} E_{30} E_{35} E_{40} E_{49} E_{51}}{E_3 E_8 E_{13} E_{17} E_{18} E_{22} E_{27} E_{32} E_{38} E_{43} E_{48} E_{52}}.$$

Using (6.3), we see that the right-hand side is a modular function for the group $\Gamma_1(105)$. We estimate the sum of the orders of the right-hand side at the cusps using the same programmes as before with modifications of the parameters. The upper bound on the order of the right-hand side at $i\infty$ is also 148, so it suffices to check that the first 148 coefficients of the right-hand side are zero in order to prove the identity, which we checked using a computer. □

Proof of Theorem 5 It is elementary to see that squares that are congruent to 1 modulo 840 are of the form S^2, where $S \equiv 1, 29, 41, 71, 139, 169, 181, 209 \pmod{210}$. If we now do a computation analogous to the one in the proof of Theorem 1, we obtain

$$\sum_{n=0}^{\infty}(-1)^{\lfloor (n+4)/8 \rfloor} q^{a_n} = (q^{105}, q^{52}, q^{53}; q^{105})_\infty + q\,(q^{105}, q^{38}, q^{67}; q^{105})_\infty + q^2\,(q^{105}, q^{32}, q^{73}; q^{105})_\infty + q^6\,(q^{105}, q^{17}, q^{88}; q^{105})_\infty. \tag{7.7}$$

Next, we observe that, using the notation (6.1) with $N = 105$, the assertion of the theorem is equivalent to

$$0 = 1 + \frac{E_{38}}{E_{52}} + \frac{E_{32}}{E_{52}} + \frac{E_{17}}{E_{52}} - \frac{E_3 E_7 E_{10} E_{17} E_{18} E_{25} E_{28} E_{32} E_{35} E_{38} E_{42} E_{45}}{E_1 E_6 E_9 E_{16} E_{19} E_{26} E_{29} E_{34} E_{36} E_{41} E_{44} E_{51}}.$$

Using (6.3), we see that the right-hand side is a modular function for the group $\Gamma_1(105)$. We estimate the sum of the orders of the right-hand side at the cusps using the same programmes as before with modifications of the parameters. The upper bound on the order of the right-hand side at $i\infty$ is also 148, so it suffices to check that the first 148 coefficients of the right-hand side are zero in order to prove the identity, which we checked using a computer. □

Proof of Theorem 6 It is elementary to see that squares that are congruent to 169 modulo 840 are of the form S^2, where $S \equiv 13, 43, 83, 97, 113, 127, 167, 197 \pmod{210}$. If we now do a computation analogous to the one in the proof of Theorem 1, we obtain

$$\sum_{n=0}^{\infty}(-1)^{t(n)} q^{a_n} = (q^{105}, q^{46}, q^{59}; q^{105})_\infty + q^2\,(q^{105}, q^{31}, q^{74}; q^{105})_\infty + q^8\,(q^{105}, q^{11}, q^{94}; q^{105})_\infty - q^{11}\,(q^{105}, q^{4}, q^{101}; q^{105})_\infty. \tag{7.8}$$

Next, we observe that, using the notation (6.1) with $N = 105$, the assertion of the theorem is equivalent to

$$0 = 1 + \frac{E_{31}}{E_{46}} + \frac{E_{11}}{E_{46}} - \frac{E_4}{E_{46}} - \frac{E_4 E_{10} E_{11} E_{14} E_{21} E_{24} E_{25} E_{31} E_{35} E_{39} E_{45} E_{49}}{E_2 E_8 E_{12} E_{13} E_{22} E_{23} E_{27} E_{33} E_{37} E_{43} E_{47} E_{48}}.$$

Using (6.3), we see that the right-hand side is a modular function for the group $\Gamma_1(105)$. We estimate the sum of the orders of the right-hand side at the cusps using the same programmes as before with modifications of the parameters. The upper bound on the order of the right-hand side at $i\infty$ is also 148, so it suffices to check that the first 148 coefficients of the right-hand side are zero in order to prove the identity, which we checked using a computer. □

Proof of Theorem 7 It is elementary to see that squares that are congruent to 1 modulo 240 are of the form S^2, where $S \equiv 1, 31, 41, 49, 71, 79, 89, 119 \pmod{120}$. If we now do a computation analogous to the one in the proof of Theorem 1, we obtain

$$\sum_{n=0}^{\infty}(-1)^{\lfloor (n+2)/4\rfloor}q^{a_n} = (q^{120}, -q^{59}, -q^{61}; q^{120})_\infty + q^4\,(q^{120}, -q^{29}, -q^{91}; q^{120})_\infty \\ - q^7\,(q^{120}, -q^{19}, -q^{101}; q^{120})_\infty - q^{10}\,(q^{120}, -q^{11}, -q^{109}; q^{120})_\infty. \tag{7.9}$$

Rewriting this expression, we see that the assertion of the theorem is equivalent to

$$0 = \frac{(q^{120}; q^{120})_\infty (q^{118}, q^{122}; q^{240})_\infty}{(q^{59}, q^{61}; q^{120})_\infty} + q^4 \frac{(q^{120}; q^{120})_\infty (q^{58}, q^{182}; q^{240})_\infty}{(q^{29}, q^{91}; q^{120})_\infty} \\ - q^7 \frac{(q^{120}; q^{120})_\infty (q^{22}, q^{218}; q^{240})_\infty}{(q^{11}, q^{109}; q^{120})_\infty} - q^{10} \frac{(q^{120}; q^{120})_\infty (q^{22}, q^{218}; q^{240})_\infty}{(q^{11}, q^{109}; q^{120})_\infty} \\ - \frac{(q, q^7, q^8; q^8)_\infty (q^6, q^{10}; q^{16})_\infty}{(q, q^4; q^5)_\infty}.$$

By dividing both sides of the identity by the first term on the right-hand side and using the notation (6.1) with $N = 240$, we obtain

$$0 = 1 + \frac{E_{58}E_{59}E_{61}}{E_{29}E_{91}E_{118}} - \frac{E_{38}E_{59}E_{61}}{E_{19}E_{101}E_{118}} - \frac{E_{22}E_{59}E_{61}}{E_{11}E_{109}E_{118}} \\ - \frac{E_7E_8E_{10}E_{15}E_{17}E_{22}E_{23}E_{25}E_{32}E_{33}E_{38}E_{40}E_{42}E_{47}E_{48}E_{55}E_{57}E_{58}}{E_4E_{11}E_{14}E_{19}E_{21}E_{29}E_{34}E_{36}E_{44}E_{46}E_{51}} \\ \times \frac{E_{63}E_{65}E_{70}E_{72}E_{73}E_{80}E_{87}E_{88}E_{90}E_{95}E_{97}E_{102}E_{103}E_{105}E_{112}E_{113}}{E_{66}E_{69}E_{76}E_{84}E_{91}E_{94}E_{99}E_{101}E_{109}E_{114}E_{116}}.$$

Each term on the right-hand side is a modular function for the group $\Gamma_1(240)$. As before, using *Magma* we compute a list of all the cusps. Here we have 448 cusps. Again, we can give an upper bound on the order of the right-hand side at $i\infty$. Running our programme, we obtain 592. We need to verify that the right-hand side has the form $0 + 0q + \cdots + 0q^{592} + \cdots$, which can be routinely done. This proves the identity. □

Proof of Theorem 8 It is elementary to see that squares that are congruent to 49 modulo 240 are of the form S^2, where $S \equiv 7, 17, 23, 47, 73, 97, 103, 113 \pmod{120}$. If we now do a computation analogous to the one in the proof of Theorem 1, we obtain

$$\sum_{n=0}^{\infty}(-1)^{\lfloor 5n/4\rfloor}q^{a_n} = (q^{120}, -q^{53}, -q^{67}; q^{120})_\infty - q\,(q^{120}, -q^{43}, -q^{77}; q^{120})_\infty \tag{7.10}$$
$$+q^2\,(q^{120}, -q^{37}, -q^{83}; q^{120})_\infty - q^9\,(q^{120}, -q^{13}, -q^{107}; q^{120})_\infty$$
$$= \frac{(q^{120}; q^{120})_\infty (q^{106}, q^{134}; q^{240})_\infty}{(q^{53}, q^{67}; q^{120})_\infty} - q\frac{(q^{120}; q^{120})_\infty (q^{86}, q^{154}; q^{240})_\infty}{(q^{43}, q^{77}; q^{120})_\infty}$$
$$+q^2\frac{(q^{120}; q^{120})_\infty (q^{74}, q^{166}; q^{240})_\infty}{(q^{37}, q^{83}; q^{120})_\infty} - q^9\frac{(q^{120}; q^{120})_\infty (q^{26}, q^{214}; q^{240})_\infty}{(q^{13}, q^{107}; q^{120})_\infty}.$$

Next, we observe that, using the notation (6.1) with $N = 240$, the assertion of the theorem is equivalent to

$$0 = 1 - \frac{E_{53}E_{67}E_{86}}{E_{43}E_{77}E_{106}} + \frac{E_{53}E_{67}E_{74}}{E_{37}E_{83}E_{106}} - \frac{E_{26}E_{53}E_{67}}{E_{13}E_{106}E_{107}}$$
$$- \frac{E_1E_6E_9E_{10}E_{15}E_{16}E_{24}E_{25}E_{26}E_{31}E_{39}E_{40}E_{41}E_{49}E_{54}E_{55}E_{56}}{E_2E_3E_{12}E_{13}E_{18}E_{27}E_{28}E_{37}E_{43}E_{52}}$$
$$\times \frac{E_{64}E_{65}E_{70}E_{71}E_{74}E_{79}E_{80}E_{81}E_{86}E_{89}E_{90}E_{95}E_{96}E_{104}E_{105}E_{111}E_{119}}{E_{62}E_{68}E_{77}E_{78}E_{82}E_{83}E_{92}E_{93}E_{98}E_{107}E_{108}E_{117}}.$$

As before, each term is a modular function for the group $\Gamma_1(240)$. The upper bound on the order of the right-hand side at $i\infty$ is 592. We verified that the right-hand side has the form $0 + 0q + \cdots + 0q^{592} + \cdots$. This proves the theorem. □

Proof of Theorem 9 It is elementary to see that squares that are congruent to 121 modulo 240 are of the form S^2, where $S \equiv 11, 19, 29, 59, 61, 91, 101, 109 \pmod{120}$. If we now do a computation analogous to the one in the proof of Theorem 1, we obtain

$$\sum_{n=0}^{\infty}(-1)^{\lfloor (n+2)/4\rfloor}q^{a_n} = (q^{120}, -q^{49}, -q^{71}; q^{120})_\infty + q\,(q^{120}, -q^{41}, -q^{79}; q^{120})_\infty$$
$$-q^3\,(q^{120}, -q^{31}, -q^{89}; q^{120})_\infty - q^{14}\,(q^{120}, -q, -q^{119}; q^{120})_\infty$$
$$= \frac{(q^{120}; q^{120})_\infty (q^{98}, q^{142}; q^{240})_\infty}{(q^{49}, q^{71}; q^{120})_\infty} + q\frac{(q^{120}; q^{120})_\infty (q^{82}, q^{158}; q^{240})_\infty}{(q^{41}, q^{79}; q^{120})_\infty}$$
$$-q^3\frac{(q^{120}; q^{120})_\infty (q^{62}, q^{178}; q^{240})_\infty}{(q^{31}, q^{89}; q^{120})_\infty} - q^{14}\frac{(q^{120}; q^{120})_\infty (q^{2}, q^{238}; q^{240})_\infty}{(q^{1}, q^{119}; q^{120})_\infty}. \tag{7.11}$$

Next, we observe that, using the notation (6.1) with $N = 240$, the assertion of the theorem is equivalent to

$$0 = 1 + \frac{E_{49}E_{71}E_{82}}{E_{41}E_{79}E_{98}} - \frac{E_{49}E_{62}E_{71}}{E_{31}E_{89}E_{98}} - \frac{E_2E_{49}E_{71}}{E_1E_{98}E_{119}}$$
$$- \frac{E_2E_3E_5E_8E_{13}E_{18}E_{27}E_{30}E_{32}E_{35}E_{37}E_{40}E_{43}E_{45}E_{48}E_{50}E_{53}}{E_1E_4E_6E_9E_{26}E_{31}E_{36}E_{39}E_{41}E_{44}E_{54}}$$
$$\times \frac{E_{62}E_{67}E_{72}E_{75}E_{77}E_{78}E_{80}E_{82}E_{83}E_{85}E_{88}E_{93}E_{107}E_{110}E_{112}E_{115}E_{117}}{E_{74}E_{76}E_{79}E_{81}E_{84}E_{86}E_{89}E_{106}E_{111}E_{116}E_{119}}.$$

As before, each term is a modular function for the group $\Gamma_1(240)$. The upper bound on the order of the right-hand side at $i\infty$ is 592. We verified that the right-hand side has the form $0 + 0q + \cdots + 0q^{592} + \cdots$. This proves the theorem. □

Proof of Theorem 10 It is elementary to see that squares that are congruent to 169 modulo 240 are of the form S^2, where $S \equiv 13, 37, 43, 53, 67, 77, 83, 107 \pmod{120}$. If we now do a computation analogous to the one in the proof of Theorem 1, we obtain

$$\begin{aligned}\sum_{n=0}^{\infty}(-1)^{\lfloor 5n/4\rfloor} q^{a_n} &= (q^{120}, -q^{47}, -q^{73}; q^{120})_\infty - q^5\,(q^{120}, -q^{23}, -q^{97}; q^{120})_\infty \\ &\quad + q^7\,(q^{120}, -q^{17}, -q^{103}; q^{120})_\infty - q^{11}\,(q^{120}, -q^{7}, -q^{113}; q^{120})_\infty \\ &= \frac{(q^{120}; q^{120})_\infty (q^{94}, q^{146}; q^{240})_\infty}{(q^{47}, q^{73}; q^{120})_\infty} - q^5 \frac{(q^{120}; q^{120})_\infty (q^{46}, q^{194}; q^{240})_\infty}{(q^{23}, q^{97}; q^{120})_\infty} \\ &\quad + q^7 \frac{(q^{120}; q^{120})_\infty (q^{34}, q^{206}; q^{240})_\infty}{(q^{17}, q^{103}; q^{120})_\infty} - q^{11} \frac{(q^{120}; q^{120})_\infty (q^{14}, q^{226}; q^{240})_\infty}{(q^{7}, q^{113}; q^{120})_\infty}. \end{aligned} \tag{7.12}$$

Next, we observe that, using the notation (6.1) with $N = 240$, the assertion of the theorem is equivalent to

$$\begin{aligned} 0 = 1 &- \frac{E_{46}E_{47}E_{73}}{E_{23}E_{94}E_{97}} + \frac{E_{34}E_{47}E_{73}}{E_{17}E_{94}E_{103}} - \frac{E_{14}E_{47}E_{73}}{E_7E_{94}E_{113}} \\ &- \frac{E_5E_{11}E_{14}E_{16}E_{19}E_{21}E_{24}E_{29}E_{30}E_{34}E_{35}E_{40}E_{45}E_{46}E_{50}E_{51}E_{56}E_{59}}{E_7E_{12}E_{17}E_{22}E_{23}E_{28}E_{33}E_{38}E_{42}E_{52}E_{57}E_{58}} \\ &\times \frac{E_{61}E_{64}E_{66}E_{69}E_{75}E_{80}E_{85}E_{91}E_{96}E_{99}E_{101}E_{104}E_{109}E_{110}E_{114}E_{115}}{E_{63}E_{68}E_{87}E_{92}E_{97}E_{102}E_{103}E_{108}E_{113}E_{118}}. \end{aligned}$$

As before, each term is a modular function for the group $\Gamma_1(240)$. The upper bound on the order of the right-hand side at $i\infty$ is 592. We verified that the right-hand side has the form $0 + 0q + \cdots + 0q^{592} + \cdots$. This proves the theorem. □

Proof of Theorem 11 This is a special case of Theorem 36. □

Proof of Theorem 12 This is a special case of Theorem 36. □

Proof of Theorem 13 This is a direct consequence of the Jacobi triple product identity (4.1): one replaces q by q^{10} and then chooses $z = -q^6$ there. □

Proof of Theorem 14 Using (4.3), we obtain

$$\sum_{n=0}^{\infty}\left(q^{n(n+1)} - q^{5n(n+1)+1}\right) = (q^2; q^2)_\infty\,(-q^2; q^2)_\infty^2 - q\,(q^{10}; q^{10})_\infty\,(-q^{10}; q^{10})_\infty^2. \tag{7.13}$$

Hence, the assertion of the theorem is equivalent to

$$0 = \frac{(q^4; q^4)_\infty^2}{(q^2; q^2)_\infty} - q\frac{(q^{20}; q^{20})_\infty^2}{(q^{10}; q^{10})_\infty} - \frac{(q, q^9, q^{10}; q^{10})_\infty\,(q^8, q^{12}; q^{20})_\infty}{(q^2, q^3; q^5)_\infty}.$$

Using the notation (6.1) with $N = 20$, we see that this is equivalent to

$$0 = 1 - \frac{E_2E_6}{E_4E_8} - \frac{E_1E_6E_9E_{10}}{E_3E_4E_7E_8}.$$

The right-hand side is a modular function for the group $\Gamma_1(20)$. The cusps for this group are computed as usual using *Magma*:

$$\left\{\infty, 0, \frac{1}{7}, \frac{3}{20}, \frac{1}{6}, \frac{2}{11}, \frac{1}{5}, \frac{1}{4}, \frac{3}{10}, \frac{1}{3}, \frac{7}{20}, \frac{11}{30}, \frac{3}{8}, \frac{2}{5}, \frac{9}{20}, \frac{1}{2}, \frac{7}{12}, \frac{3}{5}, \frac{3}{4}, \frac{4}{5}\right\}.$$

There are in total 20 cusps. Estimating the order of the right-hand side at $i\infty$, one obtains an upper bound of 4. Hence it is sufficient to show that the right-hand side has the form $0 + 0q + 0q^2 + 0q^3 + 0q^4 + \cdots$, which can be done routinely. □

Proof of Theorem 15 Using (4.2), we obtain

$$1+\sum_{n=1}^{\infty}\left(q^{n^2}+q^{5n^2}\right) = \frac{1}{2}\left((q^2;q^2)_\infty(-q;q^2)_\infty^2 + (q^{10};q^{10})_\infty(-q^5;q^{10})_\infty^2\right). \tag{7.14}$$

Hence, the assertion of the theorem is equivalent to

$$0 = \frac{1}{2}\frac{(q^2;q^2)_\infty(q^2;q^4)_\infty^2}{(q;q^2)_\infty^2} + \frac{1}{2}\frac{(q^{10};q^{10})_\infty(q^{10};q^{20})_\infty^2}{(q^5;q^{10})_\infty^2} - \frac{(q^2,q^8,q^{10};q^{10})_\infty(q^6,q^{14};q^{20})_\infty}{(q,q^4;q^5)_\infty}.$$

Using the notation (6.1) with $N = 20$, we see that this is equivalent to

$$0 = \frac{1}{2} + \frac{1}{2}\frac{E_1^2E_3^2E_7^2E_9^2}{E_2^3E_4E_6^3E_8} - \frac{E_1E_3^2E_5^2E_7^2E_9}{E_2^2E_4^2E_6^3E_{10}}.$$

The right-hand side is a modular function for the group $\Gamma_1(20)$. There are in total the 20 cusps exhibited in the previous proof. Estimating the order of the right-hand side at $i\infty$, one obtains an upper bound of 4. Hence it is sufficient to show that the right-hand side has the form $0 + 0q + 0q^2 + 0q^3 + 0q^4 + \cdots$, which can be done routinely. □

Proof of Theorem 16 This is a direct consequence of the Jacobi triple product identity (4.1): one replaces q by q^{10} and then chooses $z = -q^7$ there. □

Proof of Theorem 17 Using (4.3), we obtain

$$\sum_{n=0}^{\infty}\left(q^{n(n+1)}+q^{5n(n+1)+1}\right) = (q^2;q^2)_\infty(-q^2;q^2)_\infty^2 + q\,(q^{10};q^{10})_\infty(-q^{10};q^{10})_\infty^2. \tag{7.15}$$

Hence, the assertion of the theorem is equivalent to

$$0 = \frac{(q^4;q^4)_\infty^2}{(q^2;q^2)_\infty} + q\frac{(q^{20};q^{20})_\infty^2}{(q^{10};q^{10})_\infty} - \frac{(q^3,q^7,q^{10};q^{10})_\infty(q^4,q^{16};q^{20})_\infty}{(q,q^4;q^5)_\infty}.$$

Using the notation (6.1) with $N = 20$, we see that this is equivalent to

$$0 = 1 + \frac{E_2 E_6}{E_4 E_8} - \frac{E_2 E_3 E_7 E_{10}}{E_1 E_4 E_8 E_9}.$$

The right-hand side is a modular function for the group $\Gamma_1(20)$. There are in total the 20 cusps exhibited in the proof of Theorem 15. Estimating the order of the right-hand side at $i\infty$, one obtains an upper bound of 4. Hence it is sufficient to show that the right-hand side has the form $0 + 0q + 0q^2 + 0q^3 + 0q^4 + \cdots$, which can be done routinely. □

Proof of Theorem 18 This is a direct consequence of the Jacobi triple product identity (4.1): one replaces q by q^{10} and then chooses $z = -q^8$ there. □

Proof of Theorem 19 This is a direct consequence of the Jacobi triple product identity (4.1): one replaces q by q^{10} and then chooses $z = -q^9$ there. □

Proof of Theorem 20 Using (4.2), we obtain

$$\sum_{n=1}^{\infty} \left(q^{n^2-1} - q^{5n^2-1}\right) = \frac{1}{2q}\left((q^2;q^2)_\infty \, (-q;q^2)_\infty^2 - (q^{10};q^{10})_\infty \, (-q^5;q^{10})_\infty^2\right). \tag{7.16}$$

Hence, the assertion of the theorem is equivalent to

$$0 = \frac{1}{2q}\frac{(q^2;q^2)_\infty (q^2;q^4)_\infty^2}{(q;q^2)_\infty^2} - \frac{1}{2q}\frac{(q^{10};q^{10})_\infty (q^{10};q^{20})_\infty^2}{(q^5;q^{10})_\infty^2} - \frac{(q^4,q^6,q^{10};q^{10})_\infty (q^2,q^{18};q^{20})_\infty}{(q^2;q^3;q^5)_\infty}.$$

Using the notation (6.1) with $N = 20$, we see that this is equivalent to

$$0 = \frac{1}{2} - \frac{1}{2}\frac{E_1^2 E_3^2 E_7^2 E_9^2}{E_2^3 E_4 E_6^3 E_8} - \frac{E_1^2 E_3 E_5^2 E_7 E_9^2}{E_2^3 E_6^2 E_8^2 E_{10}}.$$

The right-hand side is a modular function for the group $\Gamma_1(20)$. There are in total the 20 cusps exhibited in the proof of Theorem 15. Estimating the order of the right-hand side at $i\infty$, one obtains an upper bound of 4. Hence it is sufficient to show that the right-hand side has the form $0 + 0q + 0q^2 + 0q^3 + 0q^4 + \cdots$, which can be done routinely. □

Proof of Theorem 21 This is a special case of Corollary 35. □

Proof of Theorem 22 This is a special case of Theorem 34. □

Proof of Theorem 23 This is a special case of Corollary 35. □

Proof of Theorem 24 This is a special case of Theorem 34. □

Proof of Theorem 25 This is a special case of Theorem 34. □

Proof of Theorem 26 It is elementary to see that squares that are congruent to 1 modulo 48 are of the form N^2, where $N \equiv 1, 7, 17, 23 \pmod{24}$. If we now do a computation analogous to the one in the proof of Theorem 1, we obtain

$$\sum_{n=0}^{\infty}(-1)^{\lfloor (n+2)/4 \rfloor} q^{a_n} = (q^{24}, q^{13}, q^{11}; q^{24})_\infty + q(q^{24}, q^{19}, q^5; q^{24})_\infty. \tag{7.17}$$

Next, we observe that, using the notation (6.1) with $N = 24$, the assertion of the theorem is equivalent to

$$0 = 1 + \frac{E_5}{E_{11}} - \frac{E_2 E_6 E_8 E_{10}}{E_1 E_7 E_9 E_{11}}.$$

The right-hand side is a modular function for the group $\Gamma_1(24)$. The cusps for this group are computed as usual using *Magma*:

$$\left\{\infty, 0, \frac{1}{8}, \frac{1}{7}, \frac{1}{6}, \frac{2}{11}, \frac{1}{5}, \frac{5}{24}, \frac{3}{14}, \frac{2}{9}, \frac{1}{4}, \frac{7}{24}, \frac{11}{36}, \frac{1}{3}, \frac{3}{8}, \frac{7}{18}, \frac{5}{12}, \frac{4}{9}, \frac{11}{24}, \frac{1}{2}, \frac{9}{16}, \frac{5}{8}, \frac{2}{3}, \frac{3}{4}\right\}.$$

There are in total 24 cusps. Estimating the order of the right-hand side at $i\infty$, one obtains an upper bound of 4. Hence it is sufficient to show that the right-hand side has the form $0 + 0q + 0q^2 + 0q^3 + 0q^4 + \cdots$, which can be done routinely. □

Proof of Theorem 27 It is elementary to see that squares that are congruent to 25 modulo 48 are of the form S^2, where $S \equiv 5, 11, 13, 19 \pmod{24}$. If we now do a computation analogous to the one in the proof of Theorem 1, we obtain

$$\sum_{n=0}^{\infty}(-1)^{\lfloor 5n/4 \rfloor} q^{a_n} = (q^{24}, q^{17}, q^7; q^{24})_\infty - q^2(q^{24}, q^{23}, q; q^{24})_\infty. \tag{7.18}$$

Next, we observe that, using the notation (6.1) with $N = 24$, the assertion of the theorem is equivalent to

$$0 = 1 - \frac{E_1}{E_7} - \frac{E_2 E_6 E_8 E_{10}}{E_3 E_5 E_7 E_{11}}.$$

The right-hand side is a modular function for the group $\Gamma_1(24)$. There are in total the 24 cusps exhibited in the previous proof. Estimating the order of the right-hand side at $i\infty$, one obtains an upper bound of 4. Hence it is sufficient to show that the right-hand side has the form $0 + 0q + 0q^2 + 0q^3 + 0q^4 + \cdots$, which can be done routinely. □

Proof of Theorem 28 This is a special case of Theorem 36. □

Proof of Theorem 29 This is a special case of Theorem 36. □

Proof of Theorem 30 This theorem is a special case of Corollary 37. □

Proof of Theorem 31 This is a special case of Theorem 38. □

Proof of Theorem 32 This is a special case of Theorem 38. □

8 Theta Function Identities

The purpose of this section is, first of all, to present Weierstraß' addition formula for theta functions, and, second, to make two special cases explicit that are particularly used in the proofs of our theorems from Sects. 2 and 3 given in the next section, and also in the proofs in Sect. 10.

In this section and the following ones, we use a different notation for the theta functions that appear in our context, namely

$$\theta(\alpha; q) := (\alpha, q/\alpha; q)_\infty.$$

It should be noted that, up to a power of q, the function $E_g(q; N)$ that we used in Sects. 6 and 7 can be expressed as $\theta(q^g; q^N)$.

Using the above notation, Weierstraß' addition formula (cf. [7, p. 451, Example 5]) reads

$$\begin{aligned}\theta(xy; q)\,\theta(x/y; q)\,\theta(uv; q)\,\theta(u/v; q) - \theta(xv; q)\,\theta(x/v; q)\,\theta(uy; q)\,\theta(u/y; q) \\ = \frac{u}{y}\,\theta(yv; q)\,\theta(y/v; q)\,\theta(xu; q)\,\theta(x/u; q).\end{aligned} \tag{8.1}$$

Two specialisations of this formula are of particular importance in our context. If, in (8.1), we replace q by q^{3N} and specialise $x = q^N$, $y = u^2/q^N$, and $v = q^N/u$, then we obtain the relation

$$\begin{aligned}&\theta(u^2; q^{3N})\,\theta(q^{2N}/u^2; q^{3N})\,\theta(q^N; q^{3N})\,\theta(u^2/q^N; q^{3N}) \\ &\quad - \theta(q^{2N}/u; q^{3N})\,\theta(u; q^{3N})\,\theta(u^3/q^N; q^{3N})\,\theta(q^N/u; q^{3N}) \\ &\qquad = \frac{q^N}{u}\theta(u; q^{3N})\,\theta(u^3/q^{2N}; q^{3N})\,\theta(q^N u; q^{3N})\,\theta(q^N/u; q^{3N}),\end{aligned}$$

or, equivalently,

$$\begin{aligned}&\theta(u^3/q^N; q^{3N}) + \frac{q^N}{u}\,\theta(u^3/q^{2N}; q^{3N}) \\ &\qquad = \frac{\theta(u^2; q^{3N})\,\theta(q^{2N}/u^2; q^{3N})\,\theta(q^N; q^{3N})\,\theta(u^2/q^N; q^{3N})}{\theta(u; q^{3N})\,\theta(q^N/u; q^{3N})\,\theta(q^N u; q^{3N})}.\end{aligned}$$

Written in alternative notation, this is

$$(u^3/q^N, q^{4N}/u^3, q^{3N}; q^{3N})_\infty + \frac{q^N}{u}(u^3/q^{2N}, q^{5N}/u^3, q^{3N}; q^{3N})_\infty \qquad (8.2)$$
$$= \frac{(u^2/q^N, q^{2N}/u^2, q^N; q^N)_\infty}{(u, q^N/u; q^N)_\infty}.$$

Similarly, if in (8.1) we replace q by q^{3N} and specialise $x = q^{2N}$, $y = u^2/q^{2N}$, and $v = q^{2N}/u$, then we obtain the relation

$$\theta(u^2; q^{3N})\,\theta(q^{4N}/u^2; q^{3N})\,\theta(q^{2N}; q^{3N})\,\theta(u^2/q^{2N}; q^{3N})$$
$$- \theta(q^{4N}/u; q^{3N})\,\theta(u; q^{3N})\,\theta(u^3/q^{2N}; q^{3N})\,\theta(q^{2N}/u; q^{3N})$$
$$= \frac{q^{2N}}{u}\theta(u; q^{3N})\,\theta(u^3/q^{4N}; q^{3N})\,\theta(q^{2N}u; q^{3N})\,\theta(q^{2N}/u; q^{3N}),$$

or, equivalently,

$$\theta(u^3/q^{2N}; q^{3N}) - \frac{q^{3N}}{u^2}\theta(u^3/q^{4N}; q^{3N})$$
$$= \frac{\theta(u^2; q^{3N})\,\theta(q^{4N}/u^2; q^{3N})\,\theta(q^{2N}; q^{3N})\,\theta(u^2/q^{2N}; q^{3N})}{\theta(u; q^{3N})\,\theta(q^{2N}/u; q^{3N})\,\theta(u/q^N; q^{3N})}.$$

Written in alternative notation, this is

$$(u^3/q^{2N}, q^{5N}/u^3, q^{3N}; q^{3N})_\infty - \frac{q^{3N}}{u^2}(u^3/q^{4N}, q^{7N}/u^3, q^{3N}; q^{3N})_\infty \qquad (8.3)$$
$$= \frac{(u^2/q^{2N}, q^{3N}/u^2, q^N; q^N)_\infty}{(u/q^N, q^{2N}/u; q^N)_\infty}.$$

9 Proofs by Using the Weierstraß relation

In this section, we provide proofs of the theorems in Sects. 2 and 3 that utilise the Weierstraß relation (8.1). Again, for the theorems which are specialisations of the parametric theorems in Sect. 10, we refer to the proofs given there (which also make use the Weierstraß relation).

Proof of Theorem 1 Our point of departure is (7.2). By (8.2) with $N = 35$ and $u = q^{26}$, respectively with $N = 35$ and $u = q^{19}$, we get

$$\sum_{n=0}^{\infty}(-1)^{t(n)}q^{a_n} = \frac{(q^{17}, q^{18}, q^{35}; q^{35})_\infty}{(q^{26}, q^9; q^{35})_\infty} + q^4\frac{(q^3, q^{32}, q^{35}; q^{35})_\infty}{(q^{19}, q^{16}; q^{35})_\infty}.$$

If we now replace q by q^{35} and choose $u = q^{10}$, $v = q^3$, $x = q^{14}$, and $y = q^6$ in (8.1), we obtain[2]

$$\theta(q^{17}; q^{35})\,\theta(q^{11}; q^{35})\,\theta(q^{16}; q^{35})\,\theta(q^4; q^{35}) + q^4\theta(q^9; q^{35})\,\theta(q^3; q^{35})\,\theta(q^{24}; q^{35})\,\theta(q^4; q^{35})$$
$$= \theta(q^{20}; q^{35})\,\theta(q^8; q^{35})\,\theta(q^{13}; q^{35})\,\theta(q^7; q^{35}),$$

and thus the above right-hand side becomes

$$\frac{\theta(q^{20}; q^{35})\,\theta(q^8; q^{35})\,\theta(q^{13}; q^{35})\,\theta(q^7; q^{35})\,(q^{35}; q^{35})_\infty}{\theta(q^{16}; q^{35})\,\theta(q^9; q^{35})\,\theta(q^{11}; q^{35})\,\theta(q^4; q^{35})},$$

which is equivalent to the right-hand side of (2.1). □

Proof of Theorem 2 Our point of departure is (7.4). By (8.2) with $N = 35$ and $u = q^{33}$, respectively with $N = 35$ and $u = q^{23}$, we get

$$\sum_{n=0}^{\infty}(-1)^{t(n)}q^{a_n} = \frac{(q^{31}, q^4, q^{35}; q^{35})_\infty}{(q^{33}, q^2; q^{35})_\infty} - q\frac{(q^{11}, q^{24}, q^{35}; q^{35})_\infty}{(q^{23}, q^{12}; q^{35})_\infty}.$$

If we now replace q by q^{35} and choose $u = q^{15}$, $v = q^3$, $x = q^{17}$, and $y = q^{14}$ in (8.1), we obtain

$$\theta(q^{31}; q^{35})\,\theta(q^3; q^{35})\,\theta(q^{18}; q^{35})\,\theta(q^{12}; q^{35}) - \theta(q^{20}; q^{35})\,\theta(q^{14}; q^{35})\,\theta(q^{29}; q^{35})\,\theta(q; q^{35})$$
$$= q\theta(q^{17}; q^{35})\,\theta(q^{11}; q^{35})\,\theta(q^{32}; q^{35})\,\theta(q^2; q^{35}),$$

and thus the above right-hand side becomes

$$\frac{\theta(q^{20}; q^{35})\,\theta(q^{14}; q^{35})\,\theta(q^{29}; q^{35})\,\theta(q; q^{35})\,(q^{35}; q^{35})_\infty}{\theta(q^{17}; q^{35})\,\theta(q^{12}; q^{35})\,\theta(q^{32}; q^{35})\,\theta(q^2; q^{35})},$$

which is equivalent to the right-hand side of (2.2). □

Proof of Theorem 3 Our point of departure is (7.5). By (8.2) with $N = 35$ and $u = q^{31}$, respectively with $N = 35$ and $u = q^{24}$, we get

$$\sum_{n=0}^{\infty}(-1)^{\lfloor (n+4)/8\rfloor}q^{a_n} = \frac{(q^{27}, q^8, q^{35}; q^{35})_\infty}{(q^{31}, q^4; q^{35})_\infty} + q\frac{(q^{13}, q^{22}, q^{35}; q^{35})_\infty}{(q^{24}, q^{11}; q^{35})_\infty}.$$

If we now replace q by q^{35} and choose $u = q^{14}$, $v = q^9$, $x = q^{15}$, and $y = q^{13}$ in (8.1), we obtain

---

[2] In case the reader wonders how we came up with these choices of u, v, x, y (and the choices in subsequent proofs): after a lot of trial and error which produced some useful choices in certain cases, but did not lead to the recognition of any underlying patterns (we doubt in fact that there are), we decided to write a *Maple* programme that goes through all possible choices of u, v, x, y in non-negative powers of q and outputs choices that are appropriate to establish our identities.

$$\theta(q^{28};q^{35})\,\theta(q^{2};q^{35})\,\theta(q^{23};q^{35})\,\theta(q^{5};q^{35}) - \theta(q^{24};q^{35})\,\theta(q^{6};q^{35})\,\theta(q^{27};q^{35})\,\theta(q;q^{35})$$
$$= q\theta(q^{2};q^{35})\,\theta(q^{4};q^{35})\,\theta(q^{29};q^{35})\,\theta(q;q^{35}),$$

and thus the above right-hand side becomes

$$\frac{\theta(q^{28};q^{35})\,\theta(q^{2};q^{35})\,\theta(q^{23};q^{35})\,\theta(q^{5};q^{35})(q^{35};q^{35})_\infty}{\theta(q^{4};q^{35})\,\theta(q^{11};q^{35})\,\theta(q^{6};q^{35})\,\theta(q;q^{35})},$$

which is equivalent to the right-hand side of (2.3). □

Proof of Theorem 4 Our point of departure is (7.6). By (8.2) with $N = 35$ and $u = q^{32}$, respectively with $N = 35$ and $u = q^{18}$, we get

$$\sum_{n=0}^{\infty}(-1)^{t(n)}q^{a_n} = \frac{(q^{29},q^{6},q^{35};q^{35})_\infty}{(q^{32},q^{3};q^{35})_\infty} - q^5\frac{(q,q^{34},q^{35};q^{35})_\infty}{(q^{18},q^{17};q^{35})_\infty}.$$

If we now replace q by q^{35} and choose $u = q^7$, $v = q^{15}$, $x = q$, and $y = q^2$ in (8.1), we obtain

$$\theta(q^{17};q^{35})\,\theta(q^{8};q^{35})\,\theta(q^{6};q^{35})\,\theta(q^{13};q^{35}) - \theta(q^{9};q^{35})\,\theta(q^{16};q^{35})\,\theta(q^{14};q^{35})\,\theta(q^{5};q^{35})$$
$$= q^5\theta(q;q^{35})\,\theta(q^{8};q^{35})\,\theta(q^{22};q^{35})\,\theta(q^{3};q^{35})$$

after little manipulation. Thus, the above right-hand side becomes

$$\frac{\theta(q^{9};q^{35})\,\theta(q^{16};q^{35})\,\theta(q^{14};q^{35})\,\theta(q^{5};q^{35})(q^{35};q^{35})_\infty}{\theta(q^{17};q^{35})\,\theta(q^{3};q^{35})\,\theta(q^{13};q^{35})\,\theta(q^{8};q^{35})},$$

which is equivalent to the right-hand side of (2.4). □

Proof of Theorem 5 Our point of departure is (7.7). By (8.2) with $N = 35$ and $u = q^{29}$, respectively with $N = 35$ and $u = q^{34}$, we get

$$\sum_{n=0}^{\infty}(-1)^{\lfloor (n+4)/8\rfloor}q^{a_n} = \frac{(q^{23},q^{12},q^{35};q^{35})_\infty}{(q^{29},q^{6};q^{35})_\infty} + q\frac{(q^{33},q^{2},q^{35};q^{35})_\infty}{(q^{34},q;q^{35})_\infty}.$$

If we now replace q by q^{35} and choose $u = q^5$, $v = q^{14}$, $x = q^2$, and $y = q^4$ in (8.1), we obtain

$$\theta(q^{18};q^{35})\,\theta(q^{3};q^{35})\,\theta(q^{10};q^{35})\,\theta(q^{7};q^{35}) - \theta(q^{12};q^{35})\,\theta(q^{9};q^{35})\,\theta(q^{16};q^{35})\,\theta(q;q^{35})$$
$$= q\theta(q^{9};q^{35})\,\theta(q^{6};q^{35})\,\theta(q^{19};q^{35})\,\theta(q^{2};q^{35})$$

after little manipulation. Thus, the above right-hand side becomes

$$\frac{\theta(q^{18};q^{35})\,\theta(q^{3};q^{35})\,\theta(q^{10};q^{35})\,\theta(q^{7};q^{35})(q^{35};q^{35})_\infty}{\theta(q^{6};q^{35})\,\theta(q;q^{35})\,\theta(q^{16};q^{35})\,\theta(q^{9};q^{35})},$$

which is equivalent to the right-hand side of (2.5). □

Proof of Theorem 6 Our point of departure is (7.8). By (8.2) with $N = 35$ and $u = q^{27}$, respectively with $N = 35$ and $u = q^{22}$, we get

$$\sum_{n=0}^{\infty}(-1)^{t(n)}q^{a_n} = \frac{(q^{19},q^{16},q^{35};q^{35})_\infty}{(q^{27},q^{8};q^{35})_\infty} + q^2\frac{(q^{9},q^{26},q^{35};q^{35})_\infty}{(q^{22},q^{13};q^{35})_\infty}.$$

If we now replace q by q^{35} and choose $u = q^7$, $v = q^{16}$, $x = q^3$, and $y = q^5$ in (8.1), we obtain

$$\theta(q^{21};q^{35})\,\theta(q^{4};q^{35})\,\theta(q^{11};q^{35})\,\theta(q^{10};q^{35}) - \theta(q^{13};q^{35})\,\theta(q^{12};q^{35})\,\theta(q^{19};q^{35})\,\theta(q^{2};q^{35})$$
$$= q^2\theta(q^{9};q^{35})\,\theta(q^{8};q^{35})\,\theta(q^{23};q^{35})\,\theta(q^{2};q^{35})$$

after little manipulation. Thus, the above right-hand side becomes

$$\frac{\theta(q^{21};q^{35})\,\theta(q^{4};q^{35})\,\theta(q^{11};q^{35})\,\theta(q^{10};q^{35})(q^{35};q^{35})_\infty}{\theta(q^{8};q^{35})\,\theta(q^{13};q^{35})\,\theta(q^{12};q^{35})\,\theta(q^{2};q^{35})},$$

which is equivalent to the right-hand side of (2.6). □

Proof of Theorem 7 Our point of departure is (7.9). By (8.2) with $N = 40$ and $u = -q^{33}$, respectively with $N = 40$ and $u = -q^{23}$, we get

$$\sum_{n=0}^{\infty}(-1)^{\lfloor (n+2)/4\rfloor}q^{a_n} = \frac{(q^{26},q^{14},q^{40};q^{40})_\infty}{(-q^{33},-q^{7};q^{40})_\infty} + q^4\frac{(q^{6},q^{34},q^{40};q^{40})_\infty}{(-q^{23},-q^{17};q^{40})_\infty}.$$

If we now replace q by q^{40} and choose $u = -q^9$, $v = q^{16}$, $x = -q$, and $y = -q^5$ in (8.1), we obtain

$$\theta(-q^{21};q^{40})\,\theta(q^{8};q^{40})\,\theta(q^{10};q^{40})\,\theta(-q^{11};q^{40})$$
$$- \theta(q^{14};q^{40})\,\theta(-q^{15};q^{40})\,\theta(-q^{17};q^{40})\,\theta(q^{4};q^{40})$$
$$= q^4\theta(q^{6};q^{40})\,\theta(-q^{7};q^{40})\,\theta(-q^{25};q^{40})\,\theta(q^{4};q^{40})$$

after little manipulation. Thus, the above right-hand side becomes

$$\frac{\theta(-q^{21};q^{40})\,\theta(q^{8};q^{40})\,\theta(q^{10};q^{40})\,\theta(-q^{11};q^{40})\theta(q;q^{40})\,(q^{40};q^{40})_\infty}{\theta(-q^{17};q^{40})\,\theta(-q^{7};q^{40})\,\theta(-q^{15};q^{40})\,\theta(q^{4};q^{40})},$$

which is equivalent to the right-hand side of (2.7). □

Proof of Theorem 8 Our point of departure is the first two lines in (7.10). By (8.2) with $N = 40$ and $u = -q^{31}$, respectively with $N = 40$ and $u = -q^{39}$, we get

$$\sum_{n=0}^{\infty}(-1)^{\lfloor 5n/4\rfloor}q^{a_n} = \frac{(q^{22}, q^{18}, q^{40}; q^{40})_\infty}{(-q^{31}, -q^{9}; q^{40})_\infty} - q\frac{(q^{38}, q^{2}, q^{40}; q^{40})_\infty}{(-q^{39}, -q; q^{40})_\infty}.$$

If we now replace q by q^{40} and choose $u = -q^8$, $v = q^5$, $x = q^{17}$, and $y = q^7$ in (8.1), we obtain

$$\begin{aligned}\theta(q^{24}; q^{40})\,\theta(q^{10}; q^{40})\,\theta(-q^{13}; q^{40})\,\theta(-q^{3}; q^{40})&\\ -\,\theta(q^{22}; q^{40})\,\theta(q^{12}; q^{40})\,\theta(-q^{15}; q^{40})\,\theta(-q; q^{40})&\\ = -q\theta(q^{12}; q^{40})\,\theta(q^{2}; q^{40})\,\theta(-q^{25}; q^{40})\,\theta(-q^{9}; q^{40}),&\end{aligned}$$

and thus the above right-hand side becomes

$$\frac{\theta(q^{24}; q^{40})\,\theta(q^{10}; q^{40})\,\theta(-q^{13}; q^{40})\,\theta(-q^{3}; q^{40})(q^{40}; q^{40})_\infty}{\theta(-q; q^{40})\,\theta(-q^{9}; q^{40})\,\theta(q^{12}; q^{40})\,\theta(-q^{15}; q^{40})},$$

which is equivalent to the right-hand side of (2.8). □

Proof of Theorem 9 The point of departure is the first two lines in (7.11). By (8.2) with $N = 40$ and $u = -q^{37}$, respectively with $N = 40$ and $u = -q^{27}$, we get

$$\sum_{n=0}^{\infty}(-1)^{\lfloor (n+2)/4\rfloor}q^{a_n} = \frac{(q^{34}, q^{6}, q^{40}; q^{40})_\infty}{(-q^{37}, -q^{3}; q^{40})_\infty} + q\frac{(q^{14}, q^{26}, q^{40}; q^{40})_\infty}{(-q^{27}, -q^{13}; q^{40})_\infty}.$$

By (8.1) with $N = 40$, $u = -q^{16}$, $v = q^{11}$, $y = q^{15}$, and $x = q^{19}$, we get

$$\begin{aligned}\theta(q^{34}; q^{40})\,\theta(q^{4}; q^{40})\,\theta(-q^{27}; q^{40})\,\theta(-q^{5}; q^{40}) - \theta(q^{30}; q^{40})\,\theta(q^{8}; q^{40})\,\theta(-q^{31}; q^{40})\,\theta(-q; q^{40})&\\ = -q\,\theta(q^{26}; q^{40})\,\theta(q^{4}; q^{40})\,\theta(-q^{35}; q^{40})\,\theta(-q^{3}; q^{40}),&\end{aligned}$$

and thus the above right-hand side becomes

$$\frac{\theta(q^{30}; q^{40})\,\theta(q^{8}; q^{40})\,\theta(-q^{31}; q^{40})\,\theta(-q; q^{40})\,(q^{40}; q^{40})_\infty}{\theta(-q^{27}; q^{40})\,\theta(-q^{37}; q^{40})\,\theta(q^{4}; q^{40})\,\theta(-q^{5}; q^{40})},$$

which is equivalent to the right-hand side of (2.9). □

Proof of Theorem 10 Our point of departure is the first two lines in (7.12). By (8.2) with $N = 40$ and $u = -q^{29}$, respectively by (8.3) with $N = 40$ and $u = -q^{59}$, we get

$$\sum_{n=0}^{\infty}(-1)^{\lfloor 5n/4\rfloor}q^{a_n} = \frac{(q^{18}, q^{22}, q^{40}; q^{40})_\infty}{(-q^{29}, -q^{11}; q^{40})_\infty} - q^5\frac{(q^{38}, q^{2}, q^{40}; q^{40})_\infty}{(-q^{19}, -q^{21}; q^{40})_\infty}.$$

If we now replace q by q^{40} and choose $u = -q^{13}$, $v = -q^{15}$, $x = -q^3$, and $y = q^8$ in (8.1), we obtain

$$\begin{aligned}\theta(-q^{23}; q^{40})\,\theta(q^{10}; q^{40})\,\theta(q^{16}; q^{40})\,\theta(-q^7; q^{40})&\\ -\theta(-q^{21}; q^{40})\,\theta(q^{12}; q^{40})\,\theta(q^{18}; q^{40})\,\theta(-q^5; q^{40})&\\ = -q^5\theta(-q^{11}; q^{40})\,\theta(q^2; q^{40})\,\theta(q^{28}; q^{40})\,\theta(-q^5; q^{40})&\end{aligned}$$

after little manipulation. Thus, the above right-hand side becomes

$$\frac{\theta(-q^{23}; q^{40})\,\theta(q^{10}; q^{40})\,\theta(q^{16}; q^{40})\,\theta(-q^7; q^{40})(q^{40}; q^{40})_\infty}{\theta(-q^{11}; q^{40})\,\theta(-q^{19}; q^{40})\,\theta(q^{12}; q^{40})\,\theta(-q^5; q^{40})},$$

which is equivalent to the right-hand side of (2.10). □

Proof of Theorem 11 This is a special case of Theorem 36. □

Proof of Theorem 12 This is a special case of Theorem 36. □

Proof of Theorem 13 As we already said, this is a direct consequence of the Jacobi triple product identity (4.1): one replaces q by q^{10} and then chooses $z = -q^6$ there. □

Proof of Theorem 14 We start again with (cf. (7.13))

$$\sum_{n=0}^{\infty}\left(q^{n(n+1)} - q^{5n(n+1)+1}\right) = (q^2; q^2)_\infty\,(-q^2; q^2)_\infty^2 - q\,(q^{10}; q^{10})_\infty\,(-q^{10}; q^{10})_\infty^2.$$

The right-hand side can be rewritten as

$$\frac{(q^4; q^4)_\infty^2}{(q^2; q^2)_\infty} - q\frac{(q^{20}; q^{20})_\infty^2}{q^{10}; q^{10})_\infty} = \frac{(q^{20}; q^{20})_\infty^2}{q^{10}; q^{10})_\infty}\left(\frac{\theta(q^4; q^{20})\,\theta(q^8; q^{20})}{\theta(q^2; q^{20})\,\theta(q^6; q^{20})} - q\right). \tag{9.1}$$

If in (8.1) we choose $u = q^5$, $v = q^2$, $x = q^{12}$, and $y = q^4$, then we obtain

$$\begin{aligned}\theta(q^{16}; q^{20})\,\theta(q^8; q^{20})\,\theta(q^7; q^{20})\,\theta(q^3; q^{20}) - \theta(q^{14}; q^{20})\,\theta(q^{10}; q^{20})\,\theta(q^9; q^{20})\,\theta(q; q^{20})&\\ = q\,\theta(q^6; q^{20})\,\theta(q^2; q^{20})\,\theta(q^{17}; q^{20})\,\theta(q^7; q^{20}).&\end{aligned}$$

If this is used in (9.1), then we obtain the right-hand side of (2.14) after little manipulation. □

Proof of Theorem 15 We start again with (cf. (7.14))

$$1 + \sum_{n=1}^{\infty}\left(q^{n^2} + q^{5n^2}\right) = \frac{1}{2}\left((q^2; q^2)_\infty\,(-q; q^2)_\infty^2 + (q^{10}; q^{10})_\infty\,(-q^5; q^{10})_\infty^2\right). \tag{9.2}$$

Now, we have the relation[3] (cf. [2, Eq. (8.10.9)])

$$(q;q^2)_\infty^{-1} = (-q;q)_\infty. \tag{9.3}$$

Upon replacement of q by $-q$, we obtain the variant

$$(-q;q^2)_\infty^{-1} = (q;q^2)_\infty\,(-q^2;q^2)_\infty. \tag{9.4}$$

We use the latter identity in (9.2) and get

$$\begin{aligned}1+\sum_{n=1}^{\infty}\left(q^{n^2}+q^{5n^2}\right) &= \frac{1}{2}\left(\frac{(q^2;q^2)_\infty\,(-q;q^2)_\infty}{(q;q^2)_\infty\,(-q^2;q^2)_\infty} + (q^{10};q^{10})_\infty\,(-q^5;q^{10})_\infty^2\right)\\ &= \frac{1}{2}\left(\frac{\theta(q^2;q^{10})\,\theta(q^4;q^{10})\,(q^{10};q^{10})_\infty\,\theta(-q;q^{10})\,\theta(-q^3;q^{10})\,(-q^5;q^{10})_\infty}{\theta(q;q^{10})\,\theta(q^3;q^{10})\,(q^5;q^{10})_\infty\,\theta(-q^2;q^{10})\,\theta(-q^4;q^{10})\,(-q^{10};q^{10})_\infty}\right.\\ &\qquad\left. + (q^{10};q^{10})_\infty\,(-q^5;q^{10})_\infty^2\right).\end{aligned}$$

If, in the first expression within parentheses we use (9.4) with q replaced by q^5, then the above becomes

$$\begin{aligned}1+\sum_{n=1}^{\infty}\left(q^{n^2}+q^{5n^2}\right) &= \frac{(q^{10};q^{10})_\infty\,(-q^5;q^{10})_\infty^2}{2\,\theta(q;q^{10})\,\theta(q^3;q^{10})\,\theta(-q^2;q^{10})\,\theta(-q^4;q^{10})}\\ &\quad\times\Big(\theta(q^2;q^{10})\,\theta(q^4;q^{10})\,\theta(-q;q^{10})\,\theta(-q^3;q^{10})\\ &\qquad + \theta(q;q^{10})\,\theta(q^3;q^{10})\,\theta(-q^2;q^{10})\,\theta(-q^4;q^{10})\Big).\end{aligned}$$

We now choose $u=-q^3$, $v=q$, $x=q^4$, and $y=q^3$ in (8.1). This gives the relation

$$\begin{aligned}\theta(q^2;q^{10})\,\theta(q^4;q^{10})\,\theta(-q;q^{10})\,\theta(-q^3;q^{10}) + \theta(q;q^{10})\,\theta(q^3;q^{10})\,\theta(-q^2;q^{10})\,\theta(-q^4;q^{10})\\ = \theta(q^3;q^{10})\,\theta(-q^4;q^{10})\,\theta(q^5;q^{10})\,\theta(-1;q^{10}).\end{aligned}$$

If this is used in the above expression, then we obtain (2.15) after some manipulation, where (9.4) is again used with q replaced by q^5. □

Proof of Theorem 16 As we already said, this is a direct consequence of the Jacobi triple product identity (4.1): one replaces q by q^{10} and then chooses $z=-q^7$ there. □

Proof of Theorem 17 We start again with (cf. (7.15))

[3] In combinatorial terms, this is Euler's theorem that the number of partitions of n into odd parts equals the number of partitions of n into distinct parts.

$$\sum_{n=0}^{\infty}\left(q^{n(n+1)}+q^{5n(n+1)+1}\right)=(q^2;q^2)_\infty\,(-q^2;q^2)_\infty^2+q\,(q^{10};q^{10})_\infty\,(-q^{10};q^{10})_\infty^2.$$

The right-hand side can be rewritten as

$$\frac{(q^4;q^4)_\infty^2}{(q^2;q^2)_\infty}+q\frac{(q^{20};q^{20})_\infty^2}{q^{10};q^{10})_\infty}=\frac{(q^{20};q^{20})_\infty^2}{q^{10};q^{10})_\infty}\left(\frac{\theta(q^4;q^{20})\,\theta(q^8;q^{20})}{\theta(q^2;q^{20})\,\theta(q^6;q^{20})}+q\right). \tag{9.5}$$

If in (8.1) we choose $u=q^5$, $v=q^2$, $x=q^6$, and $y=q^4$, then we obtain

$$\begin{aligned}\theta(q^{10};q^{20})\,\theta(q^2;q^{20})\,\theta(q^7;q^{20})\,\theta(q^3;q^{20})-\theta(q^8;q^{20})\,\theta(q^4;q^{20})\,\theta(q^9;q^{20})\,\theta(q;q^{20})\\=q\,\theta(q^6;q^{20})\,\theta(q^2;q^{20})\,\theta(q^{11};q^{20})\,\theta(q;q^{20}).\end{aligned}$$

If this is used in (9.5), then we obtain the right-hand side of (2.17) after little manipulation. □

Proof of Theorem 18 As we already said, this is a direct consequence of the Jacobi triple product identity (4.1): one replaces q by q^{10} and then chooses $z=-q^8$ there. □

Proof of Theorem 19 As we already said, this is a direct consequence of the Jacobi triple product identity (4.1): one replaces q by q^{10} and then chooses $z=-q^9$ there. □

Proof of Theorem 20 We start again with (cf. (7.16))

$$\sum_{n=1}^{\infty}\left(q^{n^2-1}-q^{5n^2-1}\right)=\frac{1}{2q}\left((q^2;q^2)_\infty\,(-q;q^2)_\infty^2-(q^{10};q^{10})_\infty\,(-q^5;q^{10})_\infty^2\right). \tag{9.6}$$

The right-hand side is almost the same expression as the one on the right-hand side of (9.2), except for a multiplicative factor of $1/q$ and a changed sign. Hence, by proceeding as in the alternative proof of Theorem 15, we arrive at

$$\begin{aligned}\sum_{n=1}^{\infty}\left(q^{n^2-1}-q^{5n^2-1}\right)&=\frac{(q^{10};q^{10})_\infty\,(-q^5;q^{10})_\infty^2}{2q\,\theta(q;q^{10})\,\theta(q^3;q^{10})\,\theta(-q^2;q^{10})\,\theta(-q^4;q^{10})}\\&\quad\times\Big(\theta(q^2;q^{10})\,\theta(q^4;q^{10})\,\theta(-q;q^{10})\,\theta(-q^3;q^{10})\\&\qquad-\theta(q;q^{10})\,\theta(q^3;q^{10})\,\theta(-q^2;q^{10})\,\theta(-q^4;q^{10})\Big).\end{aligned}$$

We now choose $u=-q^3$, $v=q$, $x=q^4$, and $y=q^3$ in (8.1). This gives the relation

$$\begin{aligned}\theta(q^4;q^{10})\,\theta(q^2;q^{10})\,\theta(-q^3;q^{10})\,\theta(-q;q^{10})-\theta(-q^4;q^{10})\,\theta(-q^2;q^{10})\,\theta(q^3;q^{10})\,\theta(q;q^{10})\\=q\,\theta(-q^2;q^{10})\,\theta(-1;q^{10})\,\theta(q^5;q^{10})\,\theta(q;q^{10}).\end{aligned}$$

If this is used in the above expression, then we obtain (2.20) after some manipulation, where (9.4) is again used with q replaced by q^5. □

Proof of Theorem 21 This is a special case of Corollary 35. □

Proof of Theorem 22 This theorem is a special case of Theorem 34. □

Proof of Theorem 23 This is a special case of Corollary 35. □

Proof of Theorem 24 This theorem is a special case of Theorem 34. □

Proof of Theorem 25 This theorem is a special case of Theorem 34. □

Proof of Theorem 26 Using (8.2) with $N = 8$ and $u = q^7$, the right-hand side of (7.17) can be simplified into one product. This gives the claimed result. □

Proof of Theorem 27 Using (8.3) with $N = 8$ and $u = q^{11}$, the right-hand side of (7.18) can be simplified into one product. This gives the claimed result. □

Proof of Theorem 28 This theorem is a special case of Theorem 36. □

Proof of Theorem 29 This theorem is a special case of Theorem 36. □

Proof of Theorem 30 This theorem is a special case of Corollary 37. □

Proof of Theorem 31 This is a special case of Theorem 38. □

Proof of Theorem 32 This is a special case of Theorem 38. □

10 Parametric Families of Generating Functions for sequences of Squares

Here we present two results on generating functions for sequences of squares that contain parameters. The first of these results consists in Theorem 34 and Corollary 35, while the second consists in Theorem 36 and Corollary 37. For each theorem-corollary pair the proof is the same, but the theorem covers a parameter range that is different from that of the corollary. We conclude the section by stating, and proving, a uniform version of Theorems 31 and 32.

The following theorem covers Theorems 22, 24 and 25.

Theorem 34 *Let P be an odd prime power different from a power of 3, and let a be an odd positive integer relatively prime to P and less than P. Let $(a_n)_{n\ge 0}$ be the sequence of non-negative integers m such that $24Pm + a^2$ is a square.*

(1) If $a \equiv P \pmod 3$, then

$$\sum_{n=0}^{\infty}(-1)^{\lfloor (n+2)/4\rfloor} q^{a_n} = \frac{(q^{(P-a)/3}, q^{(2P+a)/3}, q^P; q^P)_\infty}{(q^{(P-a)/6}, q^{(5P+a)/6}; q^P)_\infty}. \tag{10.1}$$

(2) If $a \not\equiv P$ (mod 3), *then*

$$\sum_{n=0}^{\infty}(-1)^{\lfloor (n+2)/4\rfloor} q^{a_n} = \frac{(q^{(P+a)/3}, q^{(2P-a)/3}, q^P; q^P)_\infty}{(q^{(P+a)/6}, q^{(5P-a)/6}; q^P)_\infty}. \tag{10.2}$$

Proof We have to find all S such that

$$S^2 \equiv a^2 \pmod{24P}. \tag{10.3}$$

We claim that there are the following two cases:

(C1) If $a \equiv P$ (mod 3), then $S \equiv a, 2P - a, 4P + a, 6P - a$ (mod $6P$).
(C2) If $a \not\equiv P$ (mod 3), then $S \equiv a, 2P + a, 4P - a, 6P - a$ (mod $6P$).

It should be noted that the condition $a < P$ guarantees that, in both cases, the congruence classes above are listed in increasing order.

By assumption, a is odd. Hence, also S must be odd and automatically $S^2 \equiv a^2 \equiv 1$ (mod 8). Consequently, the congruence (10.3) can be reduced to

$$(S-a)(S+a) \equiv 0 \pmod{6P}.$$

Two solutions are immediate, namely $S \equiv a$ (mod $6P$) and $S \equiv -a$ (mod $6P$). There are two further possibilities:

$$S \equiv a \text{ (mod 6)} \quad \text{and} \quad S \equiv -a \text{ (mod } P), \tag{10.4}$$

and

$$S \equiv -a \text{ (mod 6)} \quad \text{and} \quad S \equiv a \text{ (mod } P). \tag{10.5}$$

Under our assumption that a is relatively prime to P, there are no other possibilities. For, writing $P = p^e$ with p an odd prime number, the simultaneous congruences

$$S \equiv a \text{ (mod } p^\alpha) \quad \text{and} \quad S \equiv -a \text{ (mod } p^\beta)$$

with $\alpha + \beta = e$ and both α and β positive would imply that $2a$ is divisible by p, a contradiction to our assumptions.

It is straightforward to see that, depending on whether $a \equiv P$ (mod 3) or not, the solutions to (10.4) respectively to (10.5) are given by the second and third option in (C1) and (C2) above.

We now discuss Case (C1). Here, we have

$$\sum_{n=0}^{\infty}(-1)^{\lfloor(n+2)/4\rfloor}q^{a_n}=\sum_{k=0}^{\infty}(-1)^k q^{\frac{1}{24P}((6Pk+a)^2-a^2)}+\sum_{k=0}^{\infty}(-1)^k q^{\frac{1}{24P}((6Pk+2P-a)^2-a^2)}$$
$$-\sum_{k=0}^{\infty}(-1)^k q^{\frac{1}{24P}((6Pk+4P+a)^2-a^2)}-\sum_{k=0}^{\infty}(-1)^k q^{\frac{1}{24P}((6Pk+6P-a)^2-a^2)}$$
$$=\sum_{k=0}^{\infty}(-1)^k q^{\frac{1}{2}(3Pk^2+ak)}+\sum_{k=0}^{\infty}(-1)^k q^{\frac{1}{2}(3Pk^2+(2P-a)k)+\frac{1}{6}(P-a)}$$
$$-\sum_{k=0}^{\infty}(-1)^k q^{\frac{1}{2}(3Pk^2+(4P+a)k)+\frac{1}{3}(2P+a)}-\sum_{k=0}^{\infty}(-1)^k q^{\frac{1}{2}(3Pk^2+(6P-a)k)+\frac{1}{2}(3P-a)}.$$

Again, sums can be put together in pairs, so that one obtains two sums over *all* integers k:

$$\sum_{n=0}^{\infty}(-1)^{\lfloor(n+2)/4\rfloor}q^{a_n}=\sum_{k=-\infty}^{\infty}(-1)^k q^{\frac{1}{2}(3Pk^2+ak)}+\sum_{k=-\infty}^{\infty}(-1)^k q^{\frac{1}{2}(3Pk^2+(2P-a)k)+\frac{1}{6}(P-a)}.$$

Now, to each of these sums we apply the Jacobi triple product identity (4.1) to get

$$\sum_{n=0}^{\infty}(-1)^{\lfloor(n+2)/4\rfloor}q^{a_n}=(q^{3P},q^{(3P+a)/2},q^{(3P-a)/2};q^{3P})_\infty$$
$$+q^{(P-a)/6}(q^{3P},q^{(5P-a)/2},q^{(P+a)/2};q^{3P})_\infty.$$

The sum of these two products simplifies to the single product on the right-hand side of (10.1) as is seen by applying (8.2) with $N=P$ and $u=q^{(5P+a)/6}$.
On the other hand, if we are in Case (C2), then we have

$$\sum_{n=0}^{\infty}(-1)^{\lfloor(n+2)/4\rfloor}q^{a_n}=\sum_{k=0}^{\infty}(-1)^k q^{\frac{1}{24P}((6Pk+a)^2-a^2)}+\sum_{k=0}^{\infty}(-1)^k q^{\frac{1}{24P}((6Pk+2P+a)^2-a^2)}$$
$$-\sum_{k=0}^{\infty}(-1)^k q^{\frac{1}{24P}((6Pk+4P-a)^2-a^2)}-\sum_{k=0}^{\infty}(-1)^k q^{\frac{1}{24P}((6Pk+6P-a)^2-a^2)}$$
$$=\sum_{k=0}^{\infty}(-1)^k q^{\frac{1}{2}(3Pk^2+ak)}+\sum_{k=0}^{\infty}(-1)^k q^{\frac{1}{2}(3Pk^2+(2P+a)k)+\frac{1}{6}(P+a)}$$
$$-\sum_{k=0}^{\infty}(-1)^k q^{\frac{1}{2}(3Pk^2+(4P-a)k)+\frac{1}{3}(2P-a)}-\sum_{k=0}^{\infty}(-1)^k q^{\frac{1}{2}(3Pk^2+(6P-a)k)+\frac{1}{2}(3P-a)}.$$

Again, sums can be put together in pairs, so that one obtains two sums over *all* integers k:

$$\sum_{n=0}^{\infty}(-1)^{\lfloor (n+2)/4\rfloor}q^{a_n}=\sum_{k=-\infty}^{\infty}(-1)^k q^{\frac{1}{2}(3Pk^2+ak)}+\sum_{k=-\infty}^{\infty}(-1)^k q^{\frac{1}{2}(3Pk^2+(2P+a)k)+\frac{1}{6}(P+a)}.$$

Now, to each of these sums we apply the Jacobi triple product identity (4.1) to get

$$\sum_{n=0}^{\infty}(-1)^{\lfloor (n+2)/4\rfloor}q^{a_n}=(q^{3P},q^{(3P+a)/2},q^{(3P-a)/2};q^{3P})_\infty$$
$$+q^{(P+a)/6}(q^{3P},q^{(5P+a)/2},q^{(P-a)/2};q^{3P})_\infty.$$

The sum of these two products simplifies to the single product on the right-hand side of (10.2) as is seen by applying (8.2) with $N=P$ and $u=q^{(5P-a)/6}$. □

The following corollary covers Theorems 21 and 23.

Corollary 35 *Let P be an odd prime power different from a power of 3, and let a be an odd positive integer relatively prime to P with $P<a<2P$. Let $(a_n)_{n\ge 0}$ be the sequence of non-negative integers m such that $24Pm+a^2$ is a square.*

(1) If $a\equiv P\pmod 3$, then

$$\sum_{n=0}^{\infty}(-1)^{\lfloor (n+2)/4\rfloor}q^{a_n}=\frac{(q^{(P-a)/3},q^{(2P+a)/3},q^P;q^P)_\infty}{(q^{(P-a)/6},q^{(5P+a)/6};q^P)_\infty}.\tag{10.6}$$

(2) If $a\not\equiv P\pmod 3$, then

$$\sum_{n=0}^{\infty}(-1)^{\lfloor 5n/4\rfloor}q^{a_n}=\frac{(q^{(P+a)/3},q^{(2P-a)/3},q^P;q^P)_\infty}{(q^{(P+a)/6},q^{(5P-a)/6};q^P)_\infty}.\tag{10.7}$$

Proof If one looks through the arguments of the proof of Theorem 34, then one sees that everything can be copied verbatim, until it comes to the description of the two cases to be considered: we have to adapt the order of the congruence classes, as shown below.

(C1') If $a\equiv P\pmod 3$, then $S\equiv 2P-a,a,6P-a,4P+a\pmod{6P}$.
(C2') If $a\not\equiv P\pmod 3$, then $S\equiv a,4P-a,2P+a,6P-a\pmod{6P}$.

While in Case (C1'), the rest of the proof can be copied, in Case (C2') this requires a change in sign in the original sum from $(-1)^{\lfloor (n+2)/4\rfloor}$ to $(-1)^{\lfloor 5n/4\rfloor}$. □

The following theorem covers Theorems 11, 12, 28 and 29.

Theorem 36 *Let P be an odd prime power different from a power of 3, and let a be a positive integer relatively prime to P and less than $P/2$. Let $(a_n)_{n\ge 0}$ be the sequence of non-negative integers m such that $3Pm+a^2$ is a square.*

(1) If $a\equiv P\pmod 3$, then

$$\sum_{n=0}^{\infty}(-1)^{\lfloor (n+2)/4\rfloor}q^{a_n}=\frac{(q^{(4P-4a)/3},q^{(2P+4a)/3},q^{2P};q^{2P})_\infty}{(q^{(5P-2a)/3},q^{(P+2a)/3};q^{2P})_\infty}. \tag{10.8}$$

(2) If $a\not\equiv P \pmod 3$*, then*

$$\sum_{n=0}^{\infty}(-1)^{\lfloor (n+2)/4\rfloor}q^{a_n}=\frac{(q^{(4P+4a)/3},q^{(2P-4a)/3},q^{2P};q^{2P})_\infty}{(q^{(5P+2a)/3},q^{(P-2a)/3};q^{2P})_\infty}. \tag{10.9}$$

Proof We have to find all S such that

$$S^2\equiv a^2 \pmod{3P}.$$

As before, there are two cases:

(C1) If $a\equiv P \pmod 3$, then $S\equiv a,\ P+a,\ 2P-a,\ 3P-a \pmod{3P}$.
(C2) If $a\not\equiv P \pmod 3$, then $S\equiv a,\ P-a,\ 2P+a,\ 3P-a \pmod{3P}$.

We now discuss Case (C1). Here, we have

$$\begin{aligned}\sum_{n=0}^{\infty}(-1)^{\lfloor (n+2)/4\rfloor}q^{a_n}&=\sum_{k=0}^{\infty}(-1)^k q^{\frac{1}{3P}((3Pk+a)^2-a^2)}+\sum_{k=0}^{\infty}(-1)^k q^{\frac{1}{3P}((3Pk+P+a)^2-a^2)}\\&\quad-\sum_{k=0}^{\infty}(-1)^k q^{\frac{1}{3P}((3Pk+2P-a)^2-a^2)}-\sum_{k=0}^{\infty}(-1)^k q^{\frac{1}{3P}((3Pk+3P-a)^2-a^2)}\\&=\sum_{k=0}^{\infty}(-1)^k q^{3Pk^2+2ak}+\sum_{k=0}^{\infty}(-1)^k q^{3Pk^2+2(P+a)k+\frac{1}{3}(P+2a)}\\&\quad-\sum_{k=0}^{\infty}(-1)^k q^{3Pk^2+2(2P-a)k+\frac{4}{3}(P-a)}-\sum_{k=0}^{\infty}(-1)^k q^{3Pk^2+2(3P-a)k+3P-2a}.\end{aligned}$$

Again, sums can be put together in pairs, so that one obtains two sums over *all* integers k:

$$\sum_{n=0}^{\infty}(-1)^{\lfloor (n+2)/4\rfloor}q^{a_n}=\sum_{k=-\infty}^{\infty}(-1)^k q^{3Pk^2+2ak}+\sum_{k=-\infty}^{\infty}(-1)^k q^{3Pk^2+2(P+a)k+\frac{1}{3}(P+2a)}.$$

Now, to each of these sums we apply the Jacobi triple product identity (4.1) to get

$$\begin{aligned}\sum_{n=0}^{\infty}(-1)^{\lfloor (n+2)/4\rfloor}q^{a_n}&=(q^{6P},q^{3P+2a},q^{3P-2a};q^{6P})_\infty\\&\qquad+q^{(P+2a)/3}(q^{6P},q^{5P+2a},q^{P-2a};q^{6P})_\infty.\end{aligned}$$

The sum of these two products simplifies to the single product on the right-hand side of (10.8) as is seen by applying (8.2) with $N=2P$ and $u=q^{(5P-2a)/3}$.

On the other hand, if we are in Case (C2), then we have

$$\sum_{n=0}^{\infty}(-1)^{\lfloor (n+2)/4\rfloor}q^{a_n}=\sum_{k=0}^{\infty}(-1)^k q^{\frac{1}{3P}((3Pk+a)^2-a^2)}+\sum_{k=0}^{\infty}(-1)^k q^{\frac{1}{3P}((3Pk+P-a)^2-a^2)}$$
$$-\sum_{k=0}^{\infty}(-1)^k q^{\frac{1}{3P}((3Pk+2P+a)^2-a^2)}-\sum_{k=0}^{\infty}(-1)^k q^{\frac{1}{3P}((3Pk+3P-a)^2-a^2)}$$
$$=\sum_{k=0}^{\infty}(-1)^k q^{3Pk^2+2ak}+\sum_{k=0}^{\infty}(-1)^k q^{3Pk^2+2(P-a)k+\frac{1}{3}(P-2a)}$$
$$-\sum_{k=0}^{\infty}(-1)^k q^{3Pk^2+2(2P+a)k+\frac{4}{3}(P+a)}-\sum_{k=0}^{\infty}(-1)^k q^{3Pk^2+2(3P-a)k+3P-2a}.$$

Again, sums can be put together in pairs, so that one obtains two sums over *all* integers k:

$$\sum_{n=0}^{\infty}(-1)^{\lfloor (n+2)/4\rfloor}q^{a_n}=\sum_{k=-\infty}^{\infty}(-1)^k q^{3Pk^2+2ak}+\sum_{k=-\infty}^{\infty}(-1)^k q^{3Pk^2+2(P-a)k+\frac{1}{3}(P-2a)}.$$

Now, to each of these sums we apply the Jacobi triple product identity (4.1) to get

$$\sum_{n=0}^{\infty}(-1)^{\lfloor (n+2)/4\rfloor}q^{a_n}=(q^{6P},q^{3P+2a},q^{3P-2a};q^{6P})_\infty$$
$$+q^{(P-2a)/3}(q^{6P},q^{5P-2a},q^{P+2a};q^{6P})_\infty.$$

The sum of these two products simplifies to the single product on the right-hand side of (10.8) as is seen by applying (8.2) with $N=2P$ and $u=q^{(5P+2a)/3}$. □

The following corollary covers Theorem 30.

Corollary 37 *Let P be an odd prime power different from a power of* 3, *and let a be a positive integer relatively prime to P with $P/2<a<P$. Let $(a_n)_{n\ge 0}$ be the sequence of non-negative integers m such that $3Pm+a^2$ is a square.*

(1) If $a\equiv P \pmod 3$, then

$$\sum_{n=0}^{\infty}(-1)^{\lfloor 5n/4\rfloor}q^{a_n}=\frac{(q^{(4P-4a)/3},q^{(2P+4a)/3},q^{2P};q^{2P})_\infty}{(q^{(5P-2a)/3},q^{(P+2a)/3};q^{2P})_\infty}. \tag{10.10}$$

(2) If $a\not\equiv P \pmod 3$, then

$$\sum_{n=0}^{\infty}(-1)^{\lfloor (n+2)/4\rfloor}q^{a_n}=\frac{(q^{(4P+4a)/3},q^{(2P-4a)/3},q^{2P};q^{2P})_\infty}{(q^{(5P+2a)/3},q^{(P-2a)/3};q^{2P})_\infty}. \tag{10.11}$$

Proof Here, one goes through the proof of Theorem 36. Everything can be copied, except that the description of the two cases to be considered now reads as shown below, and that this requires a modification of the sign in Case (C1').

(C1') If $a \equiv P \pmod 3$, then $S \equiv a, 2P - a, P + a, 3P - a \pmod{3P}$.
(C2') If $a \not\equiv P \pmod 3$, then $S \equiv P - a, a, 3P - a, 2P + a \pmod{3P}$. □

The following theorem covers Theorems 31 and 32.

Theorem 38 *Let a be 1 or 3. Furthermore, let $(a_n)_{n\ge0}$ be the sequence of non-negative integers m such that $16m + a^2$ is a square. Then*

$$\sum_{n=0}^{\infty} q^{a_n} = (q^8;q^8)_\infty\,(-q^{4+a};q^8)_\infty\,(-q^{4-a};q^8)_\infty. \tag{10.12}$$

Proof We have to find all S such that

$$S^2 \equiv a^2 \pmod{16},$$

or, equivalently,

$$(S-a)(S+a) \equiv 0 \pmod{16}.$$

Since a is odd, only one of the factors $N - a$ and $N + a$ can be divisible by 4. Hence, either $S \equiv a \pmod 8$ or $S \equiv -a \pmod 8$. Consequently, we have

$$\begin{aligned}\sum_{n=0}^{\infty} q^{a_n} &= \sum_{k=0}^{\infty} q^{\frac{1}{16}((8k+a)^2-a^2)} + \sum_{k=1}^{\infty} q^{\frac{1}{16}((8k-a)^2-a^2)}\\ &= \sum_{k=0}^{\infty} q^{4k^2+ak} + \sum_{k=1}^{\infty} q^{4k^2-ak} = \sum_{k=-\infty}^{\infty} q^{4k^2+ak}.\end{aligned}$$

The proof is completed by applying the Jacobi triple product identity (4.1). □

11 Consequences and Open Problems

In this section, we record a consequence of Theorem 1 that is inspired by earlier work of Andrews and the second author [1]. Furthermore, we end by reminding the reader of a conjecture from [3] related to Theorems 1–6.

In [1], the function $M_k(n)$ is defined as the number of partitions of n in which k is the least positive integer that is not a part and there are more parts $> k$ than there are parts $< k$. For example, if $n = 18$ and $k = 3$ then we have $M_3(18) = 3$ because the three partitions in question are

$$5+5+5+2+1, \quad 6+5+4+2+1, \quad \text{and} \quad 7+4+4+2+1.$$

Let $A(n)$ be the number of the partitions of n into parts not congruent to 0, 7, 8, 13, 15, 20, 22, 27, 28 (mod 35) and the parts congruent to 4, 9, 11, 16, 19, 24, 26, 31 (mod 35) have two colours.

We have the following corollary of Theorem 1.

Corollary 39 *Let k and n be positive integers. With a_n and $t(n)$ as in Theorem 1, we have*

$$(-1)^{k-1}\left(\sum_{j=-(k-1)}^{k}(-1)^j A\big(n-j(3j-1)/2\big)-\delta(n)\right)=\sum_{j=0}^{n}(-1)^{t(j)}M_k(n-a_j),$$

where

$$\delta(n)=\begin{cases}(-1)^{t(m)}, & \text{if } n=a_m,\\ 0, & \text{otherwise.}\end{cases}$$

Proof Elementary generating function calculus gives

$$\sum_{n=0}^{\infty}A(n)q^n=\frac{1}{(q,q^4;q^5)_\infty\,(q^2,q^3,q^4,q^5;q^7)_\infty}. \tag{11.1}$$

Similarly, it is not difficult to see that the generating function for the numbers $M_k(n)$ is given by

$$\sum_{n=0}^{\infty}M_k(n)q^n=\sum_{n=k}^{\infty}\frac{q^{\binom{k}{2}+(k+1)n}}{(q;q)_n}\begin{bmatrix}n-1\\k-1\end{bmatrix}, \tag{11.2}$$

where

$$\begin{bmatrix}n\\k\end{bmatrix}=\begin{cases}\dfrac{(1-q)(1-q^2)\cdots(1-q^n)}{(1-q)(1-q^2)\cdots(1-q^k)\,(1-q)(1-q^2)\cdots(1-q^{n-k})}, & \text{if } 0\leqslant k\leqslant n,\\ 0, & \text{otherwise,}\end{cases}$$

is the usual q-binomial coefficient.

Andrews and the second author [1] proved the following truncated form of Euler's pentagonal number theorem (1.1):

$$\frac{(-1)^{k-1}}{(q;q)_\infty}\sum_{n=-(k-1)}^{k}(-1)^n q^{n(3n-1)/2}=(-1)^{k-1}+\sum_{n=k}^{\infty}\frac{q^{\binom{k}{2}+(k+1)n}}{(q;q)_n}\begin{bmatrix}n-1\\k-1\end{bmatrix}. \quad (11.3)$$

Multiplying both sides of (11.3) by (11.1), and using Theorem 1 and (11.2), we obtain

$$(-1)^{k-1}\left(\left(\sum_{n=1}^{\infty}A(n)q^n\right)\left(\sum_{n=-(k-1)}^{k}(-1)^n q^{n(3n-1)/2}\right)-\sum_{n=0}^{\infty}(-1)^{t(n)}q^{a_n}\right)$$
$$=\left(\sum_{n=0}^{\infty}(-1)^{t(n)}q^{a_n}\right)\left(\sum_{n=0}^{\infty}M_k(n)q^n\right).$$

The assertion of the corollary now follows by comparing coefficients of q^n on both sides of this equation. □

According to (11.3), for $k>0$, the coefficients of q^n in the series

$$(-1)^{k-1}\left(\frac{1}{(q;q)_\infty}\sum_{j=-(k-1)}^{k}(-1)^j q^{j(3j-1)/2}-1\right)$$

are all zero for $0\leqslant n<k(3k+1)/2$, and for $n\geqslant k(3k+1)/2$ all the coefficients are positive. Related to this result on truncated pentagonal number series, we remark that there is substantial numerical evidence that there is in fact a stronger result.

Conjecture 40 *For $k>0$, the coefficients of q^n in the series*

$$(-1)^{k-1}\left(\frac{1}{(q;q)_\infty}\sum_{j=-(k-1)}^{k}(-1)^j q^{j(3j-1)/2}-1\right)(q,q^6,q^7;q^7)_\infty$$
$$=\frac{(-1)^k}{(q^2,q^3,q^4,q^5;q^7)_\infty}\sum_{j=k}^{\infty}(-1)^j q^{j(3j+1)/2}\left(1-q^{2j+1}\right)$$

are all zero for $0\leqslant n<k(3k+1)/2$ and $n=k(3k+1)/2+1$. For $n=k(3k+1)/2$ and $n\geqslant k(3k+1)/2+2$ all the coefficients are positive.

Remark. The equality above follows easily from (1.1) and little manipulation.

If we assume Conjecture 40, then we immediately deduce that the partition function $A(n)$ satisfies the following infinite families of linear inequalities.

Conjecture 41 *For $k>0$, we have*

$$(-1)^{k-1}\left(\sum_{j=-(k-1)}^{k}(-1)^j A\big(n-j(3j-1)/2\big)-\delta(n)\right)\geqslant 0,$$

with strict inequalities if $n = k(3k+1)/2$ *or* $n \geqslant k(3k+1)/2+2$.

To conclude the article, we want to recall a conjecture from [3] that is very similar in appearance to Conjecture 40 and is related (again via (1.1)) to Theorems 1–6.

Conjecture 42 *For* $k > 0$ *and* $S \in \{1,2,3,4,5,6\}$, *the coefficients of* q^n *in the series*

$$\frac{(-1)^k}{(q,q^4;q^5)_\infty}\sum_{j=k}^{\infty}(-1)^j q^{7j(j+1)/2-jS}\left(1-q^{(2j+1)S}\right)$$

and

$$\frac{(-1)^k}{(q^2,q^3;q^5)_\infty}\sum_{j=k}^{\infty}(-1)^j q^{7j(j+1)/2-jS}\left(1-q^{(2j+1)S}\right)$$

are non-negative.

Acknowledgement We thank the anonymous referees for a very careful reading of the original manuscript.

References

1. G. E. Andrews and M. Merca, *The truncated pentagonal number theorem*, J. Combin. Theory Ser. A **119** (2012), 1639–1643.
2. G. Gasper and M. Rahman, *Basic Hypergeometric Series*, Encyclopedia of Mathematics And Its Applications 35, Cambridge University Press, Cambridge, 1990.
3. M. Merca, *Truncated theta series and Rogers–Ramanujan functions*, Exp. Math. (to appear), https://doi.org/10.1080/10586458.2018.1542642.
4. R. Miranda, *Algebraic Curves and Riemann Surfaces*, Graduate Studies in Mathematics, vol. 5, Amer. Math. Soc., Providence, R.I., 1995.
5. T. Miyake, *Modular Forms*, translated from the Japanese by Y. Maeda, Springer–Verlag, Berlin, 1989.
6. P. Paule and C.-S. Radu, *Partition analysis, modular functions, and computer algebra*, in: Recent Trends in Combinatorics, IMA Vol. Math. Appl., vol. 159, Springer, Cham, 2016, pp. 511–543.
7. E. T. Whittaker and G. N. Watson, *A Course of Modern Analysis*, reprint of the 4th ed. (1927), Cambridge University Press, Cambridge, 1996.
8. Y. Yang, *Defining equations of modular curves*, Adv. Math. **204** (2006), 481–508.

A Theta Identity of Gauss Connecting Functions from Additive and Multiplicative Number Theory

Mircea Merca

Abstract Let α and β be two nonnegative integers such that $\beta < \alpha$. For an arbitrary sequence $\{a_n\}_{n\geqslant 1}$ of complex numbers, we investigate linear combinations of the form $\sum_{k\geqslant 1} S(\alpha k - \beta, n)a_k$, where $S(k, n)$ is the total number of k's in all the partitions of n into parts not congruent to 2 modulo 4. The general nature of the numbers a_n allows us to provide new connections between partitions and functions from multiplicative number theory.

Keywords Theta series · Partitions · Divisors

MSC 2010 11P81 · 05A19

1 Introduction

Questions involving divisors of an integer have long been studied and they underlie the deepest unsolved problems in number theory and related fields. The study of the partitions of an integer is much younger and has been continued with great interest since the time of Euler which is considered to be the founder of the subject. The history of both subjects is rich, interesting and very appealing. The two branches of number theory, additive and multiplicative, turn out to be related in many interesting ways. Some of the fascination of these branches of number theory lies in unexpected results, including connections between concepts that appear to be unrelated.

For example, the divisors of numbers have been studied from the point of view of partitions of integers for a long time [9]. While Euler's partition function $p(n)$ is clearly increasing, there is no apparent regularity in the sum of divisors function

$$\sigma(n) = \sum_{d|n} d.$$

M. Merca (✉)
Department of Mathematics, University of Craiova, Craiova, Romania
e-mail: mircea.merca@profinfo.edu.ro

Academy of Romanian Scientists, Ilfov 3, Sector 5, Bucharest, Romania

A. Bostan and K. Raschel (eds.), *Transcendence in Algebra, Combinatorics, Geometry and Number Theory*, Springer Proceedings in Mathematics & Statistics 373,
https://doi.org/10.1007/978-3-030-84304-5_9

However, it is well-know [2] that Euler's partition function $p(n)$ and the sum of divisors function satisfy common recursive relations with only $p(0)$ different from $\sigma(0)$:

$$\sum_{k=-\infty}^{\infty} (-1)^k p\big(n - k(3k-1)/2\big) = \delta_{0,n}, \quad \text{with} \quad p(0) = 1, \tag{1}$$

and

$$\sum_{k=-\infty}^{\infty} (-1)^k \sigma\big(n - k(3k-1)/2\big) = 0, \quad \text{with} \quad \sigma(0) \text{ replaced by } n, \tag{2}$$

where $\delta_{i,j}$ is the Kronecker delta and $p(k) = \sigma(k) = 0$ if $k < 0$. Beyond these common recursive relations there are really neat formulas that combine the two functions

$$np(n) = \sum_{k=0}^{n} \sigma(k) p(n-k) \tag{3}$$

and

$$\sigma(n) + \sum_{k=-\infty}^{\infty} (-1)^k \cdot k(3k-1)/2 \cdot p\big(n - k(3k-1)/2\big) = 0, \tag{4}$$

with $\sigma(0) = 0$. Euler recognized the mystery of these relations as early as the eighteenth century. Recently, these results were extended to the number of divisors function $\tau(n)$, obtaining new connections between partitions and divisors [4, 10, 11]. For instance, according to [11, Corollaries 2.1 and 5.2] we have

$$\tau(n) = \sum_{k=1}^{n} s_{o-e}(k) p(n-k)$$

and

$$\sum_{k=-\infty}^{\infty} (-1)^k \tau\big(n - k(3k-1)/2\big) = s_{o-e}(n),$$

where $s_{o-e}(n)$ is the difference between the number of parts in all the partitions of n into odd number of distinct parts and the number of parts in all the partitions of n into even number of distinct parts. Moreover, in [12] the author considered a factorization for the partial sum of Lambert series and showed that, on the multiplicative number theory side, these connections can be extended to other important functions. A generalization of this study can be seen in [16, 17] where the authors investigated new variants of the Lambert series factorization theorems.

The primary goal of this paper is to find new connections between partitions and functions from the multiplicative number theory. To do this, we consider the total number of k's in all the partitions of n into parts not congruent to 2 modulo 4, denoted by $S(k,n)$. For example, the partitions of 7 into parts not congruent to 2 modulo 4 are: $7, 5+1+1, 4+3, 4+1+1+1, 3+3+1, 3+1+1+1+1$, and $1+1+1+1+1+1+1$. Then we have $S(1,7)=17$, $S(2,7)=0$, $S(3,7)=4$, $S(4,7)=2$, $S(5,7)=1$, $S(6,7)=0$ and $S(7,7)=1$.

Let α and β be two integer such that $0 \leqslant \beta < \alpha$. For an arbitrary sequence $\{a_n\}_{n\geqslant 1}$ of complex numbers we consider

$$A(a,\alpha,\beta;n) = \sum_{k=1}^{\infty} S(\alpha k-\beta, n) a_k. \tag{5}$$

For example, if $a_n = 1$ for any positive integer n, then $A(1,\alpha,\beta;n)$ is the number of the parts which are congruent to $-\beta$ modulo α in all the partitions of n into parts not congruent to 2 modulo 4. On the other hand, we have:

$$\begin{aligned}
A(a,1,0;7) &= 17a_1 + 4a_3 + 2a_4 + a_5 + a_7, \\
A(a,2,0;7) &= 2a_2, \\
A(a,2,1;7) &= 17a_1 + 4a_2 + a_3 + a_4, \\
A(a,3,0;7) &= 4a_1, \\
A(a,3,1;7) &= a_2, \\
A(a,3,2;7) &= 17a_1 + 2a_2 + a_3.
\end{aligned}$$

Moreover, for $\alpha, \beta \in \mathbb{Z}$ such that $0 \leqslant \beta < \alpha$, and an arbitrary sequence $\{a_n\}_{n\geqslant 1}$ of complex numbers, we consider generalized Lambert series expansions of the form [17]

$$\sum_{\substack{n=1 \\ \alpha n-\beta \not\equiv 2 \mod 4}}^{\infty} \frac{a_n q^{\alpha n-\beta}}{1-q^{\alpha n-\beta}} = \sum_{n=1}^{\infty} B(a,\alpha,\beta;n) q^n, \qquad |q|<1. \tag{6}$$

The coefficients of the generalized Lambert series expansion on the left-hand side of the previous equation are given on the right by

$$B(a,\alpha,\beta;n) = \sum_{\substack{\alpha d-\beta | n \\ \alpha d-\beta \not\equiv 2 \mod 4}} a_d .$$

Clearly this sum runs over d such that $\alpha d - \beta$ is a positive divisor of n which is not congruent to 2 modulo 4. If $a_n = 1$ for any positive integer n, then $B(1,\alpha,\beta;n)$ is the number of the positive divisors of n which are congruent to $-\beta$ modulo α and are not congruent to 2 modulo 4.

In this paper, we combine the generalized Lambert series expansion (6) to the following theta identity which is often attributed to Gauss [2, p. 23, eqs. (2.2.13)]:

$$\sum_{n=0}^{\infty}(-q)^{n(n+1)/2}=\frac{(q^2;q^2)_\infty}{(-q;q^2)_\infty}. \tag{7}$$

Here and throughout this paper, we use the following customary q-series notation:

$$(a;q)_n=\begin{cases}1, & \text{for } n=0,\\ (1-a)(1-aq)\cdots(1-aq^{n-1}), & \text{for } n>0;\end{cases}$$
$$(a;q)_\infty=\lim_{n\to\infty}(a;q)_n.$$

Considering that the infinite product $(a;q)_\infty$ diverges when $a\neq 0$ and $|q|\geqslant 1$, whenever $(a;q)_\infty$ appears in a formula, we shall assume $|q|<1$.

Following Hirschhorn and Sellers [7], we denote by $\text{pod}(n)$ the number of partitions of n into parts not congruent to 2 modulo 4. We remark that the generating function of $\text{pod}(n)$ is the reciprocal of the product in (7), namely,

$$\sum_{n=1}^{\infty}\text{pod}(n)q^n=\frac{(q^2;q^4)_\infty}{(q;q)_\infty}=\frac{(-q;q^2)_\infty}{(q^2;q^2)_\infty}.$$

Rewriting (7) as

$$\frac{(-q;q^2)_\infty}{(q^2;q^2)_\infty}\sum_{n=0}^{\infty}(-q)^{n(n+1)/2}=1,$$

we deduce the following linear recurrence relation for the partition function $\text{pod}(n)$:

$$\sum_{j=0}^{\infty}(-1)^{j(j+1)/2}\,\text{pod}\big(n-j(j+1)/2\big)=\delta_{0,n}, \tag{8}$$

where $\delta_{i,j}$ is the Kronecker delta function and $\text{pod}(n)=0$ if n is a negative integer.

Andrews and Merca [3] defined $MP_k(n)$ to be the number of partitions of n in which the first part larger than $2k-1$ is odd and appears exactly k times. All other odd parts appear at most once. For example, $MP_2(19)=10$, and the partitions in question are $9+9+1, 9+5+5, 8+5+5+1, 7+7+3+2, 7+7+2+2+1, 7+5+5+2$, $6+5+5+3$, $6+5+5+2+1$, $5+5+3+2+2+2$, $5+5+2+2+2+2+1$.

In this article, we provide the following identity.

Theorem 1.1. *Let α, k, n be positive integers and let β be a nonnegative integer such that $\beta<\alpha$. For an arbitrary sequence $\{a_m\}_{m\geqslant 1}$ of complex numbers, we have*

$$(-1)^{k-1}\left(\sum_{j=0}^{2k-1}(-1)^{j(j+1)/2}A\big(a,\alpha,\beta;n-j(j+1)/2\big)-B(a,\alpha,\beta;n)\right)$$
$$=\sum_{j=1}^{n}B(a,\alpha,\beta;j)MP_k(n-j).$$

Considering this theorem we derive the following infinite family of linear inequalities.

Corollary 1.2. *Let α, k, n be positive integers and let β be a nonnegative integer such that $\beta < \alpha$. For an arbitrary sequence $\{a_m\}_{m\geqslant 1}$ of nonnegative real numbers, we have*

$$(-1)^{k-1}\left(\sum_{j=0}^{2k-1}(-1)^{j(j+1)/2}A\big(a,\alpha,\beta;n-j(j+1)/2\big)-B(a,\alpha,\beta;n)\right)\geqslant 0,$$

with strict inequality if $n \geqslant k(2k+1)$.

For example,

$$A(a,\alpha,\beta;n)-A(a,\alpha,\beta;n-1)\geqslant B(a,\alpha,\beta;n),$$
$$A(a,\alpha,\beta;n)-A(a,\alpha,\beta;n-1)-A(a,\alpha,\beta;n-3)+A(a,\alpha,\beta;n-6)\leqslant B(a,\alpha,\beta;n).$$

On the other hand, the limiting case $k\to\infty$ of Theorem 1.1 reads as follows.

Corollary 1.3. *Let α, β, n be nonnegative integers such that $\beta < \alpha$. For an arbitrary sequence $\{a_m\}_{m\geqslant 1}$ of complex numbers, we have*

$$\sum_{j=0}^{\infty}(-1)^{j(j+1)/2}A\big(a,\alpha,\beta;n-j(j+1)/2\big)=B(a,\alpha,\beta;n).$$

The general nature of the sequence $\{a_n\}_{n\geqslant 1}$ allows for applications of our results to many important functions from multiplicative number theory. The ordinary Lambert series expansions studied in [12, 16, 18] can be converted into a series of the form in (6) as follows:

$$\sum_{n=1}^{\infty}\frac{f(n)q^{\alpha n-\beta}}{1-q^{\alpha n-\beta}}=\sum_{n=1}^{\infty}\left(\sum_{\alpha d-\beta|n}f(d)\right)q^n,$$

where f is any of the following functions: the *Möbius function* $\mu(n)$, *Euler's totient function* $\phi(n)$, the generalized *sum of divisors function* $\sigma_\alpha(n)$, *Liouville's function* $\lambda(n)$, *von Mangoldt's function* $\Lambda(n)$, and *Jordan's totient function* $J_t(n)$. On the other hand, many identities of the form

$$\sum_{k=1}^{\lfloor n/\alpha \rfloor} S(\alpha k - \beta, n) f(k) = \sum_{k=1}^{n} B(f, \alpha, \beta; k) \operatorname{pod}(n-k) \tag{9}$$

can be easily derived.

These relations are important since there are rarely such simple and universal identities expressing formulas for an entire class of special arithmetic functions considered in the context of so many applications in number theory and combinatorics.

For example, the classical Möbius function $\mu(n)$ is an important multiplicative function in number theory and combinatorics. This function is defined for all positive integers n and has its values in $\{-1, 0, 1\}$ depending on the factorization of n into prime factors:

$$\mu(n) = \begin{cases} 0, & \text{if n has a squared prime factor,} \\ (-1)^k, & \text{if n is a product of k distinct primes.} \end{cases}$$

Considering the identity [6, Theorem 263]

$$\sum_{n=1}^{\infty} \frac{\mu(n) q^n}{1-q^n} = q,$$

and the expression (10) for the generating function of $A(a, \alpha, \beta; n)$, we deduce that

$$\sum_{n=1}^{\infty} A(\mu, 4, 0; n) q^n = \frac{(-q; q^2)_\infty}{(q^2; q^2)_\infty} \cdot q^4 = \sum_{n=4}^{\infty} \operatorname{pod}(n-4) q^n.$$

For $(a, \alpha, \beta) = (\mu, 4, 0)$, we see that the statement of Corollary 1.3 reduces to the recurrence relation (8). In addition, we remark that the partition function $\operatorname{pod}(n)$ can be expressed in terms of the classical Möbius function $\mu(n)$ as follows:

$$\operatorname{pod}(n) = \sum_{k=1}^{n+4} S(k, n+4) \mu(k).$$

Clearly this relation is a particular case of the identity (9). A similar decomposition was obtained in [15] for the Euler's partition function $p(n)$: for $n \geqslant 0$, $r \in \{1, 2\}$,

$$p(n) = (-1)^{r-1} \sum_{k=r}^{n+r} S_{n+r,k}^{(r)} \mu(k),$$

where $S_{n,k}^{(r)}$ is the number of k's in all the partitions of n with the smallest part at least r.

The rest of this paper is organized as follows. We first prove Theorem 1.1 in Sect. 2. In Sect. 3, we provide connections between the number of odd unitary divisors of n and the number of squarefree parts in all the partitions of n into parts not congruent to 2 modulo 4. In Sect. 4, we provide a decomposition of the absolute difference between the sum of odd positive divisors of n and the sum of even positive divisors of n in terms of the partition function $\mathrm{pod}(n)$.

2 Proof of Theorem 1.1

First we want the generating function for partitions into parts not congruent to 2 modulo 4 where z keeps track of the number of parts equal to k. This is

$$\frac{1}{1-zq^k}\prod_{\substack{n=1\\n\neq k}}^{\infty}\frac{1}{1-q^n}\prod_{n=0}^{\infty}(1-q^{4n+2}) = \frac{(q^2;q^4)_\infty(1-q^k)}{(q;q)_\infty(1-zq^k)}$$
$$= \frac{(-q;q^2)_\infty(1-q^k)}{(q^2;q^2)_\infty(1-zq^k)}.$$

For $k \not\equiv 2 \mod 4$, the generating function of the number of k's in all partitions of n into parts not congruent to 2 modulo 4 is given by

$$\sum_{n=0}^{\infty} S(k,n)q^n = \frac{d}{dz}\bigg|_{z=1}\frac{(-q;q^2)_\infty(1-q^k)}{(q^2;q^2)_\infty(1-zq^k)}$$
$$= \frac{q^k}{1-q^k}\cdot\frac{(-q;q^2)_\infty}{(q^2;q^2)_\infty}.$$

We can write

$$\sum_{\substack{k=1\\ \alpha k-\beta\not\equiv 2 \mod 4}}^{\infty}\left(\sum_{n=1}^{\infty} S(\alpha k-\beta,n)q^n\right)a_k = \frac{(-q;q^2)_\infty}{(q^2;q^2)_\infty}\sum_{\substack{k=1\\ \alpha k-\beta\not\equiv 2 \mod 4}}^{\infty}\frac{a_k q^{\alpha k-\beta}}{1-q^{\alpha k-\beta}}.$$

Thus, considering (5), we deduce the following expression for the generating function of $A(a,\alpha,\beta;n)$:

$$\sum_{n=1}^{\infty} A(a,\alpha,\beta;n)q^n = \frac{(-q;q^2)_\infty}{(q^2;q^2)_\infty}\sum_{\substack{k=1\\ \alpha k-\beta\not\equiv 2 \mod 4}}^{\infty}\frac{a_k q^{\alpha k-\beta}}{1-q^{\alpha k-\beta}}. \tag{10}$$

In [3], the authors considered the theta identity (7) and proved the following truncated form:

$$
\begin{aligned}
&\frac{(-q;q^2)_\infty}{(q^2;q^2)_\infty}\sum_{j=0}^{2k-1}(-q)^{j(j+1)/2} \\
&\quad = 1+(-1)^{k-1}\frac{(-q;q^2)_k}{(q^2;q^2)_{k-1}}\sum_{j=0}^{\infty}\frac{q^{k(2k+2j+1)}(-q^{2k+2j+3};q^2)_\infty}{(q^{2k+2j+2};q^2)_\infty}.
\end{aligned}
\tag{11}
$$

Multiplying both sides of (11) by

$$
\sum_{n=1}^{\infty} B(a,\alpha,\beta;n)q^n
$$

we obtain

$$
\begin{aligned}
&(-1)^{k-1}\left(\left(\sum_{n=1}^{\infty}A(a,\alpha,\beta;n)q^n\right)\left(\sum_{n=0}^{2k-1}(-q)^{n(n+1)/2}\right)-\sum_{n=1}^{\infty}B(a,\alpha,\beta;n)q^n\right)\\
&=\left(\sum_{n=1}^{\infty}B(a,\alpha,\beta;n)q^n\right)\left(\sum_{n=0}^{\infty}MP_k(n)q^n\right),
\end{aligned}
$$

where we have invoked the generating function for $MP_k(n)$ [3],

$$
\sum_{n=0}^{\infty}MP_k(n)q^n=\frac{(-q;q^2)_k}{(q^2;q^2)_{k-1}}\sum_{j=0}^{\infty}\frac{q^{k(2k+2j+1)}(-q^{2k+2j+3};q^2)_\infty}{(q^{2k+2j+2};q^2)_\infty}.
$$

The assertion of the theorem now follows easily by comparing coefficients of q^n on both sides of this equation.

3 Odd Unitary Divisors and Squarefree Parts

Taking into account that

$$
|\mu(n)|=\begin{cases}0, & \text{if n has a squared prime factor,}\\ 1, & \text{otherwise,}\end{cases}
$$

we deduce that $B(|\mu|,\alpha,\beta;n)$ is the number of the divisors of n of the form $\alpha d-\beta$ where d is a squarefree positive integer and $\alpha d-\beta$ is not congruent to 2 modulo 4. In addition, $A(|\mu|,\alpha,\beta;n)$ is the number of the parts which are of the form $\alpha k-\beta$, with k a squarefree positive integer, in all the partitions of n into parts not congruent to 2 modulo 4.

Recall that a natural number d is a *unitary divisor* of a number n if d is a divisor of n and d and n/d are coprime. It is well known [6, Theorem 264] that the sum over all the positive divisors of n of the absolute value of the Möbius function is equal to the number of unitary divisors of n,

$$\sum_{d|n} |\mu(d)| = 2^{\omega(n)},$$

where $\omega(n)$ is an additive function defined as the number of distinct primes dividing n. On the other hand, we have

$$\begin{aligned}
&\sum_{n=1}^{\infty} B(|\mu|, 1, 0; n) q^n \\
&= \sum_{\substack{n=1 \\ n \not\equiv 2 \mod 4}}^{\infty} \frac{|\mu(n)| q^n}{1-q^n} \\
&= \sum_{n=1}^{\infty} \frac{|\mu(n)| q^n}{1-q^n} - \sum_{n=1}^{\infty} \frac{|\mu(4n-2)| q^{4n-2}}{1-q^{4n-2}} \\
&= \sum_{n=1}^{\infty} \frac{|\mu(2n-1)| q^{2n-1}}{1-q^{2n-1}} + \sum_{n=1}^{\infty} \frac{|\mu(2n)| q^{2n}}{1-q^{2n}} - \sum_{n=1}^{\infty} \frac{|\mu(4n-2)| q^{4n-2}}{1-q^{4n-2}} \\
&= \sum_{n=1}^{\infty} \frac{|\mu(2n-1)| q^{2n-1}}{1-q^{2n-1}} + \sum_{n=1}^{\infty} \frac{|\mu(4n)| q^{4n}}{1-q^{4n}} \\
&= \sum_{n=1}^{\infty} \left(\sum_{2d-1|n} |\mu(2d-1)| \right) q^n,
\end{aligned}$$

where we have invoked that $\mu(4n) = 0$ for any positive integer n.

It is clear that $B(|\mu|, 1, 0; n)$ is the number of the odd unitary divisors of n, while $A(|\mu|, 1, 0; n)$ is the number of the squarefree parts in all the partitions of n into parts not congruent to 2 modulo 4.

We have the following particular results.

Corollary 3.1. *For $n \geqslant 0$, the number of squarefree parts in all the partitions of n into parts congruent to 2 modulo 4 equals*

$$\sum_{j=1}^{n} 2^{\omega_o(j)} \operatorname{pod}(n-j),$$

where $\omega_o(j)$ denotes the number of distinct odd primes dividing j.

For example, the number of squarefree parts in all the partitions of 7 into parts congruent to 2 modulo 4 is equal to

$$1+3+1+3+3+5+7=23.$$

On the other hand, we have

$$\begin{aligned}&\mathrm{pod}(6)+\mathrm{pod}(5)+2\,\mathrm{pod}(4)+\mathrm{pod}(3)+2\,\mathrm{pod}(2)+2\,\mathrm{pod}(1)+2\,\mathrm{pod}(0)\\&=5+4+6+2+2+2+2=23.\end{aligned}$$

Corollary 3.2. *For $k, n > 0$, the number of squarefree parts in all the partitions of n into parts congruent to 2 modulo 4 and the number of odd unitary divisors of n satisfy the identity*

$$\begin{aligned}&(-1)^{k-1}\left(\sum_{j=0}^{2k-1}(-1)^{j(j+1)/2}A\big(|\mu|,1,0;n-j(j+1)/2\big)-2^{\omega_o(n)}\right)\\&=\sum_{j=1}^{n}2^{\omega_o(j)}MP_k(n-j).\end{aligned}$$

4 Alternating Sum of Divisors Function

For any positive integer n we define $\sigma_{o-e}(n)$ to be the difference between the sum of odd positive divisors of n and the sum of even positive divisors of n, namely

$$\sigma_{o-e}(n)=\sum_{d|n}(-1)^{1+d}d.$$

If we consider $a_n = n$, $\alpha = 1$ and $\beta = 0$ then it is clear that

$$A(a,1,0;n)=n\cdot\mathrm{pod}(n).$$

On the other hand, considering $\sigma_{odd}(n)$ to be the sum of odd positive divisors of n and $\sigma_{even}(n)$ to be the sum of even positive divisors of n, we may write:

$$\begin{aligned}\sum_{n=1}^{\infty}B(a,1,0;n)q^n&=\sum_{n=1}^{\infty}\frac{nq^n}{1-q^n}-\sum_{n=1}^{\infty}\frac{(4n-2)q^{4n-2}}{1-q^{4n-2}}\\&=\sum_{n=1}^{\infty}\big(\sigma_{even}(n)+\sigma_{odd}(n)\big)q^n-2\sum_{n=1}^{\infty}\sigma_{odd}(n)q^{2n}\\&=\sum_{n=1}^{\infty}\big(\sigma_{even}(2n)-\sigma_{odd}(2n)\big)q^{2n}+\sum_{n=1}^{\infty}\sigma_{odd}(2n-1)q^{2n-1}\end{aligned}$$

$$= \sum_{n=1}^{\infty} |\sigma_{o-e}(n)| q^n,$$

where we have invoked that $\sigma_{even}(2n-1) = 0$.

We remark the following consequence of (10).

Corollary 4.1. *Let n be a positive integer. Then*

$$n \cdot \mathrm{pod}(n) = \sum_{j=1}^{n} |\sigma_{o-e}(j)| \cdot \mathrm{pod}(n-j).$$

According to Corollary 1.3, we obtain the following decomposition of the absolute difference between the sum of odd positive divisors of n and the sum of even positive divisors of n in terms of the partition function $\mathrm{pod}(n)$.

Corollary 4.2. *Let n be a positive integer. Then*

$$|\sigma_{o-e}(n)| = \sum_{j=0}^{\infty} (-1)^{j(j+1)/2} \big(n - j(j+1)/2\big) \cdot \mathrm{pod}\left(n - j(j+1)/2\right).$$

Theorem 1.1 provides the following truncated version of this corollary.

Corollary 4.3. *Let k and n be positive integers. Then*

$$(-1)^{k-1} \left(\sum_{j=0}^{2k-1} (-1)^{j(j+1)/2} \big(n - j(j+1)/2\big) \cdot \mathrm{pod}\left(n - j(j+1)/2\right) - |\sigma_{o-e}(n)| \right)$$
$$= \sum_{j=1}^{n} |\sigma_{o-e}(j)| MP_k(n-j).$$

Relevant to this corollary, it would be very appealing to have combinatorial interpretations of

$$\sum_{j=1}^{n} |\sigma_{o-e}(j)| MP_k(n-j).$$

5 Concluding Remarks

The results proved in this article continue the spirit of [10–18] by connecting the seemingly disparate branches of additive and multiplicative number theory in new and interesting ways. Our continued aim in exploring factorization theorems for generalized Lambert series is to branch out and provide new connections between the inherently multiplicative structure of the Lambert series generating functions and the additive theory of partitions. The investigation of these topics will be a fruitful source of new identities and insights to other special multiplicative and additive functions.

For further reading at the intersection of the additive and multiplicative branches of number theory, we recommend [1], [5, Ch. VII], [8, 19, 20].

References

1. Alladi, K., Erdős, P.: On an additive arithmetic function, *Pacific J. Math.* **71**(2), 275–294 (1977)
2. G.E. Andrews, *The Theory of Partitions*, Cambridge Mathematical Library, Cambridge University Press, Cambridge, 1998. Reprint of the 1976 original.
3. G.E. Andrews, M. Merca, Truncated theta series and a problem of Guo and Zeng, *J. Combin. Theory Ser. A*, **154** (2018) 610–619.
4. C. Ballantine, M. Merca, New convolutions for the number of divisors, *J. Number Theory*, **170** (2017), 17–34.
5. T. Cai, *The book of numbers*, World Scientific Publishing Co. Pte Ltd, New Jersey, 2017.
6. G.H. Hardy, E.M. Wright: *An Introduction to the Theory of Numbers*, *5th ed.*, Clarendon Press, Oxford, 1979
7. M.D. Hirschhorn, J.A. Sellers, Arithmetic properties of partitions with odd parts distinct, *Ramanujan J.* **22** (2010) 273–284.
8. M. Jameson, R. Schneider: Combinatorial applications of Möbius inversion, *Proc. Amer. Math. Soc.* **142**(9), 2965–2971 (2014)
9. P.A. MacMahon, Divisors of numbers and their continuations in the theory of partitions, *Proc. London Math. Soc.*, **s2-19**(1) (1921), 75–113.
10. M. Merca, A new look on the generating function for the number of divisors, *J. Number Theory*, **149** (2015), 57–69.
11. M. Merca, Combinatorial interpretations of a recent convolution for the number of divisors of a positive integer, *J. Number Theory*, **160** (2016), 60–75.
12. M. Merca, The Lambert series factorization theorem, *Ramanujan J.*, 44(2) (2017) 417–435.
13. M. Merca, New connections between functions from additive and multiplicative number theory, *Mediterr. J. Math.*, **15**:36 (2018).
14. M. Merca, M.D. Schmidt, A partition identity related to Stanley's theorem, *Amer. Math. Monthly*, **125**(10) (2018) 929–933.
15. M. Merca, M.D. Schmidt, The partition function $p(n)$ in terms of the classical Möbius function, *Ramanujan J.*, **49** (2019) 87–96.
16. M. Merca, M.D. Schmidt, Generating special arithmetic functions by Lambert series factorizations, *Contrib. Discrete Math.*, **14** (2019) 31–45.
17. M. Merca, M.D. Schmidt, Factorization theorems for generalized Lambert series and applications, *Ramanujan J.*, **51** (2020) 391–419.
18. M.D. Schmidt, New recurrence relations and matrix equations for arithmetic functions generated by Lambert series, *Acta Arith.*, **181** (2017) 355–367.
19. R. Schneider: Arithmetic of partitions and the q-bracket operator, *Proc. Amer. Math. Soc.* **145**(5), 1953–1968 (2017)
20. T. Wakhare: Special classes of q-bracket operators, *Ramanujan J.* **47** (2018) 309–316.

Combinatorial Quantum Field Theory and the Jacobian Conjecture

A. Tanasa

Abstract In this short review we first recall combinatorial or (0–dimensional) quantum field theory (QFT). We then give the main idea of a standard QFT method, called the intermediate field method, and we review how to apply this method to a combinatorial QFT reformulation of the celebrated Jacobian Conjecture on the invertibility of polynomial systems. This approach establishes a related theorem concerning partial elimination of variables that implies a reduction of the generic case to the quadratic one. Note that this does not imply solving the Jacobian Conjecture, because one needs to introduce a supplementary parameter for the dimension of a certain linear subspace where the system holds.

1 Introduction

In 1939, Keller formulated in [Kel39] the **Jacobian Conjecture**, as a remarkably simple and natural conjecture on the invertibility of polynomial systems. The conjecture states that a polynomial system $F : \mathbb{C}^n \to \mathbb{C}^n$ is invertible, and its inverse is polynomial, if and only if the determinant of the Jacobian matrix $J_F(z) = \left(\mathrm{d}F_i(z)/\mathrm{d}z_j\right)_{1\leq i,j\leq n}$ is a non-zero constant.[1]

In order to illustrate this celebrated conjecture, let us give a simple example. Let $n = 2$, $F(z_1, z_2) = (z_1 + z_2^3, z_2)$. The Jacobian writes:

[1]In [Kel39], Keller formulated this conjecture for $n = 2$ and polynomials with integral coefficients, but this conjecture was shortly after generalized to the form above.

Partially supported by the PN 09370102 grant.

A. Tanasa (✉)
LaBRI, CNRS UMR 5800, Université de Bordeaux, Talence, France
e-mail: adrian.tanasa@labri.fr

Horia Hulubei National Institute of Physics and Nuclear Engineering, Magurele, Romania

I. U. F., Paris, France

A. Bostan and K. Raschel (eds.), *Transcendence in Algebra, Combinatorics, Geometry and Number Theory*, Springer Proceedings in Mathematics & Statistics 373,
https://doi.org/10.1007/978-3-030-84304-5_10

$$\det \begin{pmatrix} \frac{dF_1}{dz_1} & \frac{dF_1}{dz_2} \\ \frac{dF_2}{dz_1} & \frac{dF_2}{dz_2} \end{pmatrix} = \det \begin{pmatrix} 1 & 3z_2^2 \\ 0 & 1 \end{pmatrix} = 1. \tag{1}$$

One easily finds: $F^{-1}(z_1, z_2) = (z_1 - z_2^3, z_2)$.

Despite several efforts (of mathematicians such as Gröbner [Gro61], Oda [Oda80], to name only a couple), and various promising partial results, the Jacobian Conjecture remains unsolved. An introduction to the problem, the context, and the state of the art up to 1982, can be found in the seminal paper [BCW82].

We say that $F : \mathbb{C}^n \to \mathbb{C}^n$ is a *polynomial system* if all the coordinate functions F_j's are polynomials. Let us call $\mathcal{P}_n$ the set of such functions. For a function F, define

$$J_F(z) = \left(\frac{\mathrm{d}}{\mathrm{d}z_j} F_i(z) \right)_{1 \le i, j \le n}, \tag{2}$$

the corresponding Jacobian matrix and the two subspaces of $\mathcal{P}_n$:

Definition 1.1.

$$\mathcal{J}_n^{\mathrm{lin}} := \{F \in \mathcal{P}_n \mid \det J_F(z) = c \in \mathbb{C}^\times\}, \tag{3}$$

$$\mathcal{J}_n := \{F \in \mathcal{P}_n \mid F \textit{ is invertible}\}. \tag{4}$$

It will often be sufficient to analyse the subset of $\mathcal{J}^{\mathrm{lin}}$ such that $\det J_F(z) = 1$. The Jacobian Conjecture rewrites as

Conjecture 1.2 (Jacobian Conjecture, [Kel39]).

$$\mathcal{J}_n^{\mathrm{lin}} = \mathcal{J}_n \qquad \forall n\,. \tag{5}$$

Define the *total degree* of F, $\deg(F)$, as $\max_j \deg(F_j(z))$, and introduce the subspaces

$$\mathcal{P}_{n,d} = \{F \in \mathcal{P}_n \mid \deg(F) \le d\} \tag{6}$$

and similarly for $\mathcal{J}$ and $\mathcal{J}^{\mathrm{lin}}$.

Let us now mention two positive results on the conjecture:

1. a theorem for the quadratic case ($d = 2$), established first in [Wan80], and then in [Oda80] (see also [Wri89, Lemma 3.5] and [BCW82, Thm. 2.4, pag. 298]).

Theorem 1.3 ([Wan80]).

$$\mathcal{J}_{n,2}^{\mathrm{lin}} = \mathcal{J}_{n,2} \qquad \forall n\,. \tag{7}$$

2. a reduction theorem, from the general case to the cubic case, that one can find in Bass *et al.* [BCW82, Sec. II]).

Theorem 1.4 ([BCW82]).

$$\mathcal{J}_{n,3}^{\text{lin}} = \mathcal{J}_{n,3} \ \forall n \quad \Longrightarrow \quad \mathcal{J}_{n}^{\text{lin}} = \mathcal{J}_{n} \ \forall n \,. \tag{8}$$

The result of [dGST16] that we will review here from a quantum field theoretical (QFT) perspective is a reduction theorem, somewhat analogous to the one above, that tries to "fill the gap" between the cases of degree two and three. However, we need to introduce an adaptation of the statement of the Jacobian conjecture with one more parameter.

Definition 1.5. *For $n' \leq n$ and $F \in \mathcal{P}_{n,d}$, we write $z = (z_1, z_2)$ and $F = (F_1, F_2)$ to distinguish components in the two subspaces $\mathbb{C}^{n'} \times \mathbb{C}^{n-n'} \equiv \mathbb{C}^n$. We set $R(z_2; z_1)$ as a synonymous for $F_2(z_1, z_2)$, emphasizing that, in R, we consider z_2 as the variables in a polynomial system, and z_1 as parameters.*

The invertibility of R, denoted by $R(\cdot; z_1) \in \mathcal{J}_{n-n',d}$, for a fixed z_1, means that there exists a polynomial R^{-1} with variables $y_2 \in \mathbb{C}^{n-n'}$, and depending on z_1, such that $R^{-1}(R(z_2; z_1); z_1) = z_2$, for all $z_2 \in \mathbb{C}^{n-n'}$.

We then define the following subspaces of $\mathcal{P}_{n,d}$

$$\mathcal{J}_{n,d;n'} := \{F \in \mathcal{P}_{n,d} \mid R(\cdot; z_1) \in \mathcal{J}_{n-n',d}, \forall z_1 \in \mathbb{C}^{n'} \text{ and } F^{-1} \text{ restricted to } \mathbb{C}^{n'} \times \{0\} \text{ is in } \mathcal{P}_{n'}\}$$

$$\mathcal{J}_{n,d;n'}^{\text{lin}} := \{F \in \mathcal{P}_{n,d} \mid R \in \mathcal{J}_{n-n',d} \text{ and } (\det J_F)(z_1, R^{-1}(0, z_1)) = c \in \mathbb{C}^{\times}, \ \forall z_1 \in \mathbb{C}^{n'}\}$$

Note that the first definition is a generalisation of $\mathcal{J}_{n,d}$. Let us now state the reduction theorem we will review in this paper:

Theorem 1.6 ([dGST16]). *For $n \in \mathbb{N}$ and $d \geq 3$, there exists an injective map $\Phi : \mathcal{P}_{n,d} \mapsto \mathcal{P}_{n(n+1),d-1}$ satisfying*

$$\Phi(\mathcal{J}_{n,d}^{\text{lin}}) \equiv \mathcal{J}_{n(n+1),d-1;n}^{\text{lin}} \cap Im(\Phi)\,; \qquad \Phi(\mathcal{J}_{n,d}) \equiv \mathcal{J}_{n(n+1),d-1;n} \cap Im(\Phi)\,, \tag{9}$$

where $Im(\Phi) = \Phi(\mathcal{P}_{n,d})$.

Combining Theorem 1.4 and the theorem above, the full Jacobian Conjecture reduces to the question whether

$$\mathcal{J}_{n(n+1),2;n}^{\text{lin}} \cap \text{Im}(\Phi) = \mathcal{J}_{n(n+1),2;n} \cap \text{Im}(\Phi), \ \forall n. \tag{10}$$

2 Combinatorial (or 0–dimensional) QFT and the Intermediate Field Method

In this section we introduce combinatorial (or 0–dimensional) QFT and we then give the general idea of the so-called intermediate field method.

2.1 Combinatorial (or 0–dimensional) QFT

Usually in QFT, the scalar field φ is function of space ($\mathbb{R}^D$), see for example the review papers [Abd03b] or [Tan12], and $D = 4$.

However, one can consider the case $D = 0$. In this case the scalar field φ is not a function of space (since there is no space anymore), and φ is simply a (real or complex) variable. Some authors refer to this simpler formulation of QFT as **combinatorial QFT**, and we will do the same here.

Thus, in the real case, the *partition function* (which, from a combinatorial point of view, is a generating function) of the so-called φ^4 model writes as the integral

$$Z = \int_{\mathbb{R}} d\varphi \, e^{-\frac{1}{2}\varphi^2 + \frac{\lambda}{4!}\varphi^4}. \tag{11}$$

The constant λ is called the QFT **coupling constant**. The term φ^4 is called an interaction term of degree four.

It is worth emphasizing here that in combinatorial QFT, functional integrals (which are particularly difficult to rigorously define for $D \geq 1$) become usual (real or complex) integrals. Combinatorial QFT is thus much easier to handle from a mathematical point of view.

However, combinatorial QFT still presents a certain interest for the mathematical physicist because it can be seen as some kind of "laboratory" to test the usual QFT mathematical tools. Thus, one needs to evaluate integrals of the same type as in usual QFT, namely integrals of the type

$$\frac{\lambda^n}{n} \int d\varphi \, e^{-\varphi^2/2} \left(\frac{\varphi^4}{4!}\right)^n \tag{12}$$

coming from the so-called QFT perturbative expansion (which comes from Taylor expanding the exponential in (11) and then dealing with each term one by one, instead of dealing with the integral as a whole). In order to evaluate this type of integrals, one can use standard QFT techniques. Namely, one can define

$$Z_0(J) = \int_{\mathbb{R}} d\varphi \, e^{-\varphi^2/2 + J\phi} \, , \tag{13}$$

where the additional field J (here a real variable) is called the QFT **source**.

Using this QFT source technique, the $(2k)$–point correlation functions can be computed in the following way:

$$\int_{\mathbb{R}} d\phi \, e^{-\phi^2/2} \varphi^{2k} = \frac{\partial^{2k}}{\partial J^{2k}} \int_{\mathbb{R}} d\varphi \, e^{-\varphi^2/2 + J\varphi} |_{J=0} = \frac{\partial^{2k}}{\partial J^{2k}} e^{J^2/2} |_{J=0}. \tag{14}$$

2.2 *The Intermediate Field Method*

We give now the general idea of the so-called intermediate field method. Note that we present this method in the case of combinatorial QFT, but this idea can generalize to arbitrary D.

Thus, the intermediate field method consists of introducing a new field, σ, used to rewrite the interaction in a way that allows for the degree of the interaction to be reduced.

In order to illustrate this, let us take the example of the φ^6 model, where φ is, as above, a real 0–dimensional field (thus, a real variable).

The partition function of the φ^6 combinatorial QFT model writes:

$$Z(\lambda) = \int_{\mathbb{R}} \frac{d\varphi}{\sqrt{2\pi}} e^{-\frac{1}{2}\varphi^2} e^{-\lambda\varphi^6} . \tag{15}$$

The intermediate field model consists of rewriting this partition function with the help of a supplementary integral on the intermediate field σ in the following way:

$$Z(\lambda) = \int_{\mathbb{R}} \frac{d\varphi}{\sqrt{2\pi}} e^{-\frac{1}{2}\varphi^2} \int_{\mathbb{R}} \frac{d\sigma}{\sqrt{2\pi}} e^{-\frac{1}{2}\sigma^2} e^{\imath\sqrt{2\lambda}\varphi^3\sigma} . \tag{16}$$

Note that this allowed one to replace the interacting term φ^6 of (15) by a lower degree interacting term $\varphi^3\sigma$. This lowering of the degree of the interaction can be particularly useful in various contexts, such as the one of the Jacobian Conjecture we are dealing with here.

3 The Jacobian Conjecture as Combinatorial QFT Model (the Abdesselam-Rivasseau Model)

In [Abd03a], the Jacobian Conjecture was re-expressed as combinatorial QFT model which we call here the Abdesselam-Rivasseau model.

Let $n, d \geq 1$, and let $F \in \mathcal{P}_{n,d}$. Invertibility of $F_1(z)$ is equivalent to the invertibility of $F_2(z) := F_1(Rz + u)$, for $R \in \mathrm{GL}(n, \mathbb{C})$ and $u \in \mathbb{C}^n$, and F_1, F_2 have the same degree, thus w.l.o.g. we can assume that $F(z) = z + \mathcal{O}(z^2)$. The coordinate functions of F can thus be written as

$$F_i(z) = z_i - \sum_{k=2}^{d} \sum_{j_1,\ldots,j_k=1}^{n} w^{(k)}_{i,j_1\ldots j_k} z_{j_1}\ldots z_{j_k} =: z_i - \sum_{k=2}^{d} W_i^{(k)}(z) , \tag{17}$$

for $i \leq n$ and $w^{(k)}_{i,j_1\ldots j_k}$ some coefficients which are nothing but the combinatorial QFT coupling constants (see Sect. 2 above).

The partition function of the Abdesselam-Rivasseau combinatorial QFT model writes

$$Z(J,K)=\int_{\mathbb{C}^n} \mathrm{d}\varphi \mathrm{d}\varphi^\dagger e^{-\varphi^\dagger\varphi+\varphi^\dagger\sum_{k=2}^{d}W^{(k)}(\varphi)+J^\dagger\varphi+\varphi^\dagger K},$$

where J, K are vectors in $\mathbb{C}^n$ (and they are nothing but the combinatorial QFT sources, see again Sect. 2). The measure of the integral above is

$$\mathrm{d}\varphi \mathrm{d}\varphi^\dagger := \prod_{i=1}^{n} \frac{\mathrm{dRe}\varphi_i \ \mathrm{dIm}\varphi_i}{\pi}, \tag{18}$$

and we have used the standard QFT notations:

$$\varphi^\dagger K := \sum_{i=1}^{n} \varphi_i^\dagger K_i, \quad J^\dagger\varphi := \sum_{i=1}^{n} J_i^\dagger \varphi_i \ . \tag{19}$$

Note that the Abdesselam-Rivasseau model has very particular Feynman graphs obtained through perturbative expansion (see the original article [Abd03a] for more details).

Independently of this, setting the coupling constants to zero (the free theory), the partition function is calculated by Gaussian integration:

$$\int_{\mathbb{C}^n} \mathrm{d}\varphi \mathrm{d}\varphi^\dagger e^{-\varphi^\dagger\varphi+J^\dagger\varphi+\varphi^\dagger K} = e^{J^\dagger K}. \tag{20}$$

Let us now stress the following two crucial points (see again [Abd03a] for details):

1. The inverse G of F corresponds to the 1–point correlation function:

$$G_i(u)=\frac{\int_{\mathbb{C}^n} \mathrm{d}\varphi \mathrm{d}\varphi^\dagger \varphi_i e^{-\varphi^\dagger\varphi+\varphi^\dagger\sum_{k=2}^{d}W^{(k)}(\varphi)+\varphi^\dagger u}}{\int_{\mathbb{C}^n} \mathrm{d}\varphi \mathrm{d}\varphi^\dagger e^{-\varphi^\dagger\varphi+\varphi^\dagger\sum_{k=2}^{d}W^{(k)}(\varphi)+\varphi^\dagger u}} \ . \tag{21}$$

2. The partition function Z coincides with the inverse of the Jacobian:

$$Z(0,u)=\det(\partial G(u)) = JG(u) = \frac{1}{JF(G(u))}. \tag{22}$$

The sets of polynomial functions involved in the Jacobian Conjecture can be written in this framework:

$$\mathcal{J}_{n,d}^{\mathrm{lin}} = \{F \in \mathcal{P}_{n,d} \mid Z(0,u)=1, \ \forall u \in \mathbb{C}^n\},$$
$$\mathcal{J}_{n,d} = \{F \in \mathcal{P}_{n,d} \mid G_i(u) \text{ given by (21) is in } \mathcal{P}_n\}.$$

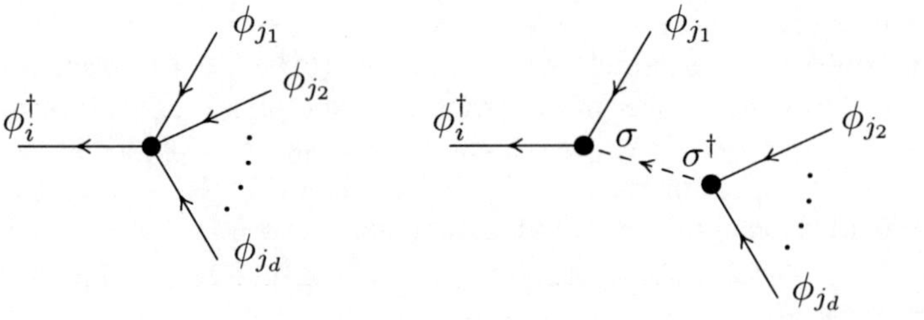

Fig. 1 Illustration of the intermediate field method. On the left hand side, the interaction term before applying the intermediate field method, and on the right hand side the interaction term after applying the method

4 The Intermediate Field Method for the Abdesselam-Rivasseau Model

As we have seen in Subsect. 2.2, applying the combinatorial QFT intermediate field method will reduce the degree d of F.

We need to add n^2 complex intermediate fields σ_{ij} $(i, j = 1, \dots, n)$ to the model. Indeed, we have, from the general formula (20) of Gaussian integration,

$$e^{(\varphi_i^\dagger \varphi_j)\left(\sum_{j_2,\dots,j_d=1}^{n} w^{(d)}_{i,j,j_2\cdots j_d}\varphi_{j_2}\cdots\varphi_{j_d}\right)}$$
$$= \int_{\mathbb{C}^{n^2}} \mathrm{d}\sigma_{i,j}\mathrm{d}\sigma_{i,j}^\dagger e^{-\sigma_{i,j}^\dagger\sigma_{i,j}+\sigma_{i,j}^\dagger\left(\sum_{j_2,\dots,j_d=1}^{n} w^{(d)}_{i,j,j_2\cdots j_d}\varphi_{j_2}\cdots\varphi_{j_d}\right)+(\varphi_i^\dagger\varphi_j)\sigma_{i,j}} . \tag{23}$$

We now use the identity (23), for each pair (i, j), in the partition function of the model with n dimensions and degree d, in order to re-express the monomials of degree d in the fields φ. This leads to:

$$Z(J, K) = \int_{\mathbb{C}^n} \mathrm{d}\varphi\mathrm{d}\varphi^\dagger \int_{\mathbb{C}^{n^2}} \mathrm{d}\sigma\mathrm{d}\sigma^\dagger e^{-\varphi^\dagger\varphi+\varphi^\dagger\sum_{k=2}^{d-1}W^{(k)}(\varphi)+J^\dagger\varphi+\varphi^\dagger K}$$
$$e^{\sum_{i,j=1}^{n}\left(-\sigma_{i,j}^\dagger\sigma_{i,j}+\sigma_{i,j}^\dagger\sum_{j_2,\dots,j_d=1}^{n} w^{(d)}_{i,j,j_2\cdots j_d}\varphi_{j_2}\cdots\varphi_{j_d}+\varphi_i^\dagger\varphi_j\sigma_{i,j}\right)}.$$

For a graphical representation of the intermediate field method leading to the model in dimension $n(n+1)$, see Fig. 1.

We define the new vector ϕ of $\mathbb{C}^{n+n^2}$ by $\phi = (\varphi_1, \dots, \varphi_n, \sigma_{1,1}, \dots, \sigma_{1,n}, \cdots, \sigma_{n,1}, \dots, \sigma_{n,n})$. We further define the interaction coupling constants $\tilde{w}$ as:

- for $k = d-1$, we set $\tilde{w}^{(d-1)}_{i,j,j_2\cdots j_d} := w^{(d-1)}_{i,j,j_2\cdots j_d}$ and $\tilde{w}^{(d-1)}_{i\cdot n+j,j_2\cdots j_d} = w^{(d)}_{i,j,j_2\cdots j_d}$ with $i, j, j_2, \dots j_n \leq n$
- for $k \in \{3, \dots, d-2\}$, we set $\tilde{w}^{(k)}_{i,j,j_2\cdots j_k} := w^{(k)}_{i,j,j_2\cdots j_k}$ with $i, j, j_2, \dots j_n \leq n$
- for $k = 2$, we set $\tilde{w}^{(2)}_{i,j,j_2} := w^{(2)}_{i,j,j_2}$ and $\tilde{w}^{(2)}_{i,j,i\cdot n+j} = 1$ with $i, j, j_2 \leq n$.

The remaining coefficients of $\tilde{w}$ are set to 0.

In the same way, the external sources are defined to be $\tilde{J} = (J, 0)$ and $\tilde{K} = (K, 0)$, where, of course, the number of extra vanishing coordinates is n^2. It is important to note that these external sources have fewer degrees of freedom than coordinates (n vs. $n(n+1)$). We also remark that, for generic d, in order to have a relation adapted to induction, it is crucial to consider an inhomogeneous model, since the intermediate field method creates terms of degrees $d-1$ (the term $\sigma^\dagger \phi_{j_2} \dots \phi_{j_d}$) and the term $\phi_i^\dagger \phi_{j_1} \sigma$ - see again Fig. 1.

The resulting QFT model, after applying the intermediate field method has partition function

$$Z(\tilde{J}, \tilde{K}) = \int_{\mathbb{C}^{n+n^2}} \mathrm{d}\phi \mathrm{d}\phi^\dagger e^{-\phi^\dagger \phi + \phi^\dagger \sum_{k=2}^{d-1} \tilde{W}^{(k)}(\phi) + \tilde{J}^\dagger \phi + \phi^\dagger \tilde{K}}$$

and 1–point correlation function

$$G_i(u) = \frac{\int_{\mathbb{C}^{n+n^2}} \mathrm{d}\phi \mathrm{d}\phi^\dagger \phi_i e^{-\phi^\dagger \phi + \phi^\dagger \sum_{k=2}^{d-1} \tilde{W}^{(k)}(\phi) + \phi^\dagger \tilde{u}}}{\int_{\mathbb{C}^{n+n^2}} \mathrm{d}\phi \mathrm{d}\phi^\dagger e^{-\phi^\dagger \phi + \phi^\dagger \sum_{k=2}^{d-1} \tilde{W}^{(k)}(\phi) + \phi^\dagger \tilde{u}}},$$

for $i \in \{1, \dots, n\}$.

5 A Particular Case: $n = 2$, $d = 3$

As an illustration of the general mechanism presented in Sections 3 and 4, we present here in detail the particular case $n = 2$, $d = 3$.

One thus has $F \in \mathcal{P}_{2,3}$ and

$$F_i(z) = z_i - \sum_{k=2}^{3} \sum_{j_1, \dots, j_k = 1}^{2} w_{i, j_1 \dots j_k}^{(k)} z_{j_1} \dots z_{j_k} =: z_i - \sum_{k=2}^{3} W_i^{(k)}(z), \tag{24}$$

for $i = 1, 2$ and $w_{i, j_1 \dots j_k}^{(k)}$ the combinatorial QFT coupling constants.

In this particular case, the partition function of the Abdesselam-Rivasseau combinatorial QFT model writes

$$Z(J, K) = \int_{\mathbb{C}^2} \mathrm{d}\varphi \mathrm{d}\varphi^\dagger e^{-\varphi^\dagger \varphi + \varphi^\dagger \sum_{k=2}^{3} W^{(k)}(\varphi) + J^\dagger \varphi + \varphi^\dagger K},$$

where the combinatorial QFT sources J, K are now vectors in $\mathbb{C}^2$.

The measure of the integral above is

$$\mathrm{d}\varphi \mathrm{d}\varphi^\dagger := \prod_{i=1}^{2} \frac{\mathrm{dRe}\varphi_i \, \mathrm{dIm}\varphi_i}{\pi}, \tag{25}$$

The standard QFT notations write explicitly:

$$\varphi^\dagger K := \sum_{i=1}^{2} \varphi_i^\dagger K_i, \quad J^\dagger \varphi := \sum_{i=1}^{2} J_i^\dagger \varphi_i \ . \tag{26}$$

One has

1. The inverse G of F corresponds to the 1–point correlation function:

$$G_i(u) = \frac{\int_{\mathbb{C}^2} \mathrm{d}\varphi \mathrm{d}\varphi^\dagger \varphi_i e^{-\varphi^\dagger\varphi + \varphi^\dagger \sum_{k=2}^{3} W^{(k)}(\varphi) + \varphi^\dagger u}}{\int_{\mathbb{C}^n} \mathrm{d}\varphi \mathrm{d}\varphi^\dagger e^{-\varphi^\dagger\varphi + \varphi^\dagger \sum_{k=2}^{3} W^{(k)}(\varphi) + \varphi^\dagger u}} \ . \tag{27}$$

2. The partition function Z coincides with the inverse of the Jacobian:

$$Z(0, u) = \det(\partial G(u)) = JG(u) = \frac{1}{JF(G(u))}. \tag{28}$$

We thus need to add 4 complex intermediate fields σ_{ij} $(i, j = 1, 2)$ to the model. One then has the identity:

$$\begin{aligned} & e^{(\varphi_i^\dagger \varphi_j)\left(\sum_{j_2, j_3=1}^{2} w^{(3)}_{i,j,j_2,j_3} \varphi_{j_2}\varphi_{j_3}\right)} \\ & = \int_{\mathbb{C}} \mathrm{d}\sigma_{i,j} \mathrm{d}\sigma_{i,j}^\dagger e^{-\sigma_{i,j}^\dagger \sigma_{i,j} + \sigma_{i,j}^\dagger \left(\sum_{j_2, j_3=1}^{2} w^{(3)}_{i,j,j_2,j_3} \varphi_{j_2}\varphi_{j_3}\right) + (\varphi_i^\dagger \varphi_j)\sigma_{i,j}} \ . \end{aligned} \tag{29}$$

We now use the identity above, for each of the four pairs (i, j), in the partition function of the model with 2 dimensions and degree 3, in order to re-express the monomials of degree 3 in the fields φ. This leads to:

$$\begin{aligned} Z(J, K) = & \int_{\mathbb{C}^2} \mathrm{d}\varphi \mathrm{d}\varphi^\dagger \int_{\mathbb{C}^4} \mathrm{d}\sigma \mathrm{d}\sigma^\dagger e^{-\varphi^\dagger\varphi + \varphi^\dagger W^{(2)}(\varphi) + J^\dagger\varphi + \varphi^\dagger K} \\ & e^{\sum_{i,j=1}^{2} \left(-\sigma_{i,j}^\dagger \sigma_{i,j} + \sigma_{i,j}^\dagger \sum_{j_2, j_3=1}^{2} w^{(d)}_{i,j,j_2 \cdots j_d} \varphi_{j_2} \cdots \varphi_{j_d} + \varphi_i^\dagger \varphi_j \sigma_{i,j}\right)} . \end{aligned}$$

This thus leads to a QFT model of dimension 6 (2 fields φ_1 and φ_2 and four intermediate fields $\sigma_{1,1}, \sigma_{1,2}, \sigma_{2,1}, \sigma_{2,2}$). We define a new vector ϕ of $\mathbb{C}^6$ by

$$\phi = (\varphi_1, \varphi_2, \sigma_{1,1}, \sigma_{1,2}, \sigma_{2,1}, \sigma_{2,2}).$$

The coupling constants $\tilde{w}$ is defined accordingly (see the previous section).

The QFT sources are also vectors of $\mathbb{C}^6$ and are defined to be

$$\tilde{J} := (J_1, J_2, 0, 0, 0, 0) \text{ and } \tilde{K} := (K_1, K_2, 0, 0, 0, 0) \tag{30}$$

The resulting QFT model, after applying the intermediate field method in this particular case writes:

$$Z(\tilde{J}, \tilde{K}) = \int_{\mathbb{C}^6} \mathrm{d}\phi \mathrm{d}\phi^\dagger e^{-\phi^\dagger\phi + \phi^\dagger \tilde{W}^{(2)}(\phi) + \tilde{J}^\dagger \phi + \phi^\dagger \tilde{K}}$$

Finally, the 1–point correlation function writes in this particular case:

$$G_i(\tilde{u}) = \frac{\int_{\mathbb{C}^6} \mathrm{d}\phi \mathrm{d}\phi^\dagger \phi_i e^{-\phi^\dagger\phi + \phi^\dagger \tilde{W}^{(2)}(\phi) + \phi^\dagger \tilde{u}}}{\int_{\mathbb{C}^{n+n^2}} \mathrm{d}\phi \mathrm{d}\phi^\dagger e^{-\phi^\dagger\phi + \phi^\dagger \tilde{W}^{(2)}(\phi) + \phi^\dagger \tilde{u}}},$$

for $i \in \{1, 2\}$.

We have thus illustrated the reduction of the QFT model with dimension 2 and degree 3 to the model with dimension 6 and degree 2.

6 Concluding Remarks

We have thus showed, by a QFT-inspired method, that the partition function (resp. the one-point correlation function) of the model with dimension $n \in \mathbb{N}$ and degree $d \in \mathbb{N} \setminus \{1, 2\}$ is equal to the partition function (resp. the n first coordinates of the one-point correlation function) of the model with dimension $n(n+1)$ and degree $d-1$, up to a redefinition of the coupling constant $w \mapsto \tilde{w}$ and a trivial redefinition of the external sources. Since, as announced above, the partition function corresponds to the inverse of the Jacobian (resp. the one-point correlation function corresponds to the formal inverse), this gives a QFT proof of Theorem 1.6. This represents a reduction theorem to the quadratic case for Jacobian Conjecture, up to the addition of a new parameter, related to the introduction of additional intermediate fields σ.

An algebraic proof (not using any QFT arguments) of Theorem 1.6 was also derived in the original article [dGST16]. This proofs rely on two intermediate lemmas:

Lemma 6.1 (Partial elimination, linearized version). *Let $N = n_1 + n_2$, and $S \in \mathcal{P}_N$. Write $z = (z_1, z_2)$ for $z \in \mathbb{K}^N = \mathbb{K}^{n_1} \times \mathbb{K}^{n_2}$ and so on. Call $R(z_2; z_1) = S_2(z_1, z_2)$ the system in $\mathcal{P}_{n_2}$, where z_1 coordinates are intended as parameters. Assume by hypothesis that $R(\cdot; z_1) \in \mathcal{J}_{n_2}$ for all $z_1 \in \mathbb{K}^{n_1}$. Define $H(z_1; y_2) = S_1(z_1, R^{-1}(y_2; z_1)) \in \mathcal{P}_{n_1}$. We have*

$$S \in \mathcal{J}^{\mathrm{lin}}_{N;n_1} \quad \textit{iff} \quad H(\cdot; 0) \in \mathcal{J}^{\mathrm{lin}}_{n_1}. \tag{31}$$

Lemma 6.2 (Partial elimination, invertible version). *Let S, R and H as in Lemma 6.1. In particular, assume by hypothesis that $R(\cdot; z_1) \in \mathcal{J}_{n_2}$ for all $z_1 \in \mathbb{K}^{n_1}$. We have*

$$S(z) \in \mathcal{J}_{N;n_1} \quad \textit{iff} \quad H(\cdot; 0) \in \mathcal{J}_{n_1}. \tag{32}$$

References

[Abd03a] A. Abdesselam. The Jacobian Conjecture as a Problem of Perturbative Quantum Field Theory. *Annales Henri Poincaré*, 4:199–215, 2003.

[Abd03b] Abdelmalek Abdesselam. Feynman diagrams in algebraic combinatorics. *Séminaire Lotharingien de Combinatoire*, B49c, 2003.

[BCW82] C. W. Bass, E. H. Connell, and D. Wright. The Jacobian conjecture: reduction of degree and formal expansion of the inverse. *Bull. Amer. Math. Soc.*, 7(2):287–330, 1982.

[dGST16] Axel de Goursac, Andrea Sportiello, and Adrian Tanasa. The Jacobian Conjecture, a Reduction of the Degree to the Quadratic Case. *Annales Henri Poincare*, 17(11):3237–3254, 2016.

[Gro61] W. Grobner. Sopra un teorema du B. Segre. *Atti. Accad. Naz. Lincei Rend. Cl. Sci. Fis. Mat. Nat.*, 31:118–122, 1961.

[Kel39] O. H. Keller. Ganze Cremona Transformations. *Monats. Math. Phys.*, 47:299–306, 1939.

[Oda80] S. Oda. The Jacobian problem and the simply-connectedness of $\mathbf{a}^n$ over a field k of characteristic zero. Osaka University preprint, 1980.

[Tan12] Adrian Tanasa. Some combinatorial aspects of quantum field theory. *Séminaire Lotharingien de Combinatoire*, B65g, 2012.

[Wan80] S. Wang. A Jacobian criterion for separability. *Journal of Algebra*, 65:453–494, 1980.

[Wri89] D. Wright. The tree formulas for reversion of power series. *Journal of Pure and Applied Algebra*, 57(2):191–211, 1989.

How Regular are Regular Singularities?

Herwig Hauser

Abstract An abstract look at the work of Fuchs and Frobenius on the solutions of ordinary differential equations at regular singularites.

The differential equation

$$x^2y'' - 3xy' + 3y = 0 \tag{1}$$

has solutions $y_1 = x$ and $y_2 = x^3$. The equation

$$x^3y''' - 4x^2y'' + 9xy' - 9 = 0 \tag{2}$$

has likewise the solutions $y_1 = x$ and $y_2 = x^3$, but there seems to hang around one more solution which we missed. Where is it?

An *Euler differential equation* $L_0y = 0$ is given by a linear differential operator of the form

$$L_0 = \sum\nolimits_{i=0}^{n} c_i x^i \partial^i,$$

with $c_i \in \mathbb{C}$ and $\partial = \partial_x$. It acts on monomials x^ρ via

$$L_0 x^\rho = \chi_0(\rho) \cdot x^\rho,$$

where

$$\chi_0(t) = \sum\nolimits_{i=0}^{n} c_i t^{\underline{i}}$$

denotes the *indicial polynomial* of L_0. The falling factorial $t^{\underline{i}} = t(t-1)\cdots(t-i+1)$ is called the *Pochhammer symbol*. The above two equations are homogeneous

Supported by the Austrian Science Fund FWF, project P-31338. We are very grateful to anonymous referees for pointing at two serious bugs in an earlier draft of the paper and for suggesting improvements in the presentation.

H. Hauser (✉)
Faculty of Mathematics, University of Vienna, Vienna, Austria
e-mail: herwig.hauser@univie.ac.at

A. Bostan and K. Raschel (eds.), *Transcendence in Algebra, Combinatorics, Geometry and Number Theory*, Springer Proceedings in Mathematics & Statistics 373,
https://doi.org/10.1007/978-3-030-84304-5_11

with indicial polynomials $\chi_0(t) = (t-1)(t-3)$ and $\chi_0(t) = (t-1)(t-3)^2$, respectively. This makes it clear that x and x^3 are solutions. But in the second equation, $\rho = 3$ is a double root of χ_0, so something special seems to happen. The classical theory of ordinary linear differential equations tells us that this peculiarity can be understood if we allow logarithms. And, indeed,

$$y_3 = x^3 \log(x)$$

is a third linearly independent solution of the second equation. Up to now, things are straightforward. Let us alter a bit the equations by adding to them a higher degree term, e.g., x^3y. Here, we take the natural grading on $\mathbb{C}[x, \partial]$ given by $\deg(x^i\partial^j) = i - j$ so that Euler operators have degree 0 and x^3y has degree 3. The modification

$$x^2y'' - 3xy' + 3y + x^3y = 0 \tag{1'}$$

of Eq. (1) has no obvious solutions. It is, however, tempting to expect that its solutions are *perturbations* of x and x^3, say, of the form $x \cdot g(x)$ and $x^3 \cdot h(x)$ for some holomorphic functions g and h not vanishing at 0. Taking an unknown ansatz $g = \sum a_k x^k$ and $h = \sum b_k x^k$ and plugging these power series into the equation yields linear recurrence equations for the coefficients a_k and b_k. They can be solved iteratively. Hence g and h exist as formal power series, and then an analysis of the growth behaviour proves their convergence. As analysis is not our main interest here, we skip it.

If we replace the perturbation term x^3y by, say, x^2y, the story makes a brusque turn. The equation

$$x^2y'' - 3xy' + 3y + x^2y = 0 \tag{1''}$$

has a solution of the form $x^3h(x)$, but no longer one of the form $xg(x)$, with g holomorphic and $g(0) \neq 0$.[1] It is not clear at that point whether this is just a coincidence or whether there lies a deeper reason behind the different behaviour of the solutions x and x^3 of Eq. (1). The situation becomes still more complicated when we perturb the second Eq. (2). We will return to it later in this text.

Two of the main protagonists in resolving the preceding quandary and in constructing a basis of solutions were Lazarus Immanuel Fuchs (1833–1902, student of Weierstrass) and Ferdinand Georg Frobenius (1849–1917, student of Weierstrass and Kummer) with articles in the Journal für Reine und Angewandte Mathematik in 1866 and 1868, respectively, 1873 [Fu1, Fu2, Fro]. They proposed smart algorithms which produced a precise description of the local solutions of a differential equation near a regular singularity: logarithms, as is now standard knowledge, play a prominent role.

To put their findings on a more conceptual basis, we will rely on a simple though fundamental result from functional analysis about the perturbation of linear operators.

[1] The linear recurrence for the coefficients of $xg(x) = \sum a_k x^{k+1}$ cannot be solved for $a_0 \neq 0$.

Theorem 1. *Let E be a Banach space and let $L_0, T : E \to F$ be continuous linear operators with values in a topological vector space F. Suppose that the image of T is contained in the image I of L_0 and that the kernel of L_0 admits a closed direct complement H in E,*

$$\mathrm{Ker}(L_0) \oplus H = E.$$

Assume that the inverse $S : I \to H$ of $L_{0|H} : H \to I$ satisfies $|S \circ T| < 1$, where $| - |$ denotes the operator norm induced by the norm of E. Set $L = L_0 + T$. Then $u := \mathrm{Id}_E + S \circ T : E \to E$ is a continuous linear automorphism of E which satisfies

$$L \circ u^{-1} = L_0.$$

Proof. The operator u is well defined since $\mathrm{Im}(T) \subseteq \mathrm{Im}(L_0)$, and continuous because of $|S \circ T| < 1$. Consider the geometric series $v = \sum_{k=0}^{\infty}(-1)^k(S \circ T)^k$. As $|S \circ T| < 1$ and the space of continuous linear operators between Banach spaces is again a Banach space, v defines a continuous linear operator on E. It is then clear that v is inverse to u. Hence u is a continuous linear automorphism of E. We are left to show that $L = L_0 \circ u$. But

$$\begin{aligned} L_0 \circ u &= L_0 \circ (\mathrm{Id}_E + S \circ T) \\ &= L_0 + L_0 \circ S \circ T \\ &= L_0 + L_0 \circ (L_{0|H})^{-1} \circ T \\ &= L_0 + T \\ &= L. \end{aligned}$$

This proves the theorem. ◯

This innocuous looking result is in fact very useful in our context (and in many others). We first apply it to the solution $y_2 = x^3$ of Eq. (1).

Proposition 1. *Let $L_0 = x^2\partial^2 - 3x\partial + 3$ be the Euler operator from above. Let $T = \sum_{ij} c_{ij}x^i\partial^j$ be a second order linear differential operator with polynomial or holomorphic coefficients for which $i - j \geq 1$ whenever $c_{ij} \neq 0$. Set $L = L_0 + T$. Then $Ly = 0$ has a solution of the form $y = x^3h(x)$ with h holomorphic and $h(0) \neq 0$.*

Example. The proposition shows that Eqs. (1') and (1") from above, that is, $x^2y'' - 3xy' + 3y + x^3y = 0$ and $x^2y'' - 3xy' + 3y + x^2y = 0$, have both a solution of the form $y = x^3h(x)$ with h holomorphic and $h(0) \neq 0$.

Proof. We first look for formal power series solutions of $Ly = 0$. The operator L_0 sends $x^3\mathbb{C}[[x]]$ onto $x^4\mathbb{C}[[x]]$ and its restriction to $x^4\mathbb{C}[[x]]$ (which is a direct complement of the kernel $x^3\mathbb{C}$ of L_0) gives an automorphism of $x^4\mathbb{C}[[x]]$. Let S be its inverse. As $x^4\mathbb{C}[[x]]$ contains $T(x^3\mathbb{C}[[x]])$ by the assumption on T the composition $S \circ T : x^3\mathbb{C}[[x]] \to x^4\mathbb{C}[[x]] \subseteq x^3\mathbb{C}[[x]]$ is well defined. It is easy to see that applying the operator $S \circ T$ increases the order of power series in $x^3\mathbb{C}[[x]]$. Therefore $v = \sum_{k=0}^{\infty}(-1)^k(S \circ T)^k$ defines an automorphism of $x^3\mathbb{C}[[x]]$ (use here that $\mathbb{C}[[x]]$ is complete with respect to the x-adic topology). It is inverse to $u = \mathrm{Id}_{x^3\mathbb{C}[[x]]} + S \circ T$. The same calculation as in the proof of Theorem 1 shows that $L \circ v = L \circ u^{-1} = L_0$ holds on $x^3\mathbb{C}[[x]]$. We conclude that $y = u^{-1}(x^3) =: x^3h(x) \in x^3\mathbb{C}[[x]]$ is a formal solution of $Ly = 0$ with $h(0) \neq 0$.

Convergence is more delicate: For a power series $h(x) = \sum_{k=0}^{\infty} a_kx^k$ and $s > 0$ we set $|h|_s = \sum_{k=0}^{\infty} |a_k|s^k$ and let $\mathbb{C}\{x\}_s$ denote the Banach space of convergent series h with finite s-norm $|h|_s < \infty$. Note that $\mathbb{C}\{x\}_s \subseteq \mathbb{C}\{x\}_{s'}$ for $0 < s' \leq s$ and $\mathbb{C}\{x\} = \bigcup_{s>0} \mathbb{C}\{x\}_s$. It is well known that differentiation $h \to h'$ sends $\mathbb{C}\{x\}_s$ into $\mathbb{C}\{x\}_{s'}$ for all $0 < s' < s$, see [GrRe, Satz 3, p. 19], but not necessarily into $\mathbb{C}\{x\}_s$, just take $h = \sum_{k=1}^{\infty} \frac{x^{k+1}}{k(k+1)} \in \mathbb{C}\{x\}_1$ with $h' \notin \mathbb{C}\{x\}_1$.

Let $E_s = x^3\mathbb{C}\{x\}_s$ for $s > 0$, $F = \mathbb{C}\{x\}$, and write $L_0, T : E_s \to F$ also for the restrictions of L_0 and T to E_s. From the first part of the proof we already dispose of the map $S \circ T : x^3\mathbb{C}[[x]] \to x^3\mathbb{C}[[x]]$ between formal power series spaces. The crucial point now is to show that $S \circ T$ sends $E_s = x^3\mathbb{C}\{x\}_s$ to itself and that the restriction $(S \circ T)_s : E_s \to E_s$ satisfies $|(S \circ T)_s| < 1$ for $s > 0$ sufficiently small. But a convergent $h = \sum_{k=0}^{\infty} a_kx^k$ is sent by $S \circ T$ to

$$(S \circ T)(h) = S\Big(\sum_{ij} c_{ij} \sum_k a_k k^{\underline{j}} x^{k+i-j}\Big) = \sum_{ij}\sum_k \frac{k^{\underline{j}}}{(k+i-j-1)(k+i-j-3)} c_{ij}a_kx^{k+i-j},$$

by the very definitions of T and S. Recall here that $\chi_{L_0}(t) = (t-1)(t-3)$ is the indicial polynomial of L_0 and that S is the inverse of $L_{0|x^4\mathbb{C}\{x\}}$. As $j \leq 2$, the ratios $\frac{k^{\underline{j}}}{(k+i-j-1)(k+i-j-3)}$ remain bounded as i and k go to ∞. Therefore $(S \circ T)(h)$ converges again. But as $i - j \geq 1$ for all i, j with $c_{ij} \neq 0$, a short computation shows that one can even achieve

$$|(S \circ T)_s(h)|_s \leq C|h|_s$$

for all $0 < s < s_0$ and some constant $C < 1$ independent of h, provided that s_0 is sufficiently small (s_0 will also be independent of h). Therefore $|(S \circ T)_s| < 1$. This shows that the restriction $u_s : E_s \to E_s$ of u is a well defined continuous linear automorphism with inverse the restriction v_s of v.

By Theorem 1, we now get $L \circ u_s^{-1} = (L_0 + T) \circ u_s^{-1} = L_0$ on E_s. This implies that $y = u_s^{-1}(x^3) = x^3h(x)$ with h holomorphic and $h(0) \neq 0$ is a solution of $Ly = 0$. ⟲

Remark. The argument applied to x^3 in the preceding examples (1') and (1") does not work equally well for the solution $y_1 = x$ of Eq. (1), even when looking just for formal solutions: We would have to let L_0 operate on the space $x\mathbb{C}[[x]]$, observing that the image of L_0 now equals $x^2\mathbb{C} \oplus x^4\mathbb{C}[[x]]$, with a *gap* at x^3. For Eq. (1'), the image of the operator T defined by $Ty = x^3y$ is $x^4\mathbb{C}[[x]]$. It is contained in the image of L_0 and one can find the solution $xg(x)$ of Eq. (1') satisfying $g(0) \neq 0$ as in the proposition. On the other hand, for Eq. (1"), the image of the operator T defined by $Ty = x^2 y$ equals $x^3\mathbb{C}[[x]]$ and is thus *not* contained in the image of L_0. The reasoning does not apply.

Lazarus Fuchs and Georg Frobenius would now potentially say the following: The roots of the indicial polynomial of (1) are 1 and 3, the it local exponents of Eq. (1) at 0. They are congruent modulo $\mathbb{Z}$. The larger one, 3, is maximal: $3 + k$ is not a root for any positive integer k. Therefore, the local solutions of (1') and (1") at 0 with respect to the maximal exponent $\rho = 3$ are of the form $x^3 h(x)$ with h holomorphic, regardless of the perturbation term.

And F. G. F. would continue: The exponent $\sigma = 1$, however, is not maximal. Extra caution has to be taken to find the solutions $xg(x)$ with $g(0) \neq 0$. A nice trick would be to consider a formal expression $x^t h(x)$, where now t is a variable, and to try to solve the equation for this term. One will fail, but may observe that differentiating the whole situation with respect to t does produce a solution, now involving a logarithm (it comes from the differentiation with respect to the exponent). This mysterious manipulation is suggested and explained in my paper [Fro]. The prospective solution of Eq. (1") corresponding to the exponent $\sigma = 1$ will be of the form $y_1 = x[g(x) + h(x)\log(x)]$. Perfect!

Let $L \in \mathbb{C}\{x\}[\partial]$ be a linear differential operator with polynomial or holomorphic coefficients, $L = \sum c_{ij} x^i \partial^j$. Consider the vector space $\mathbb{C}\{x\}[z]$ of polynomials in a new variable z with coefficients in $\mathbb{C}\{x\}$. The *extension* $\mathbb{L}$ of L to $\mathbb{C}\{x\}[z]$ is defined as

$$\mathbb{L} = \sum c_{ij} x^i \boldsymbol{\partial}^j,$$

where the derivation $\boldsymbol{\partial}$ on $\mathbb{C}\{x\}[z]$ is prescribed by the two requirements $\boldsymbol{\partial} x = 1$ and $\boldsymbol{\partial} z = x^{-1}$ (compare this with section 4 in [Hon]).

Lemma 1. *If L_0 is an n-th order Euler operator with indicial polynomial χ_0 then*

$$\mathbb{L}_0(x^t z^k) = x^t \cdot [\chi_0(t) z^k + \chi_0'(t) k z^{k-1} + \ldots + \frac{1}{n!}\chi_0^{(n)}(t) k^{\underline{n}} z^{k-n}].$$

Proof. Pastime for the dentist's waiting room.[2] ○

Lemma 2. *If L_0 is an Euler operator and $\rho \in \mathbb{C}$ is an m-fold root of χ_0, then*

$$x^\rho, x^\rho z, \ldots, x^\rho z^{m-1}$$

are solutions of $\mathbb{L}_0 y = 0$ in $\mathbb{C}\{x\}[z]$. ○

Define, for $\ell \geq 0$, linear maps $(\partial^j)^{(\ell)} : \mathbb{C}\{x\} \to \mathbb{C}\{x\}$ by

$$(\partial^j)^{(\ell)}(x^t) = (t^{\underline{j}})^{(\ell)} x^{t-j}.$$

Note that $(\partial^j)^{(0)} = \partial^j$ is the usual j-fold differentiation. For an arbitrary differential operator $L = \sum_{j=0}^n \sum_{i=0}^\infty c_{ij} x^i \partial^j$ we call

[2] It suffices to prove this for $\mathbb{L}_0 = \boldsymbol{\partial}^j$. One may replace z by $\log(x)$ and apply the usual differentiation rules, replacing afterwards all powers $\log(x)^i$ by z^i again.

$$L^{(\ell)} = \sum_{j=0}^{n} \sum_{i=0}^{\infty} c_{ij} x^i (\partial^j)^{(\ell)}$$

the *ℓ-th Pochhammer derivative* of L. This is no longer a differential operator in the classical sense, but, of course, defines again a linear map $L^{(\ell)} : \mathbb{C}\{x\} \to \mathbb{C}\{x\}$. If $L = L_0$ is an Euler operator with indicial polynomial χ_0, then

$$L_0^{(\ell)}(x^t) = \chi_0^{(\ell)}(t) x^t,$$

where $\chi_0^{(\ell)}$ denotes the ℓ-fold derivative of χ_0. This fact motivates the differentiation notation $(\partial^j)^{(\ell)}$. One may call $\chi_0^{(\ell)}$ the indicial polynomial of $L_0^{(\ell)}$.

Leo August Pochhammer (1841–1920, student of Kummer) may complain here that his name is used without having been asked. He may also request, possibly in latin, to find an axiomatic characterization of the linear maps $(\partial^j)^{(\ell)} : \mathbb{C}\{x\} \to \mathbb{C}\{x\}$.

Lemma 3. *The extension $\mathbb{L}$ to $\mathbb{C}\{x\}[z]$ of an n-th order differential operator L has "Taylor expansion"*

$$\mathbb{L} = L + L'\partial_z + \frac{1}{2} L'' \partial_z^2 + \ldots + \frac{1}{\ell!} L^{(\ell)} \partial_z^\ell + \ldots + \frac{1}{n!} L^{(n)} \partial_z^n,$$

where now L, L', ... are taken as maps on $\mathbb{C}\{x\}[z]$, acting on the coefficients in $\mathbb{C}\{x\}$ of the polynomials in z while leaving z invariant, and where ∂_z denotes the usual differentiation with respect to z.

Proof. As for Lemma 1. ◯

Lemma 4. $\mathbb{L}(x^i z^k)_{|z=\log(x)} = L(x^i \log(x)^k)$. ◯

Let $L = \sum_{j=0}^{n} \sum_{i=0}^{\infty} c_{ij} x^i \partial^j$ be an n-th order differential operator in $\mathbb{C}\{x\}[\partial]$. Its *initial form* is the homogeneous operator $L_0 = \sum_{i-j=\tau} c_{ij} x^i \partial^j$ where the *shift* τ is the minimal difference $i - j$ for which some c_{ij} is non zero. Up to multiplication of L by a Laurent monomial we may and will assume that the shift τ is 0, i.e., that L_0 is an Euler operator. The origin 0 is called a *regular singularity* of L if L_0 is again an operator of order n. It just means that $c_{nn} \neq 0$. This is an important concept. It ensures that formal power series solutions of $Ly = 0$ are actually convergent. Many characterizations exist [Sin, Proposition 1.4.2, vdPS, section 5.1]. The operator L of Proposition 1 has a regular singularity at 0 and initial form L_0.

The indicial polynomial χ_L of an operator L is defined as the indicial polynomial χ_0 of L_0, and its roots $\rho \in \mathbb{C}$ are the *local exponents* of L at 0. They may be multiple roots as in Eq. (2).

Proposition 2. *Let L be an operator with regular singularity at 0 and initial form $L_0 = x^3\partial^3 - 4x^2\partial^2 + 9x\partial - 9$ as in Eq. (2) above. Then $Ly = 0$ has two solutions of the form $y_1 = x^3 h(x)$ and $y_2 = x^3[g(x) + h(x)\log(x)]$ with g, h holomorphic and $g(0), h(0) \neq 0$.*

Proof. Write $L = L_0 + T$ with T an operator of shift ≥ 1 and denote by $\mathbb{L}_0$ and $\mathbb{T}$ the extensions of L_0 and T to the subspace

$$E = x^\rho \mathbb{C}\{x\} \oplus x^\rho \mathbb{C}\{x\} z = x^3 \mathbb{C}\{x\} \oplus x^3 \mathbb{C}\{x\} z$$

of $x^3\mathbb{C}\{x\}[z]$. The image of $\mathbb{L}_0$ equals $x^4\mathbb{C}\{x\} \oplus x^4\mathbb{C}\{x\}z$ since $\rho = 3$ is a maximal exponent of L_0. It thus contains the image of $\mathbb{T}$. Restricting to power series in x of finite s-norm the theorem applies. An appropriate operator $\mathbb{S}$ depends on the choice of a direct complement H of the kernel of $\mathbb{L}_0$ in E. One preferably takes here $H = x^{\rho+1}\mathbb{C}\{x\} \oplus x^{\rho+1}\mathbb{C}\{x\}z = x^4\mathbb{C}\{x\} \oplus x^4\mathbb{C}\{x\}z$. To establish the norm estimate $|\mathbb{S} \circ \mathbb{T}| < 1$ for the restriction to suitable Banach spaces E_s, for $s > 0$ sufficiently small, a similar estimate as in the proof of Proposition 1 works, using now that 0 is a regular singularity of L.[3] One gets $\mathbb{L} \circ u^{-1} = \mathbb{L}_0$. Now pull back with u^{-1} the solutions x^3 and $x^3 z$ of $\mathbb{L}_0 y = 0$ to get the solutions of $\mathbb{L}y = 0$ in E. Then replace there z by $\log(x)$. The resulting solutions y_1 and y_2 of $Ly = 0$ are of the form as indicated.[4] ○

At that point, the attentive reader may have observed that the second exponent $\sigma = 1$ and the solution $y_1 = x$ of the Euler equation $L_0 y = 0$ have been neglected in the statement of Proposition 2. There is a reason for this:

Let L_0 have two local exponents ρ and σ which are congruent to each other modulo $\mathbb{Z}$, that is, $\sigma = \rho - e$, $e \in \mathbb{N}_{>0}$. Consider the subspace

$$E = x^\sigma \mathbb{C}\{x\} \oplus x^\rho \mathbb{C}\{x\} z$$

of $x^\sigma \mathbb{C}\{x\}[z]$ and the extension

$$\mathbb{L}_0 : E \to E$$

of L_0 to it. What is its image?

Let $L_0 = x^2\partial^2 - 3x\partial + 3$ be the operator from Eq. (1). We have $\rho = 3$ and $\sigma = 1$. A gentle computation gives (for general ρ and σ)

$$\mathbb{L}_0(x^\sigma + x^\rho z) = [\chi_0(\sigma)x^\sigma + \chi_0'(\rho)x^\rho] + [\chi_0(\rho)x^\rho]z,$$

say,

$$\mathbb{L}_0(x^\sigma + x^\rho z) = [(\sigma - 1)(\sigma - 3)x^\sigma + 2(\rho - 2)x^\rho] + [(\rho - 1)(\rho - 3)x^\rho]z.$$

A short moment of reflection will confirm that the image $\mathbb{L}_0(E)$ of $\mathbb{L}_0$ is therefore

[3] The critical ratio in the expansion of $(\mathbb{S} \circ \mathbb{T})(h)$ is now $\frac{k^j}{\chi_{L_0}(k+i-j)}$. The assumption on the regularity of the singularity signifies that the indicial polynomial χ_{L_0} has degree equal to the order of L, hence the ratio is bounded from above as k and i tend to ∞.

[4] The reappearence of the function h in y_2 is due to the special shape of u; the proof would require some computation.

$$x^{\sigma+1}\mathbb{C}\{x\} \oplus x^{\rho+1}\mathbb{C}\{x\}z = x^2\mathbb{C}\{x\} \oplus x^4\mathbb{C}\{x\}z.$$

Both summands have *no* gaps. Sounds good!

Proposition 3. *Let L be differential operator with regular singularity at* 0 *and initial form* $L_0 = x^2\partial^2 - 3x\partial + 3$ *as in Eq. (1) above. Then* $Ly = 0$ *has a basis of local solutions of the form* $y_1 = x^3 f(x)$ *and* $y_2 = xg(x) + x^3h(x)\log(x)$ *with* f, g *and* h *holomorphic and none vanishing at* 0.

Proof. Write $L = L_0 + T$. Denote by $\mathbb{T}$ the extension of T to $x^\sigma\mathbb{C}\{x\} \oplus x^\rho\mathbb{C}\{x\}z$. Its image is contained in the image of $\mathbb{L}$ as computed before, because T has shift ≥ 1. Restricting to power series of finite s-norm, for $s > 0$ sufficiently small, the theorem applies to $\mathbb{L}$. Lift with u^{-1} the solutions x and x^3 of $\mathbb{L}_0 y = 0$ to solutions of $\mathbb{L}y = 0$. Then replace z by $\log(x)$. ◌

Lothar Heffter (1862–1962, student of Fuchs) would now possibly reply that this perturbation method already appeared in his book from 1894 [Hef]. This could to a certain extent be correct, though it is hard to compare the two approaches in detail. The reader may consult [Poo, section V.16] or [Mez, section 4.4] to get a personal opinion.

We can now understand why for Eq. (1')

$$x^2y'' - 3xy' + 3y + x^3y = 0$$

it is easier to construct a basis of solutions than for (1")

$$x^2y'' - 3xy' + 3y + x^2y = 0.$$

In the first case, the perturbation term T is defined by $Ty = x^3y$. Its restriction to $x^\sigma\mathbb{C}\{x\} = x\mathbb{C}\{x\}$ has image $x^4\mathbb{C}\{x\}$ contained in the image of $L_{|x\mathbb{C}\{x\}}$ which is $\mathbb{C}x^2 \oplus x^4\mathbb{C}\{x\}$. This does not happen for Eq. (1"), where T is defined by $Ty = x^2y$ and has image $x^3\mathbb{C}\{x\}$ not contained in the image of L_0. In the first case, we find an isomorphism u of $x\mathbb{C}\{x\}$ such that $L \circ u^{-1} = L_0$, whereas in the second case we have to resort to $E = x\mathbb{C}\{x\} \oplus x^3\mathbb{C}\{x\}z$ and the extensions $\mathbb{L}$ and $\mathbb{T}_0$ of L and T to normalize $\mathbb{L}$ to $\mathbb{L}_0$ by an automorphism of E. This creates the presence of logarithms in the respective solutions.

One may want to try to work instead with the extension $\mathbb{L}$ induced by L on the larger space $\widetilde{E} = x^\sigma\mathbb{C}\{x\} \oplus x^\sigma\mathbb{C}\{x\}z = x\mathbb{C}\{x\} \oplus x\mathbb{C}\{x\}z$ instead of $E = x\mathbb{C}\{x\} \oplus x^3\mathbb{C}\{x\}z$. This would be suggested by the description of solutions by Frobenius [Fro, p. 222]. However, one can check that for the extension of $\mathbb{L}$ to $\widetilde{E}$ the normalization procedure of the theorem does not go through.

In Proposition 1 we constructed one solution for one simple maximal exponent, in Proposition 2 two solutions for one maximal exponent of multiplicity 2, in Proposition 3 two solutions for one simple maximal and one non maximal exponent. There is one more case: one maximal exponent of multiplicity 2 and one non maximal exponent.

Proposition 4. *Let L be differential operator with regular singularity at* 0 *and initial form* $L_0 = x^3\partial^3 - 4x^2\partial^2 + 9x\partial - 9$ *as in Eq. (2) above. Then* $Ly = 0$ *has a basis of local solutions at* 0 *of the form*

$$\begin{aligned} y_1 &= x^3 f_0(x), \\ y_2 &= x^3[f_1(x) + f_0(x)\log(x)], \\ y_3 &= xg(x) + x^3h_1(x)\log(x) + x^3h_2(x)\log(x)^2, \end{aligned}$$

with f_0, f_1, g, h_1, *and* h_2 *holomorphic and none vanishing at* 0.

Proof. For the first two solutions we may work with the extensions $\mathbb{L}$ and $\mathbb{T}$ of L and T to $E = x^3\mathbb{C}\{x\} \oplus x^3\mathbb{C}\{x\}z$ and apply the same arguments as in the proof of Proposition 2. For the third solution, things are getting more complicated, since we now have to extend L and T to the space

$$E' = x\mathbb{C}\{x\} \oplus x^3\mathbb{C}\{x\}z \oplus x^3\mathbb{C}\{x\}z^2.$$

Call $\mathbb{L}$ and $\mathbb{T}$ the respective extensions. Again, the composition $\mathbb{L} \circ u^{-1} = \mathbb{L}_0$ normalizes $\mathbb{L}$ on E' to its initial form $\mathbb{L}_0$ for a suitable automorphism u of E'. To obtain the exact shape of the third solution, one has to choose $\mathbb{S}$ suitably as in the proof of Proposition 2 and then look carefully at the action of u on the three summands of E. This is a bit tedious and will be omitted.[5] ↺

At least now it should have become clear how the story continues: the local exponents of a given differential operator L with regular singularity at 0 have to be listed in sets of exponents congruent to each other modulo $\mathbb{Z}$. For the largest elements of each of these sets, say ρ, with multiplicity m as a root of χ_L, arguments as used for Propositions 1 and 2 apply, yielding solutions

$$y_k(x) = x^\rho \left[f_{k-1}(x) + f_{k-2}(x)\log(x) + \ldots + f_0(x)\log(x)^{k-1}\right],$$

for $k = 1, ..., m$ and with holomorphic functions $f_0, ..., f_{m-1}$, none vanishing at 0. The second largest exponent of the set containing ρ, say σ, of multiplicity ℓ, yields solutions of the form

$$\begin{aligned} y_k(x) = x^\sigma \quad & \left[g_{k-1}(x) + g_{k-2}(x)\log(x) + \ldots + g_0(x)\log(x)^{k-1}\right] \\ & + x^\rho \log(x)^k \left[h_{m-1}(x) + h_{m-2}(x)\log(x) + \ldots + h_0(x)\log(x)^{m-1}\right], \end{aligned}$$

for $k = 1, ..., \ell$ and with holomorphic $g_0, ..., g_{\ell-1}, h_0, ..., h_{m-1}$, none vanishing at 0.[6] And so on for the other exponents.[7]

---

[5] An comprehensive manual for this type of extension techniques is in preparation [Hau].

[6] Each y_k has $k + m$ summands, for $k = 1, ..., \ell$, but there appear, in total, only $\ell + m$ different coefficient functions.

[7] The respective formulas (11) in [Ful, p. 136], (12) in [Fro, p. 222], and in [Ince, p. 401] are somewhat cumbersome to disentangle.

Theorem 2. *Every n-th order linear differential operator L with regular singularity at* 0*, initial form* L_0*, and local exponents* $\rho \in \Omega$*, admits an extension* $\mathbb{L}$ *to a finite rank* $\mathbb{C}\{x\}$*-submodule E of* $\prod_{\rho\in\Omega} x^\rho\mathbb{C}\{x\}[z]$ *which contains a basis of solutions of* $\mathbb{L}_0$ *at* 0 *and such that*

$$\mathbb{L} \circ u^{-1} = \mathbb{L}_0,$$

where the normalizing automorphism $u = \mathrm{Id}_E + \mathbb{S} \circ \mathbb{T}$ *of E is built from an "inverse"* $\mathbb{S}$ *of* $\mathbb{L}_0$ *and the higher degree terms* $\mathbb{T} = \mathbb{L} - \mathbb{L}_0$ *of* $\mathbb{L}$*. A basis of local solutions of* $Ly = 0$ *can be computed by pulling back via* u^{-1} *the trivial solutions of* $\mathbb{L}_0$ *in E and setting* $z = \log(x)$*.*

Frobenius: So it seems that regular singularities are even more regular than my dear colleague, Herr Fuchs, may have suspected. Fuchs: Yes, lieber Herr Frobenius, this, indeed, seems to be the case.

Idea of Proof. The $\mathbb{C}\{x\}$-module E will be a cartesian product $E = \prod_{\rho\in\Omega} E_\rho$ of finite rank $\mathbb{C}\{x\}$-modules E_ρ attached to each $\rho \in \Omega$. We sketch their construction: Let Ω_1 be a set of local exponents all of which are congruent to each other modulo $\mathbb{Z}$ and such that no other exponent in Ω is congruent modulo $\mathbb{Z}$ to an element of Ω_1. This set is naturally ordered by considering differences. Let ρ be the maximal exponent in Ω_1, with multiplicity m, and let σ be the second largest exponent in Ω_1, with multiplicity ℓ. The factor E_ρ of E taking care of ρ is of the form

$$E_\rho = x^\rho\mathbb{C}\{x\} \oplus x^\rho\mathbb{C}\{x\}z \oplus \cdots \oplus x^\rho\mathbb{C}\{x\}z^{m-1}.$$

The summand E_σ of E taking care of σ is of the form

$$E_\sigma = x^\sigma\mathbb{C}\{x\} \oplus \cdots \oplus x^\sigma\mathbb{C}\{x\}z^{\ell-1} \oplus x^\rho\mathbb{C}\{x\}z^\ell \oplus \cdots \oplus x^\rho\mathbb{C}\{x\}z^{\ell+m-1}.$$

Compare these formulas with the description of the solutions y_1, y_2 and y_3 in Proposition 4. It can now be guessed how the construction of E continues for smaller exponents in Ω_1. To establish the theorem, it then suffices to show that the image of $\mathbb{L}_0$ on each E_ρ has no gaps, with the exponents of x^ρ, x^σ, ..., shifted by 1. To illustrate, one has to show that

$$\mathbb{L}_0(E_\rho) = x^{\rho+1}\mathbb{C}\{x\} \oplus \cdots \oplus x^{\rho+1}\mathbb{C}\{x\}z^{m-1},$$

$$\mathbb{L}_0(E_\sigma) = x^{\sigma+1}\mathbb{C}\{x\} \oplus \cdots \oplus x^{\sigma+1}\mathbb{C}\{x\}z^{\ell-1} \oplus x^{\rho+1}\mathbb{C}\{x\}z^\ell \oplus \cdots \oplus x^{\rho+1}\mathbb{C}\{x\}z^{\ell+m-1}.$$

Proving these equalities is a purely combinatorial task, using the formula from Lemma 1. They imply that the image of each E_ρ under $\mathbb{T}$ is contained in the image of $\mathbb{L}_0$. This is required in order to be able to apply Theorem 1 and to get $\mathbb{L} \circ u^{-1} = \mathbb{L}_0$. ↺

Remarks. (a) In the case that L has polynomial coefficients, $L \in \mathbb{C}[x, \partial]$, the automorphism u restricts to a linear endomorphism u° of $E^\circ = E \cap \sum_{\rho\in\Omega} x^\rho\mathbb{C}[x][z]$ which may be called *differentially étale*: its degree 0 term is the identity, whereas the remaining terms increase the order in x at 0; after extending E° to an analogous $\mathbb{C}[[x]]$-submodule $\widehat{E}$ of $\prod_{\rho\in\Omega} x^\rho\mathbb{C}[[x]][z]$ by allowing formal power series, the extension of u° to $\widehat{E}$ becomes an automorphism $\widehat{u} : \widehat{E} \to \widehat{E}$. It is a mystery which formal power series appear in the coefficients of z^i in the images under $\widehat{u}^{-1}$ of polynomials in E°. They must be quite special![8]

(b) The hypothesis in the theorem that L has a regular singularity at 0 is used to ensure the norm estimate required for the application of Theorem 1. This estimate was responsible to have u^{-1} map convergent series to convergent ones. Dropping the regularity condition, one still gets a normalizing automorphism u but now for formal power series with a $\mathbb{C}[[x]]$-submodule $\widehat{E}$ of $\prod_{\rho\in\Omega} x^\rho\mathbb{C}[[x]][z]$. The convergence of the geometric series to an operator defining u^{-1} is deduced in this case from the fact that free finite rank $\mathbb{C}[[x]]$-modules are complete with respect to the x-adic topology (as it was used with $\mathbb{C}[[x]]$ itself at the beginning of the proof of Proposition 1). With respect to the solutions of the equations, they have precisely the same shape as in the regular singular case, but the "coefficient functions" f_i, g_i and h_i are now just formal power series and need no longer be holomorphic.
One has to be cautious here as the order of L_0 is strictly smaller than the order of L if the singularity is irregular. Hence one does not get n linearly independent formal solutions but just as many as the order of L_0 indicates.[9]

(c) For an extension of Thm. 2 to several variables, but restricting to formal power series (no logarithms and powers with complex exponent), see the Monomialization Theorem in [GH, p. 11]. In this paper, a generalization of the concept of regular singularity for partial differential equations is proposed.

(d) It is tempting to have a glance at differential equations with polynomial or formal power series coefficients defined over a field of positive characteristic $p > 0$. Logarithms no longer exist, and already $y' = y$ has no solution. Does there exist a (significant) variant of Theorem 2 which is valid in positive characteristic?

One More Pastime. Every linear differential equation $Ly = 0$ with holomorphic coefficients at 0 is equivalent to a system of first order linear differential equations, $Y' = AY$, where $Y = (y_1, \ldots, y_n)^t$ is a column vector of unknown functions and $A \in \mathrm{M}_n(\mathrm{Quot}(\mathbb{C}\{x\}))$ is a matrix with meromorphic entries. Formulate and prove the normal form theorem from above in terms of such systems and the matrix A (you may compare your findings with Thm. 1 in [Turr], see also [Lev]).

How Regular are Regular Singularities? Even though we have not given all details (actually, we gave only very few), we hope to have conveyed the flavour of the techniques: The statement in Theorem 2 is a normal form theorem for differential operators acting on specific spaces of formal or convergent power series (namely, on the spaces $\widehat{E}_\rho$ and E_ρ). It asserts that all relevant information is contained, up to

---

[8] This comment may seem itself to be a bit mysterious. To catch its perspective requires a moment of contemplation.

[9] We are indebted to a referee for emphasizing this detail.

an automorphism of the source space, in the initial form of the operator. This holds in both the formal and the convergent setting, in the first case regardless of whether the singularity is regular or not. But if it is regular, one gets in addition that the normalizing automorphism sends convergent series to convergent series. Therefore, differential equations with regular singularities are locally equivalent to the Euler equation given by the initial form of the operator. As such, regular singularities behave also regularly in this sense.

The normal form of the operator has an immediate consequence on the description of the solutions of the differential equation defined by the operator: One solves the initial equation and lifts its solutions with the normalizing automorphism to the solutions of the original equation. This, again, works for the formal and convergent setting. And as a by-product, similar perturbation methods apply to prove more generally the Malgrange index theorem [Mal, Kom]: It counts the number of convergent solutions of irregular singular differential equations among all formal solutions. For the case of several variables, we refer to the paper [GH] where the notion of perfect operator gives a prospective generalization of regular singularities to partial differential equations.

Thanks. Being an apprentice in the field, the author has enormously profited from many conversations with insiders, among them: Alin Bostan, Michael Singer, Duco van Straten, Frits Beukers, Julien Roques, Eric Delaygue, Boris Adamczewski, Michael Wibmer, Fernando Rodriguez-Villegas, Nicholas Katz, Orlando Neto, Carlos Florentino, Francisco Castro, Luis Narváez, Jean-Marie Maillard, Anton Mellit, Pierre Lairez, and Sergey Yurkevich.

Bibliography

The article [Fro] of Frobenius is truly inspiring and cristal-clear, though compact. It takes up and extends ideas and constructions of Fuchs [Fu1, Fu2], papers which are a bit more outspread but layed the basis of the theory. The proof that one obtains indeed a basis of solutions—this is only sketched in [Fro]—is detailed in [Tho] and [Ince, p. 402]. These methods are reproduced, among many other sources and in slightly different language, in [Poo, Mez, Tou, CH, Sau]. The book [Poo] of Poole and the thesis [Mez] describe also the recursive approach of Heffter [Hef]. For more on regular singularities, see [vdPS], [Sin] and [Was]. Historical information can be found in [Gra, Schl]. Finally, the article [GH] extends the methods of the present text to the multivariate case of partial differential equations.

References

[CH] Crespo, T., Hajto, Z.: Algebraic groups and differential Galois theory. Grad. Studies Math. 122, Amer. Math. Soc. 2011.

[Fro] Frobenius, F.: Über die Integration der linearen Differentialgleichungen durch Reihen. J. Reine Angew. Math. 76 (1873), 214–235.

[Fu1] Fuchs, L.: Zur Theorie der linearen Differentialgleichungen mit veränderlichen Coefficienten. J. Reine Angew. Math. 66 (1866), 121–160.

[Fu2] Fuchs, L.: Zur Theorie der linearen Differentialgleichungen mit veränderlichen Coefficienten (Ergänzungen zu der im 66. Bande dieses Journals enthaltenen Abhandlung). J. Reine Angew. Math. 68 (1868), 354–385.

[GH] Gann, S., Hauser, H.: Perfect bases for differential equations. J. Symb. Comp. 40 (2005), 979–997.

[Gra] Gray, J.: Linear differential equations and group theory from Riemann to Poincaré. Birkhäuser, 2000.

[GrRe] Grauert, H., Remmert, R.: Analytische Stellenalgebren. Springer 1971.

[Hau] Hauser, H.: Algebraic aspects of Fuchsian differential equations. Manuscript 2020, 140 pp, forthcoming.

[Hef] Heffter, L.: Einleitung in die Theorie der linearen Differentialgleichungen. Teubner, 1894.

[Hon] Honda, T.: Algebraic differential equations. In: Symposia Math. 24. Academic Press 1981, 169–204.

[Ince] Ince, E.: Ordinary Differential Equations. Longmans 1927.

[Kom] Komatsu, H. On the index of ordinary differential operators. J. Fac. Sci. Univ. Tokyo 18 (1971), 379–398.

[Lev] Levelt, A.: Jordan decomposition for a class of singular differential operators. Ark. Mat. 13 (1975), 1–27.

[Mal] Malgrange, B.: Sur les points singuliers des équations différentielles. L'Ens. Math. 20 (1974), 147–176.

[Mez] Mezzarobba, M.: Autour de l'évaluation numérique des fonctions D-finies. Thèse École Polytechnique 2011.

[Poo] Poole, E.: Introduction to the theory of linear differential equations. Oxford 1936.

[Sau] Sauloy, J.: Differential Galois theory through Riemann-Hilbert correspondence: an elementary introduction. Amer. Math. Soc. 2016.

[Schl] Schlesinger, L.: Bericht über die Entwicklung der Theorie der linearen Differentialgleichungen seit 1856. Jahresbericht Dt. Mathematiker-Vereinigung 18 (1909), 133–266.

[Tho] Thomé, L.: Zur Theorie der linearen Differentialgleichungen. J. Reine Angew. Math. 74 (1872), 193–217.

[Tou] Tournier, É.: Solutions formelles d'équations différentielles. Thèse Université de Grenoble, 1987.

[Turr] Turrittin, H.L.: Convergent solutions of ordinary linear homogeneous differential equations in the neighborhood of an irregular singular point. Acta Math. 93 (1955), 27–66.

[Sin] Singer, M.: Introduction to the Galois theory of linear differential equations. In: Algebraic theory of differential equations. London Math. Soc. Lecture Note Ser. 357. Cambridge Univ. Press 2009, pp. 1–82.

[vdPS] van der Put, M., Singer, M.: Galois theory of linear differential equations. Springer, 2003.

[Was] Wasow, W.: Asymptotic expansions for ordinary differential equations. Wiley, 1965, Dover 1987.

Néron Desingularization of Extensions of Valuation Rings

With an Appendix by Kęstutis Česnavičius

Dorin Popescu

Abstract Zariski's local uniformization, a weak form of resolution of singularities, implies that every valuation ring containing **Q** is a filtered direct limit of smooth **Q**-algebras. Given an immediate extension of valuation rings $V \subset V'$ containing **Q** we show that V' is a filtered direct limit of smooth V-algebras. This corrects a paper of us [23] where we thought that we may reduce to the case when the value groups are finitely generated. For this correction we use an infinite tower of ultrapowers construction that rests on results from model theory.

1 A Version of Local Uniformization

Zariski proved in characteristic 0 in [34], that any integral algebraic variety X equipped with a dominant morphism $v : \mathrm{Spec}(V) \to X$ from a valuation ring V can be "desingularized along V": there exists a proper birational map $\tilde{X} \to X$ for which the lift $\tilde{v} : \mathrm{Spec}(V) \to \tilde{X}$ of v supplied by the valuative criterion of properness factors through the regular locus of $\tilde{X}$. This implies the following theorem.

Theorem 1 *(Zariski) Every valuation ring V containing a field K of characteristic zero is a filtered direct limit of smooth K-subalgebras of V(in particular they are regular rings).*

A smooth algebra is here finitely presented. A ring map $A \to A'$ is *ind-smooth* if A' is a filtered direct limit of smooth A-algebras. A filtered direct limit (in other words a filtered colimit) is a limit indexed by a small category that is filtered (see [32, 002V] or [32, 04AX]). A filtered union is a filtered direct limit in which all objects are subobjects of the final colimit, so that in particular all the transition arrows are monomorphisms. The above theorem says that $K \to V$ is ind-smooth. Actually, Zariski proved that V is a filtered union of smooth K-subalgebras of V. One goal of this paper is to show a weaker statement:

D. Popescu (✉)
Simion Stoilow Institute of Mathematics of the Romanian Academy, Research Unit 5, University of Bucharest, P.O. Box 1-764, 014700 Bucharest, Romania

A. Bostan and K. Raschel (eds.), *Transcendence in Algebra, Combinatorics, Geometry and Number Theory*, Springer Proceedings in Mathematics & Statistics 373,
https://doi.org/10.1007/978-3-030-84304-5_12

Theorem 2 *Let $V \subset V'$ be an immediate extension of valuation rings containing* **Q**. *Then $V \subset V'$ is ind-smooth. If* $\dim V = \dim V' = 1$ *and the residue field extension of $V \subset V'$ is trivial then $V \subset V'$ is ind-smooth if and only if the value group extension of $V \subset V'$ is trivial.*

The proof follows from Theorem 21, Lemma 29 and Proposition 38. The above result was stated by mistake in [23] in a more general case. A different proof of Theorem 1 is given in Theorem 35 and the method use some facts from model theory described below.

3. The Method of Proof. To achieve the desingularization claimed in Theorem 2, we replace the initial V by the limit $\tilde{V}$ of a certain countable tower of iterated ultrapowers of V, constructed in such a way that $\tilde{V}$ would, in turn, be an immediate extension of a filtered increasing union of valuation rings for which one knows local uniformization. To then conclude, we argue that large immediate extensions and ultrapowers interact well with desingularization.

The techniques we use include extensions of steps of the General Néron desingularization, notably, Lemma 7 that is also key for reductions to complete rank 1 cases. In the purely transcendental case, Kaplansky's classification [12] of Ostrowski's pseudo-convergent sequences plays an important role.

The utility of desingularizing immediate extensions is evident already in the case when V' is complete of rank 1 with a finitely generated value group Γ'. Such a V' has a coefficient field k, so, by choosing a presentation $\Gamma' \cong \mathbf{Z}\mathrm{val}(x_1) \oplus \cdots \oplus \mathbf{Z}\mathrm{val}(x_n)$, one obtains the immediate extension $V' \cap k(x_1, \ldots, x_n) \subset V'$. To show that $V = k \subset V'$ is ind-smooth, it remains to observe that a local uniformization of $V' \cap k(x_1, \ldots, x_n)$ may be constructed using Perron's algorithm in the style of Zariski.

The goal of the tower of ultrapowers argument given in the Appendix is to overcome the obstacle that in general Γ may not be finitely generated and there may not even be a group section $s : \Gamma \to K^*$ to $\mathrm{val} : K^* \to \Gamma$ (roughly, such an s suffices). Nevertheless, s can always be arranged for any finitely generated submonoid of Γ, and the idea is to then use the following fact from model theory: for a system of equations whose finite subsystems have solutions in V, the entire system has a solution in a well-chosen ultrapower of V (see the Appendix).

This fact, which rests on the Keisler–Kunen theorem about the existence of good ultrafilters, permits us to obtain s at the expense of passing to an ultrapower. However, such a passage replaces Γ by its corresponding ultrapower Γ^* and, in order to extend s to this Γ^*, we then need another ultrapower and some model-theoretic facts about algebraic compactness that ensure that $\Gamma \to \Gamma^*$ be a split injection. Even though the new ultrapower again enlarges the value group, by repeating the construction countably many times, in the limit we find our final s and can conclude.

We should mention that the proof of the main part of Theorem 2 uses only Lemma 7 and Corollary 19. The last sentence from Theorem 2 needs the method explained above together with some facts from André homology.

We owe thanks to Kęstutis Česnavičius especially for the Appendix, but also for many ideas and his great help on the presentation of the paper. Also we owe thanks to the referees who pointed out several mistakes in a previous version of the paper especially in the former Proposition 20.

2 A Reduction to the Case of Complete Valuation Rings of Rank 1

We begin by reviewing the following class of generators of the singular ideal.

For a finitely presented ring map $A \to B$, an element $b \in B$ is *standard over A* if there exists a presentation $B \cong A[X_1, \dots, X_m]/I$ and $f_1, \dots, f_r \in I$ with $r \leq m$ such that $b = b'b''$ with $b' = \det((\partial f_i/\partial X_j))_{1 \leq i, j \leq r} \in A[X_1, \dots, X_m]$ and a $b'' \in A[X_1, \dots, X_m]$ that kills $I/(f_1, \dots, f_r)$ (our standard element is a special power of the standard element from [33, Definition, page 9] given in the particular case of the valuation rings). Any multiple of an element b standard over A is standard over A. The definition is compatible with base change: more precisely, for any morphism $A \to A'$, elements of B standard over A map to elements of $B \otimes_A A'$ standard over A'.

Lemma 4 *For a finitely presented ring map $A \to B$, the loci of vanishing of standard over A elements of B cut out the locus of non-smoothness of* $\mathrm{Spec}(B) \to \mathrm{Spec}(A)$. *The radical of the ideal generated by the elements of B standard over A is $H_{B/A}$.*

Proof The argument is standard (compare with [8], [33, 4.3]) but we include it due to the lack of a convenient reference.

If $b \in B$ is standard over A, then B_b is the localization of the standard smooth A-algebra $(A[X_1, \dots, X_n]/(f_1, \dots, f_r))_{\det((\partial f_i/\partial X_j))_{1 \leq i, j \leq r}}$ (see [32, 00T8]), so is A-smooth. Conversely, if B_b is the coordinate ring of a smooth neighborhood of a fixed prime $\mathfrak{p} \subset B$, then we may choose a presentation $B[X_1, \dots, X_m]/I$ and $f_1, \dots, f_r \in I$ such that, at the expense of localizing at $\mathfrak{p}$ further. The module $(I/I^2)_b$ is a free B_b-module with a basis given by the classes of $f_1, \dots, f_r$ and $(I/I^2)_b \to \bigoplus_{i=1}^m B_b \cdot dX_i$ is a split injection such that $dX_{r+1}, \dots, dX_m$ maps to a basis for the quotient. The first condition and Nakayama's lemma [32, 00DV] then supply an $i \in I$ with $(1 + i/b^n)I_b \subset (f_1, \dots, f_r)_b$ for some $n > 0$. It follows that $b^N(b^n + i)$ for some $N > 0$ kills $I/(f_1, \dots, f_n)$ and maps to a power of b in B. The second condition implies that $b' = \det((\partial f_i/\partial X_j))_{1 \leq i, j \leq r}$ is a unit in B_b, so that b' divides some power of b in B. In conclusion, some power of b is standard over A, as desired. □

To stress the relevance of the desingularization lemma 7, we recall the following well-known lemma (see [33, (1.5)] or [32, 07C3]) and definitions, which will be crucial throughout this paper.

Lemma 5 *For a ring R and a set $\mathcal{S}$ of finitely presented R-algebras, an R-algebra R' is a filtered direct limit of elements of $\mathcal{S}$ if and only if every R-morphism $B \to R'$ with B a finitely presented R-algebra factors as $B \to S \to R'$ for some $S \in \mathcal{S}$.*

By [26, 1.8] (see also [32, 07GC]), a map of Noetherian rings is ind-smooth if and only if A' is A-flat and has geometrically regular A-fibers. In particular, a field extension K'/K is ind-smooth if and only if it is separable.

Concretely, by Lemma 5, a ring map $A \to A'$ is ind-smooth if and only if every factorization $A \to B \to A'$ with B finitely presented over A can be refined to $A \to B \to S \to A'$ with S smooth (or merely ind-smooth) over A. Thus, a finite product or a filtered direct limit of ind-smooth A-algebras is ind-smooth. Evidently, ind-smooth morphisms are stable under base change. They are also stable under compositions, in fact, we have the following slightly finer criterion.

Lemma 6 *For an ind-smooth map $A \to A'$ and a map $A' \to A''$ such that for every factorization $A \to B \to A''$ with B finitely presented over A the induced factorization $A' \to A' \otimes_A B \to A''$ can be refined to $A' \to A' \otimes_A B \to S' \to A''$ for some smooth A'-algebra S', the map $A \to A''$ is ind-smooth. In particular, the composition of ind-smooth maps is ind-smooth.*

Proof It suffices to argue that the map $A \to A' \to S'$ is ind-smooth. For this, we express A' as a filtered direct limit of smooth A-algebras S_i, note that S' descends to a smooth S_i-algebra S'_i for some i, and conclude that S' is then the filtered direct limit of the smooth A-algebras $S'_j = S_j \otimes_{S_i} S'_i$ with $j \geq i$. □

The following lemma originates in [24, (7.1)] and its variants have appeared, for instance, in [33, 18.1], [26, 7.2], [32, 07CT], [15, Proposition 3], and [28, Proposition 5]. The version below differs in two aspects: we do not assume Noetherianness and do not require the elements a or b to come from the base ring A. The latter improvement is particularly convenient for our purposes—we recall that in the General Néron desingularization arranging for b to come from A is an additional step before one can apply the desingularization lemma (compare with, for instance, [32, 07F4]).

Lemma 7 *For a commutative diagram of ring morphisms*

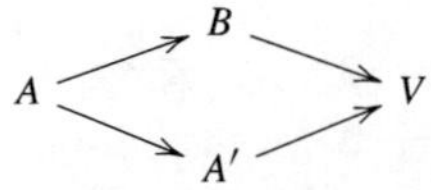

that factors as follows

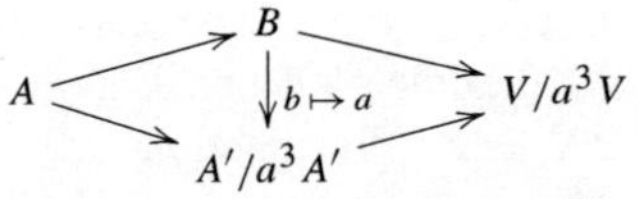

with B finitely presented over A, a $b \in B$ that is standard over A, and a nonzerodivisor $a \in A'$ that maps to a nonzerodivisor in V that lies in every maximal ideal of V, there is a smooth A'-algebra S such that the original diagram factors as follows:

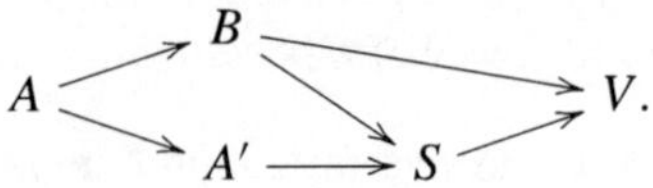

Proof The finitely presented A'-algebra $B \otimes_A A'$ comes equipped with a morphism to V and a retraction modulo b^3 to A'/a^3A' that sends b to a. Moreover, the image of b in $B \otimes_A A'$ is standard over A'. Thus, by replacing A by A' and B by $B \otimes_A A'$, we reduce to the case $A = A'$.

Since the images of a and b in V agree modulo a^3V, these images are unit multiples of each other. We write

$$B = A[X_1, \ldots, X_m]/I \quad \text{and} \quad f_1, \ldots, f_r \in I$$

and choose $b' = \det((\partial f_i/\partial X_j)_{1 \le i,\, j \le r}) \in A[X_1, \ldots, X_m]$ and a $b'' \in A[X_1, \ldots, X_m]$ that kills $I/(f_1, \ldots, f_r)$ with $b = b'b''$ in B. In these coordinates, we fix a map

$A[X_1, \ldots, X_m] \xrightarrow{f \mapsto \tilde{f}} A$ that makes the diagram

$$\begin{array}{ccc} A[X_1, \ldots, X_m] \twoheadrightarrow B & \twoheadrightarrow & B/b^3B \\ \downarrow & & \downarrow \\ A & \twoheadrightarrow & A/a^3A \end{array}$$

commute, so that $\tilde{f} \in a^3A$ for every $f \in I$. In particular, the assumption $b \mapsto a$ gives $\widetilde{b'b''} \equiv a \bmod a^3A$. It follows that

$$\widetilde{b'b''} = au \quad \text{for some } u \in 1 + a^2A,$$

so that, in particular, u maps to a unit in V.

We consider the $m \times m$ matrix Δ given by

$$\begin{pmatrix} \partial f_1/\partial X_1 & \partial f_1/\partial X_2 & \ldots & \partial f_1/\partial X_r & \partial f_1/\partial X_{r+1} & \partial f_1/\partial X_{r+2} & \ldots & \partial f_1/\partial X_m \\ \partial f_2/\partial X_1 & \partial f_2/\partial X_2 & \ldots & \partial f_2/\partial X_r & \partial f_2/\partial X_{r+1} & \partial f_2/\partial X_{r+2} & \ldots & \partial f_2/\partial X_m \\ \vdots & \vdots & \ddots & \vdots & \vdots & \vdots & \ddots & \vdots \\ \partial f_r/\partial X_1 & \partial f_r/\partial X_2 & \ldots & \partial f_r/\partial X_r & \partial f_r/\partial X_{r+1} & \partial f_r/\partial X_{r+2} & \ldots & \partial f_r/\partial X_m \\ 0 & 0 & \ldots & 0 & 1 & 0 & \ldots & 0 \\ 0 & 0 & \ldots & 0 & 0 & 1 & \ldots & 0 \\ \vdots & \vdots & \ddots & \vdots & \vdots & \vdots & \ddots & \vdots \\ 0 & 0 & \ldots & 0 & 0 & 0 & \ldots & 1 \end{pmatrix}$$

that satisfies $\det(\Delta) = b'$. We let $Ad(\Delta)$ denote the adjoint matrix, so that

$$Ad(\Delta) \cdot \Delta = \Delta \cdot Ad(\Delta) = b' \cdot Id_{m \times m}.$$

We let x_i and x_i' be the images in V of X_i and $\tilde{X}_i$, respectively, so that, by construction, $x_i - x_i' \in a^3V$. Moreover, a is a nonzerodivisor in V and there we have that

$$\begin{pmatrix} t_1 \\ \vdots \\ t_m \end{pmatrix} \tilde{\Delta} \begin{pmatrix} (x_1 - x_1')/a^2 \\ \vdots \\ (x_m - x_m')/a^2 \end{pmatrix} \quad \text{satisfies } t_i \in aV \text{ and } ab''\widetilde{Ad(\Delta)} \begin{pmatrix} t_1 \\ \vdots \\ t_m \end{pmatrix} = u \begin{pmatrix} x_1 - x_1' \\ \vdots \\ x_m - x_m' \end{pmatrix}.$$

We let $T_1, \ldots, T_m$ be new variables and set

$$\begin{pmatrix} h_1 \\ \vdots \\ h_m \end{pmatrix} = u \begin{pmatrix} X_1 - \tilde{X}_1 \\ \vdots \\ X_m - \tilde{X}_m \end{pmatrix} - ab''\widetilde{Ad(\Delta)} \begin{pmatrix} T_1 \\ \vdots \\ T_m \end{pmatrix}, \quad \text{so that } h_i \in A[X_1, \ldots, X_m, T_1, \ldots, T_m].$$

By construction, if we map T_i to t_i in V, then the h_i map to 0, so we obtain the map

$$\varphi : A_u[X_1, \ldots, X_m, T_1, \ldots, T_m]/(h_1, \ldots, h_m) \to V \text{ given by } X_i \mapsto x_i, \ T_i \mapsto t_i.$$

Since we have inverted u, the source of this map may be identified with $A_u[T_1, \ldots, T_m]$. To proceed further, we will use Taylor's formula to express each f_i in terms of this identification.

By Taylor's formula, for any ring R, any section $R[X_1, \ldots, X_m] \xrightarrow{f \mapsto \tilde{f}} R$, and any $f \in R[X_1, \ldots, X_m]$,

$$f - \tilde{f} - \sum_{i=1}^{m} \widetilde{(\partial f/\partial X_i)}(X_i - \tilde{X}_i) \in (X_1 - \tilde{X}_1, \ldots, X_m - \widetilde{X_m})^2 \subset R[X_1, \ldots, X_m].$$

In particular, by applying this with $R = A[T_1, \ldots, T_m]$ and letting d denote the maximal total degree of any monomial that appears in some f_i, we obtain

$$Q_i \in (T_1, \ldots, T_m)^2 \subset A[T_1, \ldots, T_m]$$

for which

$$u^d f_i - u^d \tilde{f}_i \equiv u^{d-1} a\tilde{b''} \ (\widetilde{(\partial f_i/\partial X_1)}, \ldots, \widetilde{(\partial f_i/\partial X_m)})\widetilde{Ad(\Delta)} \begin{pmatrix} T_1 \\ \vdots \\ T_m \end{pmatrix}$$
$$+ a^2 Q_i \bmod (h_1, \ldots, h_m) \equiv u^{d-1} a\tilde{b''}\tilde{b'} T_i + a^2 Q_i \equiv a^2 u^d T_i + a^2 Q_i \bmod (h_1, \ldots, h_m)$$

We have $\tilde{f}_i = a^2 b_i$ for some $b_i \in aA$, and for $1 \leq i \leq r$ we set

$$g_i = u^d b_i + u^d T_i + Q_i \in A[T_1, \ldots, T_m], \text{ so that } a^2 g_i \equiv u^d f_i \bmod (h_1, \ldots, h_m).$$

This achieves the promised expression of f_i in terms of the identification of the source of φ with $A[T_1, \ldots, T_m]$ and simultaneously shows that each g_i vanishes in V, so that φ induces a map

$$\varphi\colon A_u[X_1, \ldots, X_m, T_1, \ldots, T_m]/(I, g_1, \ldots, g_r, h_1, \ldots, h_m) \to V.$$

In $A[X_1, \ldots, X_m]$ the element $b'b'' - au$ lies in the ideal $(X_1 - \tilde{X}_1, \ldots, X_m - \tilde{X}_m)$, so in the quotient $A_u[X_1, \ldots, X_m, T_1, \ldots, T_m]/(h_1, \ldots, h_m)$ it lies in the ideal $a(T_1, \ldots, T_m)$. It then follows from the definition of b'' and the fact that after inverting u and modulo $(h_1, \ldots, h_m)$ the ideal $(g_1, \ldots, g_r)$ contains $(f_1, \ldots, f_r)$ that some element from the coset $a(u + (T_1, \ldots, T_m))$ kills the image of I in

$$A_u[X_1, \ldots, X_m, T_1, \ldots, T_m]/(g_1, \ldots, g_r, h_1, \ldots, h_m).$$

Setting $u' = \det((\partial g_i/\partial T_j)_{1 \leq i, j \leq r})$, we deduce that the same then holds in the localization

$$(A_u[X_1, \ldots, X_m, T_1, \ldots, T_m]/(g_1, \ldots, g_r, h_1, \ldots, h_m))_{u'} \cong (A_u[T_1, \ldots, T_m]/(g_1, \ldots, g_r))_{u'}.$$

However, the latter is smooth over A, to the effect that a is a nonzerodivisor in the ring above. It follows that even some element $u'' \in u + (T_1, \ldots, T_m)$ kills the image of I in the ring above. By construction, both u' and u'' map to units in V and φ factors through the A-smooth algebra

$$S = (A_u[X_1, \ldots, X_m, T_1, \ldots, T_m]/(g_1, \ldots, g_r, h_1, \ldots, h_m))_{u'u''}.$$

□

In some situations, when applying Lemma 7 we will not initially have a map $A' \to V$. The following lifting lemma will help to bypass this obstacle. Its key novel aspect is that the elements s, s', and v need not come from the base ring A (compare with [24, (8.1], [33, (17.1)], or [32, 07CP]).

Lemma 8 *For a ring morphism $A \to V$ with V local, a smooth A-algebra S, an element $s \in S$, a nonunit $v \in V$, and a factorization*

$$A \to S \xrightarrow{s \mapsto v} V/v^n V \text{ for some } n \geq 2,$$

there are a smooth A-algebra S', an element $s' \in S'$, and factorizations

$$A \to S' \xrightarrow{s' \mapsto uv} V \text{ with } u \in V^* \text{ and } A \to S \xrightarrow{s \mapsto s'} S'/s'^n S' \to V/v^n V;$$

if s is the image of an element $a \in A$, then one may choose $s' = a$.

Proof Due to the local structure of smooth and étale morphisms [32, 054L,00UE], by localizing S around the preimage of the maximal ideal of V, we may assume that S is standard étale over a polynomial A-algebra, that is, that

$$S \cong (A[X_1, \ldots, X_d, Y]/(f))_{g \cdot \partial f/\partial Y} \text{ for some } f, g \in A[X_1, \ldots, X_d, Y] \text{ with } f \text{ monic in } Y.$$

For a suitable $n \in \mathbf{N}$, some unit multiple of $s \in S$ of the form $(g \cdot (\partial f/\partial Y))^N \cdot s$ lifts to an $\tilde{s} \in A[X_1, \ldots, X_d, Y]$. Letting $x_1, \ldots, x_d, y$ be some lifts to V of the images of $X_1, \ldots, X_d, Y$ in $V/v^n V$, we find that the A-morphism

$$A[X_1, \ldots, X_d, Y] \xrightarrow{X_i \mapsto x_i,\, Y \mapsto y} V$$

maps $\tilde{s}$ to a unit multiple of v (as may be checked modulo v^n), so it maps f to $\tilde{s}^n w$ for some $w \in V$. Thus, we obtain the A-morphism

$$S' = (A[X_1, \ldots, X_d, Y, W])_{g \cdot \partial f/\partial Y \cdot \partial(f - \tilde{s}^n W)/\partial Y}/(f - \tilde{s}^n W) \xrightarrow{W \mapsto w} V.$$

By construction, S' is A-smooth and, setting $s' = \tilde{s} \cdot (g \cdot \partial f/\partial Y)^{-N}$ in S' we have the identification

$$S'/s'^n S' \cong (A[X_1, \ldots, X_d, Y, W])_{g \cdot \partial f/\partial Y \cdot \partial(f - \tilde{s}^n W)/\partial Y}/(f, s'^n) \cong ((S/s^n S)[W])_{\partial(f - \tilde{s}^n W)/\partial Y}$$

with s' corresponding to s and compatibly with the maps to $V/v^n V$. The main part of the claim follows, and for the remaining assertion about a note that if s is the image of an $a \in A$, then we may choose $N = 0$ and $\tilde{s} = s' = a$ above. □

For desingularizing valuation rings, the above lemmas will be useful in several different ways. We illustrate this right away with the following results that facilitate passage to completions.

Proposition 9 *For a ring A, a dense extension of valuation rings (see Sect. 3) $V \subset V'$, K the fraction field of V, a ring morphism $A \to V$, a finitely presented A-algebra B, and maps*

$$A \to B \to V \text{ such that } B \to K \text{ factors through some } A\text{-smooth localization of } B$$

suppose that there exist a smooth A-algebra S' and a factorization $A \to B \to S' \to V'$. Then there exist a smooth A-algebra S and a factorization $A \to B \to S \to V$. In particular, there exist a smooth A-algebra S and a factorization $A \to B \to S \to V$ if there exist a smooth A-algebra $\hat{S}$ and a factorization $A \to B \to \hat{S} \to \hat{V}$, $\hat{V}$ being the completion of V.

Proof By hypothesis $H_{B/A}V \neq 0$ and let $b \in H_{B/A}V, b \neq 0$. Let $B \cong A[Y]/I, Y = (Y_1 \ldots, Y_m)$, I being a finitely generated ideal. Changing A by $A[Z]$, B by $B[Z]$, the map $B[Z] \to V$ being given by $Z \to b$, we may assume that b comes in fact from A. Indeed, if S is given for B, let us say as in (1) then $S[Z]$ could be taken for $B[Z]$ as in (1). Similarly as in [15, Lemma 4] we may assume that for some polynomials $f = (f_1, \ldots, f_r)$ from I, we have $b \in NMB$ for some $N \in ((f) : I)$ and a $r \times r$-minor M of the Jacobian matrix $(\partial f_i/\partial Y_j)$. Thus we may assume b is standard for B over V, which is necessary later to apply Lemma 7. Note that the composite map $B \to V \to V/b^3 V \cong V'/b^3 V'$ factors through a smooth $A/b^3 A$-algebra. By Lemma 7 $B \to V$ factors through a smooth A-algebra as well.

Indeed, since $V/b^3 V \cong V'/b^3 V'$, Lemma 8 supplies a smooth A-algebra S'_0, an $s' \in S'$, a factorization $A \to S'_0 \to V$ that sends s' to a unit multiple of b in V, and a factorization

$$\begin{array}{ccccc} & \nearrow & B/b^3 B \longrightarrow S'/b^3 S' & \searrow & \\ A & & \quad\quad\quad\; {\scriptstyle b \mapsto s'} \downarrow & & V/b^3 V. \\ & \searrow & \quad\quad\quad\; S'_0/s'^3 S'_0 & \nearrow & \end{array}$$

The local ring of S'_0 at the preimage of the maximal ideal of V is a domain (see [32, 033C],) and s' is nonzero in this local ring, so it is a nonzerodivisor there.

Thus, Lemma 7 applies and supplies a smooth S_0'-algebra S'' with a factorization $A \to B \to S'' \to V$. Note that S'' is a smooth A-algebra. □

To draw further consequences, we will use the following well-known result of Nagata (see [20, Theorem 4] or [32, 053E]).

Lemma 10 *Any finitely generated, flat (equivalently, torsion free) algebra over a valuation ring is finitely presented.*

Corollary 11 *For a local injection $V \to V'$ of valuation rings that induces a separable extension K'/K of fraction fields, if the map $V \to \tilde{V}'$ is ind-smooth, $\tilde{V}'$ being the completion of V', then so is $V \to V'$.*

Proof The separability assumption and Lemma 10 imply that Proposition 9 applies to every finite type V-subalgebra $B \subset V'$: a limit argument reduces to showing that the smooth locus of B over V is nonempty, which follows from the separability of $\mathrm{Frac}(B)/K$ thanks to [10, (6.7.4.1) in IV2] and [10, (17.5.1) in IV4]. It then remains to apply Lemma 5. □

The work above allows us to relate certain "formal desingularization" extensions of valuation rings studied in [25, section 6] to "weak desingularization" (that is, ind-smooth) extensions as follows.

Proposition 12 *Fix a local injection $V \to V'$ of valuation rings with fraction fields $K \to K'$ such that* $\mathrm{val}(V \setminus \{0\})$ *is cofinal in* $\mathrm{val}(V' \setminus \{0\})$ *and for each $0 \neq v \in V$ the map $V/vV \to V'/vV'$ is ind-smooth, a finitely presented A-algebra B, and maps $V \to B \to V'$ such that the map $B \to K'$ factors through some V-smooth localization of B. There is a smooth V-algebra S and a factorization $V \to B \to S \to V'$. If, in addition, K'/K is separable, then V' is ind-smooth over V.*

Proof In the case when K'/K is separable, B could be any finite type V-subalgebra of V', which is finitely presented by Lemma 10. So the last assertion follows from the rest and Lemma 5. For the assertion about B, we use Lemma 4 to choose an element $b \in B$ standard over V that does not die in V'. We assume that b is not a unit in V' (or else we may set $S = B_b$) and we choose a $0 \neq v \in V$ with $\mathrm{val}(v) > \mathrm{val}(b)$, where $\mathrm{val}(b)$ denotes the valuation of b considered in V'. By our assumptions, there are a smooth V/v^3V-algebra $\bar{S}$, an $s \in \bar{S}$, and a factorization $V \to B \to \bar{S} \xrightarrow{s \mapsto v} V'/v^3V'$ such that $b \mid s$ in $\bar{S}$. Thus, since $\bar{S}$ lifts to a smooth V-algebra (see [3, (1.3.1)] or [32, 07M8]), Lemma 8 supplies a smooth V-algebra S', an $s' \in S'$, a factorization $V \to S' \to V'$ that sends s' to a unit multiple of v, and a factorization $V \to B \to \bar{S} \xrightarrow{s \mapsto s'} S'/s'^3S' \to V'/v^3V'$. Since $b \mid s'$ in S'/s'^3S', by replacing S' by its localization by an element of $1 + s'^2S'$ if necessary we may ensure that $a \mid s'$ in S' for some lift $a \in S'$ of $b \in S'/s'^3S'$. Then we have a factorization

$$V \to B \xrightarrow{b \mapsto a} S'/a^3S' \to V'/a^3V'.$$

As in the proof of Lemma 9, Lemma 7 then supplies a smooth V-algebra S. □

The following localization lemma, a variant of [27, Lemma 2], [33, (12.2)], or [32, 07F9], will permit us to localize our valuation rings when arguing their ind-smoothness.

Lemma 13 *For ring maps $A \to B \to V$ with B of finite type over A, a prime $\mathfrak{P} \subset V$ with preimage $\mathfrak{p} \subset A$, and a factorization $A \to B \to S' \to V_{\mathfrak{P}}$ for a finitely presented $A_{\mathfrak{p}}$-algebra S', there are a finitely presented A-algebra S, an $s \in S$ with $S_s \otimes_A A_{\mathfrak{p}} \simeq S'[X, X^{-1}]$, and a factorization*

$$A \to B \to S \to V \ \text{ such that } \ S \to V_{\mathfrak{P}} \ \text{ factors as } \ S \to S_s \otimes_A A_{\mathfrak{p}} \to V_{\mathfrak{P}}.$$

Proof Following the argument of [33, (12.2)], we choose a presentation

$$S' \simeq (B \otimes_A A_{\mathfrak{p}})[X_1, \ldots, X_n]/(f_1(X_1, \ldots, X_n), \ldots, f_m(X_1, \ldots, X_n))$$

(see [32, 00F4]) in which the polynomials f_i have coefficients in B, and we set

$$S = B[X_0, X_1, \ldots, X_n]/(X_0^N f_1(X_1/X_0, \ldots, X_n/X_0), \ldots, X_0^N f_m(X_1/X_0, \ldots, X_n/X_0))$$

for a large enough $N > 0$ for which each $X_0^N f_i(X_1/X_0, \ldots, X_n/X_0)$ is a (necessarily homogeneous) polynomial in $X_0, X_1, \ldots, X_n$ of positive degree and coefficients in B. We set $s = X_0$, so that a desired isomorphism $S_s \otimes_A A_{\mathfrak{p}} \simeq S'[X, X^{-1}]$ is induced by the change of variables $X_0 \mapsto X$ and $X_i \mapsto XX_i$ for $1 \leq i \leq n$. To build the map $S \to V$, we first choose $x_1, \ldots, x_n \in V$ and $t \in V \setminus \mathfrak{P}$ such that X_i maps to $x_i/t \in V_{\mathfrak{P}}$. Continuing to use abusive notation for homogeneous polynomials, we note that the ("homogeneous" in $t, x_1, \ldots, x_n$) elements $t^N f_i(x_1/t, \ldots, x_n/t)$ of V die in $V_{\mathfrak{P}}$, so they are killed by some $t' \in V \setminus \mathfrak{P}$. Thus, the B-morphism

$$B[X_0, X_1, \ldots, X_n] \to V \ \text{ given by } \ X_0 \mapsto t't, X_1 \mapsto t'x_1, \ldots, X_n \mapsto t'x_n$$

factors through S. By construction, the resulting morphism $S \to V_{\mathfrak{P}}$ factors through the localization $S_s \otimes_A A_{\mathfrak{p}}$ of S, as desired. □

We are ready for the promised reduction to complete, one-dimensional valuation rings.

Proposition 14 *Consider the following property of a valuation ring V and a subring $A \subset V$:*

(∗) *every $A \to B \to V$ with B a finite type A-algebra such that $B \to \mathrm{Frac}(V)$ factors through an A-smooth localization of B has a refinement $A \to B \to S \to V$ with S smooth over A.*

For a finite dimensional valuation ring V with a subfield $A \subset V$, if for all consecutive primes $\mathfrak{q}' \subset \mathfrak{q} \subset V$ the complete height one valuation ring $\widehat{(V/\mathfrak{q}')_{\mathfrak{q}}}$ satisfies (∗), then so does V.

Proof We fix a finite type A-algebra B equipped with a factorization $A \to B \to V$ as in $(*)$, which we need to factor further as $A \to B \to S \to V$ for some smooth A-algebra S. When $B \to V$ itself factors through an A-smooth localization of B, there is nothing to show. Otherwise, since V is of finite height, we may choose the minimal prime $\mathfrak{q} \subset V$ whose preimage in B does not lie in the A-smooth locus of $\mathrm{Spec}(B)$ and the largest prime $\mathfrak{q}' \subsetneq \mathfrak{q} \subset V$ properly contained in $\mathfrak{q}$ (the assumption in $(*)$ ensures that $\mathfrak{q}'$ exists). Thanks to Lemma 13, we may replace V by $V_{\mathfrak{q}}$ to reduce to the case when $\mathfrak{q}$ is the maximal ideal (so that $(\widetilde{V/\mathfrak{q}'})_{\mathfrak{q}} = \widetilde{V/\mathfrak{q}'}$): indeed, once we resolve this case, then, by using Lemma 13, we will be able to refine B to an A-algebra that either is smooth or for which $\mathfrak{q}$ is strictly larger, and, by iteration, we will then arrive at a desired S.

By Lemma 4, there is an element $b \in B$ standard over A that maps to $\mathfrak{q} \setminus \mathfrak{q}'$. The property from $(*)$ of $\widetilde{V/\mathfrak{q}'}$ then supplies a smooth A-algebra S', an element $s \in S'$ (the image of b), and a factorization

$$\begin{array}{ccccc} & \nearrow & B & \searrow & \\ A & & \big\downarrow{\scriptstyle b \mapsto s} & & \widetilde{V/\mathfrak{q}'}/b^3\widetilde{V/\mathfrak{q}'} \cong V/b^3V. \\ & \searrow & S'/s^3S' & \nearrow & \end{array}$$

Thanks to Lemma 8, we may change S' in order to make sure that the map $S'/s^3S' \to V/b^3V$ lifts to an A-morphism $S' \to V$. This puts us in a situation in which we may apply Lemma 7 to obtain a smooth S'-algebra S with a desired factorization $A \to B \to S \to V$. □

3 Ind-Smoothness of Large Immediate Extensions of Valuation Rings

Our next goal is to find a large class of extensions of valuation rings that are ind-smooth. The argument combines classical results from valuation theory that go back to Kaplansky, results from [23] (see Lemma 15 and its proof), and the desingularization lemmas from Sect. 2.

Consider the case when V is not noetherian and its associated valuation has rank one. In the Noetherian case a immediate extension of valuation rings $V \subset V'$ is dense, but in general case it need not be. If $V \supset \mathbf{Q}$ the problem is solved by Ostrowski's Defektsatz [22] but when the characteristic of the residue field of V is > 0 the immediate algebraic extensions present extra difficulties.

An inclusion $V \subset V'$ of valuation rings is an *immediate extension* if it is local as a map of local rings and induces isomorphisms between the value groups and the residue fields of V and V'. For such a $V \subset V'$, letting K'/K be the induced fraction field extension, we have $V = V' \cap K$ (see [5, (4.1) in VI]). Moreover, for any subextension $K'/K''/K$ and the valuation ring $V'' = V' \cap K''$, both $V \subset V''$ and

$V'' \subset V'$ are then also immediate extensions (to check the value group requirement one uses that any $v'' \in V''$ is a unit if and only if so is its image in V').

For example, for any valuation ring V, the extension $V \subset \tilde{V}$ is immediate (see [33]), $\tilde{V}'$ being the completion of V'.

For a valuation ring V with the fraction field K, a sequence $\{v_i\}_{i<\omega}$ in K indexed by the ordinals i less than a fixed limit ordinal ω is *pseudo-convergent* if

$\mathrm{val}(v_i - v_{i''}) < \mathrm{val}(v_{i'} - v_{i''})$ (that is, $\mathrm{val}(v_i - v_{i'}) < \mathrm{val}(v_{i'} - v_{i''})$) for $i < i' < i'' < \omega$ (see [12, 33]). A (possibly nonunique) *pseudo-limit* of a pseudo-convergent sequence $\{v_i\}_{i<\omega}$ is an element $\alpha \in K$ with

$\mathrm{val}(\alpha - v_i) < \mathrm{val}(\alpha - v_{i'})$ (that is, $\mathrm{val}(\alpha - v_i) = \mathrm{val}(v_i - v_{i'})$) for $i < i' < \omega$. A pseudo-convergent sequence $\{v_i\}_{i<\omega}$ in K is

(1) *algebraic* if some $f \in K[T]$ satisfies $\mathrm{val}(f(v_i)) < \mathrm{val}(f(v_{i'}))$ for large enough $i < i' < \omega$;
(2) *transcendental* if each $f \in K[T]$ satisfies $\mathrm{val}(f(v_i)) = \mathrm{val}(f(v_{i'}))$ for large enough $i < i' < \omega$.

(Here "large enough" means larger than a fixed ordinal $\omega' < \omega$ that is allowed to depend on f.) In both cases, [12, Theorems 1, 2] describe the valuation of K' that extends V of K. For instance, in the transcendental case, by [12, Theorem 2], this valuation on $K(t)$ is given by setting

$$\mathrm{val}(((f(t))/(g(t))) = \mathrm{val}(f(v_i)) - \mathrm{val}(g(v_i)) \quad \text{for large enough} \quad i < \omega.$$

These results lead to [12, Theorem 4]: a valuation ring V has no nontrivial immediate extensions if and only if each pseudo-convergent sequence in its fraction field K has a pseudo-limit in K. If for all $\gamma \in \Gamma$, the value group of V, there exists $i < i'$ sufficiently large such that $\mathrm{val}(v_i - v_{i'}) > \gamma$ then we call $\{v_i\}_{i<\omega}$ *fundamental*. As in [23, Lemma 3.2] we get the following lemma.

Lemma 15 *For an immediate extension $V \subset V'$ of valuation rings and a transcendental pseudo-convergent sequence $(v_i)_{i<\omega}$ in K, which has a pseudo-limit v' in K' but no pseudo-limit in K the valuation ring $V'' = V' \cap K(v')$ is a filtered union of smooth V-subalgebras.*

Proof For each i set $x_i = (v' - v_i)/(v_{i+1} - v_i)$, so that x_i is a unit in V'. Let $\mathfrak{m}'$ be the maximal ideal of V'. We show that for every polynomial $0 \neq f \in V[t]$ it holds

$$f(v') \in f(v_i) \cdot (1 + \mathfrak{m}' \cap V[x_i]) \quad \text{for every large enough} \quad i < \omega.$$

Since $\{v_i\}_{i<\omega}$ is transcendental, for each $g(t) \in K[t]$, the value $\mathrm{val}(g(v_i))$ is constant for large i. Moreover, for large i the values $\mathrm{val}(v' - v_i)$ are strictly increasing as i increases. Thus, in the Taylor expansion[1]

[1] The polynomials $D^{(n)} f \in R[t]$ for $f \in R[t]$ make sense for any ring R: indeed, one constructs the Taylor expansion in the universal case $R = \mathbf{Z}[a_0, \ldots, a_{\deg f}]$ by using the equality $n! \cdot (D^{(n)} f) = f^{(n)}$.

$$f(v') = \sum_{n=0}^{\deg f} (D^{(n)} f)(v_i) \cdot (v' - v_i)^n \quad \text{with} \quad D^{(n)} f \in V[t]$$

the values $\mathrm{val}((D^{(n)} f)(v_i) \cdot (v' - v_i)^n)$ are pairwise distinct for every large enough i. Consequently, since $\mathrm{val}(f(v')) = \mathrm{val}(f(v_i))$ for large i, we conclude that

$$\mathrm{val}((D^{(n)} f)(v_i) \cdot (v' - v_i)^n) > \mathrm{val}(f(v_i)) \text{ for every } n > 0 \text{ and large enough } i < \omega.$$

It remains to note that

$$(v' - v_i)^n = x_i^n \cdot (v_{i+1} - v_i)^n \text{ and } \mathrm{val}(v' - v_i) = \mathrm{val}(v_{i+1} - v_i),$$

which is enough for our claim. In particular, we get that $f(v')$ is transcendental over K. The element x_i is transcendental over K, so $V[x_i] \subset V'$ is the polynomial algebra. Moreover, for $i < i' < \omega$ we have $x_i = x_{i'} \cdot (v_{i'+1} - v_{i'})/(v_{i+1} - v_i) + (v_{i'} - v_i)/(v_{i+1} - v_i)$, so $V[x_i] \subset V[x_{i'}] \subset V'$. Consequently, we arrive at a nested sequence

$$\{V[x_i]_{\mathfrak{m}' \cap V[x_i]}\}_{i<\omega} \text{ of ind-smooth } V\text{-subalgebras of } V',$$

and, it remains to show that every element of V'' belongs to some $V[x_i]_{\mathfrak{m}' \cap V[x_i]}$. In fact, it suffices to show that each $0 \neq f \in V[t]$ satisfies

$$f(v') \in f(v_i) \cdot (1 + \mathfrak{m}' \cap V[x_i]) \quad \text{for every large enough} \quad i < \omega$$

which was done above. □

Lemma 16 *For an immediate extension $V \subset V'$ of valuation rings containing* $\mathbf{Q}$ *with value group $\Gamma \subset \mathbf{R}$ every algebraic pseudo-convergent sequence $(v_i)_{i<\alpha}$ in K, which is not a fundamental sequence but has a pseudo-limit v' in K' has also a pseudo-limit in K.*

Proof By [12, Theorem 3] there exists an immediate extension of valued fields $K \subset K(u)$ such that u is algebraic over K and it is a pseudo-limit of (v_i) in $K(u)$. As a consequence of Ostrowski's Defektsatz [22, Sect 9, No 55] (see [23, Corollary 4.2], or [25, Corollary 3.10]) we see that $K \subset K(u)$ is dense, that is, u belongs to the completion of K. Thus (v_i) has a pseudo-limit in K by [25, Lemma 2.5]. □

Remark 17 *The above lemma is false if the characteristic of the residue field of a valuation rings is > 0 (see [25, Example 3.13] inspired by [22, Sect 9, No 57]).*

Proposition 18 *For an immediate extension $V \subset V'$ of valuation rings containing* $\mathbf{Q}$ *with value group $\Gamma \subset \mathbf{R}$, V' is ind-smooth over V.*

Proof Applying Lemma 15 possible infinitely, even uncountably many, we find a pure transcendental extension $K'' \subset K'$ of K such that $V'' = V' \cap K''$ is ind-smooth over V and all transcendental pseudo-convergent sequences in V'' over V having a pseudo limit in V', which are not fundamental sequences, have pseudo-limits in V''. By Lemma 16 we see that this holds also for algebraic pseudo-convergent sequences from V'', which are not fundamental sequences. It follows that the extension $V'' \subset V'$ is dense using [12, Theorem 1]. Now, it is enough to apply Lemma 7 to see that V' is ind-smooth over V''. Indeed, let $B \subset V'$ be a finitely generated V''-subalgebra and w its inclusion. By separability, $w(H_{B/V''}) \neq 0$ and choose an element $d \in V''$, which is standard for B over V''. Then the composite map $B \to V' \to V'/d^3V' \cong V''/d^3V''$ factors obviously to a smooth V''/d^3V''-algebra. By Lemma 7 w factors through a smooth V''-algebra. □

Corollary 19 *For an extension $V \subset V'$ of valuation rings containing* **Q** *with the same value group $\Gamma \subset$* **R**, *V' is ind-smooth over V.*

Proof By Proposition 9 we may reduce to the case when V, V' are complete and so they are Henselian since $\dim V = 1$, that is, they contain their residue fields k, k'. Let K, K' be the fraction fields of V, resp. V'. Then V' is an immediate extension of $V'' = V' \cap K(k')$ and so it is ind-smooth by the above proposition. Express k' as a filtered union of some finitely generated field extensions (k_i) of k. It is enough to see that $V_i = V' \cap K(k_i)$ is an ind-smooth extension of V. But V_i is even essentially smooth over V because k_i is so over k. □

4 Extensions of Valuation Rings

The following proposition is an extension of Corollary 19.

Proposition 20 *Let $V \subset V'$ be an extension of valuation rings containing* **Q**. *Suppose* $\dim V < \infty$ *and the value group extension of $V \subset V'$ is trivial. Then V' is ind-smooth over V.*

Proof For the proof we apply Lemma 5.

Let E be a V-algebra of finite presentation, let us say $E \cong V[Y]/I$, $Y = (Y_1 \ldots, Y_m)$, I being a finitely generated ideal. Let $w : E \to V'$ be a V-morphism. We will show that w factors through a smooth V-algebra. E is finitely generated and so it is $\operatorname{Im} w$. By Lemma 10 we see that $\operatorname{Im} w$ is finitely presented. So we may replace E by $\operatorname{Im} w$, that is we may assume w injective. By separability we have $w(H_{E/V}) \neq 0$, let us assume that $w(H_{E/V})V' \supset zV'$ for some $z \in V, z \neq 0$. Replacing z by a power of it we may assume that $z = \sum_i^s b_i b_i'$ for some $b_i = \det(\partial f_{ij}/\partial Y_{j_i})$ for some systems of polynomials f_i from I and $b_i'' \in V[Y]$ which kills $I/(f_i)$. Similarly as in [15, Lemma 4] we may assume that we can take $s = 1$, that is for some polynomials $f = (f_1, \ldots, f_r)$ from I, we have $z \in NME$ for some $N \in ((f) : I)$ and a $r \times r$-minor M of the Jacobian matrix $(\partial f_i/\partial Y_j)$ (since V' is a valuation ring this reduction is much easier). Thus we may assume z is standard over V (see the beginning of Sect. 2), which is necessary later to apply Lemma 7. Let $q_2' \in \operatorname{Spec} V'$, be the minimal prime ideal of zV' and $q_2 = q_2' \cap V$. As the value group extension of $V \subset V'$ is trivial we have $q_2' = q_2V'$.

Let $q_1 \in \operatorname{Spec} V$, $q_1 \subset q_2$ be the greatest prime ideal of V not containing z. Then $q_1 \neq q_2$. The extension $V_{q_2}/q_1 V_{q_2} \subset V'_{q'_2}/q_1 V_{q'_2}$ has the trivial value group extension and so it is ind-smooth by Corollary 19. The composite map $E \xrightarrow{w} V' \to V'_{q'_2}/q_2 V'_{q'_2}$ factors by a smooth $V_{q_2}/q_1 V_{q_2}$-algebra G, let us say it is the composite map $E \xrightarrow{\alpha} G \xrightarrow{\beta} V'_{q'_2}/q_2 V'_{q'_2}$. We may assume that $G = (V_{q_2}/q_1 V_{q_2})[U]_{g'h}/(g)$, with $U = (U_1, \ldots, U_l)$, $g' = \partial g/\partial U_1$, $g, h \in V[U]$ by [33, Theorem 2.5] and let β be given by $U + (g) \to u + q_1 V'_{q'_2}$ for some $u \in (V'_{q'_2})^l$. Note that [33, Theorem 2.5] gives just that a localization of G has the form a localization of $C = (V_{q_2}/q_1 V_{q_2})[U]_{g'}/(g)$ and so the above composite map factors through a C_h for some $h \in V[Y]$. Then $g(u) \equiv 0$ modulo $q_1 V'_{q'_2}$ and in particular $g(u) \equiv 0$ modulo $z^3 V'_{q'_2}$. Then $g(u) = z^3 t$ for some $t \in V'_{q'_2}$. Note that the composite map $E \to V' \to V'_{q'_2}$ factors through the smooth V_{q_2}-algebra $D = (V_{q_2}[U, T]/(g - z^3 T))_{g'h}$ modulo z^3, where $T \to t$. By Lemma 7 we see that $E \to V'_{q'_2}$ factors through a smooth D-algebra D' which is also smooth over V_{q_2}. Using Lemma 13 we see that w factors through a finitely presented V-algebra E'', let us say through a map $w'' : E'' \to V'$ with $w''(H_{E''/V}) \not\subset q'_2$. More precisely, by Lemma 13 there exist a finitely presented V-algebra E'' and $c \in E''$ with $E''_c \otimes_V V_q \cong D'[X, X^{-1}]$ and a factorization $V \to E \to E'' \to V'$ such that $E' \to V'_{q'}$ factors through $E' \to E'_c \otimes_V V_q \to V'_{q'}$. Note that $\dim V' = \dim V < \infty$ because V, V' have the same value group. We arrive in finite steps using induction on $\dim V'/zV'$ in the case when z is a unit, that is we can embed E in a smooth V-algebra. This is enough by Lemma 5. □

Theorem 21 *Let $V \subset V'$ be an immediate extension of valuation rings containing* **Q**. *Then V' is ind-smooth over V.*

Proof Let $K \subset K'$ be the fraction field of $V \subset V'$ and $K'' \subset K'$ a pure transcendental extension of K generated by a transcendental basis of K'/K, that is K'/K is algebraic. Applying Lemma 15 possible infinitely and even uncountably many, as in Proposition 18 we see that $V'' = V' \cap K''$ is ind-smooth over V and all transcendental pseudo-convergent sequences in V'' over V having a pseudo limit in V', which are not fundamental sequences, have pseudo-limits in V''. Thus we reduce to show that V' is ind-smooth over V when K'/K is algebraic. Actually, it is enough to assume K'/K finite because V' is the filtered union of $V' \cap L$ for all subfields $L \subset K'$ which are finite extension over K.

Let $E = V[Y]/I, Y = (Y_1, \ldots, Y_n)$ be a finitely generated V-subalgebra of V' (so finitely presented by Lemma 10) and a map $w : E \to V'$. By Lemma 5 it is enough to show that w factors through a smooth V-algebra. Consider $H_{E/V}$ and a standard element $z \in V$ for E over V, so $w(z) \in w(H_{E/V})V'$, as in the proof of Proposition 20. If $V \subset V'$ is dense we may apply Proposition 9 to see that w factors through a smooth V-algebra (note that $w(H_{E/V}) \neq 0$ says that the composite map $E \to V' \to K'$ factors through a smooth V-algebra). In the remaining case the factorization is constructed in some steps: for the standard element z one chooses adjacent prime ideals $q_1 \subset q$ of V such that $w(z) \in qV' \setminus q_1 V'$ and construct a factorization $E \to E' \xrightarrow{w'} V'$ such that $w'(H_{E'/V}) \not\subset qV'$, where E' is finitely presented over V. If after finite steps we get a factorization $E \to E^{(n)} \xrightarrow{w^{(n)}} V'$ such that $w^{(n)}(H_{E^{(n)}/V})V' = V'$ the goal is reached. A hard problem is to show that we can find such $E^{(n)}$ in finite steps. For this we will consider a finite partition $\mathcal{P}_i$, $i = 1, \dots, s$ of Spec V corresponding to those $q \in \operatorname{Spec} V$ which have the same dimension $f_i = f_q \leq f = [K' : K]$ of the fraction field extension $K_q \subset K'_q$ of $V/q \subset V'/qV'$. We will see that to each construction q_1 change from one $\mathcal{P}_j$ to another one from $\mathcal{P}_i$ with $j < i$ and $f_i < f_j$. Finally we arrive in finite steps to the case $f_{q_1} = 1$, which is done easily as in the dense case.

Assume $V \subset V'$ is not dense and $q, q_1, f_{q_1}, (\mathcal{P}_i)_{1 \leq i \leq s}$ as above. More precisely, let $q' \in \operatorname{Spec} V'$ be the minimal prime ideal of $w(z)V'$ and $q = q' \cap V$. Thus $qV' = q'$ because $V \subset V'$ is immediate. Let $q'_1 \in \operatorname{Spec} V'$ be the prime ideal corresponding to the maximal ideal of the fraction ring of V' with respect to the multiplicative system generated by z. Then q'_1 is the biggest prime ideal of V' contained strictly in q' and so height$(q'/q'_1) = 1$. Set $q_1 = q'_1 \cap V$. We have $f_q \leq f_{q_1}$.

Let x_q be a primitive element of the separable finite extension K'_q/K_q and $g_q \in V/q[X]$ be a primitive polynomial multiple of $\operatorname{Irr}(x_q, K_q)$ by a nonzero constant of K. Note that if $q, \mathfrak{q} \in \mathcal{P}_i, q \subset \mathfrak{q}$ then $f_q = f_{\mathfrak{q}} = f_i$ and g_q remains irreducible over $V/\mathfrak{q}$. Clearly, $f_s = 1$ because $f_{\mathfrak{m}} = 1$ for the maximal ideal $\mathfrak{m}$ of V, the extension $V \subset V'$ being immediate. A set $\mathcal{P}_i$ has a maximum element for inclusion namely $\mathfrak{p}_i = \cup_{\mathfrak{q} \in \mathcal{P}_i} \mathfrak{q}$. Indeed, $\mathfrak{p}_i$ is clearly a prime ideal and if $f_{\mathfrak{p}_i} < f_i$ then $f_{\mathfrak{q}} < f_i$ for some $\mathfrak{q} \in \mathcal{P}_i$, which is false.

Assume $q_1 \in \mathcal{P}_j$. If $q_1 \neq \mathfrak{p}_j$ then $(V/q_1)_{\mathfrak{p}_j} \subset (V'/q_1V')_{\mathfrak{p}_jV'}$ is in fact a localization of $(V/q_1)[X]/(g_{q_1})$ because $g'_{q_1} = \partial g_{q_1}/\partial X$ corresponds to a unit in $(V'/q_1V')_{\mathfrak{p}_j}$ and so the composite map $E \to V' \to (V'/q_1V')_{\mathfrak{p}_jV'}$ factors through an etale V/q_1-algebra of the form $((V/q_1)[X]/(g_{q_1})_{g'_{q_1}h}$ for some $h \in V[X]$. In particular $E \to V' \to (V'/(z^3))_{\mathfrak{p}_jV'}$ factors through an etale $V/(z^3)$-algebra and by Lemma 7 the map $E \to V' \to V'_{\mathfrak{p}_jV'}$ factors through a smooth V-algebra. Using Lemma 13 we see that w factors through a finitely presented V-algebra E', let us say through a map $w' : E' \to V'$ with $w'(H_{E'/V}) \not\subset \mathfrak{p}_jV'$. Changing E by E' we see that the new q belongs to $\mathcal{P}_i$ for some $i > j$. Moreover, the new q_1 belongs also to $\mathcal{P}_i$ for some $i > j$, because otherwise we get $q_1 = \mathfrak{p}_j$.

If $q_1 = \mathfrak{p}_j$ then $q \in \mathcal{P}_{j+1}$ and we apply Corollary 19. Then we see that $(V/q_1)_q \subset (V'/q'_1)_{q'}$ is ind-smooth and as above we see that the composite map $E \to V' \to (V'/q'_1)_{q'}$ factors through a smooth V/q_1-algebra and finally by Lemmas 7 and 13 we get that w factors through a finitely presented V-algebra E', let us say through a map $w' : E' \to V'$ with $w'(H_{E'/V}) \not\subset qV'$. Now the new q_1, that is the old q, belongs to $\mathcal{P}_{j+1}$. In some steps (at most s) we arrive to the case when $f_{q_1} = 1$.

If $f_{q_1} = 1$ then we get $f_{q''_1} = 1$ for all $q''_1 \in \operatorname{Spec} V$ containing q_1. Actually, we get $V/q = V'/q'$ and so in particular $V/q \subset V'/q'$ is ind-smooth. Using Lemma 7 we see that w factors through a smooth (in fact etale) V-algebra. Applying Lemma 5 we are done. □

Proposition 22 *Let $V \subset V'$ be an extension of valuation rings. Suppose that*

(1) *V is a discrete valuation ring extending $\mathbf{Z}_{(p)}$ with π its local parameter, and p a prime number.*
(2) *$\pi V'$ is the maximal ideal of V',*
(3) *the residue field extension of $V \subset V'$ is separable.*

Then $V \to V'$ is ind-smooth.

Proof Let E, w, $H_{E/V}$, z be as in Proposition 20 and we may assume K' is the fraction field of $\mathrm{Im}\, w$. choose $q_2' \in \mathrm{Spec}\, V'$ a minimal prime ideal of $w(H_{E/V})V'$. If $q_2' \neq \pi V'$ then using Zariski's Uniformization Theorem we may change E, w with some E', w' such that $w'(H_{E'/V})V' \not\subset q_2'$. Step by step we arrive to the case when either $w(H_{E/V})V' = V'$, or $w(H_{E/V})V'$ is a $\pi V'$-primary ideal. In the first case, w factors through a localization of E which is smooth. In the second case, $q_1' = \cap_{i \in \mathbf{N}} \pi^i V'$ is a prime ideal and the composite map $V \to V' \to V'/q_1'$ is a regular map of discrete valuation rings and so an ind-smooth map by the classical Néron Desingularization. The proof ends by using Lemma 5. □

Corollary 23 *Let V be a discrete valuation ring extending $\mathbf{Z}_{(p)}$ with p a prime number and V' an ultrapower of V with respect to a nonprincipal ultrafilter on $\mathbf{N}$. Then $V \subset V'$ is ind-smooth.*

For the proof note that the maximal ideal of V generates the maximal ideal of V' and apply the above proposition.

Proposition 24 *Let V be a discrete valuation ring extending $\mathbf{Z}_{(p)}$ with p a prime number and $V \subset V'$ an extension of valuation rings such that*

(1) *p is a local parameter of V,*
(2) *pV' is a $\mathfrak{m}'$-primary ideal of V', where $\mathfrak{m}'$ is the maximal ideal of V',*
(3) *the residue field extension of $V \subset V'$ is separable.*

Then V' is a filtered direct limit of regular local rings essentially of finite type over V.

Proof As in Proposition 22 we may consider $w : E \to V'$ and we may reduce to the case when $w'(H_{E'/V})V'$ is $\mathfrak{m}'$-primary ideal. We may assume that p^s is a standard for E over V for some $s \in \mathbf{N}$ and as in the proof of [30, Theorem 3.6] there exists a local essentially smooth V-algebra G and $b \in G$ such that the map $E/p^{3s}E \to V'/p^{3s}V'$ factors through $G/(p^{3s}, p - b)$. Then a variant of Lemma 7 in the idea of [30, Proposition 3.4] shows that w factors through a local essentially smooth $D = G/(p - b)$-algebra D'. This D' is regular local since D is so. Now apply Lemma 5. □

5 Structure of Equicharacteristic Valuation Rings Possessing a Cross-Section

Modulo all the reductions and simplifications that go into the overall proof of Theorem 2, our ultimate source of expressions of valuation rings as filtered direct limits of smooth rings is Lemma 26 below. This lemma describes some valuations on an affine space for which local uniformizations can be constructed by successively blowing up regular centers as in [31, 4.5, 4.19] following Perron's algorithm (whose relevance to the resolution of singularities was explained already in [34]). We present a more direct argument for this uniformization that is close to [23, Lemma 4.6] and rests on the following lemma that captures the "combinatorial" part of local uniformization.

We will need the following lemma (see [7, 2.2], or [23, 4.6.1], or [9, 6.1.30]).

Lemma 25 *For a totally ordered abelian group Γ, the submonoid $\Gamma_{\geq 0} \subset \Gamma$ of non-negative elements is a filtered union of its finite free submonoids isomorphic to $\mathbf{Z}^r_{\geq 0}$, where $r \in \mathbf{Z}_{\geq 0}$ need not be constant.*

We include a mixed characteristic version of the following lemma because it requires virtually no additional effort in comparison to the equicharacteristic case that we will use below.

Lemma 26 *(1) For a field $\mathbf{F}$, a valuation ring $\mathbf{F} \subset V$ with fraction field $\mathbf{F}(x_1, \dots, x_n)$ such that* $\operatorname{val}(x_1), \dots, \operatorname{val}(x_n)$ *are $\mathbf{Z}$-linearly independent is a countable direct union of essentially smooth $\mathbf{F}$-algebras.*

(2) For a discrete valuation ring Λ with uniformizer π and fraction field $\mathbf{F}$, a valuation ring $\Lambda \subset V$ that dominates Λ and has fraction field $\mathbf{F}(x_1, \dots, x_n)$ such that $\operatorname{val}(\pi), \operatorname{val}(x_1), \dots, \operatorname{val}(x_n)$ *are $\mathbf{Z}$-linearly independent is a countable direct union of regular local Λ-algebras of the form*

$$(\Lambda[Y_1, \dots, Y_{n+1}]/(\pi - Y_1^{b_1} \cdots Y_{n+1}^{b_{n+1}}))_{(Y_1, \dots, Y_{n+1})} \text{ with } \gcd(b_1, \dots, b_{n+1}) = 1.$$

Proof To avoid repeating the argument, we will prove both claims simultaneously, so in (1) we set $\Lambda = \mathbf{F}$ and $\pi = 0$ and in both parts we set $p = \operatorname{Char}(\Lambda/(\pi))$. By [5, Theorem 1 in VI (10.3)],

$$\gamma_1 = \operatorname{val}(x_1), \quad \dots, \quad \gamma_n = \operatorname{val}(x_n), \quad \gamma_{n+1} = \operatorname{val}(\pi) \text{ (resp. } \gamma_{n+1} = 0 \text{ if } \pi = 0)$$

satisfy $\Gamma \cong \mathbf{Z}\gamma_1 \oplus \cdots \oplus \mathbf{Z}\gamma_{n+1}$, where Γ is the value group of V. We set $N = n+1$ (resp., $N = n$ if $\pi = 0$) and use Lemma 25 to find a countable sequence $\Gamma_0 \subset \Gamma_1 \subset \dots$ of submonoids of $\Gamma_{\geq 0}$ with $\Gamma_i \simeq \mathbf{Z}^N_{\geq 0}$ for each i and $\Gamma_{\geq 0} = \bigcup_{i \geq 0} \Gamma_i$. We fix a $\mathbf{Z}_{\geq 0}$-basis $\nu_{i1}, \dots, \nu_{iN}$ of Γ_i with $(\nu_{01}, \dots, \nu_{0N}) = (\gamma_1, \dots, \gamma_N)$, so that the elements $\nu_{i1}, \dots, \nu_{iN}$ are $\mathbf{Z}$-linearly independent in Γ, and we express them in terms of the fixed $\mathbf{Z}$-basis:

$$\nu_{ij} = d_{ij1}\gamma_1 + \cdots + d_{ijN}\gamma_N \text{ for unique } d_{ij1}, \dots, d_{ijN} \in \mathbf{Z} \text{ and every } j = 1, \dots, N.$$

We set $x_{n+1} = \pi$ and note that, by construction, for each $i \geq 0$ and $1 \leq j \leq N$, the element
$y_{ij} = x_1^{d_{ij1}} \cdots x_N^{d_{ijN}} \in \mathbf{F}(x_1, \dots, x_n)$ has valuation ν_{ij}.

Since $\Gamma_{i'} \subset \Gamma_i$ for $i' < i$, each $y_{i'j}$ is in a unique way a monomial in the elements $y_{i1}, \dots, y_{iN}$:

$$\text{if we express } \nu_{i'j} = b_{i'i1}\nu_{i1} + \cdots + b_{i'iN}\nu_{iN} \text{ with } b_{i'ij} \in \mathbf{Z}_{\geq 0}, \text{ then } y_{i'j} = y_{i1}^{b_{i'i1}} \cdots y_{iN}^{b_{i'iN}}.$$

Since the valuations of $y_{i1}, \dots, y_{iN}$ are $\mathbf{Z}$-linearly independent, the Λ-subalgebra $\Lambda[y_{i1}, \dots, y_{iN}]$ of $\mathbf{F}(x_1, \dots, x_n)$ is the regular ring

$$\Lambda[Y_{i1}, \dots, Y_{iN}]/(\pi - Y_{i1}^{b_1} \cdots Y_{iN}^{b_N}) \text{ with } b_i = b_{0Ni} \text{ (resp. } \Lambda[Y_{i1}, \dots, Y_{iN}] \text{ if } \pi = 0),$$

where $\gcd(b_1, \dots, b_N) = 1$ because $\mathrm{val}(\pi)$ is assumed to be a primitive element of Γ when $\pi \neq 0$. In particular, we obtain a nested sequence of Λ-subalgebras

$$R_i = \Lambda[y_{i1}, \dots, y_{iN}]_{(y_{i1}, \dots, y_{iN})} \subset V$$

that are regular (resp., essentially smooth if $\pi = 0$) and it remains to argue that every $f \in V$ belongs to some R_i. For this, we first express f as a rational function as follows:

$$f = \left(\sum \lambda_{s_1, \dots, s_N} x_1^{s_1} \cdots x_N^{s_N}\right) / \left(\sum \lambda'_{r_1, \dots, r_N} x_1^{r_1} \cdots x_N^{r_N}\right) \text{ with } \lambda_{s_1, \dots, s_N}, \lambda'_{r_1, \dots, r_N} \in \Lambda^\times \cup \{0\}.$$

The linear independence of the γ_i ensures that the valuations of the monomials that appear in the numerator (resp., denominator) are all distinct. Thus, by taking out the monomials with minimal valuations, we reduce to showing that every $x_1^{\alpha_1} \cdots x_N^{\alpha_N}$ with $\alpha_j \in \mathbf{Z}$ and $\alpha_1\gamma_1 + \cdots + \alpha_N\gamma_N > 0$ is a product of nonnegative powers of the elements $y_{i1}, \dots, y_{iN}$ for some $i \geq 0$. For this, it suffices to note that $\alpha_1\gamma_1 + \cdots + \alpha_n\gamma_n$ lies in some Γ_i, and then to express it as a $\mathbf{Z}_{\geq 0}$-linear combination of the $\nu_{i1}, \dots, \nu_{iN}$: more precisely, if $\alpha_1\gamma_1 + \cdots + \alpha_N\gamma_N = c_1\nu_{i1} + \cdots + c_N\nu_{iN}$ with $c_j \in \mathbf{Z}_{\geq 0}$, then

$$x_1^{\alpha_1} \cdots x_N^{\alpha_N} = y_{i1}^{c_1} \cdots y_{iN}^{c_N}.$$

□

For a valuation ring V with the value group Γ and the fraction field K, a *cross-section* of V is a section

$$s\colon \Gamma \to K^* \text{ in the category of abelian groups of the valuation map } \mathrm{val}\colon K^* \to \Gamma$$

(for more details, see the Appendix).

Proposition 27 *An equicharacteristic valuation ring V that has a cross-section $s\colon \Gamma \to K^*$ and a subfield $k \subset V$ lifting the residue field is an immediate extension*

$$V_0 = \bigcup_i V_i \subset V$$

of a filtered union of valuation subrings $V_i \subset V$ dominated by V such that each V_i has a finitely generated value group, is a countable increasing union of localizations of smooth k-subalgebras of V so V_0 is ind-smooth over k, and has the restriction of s as a cross-section.

Proof By Lemma 25, the submonoid $\Gamma_{\geq 0} \subset \Gamma$ of positive elements is a filtered union $\Gamma_{\geq 0} = \bigcup_i \Gamma_i$ of submonoids $\Gamma_i \simeq \mathbf{Z}_{\geq 0}^{d_i}$ with $d_i \geq 0$. Thus, the cross-section s gives rise to the filtered system of subfields $k_i = k(s(\gamma) \mid \gamma \in \Gamma_i)$ of the field of fractions K of V. By choosing a $\mathbf{Z}_{\geq 0}$-basis for Γ_i and applying [5, Theorem 1, in VI Sect. 10] we see that each k_i is a purely transcendental extension of k and that the value group of the valuation subring $V_i = V \cap k_i$ of V is $\mathbf{Z}^{d_i} \simeq \mathbf{Z}\Gamma_i \subset \Gamma$. By construction, s restricts to a cross-section of V_i and, by Lemma 26, each V_i is a filtered union of localizations of k-subalgebras. The construction ensures that V is an immediate extension of the resulting V_0. □

6 Counterexamples When the Value Groups Are Finitely Generated

Lemma 28 *Let $V \subset V'$ be an extension of valuation rings which is ind-smooth. Then $\Omega_{V'/V}$, that is $H_0(V, V', V')$ in terms of Andre-Quillen homology, is a flat V'-module and $H_1(V, V', V') = 0$ (the last homology is usually denoted by $\Gamma_{V'/V}$).*

Proof Assume that V' is the filtered direct limit of some smooth V -algebras B_i, $i \in I$. Then $\Omega_{B_i/V}$ is projective over B_i and $H_1(V, B_i, B_i) = 0$ by e.g. [33, Theorem 3.4]. But $\Omega_{V'/V}$, and $H_1(V, V', V')$ are filtered direct limits of $V' \otimes_{B_i} \Omega_{B_i/V}$ resp. $V \otimes_{B_i} H_1(V, B_i, B_i)$ by [33, Lemma 3.2], which is enough. □

Lemma 29 *Let $V \subset V'$ be an extension of valuation rings of dimension one with the same residue field and let $\Gamma \subsetneq \Gamma'$ be their value group extension. Assume that Γ'/Γ has torsion. Then the extension $V \subset V'$ is not ind-smooth.*

Proof Let $\gamma \in \Gamma' \setminus \Gamma$ be such that $n\gamma \in \Gamma$ for some positive integer n. Choose an element $x \in V'$ such that $\mathrm{val}(x) = \gamma$. Then $x^n = zt$ for some $z \in V$ and an unit $t \in V'$. Thus the system S of polynomials $X^n = zT$, $TT' = 1$ over V has a solution in V'. If V' is ind-smooth over V then S has a solution in a smooth V-algebra and so one $(\tilde{x}, \tilde{t}, \tilde{t}')$ in the completion of V. But then $\gamma = \mathrm{val}(z)/n = \mathrm{val}(\tilde{x})$ must be in Γ which is false. □

Lemma 30 *Let $V \subset V'$ be an extension of valuation rings of dimension one containing* **Q** *having the same residue field k. Assume that V contains k and its value group $\Gamma \subset \mathbf{R}$ is dense in* **R**. *Also assume that the value group $\Gamma' \subset \mathbf{R}$ of V' is finitely generated (that is finitely generated over 0), $\Gamma \neq \Gamma'$ and Γ'/Γ has no torsion. Then the extension $V \subset V'$ is not ind-smooth.*

Proof Since Γ is free over **Z** we may take a basis of positive elements $\gamma_1, \ldots, \gamma_m$ of Γ which may be completed with some positive elements $\gamma_{m+1}, \ldots, \gamma_n \in \Gamma'$ to a basis of Γ'. Choose $x_1, \ldots, x_m \in V$ and $x_{m+1}, \ldots, x_n \in V'$ such that $\mathrm{val}(x_i) = \gamma_i$. Let $V_0 = V \cap k(x_1, \ldots, x_m)$ and $V_0' = V' \cap k(x_1, \ldots, x_n)$.

We will show that Ω_{V_0'/V_0} has torsion. **First assume** that $n = m + 1$. We will use the proof of [23, Lemma 7.2]. By Lemma 25 $\Gamma_+ = \cup_{j \in \mathbf{N}} \Gamma_j$ for some monoids $\Gamma_j \subset \Gamma_+$ generated by bases of Γ, the union being filtered. We consider as in the quoted lemma two real sequences (u_i), (v_i) which converge in **R** to γ_n and such that

1) $u_j, v_j \in \Gamma_j$ and $u_{j+1} - u_j, v_j - v_{j+1} \in \Gamma_{j+1}$,

2) $v_j - u_j$ is an element of the basis ν_j of Γ generating Γ_j, we may assume $v_j - u_j = \nu_{j1}$.

3) $u_j < \gamma_n < v_j$ for all j.

We may also suppose that $u_{j+1} - u_j = v_j - v_{j+1}$ if necessary restricting to a subsequence of (Γ_j). Let a_j, b_j be in V with values u_j, resp. v_j, and take $y_{jn} = x_n/a_j$ and $z_{jn} = b_j/x_n$ in V'. As in the proof of Lemma 4.2 a), we have $\nu_{ji} = d_{ji1}\gamma_1 + \ldots + d_{jim}\gamma_m$ and set $y_{ji} = x_1^{d_{ji1}} \cdots x_m^{d_{jim}} \in V$ which has valuation $\nu_{ji}, i \in [m]$. Then V_0 is a filtered union of localizations B_j of $k[y_{j1}, \ldots, y_{jm}]$ and V_0' is a filtered union of localizations C_j of $B_j[Z_j, Z_j']/(Z_j Z_j' - y_{j1}) \cong B_j[z_j, z_j']$, where $z_j = x_n/a_j$ and $z_j' = b_j/x_n$ in V'. Note that the map $C_j \to C_{j+1}$ is given by $Z_j \to (a_{j+1}/a_j)Z_{j+1}$, $Z_j' \to (b_j/b_{j+1})Z_{j+1}'$.

We claim that the map $f_j : C_{j+1} \otimes_{C_j} \Omega_{C_j/B_j} \to \Omega_{C_{j+1}/B_{j+1}}$ given by $dz_j \to (a_{j+1}/a_j)dz_{j+1}, dz_j' \to (b_j/b_{j+1})dz_{j+1}'$ is injective. Indeed, an element from Ker f_j induced by $w = \alpha \otimes dz_j + \beta \otimes dz_j'$, $\alpha, \beta \in C_{j+1}$ must go by f_j in

$$\alpha(a_{j+1}/a_j)dz_{j+1} + \beta(b_j/b_{j+1})dz_{j+1}' \in < z_{j+1}'dz_{j+1} + z_{j+1}dz_{j+1}' >$$

in $C_{j+1}dz_{j+1} \oplus C_{j+1}dz_{j+1}'$. So $\alpha(a_{j+1}/a_j) = \mu z_{j+1}'$ and $\beta(b_j/b_{j+1}) = \mu z_{j+1}$ for some $\mu \in C_{j+1}$. It follows that

$$w = (\mu(b_{j+1}/b_j)(a_{j+1}/a_j)) \otimes z_j'dz_j + (\mu(b_{j+1}/b_j)(a_{j+1}/a_j)) \otimes z_j dz_j'$$

belongs to $< z_j'dz_j + z_j dz_j' >$, which shows our claim.

We may assume that $\mathrm{val}(z_j) \leq \mathrm{val}(z'_j)$. For some $j' \geq j$ we have $z'_j = t_j z_j$ in $C_{j'}$ for some t_j in $C_{j'}$. We have dz_j, dz'_j in $\Omega_{C_{j'}/B_{j'}}$ and $z_j w'_j = 0$, for $w'_j = dz'_j + t_j dz_j$. So $C_{j'} \otimes_{C_j} \Omega_{C_j/B_j}$ is not torsion free. Here we should point out that the localizations are given by elements from $\rho + ((y_{j'i})_i, z_{j'}, z'_{j'})$, $\rho \in k$, $\rho \neq 0$, which cannot kill w'_j. Since f_j are injective we see that $\Omega_{V'_0/V_0}$ has torsion.

Now **assume** that $n > m+1$ and consider $V''_0 = V' \cap k(x_1, \ldots, x_{n-1})$. Apply induction on $n-m$, the case $n-m=1$ having been considered above. By induction hypothesis, we assume that $\Omega_{V''_0/V_0}$ has torsion. As above V''_0 is a filtered direct limit of some localizations $\tilde{B}_j$ of $k[\tilde{y}_{j1}, \ldots, \tilde{y}_{j,n-1}]$ and V'_0 is the filtered direct limit of some localizations $\tilde{C}_j$ of $\tilde{B}_j[\tilde{z}_j, \tilde{z}'_j]$. Set $I_j = (Z_j Z'_j - \tilde{y}_{j1}) \subset \tilde{B}_j[Z_j, Z'_j]$. By definition we have the following exact sequence

$$0 \to H_1(\tilde{B}_j, \tilde{C}_j, \tilde{C}_j) \to I_j/I_j^2 \xrightarrow{d} \tilde{C}_j d\tilde{z}_j \oplus \tilde{C}_j d\tilde{z}'_j \to \Omega_{\tilde{C}_j/\tilde{B}_j} \to 0.$$

If $h \in I_j$ induces an element in $\mathrm{Ker}\, d$ then we get

$$(\partial h/\partial Z_j)(\tilde{z}_j, \tilde{z}'_j) d\tilde{z}_j \oplus (\partial h/\partial Z'_j)(\tilde{z}_j, \tilde{z}'_j) d\tilde{z}'_j = 0.$$

But $h = \tilde{h}(Z_j Z'_j - \tilde{y}_{j1})$ for some $\tilde{h} \in \tilde{B}_j[Z_j, Z'j]$ and it follows $\tilde{h}(\tilde{z}_j, \tilde{z}'_j)\tilde{z}'_j = 0$, that is $\tilde{h}(\tilde{z}_j, \tilde{z}'_j) = 0$ and so $\tilde{h} \in I_j$. Hence d is injective and $H_1(\tilde{B}_j, \tilde{C}_j, \tilde{C}_j) = 0$.

In the Jacobi-Zariski sequence ([33, Theorem 3.3] applied to $V_0 \to V''_0 \to V'_0$

$$0 = H_1(V''_0, V'_0, V'_0) \to V'_0 \otimes_{V''_0} \Omega_{V''_0/V_0} \xrightarrow{\lambda} \Omega_{V'_0/V_0} \to \Omega_{V'_0/V''_0} \to 0$$

we see that the map λ is injective. It follows that $\Omega_{V'_0/V_0}$ has torsion which proves our claim.

By Proposition 18 the immediate extension $V_0 \subset V$ is ind-smooth. Assume, aiming for contradiction, that V' is ind-smooth over V. Then V' is ind-smooth over V_0 and by the above lemma we get Ω_{V'/V_0} flat over V'. Again by Proposition 18 we have V' ind-smooth over V'_0. As in the above lemma we obtain that Ω_{V'/V'_0} is a flat module over V'. In the Jacobi-Zariski sequence applied to $V_0 \to V'_0 \to V'$

$$H_1(V'_0, V', V') \to V' \otimes_{V'_0} \Omega_{V'_0/V_0} \to \Omega_{V'/V_0} \to \Omega_{V'/V'_0} \to 0$$

we have $H_1(V'_0, V', V') = 0$ and the last two modules are flat by the above lemma. We obtain that $V' \otimes_{V'_0} \Omega_{V'_0/V_0}$ is flat also over V' and so $\Omega_{V'_0/V_0}$ is also flat, which is not possible because it has torsion. Thus V' is not ind-smooth over V. □

Remark 31 *In the above proof the main point was to show that when* $\Gamma'/\Gamma \neq 0$ *has no torsion then* $\Omega_{V'_0/V_0}$ *has torsion.*

Lemma 32 *Let* $V \subset V'$ *be an extension of valuation rings of dimension one containing* $\mathbf{Q}$. *Assume that* V *contains its residue field and its value group* $\Gamma \subset \mathbf{R}$ *is dense in* $\mathbf{R}$. *Also assume that the value group* $\Gamma' \subset \mathbf{R}$ *of* V' *is finitely generated and* Γ'/Γ, $\Gamma \neq \Gamma'$ *has no torsion. Then the extension* $V \subset V'$ *is not ind-smooth.*

Proof In the proof of Lemma 30 take $W = V' \cap \mathrm{Frac}(V)(x_{m+1}, \dots, x_n)$. Then the extension $V \subset W$ has the same residue field but the value group extension is $\Gamma \subset \Gamma'$. Then $\Omega_{W/V}$ has torsion as in the proof of the quoted lemma. In the Jacobi-Zariski sequence applied to V, W, V'

$$H_1(W, V', V') \to V' \otimes_W \Omega_{W/V} \to \Omega_{V'/V}$$

we see that the left module is zero because the valuation extension $W \subset V'$ is ind-smooth (see Lemma 28), having the same value group (see Corollary 19). It follows that $\Omega_{V'/V}$ has torsion. But if $V \subset V'$ is ind-smooth then $\Omega_{V'/V}$ is torsion free. So $V \subset V'$ is not ind-smooth. □

Lemma 33 *Let $V \subset V'$ be an extension of valuation rings of dimension one containing $\mathbf{Q}$ with value groups $\Gamma \subset \Gamma'$ and having the same residue field k. Assume that V contains k and its value group $\Gamma \subset \mathbf{R}$ is dense in $\mathbf{R}$. Also assume that the value group $\Gamma'/\Gamma \neq 0$ has no torsion and V' has a cross-section $s : \Gamma' \to K'^*$ such that $s(\Gamma) \subset K^*$, K, K' being the fraction fields of V, resp. V'. Then the extension $V \subset V'$ is not ind-smooth.*

Proof We follow the proof of the above lemma. Take $V_0 = V \cap k(s(\Gamma))$ and $V'_0 = V' \cap k(s(\Gamma'))$. For every finitely generated $\Gamma_1 \subset \Gamma$ and $\Gamma'_1 \subset \Gamma'$ such that $\Gamma_1 \subset \Gamma'_1 \not\subset \Gamma$ we see as in the above lemma that for $V_{10} = V \cap k(s(\Gamma_1))$, $V'_{10} = V' \cap k(s(\Gamma'_1))$ we have a torsion in $\Omega_{V'_{10}/V_{10}}$. Then as in the above lemma we get a torsion in $\Omega_{V'_0/V_0}$ by [33, Lemma 3.2] and so in $\Omega_{V'/V}$ which implies that V' is not ind-smooth over V by Lemma 29. □

Lemma 34 *Let $V \subset V'$ be an extension of valuation rings of dimension one containing $\mathbf{Q}$ with value groups $\Gamma \subset \Gamma'$. Assume that V contains its residue field and its value group $\Gamma \subset \mathbf{R}$ is dense in $\mathbf{R}$. Also assume that the group $\Gamma'/\Gamma \neq 0$ has no torsion and V' has a cross-section $s : \Gamma' \to K'^*$ such that $s(\Gamma) \subset K^*$, K, K' being the fraction fields of V, resp. V'. Then the extension $V \subset V'$ is not ind-smooth.*

The proof follows as in Lemma 33 using now Lemma 32.

7 The Case When the Value Group Is Not Finitely Generated

A weaker form of Theorem 1 is given below with an independent proof (note that in the proof of Theorem 21 we do not use the Zariski's Uniformization Theorem.

Theorem 35 *Every valuation ring V containing its residue field k of characteristic zero is a filtered direct limit of smooth k-algebras $(R_i)_i$, that is the inclusion $k \subset V$ is ind-smooth (in particular, all the R_i are regular rings).*

Proof Let Γ be the value group of V, K the fraction field of V and $\tilde{k}$, $\tilde{\Gamma}$, $\tilde{V}$, $\tilde{K}$, $\tilde{s}:\tilde{\Gamma}\to\tilde{K}^*$ be given as in Theorem A.10. Note that in the proof of Theorem A.10 the ultrapoduct $k_1\subset V_1$ of $k\subset V$ gives an inclusion in V_1 of its residue field. More precisely, given a map $f:A\to B$ the ultrapower of f is the map between the ultrapowers of A and B given by $[a_i]\to[f(a_i)]$. By induction we see that the ultraproduct $k_n\subset V_n$ of $k_{n-1}\subset V_{n-1}$ gives an inclusion in V_n of its residue field. Thus $\tilde{V}=\cup_{n\in\mathbf{N}}V_n$ contains its residue field $\tilde{k}=\cup_{n\in\mathbf{N}}k_n$.

By Proposition 27 we see that $\tilde{V}$ is an immediate extension of a valuation ring $\tilde{V_0}$ which is a filtered union of localizations of smooth $\tilde{k}$-algebras. By Theorem 21 we get $\tilde{V_0}\subset\tilde{V}$ ind-smooth. Hence $k\subset\tilde{V}$ is ind-smooth because $k\subset\tilde{k}$ is separable and so ind-smooth (see Lemma 6). It follows that $k\subset V$ is ind-smooth too. Indeed, let $E=k[Y]/I$, $Y=(Y_1,\dots,Y_n)$ be a finitely generated k-algebra and $w:E\to V$ be a morphism of k-algebras. For the result we will apply Lemma 5 if we show that w factors through a smooth k-algebra. But the composite map $E\to V\to\tilde{V}$ factors through a smooth k-algebra $S=k[Z]/(h)$, $Z=(Z_1,\dots,Z_s)$ for a system of polynomials h from $k[Z]$, thus it is the composite map $E\xrightarrow{t}S\xrightarrow{\tilde{w}}\tilde{V}$, where $\tilde{w}$ is given by $Z\to\tilde{z}\in\tilde{V}^s$ and t is induced by $Y\to g\in k[Z]^n$. Then $\tilde{z}$ is a solution of h and of the system P of polynomial equations $g(Z)=w(Y)$ over V. Actually, $\tilde{z}$ belongs to some V_n and so h, P has also a solution z_{n-1} in V_{n-1} because V_n is an ultrapower of V_{n-1}. By induction we get a solution $z\in V$ of h, P. Therefore, w factors through the map $S\to V$, $Z\to z$. □

Lemma 36 *Let B be an A-algebra and A^n, B^n be the product of n-copies of A, resp. B. Then $\Omega_{B^n/A^n}\cong\Omega^n_{B/A}$ and $H_1(A^n,B^n,B^n)\cong H_1(A,B,B)^n$.*

Proof We treat only the case $n=2$. If $B=A[X]/I$, $X=(X_i)_i$ then $B^2=A^2[X]/J$, where J is given by polynomials of the form $h_{f,g}=\sum_{j=(j_1,\dots,j_s)}(a_j,b_j)X^j$ for some polynomials $f=\sum a_jX_j$, $g=\sum b_jX^j$ from I. Then Ω_{B^2/A^2} is the cokernel of the map $d:J/J^2\to\oplus_iB^2dX_i$ given by $h_{f,g}\to\sum_i(\partial f/\partial X_i,\partial g/\partial X_i)dX_i=\sum_i(\partial f/\partial X_idX_i,\partial g/\partial X_idX_i)$. Also $H_1(A^2,B^2,B^2)$ is the kernel of d and note that $d(h_{(f,g)})=0$ if and only if $d_0(f)=0$ and $d_0(g)=0$, d_0 being the map $I/I^2\to\oplus_iBdX_i$. Hence $\Omega_{B^2/A^2}\cong\Omega^2_{B/A}$ and $H_1(A^2,B^2,B^2)\cong H_1(A,B,B)^2$. □

Lemma 37 *Let B be an A-algebra, $\mathcal{U}$ and ultrafilter on a set U and $\tilde{B}$, resp. $\tilde{A}$ the ultrapowers of B (see the Appendix for the details), resp. A with respect to $\mathcal{U}$. Then $\Omega_{\tilde{B}/\tilde{A}}$ (resp. $H_1(\tilde{A},\tilde{B},\tilde{B})$) is the corresponding ultrapower of $\Omega_{B/A}$ (resp. $H_1(A,B,B)$) with respect to $\mathcal{U}$. In particular, $\Omega_{B/A}$ has torsion if and only if $\Omega_{\tilde{B}/\tilde{A}}$ has torsion and $H_1(\tilde{A},\tilde{B},\tilde{B})=0$ if and only if $H_1(A,B,B)=0$.*

For the proof note for example that $\Omega_{\tilde{B}/\tilde{A}}$ is the filtered direct limit of $(\Pi_{u\in U'}(\Omega_{B/A})_u)_{U'\in\mathcal{U}}$, where $(\Omega_{B/A})_u=\Omega_{B/A}$ (see the Appendix).

Proposition 38 *Let $V \subset V'$ be an extension of valuation rings of dimension one containing* $\mathbf{Q}$ *such that its residue field extension is trivial. Assume that the value groups $\Gamma, \Gamma' \subset \mathbf{R}$ of V respectively V' are dense in* $\mathbf{R}$ *and the factor of the value groups $\Gamma'/\Gamma \neq 0$ has no torsion. Then the extension $V \subset V'$ is not ind-smooth.*

Proof Using Proposition 9 we may suppose that V, V' are complete. So V contains its residue field. By Variant A.11 we find an extension of valuation rings $\tilde{V} \subset \tilde{V}'$ such that there exists a cross-section $\tilde{s} : \tilde{\Gamma}' \to (\tilde{K}')^*$ such that $\tilde{s}(\tilde{\Gamma}) \subset \tilde{K}$. We remind that we wrote Γ' as a filtered union of finitely generated subgroups $(\Gamma'_i)_{i \in I}$ and set $\Gamma_i = \Gamma'_i \cap \Gamma$. Set $V_{0i} = V \cap s'(\Gamma_i) = V' \cap s'(\Gamma_i)$, $V'_{0i} = V' \cap s'(\Gamma'_i)$. By Remark 31 the modules $\Omega_{V'_{0i}/V_{0i}}, i \in I$ have torsion. Note that the filtered union V_{01} of $V_{0i}, i \in I$ is a valuation ring with value group Γ, and similarly consider V'_{01} which has the value group Γ'. Clearly, $\Omega_{V'_{01}/V_{01}}$ has torsion since it is the limit of $V'_{01} \otimes_{V'_{01}} \Omega_{V'_{0i}/V_{0i}}$. Set $\tilde{V}_0 = \tilde{V} \cap s'(\tilde{\Gamma})$, $\tilde{V}'_0 = \tilde{V}' \cap s'(\tilde{\Gamma}')$. By iteration we define the extensions $V_{0n} \subset V_n$ and $V'_{0n} \subset V'_n$ with the same value group Γ_n, Γ'_n obtained taking n-ultrapowers of Γ, resp. Γ' and we see that $\Omega_{V'_{0n}/V_{0n}}$ has torsion by Lemma 37. Then $\Omega_{\tilde{V}'_0/\tilde{V}_0}$ has torsion since it is the limit of $\tilde{V}'_0 \otimes_{V'_{0n}} \Omega_{V'_{0n}/V_{0n}}$.

Assume that $V \subset V'$ is ind-smooth. In the Jacobi-Zariski sequence applied to $\tilde{V}_0, \tilde{V}, \tilde{V}'$

$$H_1(\tilde{V}, \tilde{V}', \tilde{V}') \to \tilde{V}' \otimes_{\tilde{V}} \Omega_{\tilde{V}/\tilde{V}_0} \to \Omega_{\tilde{V}'/\tilde{V}_0} \to \Omega_{\tilde{V}'/\tilde{V}} \to 0$$

we claim that the left module is zero and the last module has no torsion by Lemmas 28, 37 because $\Omega_{V'/V}$ has no torsion and $H_1(V, V', V') = 0$, $V \subset V'$ being ind-smooth. More precisely, we see that $\Omega_{V'_n/V_n}$ has no torsion and $H_1(V_n, V'_n, V'_n) = 0$ for all n using Lemma 28 and by iteration Lemma 37. At the limit we get our claim.

Also $\Omega_{\tilde{V}/\tilde{V}_0}$ is flat (so it has no torsion) since the extension $\tilde{V}_0 \subset \tilde{V}$ is ind-smooth having the same value group (see Proposition 20). It follows that $\Omega_{\tilde{V}'/\tilde{V}_0}$ has also no torsion.

Now, in the Jacobi-Zariski sequence applied to $\tilde{V}_0, \tilde{V}'_0, \tilde{V}'$

$$H_1(\tilde{V}_0, \tilde{V}', \tilde{V}') \to \tilde{V}' \otimes_{\tilde{V}'_0} \Omega_{\tilde{V}'_0/\tilde{V}_0} \to \Omega_{\tilde{V}'/\tilde{V}_0}$$

we see that the left module is zero by Lemma 28 because $\tilde{V}'_0 \subset \tilde{V}'$ is ind-smooth by Proposition 20. As above $\Omega_{\tilde{V}'_0/\tilde{V}_0}$ has torsion and so $\Omega_{\tilde{V}'/\tilde{V}_0}$ has torsion too, which is false. □

Acknowledgements This work has been partially elaborated in the frame of the International Research Network ECO-Math.

Appendix. Cross-sections via Infinite Towers of Ultrapowers by Kęstutis Česnavičius[2]

The goal of this Appendix is to show that by replacing a valuation ring by the limit of an infinite tower of its suitable ultrapowers one may arrange the valuation map $\mathrm{val}\colon V\setminus\{0\}\to\Gamma$ to admit a multiplicative section (see Theorem A.10). For this, we use techniques from model theory, specifically, the Keisler–Kunen theorem about the existence of good ultrafilters[3]: the idea is that constructing a section amounts to solving a system of equations for which any finite subsystem has a solution, and such systems always have solutions in well-chosen ultrapowers. For instance, if the system is countable, then solutions exists in any nonprincipal ultrafilter on $\mathbf{N}$, and in general the main subtlety is in constructing the ultrafilter (within ZFC).

A.1. Cross-Sections of Valuation Rings. For a valuation ring V with the value group Γ and the fraction field K, a *cross-section* of V is a section

$$s\colon \Gamma\to K^*\ \text{ in the category of abelian groups to the valuation map }\ \mathrm{val}\colon K^*\to\Gamma.$$

Concretely, s is a group homomorphism such that $\mathrm{val}(s(\gamma))=\gamma$ for $\gamma\in\Gamma$. For a submonoid $M\subset\Gamma$, a *partial cross-section* defined on M is a monoid morphism $s\colon M\to K^*$ (concretely, $s(0)=1$ and $s(m+m')=s(m)s(m')$) with $\mathrm{val}(s(m))=m$ for $m\in M$. Evidently, partial cross-sections exist for $M\simeq\mathbf{Z}^r_{\geq 0}$ and correspond to choices of tuples of elements of K^* whose valuations form the standard basis of $\mathbf{Z}^r_{\geq 0}$.

Example A.2 Cross-sections exist when Γ is free as a $\mathbf{Z}$-module, for instance, when it is finitely generated. As we now explain, they also exist when V is strictly Henselian of residue characteristic p and there is a free subgroup $\Gamma_0\subset\Gamma$ such that Γ/Γ_0 is torsion with $(\Gamma/\Gamma_0)[p^\infty]=0$. Indeed, we first define s on Γ_0 and then, by Zorn's lemma, reduce to the situation in which s is already defined on some subgroup $\Gamma'\supset\Gamma_0$ and needs to be extended to a $\Gamma''\supsetneq\Gamma'$ with Γ''/Γ' cyclic of order n prime to p. For the latter, we first choose an $x\in V$ such that $\mathrm{val}(x)$ lies in Γ'' and generates the quotient Γ''/Γ', which gives the following equation in V:

$$x^n=u\cdot s(n\cdot\mathrm{val}(x))\quad\text{for some}\quad u\in V^*.$$

[2] CNRS, UMR 8628, Laboratoire de Mathématiques d'Orsay, Univ. Paris-Sud, Université Paris-Saclay, 91405 Orsay, France. Email: kestutis@math.u-psud.fr

I thank Dorin Popescu for helpful discussions and for hosting me during a visit to Romania in March 2019. I thank Matthias Aschenbrenner for helpful comments. This project has received funding from l'Agence Universitaire de la Francophonie and the European Research Council (ERC) under the European Union's Horizon 2020 research and innovation programme (grant agreement No. 851146).

[3] After this appendix was written, we learned of a much simpler way to deduce sharper versions of Theorem A.10 and Variant A.11 from the results reviewed in §A.8, see [1, 3.3.39, 3.3.40]. We left this appendix in place in case the method used here would prove useful for other purposes.

Since $p \nmid n$, Hensel's Lemma [10, IV, 18.5.17] (which is the Implicit Function Theorem in this context) now implies that the equation $X^n = u$ has a solution in V, so we may adjust x to assume that $u = 1$. Granted this, s then extends to Γ'' by setting $s(\mathrm{val}(x)) = x$: indeed, any relation $N \cdot \mathrm{val}(x) = \gamma$ with $N \in \mathbf{Z}$ and $\gamma \in \Gamma'$ must be a multiple of such a relation with $N = n$, so $s(N \cdot \mathrm{val}(x)) = s(\gamma)$.

A.3. Ultrafilters and Ultraproducts. We recall that an *ultrafilter* on a nonempty set U is a set $\mathscr{U}$ of subsets of U that is closed under finite intersections, closed under taking supersets, does not contain the empty set, and for every $U' \subset U$ contains either U' or $U \setminus U'$. Such a $\mathscr{U}$ is *principal* if it consists of all the subsets containing some fixed $u \in U$, and is *nonprincipal* otherwise. An ultrafilter $\mathscr{U}$ is *countably incomplete* if some countable collection of elements of $\mathscr{U}$ has an empty intersection. Such a $\mathscr{U}$ is also nonprincipal and it exists whenever U is infinite (see [4, §A.3, 8.4]). We view any ultrafilter $\mathscr{U}$ as a partially ordered set, where $U' \leq U''$ if $U' \supseteq U''$.

For any category $\mathcal{C}$ that has small products and filtered direct limits, the *ultraproduct* of a set $\{C_u\}_{u \in U}$ of objects of $\mathcal{C}$ with respect to an ultrafilter $\mathscr{U}$ on U, which we denote abusively by $\prod_{\mathscr{U}} C_u$, is

$$\varinjlim_{U' \in \mathscr{U}} \left(\textstyle\prod_{u \in U'} C_u\right) \quad \text{where transition maps are projections onto partial products}$$

(the limit is filtered because $\mathscr{U}$ is closed under finite intersections). In the case when all the C_u are the same object $C \in \mathcal{C}$, we call $\prod_{\mathscr{U}} C$ an *ultrapower* of C.

A.4. Ultraproducts of Valuation Rings. We will work with ultraproducts of rings or modules. For instance, an ultraproduct of fields is again a field: every nonzero element is invertible (thanks to the axiom that $U' \in \mathscr{U}$ or $U \setminus U' \in \mathscr{U}$). Likewise, an ultraproduct $\prod_{\mathscr{U}} V_u$ of valuation rings $\{V_u\}_{u \in U}$ with fraction fields $\{K_u\}_{u \in U}$ is a valuation ring with fraction field $\prod_{\mathscr{U}} K_u$: for any nonzero element v of the latter, either v or v^{-1} lies in $\prod_{\mathscr{U}} V_u$. We see similarly that

1. the maximal ideal of $\prod_{\mathscr{U}} V_u$ is the ultraproduct $\prod_{\mathscr{U}} \mathfrak{m}_u$ of the maximal ideals;
2. the residue field of $\prod_{\mathscr{U}} V_u$ is the ultraproduct $\prod_{\mathscr{U}} k_u$ of the residue fields;
3. the value group of $\prod_{\mathscr{U}} V_u$ is the ultraproduct $\prod_{\mathscr{U}} \Gamma_u$ of the value groups;
4. the monoid of nonnegative elements $(\prod_{\mathscr{U}} \Gamma_u)_{\geq 0}$ is identified with $\prod_{\mathscr{U}} (\Gamma_u)_{\geq 0}$.

The existence of "well-chosen" ultrapowers mentioned above rests on the Keisler–Kunen theorem from model theory that we recall in the following lemma. Keisler proved it in [13] assuming the Generalized Continuum Hypothesis and Kunen gave an unconditional proof in [16, Theorem 3.2].

Lemma A.5 ([6], Theorem 6.1.4). *For an infinite set U, there is a countably incomplete ultrafilter $\mathscr{U}$ on U such that for any inclusion-reversing function*

$$f\colon \{\textit{finite subsets of } U\} \to \mathscr{U}, \quad \textit{there is a function} \quad f_0\colon \{\textit{finite subsets of } U\} \to \mathscr{U}$$

that is also inclusion-reversing and satisfies

$$f_0(U') \subset f(U') \text{ and } f_0(U' \cup U'') = f_0(U') \cap f_0(U'') \text{ for all finite subsets } U', U'' \subset U.$$

Of course, the requirement that f_0 be inclusion-reversing is superfluous: it is a special case of the requirement that f_0 transform finite unions into intersections.

We now verify that the ultrapowers that result from the ultrafilters supplied by Lemma A.5 have the promised property of solvability of systems of equations.

Proposition A.6. *For an infinite cardinal κ, every ultrafilter $\mathscr{U}$ supplied by Lemma* A.5 *for a set U of cardinality κ is such that: for any ring R (resp., and any left R-module M), any polynomial (resp., linear) system of equations*

$$\{g_i(\{X_\sigma\}_\sigma) = 0\}_{i\in I} \quad (\textit{resp.,} \quad \{\textstyle\sum_\sigma r_{i,\sigma} X_\sigma = m_i\}_{i\in I}) \quad \textit{with } \#I \le \kappa$$

in variables $\{X_\sigma\}_\sigma$ and coefficients in $\prod_{\mathscr{U}} R$ (resp., $r_{i,\sigma} \in \prod_{\mathscr{U}} R$) and $m_i \in \prod_{\mathscr{U}} M$ has a solution in $\prod_{\mathscr{U}} R$ (resp., $\prod_{\mathscr{U}} M$) as soon as so do all its finite subsystems.

Proof. The assertion is a concrete case of the model-theoretic [6, Theorem 6.1.8], and the latter is sharper in multiple aspects. For convenience, we recall the argument.

For brevity, we denote the system in question by $\{g_i = 0\}_{i\in I}$ and we lift it to a system $\{\widetilde{g}_i = 0\}_{i\in I}$ with coefficients in $\prod_{u\in U} R$ (resp., and in $\prod_{u\in U} M$) and the same variables $\{X_\sigma\}_\sigma$ by lifting the nonzero coefficients along the surjection

$$\textstyle\prod_{u\in U} R \twoheadrightarrow \prod_{\mathscr{U}} R \quad \text{(resp., and along} \quad \prod_{u\in U} M \twoheadrightarrow \prod_{\mathscr{U}} M \quad \text{for the } m_i).$$

Since $\mathscr{U}$ is countably incomplete, we may fix a decreasing sequence

$$U \supset U_0 \supset U_1 \supset U_2 \supset \cdots \quad \text{of sets in } \mathscr{U} \text{ with } \textstyle\bigcap_{n\ge 0} U_n = \emptyset.$$

We then define a function

$$f\colon \{\text{finite subsets of } I\} \to \mathscr{U}$$

by letting pr_u denote the projection onto the u-th factor of $\prod_{u\in U}$ and setting

$$f(I') := U_{\#I'} \cap \{u \in U \mid \text{the system } \{\mathrm{pr}_u(\widetilde{g}_i) = 0\}_{i\in I'} \text{ is solvable in } R \text{ (resp., } M)\}.$$

The well-definedness of f follows from the solvability of the subsystem $\{g_i = 0\}_{i\in I'}$ in $\prod_{\mathscr{U}} R$ (resp., $\prod_{\mathscr{U}} M$) and from the stability of $\mathscr{U}$ under supersets. By construction, $f(I') \supset f(I'')$ whenever $I' \subset I''$, so, since $\#I \le \#U$, Lemma A.5 supplies a function

$$f_0\colon \{\text{finite subsets of } I\} \to \mathscr{U} \text{ such that } f_0(I') \subset f(I'),\ f_0(I' \cup I'') = f_0(I') \cap f_0(I'')$$

for all finite subsets $I', I'' \subset I$ (technically, to apply Lemma A.5 we first embed I into U as a subset and then extend f to finite subsets $U' \subset U$ by the rule $U' \mapsto f(U' \cap I)$).

For each $u \in U$, we set

$$I_u := \{i \in I \mid u \in f_0(\{i\})\}.$$

Whenever, $i_1, \dots, i_n \in I_u$ are pairwise distinct, we have

$$u \in f_0(\{i_1\}) \cap \dots \cap f_0(\{i_n\}) = f_0(\{i_1, \dots, i_n\}) \subset f(\{i_1, \dots, i_n\}) \subset U_n,$$

so, since the U_n have empty intersection, each I_u is finite. Then the preceding display applied to an enumeration of I_u shows that $u \in f(I_u)$, to the effect that the system $\{\mathrm{pr}_u(\widetilde{g}_i) = 0\}_{i \in I_u}$ has a solution $\{x_{\sigma,u}\}_\sigma$ in R (resp., M).

We claim that $\{x_\sigma := (x_{\sigma,u})_{u \in U}\}_\sigma$ gives a solution in $\prod_{\mathscr{U}} R$ (resp., $\prod_{\mathscr{U}} M$) to the system $\{g_i = 0\}_{i \in I}$. Indeed, for every $i \in I$ we have $f_0(\{i\}) \in \mathscr{U}$ and for every $u \in f_0(\{i\})$ we have $i \in I_u$, so $\widetilde{g}_i(\{x_\sigma\}_\sigma) = 0$ in the projection on $\prod_{u \in f_0(\{i\})}$. □

The argument is not specific to rings or modules, and it also shows the following.

Variant A.7. *For an infinite cardinal κ, every ultrafilter $\mathscr{U}$ supplied by Lemma* A.5 *for a set U of cardinality κ is such that: for any monoid G, any system*

$$\{g_i(\{X_\sigma\}_\sigma) = g_i'(\{X_\sigma\}_\sigma)\}_{i \in I} \quad \textit{with} \quad \#I \le \kappa$$

of monomial equations in variables $\{X_\sigma\}_\sigma$ and coefficients in $\prod_{\mathscr{U}} G$ has a solution in $\prod_{\mathscr{U}} G$ as soon as so does each of its finite subsystems.

As we now review, Proposition A.6 supplies algebraically compact ultrapowers.

A.8. Algebraic Compactness. We fix a ring R and recall that a map $M \to M'$ of left R-modules is *pure* if the map $M'' \otimes_R M \to M'' \otimes_R M'$ is injective for every right R-module M''. An R-module M is *algebraically compact* (or *pure-injective*) if every pure map $M \to M'$ of R-modules is a split injection. For example, if M is an algebraically compact abelian group (so $R = \mathbf{Z}$), then every short exact sequence

$$0 \to M \to M' \to M'/M \to 0 \quad \text{of abelian groups with} \quad (M'/M)_{\mathrm{tors}} = 0 \quad \text{splits.}$$

A concrete criterion for algebraic compactness is given by [11, 7.1 (with 6.5)]: a left R-module M is algebraically compact if every system of equations

$$\{\textstyle\sum_\sigma r_{i,\sigma} X_\sigma = m_i\}_{i \in I} \quad \text{with} \quad r_{i,\sigma} \in R \quad \text{and} \quad m_i \in M$$

has a solution in M as soon as so do all its finite subsystems. Moreover, by [11, 7.28, 7.29], it suffices to consider systems with cardinality $\#I \le \max(\#R, \#\mathbf{Z})$. In particular, thanks to Proposition A.6, there is an ultrafilter $\mathscr{U}$ such that for any R-module M, the R-module $\prod_{\mathscr{U}} M$ is algebraically compact.

With model-theoretic input in place, we turn to the tower of ultrapowers argument in Theorem A.10. The final input is the following lemma proved in [7, 2.2], [23, 4.6.1], or [9, 6.1.30] that captures the "combinatorial" part of local uniformization.

Lemma A.9. *For a totally ordered abelian group Γ, the submonoid $\Gamma_{\geq 0} \subset \Gamma$ of nonnegative elements is a filtered increasing union of its finite free submonoids isomorphic to $\mathbf{Z}^r_{\geq 0}$ (where $r \in \mathbf{Z}_{\geq 0}$ need not be constant).*

Theorem A.10. *For a valuation ring V with value group Γ, there is a countable sequence of ultrafilters $\mathscr{U}_1, \mathscr{U}_2, \ldots$ on some respective sets $U_1, U_2, \ldots$ for which the valuation rings $\{V_n\}_{n\geq 0}$ defined inductively by $V_0 := V$ and $V_{n+1} := \prod_{\mathscr{U}_{n+1}} V_n$ are such that the valuation ring*

$$\widetilde{V} := \varinjlim_{n\geq 0} V_n \quad \textit{has a cross-section} \quad \widetilde{s} : \widetilde{\Gamma} \to \widetilde{K}^*,$$

where $\widetilde{K}$ and $\widetilde{\Gamma}$ are the fraction field and the value group of $\widetilde{V}$.

Proof. We let K_n and Γ_n denote the fraction field and the value group of V_n, so that $\Gamma_{n+1} \cong \prod_{\mathscr{U}_{n+1}} \Gamma_n$ and $K_{n+1} = \prod_{\mathscr{U}_{n+1}} K_n$ (see §A.4) with

$$\widetilde{\Gamma} \cong \varinjlim_{n\geq 0} \Gamma_n \text{ and } \widetilde{K} \cong \varinjlim_{n\geq 0} K_n.$$

The idea is to build ultrafilters $\mathscr{U}_n$ one by one using Lemma A.5 in such a way that a desired cross-section

$\widetilde{s}\colon \widetilde{\Gamma} \to \widetilde{K}^*$ would be the limit of compatible partial cross-sections $s_n\colon \Gamma_n \to K^*_{n+1}$.

For this, as an initial step, we replace V by a suitable ultrapower to ensure that the abelian group Γ is algebraically compact (see §A.8). Granted this, it suffices to carry out the inductive step: setting $\Gamma_{-1} := 0$ for convenience and assuming that we have already constructed s_{n-1} and V_n for some $n \geq 0$ in such a way that the abelian groups Γ_{n-1} and Γ_n are algebraically compact, it suffices to construct V_{n+1} with Γ_{n+1} algebraically compact in such a way that s_{n-1} extends to an s_n.

The role of algebraic compactness is to split the map $\Gamma_{n-1} \hookrightarrow \Gamma_n \cong \prod_{\mathscr{U}_n} \Gamma_{n-1}$ whose cokernel is torsion free:

$$\Gamma_n \cong \Gamma_{n-1} \oplus G \text{ for some subgroup } G \subset \Gamma_n.$$

Thanks to this splitting, we only need to build an ultrafilter $\mathscr{U}_{n+1}$ and a partial cross-section $s_G : G \to (\prod_{\mathscr{U}_{n+1}} K_n)^*$ such that $\prod_{\mathscr{U}_{n+1}} \Gamma_n$ is algebraically compact. In fact, we let $\mathscr{U}_{n+1}$ be any ultrafilter as in Lemma A.5 applied to the cardinal $\max(\#\Gamma_n, \#\mathbf{Z})$. Then $\prod_{\mathscr{U}_{n+1}} \Gamma_n$ is necessarily algebraically compact by the criterion reviewed in §A.8 and Proposition A.6.

The subgroup G inherits a total order from Γ_n, and any partial cross-section

$$s_{G_{\geq 0}} : G_{\geq 0} \to (\prod_{\mathscr{U}_{n+1}} V_n) \setminus \{0\} \quad \text{will give rise to a desired} \quad s_G.$$

For each $g \in G_{>0}$, we fix a $v_g \in V_n$ with $\mathrm{val}(v_g) = g$. Then $s_{G_{\geq 0}}$ amounts to a solution in $\prod_{\mathscr{U}_{n+1}} V_n$ of the following system of equations in variables $\{X_g, U_g, U'_g\}_{g \in G_{>0}}$:

$$\{X_{g+g'} = X_g X_{g'}, \quad X_g U_g = v_g, \quad U_g U'_g = 1\}_{g,\, g' \in G_{>0}}.$$

Likewise, for any submonoid $G' \subset G_{\geq 0}$, the restriction of $s_{G_{\geq 0}}|_{G'}$, that is, a partial cross-section defined on G', amounts to a solution in $\prod_{\mathscr{U}_{n+1}} V_n$ of the subsystem consisting of those equations that only involve the variables $\{X_g, U_g, U'_g\}_{g \in G'}$. However, a partial cross-section $G' \to \prod_{\mathscr{U}_{n+1}} V_n$ (and even $G' \to V_n$) certainly exists if $G' \simeq \mathbf{Z}^d_{\geq 0}$, and, by Lemma A.9, the monoid $G_{\geq 0}$ is a filtered increasing union of such G'. This implies that every finite subsystem of the above system has a solution in $\prod_{\mathscr{U}_{n+1}} V_n$ (and even in V_n). Then, by Proposition A.6, the entire system has a solution in $\prod_{\mathscr{U}_{n+1}} V_n$, which completes the inductive step. □

Variant A.11. *For every faithfully flat map $V \subset V'$ of valuation rings with value groups $\Gamma \subset \Gamma'$ such that Γ'/Γ is torsion free, there is a countable sequence of ultrafilters $\mathscr{U}_1, \mathscr{U}_2, \ldots$ on some respective sets $U_1, U_2, \ldots$ for which the valuation rings $\{V_n\}_{n \geq 0}$ and $\{V'_n\}_{n \geq 0}$ defined inductively by $V_0 := V$ and $V'_0 := V'$ with $V_{n+1} := \prod_{\mathscr{U}_{n+1}} V_n$ and $V'_{n+1} := \prod_{\mathscr{U}_{n+1}} V'_n$ are such that the valuation ring*

$$\widetilde{V}' = \varinjlim_{n \geq 0} V'_n \quad \textit{with} \quad \widetilde{K}' := \mathrm{Frac}(\widetilde{V}') \quad \textit{has a cross-section} \quad \widetilde{s} : \widetilde{\Gamma}' \to \widetilde{K}'^{*}$$

whose restriction to the value group $\widetilde{\Gamma}$ of $\widetilde{V} := \varinjlim_{n \geq 0} V_n$ lands in $\widetilde{K} := \mathrm{Frac}(\widetilde{V})$.

Proof. An ultrapower of an ultrapower is itself an ultrapower [6, 6.5.2], so we may make an initial replacement of V and V' by suitable large ultrapowers and use §A.8 with Proposition A.6 to ensure that Γ is algebraically compact and later absorb the appearing initial ultrafilter into $\mathcal{U}_1$ (alternatively, we could simply insert this initial ultrafilter as $\mathcal{U}_1$ without using loc. cit.). Then, thanks to the torsion-freeness assumption on Γ'/Γ, the inclusion $\Gamma \subset \Gamma'$ splits. A choice of a splitting induces a compatible splitting on any ultrapower, so the proof of Theorem A.10 continues to give the claimed variant granted that we take advantage of the splitting to build the cross-section in such a way that its restriction to $\tilde{\Gamma}$ lands in $\tilde{K}$. □

References

1. M. Aschenbrenner, L. van den Dries, J. van der Hoeven, *Asymptotic differential algebra and model theory of transseries*, Annals of Mathematics Studies, **195**, Princeton University Press, Princeton, NJ, 2017.

2. V. Alexandru, N. Popescu, A. Zaharescu, *All valuations on $K(X)$*, J. Math. Kyoto Univ., **30**, (1990), 281-296.
3. A. Arabia, *Relèvements des algèbres lisses et de leurs morphismes*, Comment. Math. Helv. 76 (2001), no. 4, 607–639.
4. J. Barwise, *Handbook of mathematical logic*, Studies in Logic and the Foundations of Mathematics, **90**, North-Holland Publishing Co., Amsterdam, 1977. With the cooperation of H. J. Keisler, K. Kunen, Y. N. Moschovakis and A. S. Troelstra.
5. N. Bourbaki, Éléments de mathématique. Algèbre commutative, chap. I-VII, Hermann (1961, 1964, 1965); chap. VIII-X, Springer (2006, 2007) (French).
6. C. C. Chang, H. J. Keisler, Model theory, 3rd ed., Studies in Logic and the Foundations of Mathematics, vol. 73, North-Holland Publishing Co., Amsterdam, 1990.
7. G. A. Elliott, *On totally ordered groups, and K_0*, in Ring theory (Proc. Conf., Univ. Waterloo, Waterloo, 1978), Lecture Notes in Math., vol. 734, Springer, Berlin, 1979, pp. 1–49.
8. R. Elkik, Solutions d'équations à coéfficients dans un anneaux hensélien, Ann. Sci. Ecole Normale Sup., **6** (1973), 553-604.
9. O. Gabber, L. Ramero, *Almost ring theory*, Lecture Notes in Mathematics, vol. 1800, Springer-Verlag, Berlin, 2003.
10. A. Grothendieck, J. Dieudonné, Éléments de géométrie algébrique. IV. Étude locale des schémas et des morphismes de schémas IV, Inst. Hautes Études Sci. Publ. Math. 32 (1967), 361 (French).
11. C. U. Jensen, H. Lenzing, *Model-theoretic algebra with particular emphasis on fields, rings, modules*, Algebra, Logic and Applications, vol. 2, Gordon and Breach Science Publishers, New York, 1989.
12. I. Kaplansky, *Maximal fields with valuations, Duke Math. J. 9 (1942), 303–321.*
13. H. J. Keisler, *Good ideals in fields of sets*, Ann. of Math. **(2) 79** (1964), 338–359.
14. A. Khalid, A. Popescu, D. Popescu, *Algorithms in the classical Néron Desingularization*, Bull. Math. Soc. Sci. Math. Roumanie, **61(109)**, (2018), 73-83, arXiv:AC/1702.01445.
15. Z. Kosar, G. Pfister, D. Popescu, *Constructive Néron Desingularization of algebras with big smooth locus*, Communications in Algebra, **46**, (2018), 1902-1911, arXiv:1702.01867.
16. K. Kunen, *Ultrafilters and independent sets*, Trans. Amer. Math. Soc. 172 (1972), 299–306,
17. Y. Lequain, A. Simis, *Projective modules over $R[X_1, \ldots, X_n]$*, R a Prüfer domain, J. Pure Appl. Algebra **18**, (1980), no. 2, 165–171.
18. L. Moret-Bailly, *An extension of Greenberg's theorem to general valuation rings*, Manuscripta Math. **139** (2012), no. 1-2, 153–166.
19. H. Matsumura, *Commutative Algebra*, Benjamin, New-York, 1980.
20. M. Nagata, *Finitely generated rings over a valuation ring*, J. Math. Kyoto Univ. 5 (1966), 163–169,
21. A. Néron, *Modèles minimaux des varietés abéliennes sur les corps locaux et globaux*, Publ. Math. IHES, **21**, 1964, 5-128.
22. A. Ostrowski, *Untersuchungen zur arithmetischen Theorie der Körper*, Math. Z. 39 (1935), no. 1, 321–404.
23. D. Popescu, *On Zariski's uniformization theorem*, in Algebraic geometry, Bucharest 1982 (Bucharest, 1982), Lecture Notes in Math., vol. 1056, Springer, Berlin, 1984, 264–296.
24. D. Popescu, *General Neron Desingularization*, Nagoya Math. J., **100** (1985), 97-126.
25. D. Popescu, Algebraic extensions of valued fields, J. Algebra 108 (1987), no. 2, 513–533.
26. D. Popescu, General Neron Desingularization and approximation, Nagoya Math. J., **104**, (1986), 85-115.
27. D. Popescu, *Letter to the Editor, General Neron Desingularization and approximation*, Nagoya Math. J., **118**, (1990), 45-53.
28. D. Popescu, *Artin Approximation*, in "Handbook of Algebra", vol. 2, Ed. M. Hazewinkel, Elsevier, 2000, 321-355.
29. D. Popescu, Simple General Neron Desingularization in local **Q**-algebras, Commun. Algebra 47 (2019), 923–929, arXiv:AC/1802.05109.

30. D. Popescu, *On a question of Swan* , Algebraic Geometry, **6(6)**, (2019), 716-729, arXiv:AC/1803.06956.
31. D. L. Shannon, *Monoidal transforms of regular local rings*, Amer. J. Math. 95 (1973), 294–320.
32. A. J. de Jong et al., The Stacks Project. Available at http://stacks.math.columbia.edu.
33. R. Swan, *Neron-Popescu desingularization*, in "Algebra and Geometry", Ed. M. Kang, International Press, Cambridge, (1998), 135-192.
34. O. Zariski, *Local uniformization on algebraic varieties*, Ann. of Math. **41** (1940), 852–896.

Diagonal Representation of Algebraic Power Series: A Glimpse Behind the Scenes

Sergey Yurkevich

Abstract There are many viewpoints on algebraic power series, ranging from the abstract ring-theoretic notion of Henselization to the very explicit perspective as diagonals of certain rational functions. To be more explicit on the latter, Denef and Lipshitz proved in 1987 that any algebraic power series in n variables can be written as a diagonal of a rational power series in one variable more. Their proof uses a lot of involved theory and machinery which remains hidden to the reader in the original article. In the present work we shall take a glimpse on these tools by motivating while defining them and reproving most of their interesting parts. Moreover, in the last section we provide a new significant improvement on the Artin-Mazur lemma, proving the existence of a 2-dimensional code of algebraic power series.

1 Introduction

1.1 Basic Notions and Motivation

In all this text, K will denote a field of characteristic zero, even though most presented results are known to work even for excellent local integral domains. By default, $\mathbb{N}$, $\mathbb{Q}$, $\mathbb{R}$ and $\mathbb{C}$ are sets (equipped with the appropriate algebraic structure) of natural numbers (including 0), rationals, reals and complex numbers respectively. A ring is always commutative with 1 and a ring homomorphism is always unital. By R^* we denote the set of units of a ring R, and $\widehat{R}$ stands for the algebraic completion of a local ring with respect to its maximal ideal. If not indicated otherwise, $x = (x_1, \ldots, x_n)$ is a vector of n variables and $x' = (x_1, \ldots, x_{n-1})$. In contrast, t is always one variable and when we write xt, we mean $(x_1 t, \ldots, x_n t)$. Given an n-dimensional index $\alpha \in \mathbb{N}^n$ we write $|\alpha|$ for $\alpha_1 + \cdots + \alpha_n$ and x^α for $x_1^{\alpha_1} \cdots x_n^{\alpha_n}$.

This work is based on the author's master's thesis (U. of Vienna, 2020), supervised by H. Hauser.
S. Yurkevich—Supported by the Austrian Science Fund (FWF): P-31338.

S. Yurkevich (✉)
University of Vienna, Schüttaustraße 48-6a, 1220 Vienna, Austria
e-mail: sergey.yurkevich@unvie.ac.at

A. Bostan and K. Raschel (eds.), *Transcendence in Algebra, Combinatorics, Geometry and Number Theory*, Springer Proceedings in Mathematics & Statistics 373,
https://doi.org/10.1007/978-3-030-84304-5_13

Definition A formal power series $h(x) \in K[\![x]\!]$ is called *algebraic* if there exists a non-zero polynomial $P(x, t) \in K[x, t]$ such that $P(x, h(x)) = 0$. Such a polynomial with minimal degree in t is called a *minimal polynomial of* $h(x)$. The set of algebraic power series is denoted by $K\langle x\rangle$. A power series which is not algebraic is called *transcendental.*

Consider $(x) \subseteq K[x]$, the maximal ideal inside $K[x]$ generated by the elements $x_1, \ldots, x_n$. Then, algebraically speaking, $K\langle x\rangle$ is the algebraic closure of the localization of $K[x]$ with respect to this ideal, $K[x]_{(x)}$, inside $K[\![x]\!]$. Hence the ring of algebraic power series is a subring of formal power series. Moreover it is easy to see that $K\langle x\rangle^* = K\langle x\rangle \cap K[\![x]\!]^*$. Here are several examples in one variable, $x = x_1$.

Example 1 Any polynomial $p(x) \in K[x]$ is an algebraic power series since we may chose $P(x, t) := t - p(x)$.

Example 2 The power series given by

$$(1+x)^r = \sum_{k\geq 0} \binom{r}{k} x^k,$$

for some rational number $r \in \mathbb{Q}$ is algebraic[1]. This holds true, because when $r = p/q$ for non-zero integers p, q, we may choose $P(x, t) = t^q - (1 + x)^p$ if $p, q > 0$ and $P(x, t) = t^q(1 + x)^{-p} - 1$ if p happens to be negative. We obtain again that $P(x, (1 + x)^r) = 0$.

Example 3 Consider the exponential power series:

$$\exp(x) = \sum_{k\geq 0} \frac{x^k}{k!}.$$

We claim that it is transcendental: assume it was algebraic, then after dividing the minimal polynomial by the coefficient of the leading term in t, we would find a $Q(x, t) = q_0(x) + \cdots + q_{m-1}(x)t^{m-1} + t^m \in K(x)[t]$ with $Q(x, \exp(x)) = 0$. Now, taking the derivative of $Q(x, \exp(x)) = 0$ with respect to x and using $\exp'(x) = \exp(x)$, it follows that

$$q_0'(x) + \cdots + (q_{m-1}'(x) + (m-1)q_{m-1}(x)) \exp(x)^{m-1} + m \exp(x)^m = 0.$$

Subtracting this from $m\,Q(x, \exp(x)) = 0$ and using the simple fact that no rational function $q(x)$ can satisfy $q(x) = cq'(x)$ for $c \in K^*$, we can find a non-zero polynomial of lower degree than m also annihilating $\exp(x)$. This is a contradiction with the minimality of m.

Example 4 The function $f(x) = \sqrt{x}$ is not an algebraic power series, because it is not a formal power series.

---

[1] The function $(1 + x)^r$ exists for any $r \in \mathbb{Q}$, because the characteristic of K is assumed to be zero.

Example 5 Set $f(x) = \sqrt{x+1}$ and $g(x) = \sqrt[3]{x+1}$; we already saw in Example 2 that both $f, g \in K\langle x\rangle$. To see that $f(x) + g(x) = \sqrt{x+1} + \sqrt[3]{x+1}$ is algebraic as well, just consider the polynomial

$$P(x,t) = t^6 - 3(x+1)t^4 - 2(x+1)t^3 + 3(x+1)^2t^2 - 6(x+1)^2t - x(x+1)^2$$

and verify that it indeed satisfies $P(x, f(x) + g(x)) = 0$. Finding such a $P(x,t)$ is not straightforward and may require some work.

Algebraic power series appear in many mathematical areas, such as combinatorics, algebraic geometry and number theory. In order to motivate their deep ring-theoretic study in the next sections, we first introduce a very explicit viewpoint.

Definition Let $g(x)$, $f(x,t)$ be formal power series:

$$g(x) = \sum_{i_1,\dots,i_n} g_{i_1,\dots,i_n} x_1^{i_1} \cdots x_n^{i_n} \in K[\![x]\!],$$

$$f(x,t) = \sum_{i_1,\dots,i_n,j} f_{i_1,\dots,i_n,j} x_1^{i_1} \cdots x_n^{i_n} t^j \in K[\![x,t]\!].$$

Then the *small diagonal* $\Delta(g)$ of $g(x)$ is the (univariate) formal power series given by:

$$\Delta(g(x)) = \Delta(g(x))(t) := \sum_{j\geq 0} g_{j,\dots,j} t^j \in K[\![t]\!].$$

The *big diagonal* $\mathcal{D}(f)$ of $f(x,t)$ is given by the (multivariate) formal power series:

$$\mathcal{D}(f(x,t)) = \mathcal{D}(f(x,t))(x) := \sum_{i_1+\dots+i_n=j} f_{i_1,\dots,i_n,j} x_1^{i_1} \cdots x_n^{i_n} \in K[\![x]\!].$$

Clearly, for $n = 2$ it holds that $\Delta(g(x_1, x_2)) = \mathcal{D}(g(x_1, t))$. We shall always refer to the big diagonal whenever we do not specify which diagonal we use.

Example 1 Let $x = x_1$ be one variable and $f(x,t) = 1/(1 - x - t)$. Then we obtain

$$\mathcal{D}(f(x,t))(x) = \mathcal{D}\Big(\sum_{i,j\geq 0} \binom{i+j}{i} x^i t^j\Big)(x).$$

Therefore it follows that $\mathcal{D}(f(x,t))(x) = \sum_{n\geq 0} \binom{2n}{n} x^n = (1-4x)^{-1/2}$. This function is an algebraic power series with minimal polynomial $P(x,t) = (1-4x)t^2 - 1$.

Example 2 Define the Hadamard product of two power series $f(x) = \sum_{\alpha\in\mathbb{N}^n} f_\alpha x^\alpha$ and $g(x) = \sum_{\alpha\in\mathbb{N}^n} g_\alpha x^\alpha$ to be the series $(f * g)(x) := \sum_{\alpha\in\mathbb{N}} f_\alpha g_\alpha x^\alpha$. Now let again $x = x_1$ and $f(x,t) = \sum_{i,j\geq 0} c_{i,j} x^i t^j$. Define $D := \{(i,j) \in \mathbb{N}^2 : i = j\}$ and its indicator function $\mathbb{1}_D : \mathbb{N}^2 \to \{0,1\}$, then we obtain:

$$\Big(f(x,t) * \frac{1}{1-xt}\Big)(x,t) = \Big(\sum_{i,j\geq 0} c_{i,j}x^i t^j * \sum_{i,j\geq 0} \mathbb{1}_D x^i t^j\Big)(x,t) = \sum_{i,j\geq 0} c_{i,j}\mathbb{1}_D x^i t^j$$
$$= \sum_{n\geq 0} c_{n,n}(xt)^n = \mathcal{D}(f(x,t))(xt),$$

the diagonal of $f(x,t)$, with xt substituted by x. This gives another viewpoint on the diagonal operator and was one historic reason for its definition.

Example 3 Consider the well-known power series

$$f(t) = \sum_{n\geq 0}\sum_{k=0}^{n} \binom{n}{k}^2 \binom{n+k}{k}^2 t^n \in \mathbb{Z}[\![t]\!] \subseteq \mathbb{C}[\![t]\!].$$

This series appears in Apéry's proof of the irrationality of $\zeta(3)$. It is known to be transcendental and to satisfy a Picard-Fuchs differential equation, see [ABD19, AB13]. A lengthy but simple computation shows that $f(t)$ is the small diagonal of the rational power series

$$\frac{1}{1-x_1} \cdot \frac{1}{(1-x_2)(1-x_3)(1-x_4)(1-x_5) - x_1x_2x_3} \in \mathbb{Z}[\![x_1, x_2, x_3, x_4, x_5]\!].$$

In [Str14] a diagonal representation with four variables is shown:

$$f(t) = \Delta\left(\frac{1}{(1-x_1-x_2)(1-x_3-x_4) - x_1x_2x_3x_4}\right).$$

Is is not known whether a similar rational expression using only 3 variables exists; in fact no power series is known for which 4 is provably the least number such that it is possible to write it as a small diagonal of some rational power series in that many variables [BLS17].

There are many theorems in the literature connecting algebraic power series and diagonals of rational functions. For example, Pólya observed already in 1922 that the diagonal of any rational power series in two variables is necessarily algebraic [Pó22]. Furstenberg's trick from 1967 implies that if $f \in K[\![x]\!]$ is algebraic and $x = x_1$ one variable, then there exists a rational power series $R(x,t)$ with $\mathcal{D}(R(x,t)) = f(x)$ [Fur67]; see also [Saf87], [AB13, Section 6] and [Dum16, Lemma 87]. In the same paper, Furstenberg proved that the small diagonal of any rational power series with coefficients in a field of positive characteristic is algebraic. In 1984 Deligne improved on the second result: the small diagonal of any *algebraic* power series over a field of positive characteristic is algebraic [Del84]. Note that for fields of characteristic 0 neither Deligne's nor Furstenberg's statements hold (cf. Example 3 above). Some elementary proofs of Deligne-Furstenberg's theorem have been found later by Harase [Har88], as well as by Sharif and Woodcock [SW88]. More recent and quantitative progress on this theorem is done by Adamczewski and Bell in [AB13]. Denef and

Lipshitz gave a simpler proof already in 1987 and generalized the first theorem of Furstenberg to several variables [DL87]. This generalization stated below uses very abstract theory about the ring $K\langle x\rangle$ and we will present its proof in the last section using the discussed theorems of previous sections. A recent algorithmic confrontation to the viewpoint of algebraic power series as diagonals of rational functions is explained in [BDS17]. Finally, the reader can find Christol's survey about diagonals of rational functions in [Chr15].

Theorem 1.1 (Denef and Lipshitz) *Let $f(x) \in K\langle x\rangle$ be an algebraic power series in n variables over a field K of characteristic zero*[2]*. Then there exists a rational power series in $n+1$ variables $R(x,t) \in K(x,t) \cap K[\![x,t]\!]$ such that $f(x) = \mathcal{D}(R(x,t))$*[3]*.*

While the statement of this theorem is completely elementary, its proof uses quite involved techniques of commutative algebra and algebraic geometry. The goal of this text is not only to motivate and explain them to a non-expert reader, but also to provide intuition for the main steps of the original proof. After the prefatory discussion about the ring of algebraic power series in Sect. 1, we advance to Sect. 2 which is devoted to the notion of Henselization and its connection to $K\langle x\rangle$. In Sect. 3 we introduce étale ring maps and use this tool to prove an important fact about the Henselization of certain rings. Finally, in Sect. 4 we demonstrate the proof of Theorem 1.1 and present a new application of it, Theorem 4.3, in which we improve on the so-called Artin-Mazur lemma. We will recall all non-trivial definitions and try to be self-contained throughout the whole article.

Even though the proof of Theorem 1.1 is quite difficult, we can easily show it by the same trick as Furstenberg for a subclass of algebraic power series, which we call *étale-algebraic*. As we will see, the difficulty is then reducing the general case to the étale-algebraic one. This is done by proving that any algebraic power series can be represented as a rational function in an étale-algebraic series.

Definition An algebraic power series $h(x) \in K\langle x\rangle$ with minimal polynomial $P(x,t) \in K[x,t]$ is called *étale-algebraic* if $h(0) = 0$ and $\partial_t P(0,0) \neq 0$.

Note that it immediately follows from the implicit function theorem that the minimal polynomial (as a function in t) of an étale-algebraic power series has a unique power series root vanishing at 0, which must be the étale-algebraic series itself. Therefore, étale-algebraic power series are exactly those series which are encoded uniquely by their minimal polynomial.

Example Take $x = x_1$; the algebraic power series $\tilde{h}(x) = x\sqrt{1-x}$ has minimal polynomial $\tilde{P}(x,t) = t^2 + x^3 - x^2$ which admits $\partial_t \tilde{P}(0,0) = 0$ and therefore $\tilde{h}(x)$ is

[2] In the original paper [DL87] the statement is more general, allowing for excellent local integral domains instead of only fields of characteristic 0, but the ideas of the proof are the same in the special case we consider.

[3] For completeness we mention that Denef and Lipshitz also proved another similar theorem in their paper. Using a slightly different notion of diagonal, they showed that an algebraic power series in n variables is the diagonal of a rational function in $2n$ variables, see [DL87, Theorem 6.2 (ii)].

not étale-algebraic. Clearly, $\tilde{P}(x,t) = (t - x\sqrt{1-x})(t + x\sqrt{1-x})$ has two power series solutions, both vanishing at 0. However, consider $h(x) = \sqrt{1-x} - 1$. Then still $h(0) = 0$ and for the new minimal polynomial we find $\partial_t P(0,0) = 2 \neq 0$, meaning that $h(x)$ is étale-algebraic.

Lemma 1.2 *Let $h \in K\langle x\rangle$ be étale-algebraic with minimal polynomial $P(x,t) \in K[x,t]$. Then the following rational function is a power series*

$$F := t\frac{\partial_t P(xt,t)}{P(xt,t)} \in K[\![x,t]\!].$$

Moreover, for any $i \in \mathbb{N}^n$ and $j \in \mathbb{N}$, the series $x^i h(x)^j$ is the diagonal of $(xt)^i t^{j+1} F$. In particular, Theorem 1.1 holds if f is étale-algebraic.

Proof Since $h(x)$ is a root of $P(x,t)$, we can write $P(x,t) = (t - h(x))Q(x,t)$ for some $Q(x,t) \in K[\![x]\!][t]$. Differentiating both sides with respect to t gives

$$\partial_t P(x,t) = Q(x,t) + (t - h(x))\partial_t Q(x,t), \tag{1}$$

hence, after dividing through by $P(x,t)/t$, we obtain

$$t\frac{\partial_t P(x,t)}{P(x,t)} = \frac{t}{t-h(x)} + t\frac{\partial_t Q(x,t)}{Q(x,t)}. \tag{2}$$

Equation (1) implies that $Q(0,0) \neq 0$, therefore $Q(x,t) \in K[\![x,t]\!]^*$ and consequently the second summand in the equation above is a power series. Furthermore, since $h(0) = 0$, we have

$$\frac{t}{t-h(xt)} = \frac{1}{1-t^{-1}h(xt)} \in K[\![x,t]\!].$$

This concludes, using Eq. (2), the proof of the first part of the Lemma. For the second part, consider

$$R(x,t) = (xt)^i t^{j+1}\frac{\partial_t P(xt,t)}{P(xt,t)}.$$

A straightforward computation together with Eq. (2) yields

$$\begin{aligned}\mathcal{D}(R(x,t)) &= \mathcal{D}\Big((xt)^i t^j \frac{1}{1-t^{-1}h(xt)}\Big) = \mathcal{D}\Big((xt)^i \sum_{k\geq 0} t^{j-k} h(xt)^k\Big)\\ &= \mathcal{D}\big((xt)^i h(xt)^j\big) = x^i h(x)^j.\end{aligned} \tag{3}$$

Finally, by setting $i = 0$ and $j = 1$ in (3), we obtain the theorem of Denef and Lipshitz for étale-algebraic power series. □

Now we dive into a more fundamental and basic study of our ring of interest $K\langle x\rangle$.

1.2 Weierstrass Theorems

The Weierstrass division theorem (WDT) and the preparation theorem (WPT) are fundamental classical results about the rings of formal and convergent power series. WDT is in some sense a form of the Euclidean division algorithm, but requires an extra property on the divisor. Often, applications and implications of the division algorithm for the polynomial ring can be translated via the Weierstrass division theorem to $K[\![x]\!]$. For example, it implies that the ring of formal power series is Noetherian, Henselian and a unique factorization domain. However, our main interest lies in the fact that both WDT and WPT hold true for algebraic power series and yield the same implications also for this ring. This fact was first proven by Lafon in 1965 [Laf65] and is much less known. We shall reprove it following the ideas from [LT70].

Definition We introduce the following notation and object:

(1) The *order* of a non-zero formal power series $f = \sum_{\alpha\in\mathbb{N}^n} a_\alpha x^\alpha$, denoted by $\mathrm{ord}(f)$, is the smallest integer $d \geq 0$ such that $a_\alpha \neq 0$ for some $\alpha \in \mathbb{N}^n$ with $|\alpha| = d$. For $f = 0$ we say that $\mathrm{ord}(f) = +\infty$.
(2) A power series $g \in K[\![x]\!]$ is called *x_n-regular of order d* if $g(0, \dots, 0, x_n) = x_n^d f(x_n)$ for some power series $f(x_n) \in K[\![x_n]\!]$ with $f(0) \neq 0$.
(3) A polynomial $p \in K[\![x']\!][x_n]$ is called *distinguished* if it is of the form $p = x_n^d + a_{d-1}(x')x_n^{d-1} + \cdots + a_0(x')$ for some power series $a_i(x') \in K[\![x']\!]$ with $a_i(0) = 0$ for $i = 0, \dots, d-1$.

Theorem 1.3 (WPT) *Let $g \in K[\![x]\!]$ be an x_n-regular power series of order d. Then there exist a unique $p \in K[\![x']\!][x_n]$ which is a distinguished polynomial in x_n of degree d and a unique unit $u \in K[\![x]\!]^*$, such that $g = up$.*

Theorem 1.4 (WDT) *Let $g \in K[\![x]\!]$ be an x_n-regular power series of order d. For any $f \in K[\![x]\!]$ there exist uniquely a power series $q \in K[\![x]\!]$ and an $r \in K[\![x']\!][x_n]$ which is a distinguished polynomial in x_n of degree less than d, such that $f = qg + r$.*

There are several different known ways to prove these classical theorems. The standard literature for their proofs and implications is the book "The Basic Theory of Power Series" by Ruiz [Rui93], a more recent and very short proof can be found in [Hau17]. A very explicit approach is followed by Lang in [Lan84]. Now we state and prove Lafon's algebraic versions of these theorems.

Theorem 1.5 (Algebraic WPT) *Let $g \in K\langle x\rangle$ be an x_n-regular algebraic power series of order d. Then there exist a unique $p \in K\langle x'\rangle[x_n]$ which is a distinguished polynomial in x_n of degree d with coefficients given by algebraic power series in x' and a unique unit $u \in K\langle x\rangle^*$, such that $g = up$.*

Proof First note that we may assume without loss of generality that g is irreducible as a power series. Write $g = up$ with $u \in K[\![x]\!]^*$ and $p \in K[\![x']\!][x]$ a distinguished polynomial of degree d. It follows that p has d distinct roots in an algebraic closure of the quotient ring of the formal power series $\Omega = \overline{\mathrm{Frac}(K[\![x']\!])}$, say $\alpha_1, \dots, \alpha_d$. Now let $G(x,t) = G_0(x) + G_1(x)t + \cdots + G_e(x)t^e \in K[x,t]$, $G_0 \neq 0$ be a minimal polynomial of g, i.e. we have

$$0 = G(x, g(x)) = G_0(x) + G_1(x)g(x) + \cdots + G_e(x)g(x)^e.$$

For every $i = 1, \dots, d$ we can replace x by (x', α_i) and, using $g(x', \alpha_i) = 0$, we obtain for every of those i's

$$0 = G(x', \alpha_i, g(x', \alpha_i)) = G_0(x', \alpha_i).$$

As $0 \not\equiv G_0(x', x_n) \in K[x', x_n] \subseteq K(x', x_n)$ and annihilates α_i, we get that α_i is algebraic over $K(x')$. It follows that $x_n - \alpha_i$ is algebraic over $K(x', x_n) = K(x)$. Therefore, $p = \prod_{j=1}^{d}(x_n - \alpha_j)$ is an algebraic power series and the same holds for $u = g/p$. Uniqueness is guaranteed by uniqueness of Weierstrass formal preparation (Theorem 1.3). □

Theorem 1.6 (Algebraic WDT) *Let $g \in K\langle x\rangle$ be an x_n-regular algebraic power series of order d. For any $f \in K\langle x\rangle$ there exist uniquely an algebraic power series $q \in K\langle x\rangle$ and an $r \in K\langle x'\rangle[x_n]$ which is a distinguished polynomial in x_n of degree less than d with coefficients given by algebraic power series in x', such that $f = qg + r$.*

Proof Again, we may assume that g is irreducible and by the algebraic Weierstrass preparation theorem, we may also assume without loss of generality that $g \in K\langle x'\rangle[x_n]$ is a distinguished polynomial of degree d. We can divide formally:

$$f = qg + r = qg + \sum_{j=0}^{d-1} b_j(x')x_n^j, \tag{4}$$

for $q \in K[\![x]\!]$ and $b_0(x'), b_1(x'), \dots, b_{d-1}(x') \in K[\![x']\!]$ formal power series. Because g is assumed to be a distinguished polynomial, we may write $g = \sum_{j=0}^{d} c_j(x')x_n^j$ for some $c_0(x'), \dots, c_d(x') \in K\langle x'\rangle$ algebraic power series. We get that g has d distinct roots in $\Omega = \overline{\mathrm{Frac}(K[\![x']\!])}$, say $\alpha_1, \dots, \alpha_d$. From (4), by replacing x_n with α_i, we get for every $i = 1, \dots, d$ that $f(x', \alpha_i) = \sum_{j=0}^{d-1} b_j(x')\alpha_i^j$. This can be rephrased in terms of a matrix multiplication:

$$\begin{pmatrix} f(x', \alpha_1) \\ \vdots \\ f(x', \alpha_d) \end{pmatrix} = \begin{pmatrix} 1 & \alpha_1 & \cdots & \alpha_1^{d-1} \\ 1 & \alpha_2 & \cdots & \alpha_2^{d-1} \\ \vdots & \vdots & \ddots & \vdots \\ 1 & \alpha_d & \cdots & \alpha_d^{d-1} \end{pmatrix} \begin{pmatrix} b_0(x') \\ \vdots \\ b_{d-1}(x') \end{pmatrix}.$$

The matrix above is the Vandermonde matrix and it is invertible since the α_i's are pairwise different. Therefore, each $b_i(x')$ is uniquely given by some rational expression in the $f(x', \alpha_j)$'s and α_k's for $j, k \in \{1, \ldots, d\}$. On the other hand, by the same argument as in the proof of Theorem 1.5, we obtain that each α_i is algebraic over $K(x')$. Moreover, by assumption $f(x', x_n)$ is algebraic over $K(x', x_n)$. It follows that both field extensions $K(x') \subseteq K(x', \alpha_i)$ and $K(x', \alpha_i) \subseteq K(x', \alpha_i, f(x', \alpha_i))$ are finite. Hence, $K(x') \subseteq K(x', f(x', \alpha_i))$ is finite and consequently $f(x', \alpha_i)$ is algebraic over $K(x')$. As this holds for every $i = 1, \ldots, d$, it follows that also any rational expression in the $f(x', \alpha_j)$'s and α_k's for $j, k \in \{1, \ldots, d\}$ is algebraic over $K(x')$. Recall that each $b_i(x')$ is such an expression, therefore each $b_i(x')$ is algebraic over $K(x')$ and hence an algebraic power series. It follows that $r = \sum_{j=0}^{d-1} b_j(x')x_n^j$ is an algebraic power series and finally the same holds for $q = (f - r)/g$. Uniqueness is clear by the formal WDT (Theorem 1.4). □

As already mentioned, this theorem has many implications. For example, the proofs from [Rui93] and [Lan84] that $K[\![x]\!]$ is Noetherian and a UFD can be directly carried out for $K\langle x\rangle$. Another corollary is the following:

Theorem 1.7 (Hensel's Lemma) *Let $f \in K\langle x\rangle[t]$ be a monic polynomial in t over $K\langle x\rangle$. Assume $\alpha \in K$ is a root of multiplicity d of the polynomial $f(0, t) \in K[t]$. Then there exist unique monic polynomials $p, u \in K\langle x\rangle[t]$ with $u(0, \alpha) \neq 0$, p of degree d in t, $p(0, t) = (t - \alpha)^d$ and $f = up$.*

Proof After the change of the variable t to $t + \alpha$, we may assume that $\alpha = 0$. Since α is a d-multiple root of $f(0, t)$, it follows that $f(x, t)$ is t-regular of order d. By the algebraic WPT in $n + 1$ variables we may write uniquely $f = up$ where $p \in K\langle x\rangle[t]$ is a distinguished polynomial in t of degree d and $u \in K\langle x, t\rangle^*$ a unit, hence $u(0, \alpha) = u(0, 0) \neq 0$. Moreover, since $p(x, t)$ is distinguished of degree d it follows by definition that $p(0, t) = t^d = (t - \alpha)^d$. $u(x, t) \in K\langle x\rangle[t]$ because of polynomial division in this ring. Uniqueness follows from the uniqueness of the algebraic Weierstrass division theorem. □

The statement above ensures that a root $\alpha \in K$ of $f(0, t)$ gives rise to a factorization $f = up$. This factorization is called the *lifting* of α. Lifting all roots separately, one by one, it is possible to prove that coprime factorizations can be lifted as well.

Theorem 1.8 (Hensel's Lemma) *Let $f \in K\langle x\rangle[t]$ be a monic polynomial in t over $K\langle x\rangle$. Assume $f(0, t) = \bar{p}(t)\bar{q}(t)$ factors into two monic coprime polynomials. Then there exist two unique monic polynomials $p, q \in K\langle x\rangle[t]$ with $p(0, t) = \bar{p}(t)$, $q(0, t) = \bar{q}(t)$ and $f = qp$.*

We omit the detailed proof here, because the equivalence on these two versions of Hensel's lemma will be justified in Sect. 3 in a more general setting. Instead, we explain the importance of these theorems following the motivation in [Eis95, pp. 185].

1.3 The Importance of Hensel's Lemma

Consider the nodal plane cubic curve over a field K (of characteristic 0 as always or positive but not equal to 2) given by the equation $t^2 - x^2(1 + x) = 0$ for $x = x_1$.

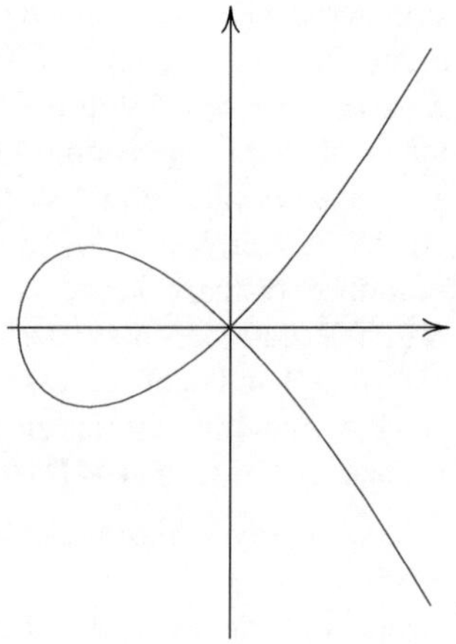

Fig. 1 The node: $\mathbb{R}[x,t]/(t^2 - x^2(1+x))$

The associated affine coordinate ring is $S = K[x,t]/(t^2 - x^2(1+x))$. Of course, the curve is irreducible and S is a domain. When looking at the picture over $\mathbb{R}$ (Fig. 1), one may think that localizing S at the maximal ideal $\mathfrak{m} = (\bar{x}, \bar{t})$ will make the ring have zero divisors, however this is not the case: every Zariski neighborhood of 0 of the node is irreducible. The reason is that over the complex numbers a neighborhood of the omitted origin is a punctured disc and therefore the curve remains irreducible. We would still like to factor $t^2 - x^2(1+x)$ somehow, in order to study the easier rings into which S will decompose. Examining a "really small neighborhood" of the node, we would expect the curve to become reducible there: for example over the ring of formal power series the expression $t^2 - x^2(1+x)$ is in fact reducible. This comes from the fact that $1+x$ has a square root in $K[\![x]\!]$ and we may therefore write $t^2 - x^2(1+x) = (t - x\sqrt{1+x})(t + x\sqrt{1+x})$. One can argue the reason why it is immediately clear that $1+x$ is a square over $K[\![x]\!]$ is that this ring satisfies Hensel's lemma! More precisely, take the polynomial $f(x,t) = t^2 - (1+x)$, then $f(0,t) = (t-1)(t+1) = \bar{p}(t)\bar{q}(t)$ and these polynomials are coprime. Therefore, by Hensel's lemma, this factorization must admit a lifting and therefore $\sqrt{1+x} \in K[\![x]\!]$. This is also the reason why we explicitly do not allow the characteristic of K to be 2: in this case $\bar{p}(t)$ and $\bar{q}(t)$ would not be relatively prime and the lifting would not be guaranteed, in fact it would not exist. However, we also see that in order to make the node reducible, we do not have to go from $K[x,t]$ all the way up to the polynomial ring over the completion $K[\![x]\!][t]$: it suffices to take *any* Henselian ring extension of S or of $K[x]_{(x)}$. This is exactly the idea and motivation for defining the Henselization.

2 Algebraic Power Series and Henselization

The main goal of this section is to stress the connection between the algebraic closure in the completion of a ring and the property of being Henselian, that is to satisfy Hensel's lemma. We will be able to prove that, under certain conditions, any Henselian ring is algebraically closed in its completion, that is, if $a \in \widehat{R}$ is algebraic over a Henselian R then it must already hold that $a \in R$. This will allow us to look at the ring of algebraic power series from a different viewpoint. The conditions may appear technical at first sight, however they have the purpose of excluding pathologies while still allowing for a large class of rings. A considerably different approach to this theory can be found in the book [KPR75] by Kurke, Pfister and Roczen.

We will work with *local rings*, i.e. with those rings R, which have exactly one maximal ideal. Usually, we will denote this maximal ideal by $\mathfrak{m}$ and let $K := R/\mathfrak{m}$ be the residue field with respect to $\mathfrak{m}$. Sometimes, we write triples $(R, \mathfrak{m}, K)$ when talking about local rings, combining these three objects. It is immediate to see that R has only one maximal ideal $\mathfrak{m}$ if and only if $R^* = R\backslash\mathfrak{m}$. Recall that given two local rings $(R, \mathfrak{m}, K)$, $(S, \mathfrak{n}, L)$, a homomorphism $\phi : R \to S$ is called *local* if $\phi(\mathfrak{m}) \subseteq \mathfrak{n}$ holds and this condition is equivalent to $\phi^{-1}(\mathfrak{n}) = \mathfrak{m}$. Note that both, the rings of formal and algebraic power series, are local with maximal ideal $\mathfrak{m} = (x) = (x_1, \ldots, x_n)$. It is also obvious that the residue field $K[\![x]\!]/(x) \cong K\langle x\rangle/(x)$ is (isomorphic to) K. Recall that the *completion* of a local ring $(R, \mathfrak{m}, K)$ is defined as the inverse limit $\varprojlim R/\mathfrak{m}^i$. A local ring $(R, \mathfrak{m}, K)$ is called *complete* if the canonical map to its completion $R \to \widehat{R}$ is an isomorphism. Note that $\bigcap_i \mathfrak{m}^i$ goes to zero under this mapping, therefore completeness implies $\bigcap_i \mathfrak{m}^i = 0$. Krull's intersection theorem justifies that the completion of a *Noetherian* ring (i.e. a ring with only finitely generated ideals) is complete, however this is false in general. Finally, recall that the completion $\widehat{R}$ of R satisfies the universal property that for any local $(R, \mathfrak{m}) \to (S, \mathfrak{n})$ with $(S, \mathfrak{n})$ complete there exists a unique factorization $R \to \widehat{R} \to S$.

2.1 *Henselian Rings*

Given a $p \in R[t]$, we will denote by $\bar{p} \in K[t]$ the reduction of p mod $\mathfrak{m}R[t]$, given by reducing all coefficients of p mod $\mathfrak{m}$.

Definition A local ring $(R, \mathfrak{m}, K)$ is called *Henselian* if the following property holds: Let $f(t) \in R[t]$ be a monic polynomial. Assume that $\bar{f}(t) = p_0(t)q_0(t)$ holds for two monic coprime polynomials $p_0(t), q_0(t) \in K[t]$. Then there exist two unique monic polynomials $p(t), q(t) \in R[t]$ satisfying $\bar{p}(t) = p_0(t), \bar{q}(t) = q_0(t)$, $\deg p(t) = \deg p_0(t)$, $\deg q(t) = \deg q_0(t)$ and $f(t) = p(t)q(t)$.

The notion of Henselian rings (or Hensel rings) was first introduced by Azumaya in [Azu51]. The property above is usually referred to as "Hensel's lemma" even though it is used as a definition. One often reads in the literature "A ring is called Henselian, if Hensel's lemma holds [in this ring]". To avoid confusion, we will call the statement above Hensel's property. The actual "lemma" of Hensel is the following classical theorem (see for example [Eis95, Chapter 7]):

Theorem 2.1 (Hensel's lemma for complete rings) *Let* $(R, \mathfrak{m}, K)$ *be a complete local ring. Then* R *is Henselian.*

Standard references for Henselian rings are [Nag62, Ray70, Gro67]. For the purpose of this work, a very significant fact is that Theorem 1.8 implies the following:

Theorem 2.2 *The ring of algebraic power series* $K\langle x\rangle$ *is Henselian.*

Another key fact about Henselian rings is the following lemma. Its statement and proof appear in [Nag62] and give an introductory flavor to this section. Recall that given an extension of rings $R \subseteq S$, an element $s \in S$ is called *integral over* R if there exists a monic polynomial $P(t) \in R[t]$ with $P(s) = 0$. The extension is said to be *integral* if this holds for any $s \in S$.

Lemma 2.3 *Let* R *be a Henselian integral domain and* R' *an integral extension of* R*. Then* R' *is a local ring.*

Proof Let $\mathfrak{m}$ be the maximal ideal of R and assume that R' has two maximal ideals $\mathfrak{m}'_1 \neq \mathfrak{m}'_2$. Take some $a \in \mathfrak{m}'_1$ which is not in $\mathfrak{m}'_2$. We have an irreducible monic polynomial $f(t) = t^n + c_{n-1}t^{n-1} + \cdots + c_0 \in R[t]$ which has a as root. Now, as $a \in \mathfrak{m}'_1$, we must have $c_0 \in \mathfrak{m}'_1 \cap R \subseteq \mathfrak{m}$. We also have that $a^n \notin \mathfrak{m}R[a]$, because $a \notin \mathfrak{m}'_2$, hence there must be a c_i which is not in $\mathfrak{m}$. Take $j \in \mathbb{N}$ such that $c_j \notin \mathfrak{m}$ but $c_{j-s} \in \mathfrak{m}$ for $0 < s \leq j$. Clearly $1 \leq j \leq n-1$ and we have

$$f(t) \equiv (t^j + c_{n-1}t^{j-1} + \cdots + c_{n-j})t^{n-j} \mod \mathfrak{m}R[t].$$

But this means that the image of f is reducible mod $\mathfrak{m}R[t]$ and using that R is Henselian we obtain that f must be reducible in $R[t]$. This is a contradiction and the assertion is proved. □

To state the main theorem of this section we will need some conditions on our local ring R. Therefore we define the necessary terms:

Definition Let R be a ring.

(1) R is called *analytically irreducible* if it is local and its completion $\widehat{R}$ is a domain. R is called *analytically normal* if it is local and $\widehat{R}$ is normal[4].
(2) Assume R be an integral domain with quotient field L. R is called *Japanese*[5] if it satisfies the so-called finiteness condition for integral extensions. This means, for every finite extension L' of the quotient field L, the integral closure of R in L' is a finitely generated R-module.

[4] Recall that a local ring is called normal if is integrally closed in its quotient field.

[5] According to [Sta20] this name was first used by Grothendieck in order to contribute to Nakayama, Takagi, Nagata and many others.

(3) R is called a *Nagata ring*[6] if R is Noetherian and for every prime ideal $\mathfrak{p} \subseteq R$, the ring $R/\mathfrak{p}$ is Japanese.

Lemma 2.4 *If R is a Nagata ring, then every ring which is a finite module over R or a ring of quotients of R is also Nagata. Any localization of R is also Nagata. Any finitely generated R-algebra is Nagata.*

We see that the category of Nagata rings is reasonably large and closed under many operations. Good references for the proofs of the lemma above are [Nag62, Section 36] and [Mat80, Section 31]. Of course, most facts can also be found in [Sta20, Section 032E].

Finally, for our purposes we need the following version of the Zariski Main Theorem, which is Theorem 37.8 in [Nag62]. Recall that S is said to be of *finite type* over R if S is isomorphic to a quotient of $R[x_1, \ldots, x_n]$ as an R-algebra.

Lemma 2.5 *Let R be an analytically normal ring. If a normal and local Nagata ring S is of finite type over R, then S analytically irreducible.*

We are ready to prove one central theorem of this section, connecting Henselian rings with the property of being algebraically closed in the completion, and which will be a crucial ingredient in the proof of Theorem 2.7. This theorem explains why the study of algebraic power series essentially comes down to studying Henselian rings. Its statement can be found in [Nag62, (44.1)], however the proof given there is very concise and in some places unclear. Therefore we shall reprove the theorem here.

Theorem 2.6 *Let R be a Henselian, analytically normal Nagata ring. Then R is algebraically closed in its completion, i.e. if $a \in \widehat{R}$ algebraic over R, then $a \in R$.*

Proof Let a be an element of $\widehat{R}$ which is algebraic over R. Then we can find $b \neq 0$ in R such that $c := ab$ is integral over R. We claim that $R[c] = R$. Assume otherwise and let $f \in R[t]$ be the minimal polynomial of c. We wish to use the lemma above on $R[c]$, but we lack the assumption of normality. So we define R' to be the integral closure of $R[c]$ in $L[c]$, where $L = \operatorname{Frac}(R)$, so R' is normal by definition. By an easy observation it follows that R' is also the integral closure of R in $L[c]$. Since R is analytically normal, it is a domain and therefore the ideal (0) is prime. By the definition of a Nagata ring, it follows that the integral closure of $R = R/(0)$ in any finite extension of L is a finitely generated R-module. Since c is integral and in particular algebraic, it follows that $L[c]$ is a finite extension of L and therefore R' is a finitely generated R-module. Furthermore, R', being finitely generated over a Nagata ring, is still Nagata by Proposition 2.4 and also R' is indeed local by Lemma 2.3,[7] hence we may apply Lemma 2.5 to get that R' is analytically irreducible. However, we have that $\widehat{R'} = R' \otimes_R \widehat{R}$ and this must be a domain. Now look at the completion

[6] In the book "Local Rings" Nagata calls these rings pseudo-geometric.

[7] This is where we use the Henselian assumption.

of $R[c]$, which is given by $R[c] \otimes_R \widehat{R}$. Now, $R \to \widehat{R}$ is flat, meaning that we also have the inclusion $R[c] \otimes_R \widehat{R} \subseteq R' \otimes_R \widehat{R}$ and hence $\widehat{R[c]}$ must be a domain as well. On the other hand, we have $\widehat{R[c]} = R[c] \otimes_R \widehat{R} = \widehat{R}[t]/(f)$, identifying f with its image in $\widehat{R}[t]$. However $c \in \widehat{R}$ is a root of f, hence $\widehat{R}[t]/(f)$ cannot be a domain: a contradiction. So $c \in R$ and hence $a \in \mathrm{Frac}(R)$. Because $R = \mathrm{Frac}(R) \cap \widehat{R}$ and since a is in both rings, we get that $a \in R$ as wanted. □

2.2 Henselian Characterization of Algebraic Power Series

We saw that Henselian rings are closely connected to algebraic closures in the completion. In particular, at this point, one may conjecture that for some, not necessarily Henselian, ring R, if we can define the "smallest" Henselian extension of R, it will be exactly the algebraic closure of R in $\widehat{R}$. Since algebraic power series are by definition the algebraic closure of $K[x]_{(x)}$ in its completion, this approach will also give a different viewpoint on our main ring of interest. Note that obviously any field is Nagata, therefore by Lemma 2.4 it follows that $K[x]$ is also a Nagata ring. Then $K[x]_{(x)}$ is again Nagata, since it is a localization.

Definition Let $(R, \mathfrak{m}, K)$ be a local ring. We say a Henselian ring R^{h} together with a local homomorphism $i : R \to R^{\mathrm{h}}$ is the *Henselization* of R, if any local homomorphism from R to a Henselian ring factors uniquely through i.

In other words, the Henselian ring R^{h} together with $i : R \to R^{\mathrm{h}}$ is the Henselization of R, if for any Henselian ring H and local $\psi : R \to H$ there exists a unique local ϕ such that the following diagram commutes:

$$\begin{array}{ccc} R & \xrightarrow{\ i\ } & R^{\mathrm{h}} \\ {\scriptstyle\psi}\downarrow & \swarrow {\scriptstyle\phi} & \\ H & & \end{array}$$

This notion was first introduced by Nagata in the article "On the theory of Henselian rings", which became the first of a trilogy [Nag53, Nag54, Nag59]. Since then, the Henselization of a ring became a very well studied object; we shall only explain those facts which are of importance for our purpose.

Note that from the definition it follows that if R^{h} exists, then it must be unique up to isomorphism. For Noetherian rings one has the inclusion $R \hookrightarrow \widehat{R}$; this together with Hensel's lemma immediately implies that i must be injective as well in this case. Moreover, the following fact follows also easily from the universal property: Assume the existence of a Henselian ring R' such that $R \subseteq R' \subseteq R^{\mathrm{h}}$, then $R' = R^{\mathrm{h}}$. In this sense we can view the Henselization as the "smallest" Henselian ring extension of R. Note that it is not obvious that R^{h} exists for any local R, however this is true and we will prove this in the next section (Sect. 3.3).

The following theorem allows a purely ring-theoretic viewpoint on $K\langle x\rangle$.

Theorem 2.7 *Let $R = K[x]_{(x)}$ be the localization of $K[x]$ at the maximal ideal (x). Assume that the Henselization of R exists*[8]*. Then it is isomorphic to the ring of algebraic power series: $R^{\mathrm{h}} \cong K\langle x\rangle$.*

For the proof we need two lemmas: we provide a proof for the first, and a reference for the second.

Lemma 2.8 *Let R^{h} be the Henselization of a Noetherian local ring R. Then $\widehat{R^{\mathrm{h}}} = \widehat{R}$.*

Proof By Hensel's lemma $\widehat{R}$ is Henselian, thus there exists a unique factorization $R \to R^{\mathrm{h}} \to \widehat{R}$. Let S be a complete local ring with maximal ideal $\mathfrak{n}$ and assume $R^{\mathrm{h}} \to S$ is a local map. Precomposing with i gives $R \to R^{\mathrm{h}} \to S$. By the universal property of the completion and then by the factorization of $R \to \widehat{R}$ we also find $R \to R^{\mathrm{h}} \to \widehat{R} \to S$. Now, since S being complete is also Henselian, the uniqueness in the universal property of R^{h} implies that these factorizations are equal. Hence, for every $R \to S$ with S complete we find $R^{\mathrm{h}} \to \widehat{R} \to S$. The universal property of the completion forces $\widehat{R^{\mathrm{h}}} = \widehat{R}$. □

Lemma 2.9 *Let R be a local Nagata ring. Then its Henselization R^{h} is also Nagata. Moreover, if R is also analytically normal then so is R^h.*

For the proof see [Nag62, (44.2,44.3)].

Proof of Theorem 2.7 By Lemma 2.9 it follows that $K[x]^{\mathrm{h}}_{(x)}$ is both analytically normal and Nagata. Because the ring of algebraic power series is Henselian (Theorem 2.2) and the Henselization is the smallest Henselian ring extension of a given ring, it suffices to show $K\langle x\rangle \subseteq K[x]^{\mathrm{h}}_{(x)}$. Let $f \in K\langle x\rangle \subseteq K[\![x]\!] = \widehat{K[x]_{(x)}}$. Then obviously f is algebraic over $K[x]^{\mathrm{h}}_{(x)}$ and by Lemma 2.8 we must have $f \in \widehat{K[x]^{\mathrm{h}}_{(x)}}$. We can apply Theorem 2.6 to see that $f \in K[x]^{\mathrm{h}}_{(x)}$. □

3 Étale Ring Maps and Henselization

3.1 Motivation for Étale Ring Maps

Before giving the rigorous definition of an étale map $R \to S$ between two rings R, S, we will try to explain the motivation behind it. Milne writes in his lecture notes [Mil13]:

> *"An étale morphism is the analogue in algebraic geometry of a local isomorphism of manifolds in differential geometry, a covering of Riemann surfaces with no branch points in complex analysis, and an unramified extension in algebraic number theory."*

---

[8] In the next section (Sect. 3.3) we will prove that the Henselization of a local ring always exists.

Of course, the importance of these objects makes it clear that one needs a definition in the setting of algebraic geometry and that this definition might be involved. There are many equivalent ways to define this analogue and we will try to motivate the one that is mostly geometric and closest to a universal property.

Consider the case of two affine algebraic varieties $X = V(f_1, \dots, f_r) \subseteq K^n$, $Y = V(g_1, \dots, g_s) \subseteq K^m$ and a morphism $f_\phi : X \to Y$ coming from $\phi : R \to S$, where

$$R := K[Y] = K[y_1, \dots, y_m]/(g_1, \dots, g_s) \text{ and}$$
$$S := K[X] = K[x_1, \dots, x_n]/(f_1, \dots, f_r)$$

are the corresponding coordinate rings. Recall that a local diffeomorphism is characterized by its bijective differential. We want to achieve an analogous property for f_ϕ by putting only algebraic conditions on ϕ.

By definition, f_ϕ maps any K-point $a := (a_1, \dots, a_n) \in X$ to a K-point $b := (b_1, \dots, b_m) \in Y$. To formulate this in an algebraic way, we can require the following diagram to commute:

$$\begin{array}{ccc} S = K[X] & \longrightarrow & K \\ \phi\uparrow & \nearrow & \\ R = K[Y] & & \end{array}$$

To see that this algebraic formulation indeed corresponds to the geometric viewpoint of sending $a \in X$ to some $b \in Y$, note that the map $K[X] \to K$ defines a K-point of X, since it maps each x_i to a_i for some $a := (a_1, \dots, a_n) \in K^n$ with the condition that each $f_j(a_1, \dots, a_n) = 0$, $1 \le j \le r$, hence, by definition, $a \in X$. Similarly, $K[Y] \to K$ is a K-point, say $b = (b_1, \dots, b_m) \in Y$, because $g_j(b_1, \dots, b_m) = 0$ for $j = 1, \dots, s$. The commutativity of the diagram means that sending $(y_1, \dots, y_m) \mapsto (b_1, \dots, b_m)$ by the diagonal map is the same as sending $(y_1 \dots, y_m) \mapsto (\phi_1(x_1, \dots, x_n), \dots, \phi_m(x_1, \dots, x_n)) \mapsto (\phi_1(a), \dots, \phi_m(a))$: K-points are sent to K-points.

Now we want to describe the behavior of f_ϕ on tangent vectors. We can formulate this in an algebraic way, by requiring the commutativity of the following diagram, adding the ring $K[\varepsilon]/(\varepsilon^2)$ to the above:

$$\begin{array}{ccc} S = K[X] & \longrightarrow & K \\ \phi\uparrow & & \uparrow \\ R = K[Y] & \longrightarrow & K[\varepsilon]/(\varepsilon^2) \end{array}$$

Since

$$\begin{array}{ccccc} K[Y] & \longrightarrow & K[\varepsilon]/(\varepsilon^2) & \longrightarrow & K \\ (y_1, \dots, y_m) & \mapsto & (b_1 + \varepsilon c_1, \dots, b_m + \varepsilon c_m) & \mapsto & (b_1, \dots, b_m), \end{array}$$

we see that this intermediate ring does not destroy the considerations above. Moreover, we claim that the map $K[Y] \to K[\varepsilon]/(\varepsilon^2)$ corresponds to a tangent vector of Y: say, we have

$$\begin{aligned} K[Y] = K[y_1, \ldots, y_m]/(g_1, \ldots, g_s) &\to K[\varepsilon]/(\varepsilon^2) \\ y_i &\mapsto b_i + \varepsilon c_i, \qquad 1 \le i \le m, \end{aligned}$$

for some $b \in K^m$ and $c := (c_1, \ldots, c_m) \in K^m$. Then it must hold that $g_j(b_1 + \varepsilon c_1, \ldots, b_m + \varepsilon c_m) = 0$ for $1 \le j \le s$. Using Taylor expansion and the fact that $\varepsilon^2 = 0$ in $K[\varepsilon]/(\varepsilon^2)$, we obtain:

$$0 = g_j(b_1 + \varepsilon c_1, \ldots, b_m + \varepsilon c_m) = g_j(b) + \sum_{i=1}^{m} \frac{\partial g_j}{\partial y_i}(b) c_i \varepsilon.$$

Comparison of the coefficients in ε gives that $g_j(b_1, \ldots, b_n) = 0$ for each j, i.e. b is a K-point of Y (what we already knew), and that

$$\sum_{i=1}^{m} \frac{\partial g_j}{\partial y_i}(b) c_i = 0, \qquad 1 \le j \le s.$$

This is of course equivalent to $c \cdot \nabla g_j(b) = 0$, i.e. c is a tangent vector of Y at b and we may say $c \in T_b Y$, the tangent space of Y at b.

Up to now, we have reformulated the property of f_ϕ to map K-points to K-points and added the potential of considering tangent vectors in terms of a commutative diagram. We can now add the final requirement to ϕ, making it the analogue of a local diffeomorphism: we want its "differential" $T_a X \to T_{f_\phi(a)} Y = T_b Y$ to be bijective. Surprisingly, this condition is very easy to add in our commutative diagram formalism: we require additionally the existence and uniqueness of the diagonal arrow, preserving commutativity:

$$\begin{array}{ccc} S = K[X] & \longrightarrow & K \\ \phi\uparrow & \searrow & \uparrow \\ R = K[Y] & \longrightarrow & K[\varepsilon]/(\varepsilon^2) \end{array}$$

By the same argument as above, we can easily convince ourselves that this diagonal map writes $K[X] \to K[\varepsilon]/(\varepsilon^2) : x_i \mapsto a_i + \varepsilon d_i$ for $i = 1, \ldots, n$ and some $d := (d_1, \ldots, d_n)$ which corresponds to a tangent vector of X. The commutativity of the upper-right triangle just means that this vector is in the tangent space $T_a X$. Finally, consider the commutativity of the lower triangle. On the one hand, we can map by the horizontal homomorphism $y_j \mapsto b_j + \varepsilon c_j$, $1 \le j \le m$ as we already saw. On the other hand, going the other path, we have again by Taylor's expansion for $j = 1, \ldots, m$:

$$y_j \mapsto \phi_j(x_1, \ldots, x_n) \mapsto \phi_j(a_1 + \varepsilon d_1, \ldots, a_n + \varepsilon d_n) = \phi_j(a) + \sum_{i=1}^{n} \frac{\partial \phi_j}{\partial x_i}(a) d_i \varepsilon.$$

Since the lower-left triangle commutes, we have by comparison of the coefficient of ε that

$$c_j = \sum_{i=1}^{n} \frac{\partial \phi_j}{\partial x_i}(a) d_i, \qquad 1 \leq j \leq m.$$

Putting these m equations together, we define the Jacobian matrix

$$J_\phi(a) := \left(\frac{\partial \phi_j}{\partial x_i}(a) \right)_{\substack{1 \leq i \leq n \\ 1 \leq j \leq m}}.$$

Then, the equation above is, of course, equivalent to $J_\phi(a)d = c$.

Hence, the existence of the diagonal arrow makes sure that for any tangent vector at $b \in Y$, we have at least one tangent vector at $a \in X$ mapping to it, in other words it ensures the surjectivity of $J_\phi(a)$. Analogously, the uniqueness of the diagonal map translates into injectivity of the differential. Equipped with this good understanding of what it means to define the algebraic analogue of a local diffeomorphism, we can step forward to its rigorous definition.

3.2 *Étale Ring Maps*

We will present only those results about étale ring maps that are important for the construction of the Henselization. For other statements and some omitted proofs we refer to standard literature such as [Mil80, Gro67, Ray70] and of course [Sta20, Section 00U0].

Definition Given an R-algebra S with the homomorphism $\phi : R \to S$, we call S *formally étale* if the following condition is satisfied:

Suppose that T is some R-algebra, $\mathfrak{n} \subseteq T$ some ideal with $\mathfrak{n}^2 = 0$ and the following diagram of R-algebra maps commutes:

$$\begin{array}{ccc} S & \xrightarrow{\bar{u}} & T/\mathfrak{n} \\ {\scriptstyle \phi}\uparrow & & \uparrow{\scriptstyle \pi} \\ R & \longrightarrow & T \end{array}$$

Then there is a unique R-algebra morphism $u : S \to T$, which lifts $\bar{u}$, i.e. the following diagram also commutes:

$$\begin{array}{ccc} S & \xrightarrow{\bar{u}} & T/\mathfrak{n} \\ {\scriptstyle\phi}\uparrow & \overset{u}{\searrow} & \uparrow{\scriptstyle\pi} \\ R & \longrightarrow & T \end{array}$$

This property is known under the name *infinitesimal lifting*. As we saw above, it reflects the definition of a local diffeomorphism inside of algebraic geometry. When dealing with a formally étale S, we will often refer to the map $\phi : R \to S$ as formally étale rather than to the R-algebra itself.

To go from *formally* étale to étale ring maps, we need to recall the notion of finitely presented algebras. It is evident that an R-algebra S is always of the form $S \cong R[x_i : i \in \mathcal{I}]/\mathfrak{a}$ for some index set $\mathcal{I}$ and an ideal $\mathfrak{a} \subseteq R[x_i : i \in \mathcal{I}]$. In practice we are often interested in a finite number of generators and a finitely generated ideal, hence we define:

Definition Let R be a ring. We say an R-algebra S is *finitely presented* if it is of the form $S \cong R[x_1, \ldots, x_n]/(f_1, \ldots, f_m)$ for some $f_i \in R[x_1, \ldots, x_n]$, $i = 1, \ldots, m$.

Naturally, we have immediately the following fact about transitivity: If S if finitely presented over R and T is finitely presented over S, then T is finitely presented over R. Note that a localization of R at one element, say $a \in R$, is finitely presented, since $R_a \cong R[t]/(at - 1)$. Therefore a localization at finitely many elements is still finitely presented, but this does not have to be true for any multiplicative system.

Definition Let S be an R-algebra. S is called *étale* if it is formally étale and finitely presented.

The following lemma is a direct consequence of our definitions and observations.

Lemma 3.1 *Let S be an R-algebra and S' an S-algebra. Assume that $R \to S$ and $S \to S'$ are (formally) étale, then the induced map $R \to S'$ is also (formally) étale.*

Another important fact which again follows easily from chasing the correct diagram states that the property of being étale is stable under base change. Before proving this statement, we want to recall the definition and add a simple remark: Let S be an R-algebra with $\phi : R \to S$ the corresponding map and let $R \to R'$ be any ring homomorphism. Then the *base change of ϕ by $R \to R'$* is the ring map $R' \to R' \otimes_R S =: S'$.

$$\begin{array}{ccc} S & \longrightarrow & S' = R' \otimes_R S \\ {\scriptstyle\phi}\uparrow & & \uparrow{\scriptstyle\text{base change of }\phi} \\ R & \longrightarrow & R' \end{array}$$

Note that the explicit description of a base change is very natural when a presentation is given: We already saw that S, being an R-algebra, is of the form

$$S \cong R[x_i : i \in \mathcal{I}]/(f_j : j \in \mathcal{J}),$$

for some index sets $\mathcal{I}, \mathcal{J}$ and polynomials $f_j \in R[x_i : i \in \mathcal{I}]$. Then, for the base change one has

$$R' \otimes_R S = R'[x_i : i \in \mathcal{I}]/(f'_j : j \in \mathcal{J}),$$

where each f'_j is the image of f_j under the map $R[x_i : i \in \mathcal{I}] \to R'[x_i : i \in \mathcal{I}]$ induced by the map $R \to R'$. In [Sta20, Tag 05G3] this fact is described as "the key to understanding base change".

Lemma 3.2 *Let $R \to S$ be étale and $R \to R'$ be arbitrary. Then $R' \to R' \otimes_R S$ is étale.*

Corollary 3.3 *Let $R \to S$ and $R \to S'$ be étale. Then $R \to S \otimes_R S'$ is étale.*

The following proposition has a more involved proof.

Proposition 3.4 *Let R be a ring, $S = R[x_1, \dots, x_n]_g/(f_1, \dots, f_n)$ for $g \in R[x_1, \dots, x_n]$ and $f_1, \dots, f_n \in R[x_1, \dots, x_n]_g$. If the image of the Jacobian determinant $\det(\frac{\partial f_j}{\partial x_i})_{1 \le i,j \le n}$ is invertible in S, then S is étale over R.*
Conversely, if $R \to S$ is étale, then there exists a presentation $S = R[x_1, \dots, x_n]/(f_1, \dots, f_n)$ such that the image of $\det(\frac{\partial f_j}{\partial x_i})_{1 \le i,j \le n}$ is invertible in S.

For the proof we refer to standard literature [Ray70], [Mil80, Corollary 3.16], to well-written lecture notes [Mil13, Hoc17] and to the Stacks Project [Sta20, Section 00U0].

Definition A finitely presented R-algebra S is called *standard étale* if it is of the form $S = R[t]_g/(f)$ for some polynomials $f, g \in R[t]$, such that f is monic and its derivative f' is invertible in S.

Note that by Proposition 3.4, it follows that a standard étale algebra is indeed étale.

There exists a structure theorem of étale algebras, making sure that any étale algebra is *locally* standard étale. In [Gro67, p. 120] Grothendieck attributes this fact to Chevalley and so shall we.

Theorem 3.5 (Chevalley) *Let S be a finitely presented R-algebra. Then S is étale over R if and only if for every prime ideal $\mathfrak{q}$ of S with contraction $\mathfrak{p}$ to R there exist $b \in S \backslash \mathfrak{q}$ and $a \in R \backslash \mathfrak{p}$ such that S_b is isomorphic to a standard étale algebra over R_a.*

The proof is an application of Zariski's main theorem, a form of which we already mentioned in Lemma 2.5. Also Nakayama's lemma and the primitive element theorem for separable field extensions play a role in the proof. In his lecture notes [Hoc17] Hochster points out that "additional trickery" is required as well. Therefore the proof is lengthy and technical and shall not be provided here. We refer to [Ray70, pp. 51], [Ive73, pp. 63] [Mil80, Theorem 3.14] as well as [Sta20, Tag 00UE], [Gro67, pp. 120] and [Hoc17, pp. 27].

3.3 Construction of the Henselization

We want to construct the Henselization of a local ring $(R, \mathfrak{m}, K)$ and consequently prove its existence and some desirable properties. First, we define the notion of étale neighborhoods like Milne in [Mil80]:

Definition Let $(R, \mathfrak{m}, K)$ be local. A pair $(S, \mathfrak{q})$ is called an *étale neighborhood* of R if S is an étale R-algebra and $\mathfrak{q}$ is a prime of S lying over $\mathfrak{m}$, such that the induced map between the residue fields $K = R/\mathfrak{m} \to S_\mathfrak{q}/\mathfrak{q}S_\mathfrak{q}$ is an isomorphism.

In order to save notation in our setting, it is more useful to work *locally* and to use the notion of pointed étale extensions, as does Hochster in his lecture notes [Hoc17]:

Definition A local ring T is called *pointed étale extension* of $(R, \mathfrak{m}, K)$ if $T = S_\mathfrak{q}$ for some étale neighborhood $(S, \mathfrak{q})$.

Before stating and proving a theorem which connects étale ring maps and Henselian rings, we need to state the following lemma.

Lemma 3.6 *Assume T and T' are pointed étale extensions of R. Then there is at most one local R-algebra homomorphism from T to T'.*

The proof of this lemma requires the study the multiplication map, given by the linear extension of $\mu : S \otimes_R S \twoheadrightarrow S$ sending $s \otimes s' \mapsto ss'$, and its kernel $\mathfrak{a} := \ker(\mu)$. We omit the details and refer to [Hoc17, pp. 45], [Ive73, p. 71], [Mil80, p. 36] and [Mil13, Section 4]. Now we are ready for the central theorem of this section. Its statement and proof can be found in the references just mentioned, however we shall reprove it again following the notes of Hochster.

Theorem 3.7 *Let $(R, \mathfrak{m}, K)$ a be local ring. The following conditions are equivalent:*

(1) R is Henselian.
(2) If $f \in R[t]$ is a monic polynomial whose reduction mod $\mathfrak{m}$, $\bar{f} \in K[t]$, *has a simple root $\lambda \in K$, then there exists an element $r \in R$ such that $r \equiv \lambda \mod \mathfrak{m}$ and $f(r) = 0$.*
(3) If $R \to T$ is a pointed étale extension, then $R \cong T$.
(4) If $f_1, \dots, f_n \in R[x_1, \dots, x_n]$ are n polynomials in n variables whose images $\overline{f_j} \mod \mathfrak{m}$ vanish simultaneously at $(\lambda_1, \dots, \lambda_n) \in K^n$ and the Jacobian determinant $\det(\frac{\partial f_j}{\partial x_i})$ does not vanish $\mod \mathfrak{m}$ at $x_1 = \lambda_1, \dots, x_n = \lambda_n$, then there are unique elements $r_1, \dots, r_n \in R$ such that for all i, we have $r_i \equiv \lambda_i \mod \mathfrak{m}$ and $f_j(r_1, \dots, r_n) = 0, 1 \leq j \leq n$.

This theorem gives a deep insight into Henselian rings. In particular, the equivalence of conditions (1) and (2) implies that it suffices to lift only simple roots in order to be able to lift coprime factorizations. Applying this for the ring of algebraic power series and Theorem 1.7 we obtain a proof for Theorem 1.8. We see that Henselian rings are connected to the theory of étale ring maps via condition (3). Moreover, (4) is a multidimensional version of Hensel's lemma for n polynomials and n variables; if the f_i's were also allowed to be power series, one would recognize the implicit function theorem. The equivalence (1) $\Leftrightarrow$ (4) states that a ring is Henselian if and only if the algebraic version of this analytic theorem holds in this ring.

Proof We will show that (1) $\Rightarrow$ (2) $\Rightarrow$ (3) $\Rightarrow$ (4) $\Rightarrow$ (1).

(1) $\Rightarrow$ (2): Suppose R is Henselian and we have a monic $f \in R[t]$ such that $\bar{f}$ has a simple root $\lambda \in K$. We may factor $\bar{f}(t) = (t-\lambda)\bar{g}(t)$ for some $\bar{g} \in K[t]$ with $\bar{g}(\lambda) \neq 0$. The polynomials $t-\lambda$ and $\bar{g}(t)$ are relatively prime. Using the assumption we find a lifting of the factorization to $f(t) = (t-r)g(t)$ for some $r \equiv \lambda \mod \mathfrak{m}$ and $g \in R[t]$. Clearly $f(r) = 0$.

(2) $\Rightarrow$ (3): Let $\phi : R \to T$ be a pointed étale extension, so a localization of an étale neighborhood. By Theorem 3.5 it follows that the étale neighborhood is locally standard étale, hence we may write $T \cong (R[t]_g/(f))_\mathfrak{q}$ for a prime ideal $\mathfrak{q} \subseteq R[t]_g/(f)$ lying over $\mathfrak{m}$ and $g, f \in R[t]$ such that f is monic and f' is invertible in T. Denoting by λ the image of $t \in T$ in K, it follows that $\bar{f}(\lambda) = 0$. Moreover, because f' is invertible in T, we must have that $\bar{f}'(\lambda) \neq 0$, hence λ is a simple root of $\bar{f}$. Using (2), we can find an $r \in R$ for which $f(r) = 0$. Therefore there exists an $h \in R[t]$ such that $f(t) = (t-r)h(t)$ and h is invertible in T, because λ is a simple root of $\bar{f}$. It follows that

$$T \cong (R[t]_g/(f))_\mathfrak{q} \cong \left(R[t]_g/((t-r)h(t))\right)_\mathfrak{q} \cong (R[t]_g/(t-r))_\mathfrak{q} \cong R_{\phi^{-1}(\mathfrak{q})}.$$

However, since ϕ and R are local, we have that $T \cong R_{\phi^{-1}(\mathfrak{q})} \cong R$, what was to be shown.

(3) $\Rightarrow$ (4): Assume we have a system of equations $f_1, \ldots, f_n \in R[x_1, \ldots, x_n]$ like in (4) with $(\lambda_1, \ldots, \lambda_n) \in K^n$ solution of $(\bar{f}_1, \ldots, \bar{f}_n) = 0$ and suppose that (3) holds. Let Q be the kernel of $\pi' : R[x_1, \ldots, x_n] \to K$, where we choose π' such that $\pi'(x_i) = \lambda_i$. By Proposition 3.4 and the assumption on the Jacobian of the $f_1, \ldots, f_n$ in (4), we have that $T := R[x_1, \ldots, x_n]_Q/(f_1, \ldots, f_n) \cong (R[x_1, \ldots, x_n]/(f_1, \ldots, f_n))_{\bar{Q}}$ is a pointed étale extension of R. Because of (3) we must have that $R \cong T$. However, solving the equations $f_1, \ldots, f_n$ and lifting the λ_i's is equivalent to giving an R-algebra map $R[x_1, \ldots, x_n]/(f_1, \ldots, f_n) \to R$ such that under the composite $R[x_1, \ldots, x_n]/(f_1, \ldots, f_n) \to R \twoheadrightarrow K$ the elements x_i map to λ_i. This is in turn equivalent to giving a map that sends Q to $\mathfrak{m}$, hence giving a local R-algebra map $T \to R$. But we have that $R \cong T$, hence the local map exists and is unique by Lemma 3.6. It provides us with a unique solution to the equations.

(4) $\Rightarrow$ (1): Let $f = t^n + c_{n-1}t^{n-1} + \cdots + c_1 t + c_0 \in R[t]$ be a monic polynomial of degree n and suppose that we have a factorization $\bar{f} = \bar{g}\bar{h}$ for some monic coprime polynomials $\bar{g}, \bar{h} \in K[t]$ of degrees d and e respectively. Let $\bar{g} = \sum_{i=0}^{d} \alpha_i t^i$ and $\bar{h} = \sum_{i=0}^{e} \beta_i t^i$ for some $\alpha_i, \beta_i \in K$ and $\alpha_d = \beta_e = 1$. We seek a lifting of the factorization to $f = gh$ for monic polynomials $g, h \in R[t]$. Let the coefficients of g and h be unknowns $y_0, \ldots, y_{d-1}$ and $z_0, \ldots, z_{e-1}$, henceforth we want to solve the equation

$$t^n + c_{n-1}t^{n-1} + \cdots + c_1 t + c_0 = (t^d + y_{d-1}t^{d-1} + \cdots + y_1 t + y_0)(t^e + z_{e-1}t^{e-1} + \cdots + z_1 t + z_0),$$

for the unknowns over R such that the residue classes of the polynomials g and h agree with $\bar{g}$ and $\bar{h}$. Comparing coefficients leads to a system of $n = d + e$ polynomial equations in as many variables:

$$\begin{cases} y_0 z_0 & = c_0, \\ y_0 z_1 + y_1 z_0 & = c_1, \\ \vdots & \\ y_{d-1} z_e + y_d z_{e-1} & = c_{n-1}. \end{cases}$$

This system has a solution mod $\mathfrak{m}$ coming from the factorization $\bar{f} = \bar{g}\bar{h}$ given by $\alpha_0, \ldots, \alpha_{d-1}, \beta_0, \ldots, \beta_{e-1} =: (\alpha, \beta)$. In order to use (4) and to lift this solution to R we have to verify that the Jacobian determinant of this system of equations does not vanish, i.e. that the matrix

$$J(y,z) := \begin{pmatrix} z_0 & z_1 & z_2 & \cdots & z_{e-1} & 1 & 0 & \cdots & 0 \\ 0 & z_0 & z_1 & \cdots & z_{e-2} & z_{e-1} & 1 & \cdots & 0 \\ \vdots & & \ddots & & & & \ddots & & \vdots \\ 0 & \cdots & 0 & z_0 & z_1 & \cdots & z_{e-2} & z_{e-1} & 1 \\ y_0 & y_1 & y_2 & \cdots & y_{d-1} & 1 & 0 & \cdots & 0 \\ 0 & y_0 & y_1 & \cdots & y_{d-2} & y_{d-1} & 1 & \cdots & 0 \\ \vdots & & \ddots & & & & \ddots & & \vdots \\ 0 & \cdots & 0 & y_0 & y_1 & \cdots & y_{d-2} & y_{d-1} & 1 \end{pmatrix}$$

is invertible at $(y, z) = (\alpha, \beta)$. However, $J(\alpha, \beta)$ is the (transpose of the) Sylvester matrix of the polynomials $\bar{g}$ and $\bar{h}$. Since the polynomials are relatively prime by assumption, we obtain that $J(\alpha, \beta)$ is invertible. This shows that the assumptions of (4) are satisfied and hence we find a unique solution for the unknowns $y_0, \ldots, y_{d-1}, z_0, \ldots, z_{e-1}$. This gives the unique factorization $f = gh$ we were looking for. □

Having in mind that (1) ⇔ (3) in this theorem, we come back to our goal of constructing the Henselization. We see that it may be a good idea to combine all possible pointed étale extensions of R into one large ring. If we can do this rigorously, then we might argue that this ring does not have any proper pointed étale extensions anymore, which will mean that it will be Henselian. Finally, we might be able to verify the universal property of the Henselization and conclude that we indeed found the correct object. Let us start executing this plan.

Given a local ring R, we wish to define a *set* of pointed étale algebras of R, say $\mathcal{R}$, that contains exactly one representative from each isomorphism class of pointed étale extensions. It is not trivial that $\mathcal{R}$ is a set, since it might turn out "too large".

However, the following result bounds the cardinality of a pointed étale extension from above and allows us to define $\mathcal{R}$ properly.

Lemma 3.8 *Let R be a local ring and T a pointed étale extension. Then T is finite if R is finite. In the other case, the cardinalities of R and T agree.*

Proof By definition, $T = S_{\mathfrak{q}}$ for some prime $\mathfrak{q} \subseteq S$ and an étale R-algebra S. Since S is finitely presented over R, we have $|S| \leq |R|^n$ for some $n \in \mathbb{N}$, where $|\cdot|$ denotes cardinality. The localization is parametrized by pairs in $(S \backslash \mathfrak{q}) \times S$ and therefore $|S_{\mathfrak{q}}| \leq |S|^2$. We have

$$|R| \leq |T| = |S_{\mathfrak{q}}| \leq |R|^{2n},$$

proving the assertion. □

Now, from the axiom of choice, it follows that the *set* $\mathcal{R}$ exists, since it is a subset of the *set* of all ring structures on a set with similar cardinality as R. Let $\mathcal{A}$ be an index set of $\mathcal{R}$, whereby *index set* means that each $i \in \mathcal{A}$ corresponds bijectively to a $T_i \in \mathcal{R}$ and we can write therefore $\mathcal{R} = (T_i)_{i \in \mathcal{A}}$.

Proposition 3.9 *For $i, j \in \mathcal{A}$ define $i \leq j$ if and only if there exists a local R-algebra map $\phi_{i,j} : T_i \to T_j$. Then $(\mathcal{A}, \leq)$ is a directed set.*

Proof Obviously, $\mathcal{A}$ is not empty since $\mathcal{R}$ contains R. Clearly, $\leq$ is reflexive, as one always has the identity map $\mathrm{id} : T_i \to T_i$ for all i. Moreover, if we have $\phi : T_i \to T_j$ and $\psi : T_j \to T_k$ both local R-algebra maps then $\psi \circ \phi : T_i \to T_k$ is a local R-algebra map. This implies that $\leq$ is transitive. Finally, we have to prove that for any two $T_i, T_j \in \mathcal{R}$, there exist $T_k \in \mathcal{R}$ and two local R-algebra maps $T_i \to T_k$ and $T_j \to T_k$. As T_i, T_j are pointed étale, they are localizations of some étale R-algebras S_i, S_j. By Corollary 3.3 we immediately have that $R \to S_i \otimes_R S_j$ is étale. Consider the composite map

$$R \to S_i \otimes_R S_j \twoheadrightarrow K \otimes_K K \xrightarrow{\cong} K,$$

which sends $r \mapsto r \cdot (1_{S_i} \otimes 1_{S_j}) \mapsto \bar{r} \cdot (1_K \otimes 1_K) \mapsto \bar{r}$ and is thus precisely the quotient map $R \twoheadrightarrow R/\mathfrak{m} \cong K$. It follows that by letting Q be the kernel of the map $S_i \otimes_R S_j \twoheadrightarrow K \otimes_K K$, we must have that $R \to (S_i \otimes_R S_j)_Q$ is local. The residue class field of $(S_i \otimes_R S_j)_Q$ is K. Set $T_k = (S_i \otimes_R S_j)_Q$ which is now by definition a pointed étale extension of R and we have maps $T_i \to T_k$ and $T_j \to T_k$. This shows the existence of a $k \in \mathcal{A}$ for given $i, j \in \mathcal{A}$ such that $i, j \leq k$ and finishes the proof. □

This proposition shows that $(\mathcal{R}, \{\phi_{i,j} : T_i \to T_j\}_{i,j \in \mathcal{A}, i \leq j})$ forms a direct system of rings. Note that because of Lemma 3.6, we know that the $\phi_{i,j}$'s are actually unique, justifying that the construction is canonical. The fact that $\mathcal{R}$ together with these maps forms a direct system of rings allows us to define the direct limit:

Definition For a local ring $(R, \mathfrak{m}, K)$ we denote $R^{\mathrm{e}} := \varinjlim_{T \in \mathcal{R}} T$.

Given a local ring R, we combine all pointed étale extensions of it to the ring R^{e} in a rigorous way using the direct limit in the definition above. Therefore, it is natural to expect that R^{e} does not have any proper pointed étale extensions anymore, which means that R^{e} is Henselian by Theorem 3.7. It is also intuitively clear that R^{e} the "smallest" extension of R that admits this property. We prove both statements below using ideas from [Ive73] and [Hoc17].

Lemma 3.10 *Let $(R, \mathfrak{m}, K)$ be a local ring. Then R^e is local with maximal ideal $\mathfrak{m}R$ and residue field K. Moreover, R^e is Henselian.*

Proof Locality, the statement about the maximal ideal and the condition on the residue field follow by construction, since every pointed étale R-algebra T is local with maximal ideal $\mathfrak{m}T$ and residue field K.

By Theorem 3.7 we only have to check the lifting of simple roots in order to verify the Henselian property. Let $f \in R^{\mathrm{e}}[t]$ be monic and $\lambda \in K$ a simple root of $\bar{f} \in K[t]$. Since $R^{\mathrm{e}} = \varinjlim_{T \in \mathcal{R}} T$, there exists some pointed étale R-algebra T such that all coefficients of f lie in T. We define $T' := (T[t]/(f))_{\mathfrak{q}}$, where $\mathfrak{q} := (\bar{t} - \lambda)$. The residue field of T' is K and because λ is a simple root, it follows that f' is invertible in T' and therefore T' is a pointed étale extension of R by Lemma 3.1 and Proposition 3.4. However, f has a root in T' and it lifts λ. This gives rise to an element $r \in R^{\mathrm{e}}$ such that $f(r) = 0$ and $\bar{r} = \lambda$. □

Theorem 3.11 *Let $(R, \mathfrak{m}, K)$ be a local ring. The Henselization of R is given by the direct limit as in Definition 3.3: $R^h = R^e$.*

Proof We will verify the universal property. From the lemma above we already have that R^{e} is local and Henselian. Let $\psi : R \to H$ be a local map from R to a Henselian ring $(H, \mathfrak{m}_H, L)$. To show that this map factors uniquely through R^{e}, it suffices to show that it factors uniquely through every $(T, \mathfrak{q}T, K)$, where $T = S_{\mathfrak{q}}$ is a pointed étale extension of R. Consider the commutative diagram of the base change:

$$\begin{array}{ccc} R & \xrightarrow{\text{étale}} & S \\ \downarrow{\scriptstyle\psi} & & \downarrow \\ H & \longrightarrow & S \otimes_R H \end{array}$$

Since $R \to S$ is étale, we obtain by Lemma 3.2 that $H \to S \otimes_R H$ is also étale. Moreover, there exists a canonical map $S \otimes_R H \to K \otimes_K L \cong L$. Denote its kernel by Q. It follows that $H \to (S \otimes_R H)_Q$ is a localization of an étale extension. Since $L \cong K \otimes_K L$, we obtain that the residue fields agree and hence this extension is pointed étale. But H is Henselian, hence $H \cong (S \otimes_R H)_Q$ by (1) $\Rightarrow$ (3) of Theorem 3.7 and therefore we found a local map $\phi : S \to (S \otimes_R H)_Q \cong H$, the map we were looking for:

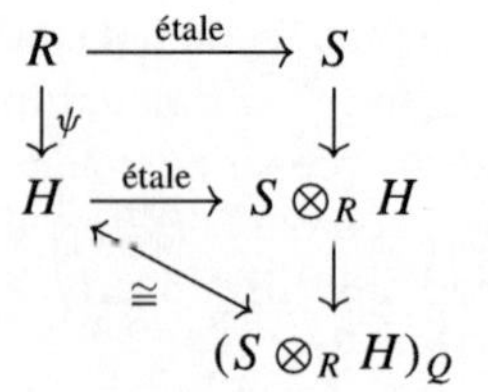

Finally, because H is pointed étale over itself as well as over $(S \otimes_R H)_Q$, we obtain that this map is unique by Lemma 3.6. □

The characterization of the Henselization as a direct limit of pointed étale extensions not only proves its existence, but also led to remarkable mathematical discoveries in this area in the second half of the last century. The theory of approximation rings and consequently the algebraic version of Artin's Approximation [Art69] use exactly this fact (amongst other). This and other related results are contained in the recent survey [Hau17] by Hauser. The Henselization of non-local rings with respect to ideals is investigated in [Gre69]. An exposition of various versions of the Henselian property can be found in [Rib85]. The connection of Henselian rings to rings satisfying Weierstrass preparation theorem was first established by Lafon in 1967 [Laf67]. Probably the most explicit application of the Theorems 3.5 and 3.11 was found by Denef and Lipshitz in 1985 and we shall explain their ideas in the next section.

4 Explicit Implications

4.1 *Proof of Theorem 1.1*

Theorem 4.1 (Denef and Lipshitz) *Let $f \in K\langle x\rangle$ be an algebraic power series. Then there exist an étale-algebraic power series h and polynomials $a_i, b_j \in K[x]$ for $0 \le i \le r$, $0 \le j \le s$, $r, s \in \mathbb{N}$, where $b_0(0) \ne 0$, such that*

$$f = \frac{a_0 + a_1 h + \cdots + a_r h^r}{b_0 + b_1 h + \cdots + b_s h^s}. \tag{5}$$

Proof Set $R := K[x]_{(x)}$. We have seen that $K\langle x\rangle = R^{\mathrm{h}} = \varinjlim_{T\in\mathcal{R}} T$, where the limit is taken over all pointed étale extensions up to isomorphism. It follows that there exists a ring $T \subseteq K\langle x\rangle$ which is a pointed étale extension of R and which contains f. Hence, $T = S_{\mathfrak{q}}$ for an étale R-algebra S and a prime ideal $\mathfrak{q} \subseteq S$ lying over $\mathfrak{m} \subseteq R$. We know furthermore by Theorem 3.5 that S is locally standard étale over R; since R is local, this means that we have an isomorphism

$$\alpha : S_b \overset{\cong}{\to} R[t]_g/(p)$$

for some $b \in S\backslash\mathfrak{q}, g \in R[t]$ and $p \in R[t]$ monic such that its derivative p' is invertible in S_b. Localizing in $\mathfrak{q}$ and $\alpha(\mathfrak{q})$, respectively, yields $T \cong \left(R[t]/(\tilde{P})\right)_{\alpha(\mathfrak{q})}$ for some $\tilde{P} \in R[t]$ such that $\tilde{P}' \notin \alpha(\mathfrak{q})$. We can rephrase the isomorphism above as $T \cong (R[\tilde{h}])_{\alpha(\mathfrak{q})}$, where $\tilde{h} \in \widehat{R} = K[\![x]\!]$ is an algebraic element over R whose minimal polynomial is exactly $\tilde{P}$. Because $\tilde{P}' \notin \tilde{\alpha}(\mathfrak{q})$, we have $\partial_t \tilde{P}(0, \tilde{h}(0)) \neq 0$. Now, any element $f \in (R[\tilde{h}])_{\alpha(\mathfrak{q})} \cong T$ is of the form a/b, for $a, b \in R[\tilde{h}]$ and $b \notin \tilde{\alpha}(\mathfrak{q})$, hence we have

$$f(x) = \frac{\tilde{a}(x, \tilde{h}(x))}{\tilde{b}(x, \tilde{h}(x))},$$

for $\tilde{a}, \tilde{b} \in K[x]_{(x)}[t]$ such that $\tilde{b}(0, \tilde{h}(0)) \neq 0$. Finally, to achieve the condition $h(0) = 0$ as in the definition of étale-algebraic, we define $h(x) = \tilde{h}(x) - \tilde{h}(0)$. It is easy to verify that the derivative of the minimal polynomial $P(x, t)$ of h does not vanish at the origin, $\partial_t P(0, 0) = \partial_t \tilde{P}(0, \tilde{h}(0)) \neq 0$, and that we have again

$$f = \frac{a_0 + a_1 h + \cdots + a_r h^r}{b_0 + b_1 h + \cdots + b_s h^s},$$

for polynomials $a_i, b_j \in K[x]$ such that $b_0(0) = \tilde{b}(0, \tilde{h}(0)) \neq 0$. □

Now we can finally prove Theorem 1.1.

Proof of Theorem 1.1 Using the theorem above it follows that there exist $r, s \in \mathbb{N}$ and $a_i(x), b_j(x) \in K[x]$ for $0 \leq i \leq r, 0 \leq j \leq s$ with $b_0(0) \neq 0$ such that

$$f(x) = \frac{a_0(x) + a_1(x)h(x) + \cdots + a_r(x)h(x)^r}{b_0(x) + b_1(x)h(x) + \cdots + b_s(x)h(x)^s}, \tag{6}$$

where $h(x) \in K\langle x\rangle$ is étale-algebraic. Define

$$W(x, t) := \frac{a_0(x) + a_1(x)t + \cdots + a_r(x)t^r}{b_0(x) + b_1(x)t + \cdots + b_s(x)t^s} \in K[x, t]_{(x,t)},$$

and let

$$R(x, t) := W(xt, t)t \frac{\partial_t P(xt, t)}{P(xt, t)}.$$

Using Lemma 1.2 and the same computation as in Eq. (3) we verify that we found the correct rational function:

$$\mathcal{D}(R(x, t)) = W(x, h(x)) = h(x).$$

□

4.2 Codes of Algebraic Power Series

Finally, we introduce a new result which can be seen as a corollary of Theorem 4.1. First, we remark that the following fact was explained in [AM65, pp. 88] and became later known under the name Artin-Mazur lemma, see [BCR98, AMR92]. In [Ron18, Proposition 9.3] a more general version of the statement, allowing for K to be any complete normal local domain (with appropriate changes to the assumptions), is presented.

Theorem 4.2 (Artin and Mazur) *Let $f \in K\langle x_1, \ldots, x_n\rangle = K\langle x\rangle$ be an algebraic power series with $f(0) = 0$. Then there exist $k \in \mathbb{N}$ and a vector of k polynomials $P(x, y_1, \ldots, y_k) \in K[x][y_1, \ldots, y_k]^k$ with the following properties:*

(1) $P(x, f, h_2, \ldots, h_k) = 0$ for algebraic power series $h_2, \ldots, h_k \in K\langle x\rangle$ with $h_i(0) = 0$ for $i = 2, \ldots, k$.
(2) The Jacobian matrix $J_P(x, y_1, \ldots, y_k)$ of $P(x, y_1, \ldots, y_n)$ with respect to the variables $y_1, \ldots, y_k$ at $x = y = 0$ is invertible: $J_P(0, 0) \in \mathrm{GL}_k(K)$.

In other words, given $f \in K\langle x\rangle$, one can find $k - 1$ algebraic power series $h_2, \ldots, h_k \in K\langle x\rangle$ and a k-dimensional vector of polynomials $P(x, y_1, \ldots, y_k) \in K[x, y]^k$, such that $P(x, f(x), h_2(x), \ldots, h_k(x)) = 0$ and the Jacobian of $P(x, y)$ with respect to y at $x = y = 0$ is invertible. Similarly to Theorem 4.1, this implies that one can repair the problem of an algebraic power series of not being étale-algebraic, now by appending $k - 1$ new power series and considering the k-dimensional analogue of the definition of étale-algebraicity. This polynomial vector $P(x, y) \in K[x, y]^k$ is referred to as *a (mother) code of the algebraic series f* in [ACJH18, Hau17, AMR92]. The authors Alonso, Castro-Jimenez and Hauser of the first reference point out that "The advantage of this code in comparison with taking the minimal polynomial lies in the fact that the latter determines the algebraic series only up to conjugation, so that extra information is necessary to specify the series, typically a sufficiently high truncation of the Taylor expansion. In contrast, the polynomial code determines the series completely and is easy to handle algebraically".

With the help of the theorem of Denef and Lipshitz we can improve on the Artin-Mazur lemma, proving that it is always possible to choose $k = 2$. Note that since Theorem 1.1 is known to hold in more generality, it is also natural that also the general version by Rond can be covered and improved by our approach.

Theorem 4.3 *Let $f \in K\langle x_1, \ldots, x_n\rangle = K\langle x\rangle$ be an algebraic power series with $f(0) = 0$. Then there exists a vector of two polynomials $P(x, y_1, y_2) \in K[x][y_1, y_2]^2$ with the following properties:*

(1) $P(x, f, h) = 0$ for some étale-algebraic power series $h \in K\langle x\rangle$.
(2) The Jacobian matrix $J_P(x, y_1, y_2)$ of $P(x, y_1, y_2)$ with respect to y_1 and y_2 at 0 is invertible: $J_P(0, 0, 0) \in \mathrm{GL}_2(K)$.

Note that in the two-dimensional square matrix $J_P(0, 0, 0)$, the first 0 means setting the variables $x_1, \ldots, x_n$ all to 0 in $J_P(x, y_1, y_2)$, whereas the other two zeros are both one-dimensional and advert to y_1 and y_2.

Proof Let $Q(x, y_1)$ be the minimal polynomial of f. If $\partial_{y_1} Q(0, 0) \neq 0$ then we can simply choose $P(x, y_1, y_2) = (Q(x, y_1), y_2)$ and the assertion follows in this case.

We are left with the more challenging case $\partial_{y_1} Q(0, 0) = 0$. By Theorem 4.1, we may write for some étale-algebraic power series $h \in K\langle x\rangle$

$$f = \frac{a_0 + a_1 h + \cdots + a_r h^r}{b_0 + b_1 h + \cdots + b_s h^s}, \tag{7}$$

for $r, s \in \mathbb{N}$ and $a_i(x), b_j(x) \in K[x]$ with $0 \le i \le r$, $0 \le j \le s$ and $b_0(0) \neq 0$. Define the polynomials

$$T_1(x, y_2) := a_0(x) + a_1(x)y_2 + \cdots + a_r(x)y_2^r,$$
$$T_2(x, y_2) := b_0(x) + b_1(x)y_2 + \cdots + b_s(x)y_2^s,$$

and get the relationship $T_1(x, h(x)) = f(x)T_2(x, h(x))$ from identity (7). Let $S(x, y_2)$ be the minimal polynomial of the étale-algebraic $h(x)$, so that $\partial_{y_2} S(0, 0) \neq 0$. Now we put

$$P(x, y_1, y_2) := \begin{pmatrix} y_1 T_2(x, y_2) - T_1(x, y_2) \\ S(x, y_2) \end{pmatrix}.$$

A simple computation confirms that this choice of P satisfies all required properties:

$$P(x, f(x), h(x)) = 0 \quad \text{and}$$
$$J_P(0, 0, 0) = \begin{pmatrix} T_2(x, y_2) & * \\ 0 & \partial_{y_2} S(x, y_2) \end{pmatrix}\Bigg|_{(0,0,0)} = \begin{pmatrix} T_2(0, 0) & * \\ 0 & \partial_{y_2} S(0, 0) \end{pmatrix}.$$

Clearly, $\det(J_P(0, 0, 0)) = T_2(0, 0)\partial_{y_2} S(0, 0) \neq 0$, because both factors are different from 0. □

Acknowledgments First of all, the author wants to thank Herwig Hauser for the supervision and correction of the master's thesis version of this work. The author is also grateful to Christoper Chiu and Giancarlo Castellano for patiently explaining to him central concepts related to this work, reading its early versions and suggesting improvements. Moreover, we thank the anonymous referee for helpful comments and finally Alin Bostan and Kilian Raschel for organizing the wonderful conference "Transient Transcendence In Transylvania" in 2019.

References

[AB13] B. Adamczewski and J. P. Bell. Diagonalization and rationalization of algebraic Laurent series. *Annales scientifiques de l'École Normale Supérieure*, Ser. 4, 46(6):963–1004, 2013

[ABD19] B. Adamczewski, J. P. Bell, and E. Delaygue. Algebraic independence of G-functions and congruences "à la Lucas". *Ann. Sci. Éc. Norm. Supér. (4)*, 52(3):515–559, 2019

[ACJH18] M. E. Alonso, F. J. Castro-Jiménez, and H. Hauser. Encoding algebraic power series. *Found. Comput. Math.*, 18(3):789–833, 2018
[AM65] M. Artin and B. Mazur. On periodic points. *Ann. of Math. (2)*, 81:82–99, 1965
[AMR92] M. E. Alonso, T. Mora, and M. Raimondo. A computational model for algebraic power series. *J. Pure Appl. Algebra*, 77(1), 1–38, 1992
[Art69] M. Artin. Algebraic approximation of structures over complete local rings. *Inst. Hautes Études Sci. Publ. Math.*, (36):23–58, 1969
[Azu51] G. Azumaya. On maximally central algebras. *Nagoya Math. J.*, 2:119–150, 1951
[BCR98] J. Bochnak, M. Coste, and M.-F. Roy. *Real algebraic geometry*, volume 36 of *Ergebnisse der Mathematik und ihrer Grenzgebiete (3)*. Springer-Verlag, Berlin, 1998. Translated from the 1987 French original, Revised by the authors
[BDS17] A. Bostan, L. Dumont, and B. Salvy. Algebraic diagonals and walks: algorithms, bounds, complexity. *J. Symbolic Comput.*, 83:68–92, 2017
[BLS17] A. Bostan, P. Lairez, and B. Salvy. Multiple binomial sums. *Journal of Symbolic Computation*, 80(2), 351–386, 2017
[Chr15] G. Christol. Diagonals of rational fractions. *Eur. Math. Soc. Newsl.*, (97):37–43, 2015
[Del84] P. Deligne. Intégration sur un cycle évanescent. *Invent. Math.*, 76(1):129–143, 1984
[DL87] J. Denef and L. Lipshitz. Algebraic power series and diagonals. *J. Number Theory*, 26(1), 46–67, 1987
[Dum16] L. Dumont. *Algorithmes rapides pour le calcul symbolique de certaines intégrales de contour à paramètre*. Theses, Université Paris-Saclay, December 2016
[Eis95] D. Eisenbud. *Commutative algebra*, volume 150 of *Graduate Texts in Mathematics*. Springer-Verlag, New York, 1995. With a view toward algebraic geometry
[Fur67] H. Furstenberg. Algebraic functions over finite fields. *J. Algebra*, 7:271–277, 1967
[Gre69] S. Greco. Henselization of a ring with respect to an ideal. *Trans. Amer. Math. Soc.*, 144:43–65, 1969
[Gro67] A. Grothendieck. Éléments de géométrie algébrique. IV. Étude locale des schémas et des morphismes de schémas IV. *Inst. Hautes Études Sci. Publ. Math.*, (32):361, 1967
[Har88] T. Harase. Algebraic elements in formal power series rings. *Israel J. Math.*, 63(3):281–288, 1988
[Hau17] H. Hauser. The classical Artin approximation theorems. *Bull. Amer. Math. Soc. (N.S.)*, 54(4):595–633, 2017
[Hoc17] M. Hochster. Math 615 Lecture Notes, 2017. Available at http://www.math.lsa.umich.edu/~hochster/615W17/615.pdf
[Ive73] B. Iversen. *Generic local structure of the morphisms in commutative algebra*. Lecture Notes in Mathematics, Vol. 310. Springer-Verlag, Berlin-New York, 1973
[KPR75] H. Kurke, G. Pfister, and M. Roczen. *Henselsche Ringe und algebraische Geometrie*. VEB Deutscher Verlag der Wissenschaften, Berlin, 1975. Mathematische Monographien, Band II
[Laf65] J.-P. Lafon. Séries formelles algébriques. *C. R. Acad. Sci. Paris*, 260:3238–3241, 1965
[Laf67] J.-P. Lafon. Anneaux henséliens et théorème de préparation. *C. R. Acad. Sci. Paris Sér. A-B*, 264:A1161–A1162, 1967
[Lan84] S. Lang. *Algebra*. Addison-Wesley Publishing Company, Advanced Book Program, Reading, MA, second edition, 1984
[LT70] F. Lazzeri and A. Tognoli. Alcune proprietà degli spazi algebrici. *Ann. Scuola Norm. Sup. Pisa Cl. Sci. (3)*, 24:597–632, 1970
[Mat80] H. Matsumura. *Commutative algebra*, volume 56 of *Mathematics Lecture Note Series*. Benjamin/Cummings Publishing Co., Inc, Reading, Mass., second edition, 1980
[Mil80] J. S. Milne. *Étale cohomology*, volume 33 of *Princeton Mathematical Series*. Princeton University Press, Princeton, N.J., 1980
[Mil13] J. S. Milne. Lectures on Etale Cohomology (v2.21), 2013. Available at www.jmilne.org/math/
[Nag53] M. Nagata. On the theory of Henselian rings. *Nagoya Math. J.*, 5:45–57, 1953
[Nag54] M. Nagata. On the theory of Henselian rings. II. *Nagoya Math. J.*, 7:1–19, 1954

[Nag59] M. Nagata. On the theory of Henselian rings. III. *Mem. Coll. Sci. Univ. Kyoto Ser. A. Math.*, 32:93–101, 1959

[Nag62] M. Nagata. *Local rings*. Interscience Tracts in Pure and Applied Mathematics, No. 13. Interscience Publishers a division of John Wiley & Sons New York-London, 1962

[Pó22] G. Pólya. Sur les séries entières, dont la somme est une fonction algébrique. *L'Enseign. Math.*, pages 38–47, 1921–1922

[Ray70] M. Raynaud. *Anneaux locaux henséliens*. Lecture Notes in Mathematics, Vol. 169. Springer-Verlag, Berlin-New York, 1970

[Rib85] P. Ribenboim. Equivalent forms of Hensel's lemma. *Exposition. Math.*, 3(1):3–24, 1985

[Ron18] G. Rond. Artin Approximation. *Journal of Singularities*, 17:108–192, 2018. 108 pages

[Rui93] J. M. Ruiz. *The basic theory of power series*. Advanced Lectures in Mathematics. Friedr. Vieweg & Sohn, Braunschweig, 1993

[Saf87] K. V. Safonov. On conditions for the sum of a power series to be algebraic and rational. *Mat. Zametki*, 41(3), 325–332, 457, 1987

[Sta20] T. Stacks Project Authors. *Stacks Project*. https://stacks.math.columbia.edu, 2020

[Str14] A. Straub. Multivariate Apéry numbers and supercongruences of rational functions. *Algebra Number Theory*, 8(8), 1985–2007, 2014

[SW88] H. Sharif and C. F. Woodcock. Algebraic functions over a field of positive characteristic and Hadamard products. *J. London Math. Soc.* (2), 37(3):395–403, 1988

Proof of Chudnovskys' Hypergeometric Series for $1/\pi$ Using Weber Modular Polynomials

Jesús Guillera

Abstract We prove rational alternating Ramanujan-type series of level 1 discovered by the brothers David and Gregory Chudnovsky, by using a method of the author. We have carried out the computations with Maple (a symbolic software for mathematics).

Keywords Hypergeometric series · Ramanujan-type series for $1/\pi$ · Chudnovskys' series for $1/\pi$ · Elliptic modular functions · Weber modular polynomials · Modular equations · Maple program

2010 Mathematics Subject Classification: 33E05 · 33C05 · 33C20 · 11F03

1 Introduction

In his famous paper [12] of 1914 Ramanujan gave a list of 17 extraordinary formulas for the number $1/\pi$, which are of the following form

$$\sum_{n=0}^{\infty} \frac{\left(\frac{1}{2}\right)_n \left(\frac{1}{s}\right)_n \left(1-\frac{1}{s}\right)_n}{(1)_n^3} (a+bn)\, z^n = \frac{1}{\pi}, \qquad (c)_0 = 1, \quad (c)_n = \prod_{j=1}^{n} (c+j-1), \tag{1}$$

where $s \in \{2, 3, 4, 6\}$, and z, b, a are algebraic numbers. Instead of using s to classify them, we will use the level ℓ of the family (the level of the modular forms that parametrize it). It is known that

$$\ell = 4 \sin^2 \frac{\pi}{s}.$$

J. Guillera (✉)
Department of Mathematics, University of Zaragoza, 50009 Zaragoza, Spain
e-mail: jguillera@protonmail.com

A. Bostan and K. Raschel (eds.), *Transcendence in Algebra, Combinatorics, Geometry and Number Theory*, Springer Proceedings in Mathematics & Statistics 373,
https://doi.org/10.1007/978-3-030-84304-5_14

The only formulas of level $\ell = 1$ ($s = 6$) in the list recorded by Ramanujan are

$$\sum_{n=0}^{\infty} \frac{\left(\frac{1}{2}\right)_n \left(\frac{1}{6}\right)_n \left(\frac{5}{6}\right)_n}{(1)_n^3} (11n+1) \left(\frac{4}{125}\right)^n = \frac{5\sqrt{15}}{6\pi}, \tag{2}$$

and

$$\sum_{n=0}^{\infty} \frac{\left(\frac{1}{2}\right)_n \left(\frac{1}{6}\right)_n \left(\frac{5}{6}\right)_n}{(1)_n^3} (133n+8) \left(\frac{4}{85}\right)^{3n} = \frac{85\sqrt{255}}{54\pi}, \tag{3}$$

see [12, eq. 33 and 34]. However the most interesting series in this level are the alternating ones, which were discovered by the brothers David and Gregory Chudnovsky in 1987 [5]. The most impressive is

$$\sum_{n=0}^{\infty} \frac{\left(\frac{1}{2}\right)_n \left(\frac{1}{6}\right)_n \left(\frac{5}{6}\right)_n}{(1)_n^3} (545140134n + 13591409) \left(\frac{-1}{53360}\right)^{3n} = \frac{\sqrt{640320^3}}{12\pi}. \tag{4}$$

The papers [5] and [12] are reprinted in [1]: A book collecting works on the number π.

In this paper we will prove alternating Ramanujan-type series for $1/\pi$ of level 1 discovered by David and Gregory Chudnovsky, by using the formulas obtained by the author in [7]. The fastest of all rational series for $1/\pi$ (not only for level 1) is (4), which provides approximately $\log_{10}(53360^3) \simeq 14.18$ correct digits of π per term. In [8], we applied our method to prove the fastest series of level 3, an alternating one discovered by Chan, Liaw and Tan [3], and in [9] we proved the fastest series due to Ramanujan [12, eq. 44].

2 Elliptic Modular Functions

According to the papers [7] and [8], we let

$$F_\ell(x) = {}_2F_1\left(\begin{matrix} \frac{1}{s}, 1-\frac{1}{s} \\ 1 \end{matrix} \,\middle|\, x \right), \qquad \ell = 4\sin^2\frac{\pi}{s},$$

where $s = 2, 3, 4, 6$, and $\ell = 4, 3, 2, 1$ is the corresponding level. The following functions $x_\ell(q)$ are modular functions (in a wide sense) of levels $\ell = 4, 2, 3$, respectively [6, p. 244 and p. 261], that parametrize x in such a way that $z_\ell(q) = 4x_\ell(q)(1 - x_\ell(q))$ are modular functions, and $F_\ell(x_\ell(q))$ modular forms of weight 2:

$$x_4(q) = 16q \prod_{n=1}^{\infty} \left(\frac{1+q^{2n}}{1+q^{2n-1}} \right)^8, \qquad x_2(q) = \frac{64q}{64q + \prod_{n=1}^{\infty}(1+q^n)^{-24}},$$

and

$$x_3(q) = \frac{27q}{27q + \prod_{n=1}^{\infty}(1+q^n+q^{2n})^{-12}}.$$

It is a well known theorem that all the elliptic modular functions are algebraically related. For example, one has [6, p. 274]:

$$J = \frac{1728}{4x_1(1-x_1)} = \frac{64(1+3x_2)^3}{x_2(1-x_2)^2} = \frac{27(1+8x_3)^3}{x_3(1-x_3)^3} = \frac{16(1+14x_4+x_4^2)^3}{x_4(1-x_4)^4}, \tag{5}$$

where J is the modular invariant:

$$J(q) = q^{-1} + 744 + 196884\,q + 21493760\,q^2 + \cdots, \tag{6}$$

that can be obtained from any of the functions $x_2(q), x_3(q)$ or $x_4(q)$. Hence, we have

$$x_1(q) = \frac{1}{2} - \frac{1}{2}\sqrt{1 - \frac{12^3}{J(q)}} = 432(q - 312q^2 + 87084q^3 - 23067968q^4 + \cdots),$$

We also know that

$$J = \frac{1728}{4x_1(1-x_1)} = \frac{(u^{24}-16)^3}{u^{24}}, \tag{7}$$

where u is related to the Dedekind η function in the following way:

$$u(q) = \frac{\eta^2(q)}{\eta(\sqrt{q})\eta(q^2)}, \quad \eta(q) = q^{1/24}\prod_{n=1}^{\infty}(1-q^n), \tag{8}$$

The importance of seeing this kind of functions as functions of τ with $q = e^{2\pi i\tau}$ and $Im(\tau) > 0$ is well known.

If we let $\beta = x_\ell(q)$, $\alpha = x_\ell(q^d)$, then a modular equation of level ℓ and degree $1/d$ (of β with respect to α) or d (of α with respect to β), is an algebraic relation $A(\alpha, \beta) = 0$. It is important to observe that in [7] we showed that all we need to know, in order to prove the Ramanujan-type series for $1/\pi$, are the modular equations satisfied by $x_\ell(q)$.

3 Weber Modular Equations

Instead of using modular equations in the R. Russel form [4] as we did in [8] and [9], we will use Weber modular equations for proving the Ramanujan-type series of level 1. If we let $\beta = x_1(q)$ and $\alpha = x_1(q^d)$, and $\Phi(u, v)$ is the Weber modular polynomial [13] of degree d, then

$$\alpha(1-\alpha) = \frac{432\, u^{24}}{(u^{24}-16)^3}, \quad \beta(1-\beta) = \frac{432\, v^{24}}{(v^{24}-16)^3}, \qquad \Phi(u, v) = 0,$$

is a modular equation of level 1 and degree d. In this paper we will apply our method to prove the alternating series of level 1 and degrees 5, 7, 11, 17 and 41. Our proofs of $1A5$, $1A11$, $1A17$ and $1A41$ (ℓAd), where A means alternating, are completely analogues, and for proving them we will suitably modify the polynomial $\Phi(u, v)$ into another polynomial $P(u, v)$ in order to have a modular equation of the form

$$\alpha(1-\alpha) = \frac{432\, u^{12}}{(u^{12}-16)^3}, \quad \beta(1-\beta) = \frac{432\, v^{12}}{(v^{12}-16)^3}, \quad P(u, v) = 0,$$

because we have observed that by doing it the computations are simpler. For proving $1A7$ we do not modify the Weber polynomial.

4 The Formulas of Our Method

From a modular equation of level ℓ and degree d (α respect to β), we can derive two real Ramanujan-type series for $1/\pi$:

$$\sum_{n=0}^{\infty} \frac{\left(\frac{1}{2}\right)_n \left(\frac{1}{s}\right)_n \left(1-\frac{1}{s}\right)_n}{(1)_n^3} (a+bn)\, z^n = \frac{1}{\pi}, \qquad \ell = 4 \sin^2 \frac{\pi}{s},$$

one of positive terms $z > 0$ and the other one being an alternating series $z < 0$. In [7] we proved that they correspond respectively to the following sets of formulas:

$$q = e^{-\pi\sqrt{\frac{4d}{\ell}}}, \quad z = 4\alpha_0\beta_0, \quad b = (1-2\alpha_0)\sqrt{\frac{4d}{\ell}}, \quad a = -2\alpha_0\beta_0 \frac{m_0'}{\alpha_0'} \frac{d}{\sqrt{\ell}}, \tag{9}$$

and

$$q = -e^{-\pi\sqrt{\frac{4d}{\ell}-1}}, \quad z = 4\alpha_0\beta_0, \quad b = (1-2\alpha_0)\sqrt{\frac{4d}{\ell}-1}, \quad a = -2\alpha_0\beta_0 \frac{m_0'}{\alpha_0'} \frac{d}{\sqrt{\ell}}, \tag{10}$$

where the multiplier $m(\alpha, \beta)$ is given by the Ramanujan formula:

$$m^2 = \frac{1}{d} \frac{\beta(1-\beta)}{\alpha(1-\alpha)} \frac{\alpha'}{\beta'}, \tag{11}$$

Taking logarithms in (11) and differentiating, we get

$$\frac{m'}{\alpha'} = \frac{m}{2\alpha'} \left(\frac{\beta'}{\beta} - \frac{\beta'}{1-\beta} - \frac{\alpha'}{\alpha} + \frac{\alpha'}{1-\alpha} + \frac{\alpha''}{\alpha'} - \frac{\beta''}{\beta'} \right), \tag{12}$$

From (11) and (12), we obtain the following formulas:

$$\frac{\beta_0'}{\alpha_0'} = \frac{1}{dm_0^2}, \qquad \frac{m_0'}{\alpha_0'} = \frac{1}{2} \left(m_0 + \frac{1}{dm_0} \right) \frac{\alpha_0 - \beta_0}{\alpha_0 \beta_0} + \frac{m_0}{2\alpha_0'} \left(\frac{\alpha_0''}{\alpha_0'} - \frac{\beta_0''}{\beta_0'} \right). \tag{13}$$

Hence for proving a Ramanujan-type series for $1/\pi$ of degree d one only needs to know a modular equation of that degree. When we apply our method we begin making $\beta = 1 - \alpha$ in the modular equation and choose a solution α_0. If with that solution we get $|m_0| \neq 1/\sqrt{d}$ then it is not of degree d and we have to try another solution. A good test to select the correct solution α_0 was explained in [7] and used in [8] and [9].

5 Proofs of Chudnovskys' Series for $1/\pi$

First we will prove the alternating series of degree 17. The proofs of the alternating series of degrees $d = 5, 11, 41$ are completely similar. Finally we will prove the series $1A7$. The paper [14] provides a proof of the Chudnovskys' series by a different method using modular forms. Another type of proof, based on proving that certain parameters are integers, is in [11].

5.1 *Proof of the Formula* **1A17**

We see in the tables of [7] that the alternating Ramanujan-type series for $1/\pi$ of level 1 and degree 17 is

$$\sum_{n=0}^{\infty} \frac{\left(\frac{1}{2}\right)_n \left(\frac{1}{6}\right)_n \left(\frac{5}{6}\right)_n}{(1)_n^3} (261702n + 10177) \left(\frac{-1}{440} \right)^{3n} = \frac{3 \cdot 440^2}{\sqrt{330}\,\pi}. \tag{14}$$

With other methods one needs to use a modular equation of degree $4d - 1 = 67$ to prove it. Here we will show how to prove it from the Weber modular equation of degree 17:

$$\alpha(1-\alpha) = \frac{432\, u^{24}}{(u^{24}-16)^3}, \quad \beta(1-\beta) = \frac{432\, v^{24}}{(v^{24}-16)^3}, \quad \Phi_{17}(u, v) = 0, \tag{15}$$

see $\Phi_{17}(u, v)$ at [13, second web-page]. However, we prefer to transform it in the following way: First write $\Phi_{17}(u, v) = 0$ as

$$Q(u, v) = uvR(u, v),$$

where

$$Q(u, v) = (u^{18} + v^{18}) + 17(u^{16}v^{10} + u^{10}v^{16}) + 119(u^{12}v^6 + u^6v^{12}) + 272(u^8v^2 + u^2v^8),$$
$$R(u, v) = -u^{16}v^{16} - 34(u^{14}v^2 + u^2v^{14}) + 34u^{12}v^{12} + 340u^8v^8 + 544u^4v^4 - 256.$$

Squaring we have $Q^2(u, v) = u^2v^2R^2(u, v)$. Finally, replacing u with $\sqrt{u}$ and v with $\sqrt{v}$ we obtain $Q^2(\sqrt{u}, \sqrt{v}) - uvR^2(\sqrt{u}, \sqrt{v}) = 0$. It is clear that the left hand side is a polynomial $P(u, v)$, and we obtain the following modular equation:

$$\alpha(1-\alpha) = \frac{432\, u^{12}}{(u^{12}-16)^3}, \quad \beta(1-\beta) = \frac{432\, v^{12}}{(v^{12}-16)^3}, \quad P(u, v) = 0. \tag{16}$$

We will use (16) instead of (15) because the calculations are simpler, and we will use Maple (a symbolic software for mathematics) to make those computations.

Our method begins taking $\beta = 1 - \alpha$. Hence, we have to find a solution of the system

$$\frac{u^{12}}{(u^{12}-16)^3} = \frac{v^{12}}{(v^{12}-16)^3}, \qquad P(u, v) = 0. \tag{17}$$

To discover it, we use the following trick. As

$$J = \frac{(u^{12}-16)^3}{u^{12}},$$

and we already know the value $J_0 = -12^3 \cdot 440^3$ of J of the formula that we want to prove, we solve first the easier equation

$$\frac{(u^{12}-16)^3}{u^{12}} = -12^3 \cdot 440^3.$$

The values of u_0 and v_0 will be two solutions of this equation. Thus, we check all possible couples of solutions to see which one satisfies the modular equation. But, as the equation seems still too complicated, we use an experimental mathematics trick. Instead of getting the exact solutions, we solve the equation numerically. In this way, we discover that

$$u_0 \simeq 0.23462350310326835353722795020707017833307475126546452629 5.$$

and

$$v_0 \simeq -1.1173117515551634176768613975103535089166537375632732263 14 - 2.6973894127951947206551153810493379348565675486560901 6393\, i.$$

Then, we guess the exact values of u_0 and v_0 by finding their corresponding minimal polynomial, and for both values we get the same minimal polynomial, namely $x^3 + 2x^2 + 8x - 2$. As

$$x^3 + 2x^2 + 8x - 2 = 0,$$

is a polynomial equation of degree 3, we can get its three exact solutions, and we easily recognize that

$$u_0 = \left(\frac{91}{1200} - \frac{3\sqrt{201}}{400}\right) H^2 + \frac{1}{3} H - \frac{2}{3},$$
$$v_0 = \left\{\left(\frac{3\sqrt{201}}{800} - \frac{91}{2400}\right) H^2 - \frac{1}{6} H - \frac{2}{3}\right\} + \left\{\left(\frac{-9\sqrt{67}}{800} + \frac{91\sqrt{3}}{2400}\right) H^2 - \frac{\sqrt{3}}{6} H\right\} i,$$

where

$$H = \left(91 + 9\sqrt{201}\right)^{\frac{1}{3}}.$$

In fact, the two imaginary solutions are conjugate, and either of them works. Finally to be sure that everything is correct, we substitute these values in the system (17), and operating symbolically with Maple we see that they are correct (Check!). At this point, we are ready to prove the formula (14). Indeed, substituting in (16), we get

$$\alpha_0 = \frac{1}{2} - \frac{651}{193600}\sqrt{22110}, \qquad \beta_0 = \frac{1}{2} + \frac{651}{193600}\sqrt{22110}.$$

Hence

$$z_0 = 4\alpha_0\beta_0 = \frac{-1}{440^3}.$$

Then, from the formula for b in (10), we get

$$b_0 = \frac{43617}{96800}\sqrt{330}.$$

Computing a_0 is more difficult. We choose u as the independent variable, and replace v with $v(u)$ in the modular equation. Differentiating $P(u, v(u)) = 0$ with respect to u we obtain the function $v'(u)$ as a quotient of polynomials in u and $v(u)$. Then differentiating $v'(u)$ we obtain the function $v''(u)$. Once we have these functions, we evaluate them at $u = u_0$ to obtain v_0' and v_0''. Differentiating

$$\alpha(1-\alpha) = \frac{432\, u^{12}}{(u^{12}-16)^3}, \quad \beta(1-\beta) = \frac{432\, v^{12}}{(v^{12}-16)^3}, \tag{18}$$

with respect to u at $u = u_0$, we obtain α'_0 and β'_0. Then, using the first identity in (13), we get

$$m_0 = m(u_0) = \sqrt{\frac{1}{d}\frac{\alpha'_0}{\beta'_0}} = \frac{\sqrt{67}}{34} + \frac{1}{34}i, \quad |m_0| = \frac{1}{\sqrt{17}}.$$

Then, differentiating (18) twice, we obtain α''_0 and β''_0. Finally, from the formula for a in (10) and the formulas (11) and (13), we obtain

$$a_0 = \frac{10177}{580800}\sqrt{330}.$$

5.2 Proof of the Formulas 1*A*5, 1*A*11, 1*A*41 and 1*A*7

The proofs of the formulas 1*A*5, 1*A*11, 1*A*41 are completely similar to the proof of 1*A*17: Modify the Weber polynomial $\Phi_d(u, v)$ in the same way that we have done in the case of degree $d = 17$. For proving 1*A*7 do not modified the Weber polynomial. Then, continue choosing the values of u_0 and v_0 that we indicate below.

5.2.1 Proof of the Formula 1*A*5

Choose

$$u_0 = \left(\frac{-1}{192} + \frac{\sqrt{57}}{64}\right)H^2 - \frac{1}{3}H + \frac{2}{3},$$

$$v_0 = \left\{\left(\frac{-\sqrt{57}}{128} + \frac{1}{384}\right)H^2 + \frac{1}{6}H + \frac{2}{3}\right\} - \left\{\left(\frac{\sqrt{3}}{384} - \frac{\sqrt{171}}{128}\right)H^2 - \frac{\sqrt{3}}{6}H\right\} i,$$

where

$$H = \left(1 + 3\sqrt{57}\right)^{\frac{1}{3}}.$$

The minimal polynomial for u_0 and v_0 is $x^3 - 2x^2 + 4x - 2$. Then follow the steps of Sect. 5.1.

5.2.2 Proof of the Formula 1*A*11

Choose

$$u_0 = \left(\frac{-35}{48} + \frac{\sqrt{129}}{16}\right)H^2 - \frac{1}{3}H + \frac{4}{3},$$

$$v_0 = \left\{\left(\frac{35}{96} - \frac{\sqrt{129}}{32}\right)H^2 + \frac{1}{6}H + \frac{4}{3}\right\} + \left\{\left(\frac{-35\sqrt{3}}{96} + \frac{3\sqrt{43}}{32}\right)H^2 + \frac{\sqrt{3}}{6}H\right\} i,$$

where

$$H = \left(35 + 3\sqrt{129}\right)^{\frac{1}{3}}.$$

The minimal polynomial for u_0 and v_0 is $x^3 - 4x^2 + 4x + 2$. Then follow the steps of Sect. 5.1.

5.2.3 Proof of the Formula $1A41$

Choose

$$u_0 = \left(\frac{-467}{13872} + \frac{11\sqrt{489}}{4624}\right) H^2 - \frac{1}{3}H + \frac{4}{3},$$

$$v_0 = \left\{\left(\frac{467}{27744} - \frac{11\sqrt{489}}{9248}\right) H^2 + \frac{1}{6}H + \frac{4}{3}\right\} + \left\{\left(\frac{467\sqrt{3}}{27744} - \frac{33\sqrt{163}}{9248}\right) H^2 - \frac{\sqrt{3}}{6}H\right\} i,$$

where

$$H = \left(467 + 33\sqrt{489}\right)^{\frac{1}{3}}.$$

The minimal polynomial for u_0 and v_0 is $x^3 - 4x^2 + 28x + 2$. Then follow the steps of Sect. 5.1.

5.2.4 Proof of the Formula $1A7$

Take $P(u, v) = \Phi_7(u, v)$ (that is, do not modify the Weber polynomial), and choose

$$u_0 = \sqrt[6]{2}\left(\sqrt[3]{2} - 1\right)^{\frac{1}{3}},$$

$$v_0 = \sqrt[6]{2}\left(\sqrt[3]{2} - 1\right)^{\frac{1}{3}} \left(\frac{1}{2}(\sqrt[3]{4} + 1) - \frac{\sqrt{3}}{6}(1 + \sqrt[3]{4} + 2\sqrt[3]{2})\, i\right).$$

Then follow the steps of Sect. 5.1.

Acknowledgements I am very grateful to Alin Bostan for giving to me the idea of using the minimal polynomials to simplify the computations of the Maple program [2]. This great idea accelerates the execution giving the outputs in a few seconds, and using only a few megabytes of ram memory, in many nowadays computers.

Appendix: A Maple Program

Below is a Maple program which automatically proves the series $1A5$, $1A11$, $1A17$ and $1A41$. The procedure `chud(·)` does it. For example `chud(41);` automatically proves the Chudnovskys' series (4). The interested reader can copy and paste in Maple, the procedures below from the paper on line. The program is also available at the web-site of the author in a text file [10]. The program run fast: chud(5), chud(11) and chud(17) take around 5 seconds each, and chud(41) approximately 20 seconds, on a modern laptop.

```
fullsimplify:=proc(expression)
combine(evalc(simplify(expand(combine(rationalize(radnormal(
expand(simplify(rationalize(simplify(
combine(radnormal(expand(expression)),radicals))))))),
radicals)))),radicals):
end:

dat5:=proc()
global Q,R,P,H,u0,v0,dec,rulepol: dec:=5:
Q:=(u,v)->u^3+v^3:
R:=(u,v)->-u^2*v^2+4:
P:=(u,v)->Q(u,v)^2-u*v*R(u,v)^2:
H:=(1+3*sqrt(57))^(1/3):
u0:=(-1/192+sqrt(57)/64)*H^2-1/3*H+2/3:
v0:=((-sqrt(57)/128+1/384)*H^2+1/6*H+2/3)-
((sqrt(3)/384-sqrt(171)/128)*H^2-sqrt(3)/6*H)*I:
rulepol:={u^3=2*u^2-4*u+2,v(u)^3=2*v(u)^2-4*v(u)+2}:
end:

dat11:=proc()
global Q,R,P,H,u0,v0,dec,rulepol: dec:=11:
Q:=(u,v)->u^6+v^6:
R:=(u,v)->u^5*v^5-11*u^4*v^4+44*u^3*v^3-88*u^2*v^2+88*u*v-32:
P:=(u,v)->Q(u,v)^2-u*v*R(u,v)^2:
H:=(35+3*sqrt(129))^(1/3):
u0:=(-35/48+sqrt(129)/16)*H^2-1/3*H+4/3:
v0:=((35/96-sqrt(129)/32)*H^2+1/6*H+4/3)+
((-35*sqrt(3)/96+3*sqrt(43)/32)*H^2+sqrt(3)/6*H)*I:
rulepol:={u^3=4*u^2-4*u-2,v(u)^3=4*v(u)^2-4*v(u)-2}:
end:


dat17:=proc() global Q,R,P,H,u0,v0,dec,rulepol: dec:=17:
Q:=(u,v)->(u^9+v^9)+17*(u^8*v^5+u^5*v^8)+
119*(u^6*v^3+u^3*v^6)+272*(u^4*v+u*v^4):
R:=(u,v)->-(u^8*v^8)-34*(u^7*v+u*v^7)+
34*(u^6*v^6)+340*(u^4*v^4)+544*(u^2*v^2)-256:
P:=(u,v)->Q(u,v)^2-(u*v)*R(u,v)^2:
```

```
H:=(91+9*sqrt(201))^(1/3):
u0:=(91/1200-3*sqrt(201)/400)*H^2+1/3*H-2/3:
v0:=((3*sqrt(201)/800-91/2400)*H^2-1/6*H-2/3)+
((-9*sqrt(67)/800+91*sqrt(3)/2400)*H^2-sqrt(3)/6*H)*I:
rulepol:={u^3=-2*u^2-8*u+2,v(u)^3=-2*v(u)^2-8*v(u)+2}:
end:

dat41:=proc() global Q,R,P,H,u0,v0,dec,rulepol: dec:=41:
Q:=(u,v)->u^(21)+v^(21)+943*(u^(20)*v^(5)+u^(5)*v^(20))+
123*(u^(20)*v^(17)+u^(17)*v^(20))+
40713*(u^(19)*v^(10)+u^(10)*v^(19))+72939*(u^18*v^3+u^3*v^18)+
3772*(u^18*v^15+u^15*v^18)+
733531*(u^17*v^8+u^8*v^17)+15088*(u^16*v+u*v^16)+
339111*(u^16*v^13+u^13*v^16)-
6494359*(u^15*v^6+u^6*v^15)+3112310*(u^14*v^11+u^11*v^14)+
11736496*(u^13*v^4+u^4*v^13)
-36004437*(u^12*v^9+u^9*v^12)+10422528*(u^11*v^2+u^2*v^11)+
49796960*(u^10*v^7+u^7*v^10)
+86812416*(u^8*v^5+u^5*v^8)+15450112*(u^6*v^3+u^3*v^6)+
8060928*(u^4*v+u*v^4):
R:=(u,v)->41*(u^(20)*v^(8)+u^(8)*v^(20))-
u^(20)*v^(20)+574*(u^(19)*v+u*v^(19))-
4059*(u^(19)*v^(13)+u^(13)*v^(19))-155554*(u^(18)*v^(6)+
u^(6)*v^(18))+574*u^(18)*v^(18)-
160310*(u^17*v^11+u^11*v^17)+701100*(u^16*v^4+u^4*v^16)-
2050*u^16*v^16-1753160*(u^15*v^9+u^9*v^15)-
2488864*(u^14*v^2+u^2*v^14)-462726*u^14*v^14+
20156994*(u^13*v^7+u^7*v^13)+10496*(u^12+v^12)-
3571756*u^12*v^12-28050560*(u^11*v^5+u^5*v^11)+
45567400*u^10*v^10-41039360*(u^9*v^3+u^3*v^9)-
57148096*u^8*v^8-16625664*(u^7*v+u*v^7)-118457856*u^6*v^6-
8396800*u^4*v^4+37617664*u^2*v^2-1048576:
P:=(u,v)->Q(u,v)^2-(u*v)*R(u,v)^2:
H:=(467+33*sqrt(489))^(1/3):
u0:=(-467/13872+11*sqrt(489)/4624)*H^2-1/3*H+4/3:
v0:=((467/27744-11*sqrt(489)/9248)*H^2+1/6*H+4/3)+
((467*sqrt(3)/27744-33*sqrt(163)/9248)*H^2
-sqrt(3)/6*H)*I:
rulepol:={u^3=4*u^2-28*u-2,v(u)^3=4*v(u)^2-28*v(u)-2}:
end:


chud:=proc(dd)
local T,w0,dv0,alpha0,beta0,z0,dalpha0,dbeta0,m0,ddv0,ddalpha0,
```

```
ddbeta0,b0,a0,J0, atu0,w,y,y0,dv,ddv,dw,ddw,dy,ddy,a1,a2,o:
o:=time():
if dd=5 then dat5() fi: if dd=11 then dat11() fi:
if dd=17 then dat17() fi: if dd=41 then dat41() fi:
atu0:={u=u0,v(u)=v0,diff(v(u),u)=dv0,diff(diff(v(u),u),
u)=ddv0}:
print(Modified-Weber-Polynomial=P(u,v)):
T:=u->P(u,v(u)):
print(u[0]=u0); print(v[0]=v0):
dv:=solve(diff(T(u),u)=0,diff(v(u),u)): ddv:=diff(dv,u):
w:=u^(12)/(u^(12)-16)^3: y:=v(u)^(12)/(v(u)^(12)-16)^3:
dw:=diff(w,u): ddw:=diff(dw,u):
dy:=subs(diff(v(u),u)=dv,diff(y,u)):
ddy:=subs(diff(v(u),u)=dv,diff(dy,u)):
w0:=fullsimplify(subs(atu0,simplify(w,rulepol)));
simplify(expand(P(u0,v0))):
T:=u->P(u,v(u)): print(): print():
alpha0:=solve(al*(1-al)=simplify(432*w0),al)[1]:
beta0:=1-alpha0; print(alpha[0]=alpha0): print(beta[0]=beta0):
z0:=expand(4*alpha0*beta0): print(z[0]=z0): J0:=12^3/z0:
b0:=fullsimplify(sqrt(4*dec-1)*(1-2*alpha0)):
print(b[0]=b0): print(): print():
dv0:=fullsimplify(subs(atu0,simplify(dv,rulepol))):
print(diff(v(u),u)(u[0])=dv0):
dalpha0:=fullsimplify(432*subs(atu0,simplify(dw,rulepol))/
(1-2*alpha0)):
print(diff(alpha(u),u)(u[0])=dalpha0):
dbeta0:=fullsimplify(432*subs(atu0,simplify(dy,rulepol))/
(1-2*beta0)):
print(diff(beta(u),u)(u[0])=dbeta0):
m0:=fullsimplify(sqrt(fullsimplify(1/dec*dalpha0/dbeta0))):
print(m[0]=m0):
ddv0:=fullsimplify(subs(atu0,simplify(ddv,rulepol))):
print(diff(diff(v(u),u),u)(u[0])=ddv0):
ddalpha0:=fullsimplify(2*dalpha0^2+
432*subs(atu0,simplify(ddw,rulepol)))/(1-2*alpha0):
print(diff(diff(alpha(u),u),u)(u[0])=ddalpha0):
ddbeta0:=fullsimplify(2*dbeta0^2+432*fullsimplify(
subs(atu0,simplify(ddy,rulepol))))/(1-2*beta0):
print(diff(diff(beta(u),u),u)(u[0])=ddbeta0):
a1:=fullsimplify(m0/2*(ddalpha0/dalpha0^2-ddbeta0/
(dalpha0*dbeta0))):
a2:=fullsimplify(1/2*(m0+1/(dec*m0))*(alpha0-beta0)/
(alpha0*beta0)):
a0:=expand(-2*alpha0*beta0*dec*(a1+a2)):
```

```
print(z[0]=z0,b[0]=b0,a[0]=a0,J[0]=J0):
print(Sum(((6*n)!/((3*n)!*(n!)^3)*(b0*n+a0)/J0^n,n=0..infinity)=
1/Pi):
print(seconds=time()-o):
end:
```

Explanation of the Program

The program obtains dv as a quotient of polynomials in u and $v(u)$ by differentiating $T(u) = P(u, v(u)) = 0$ with respect to u:

```
dv:=solve(diff(T(u),u)=0,diff(v(u),u));
```

It also obtains ddv, dw, dy, etc., which correspond respectively to the following mathematical functions dv $= dv/du$, ddv $= d^2v/du^2$, etc. Before substituting in dv, ddv, etc., the values at $u = u_0$, it uses the corresponding minimal polynomial for u_0 and v_0, to simplify the expressions (this accelerates the running of the program), and is done with

```
simplify(expression, rulepol);
```

It simplifies completely the expression according to the rule `rulepol`. Then, the substitution of the values at u_0 is made with

```
subs(atu0,expression);
```

Finally, the program simplifies the results. When the command `simplify` is not enough it uses the procedure

```
fullsimplify(expression),
```

which combines all Maple commands defined to simplify radicals. This assures that expansions, rationalization, combination of radicals, etc., are made in order to have a full simplification. This procedure has been achieved splitting all the lines of the procedure chud(·), and running line by line for deducing which sequence of Maple commands for the simplification of the radicals expressions was needed.

References

1. L. Berggren, J. Borwein and P. Borwein, Pi: A source book, Springer-Verlag, New York, Inc., 1997, 2000.

2. A. Bostan, P. Flajolet, B. Salvy and E. Schost, Fast computation of special resultants, J. Symbolic Comput. (1) **41** (2006), 1–29.
3. H.H. Chan, W.-C. Liaw and V. Tan, Ramanujan's class invariant λ_n and a new class of series for $1/\pi$, J. London Math Soc. (2) **64** (2001), 93–106.
4. H.H. Chan and W.-C. Liaw, On Russell-Type Modular Equations, Canad. J. Math. **52**, 31–36 (2000).
5. D. Chudnovsky and G. Chudnovsky, Approximations and complex multiplication according to Ramanujan, in Ramanujan Revisited, G.E. Andrews, R.A. Askey, B.C. Berndt, K.G. Ramanathan, and R.A. Rankin, eds., Academic Press, Boston, 1988, pp. 375–472.
6. S. Cooper, Ramanujan's Theta Functions, (Springer International Publishing, 2017).
7. J. Guillera, A method for proving Ramanujan's series for $1/\pi$, Ramanujan J. **52**, 421–431 (2020).
8. J. Guillera, Proof of a rational Ramanujan-type series. The fastest one in level 3, in Analytic and Combinatorial Number Theory: The Legacy of Ramanujan, in honor of Bruce Berndt in his 80th birthday, G. Andrews, M. Filaseta and Ae Ja Yee (eds.), Int. J. Number Theory, **17**(2), 473–477 (2021)
9. J. Guillera, The fastest series for $1/\pi$ due to Ramanujan. (A complete proof using Maple). Available at (https://arxiv.org/abs/1911.03968).
10. J. Guillera, Maple program for proving automatically the Chudnovskys' series for $1/\pi$, https://anamat.unizar.es/jguillera/other.html (June, 2020).
11. L. Milla, An efficient determination of the coefficients in Chudnovskys' series for $1/\pi$. Ramanujan J., (to appear).
12. S. Ramanujan, Modular equations and approximations to π. Quarterly Journal of Mathematics **45** (1914), 350-372.
13. A. Sutherland, Home page with many useful links at https://math.mit.edu/%7edrew/. List of Weber modular polynomials at https://math.mit.edu/%7edrew/WeberModPolys.html
14. Y. Zhao, Chudnovsky's formula for $1/\pi$ revisited, Available at (https://arxiv.org/abs/1807.10125).

Computing an Order-Complete Basis for $M^\infty(N)$ and Applications

Mark van Hoeij and Cristian-Silviu Radu

Abstract This paper gives a quick way to construct all modular functions for the group $\Gamma_0(N)$ having only a pole at $\tau = i\infty$. We assume that we are given two modular functions for $\Gamma_0(N)$ with poles only at $i\infty$ and coprime pole orders. As an application we obtain two new identities from which one can derive that $p(11n + 6) \equiv 0 \pmod{11}$, where $p(n)$ is the usual partition function.

1 Description of the Problem

For basic notions about modular functions used in this paper we refer to [14]. In this paper we show how to obtain an order-complete basis for $M^\infty(N)$ with an application to the case $N = 11$. We use such a basis to obtain two new Ramanujan type identities for $\sum_{n=0}^{\infty} p(11n+6)q^n$. The reason of constructing order-complete bases is that it is easy to express any modular function in $M^\infty(N)$ in terms of an order-complete basis. Examples of modular functions that have been considered also by other authors have the form:

$$F := q^\alpha \prod_{n=1}^{\infty}(1-q^{a_1 n})^{r_1}\dots(1-q^{a_k n})^{r_k}\sum_{n=0}^{\infty} a(mn+t)q^n.$$

and

$$\sum_{n=0}^{\infty} a(n)q^n = \prod_{n=1}^{\infty}(1-q^{a_1 n})^{s_1}\dots(1-q^{a_k n})^{s_k}.$$

M. van Hoeij—Supported by NSF 1618657.
C.-S. Radu—Supported by grant SFB F50-06 of the Austrian Science Fund (FWF).

M. van Hoeij · C.-S. Radu (✉)
Altenbergerstrasse 69, 4040 Linz, Austria
e-mail: sradu@risc.jku.at

A. Bostan and K. Raschel (eds.), *Transcendence in Algebra, Combinatorics, Geometry and Number Theory*, Springer Proceedings in Mathematics & Statistics 373,
https://doi.org/10.1007/978-3-030-84304-5_15

To prove that $a(mn+t) = 0 \pmod{p}$ for some prime one can express F in terms of a basis, as in this example:

$$q\prod_{n=1}^{\infty}(1-q^{7n})\sum_{n=0}^{\infty}p(7n+5)q^n = 7q\prod_{n=1}^{\infty}\frac{(1-q^{7n})^4}{(1-q^n)^4} + 49q^2\prod_{n=1}^{\infty}\frac{(1-q^{7n})^8}{(1-q^n)^8}.$$

This expression implies $p(7n+5) \equiv 0 \pmod{7}$.

We only need a single function $t := q\prod_{n=1}^{\infty}\frac{(1-q^{7n})^4}{(1-q^n)^4}$ to express any $F \in M^{\infty}(7)$ because $\Gamma_0(7)$ has genus 0. If the genus is greater than 1, more functions are needed to construct a basis. This occurs for example for identities involving $p(11n+6)$ as the genus of $\Gamma_0(11)$ is 1. Bases have also been constructed by other authors [1, 2, 4, 7–9, 11, 12] which use various tricks to produce sufficiently many new modular functions $f_1, f_2, \ldots \in M^{\infty}(N)$ until $\mathbb{C}[f_1, f_2, \ldots]$ becomes equal to $M^{\infty}(N)$. The advantage in our approach is that we only need to find two functions in $t, f \in M^{\infty}(N)$ with coprime orders at ∞ in order to construct an order-complete basis. In general $\mathbb{C}[t, f]$ will be a proper subset of $M^{\infty}(N)$, but instead of searching for more modular functions as in prior work, we fill this gap by using a *normalized integral basis*.

Let t and f be modular functions for the group $\Gamma_0(N)$ with poles only at $\tau = i\infty$, in other words, let $t, f \in M^{\infty}(N)$. Suppose that the pole orders are n and m respectively, and that $\gcd(n, m) = 1$; such functions always exist [13, Example 2.3]. Then there exists an irreducible polynomial $p = p(x, y) \in \mathbb{C}[x, y]$ with $p(t, f) = 0$, $\deg_x(p) = m$, and $\deg_y(p) = n$ by [21, Lemma 1]. One can compute p from the q-expansions of t and f by making an Ansatz for the unknown coefficients of p and solving a system of equations where each equation is a coefficient in the q-expansion of $p(t, f)$. We use p to compute in the function field $\mathbb{C}(t, f) \cong \mathbb{C}(x)[y]/(p)$.

The function field $\mathbb{C}(t, f)$ contains $M^{\infty}(N)$ see [13, Prop 4.3], here $M^{\infty}(N)$ is the set of all modular functions for the group $\Gamma_0(N)$ with no poles other than $i\infty$. Obtaining $M^{\infty}(N)$ is equivalent to finding all modular functions $h \in \mathbb{C}(t, f)$ that are *integral* over $\mathbb{C}[t]$ (which means there is a *monic* polynomial $g(X) \in \mathbb{C}[t][X]$ for which $g(h) = 0$). Thus, one starts by computing an *integral basis*, which is a basis $b_1, \ldots, b_n \in \mathbb{C}(t, f)$ of the $\mathbb{C}[t]$-module of all $h \in \mathbb{C}(t, f)$ that are integral over $\mathbb{C}[t]$. There are several algorithms to compute an integral basis [5, 20] and implementations in several computer algebra systems. Then every $h \in M^{\infty}(N)$ can be written as $h = p_1(t)b_1 + \cdots + p_n(t)b_n$ for some polynomials $p_1, \ldots, p_n$.

Given the q-expansions of h and $b_1, \ldots, b_n$ the algorithm described in [16, Alg. MW], illustrated in Example 1.1 below, can find $p_1, \ldots, p_n$ provided that the integral basis $b_1, \ldots, b_n$ is *order-complete*, which means: $\mathrm{ord}_{i\infty}(b_1) < \mathrm{ord}_{i\infty}(b_2) < \cdots < \mathrm{ord}_{i\infty}(b_n)$ and for each $i \neq j$, $\mathrm{ord}_{i\infty}(b_i) \not\equiv \mathrm{ord}_{i\infty}(b_j) \pmod{\mathrm{ord}_{i\infty}(t)}$.

Given an integral basis $b_1, \ldots, b_n$, an intermediate step towards constructing an order-complete basis is *normalization at infinity*, described Trager's PhD thesis [19, Chapter 2, Section 3], as well as in Sect. 1.2 below.

Set $q := q(\tau) = e^{2\pi i\tau}$ and consider the Atkin modular functions $G_2(\tau), G_3(\tau) \in M^{\infty}(11)$ see [1, p. 29]:

$$G_2(\tau) = q^{-2} + 2q^{-1} - 12 + 5q + 8q^2 + q^3 + 7q^4 - 11q^5 + 10q^6 - 12q^7 + \cdots$$

and

$$G_3(\tau) = q^{-3} - 3q^{-2} - 5q^{-1} + 24 - 13q - 22q^2 + 13q^3 - 5q^4 + 51q^5 + \cdots$$

Then we have

$$G_3^2 - G_2^3 + 12G_3G_2 - 11G_2^2 + 121G_3 = 0.$$

Example 1.1 Set $t := G_2, b_1 := 1$ and $b_2 := G_3$. Then $\{b_1, b_2\}$ is an order-complete integral basis of $M_\infty(11)$. Every $f \in M_\infty(11)$ can be expressed in the form $f = p_1(t)b_1 + p_2(t)b_2$. This is done by increasing the order (i.e. decreasing the pole order) of f at $i\infty$. If the $\text{ord}_{i\infty}(f)$ is even, then $\text{ord}_{i\infty}(f - cb_1t^k) > \text{ord}_{i\infty}(f)$ for appropriate $c \in \mathbb{C}$ and $k \in \mathbb{N}$, if $\text{ord}_{i\infty}(f)$ is odd, then $\text{ord}_{i\infty}(f - cb_2t^k) > \text{ord}_{i\infty}(f)$ for appropriate $c \in \mathbb{C}$ and $k \in \mathbb{N}$. This strategy increases the order of f step by step until f vanishes, obtaining $p_1(t), p_2(t)$ by keeping track of each cb_it^k.

Example 1.2 Set $t := G_2$, $b_1 := 1$ and $b_2 := G_3 + t^2$. This basis is not order-complete, however, it is normalized, see the Appendix. The process in the previous example no longer works because if f has an odd pole order, then there is no i for which b_it^k has the same pole order as f. So we need a method to turn a "bad" basis b_1, b_2 (not order-complete) into a "good" basis like the one in Example 1.1.

Example 1.3 Set $t := G_2, b_1 := 1$ and $b_2 := G_3 + t^u$ for $u \geq 3$. Then $\{b_1, b_2\}$ is an integral basis, however it is neither order-complete nor normalized, see the Appendix.

1.1 Notations

$K = \mathbb{C}(x)[y]/(p)$ where $p \in \mathbb{C}[x, y]$ is irreducible.
O_K is the ring of all elements of K that are integral over $\mathbb{C}[x]$.
R_∞ is the ring of all $h \in \mathbb{C}(x)$ that have no pole at $x = \infty$.
O_∞ is ring of all elements of K that are integral over R_∞.

To compute a basis of O_∞ as an R_∞-module, first substitute $x \mapsto 1/\tilde{x}$, then compute a *local integral basis* at $\tilde{x} = 0$ (most integral basis implementations allow the option of computing a local integral basis). After that, replace $\tilde{x}$ by $1/x$.

1.2 Normalize an Integral Basis at Infinity

The process of normalizing an integral basis at infinity was introduced in [19] in order to compute a Riemann-Roch space that was needed for integrating algebraic functions. For completeness we will describe the algorithm here. For technical properties see the Appendix.

Algorithm: Normalize an integral basis at infinity.

(1) Let $b_1, \ldots, b_n$ be a basis of O_K as $\mathbb{C}[x]$-module.
(2) Let $b'_1, \ldots, b'_n$ be a basis of O_∞ as R_∞-module.
(3) Write $b_i = \sum_{j=1}^n r_{ij} b'_i$ with $r_{ij} \in \mathbb{C}(x)$.
(4) Let $D \in \mathbb{C}[x]$ be a non-zero polynomial for which $a_{ij} := Dr_{ij} \in \mathbb{C}[x]$ for all i, j. Now $Db_i = \sum_{j=1}^n a_{ij} b'_i$.
(5) For each $i \in \{1, \ldots, n\}$, let m_i be the maximum of the degrees of $a_{i1}, \ldots, a_{in}$. Now let $V_i \in \mathbb{C}^n$ be the vector whose j'th entry is the x^{m_i}-coefficient of a_{ij}. Let $d_i := m_i - \deg_x(D)$.
(6) If $V_1, \ldots, V_n$ are linearly independent, then return $b_1, \ldots, b_n$ and $d_1, \ldots, d_n$ and stop.
Otherwise, take $c_1, \ldots, c_n \in \mathbb{C}$, not all 0, for which $c_1 V_1 + \cdots c_n V_n = 0$.
(7) Among those $i \in \{1, \ldots, n\}$ for which $c_i \neq 0$, choose one for which d_i is maximal. For this i, do the following
    (a) Replace b_i by $\sum_{k=1}^n c_k x^{d_i - d_k} b_k$.
    (b) Replace a_{ij} by $\sum_{k=1}^n c_k x^{d_i - d_k} a_{kj}$ for all $j \in \{1, \ldots, n\}$.
(8) Go back to step 5.

The $b_1, \ldots, b_n$ remain a basis of O_K throughout the algorithm because the new b_i in step 7a can be written as a nonzero constant times the old b_i plus a $\mathbb{C}[x]$-linear combination of the b_j, $j \neq i$. When we go back to step 5 the non-negative integer d_i decreases while the d_j, $j \neq i$ stay the same. Hence the algorithm must terminate.

Let $b_1, \ldots, b_n$ and $d_1, \ldots, d_n$ be the output of the algorithm. By construction, the number d_i in the algorithm is the smallest integer for which $b_i \in x^{d_i} O_\infty$. If $\beta \in O_K$ with $\beta \neq 0$ then we can write $\beta = c_1 b_1 + \cdots + c_n b_n$ for some $c_1, \ldots, c_n \in \mathbb{C}[x]$. Denote d_β as the maximum of $\deg_x(c_j) + d_j$ taken over all j for which $c_j \neq 0$. Then $\beta \in x^{d_\beta} O_\infty$ by construction. Since the vectors $V_1, \ldots, V_n$ in the algorithm are linearly independent when the algorithm terminates, there can not be any cancellation, which means that d_β is the smallest integer for which $\beta \in x^{d_\beta} O_\infty$. Because of this, we get the following:
If d is a positive integer, then the set $B_d := \{x^j b_i \mid 0 \leq j \leq d - d_i,\ 1 \leq i \leq n\}$ is a basis of $O_K \bigcap x^d O_\infty$ as $\mathbb{C}$-vector space.

Note that B_d is a basis of the Riemann-Roch space of the pole-divisor of x^d. So computing B_d can be interpreted as (i): a direct application of a normalized integral basis, or (ii): a special case of algorithms [3, 6] for Riemann-Roch spaces. The two interpretations are equivalent because the first step in computing Riemann-Roch spaces is to compute a normalized integral basis.

We can take q-expansions for each of the elements of B_d, and then make a change of basis so that the new basis B_d^{REF} will have q-expansions in Reduced Echelon Form. This means that if $b \in B_d^{\mathrm{REF}}$ and $b = a_r q^r + a_{r+1} q^{r+1} + \cdots$ with $a_r \neq 0$ then $a_r = 1$ and all other basis elements have a zero coefficient at q^r. Then B_d^{REF}, for suitable d, is an order-complete basis. For an implementation and two examples see: www.math.fsu.edu/~hoeij/files/OrderComplete.

1.3 Background on Modular Functions

Following [14] we define here a modular function for the group $\Gamma_0(N)$, for N a positive integer.

$$\Gamma_0(N) := \left\{ \begin{pmatrix} a & b \\ c & d \end{pmatrix} \in \mathrm{SL}_2(\mathbb{Z}) : N|c \right\}$$

and

$$\mathrm{SL}_2(\mathbb{Z}) := \left\{ \begin{pmatrix} a & b \\ c & d \end{pmatrix} : a, b, c, d \in \mathbb{Z}, ad - bc = 1 \right\}.$$

A holomorphic function g defined on $\mathbb{H}$ is called a modular function for $\Gamma_0(N)$ if for all $\begin{pmatrix} a & b \\ c & d \end{pmatrix} \in \mathrm{SL}_2(\mathbb{Z})$

$$g\left(\frac{a\tau + b}{c\tau + d}\right) = g(\tau), \quad \tau \in \mathbb{H}, \tag{1}$$

and if for $\gamma = \begin{pmatrix} a & b \\ c & d \end{pmatrix} \in \mathrm{SL}_2(\mathbb{Z})$ there exists an expansion of the form

$$g(\tau) = \sum_{n=m(\gamma)}^{\infty} c_n(\gamma) e^{\frac{2\pi i n \gamma^{-1}\tau}{w_\gamma}}, \tag{2}$$

for all $\tau \in \mathbb{H}$ sufficiently close to $\frac{a}{c} \in \mathbb{Q} \cup \{\infty\}$. Here $\frac{1}{0}$ should be interpreted as ∞ and

$$w_\gamma := \min\left\{ h \in \mathbb{N}\backslash\{0\} : \begin{pmatrix} 1 & h \\ 0 & 1 \end{pmatrix} \in \gamma^{-1}\Gamma_0(N)\gamma \right\}.$$

We define

$$M^\infty(N) := \left\{ f : \begin{array}{c} f \text{ is a modular function for } \Gamma_0(N), \\ \text{for all } \gamma \notin \Gamma_0(N), m(\gamma) \text{ in (2) is nonnegative} \end{array} \right\}.$$

2 New Identities

We will give two identities of Ramanujan type found using our algorithm (the second one is only on our website). Let $p(n)$ be the partition function. Define

$$t := q^{-5} \prod_{n=1}^{\infty} \left(\frac{1 - q^n}{1 - q^{11n}} \right)^{12} = q^{-5} - 12q^{-4} + 54q^{-3} - 88q^{-2} - 99q^{-1} + 540 - 418q + \cdots .$$

and

$$h := t \cdot q \prod_{k=1}^{\infty}(1-q^{11k}) \sum_{n=0}^{\infty} p(11n+6)q^n = 11q^{-4} + 165q^{-3} + 748q^{-2} + 1639q^{-1} + 3553 + \cdots$$

and

$$f := (dt/dq) \prod_{n=1}^{\infty}(1-q^n)^{-2}(1-q^{11n})^{-2} = -5q^{-6} + 38q^{-5} - 91q^{-4} + 42q^{-3} + 21q^{-2} + \cdots .$$

The goal is to prove

$$h = t \cdot q \prod_{k=1}^{\infty}(1-q^{11k}) \sum_{n=0}^{\infty} p(11n+6)q^n = 11^4 + 55t\left(5\frac{t-11^3}{f+47t} - \frac{2(71t+3f)(t+11^3)}{f^2+89ft+1424t^2}\right). \tag{3}$$

Both h and t are modular functions in $M^{\infty}(11)$, see [14, Lemma 3.1]. To prove that f is in $M^{\infty}(11)$ as well, first note that by [10, Prop. 3.1.1]

$$b(\tau) := q \prod_{n=1}^{\infty}(1-q^n)^2(1-q^{11n})^2, \quad q = e^{2\pi i \tau}$$

satisfies

$$b\left(\frac{a\tau+b}{c\tau+d}\right) = (c\tau+d)^2 b(\tau) \tag{4}$$

for all $\begin{pmatrix} a & b \\ c & d \end{pmatrix} \in \Gamma_0(11)$. Since $t \in M^{\infty}(11)$, we have

$$t\left(\frac{a\tau+b}{c\tau+d}\right) = t(\tau).$$

The derivative with respect to τ is:

$$(c\tau+d)^{-2} t'\left(\frac{a\tau+b}{c\tau+d}\right) = t'(\tau) \tag{5}$$

Multiplying (5) by $(c\tau+d)^2$ and dividing by (4) gives

$$(t'/b)\left(\frac{a\tau+b}{c\tau+d}\right) = (t'/b)(\tau).$$

Since $\frac{d}{d\tau} = 2\pi i q \frac{d}{dq}$, it follows that $t'/b = 2\pi i f$. Therefore

$$f\left(\frac{a\tau+b}{c\tau+d}\right) = f(\tau)$$

for all $\begin{pmatrix} a & b \\ c & d \end{pmatrix} \in \Gamma_0(11)$. Furthermore, since $b(\tau)$ has no zeros in the upper half plane and $t(\tau)$ is holomorphic in the upper half plane it follows that f is holomorphic in the upper half plane. Hence condition (1) of being a modular function for $\Gamma_0(11)$ is satisfied. The condition (2) is equivalent to showing that for $\gamma = \begin{pmatrix} a & b \\ c & d \end{pmatrix} \in \mathrm{SL}_2(\mathbb{Z})$ we have an expansion of the form

$$f\left(\frac{a\tau + b}{c\tau + d}\right) = \sum_{n=m(\gamma)}^{\infty} a_{\gamma}(n) q^{\frac{\gcd(c^2,n)n}{N}}. \tag{6}$$

As seen in [14], if this property hold for $\begin{pmatrix} a & b \\ c & d \end{pmatrix}$, then it also holds for $\begin{pmatrix} a' & b' \\ c' & d' \end{pmatrix}$, if there exists $\begin{pmatrix} A & B \\ C & D \end{pmatrix} \in \Gamma_0(11)$ such that $\frac{A\frac{a}{c}+B}{C\frac{a}{c}+D} = \frac{a'}{c'}$. So we need to find representatives of the orbits of the action of $\Gamma_0(11)$ on $\mathbb{Q} \cup \{i\infty\}$, that is, the cusps of $\Gamma_0(N)$. From [17] we find that these representatives are 0 and $i\infty$. Then it suffices to show (6) for two cases: $\begin{pmatrix} 1 & 0 \\ 0 & 1 \end{pmatrix}$ and $\begin{pmatrix} 0 & -1 \\ 1 & 0 \end{pmatrix}$. The first case holds because f is a q-series. For the second case we need to show that $f(-1/\tau)$ is a Laurent series in $q^{1/11}$ with finite principal part. By [15] we have

$$\eta(-1/\tau) = (-i\tau)^{1/2}\eta(\tau)$$

where

$$\eta(\tau) := e^{\frac{\pi i \tau}{12}} \prod_{n=1}^{\infty}(1 - q^n).$$

This implies

$$t(-1/\tau) = t^{-1}\left(\frac{\tau}{11}\right) \tag{7}$$

and

$$b(-1/\tau) = -\frac{1}{11} b\left(\frac{\tau}{11}\right)\tau^2.$$

The derivative of (7) is

$$\tau^{-2} t'(-1/\tau) = \frac{1}{11} t^{-2}\left(\frac{\tau}{11}\right) t'\left(\frac{\tau}{11}\right)$$

which is equivalent to

$$t'(-1/\tau) = \frac{1}{11}\tau^2 t^{-2}\left(\frac{\tau}{11}\right) t'\left(\frac{\tau}{11}\right).$$

This implies

$$(t'/b)(-1/\tau) = -(t'/b)(\frac{\tau}{11})t^{-2}(\frac{\tau}{11}).$$

Hence

$$f(-1/\tau) = -f(\tau/11)t^{-2}(\tau/11) = 5q^{4/11} + O(q^{5/11}).$$

So the last condition for f being a modular function for $\Gamma_0(11)$ is verified. In order for f to be in $M^\infty(11)$ we need the order of f to be nonnegative at all cusps except $i\infty$. That only leaves the cusp 0 where the order is 4. This shows $f \in M^\infty(11)$.

We want to express h as an element of $\mathbb{C}(t, f)$. The pole orders of t and f are 5 and 6 so $p(x, y) = \sum_{i=0}^{6}\sum_{j=0}^{5} a_{ij}x^i y^j$ is an Ansatz for the algebraic relation $p(t, f) = 0$. Solving linear equations coming from q-expansions gives

$$p(x, y) = y^5 + 170xy^4 + 9345x^2y^3 + 167320x^3y^2 + (5^5x^2 - 7903458x + 5^5 11^6)x^4.$$

We use $p(x, y)$ to compute in $\mathbb{C}(t, f) \cong \mathbb{C}(x)[y]/(p)$. We compute B_d^{REF} from the previous section with $d = 1$ and obtain b_0, b_2, b_3, b_4, b_5 where $b_0 = 1$ and $b_i = q^{-i} + c_i q^{-1} + O(q^1)$ for $i = 2, \ldots, 5$ for some constants c_i. Since h has a pole of order 4, we can write it as a linear combination of b_0, b_2, b_3, b_4. We have $b_0 = 1$ and

$$\begin{aligned}
b_2 &= 12 + \tfrac{5t}{22}\left(\tfrac{t-11^3}{f+47t} - \tfrac{(42t+f)(t+11^3)}{f^2+89ft+1424t^2}\right) &&= q^{-2} + 2q^{-1} + 5q + 8q^2 + O(q^3)\\
b_3 &= 12 + \tfrac{5t}{22}\left(3\tfrac{t-11^3}{f+47t} - \tfrac{(16t+3f)(t+11^3)}{f^2+89ft+1424t^2}\right) &&= q^{-3} + q^{-1} + 2q + 2q^2 + O(q^3)\\
b_4 &= 12 + \tfrac{5t}{22}\left(-3\tfrac{t-11^3}{f+47t} - \tfrac{(28t+19f)(t+11^3)}{f^2+89ft+1424t^2}\right) &&= q^{-4} - 2q^{-1} + 6q + 3q^2 + O(q^3).
\end{aligned}$$

Like in [16, Alg. MW], we use

$$h = 11q^{-4} + 165q^{-3} + 748q^{-2} + 1639q^{-1} + 3553 + O(q)$$

to find

$$h - 11b_4 - 165b_3 - 748b_2 - 3553b_0 = O(q).$$

This expression in $M^\infty(11)$ has no poles and a root at $\tau = i\infty$ (at $q = 0$) hence it is the zero function. Therefore

$$h = 11b_4 + 165b_3 + 748b_2 + 3553b_0.$$

Replacing b_0, b_2, b_3, b_4 with their corresponding expressions in terms of t and f gives (3).

This implies $p(11n + 6) \equiv 0 \pmod{11}$. Other expressions for h that prove this congruence were already in [1, 9], however, our expression in terms of t, f is novel.

For our second example, take t and h be as before and let

$$E_4 := 1 + 240 \sum_{n=1}^{\infty} \frac{n^3 q^n}{1 - q^n}$$

be the usual Eisenstein series. Let

$$\Delta := q \prod_{n=1}^{\infty} (1 - q^n)^{24}.$$

Let $J := E_4^3/\Delta = q^{-1} + \cdots$ and

$$f := J t^3.$$

Our goal is to find an identity of the form

$$h = p(f, t),$$

where $p(X, Y) \in \mathbb{Q}(X, Y)$.

Next we show that $f \in M^\infty(11)$. From the last chapter of [18] we find

$$J\left(\frac{a\tau + b}{c\tau + d}\right) = J(\tau)$$

for all $\begin{pmatrix} a & b \\ c & d \end{pmatrix} \in \mathrm{SL}_2(\mathbb{Z})$. Since $\mathrm{SL}_2(\mathbb{Z})$ has only one cusp, $i\infty$, and since J is a q-series it follows that J is a modular function on $\mathrm{SL}_2(\mathbb{Z})$ and thus on $\Gamma_0(11)$.

Since $t(\tau)$ is already a modular function on $\Gamma_0(11)$, it follows that f is a modular function on $\Gamma_0(11)$. To show that f is in $M^\infty(11)$ it suffices to show that the order of f at the cusp 0 is nonnegative. Since $J(-1/\tau) = (q^{-1/11})^{11} + O(1)$ the order of J at 0 is -11. The order of t at 0 is 5, so the order of f at the cusp 0 is $-11 + 3 \cdot 5 = 4 \geq 0$. This shows $f \in M^\infty(11)$.

The only pole of f is at $i\infty$, it has order 16. We compute the algebraic relation $p(t, f) = 0$ with the Ansatz method, and use p to compute B_d^{REF}. Then we express h in terms of the t and the new f. This relation, and the Maple file that computes it, are given at www.math.fsu.edu/~hoeij/files/OrderComplete.

3 Appendix

Definition 3.1 *An integral* $\mathbb{C}[t]$*-basis* $b_1, \ldots, b_n \in M^\infty(N)$ *is* normalized *if for any* $h = p_1(t)b_1 + \cdots + p_n(t)b_n$*, we have for all* i: $\mathrm{ord}_{i\infty}(p_i(t)b_i) - v_h \,\mathrm{ord}_{i\infty}(t) \geq 0$*, or equivalently* $\mathrm{ord}_{i\infty}(p_i(t)b_i/t^{v_h}) \geq 0$*, where* $v_h := \lceil \mathrm{ord}_{i\infty}(h)/\,\mathrm{ord}_{i\infty}(t) \rceil$.

Remark 3.2 Note that $\mathrm{ord}_{i\infty}(t) < 0$. An order-complete integral basis is also a normalized integral basis.

Proof that the basis of Example 1.2 *is normalized:* Let $\tilde{b}_1 := b_1$ and $\tilde{b}_2 := b_2 - t^2$. Then $\tilde{b}_1, \tilde{b}_2$ is just the basis from Example 1.1.

Now take $f = p_1(t)\tilde{b}_1 + p_2(t)\tilde{b}_2$, then

$$\mathrm{ord}_{i\infty}(f) = \min(\mathrm{ord}_{i\infty}(p_1(t)\tilde{b}_1), \mathrm{ord}_{i\infty}(p_2(t)\tilde{b}_2)).$$

Hence either $\mathrm{ord}_{i\infty}(f) = \mathrm{ord}_{i\infty}(p_1(t)\tilde{b}_1)$ or $\mathrm{ord}_{i\infty}(f) = \mathrm{ord}_{i\infty}(p_1(t)\tilde{b}_2)$. Assume $\mathrm{ord}_{i\infty}(f) = \mathrm{ord}_{i\infty}(p_1(t)\tilde{b}_1)$. Let $v_f := \lceil \mathrm{ord}_{i\infty}(f)/\mathrm{ord}_{i\infty}(t)\rceil$. Now $f = (p_1(t) - p_2(t)t^2)b_1 + p_2(t)b_2$. Hence by Definition 3.1 we have to prove that $\mathrm{ord}_{i\infty}(p_1(t) - p_2(t)t^2) \geq v_f \,\mathrm{ord}_{i\infty}(t)$ and $\mathrm{ord}_{i\infty}(p_2(t)b_2) \geq v_f \,\mathrm{ord}_{i\infty}(t)$. Since the order of f is even by assumption it follows that $v_f = \mathrm{ord}_{i\infty}(f)/\mathrm{ord}_{i\infty}(t) = \deg(p_1)$. Since

$$\begin{aligned} &-2\deg(p_1(t)b_1) = -2\deg(p_1(t)) \\ &= \mathrm{ord}_{i\infty}(p_1(t)) < \mathrm{ord}_{i\infty}(p_2(t)b_2) \\ &= \mathrm{ord}_{i\infty}(p_2(t)) - 3 = -2\deg(p_2(t)) - 3, \end{aligned}$$

it follows that

$$\deg(p_1) \geq \deg(p_2) + \lceil 3/2 \rceil. \tag{8}$$

Hence $-2\deg(p_1(t) - p_2(t)t^2) = \mathrm{ord}_{i\infty}(p_1(t) - p_2(t)t^2) \geq v_f \,\mathrm{ord}_{i\infty}(t) = -2p_1(t)$ is equivalent to $\deg(p_1(t) - p_2(t)t^2) \leq \deg(p_1(t))$, which follows by (8). Furthermore $\mathrm{ord}_{i\infty}(p_2(t)b_2) \geq v_f \,\mathrm{ord}_{i\infty}(t)$ is equivalent to (8).

Now assume $\mathrm{ord}_{i\infty}(f) = \mathrm{ord}_{i\infty}(p_2(t)\tilde{b}_2)$, now

$$v_f = \lceil (\mathrm{ord}_{i\infty}(p_2(t)\tilde{b}_2))/\mathrm{ord}_{i\infty}(t)\rceil = \lceil(-2\deg(p_2(t)) - 3)/(-2)\rceil = \deg(p_2(t)) + 2$$

Because $f = (p_1(t) - t^2 p_2(t))b_1 + p_2(t)b_2$ by Definition 3.1 we have to show

$$-2\deg(p_1 - t^2 p_2) = \mathrm{ord}_{i\infty}(p_1(t) - t^2 p_2(t)) \geq v_f \,\mathrm{ord}_{\infty}(t) = -2\deg(p_2) - 4. \tag{9}$$

First note that $\mathrm{ord}_{i\infty}(p_2(t)\tilde{b}_2) < \mathrm{ord}_{i\infty}(p_1(t)\tilde{b}_1)$ is equivalent to $-2\deg(p_2) - 3 < -2\deg(p_1)$ which implies $-2\deg(p_2) - 4 \leq -2\deg(p_1)$, this implies that $\deg(p_2) + 2 \geq \deg(p_1)$ and finally $\deg(p_1 - t^2 p_2) \leq 2 + \deg(p_2)$ which implies (9).

*Proof that the basis of Example*1.3 *is not normalized:*

To prove this assume that it is normalized. By Definition 3.3 below we have $\mathrm{ord}(\{b_1, b_2\}) = 2u$. Now $\{1, G_3\}$ is an integral basis which is order-complete and $\mathrm{ord}(\{1, G_3\}) = 3$. By Lemma 3.4 we have

$$2u = \mathrm{ord}(\{b_1, b_2\}) < |\mathrm{ord}_{i\infty}(t)| + \mathrm{ord}(\{1, G_3\}) = 2 + 3 = 5$$

hence if $u \geq 3$ this inequality can not be true giving a contradiction to the assumption that $\{b_1, b_2\}$ is normalized.

From the Examples 1.1–1.3 we see that the orders of the basis elements of a normalized integral basis are close to the orders of the basis elements of an order-complete integral basis. This statement is made precise below.

Definition 3.3 *We define the order of an integral basis $b_1, \ldots, b_n$ to be*

$$\max(|\operatorname{ord}_{i\infty}(b_1)|, \ldots, |\operatorname{ord}_{i\infty}(b_n)|).$$

Lemma 3.4 *Let $B := \{b_1, b_2, \ldots, b_n\}$ be a $\mathbb{C}[t]$ normalized integral basis of $M_{\infty}(N)$ and $\tilde{B} := \{\tilde{b}_1, \tilde{b}_2, \ldots, \tilde{b}_n\}$ be a $\mathbb{C}[t]$ order-complete integral basis of $M_{\infty}(N)$. Then* $\operatorname{ord}(\tilde{B}) + |\operatorname{ord}_{i\infty}(t)| > \operatorname{ord}(B)$.

Proof W.l.o.g assume $\operatorname{ord}(B) = -\operatorname{ord}_{i\infty}(b_n)$. Then since B and $\tilde{B}$ are both integral basis there exists an i such that

$$\tilde{b}_i = p_1(t)b_1 + \cdots + p_n(t)b_n$$

with $p_n(t) \neq 0$. Then $\operatorname{ord}_{i\infty}(p_n(t)b_n) - v \operatorname{ord}_{i\infty}(t) \geq 0$ and $v := \lceil \operatorname{ord}_{i\infty}(\tilde{b}_i)/\operatorname{ord}_{i\infty}(t)\rceil \leq \lceil -\operatorname{ord}(\tilde{B})/\operatorname{ord}_{i\infty}(t)\rceil$. Hence

$$\begin{aligned}
&\operatorname{ord}_{i\infty}(p_n(t)b_n) - v \operatorname{ord}_{i\infty}(t) \geq 0 \\
&\Rightarrow \operatorname{ord}_{i\infty}(p_n(t)b_n) \geq v \operatorname{ord}_{i\infty}(t) \geq \lceil -\operatorname{ord}(\tilde{B})/\operatorname{ord}_{i\infty}(t)\rceil \operatorname{ord}_{i\infty}(t) \\
&> (-\operatorname{ord}(\tilde{B})/\operatorname{ord}_{i\infty}(t) + 1)\operatorname{ord}_{i\infty}(t) = -\operatorname{ord}(\tilde{B}) + \operatorname{ord}_{i\infty}(t) \\
&\Rightarrow \operatorname{ord}_{i\infty}(b_n) > -\operatorname{ord}(\tilde{B}) + \operatorname{ord}_{i\infty}(t) \\
&\Rightarrow \operatorname{ord}(B) = -\operatorname{ord}_{i\infty}(b_n) < \operatorname{ord}(\tilde{B}) - \operatorname{ord}_{i\infty}(t) = \operatorname{ord}(\tilde{B}) + |\operatorname{ord}_{i\infty}(t)|
\end{aligned}$$

Acknowledgment We thank both referees for the careful reading of our paper and for their suggestions to improve our paper.

References

1. A. O. L. Atkin. Proof of a Conjecture of Ramanujan. *Glasgow Mathematical Journal*, 8:14–32, 1967.
2. F. G. Garvan. Some Congruences for Partitions that are p-Cores. *Proceedings of the London Mathematical Society*, 66:449–478, 1993.
3. F. Hess. Computing riemann–roch spaces in algebraic function fields and related topics. *Journal of Symbolic Computation*, 33:425–445, 2002.
4. K. Hughes. Ramanujan Congruences for $p_{-k}(n)$ Modulo Powers of 17. *Canadian Journal of Mathematics*, 43:506–525, 1991.
5. E. Nart J. Guàrdia, J. Montes. Higher newton polygons in the computation of discriminants and prime ideal decomposition in number fields. *J. Théor. Nombres Bordeaux*, 23:667–696, 2011.
6. K. Khuri-Makdisi. Linear algebra algorithms for divisors on an algebraic curve. *Mathematics of Computation*, 73:333–357, 2004.
7. O. Kolberg. An Elementary discussion of Certain Modular Forms. *UNIVERSITET I BERGEN ÅRBOK Naturvitenskapelig rekke*, 16, 1959.

8. O. Kolberg. Congruences Involving the Partition Function for the Moduli 17, 19, and 23. *UNIVERSITET I BERGEN ÅRBOK Naturvitenskapelig rekke*, 15, 1959.
9. J. Lehner. Ramanujan Identities Involving the Partition Function for the Moduli 11^α. *American Journal of Mathematics*, 65:492–520, 1943.
10. G. Ligozat. Courbes modulaires de genre 1. *Mémoires de la S.M.F.*, 43:5–80, 1975.
11. M. Newman. Construction and Application of a Class of Modular Functions. *Proceedings London Mathematical Society*, 3(7), 1957.
12. M. Newman. Construction and Application of a Class of Modular Functions 2. *Proceedings London Mathematical Society*, 3(9), 1959.
13. P. Paule and C.-S. Radu. A Proof of the Weierstrass Gap Theorem not using the Riemann-Roch Formula. *Annals of Combinatorics*, 23(3–4), 963–1007, 2019.
14. P. Paule and C.-S. Radu Radu. A new witness identity for $11 \mid p(11n + 6)$. In *Analytic number theory, modular forms and q-hypergeometric series*, volume 221 of *Springer Proc. Math. Stat.*, pages 625–639. Springer, Cham, 2017.
15. H. Rademacher. The Ramanujan Identities Under Modular Substitutions. *Transactions of the American Mathematical Society*, 51(3), 609–636, 1942.
16. C.-S. Radu. An Algorithmic Approach to Ramanujan-Kolberg Identities. *Journal of Symbolic Computations*, 68:225–253, 2015.
17. C.-S. Radu. An algorithm to prove algebraic relations involving eta quotients. *Annals of Combinatorics*, 22:377–391, 2018.
18. J. P. Serre. *A Course in Arithmetic*. Springer, 1996.
19. B. Trager. *Integration of algebraic functions*. PhD thesis, Dept. of EECS, MIT, 1984.
20. M. van Hoeij. An algorithm for computing an integral basis in an algebraic function field. *J. Symbolic Comput.*, 18(4):353–363, 1994.
21. Y. Yang. Defining Equations of Modular Curves. *Advances in Mathematics*, 204:481–508, 2006.

An Algorithm to Prove Holonomic Differential Equations for Modular Forms

Peter Paule and Cristian-Silviu Radu

Abstract Express a modular form g of positive weight locally in terms of a modular function h as $y(h)$, say. Then $y(h)$ as a function in h satisfies a holonomic differential equation; i.e., one which is linear with coefficients being polynomials in h. This fact traces back to Gauß and has been popularized prominently by Zagier. Using holonomic procedures, computationally it is often straightforward to derive such differential equations as conjectures. In the spirit of the "first guess, then prove" paradigm, we present a new algorithm to prove such conjectures.

Keywords Modular functions · Modular forms · Holonomic differential equations · Holonomic functions and sequences · Fricke-Klein relations · q-series · Partition congruences

2010 Mathematics Subject Classification Primary 05A30, 11F03, 68W30 · Secondary 11F33, 11P83

1 Description of Contents

The study of holonomic functions and sequences satisfying linear differential and difference equations, respectively, with polynomial coefficients has roots tracing back (at least) to the time of Gauß.

Besides holonomic functions and sequences, the second major class of objects considered in this article are modular forms and functions which are non-holonomic:

Date: April 19, 2021 (final version).
Both authors were supported by grant SFB F50-06 of the Austrian Science Fund (FWF).

P. Paule · C.-S. Radu (✉)
Research Institute for Symbolic Computation (RISC), Johannes Kepler University, 4040 Linz, Austria
e-mail: Silviu.Radu@risc.uni-linz.ac.at

P. Paule
e-mail: Peter.Paule@risc.uni-linz.ac.at

A. Bostan and K. Raschel (eds.), *Transcendence in Algebra, Combinatorics, Geometry and Number Theory*, Springer Proceedings in Mathematics & Statistics 373,
https://doi.org/10.1007/978-3-030-84304-5_16

any modular form satisfies a *non-linear* third order differential equation with constant coefficients; see, for instance, [19, Prop. 16].

Nevertheless, there is a connection between holonomic functions and modular forms which also traces back to Gauß. Namely, express a modular form g of positive weight locally in terms of a modular function h as $y(h)$, say; then $y(h)$ as a function in h satisfies a holonomic differential equation.

Zagier in his classical exposition [19, Prop. 21] introduces to this fact as follows: "... it is at the heart of the original discovery of modular forms by Gauss and of the later work of Fricke and Klein and others, and appears in modern literature as the theory of Picard-Fuchs differential equations or of the Gauss-Manin connection—but it is not nearly as well known as it ought to be."

In [12] we began an algorithmic study of this connection which, on the holonomic side, utilizes aspects of Zeilberger's "holonomic systems approach" to special functions identities [20]. Ibid., on the modular functions and forms side, we sketched a contribution to the theme of differential equations and modular forms, which follows the "first guess, then prove" paradigm. In this article we present the full mathematical details and derivations leading up to this new algorithmic tool. The algorithm, ModFormDE, provides non-trivial computer support to prove claims of the following kind: Given a modular function h, a modular form g of positive weight, both for a fixed congruence subgroup, and a linear differential equation in $y(h)$ with polynomial coefficients in h, where y as a function in h is induced by the local expansion $g = y(h)$. Prove that this particular holonomic differential equation is indeed satisfied by $y(h)$.

This article can be read completely independently from [12]; there are some natural overlaps, but those are kept to a minimum. The content is structured as follows. In Sect. 2 we present two examples to introduce in a concrete fashion to the holonomic paradigm in connection with modular forms and functions. Section 3 is a condensed listing of basic notions and facts about modular forms and functions needed; a more detailed summary is given in [12, Sec. 2]. Readers familiar with these notions will skip this section anyway. In Sect. 4 we describe the input/output specification and the steps of our algorithm ModFormDE which is based on work of Yifan Yang [18]. In Sect. 5, two illustrating examples trace through its steps. In addition, we present a family of identities, derived with ModFormDE, which we have not found in the literature. In Sect. 6 we present the two main theorems of the paper, Theorem 6.2 and Theorem 6.3; they specify bounds for the total number of poles of modular functions which are essential for the ModFormDE algorithm. In Sect. 7 we introduce local expansions; they give rise to a notion of orders of modular forms of even weight, which will be used in a crucial way. The Sects. 8, 9, and 10 derive and prove the bounds given in the main theorems; Sect. 11 gives a summary of how these things are related. The Appendix Sect. 13 contains proofs, computational aspects, and basic facts of meromorphic functions on Riemann surfaces. All this material is of relevance, but if presented within the main text, it would disturb the flow of the presentation.

Conventions used throughout this paper: N denotes a positive integer, k is a fixed non-negative integer (the weight of a modular form),

$$\mathbb{H} := \{z \in \mathbb{C} : \mathrm{Im}(z) > 0\},\ \hat{\mathbb{C}} := \mathbb{C} \cup \{\infty\},\ \text{and}\ \hat{\mathbb{Q}} := \mathbb{Q} \cup \{\infty\}.$$

The ring of univariate polynomials with complex coefficients is denoted by $\mathbb{C}[X]$, its quotient field, the field of rational functions, is $\mathbb{C}(X)$.

Throughout, Γ stands for a congruence subgroup of $\mathrm{SL}_2(\mathbb{Z})$; for definitions of congruence subgroups such as $\Gamma(N, M, P)$ see [12, Sec. 2].

2 Introductory Examples

Holonomic differential equations are linear with polynomial coefficients. To illustrate their fundamental role for this paper, we consider concrete examples.

Example 2.1 Given

$$G(t) := {}_2F_1\left({\frac{1}{2}\ \frac{1}{2} \atop 1}; t\right) = \sum_{n=0}^{\infty} \frac{(1/2)_n (1/2)_n}{(1)_n} \frac{t^n}{n!} = 1 + \frac{t}{4} + \frac{9t^2}{64} + \frac{25t^3}{256} + \frac{1225t^4}{16384} + \ldots, \quad (1)$$

where $(a)_n := a(a+1)\ldots(a+n-1)$, $n \geq 1$, and $(a)_0 = 1$.

Problem. Determine coefficients $c(n)$ such that

$$G(t) = \sum_{n=0}^{\infty} c(n) H(t)^n \text{ where } H(t) := 4t(1-t). \quad (2)$$

Using the holonomic tool-box, e.g., the RISC package `GeneratingFunctions` as described in more detail in [12], one can solve this problem as follows[1]:

Step 1: Take as input sufficiently many coefficients in the expansion (1) of $G(t)$. It turns out that in this case taking the first 12 coefficients is sufficient.

Step 2: With this input, compute sufficiently many values of the $c(n)$:

$$c(0) = 1, c(1) = \frac{1}{16}, c(2) = \frac{25}{1024}, \ldots, c(11) = \frac{2363152308430225}{1152921504606846976}.$$

"Sufficiently many" is meant with regard to the next step.

Step 3: Using a package like `GeneratingFunctions`, guess a recurrence for the sequence $(c(n))_{n\geq 0}$:

In[1]:= << **RISC`GeneratingFunctions`**

In[2]:= **cRec = GuessRE[{1,** $\frac{1}{16}, \frac{25}{1024}, \ldots, \frac{2363152308430225}{1152921504606846976}$**}, c[n]][[1]]**

Out[2]= $\{16(n+1)^2 c(n+1) - (4n+1)^2 c(n) = 0, c(0) = 1\}$

[1] The package, written in the Mathematica system by Christian Mallinger, is available upon password request to the first-named author.

In other words, we have guessed that

$$G(t) = Y(H(t)) \text{ where } Y(t) := \sum_{n=0}^{\infty} c(n)t^n \text{ with } c(n) = \frac{(1/4)_n(1/4)_n}{(1)_n n!}. \quad (3)$$

Step 4: To prove (3), we derive holonomic differential equations satisfied by $G(t)$, respectively by $Y(H(t))$. To derive the differential equation for $G(t)$, we input the first 12 coefficients of the power series expansion (1):

In[3]:= **GDE = GuessDE[{1, $\frac{1}{4}$, $\frac{9}{64}$, $\frac{25}{256}$, $\frac{1225}{16384}$, ..., $\frac{7775536041}{274877906944}$}, G[t]]**

Out[3]= $\{4(t^2-t)G''(t)+4(2t-1)G'(t)+g(t)=0, G(0)=1, G'(0)=\frac{1}{4}\}$

This was derived as a guess. But using the power series expansion (1), one can easily verify that this equation is indeed satisfied by $G(t)$.

To derive a differential equation for $Y(H(t))$, we first derive a differential equation for $Y(t)$ by converting the recurrence for the $c(n)$ into a differential equation for their generating function $Y(t) = \sum_{n\geq 0} c(n)t^n$:

In[4]:= **YDE = RE2DE[cRec, c[n], Y[t]]**

Out[4]= $\{-16(t^2-t)Y''(t)-8(3t-2)Y'(t)-Y(t)=0, Y(0)=1, Y'(0)=\frac{1}{16}\}$

Finally the differential equation for $Y(H(t))$ can be computed by exploiting the holonomic closure property of algebraic composition; see [9, Thm. 7.2.5]:

In[5]:= **ACompose[YDE, Y[t] == 4t(1 − t), Y[t]]**

Out[5]= $\{4(t^2-t)Y''(t)+4(2t-1)Y'(t)+Y(t)=0, Y(0)=1, Y'(0)=\frac{1}{4}\}$

This differential equation for $Y(H(t))$ is the same as GDE in `Out[3]` for $G(t)$; also the initial values coincide, which proves (3).

Remark 2.2 The identity in (3) is the special case $a = b = 1/4$ of

$${}_2F_1\left(\begin{matrix}2a\ 2b\\ a+b+\frac{1}{2}\end{matrix}; t\right) = {}_2F_1\left(\begin{matrix}a\ b\\ a+b+\frac{1}{2}\end{matrix}; 4t(1-t)\right), \quad (4)$$

a classical identity in the theory of hypergeometric series; e.g., [1, (3.1.3)].

Example 2.3 Given

$$g(\tau) := \theta_3(\tau)^2 = 1 + 4x + 4x^2 + 4x^4 + 8x^5 + 4x^8 + 4x^9 + \ldots \text{ with } x = e^{\pi i \tau}, \quad (5)$$

where $\tau \in \mathbb{H}$. In addition to $\theta_3(\tau)$, we also need another Jacobi function $\theta_2(\tau)$:

$$\theta_2(\tau) := \sum_{n\in\mathbb{Z}+1/2} e^{\pi i n^2 \tau} \text{ and } \theta_3(\tau) := \sum_{n\in\mathbb{Z}} e^{\pi i n^2\tau} = 1 + 2\sum_{n=1}^{\infty} e^{\pi i n^2 \tau}, \quad \tau \in \mathbb{H}. \quad (6)$$

Problem. Determine coefficients $c(n)$ such that for all $\tau \in \mathbb{H}$ with $\mathrm{Im}(\tau)$ sufficiently big:

$$g(\tau) = \sum_{n=0}^{\infty} c(n)h(\tau)^n \text{ where } h(\tau) := \frac{1}{16}\lambda(\tau)(1-\lambda(\tau)) = x - 24x^2 + \dots, \quad (7)$$

where

$$\lambda(\tau) := \frac{\theta_2(\tau)^4}{\theta_3(\tau)^4} = 16x(1 - 8x + 44x^2 + \dots) \text{ with } x = e^{\pi i\tau} \text{ and } \tau \in \mathbb{H}.$$

As explained in [12, Ex. 2.5], one can verify that for all $\gamma \in \Gamma(2,4,4)$, a congruence subgroup defined in [12, Sec. 2],

$$\theta_2(\gamma\tau)^2 = (c\tau + d)\,\theta_2(\tau)^2,$$

and for all $\gamma \in \Gamma(2,4,2)$, another congruence subgroup defined in [12, Sec. 2],

$$\theta_3(\gamma\tau)^2 = (c\tau + d)\,\theta_3(\tau)^2.$$

In other words, θ_2^2 and θ_3^2 are modular forms of weight 1 for $\Gamma(2,4,4) \subseteq \Gamma(2,4,2)$; consequently, λ is a modular function for $\Gamma(2,4,4)$. But λ is a modular function also for the bigger group $\Gamma(2,4,2)$, because

$$\theta_2(\gamma\tau)^4 = (c\tau + d)^2\,\theta_2(\tau)^4$$

for all $\gamma \in \Gamma(2,4,2)$.

The problem to find an expansion as in (7) is similar to the expansion problem (3). Differences are: in (3), $G(t)$ is a hypergeometric function, in contrast to a modular form $g(\tau)$ in (7); in addition, in (3), $H(t)$ is a rational function (actually, a polynomial) in contrast to a modular function $h(\tau)$ in (7).

Power series expansion (and, more generally, Puiseux series expansion) of modular forms in terms of modular functions is the central theme in [12]. Namely, one has the crucial fact, Proposition 4.2 below, that in expansions like (7) the coefficients $c(n)$ constitute a holonomic sequence. As a consequence, holonomic tools as in the situation of (3) can be applied. More concretely, the holonomic approach to solve problem (7), can be summarized as follows:

- By Proposition 4.2 we know that there exists a power series,

$$y(z) := \sum_{n=0}^{\infty} c(n)z^n, \quad (8)$$

with a *holonomic* coefficient sequence $(c(n))_{n\geq 0}$, such that locally

$$g(\tau) = y(h(\tau)). \tag{9}$$

- Also by Proposition 4.2, $y(h)$ must satisfy a *holonomic* differential equation of the form

$$P_m(h)y^{(m)}(h) + P_{m-1}(h)y^{(m-1)}(h) + \cdots + P_0(h)y(h) = 0, \tag{10}$$

with polynomials $P_j(X) \in \mathbb{C}[X]$ with $P_m(X) \neq 0$.
- A fundamental holonomic fact says[2]: the differential equation (10) can be converted into a recurrence for $(c(n))_{n\geq 0}$, and vice versa.
- Our algorithm ModFormDE, described in Sect. 4, can be used to prove conjectured differential equations of the form (10).

Consequently, to determine the coefficients $c(n)$ in (7), one can proceed as follows:

- First use holonomic tools to *guess* a differential equation of the form (10), then *prove* the differential equation using the algorithm ModFormDE. Finally, to determine the desired coefficient sequence $c(n)$, convert the proven differential equation into a recurrence for the $c(n)$.

To put this strategy into action, we begin by holonomic *guessing* of a recurrence for the $(c(n))_{n\geq 0}$. This way we already can see which answer for the coefficients in (7) is expected.

In the next step, we convert this recurrence into a differential equation for y, which we then prove using the algorithm ModFormDE. Finally, the proven holonomic differential equation is converted back into the recurrence for the $c(n)$, which then gives a valid and proven specification.

The computational steps are as follows:

Step 1: In the expansion (5) of $g(\tau)$, take as input sufficiently many coefficients.[3]

Step 2: With this input, as explained in more detail in [12, 5.3], compute sufficiently many values of the $c(n)$:

$$c(0) = 1, c(1) = 4, c(2) = 100, c(3) = 3600, \ldots, c(8) = 924193822500. \tag{11}$$

"Sufficiently many" is meant with regard to the next step.

Step 3: Using the `GeneratingFunctions` package, guess a recurrence for the sequence $(c(n))_{n\geq 0}$:

```
In[6]:= cList = {1, 4, 100, 3600, 152100, 7033104, 344622096, 17582760000, 924193822500};
In[7]:= yRec = GuessRE[cList, c[n]][[1]]
Out[7]= {-4(1 + 4n)^2 c[n] + (1 + n)^2 c[1 + n] = 0, c[0] = 1}
```

[2] E.g., in the given context, used systematically in [12].

[3] It turns out that taking the first 10 coefficients as given in (5) is sufficient.

In other words, expressing the solution to this recurrence (of order 1) in terms of rising factorials, we algorithmically derived the following conjecture for $c(n)$ such that (7):

$$c(n) = \frac{(1/4)_n (1/4)_n}{(1)_n} \frac{4^{3n}}{n!}. \tag{12}$$

Step 4: To prove (12), the first step is to transform the recurrence `yRec` from `Out[7]` into a holonomic differential equation satisfied by $y(z) = \sum_{n=0}^{\infty} c(n) z^n$. This is done by using the procedure call `RE2DE` as above:

In[8]:= **yDE = RE2DE[{yRec, c[n], y[z]]**

Out[8]= $\{-4\mathtt{y[z]} - (-1 + 96\mathtt{z})\mathtt{y'[z]} - (-\mathtt{z} + 64\mathtt{z}^2)\mathtt{y''[z]} = 0, \mathtt{y[0]} = 1, \mathtt{y'[0]} = 4\}$

In view of (9), this differential equation rewrites into

$$(64\,h^2 - h)\frac{d^2y}{dz^2}(h) + (96\,h - 1)\frac{dy}{dz}(h) + 4y(h) = 0 \tag{13}$$

with

$$y(0) = c_0 = 1 \text{ and } \frac{dy}{dz}(0) = c_1 = 4. \tag{14}$$

The verification of (14) is straightforward from the x-expansions (5) and (7) of g and h.

Using these x-expansions, also (13) can be verified up to a desired precision; i.e., by checking that the coefficients of x^n in the x-expansion of the left side are zero up to a certain power. But, needless to say, this gives no proof!

Step 5. To prove the correctness of (13), which is the conjectured differential equation of the form (10), we use the algorithm ModFormDE as detailed out in Sect. 5.1

Step 6. After having proved that (13), resp. `yDE` in `Out[8]`, is correct, in view of (8) we translate it back to a recurrence for the $c(n)$. Using the holonomic tool-box [12, Prop. 3.1], this can be done as follows:

In[9]:= **DE2RE[yDE, y[z], c[n]]**

Out[9]:= $\{-4(1 + 4\mathtt{n})^2\mathtt{c[n]} + (1 + \mathtt{n})^2\mathtt{c[1+n]} = 0, \mathtt{c[0]} = 1, \mathtt{c[1]} = 4\}$

As expected, this recurrence is nothing but recurrence `yRec` from `Out[7]` which we had guessed. But now it comes as a consequence of a *proven* differential equation, consequently we *proved* that (12) is indeed the answer to problem (7).

Remark 2.4 (Existence of (13)) As explained in Sect. 4, Proposition 4.1 and Example 4.4, a holonomic differential equation of order 2,

$$p_2(h)\frac{d^2y}{dz^2}(h) + p_1(h)\frac{dy}{dz}(h) + p_0(h)y(h) = 0, \tag{15}$$

with $p_j(X) \in \mathbb{C}[X]$ is guaranteed to exist.

Remark 2.5 Using holonomic tools has led us to guess that

$$\theta_3(\tau)^2 = {}_2F_1\left(\begin{matrix}\frac{1}{4}\ \frac{1}{4}\\ 1\end{matrix}; 4\lambda(\tau)(1-\lambda(\tau))\right); \tag{16}$$

using our algorithm ModFormDE, this relation is proved algorithmically. In exactly the same manner one can algorithmically derive and prove that

$$\theta_3(\tau)^2 = {}_2F_1\left(\begin{matrix}\frac{1}{2}\ \frac{1}{2}\\ 1\end{matrix}; \lambda(\tau)\right); \tag{17}$$

details are given in [12, Ex. 2.6 and Sec. 6.3]. Combining these two facts gives an alternative proof of the special instance $a = b = 1/4$ of (4).

Remark 2.6 We also note the following connection to the complete elliptic integral $K = K(\tau)$ of the first kind with modulus $k = k(\tau)$,

$$K(k(\tau)) := \int_0^{\pi/2} \frac{d\varphi}{\sqrt{1-k(\tau)^2\sin(\varphi)^2}} = \frac{\pi}{2}\theta_3(\tau)^2 \text{ where } k(\tau)^2 = \lambda(\tau). \tag{18}$$

This equality involving the theta function is immediate from (17) by series expansion of the integrand in terms of powers of $\lambda(\tau) = k(\tau)^2$. Besides other applications, the Borweins in their famous monograph [3] used this identity together with 11 similar ones [3, Thm. 5.6 and Thm. 5.7] to derive and explain identities which Ramanujan [14] gave (without too many details) to establish formulas to approximate π, respectively $1/\pi$.

3 Modular Functions and Forms: Basic Facts and Notions

In this section we list in a very condensed fashion the basic notions and facts about modular forms and functions needed. Section 2 of [12] is a less condensed, still short but complete collection of all the definitions and notions needed for the understanding of this article.

3.1 *Modular Forms and Modular Functions*

We write $M_k(\Gamma)$ to denote the $\mathbb{C}$-vector space of meromorphic modular forms of weight $k \in \mathbb{Z}_{\geq 0}$ for the congruence subgroup Γ. These functions are meromorphic functions on $\mathbb{H}$ with properties discussed in more detail in [12]. Here we restrict to those notions which are most fundamental for this article.

For $f \in M_k(\Gamma)$ and $\gamma = \left(\begin{smallmatrix} a & b \\ c & d \end{smallmatrix}\right) \in \mathrm{SL}_2(\mathbb{Z})$ the weight k operator is defined as usual by $(f|_k\gamma)(\tau) := (c\tau + d)^{-k} f(\gamma\tau)$, $\tau \in \mathbb{H}$, where $\gamma\tau := \frac{a\tau+b}{c\tau+d}$.

For each $\gamma_0 \in \mathrm{SL}_2(\mathbb{Z})$ there exist $w_0 = w_0(\gamma_0) \in \mathbb{Z}_{>0}$ and $n_0 = n_0(\gamma_0) \in \mathbb{Z}$ such that $f|_k\gamma_0$ admits a Fourier expansion (with coefficients in $\mathbb{C}$) of the form,

$$(f|_k\gamma_0)(\tau) = \sum_{n\geq n_0} a_{\gamma_0}(n) q_{w_0}^n, \quad \tau \in \mathbb{H} \text{ such that } \mathrm{Im}(\tau) \text{ is sufficiently big,} \tag{19}$$

where $q_{w_0} := e^{2\pi i\tau/w_0}$ and $a_{\gamma_0}(n_0) \neq 0$. One can show that $w_0 = w_0(\gamma_0) = w_{\gamma_0}(\Gamma)$, where[4]

$$w_{\gamma_0}(\Gamma) := \min_{m\in\mathbb{Z}_{>0}} \left\{ \begin{pmatrix} 1 & m \\ 0 & 1 \end{pmatrix} \in \gamma_0^{-1}\Gamma\gamma_0 \text{ or } \begin{pmatrix} -1 & m \\ 0 & -1 \end{pmatrix} \in \gamma_0^{-1}\Gamma\gamma_0 \right\}. \tag{20}$$

If $\gamma_0 = \left(\begin{smallmatrix} a & b \\ c & d \end{smallmatrix}\right)$, then $\gamma_0\infty = a/c$, and the order ("of vanishing") of a modular form f of weight k at $a/c \in \hat{\mathbb{Q}}$ is defined as follows:

$$\mathrm{ord}_{a/c} f := n_0, \text{ where } n_0 \text{ is taken as in the expansion (19).} \tag{21}$$

We write $M(\Gamma) := M_0(\Gamma)$ to denote the field of meromorphic modular functions for the congruence subgroup Γ. The fact that the weight $k = 0$ allows to extend $f \in M(\Gamma)$ meromorphically from $\mathbb{H}$ to all the points $a/c \in \hat{\mathbb{Q}}$; see [12, Sec. 2]. This way, each $f \in M(\Gamma)$ becomes invariant on the Γ-orbits $[\tau]_\Gamma := \{\gamma\tau : \gamma \in \Gamma\}$ where $\tau \in \mathbb{H} \cup \hat{\mathbb{Q}}$.

The set of all such orbits, denoted by $X(\Gamma)$, can be equipped with the structure of a compact Riemann surface.[5] Hence a modular function f with respect to Γ can be interpreted as a function $\hat{f} : X(\Gamma) \to \hat{\mathbb{C}}$; in fact, such $\hat{f}$ are meromorphic functions on $X(\Gamma)$.

If $\tau = a/c \in \hat{\mathbb{Q}}$ the orbits $[\tau]_\Gamma \in X(\Gamma)$ are called cusps. A Fourier expansion for $f \in M(\Gamma)$ as in (19) is called an expansion of f at the cusp $[a/c]_\Gamma$, or simply at a/c. One can define an expansion at infinity[6] by singling out the Fourier expansion (19) with the choice $\gamma_0 := \left(\begin{smallmatrix} 1 & 0 \\ 0 & 1 \end{smallmatrix}\right) = I$: in other words, for all $\tau \in \mathbb{H}$ with $\mathrm{Im}(\tau)$ sufficiently large,

$$f(\tau) = \sum_{n\geq n_0} a_I(n) q^{n/w_0}. \tag{22}$$

If $\gamma_0 = \left(\begin{smallmatrix} a & b \\ c & d \end{smallmatrix}\right)$ then $\gamma_0\infty = a/c$, and an expansion as in (19) is called expansion of f a the cusp $[a/c]_\Gamma$ or, in short, at a/c. With $\gamma_0 := \left(\begin{smallmatrix} 1 & 0 \\ 0 & 1 \end{smallmatrix}\right) = I$ this includes the expansion (22) at the cusp $[\infty]_\Gamma$ or, in short, at ∞ $(= 1/0)$. As a note, the minimal period $w_0 = w_{\gamma_0}(\Gamma)$ is independent from the choice of the representative $\gamma_0\infty$ of the cusp $[a/c]_\Gamma = [\gamma_0\infty]_\Gamma$; it is called the width of the cusp $[a/c]_\Gamma$.

---

[4] See, e.g., [6, Sec. 3.2].

[5] Charts are given explicitly in Sect. 7.1.

[6] I.e., at the point $\infty \in \hat{\mathbb{Q}}$.

For $f \in M(\Gamma)$ the order $\mathrm{ord}_{a/c} f$, as defined in (21), is also called the order of f at the cusp $[a/c]_\Gamma$. For the order of f at the cusp $[\infty]_\Gamma$ (in short, at infinity) one often uses the short hand notation,

$$\mathrm{ord}\, f := \mathrm{ord}_\infty f.$$

For orbits $[\tau]_\Gamma$, $\tau \in \mathbb{H} \cup \hat{\mathbb{Q}}$, we write in short $[\tau]$, if Γ is clear from the context.

3.2 Zero Recognition of Modular Functions

To recognize whether a modular function $t \in M(\Gamma)$ is zero, one exploits its extension to a meromorphic function $\hat{t} : X(\Gamma) \to \hat{\mathbb{C}}$ on the compact Riemann surface $X(\Gamma)$. Namely, if such functions are non-constant they have the property that

$$\text{number of poles of } \hat{t} = \text{number of zeros of } \hat{t}, \tag{23}$$

counting multiplicities. This follows immediately from the following fact (e.g., [10, Prop. 4.12]):

Lemma 3.1 *Let f be a non-constant meromorphic function on a compact Riemann surface X. Then*

$$\sum_{x \in X} \mathrm{Ord}_x f = 0. \tag{24}$$

Here the order of f at $x_0 \in X$, $\mathrm{Ord}_{x_0} f$, is defined as follows.

Definition 3.2 *Suppose*

$$f(x) = \sum_{n=m}^{\infty} c_n(\phi(x) - \phi(x_0))^n, \quad c_m \neq 0,$$

is the local Laurent expansion of f at $x_0 \in X$ using the local coordinate chart $\phi : U_0 \to \mathbb{C}$ which homeomorphically maps a neighborhood U_0 of x_0 to an open set $V_0 \subseteq \mathbb{C}$. Then,

$$\mathrm{Ord}_{x_0} f := m.$$

In our context, $X = X(\Gamma)$ and $f = \hat{t} : X(\Gamma) \to \hat{\mathbb{C}}$ where $\hat{t}$ is induced by the modular function $t \in M(\Gamma)$; moreover, $\phi = z_p$ as described in Sect. 7.1 serve as the local charts at $x_0 = P = [p] \in X(\Gamma)$.

Related to the order of a non-constant meromorphic function $f : X \to \hat{\mathbb{C}}$ on a Riemann surface X, we need the notion of multiplicity. Let $x \in X$: then for every neighborhood U of x, there exist neighborhoods $U_x \subseteq U$ of x and V of $f(x)$ such that the set $f^{-1}(v) \cap U_x$ contains exactly ℓ elements for every $v \in V \setminus \{f(x)\}$. This number ℓ is called the multiplicity of f at x; notation: $\ell = \text{mult}_x\, f$.[7]

In practice, there are various ways to bring the fact (23) into action. With regard to our algorithm ModFormDE, the strategy for zero recognition will be this.

Given $t \in M(\Gamma)$, decide whether $t = 0$. It is important to specify what "given" for a modular function $t \in M(\Gamma)$ means in our context. Namely, by this we assume that t is given in a way that allows to compute "sufficiently many" coefficients $a(n)$ in its expansion at infinity,[8]

$$t(\tau) = \sum_{n \geq n_0} a(n) q_{w_0}^n \text{ with } q_{w_0} := e^{2\pi i \tau / w_0}, \tag{25}$$

where n_0 is some integer less than $\text{ord}\, t = \text{ord}_\infty\, t$. "Sufficiently many" depends on the context. For example, to do zero recognition, we usually have a bound N on the number of poles of t,

$$\text{NofPoles}(t) \leq N.$$

As a consequence of (23), in order to prove that t is identically zero, we need to check that the coefficients $c(n_0), \ldots, c(N)$ are all 0.

It is important to notice, that "number of poles/zeros" has to be taken in the interpretation of t as the induced meromorphic function $\hat{t} : X(\Gamma) \to \mathbb{C}$; in other words, NofPoles(t) is the number of poles of $\hat{t}$, multiplicities counted.

Remark 3.3 Notions like NofPoles will be used heavily when describing the mathematical fundament of the algorithm ModFormDE. They are defined explicitly in (60) for modular functions, and in (86) for modular forms of even weight.

4 The Algorithm ModFormDE: Specification and Steps

4.1 Existence of Holonomic Differential Equations for Modular Forms

In Sect. 2 we showed how holonomic differential equations like (13) can be derived, as a *guess*, in computer-supported fashion. Modular form theory guarantees the existence of such differential equations. We continue Zagier's quote stated in the Introduction: "... it is at the heart of the original discovery of modular forms by Gauss and [...] but it is not nearly as well known as it ought to be. Here is a precise statement:"

---

[7] If x is a pole of f: $\text{mult}_x\, f = -\text{Ord}_x\, f$; otherwise, $\text{mult}_x\, f = \text{Ord}_x(f - f(x))$.

[8] Recall that w_0 is the width of the cusp $[\infty]$; in terms of charts (71): $q_{w_0} = z_\infty(\tau)$ when $\gamma_0 = \left(\begin{smallmatrix}1&0\\0&1\end{smallmatrix}\right)$.

Proposition 4.1 *Let $g(\tau)$ be a modular form of weight $k > 0$ and $h(\tau)$ a modular function, both with respect to the congruence subgroup Γ. Express $g(\tau)$ locally as $y(h(\tau))$. Then the function $y(h)$ satisfies a linear differential equation of order $k + 1$ with algebraic coefficients, or with polynomial coefficients if the compact Riemann surface $X(\Gamma)$ has genus 0 and h generates the field of modular functions on Γ (i.e., h is a Hauptmodul).*

An important fact in the light of the holonomic paradigm: if one drops to require minimality of the order of the differential equation, $y(h)$ always satisfies a *holonomic* differential equation:

Proposition 4.2 ([12], Prop. 6.2) *In the setting of Proposition 4.1, the function $y(h)$ satisfies a linear differential equation with polynomial coefficients also when the genus of $X(\Gamma)$ is non-zero or when h is not a Hauptmodul.—In these cases, the order of the differential equation in general will be larger than $k + 1$.*

Remark 4.3 In fact, one can prove even more: Regardless whether $h \in M(\Gamma)$ is a Hauptmodul or not, a holonomic differential equation for $y(h)$ of order

$$(k + 1)\,\mathrm{NofPoles}(h) \tag{26}$$

always exists; $\mathrm{NofPoles}(h)$ is the number of poles of $\hat{h}$, multiplicities counted.

Example 4.4 The *existence* of the differential equation (13) is owing to the following facts: g is a modular form of weight 1 for $\Gamma(2, 4, 2)$, h is a modular function for $\Gamma(2, 4, 2)$, and $X(\Gamma(2, 4, 2))$ has genus 0. Despite $\mathrm{ord}(h) = 1$, we have $\mathrm{NofPoles}(h) = 2$ as a modular function in $M(\Gamma(2, 4, 2)$. This means, h is not a Hauptmodul for $\Gamma(2, 4, 2)$, and despite falling under the scope of Proposition 4.2, in this case the order of the differential equation (13) stays at $k + 1 = 2$. As noted in Remark 5.1, h is a Hauptmodul for a bigger congruence subgroup. In Sect. 5.2 we present another example with an $h \in M(\Gamma)$ being no Hauptmodul, and where $X(\Gamma)$ has genus 1, and where we have no evidence for the existence of a bigger group turning h into a Hauptmodul.

After stating Proposition 4.1, Zagier [19, p. 21] continues: "This proposition is perhaps the single most important source of applications of modular forms in other branches of mathematics, so with no apology we sketch three different proofs, ..."

Zagier's third proof is constructive; i.e., given g and h, it constructs the corresponding differential equation. Following the holonomic paradigm, we take a different approach: we first *guess* the corresponding holonomic differential equation algorithmically, and then prove it using our algorithm ModFormDE.

4.2 The Algorithm ModFormDE and Local Expansions

The input and output specification of our algorithm ModFormDE is presented in Sect. 4.4. As illustrated in Sect. 2, a major application domain of the algorithm concerns local expansions, and we want to summarize briefly how these applications are connected to ModFormDE.

To this end, suppose we are given a modular form $g \in M_k(\Gamma)$ with weight $k \in \mathbb{Z}_{\geq 1}$ for the congruence subgroup Γ, and a modular function $h \in M(\Gamma)$ such that $\operatorname{ord} h = \ell \geq 1$, and both g and h have $q_{w_0} = q^{1/w_0}$ with $q = e^{2\pi i\tau}$ as the local expansion variable at infinity. Moreover, suppose we are interested in a local expansion of g of the form,[9]

$$g(\tau) = y\,(h(\tau)) \text{ where } y(z) := \sum_{n=\operatorname{ord} g}^{\infty} c(n) z^{\frac{n}{\ell}}. \tag{27}$$

According to Proposition 4.1 and Proposition 4.2 such an expansion exists with a uniquely determined holonomic coefficient sequence $(c(n))$. As described in Sect. 2, the holonomic recurrence for the $c(n)$ can be converted into a differential equation,

$$P_m(h)y^{(m)}(h) + P_{m-1}(h)y^{(m-1)}(h) + \cdots + P_0(h)y(h) = 0, \tag{28}$$

with polynomials $P_j(X) \in \mathbb{C}[x]$, not all 0. Notice that $y^{(n)}(h) := \frac{d^n y}{dz^n}(z)|_{z=h}$.

Our algorithm ModFormDE then proves whether $y(h)$ indeed satisfies a conjectured holonomic differential equation of a form as in (28). In case this is not true, the algorithm will detect this.

4.3 The Mathematical Fundament of Algorithm ModFormDE

Our algorithm ModFormDE is based on work of Yifan Yang [18]. So, before describing the steps of ModFormDE, we recall several notions used there. To adapt to applications, we extend the setting in [18] from $q = e^{2\pi i\tau}$ to $x = q^{1/w_0}$.

- Differential operators [18, p. 4]: Let $\varphi(\tau)$ be a function defined on $\mathbb{H}$ having an x-expansion $\varphi(\tau) = \tilde{\varphi}(x) := \sum_{n\geq n_0} a(n)x^n$ with $x = q^{1/w_0}$ where $q = e^{2\pi i\tau}$:

$$D_x\varphi = D_x\varphi(\tau) := \frac{w_0}{2\pi i} \cdot \frac{d\varphi}{d\tau}(\tau) = \frac{w_0}{2\pi i} \cdot \varphi'(\tau) = x\,\tilde{\varphi}'(x). \tag{29}$$

  Let $\psi = \psi(z)$ be a function meromorphic in a neighborhood of 0 punctured at 0, let $h = h(\tau) \in M(\Gamma)$:

$$D_h\psi = D_h\psi(h) := h\frac{d\psi}{dz}(h) = h\psi'(h).$$

- Fundamental functions on $\mathbb{H}$ [18, Thm. 1]:

[9] Details on Puiseux series expansion such as for y in (27) are given in [12, Sec. 4].

Let Γ be a congruence subgroup, and let $g \in M_k(\Gamma)$, $k \geq 1$, and $h \in M(\Gamma)$ be fixed:

$$G_1 := \frac{D_x h}{h} = \frac{w_0}{2\pi i} \cdot \frac{h'}{h} \text{ and } G_2 := \frac{D_x g}{g} = \frac{w_0}{2\pi i} \cdot \frac{g'}{g}; \tag{30}$$

notice that $h' = h'(\tau) = \frac{dh(\tau)}{d\tau}$ and $g' = g'(\tau) = \frac{dg(\tau)}{d\tau}$.

- Fundamental modular functions [18, Thm. 1]:

$$p_1 := \frac{D_x G_1 - 2G_1 G_2/k}{G_1^2} \text{ and } p_2 := -\frac{D_x G_2 - G_2^2/k}{G_1^2}. \tag{31}$$

  As proved in [18, Lemma 1], the p_j are modular functions in $M(\Gamma)$. Moreover, they are also algebraic functions in $h \in M(\Gamma)$. This means, for fixed $j \in \{1, 2\}$, p_j and h satisfy an algebraic relation; see the remark after Thm. 8.1 in [12].

  Because of the chain rule we have,

$$D_h y = D_h y(h) = h y'(h) = h \frac{g'}{h'} = g \frac{G_2}{G_1}. \tag{32}$$

  Yang [18, p. 9] also computed that

$$D_h^2 y = \left(1 - \frac{1}{k}\right) \cdot g \frac{G_2^2}{G_1^2} + (-p_1) \cdot g \frac{G_2}{G_1} + (-p_2) \cdot g, \tag{33}$$

and

$$\begin{aligned} D_h^3 y = {} & \left(1 - \frac{1}{k}\right)\left(1 - \frac{2}{k}\right) \cdot g \frac{G_2^3}{G_1^3} - 3\left(1 - \frac{1}{k}\right) p_1 \cdot g \frac{G_2^2}{G_1^2} \\ & + \left(p_1^2 - \left(3 - \frac{2}{k}\right) p_2 - D_h p_1\right) \cdot g \frac{G_2}{G_1} + (p_1 p_2 - D_h p_2) \cdot g, \end{aligned} \tag{34}$$

where the $p_j \in M(\Gamma)$ are the fundamental modular functions defined in (31) and $D_h p_j = h \frac{dp_j}{dh}$. Mathematical induction [18, p. 10] on $m \geq 0$ leads to,[10]

$$D_h^m y = \prod_{j=0}^{m-1} (1 - \frac{j}{k}) \cdot g \frac{G_2^m}{G_1^m} + a_{m,m-1} \cdot g \frac{G_2^{m-1}}{G_1^{m-1}} + \cdots + a_{m,1} \cdot g \frac{G_2}{G_1} + a_{m,0} \cdot g, \tag{35}$$

with the $a_{m,j}$ being multivariate polynomials from the polynomial ring

$$R := \mathbb{C}\left[h, p_1, p_2, \frac{dp_1}{dh}, \frac{dp_2}{dh}, \ldots, \frac{d^m p_1}{dh^m}, \frac{d^m p_2}{dh^m}, \ldots\right]. \tag{36}$$

[10] Notice that here we assume $D_h^0 y = y(h) = g$.

Lemma 4.5 *Let Γ be a congruence subgroup. Let $g \in M_k(\Gamma)$ with $k \geq 1$, and $h \in M(\Gamma)$. Then the elements of R are modular functions in $M(\Gamma)$.*

Proof Let $p \in \{p_1, p_2\}$. By [18, Lemma 1], $p \in M(\Gamma)$. In addition, p is an algebraic function in h; see, e.g., the remark after Thm. 8.1 in [12]. This means, there exists a polynomial

$$R(X, Y) := Y^n + c_1(X)Y^{n-1} + \cdots + c_n(X)$$

with rational function coefficients $c_j(X) \in \mathbb{C}(X)$ such that $R(h, p) = 0$. By the implicit function theorem one has that locally there exists an meromorphic function $r(z)$ such that $R(h, p) = 0$ iff $p = r(h)$. Moreover,

$$\frac{dp}{dh} = r'(h) = -\frac{\partial R}{\partial X}(h, p) / \frac{\partial R}{\partial Y}(h, p),$$

which, as a rational function in h and p, is in $M(\Gamma)$. Applying the same argument to p' and h, etc., completes the proof also for the higher derivatives of p. □

Remark 4.6 In the proof we introduced a new function symbol r when writing p as a function in h; i.e., $p = r(h)$. However, in order to keep notation as lean as possible, whenever things are clear from the context we will follow Yang, and write $p = p(h)$ instead of $p = r(h)$ when referring to p as a function in h.

By relation (35) we are led to the following fact.

Lemma 4.7 *Let Γ be a congruence subgroup. Let $g \in M_k(\Gamma)$ with $k \geq 1$, and $h \in M(\Gamma)$. Then: (1) any expression of the form,*

$$Y := Q_m(h) D_h^m y + Q_{m-1}(h) D_h^{m-1} y + \cdots + Q_0(h) y, \tag{37}$$

with polynomials $Q_j(X) \in \mathbb{C}[X]$ can be written into "Yang form" as

$$Y = \alpha_m \cdot g \frac{G_2^m}{G_1^m} + \alpha_{m-1} \cdot g \frac{G_2^{m-1}}{G_1^{m-1}} + \cdots + \alpha_0 \cdot g \text{ with } \alpha_j \in R; \tag{38}$$

(2) these coefficients α_j are uniquely determined.

Proof Part (1) is immediate from (35). Part (2) is a consequence of the fact that the $\frac{G_2^m}{G_1^m}$ are linearly independent over $M(\Gamma)$; this is proved in Proposition 13.1 in the Appendix Sect. 13. □

4.4 Input, Output, and Steps of the Algorithm ModFormDE

INPUT. (I1) $g \in M_k(\Gamma)$ and $h \in M(\Gamma)$ such that $\operatorname{ord} h = \ell \geq 1$; both functions are given in the form of their x-expansions, where $x = q^{1/w_0}$ with $q = e^{2\pi i \tau}$, $\tau \in \mathbb{H}$, is the local expansion variable at infinity. More precisely, we assume that sufficiently many coefficients of

$$g(\tau) = \sum_{n=\text{ord}\, g}^{\infty} g(n)x^n,$$

and

$$h(\tau) = x^{\ell}(1 + h_1 x + h_2 x^2 + \dots)$$

can be computed.

(I2) Polynomials $P_0(X), \dots, P_m(X)$ in $\mathbb{C}[X]$ with $P_m(X) \neq 0$.

(I3) NofPoles(h): the number of poles of $\hat{h}$, defined in (60).

(I4) If k is even, NofPoles(g): a pole number defined in (86); if k is odd, NofPoles(g^2).

(I5) If k is odd, NofCusps(Γ) and NofElliptic(Γ): the number of cusps and of elliptic points, defined in (58) and (59).

OUTPUT. Bounds for

$$\text{NofPoles}(p_1), \text{NofPoles}(p_2), \text{ and NofPoles}\left(\frac{d^j p_i}{dh^j}\right), \ i = 1, 2, \ j \geq 1. \tag{39}$$

As a consequence of the steps of the algorithm, these bounds as part of the strategy described in Sect. 3.2, enable a proof of the correctness of the differential relation,

$$P_m(h)y^{(m)}(h) + P_{m-1}(h)y^{(m-1)}(h) + \dots + P_0(h)y(h) = 0. \tag{40}$$

In case (40) is not valid, the algorithm detects this. The output bounds for (39) are specified in the Theorems 6.2 and 6.3 in Sect. 6.

THE STEPS OF THE ALGORITHM "ModFormDE":

Step 0: Rewrite the left side of (40) into the form (37).—This is done by using the relations $hy'(h) = D_h y$,

$$h^2 y''(h) = D_h^2 y - D_h y, \ h^3 y^{(3)}(h) = D_h^3 y - 3\,h D_h^2 y + (3h - 1) D_h y, \text{ a.s.o.},$$

which, for example, can be precomputed.

Step 1: Transform the expression (37) into Yang form (38).—This is done by using the relations (32), (33), (34), and (35) for $m \geq 4$, which, for example, can be precomputed.

Step 2: Owing to the uniqueness of the coefficients α_j in (38), the proof of (40) finally is reduced to prove that

$$\alpha_m = 0, \alpha_{m-1} = 0, \dots, \alpha_0 = 0. \tag{41}$$

Since the α_j are modular functions in $M(\Gamma)$, this task, owing to (23), reduces to determine upper bounds for the number of poles of each α_j. Because of $\alpha_j \in R$, by definition (36) it is sufficient to provide such bounds for h and for the $\frac{d^j p_j}{dh^j}$, $j \geq 0$, which is done in the Sects. 8, 9, and 10, together with the summary given in Sect. 11.

Finally, each zero test, $\alpha_j = 0$, is completed by computing sufficiently many coefficients in the x-expansion of α_j. This is done by using the coefficients of the x-expansions of g and h.[11]

To get an impression of the algorithm ModFormDE in action, see Sect. 5.

5 The Algorithm ModFormDE: Illustrating Applications

In this section we present non-trivial applications to illustrate the steps and various aspects of the algorithm ModFormDE. In Sect. 5.1 we exemplify the steps of the algorithm by proving (13). In Sect. 5.2 we demonstrate the behaviour of the ModFormDE algorithm in a computationally more challenging setting, in particular, where, in contrast to the example in Sect. 5.1, it needs to deal with derivatives of p_1 and p_2.

5.1 Proving (13) *with the ModFormDE Algorithm*

As a first illustration of the ModFormDE algorithm in action, we prove the validity of (13).

As in (5) and (7), we are given[12] $g \in M_1(\Gamma)$ and $h \in M(\Gamma)$ with $\operatorname{ord} h = 1$; here $\Gamma = \Gamma(2, 4, 2)$, and we note that $X(\Gamma(2, 4, 2))$ has genus $g_\Gamma = 0$. Noticing that $\operatorname{ord} g = 0$, by Proposition 4.1 and Proposition 4.2 we know that g has a local expansion of the form

$$g(\tau) = y(h(\tau)) \text{ where } y(z) := \sum_{n=0}^{\infty} c(n) z^n,$$

where $y(z)$ has a uniquely determined *holonomic* coefficient sequence $(c(n))$. In Example 2.3 we computed a list (11) of these coefficients; in addition, using software we conjectured the holonomic differential equation (13),

$$(64\,h^2 - h)\frac{d^2 y}{dz^2}(h) + (96\,h - 1)\frac{dy}{dz}(h) + 4y(h) = 0.$$

Using ModFormDE, its validity is proved as follows.

---

[11] Recall $x = q^{1/w_0}$ with $q = e^{2\pi i \tau}$.

[12] Given, in the sense of (25) in Sect. 3.2.

Concerning the input data (I1), the x-expansions, $x = q^{\pi i \tau}$, are immediate from (6).

The polynomials in (I2) are read off from the differential equation: $P_0(X) = 4$, $P_1(X) = 96X - 1$, and $P_2(X) = X(64X - 1)$.

Concerning (I3) one has,

$$\mathrm{NofPoles}(h) = 2, \tag{42}$$

which is implied by the classical fact, $\lambda(-1/\tau) = 1 - \lambda(\tau)$; see, e.g., [3, (4.3.4)].

Remark 5.1 As remarked in [12, 5.4], for the bigger group $\hat{\Gamma} := \Gamma \cup \left(\begin{smallmatrix} 0 & -1 \\ 1 & 0 \end{smallmatrix}\right) \Gamma$ the modular function h is a Hauptmodul; i.e., $\mathrm{NofPoles}(h) = 1$ with respect to $\hat{\Gamma}$. However, according to our definition, g is not a modular form for $\hat{\Gamma}$—in contrast to our incorrect claim made in [12, 5.3(ii)].

Concerning (I4): Since $k = 1$, we have to determine $\mathrm{NofPoles}(g^2)$. The product expansion of g is folklore [3, (3.16)]; it gives,

$$g(\tau)^2 = \frac{\eta(\tau)^{20}}{\eta(\tau/2)^8 \eta(2\tau)^8}, \tag{43}$$

using the Dedekind eta function $\eta(\tau) = e^{\frac{\pi i \tau}{12}} \prod_{j=1}^{\infty}(1 - q^j)$, $q = e^{2\pi i \tau}$. This representation tells that g^2 has no pole in $\mathbb{H}$. So we need to inspect the cusps of $X(\Gamma)$, which are 3 in total; namely $[0]$, $[1]$ and $[\infty]$. To see this, and to find the widths of each of these cusps, which is 2 in each instance, one can, e.g., run Magma as described in Sect. 13.2.1.

Since $g^2 \in M_2(\Gamma)$, according to Definition 7.12 we have,

$$\begin{aligned} \mathrm{NofPoles}(g^2) &= - \sum_{\substack{\text{cusps } [a/c] \in X(\Gamma) \\ \mathrm{ord}\, \tilde{F}_{a/c}(z) < 0}} \mathrm{ord}\, \tilde{F}_{a/c}(z) \\ &= - \sum_{\substack{\text{cusps } [a/c] \in X(\Gamma) \\ \mathrm{ord}_{a/c}(g^2) \leq 0}} \Big(\mathrm{ord}_{a/c}(g^2) - 1 \Big), \end{aligned} \tag{44}$$

where $\tilde{F}_p(z)$ is the Laurent series for $F := g^2$ defined in Lemma 7.6; the last equality is by (87). Hence, in view of the three cusps $[\infty]$, $[0]$, and $[1]$, we have to determine the orders of the x-expansions, $x = e^{\pi i \tau}$, of

$$(g^2)(\tau), \; (g^2|_2\gamma_0)(\tau), \; \text{and} \; (g^2|_2\gamma_1)(\tau),$$

with $\gamma_0 := \left(\begin{smallmatrix} 0 & -1 \\ 1 & 0 \end{smallmatrix}\right)$ and $\gamma_1 := \left(\begin{smallmatrix} 1 & -1 \\ 1 & 0 \end{smallmatrix}\right)$ such that $\gamma_0 \infty = 0$ and $\gamma_1 \infty = 1$.

At the cusp $[\infty]$:

$$g(\tau)^2 = 1 + 8x + 24x^2 + O(x^3), \; \text{hence } \mathrm{ord}_\infty(g^2) = 0.$$

At the cusp [0]:

$$(g^2|_2\gamma_0)(\tau) = (1 \cdot \tau + 0)^{-2} g(-1/\tau)^2 = g(\tau)^2, \quad \text{hence } \mathrm{ord}_0(g^2) = 0;$$

here one applies to (43) the transformation formula $\eta(-1/\tau) = (-i\tau)^{1/2}\eta(\tau)$.

At the cusp [1]: as explained in Sect. 13.2.2, one finds that

$$(g^2|_2\gamma_1)(\tau) = u \cdot x + O(x^2) \text{ for some non-zero } u \in \mathbb{C};$$

hence $\mathrm{ord}_1(g^2) = 1$. Summarizing, using the information in (44), gives,

$$\mathrm{NofPoles}(g^2) = -\Big(\mathrm{ord}_\infty(g^2) - 1\Big) - \Big(\mathrm{ord}_0(g^2) - 1\Big) = 1 + 1 = 2. \tag{45}$$

Since the given k is odd, we have to consider the input data (I5) instead of (I4), and note that

$$\mathrm{NofCusps}(\Gamma) = 3 \text{ and } \mathrm{NofElliptic}(\Gamma) = 0. \tag{46}$$

To obtain this information is routine and can be left to software. For instance, with the Magma package we obtained that there are three cusps $[\infty]$, [0], and [1], each of width 2. From this information one can extract (algorithmically) the data (46), and also that $g_\Gamma = 0$. This information extraction works in general.

Remark 5.2 We have seen that the non-routine part in providing the required input data (I1) to (I5) to algorithm ModFormDE, is to determine $\mathrm{NofPoles}(g^2)$. Nevertheless, for certain classes of modular functions and modular forms (e.g., those representable by eta quotients), also this step can be made algorithmic.

After providing all the required input data (I1) to (I5), we turn to the steps of algorithm ModFormDE:

Step 0: The conjectured differential equation rewrites into,

$$Y := (64\,h - 1)D_h^2 y + 32\,h D_h y + 4\,h\,y = 0. \tag{47}$$

Step 1: Owing to $g \in M_1(\Gamma)$ we have $k = 1$, and the Yang form of Y becomes

$$Y = \alpha_1 \cdot g \frac{G_2}{G_1} + \alpha_0 \cdot g, \tag{48}$$

with

$$\alpha_1 = 32\,h - (64\,h - 1)p_1 \in M(\Gamma) \text{ and } \alpha_0 = 4\,h - (64\,h - 1)p_2 \in M(\Gamma). \tag{49}$$

Step 2: For this step, since $k = 1$, we need to use Theorem 6.3. This theorem requires the notion $\mathrm{NofPoles}(f)$ for $f \in M(\Gamma)$, defined in (60), and the extended notion (86) for modular forms $f \in M_{2k}(\Gamma)$ defined in Sect. 7.

Step 2a. We first prove $\alpha_1 = 0$. By (65) of Theorem 6.3 with (42) and (45),

$$\begin{aligned}
\mathrm{NofPoles}(\alpha_1) &\leq \mathrm{NofPoles}(h) + \mathrm{NofPoles}(h) + \mathrm{NofPoles}(p_1)\\
&= 2\,\mathrm{NofPoles}(h) + \mathrm{NofPoles}(p_1)\\
&\leq 2\,\mathrm{NofPoles}(h) + (2k+4)(g_\Gamma - 1) + 4\,\mathrm{NofPoles}(h) + 2\,\mathrm{NofPoles}(g^2)\\
&= (2k+4)(g_\Gamma - 1) + 6\,\mathrm{NofPoles}(h) + 2\,\mathrm{NofPoles}(g^2)\\
&= -6 + 6\cdot 2 + 2\cdot 2 = 8.
\end{aligned}$$

Computing the x-expansion up to the power of x^8 shows that $\alpha_1(\tau) = 0 + 0x + 0x^2 + \cdots + 0x^8 + \ldots$ This implies that α_1 has at least 9 zeros and 8 poles or less; so α_1 has to be 0.

Step 2b. Second, we prove $\alpha_2 = 0$. By (66) of Theorem 6.3 with (42), (45), and (46),

$$\begin{aligned}
\mathrm{NofPoles}(\alpha_2) &\leq 2\,\mathrm{NofPoles}(h) + \mathrm{NofPoles}(p_2)\\
&\leq 2\,\mathrm{NofPoles}(h) + (6k+4)(g_\Gamma - 1) + 2\,\mathrm{NofPoles}(h) + 7\,\mathrm{NofPoles}(g^2)\\
&\quad + 2\,\mathrm{NofCusps}(\Gamma) + 2\,\mathrm{NofElliptic}(\Gamma)\\
&= (6k+4)(g_\Gamma - 1) + 4\,\mathrm{NofPoles}(h) + 7\,\mathrm{NofPoles}(g^2) + 2\cdot 3 + 2\cdot 0\\
&= -10 + 4\cdot 2 + 7\cdot 2 + 6 = 18.
\end{aligned}$$

Computing the x-expansion up to the power of x^{18} shows that $\alpha_2(\tau) = 0 + 0x + 0x^2 + \cdots + 0x^{18} + \ldots$ This implies that α_2 has at least 19 zeros and 18 poles or less; so α_2 has to be 0.

5.2 *Another Application of the ModFormDE Algorithm*

In general the ModFormDE algorithm needs to deal not only with p_1 and p_2, but also with derivatives of these modular functions. To illustrate its behavior in such a situation, we consider the following problem. Let

$$g(\tau) := \eta(\tau)^2\eta(11\tau)^2 = q - 2q^2 - q^3 + 2q^4 + q^5 + 2q^6 - 2q^7 - 2q^9 + O(q^{10})$$

and $h(\tau) := g_2(\tau)$, where g_2 was introduced by Atkin [2] in his celebrated proof of Ramanujan's observation that $11|p(11n+6)$, $11^2|p(11^2n+116)$, and so forth for all powers of 11. Here $p(n)$ denotes the number of partitions of n. The facts that $g \in M_2(\Gamma_0(11))$ and $\mathrm{NofPoles}(g) = 0$ are straight-forward consequences of transformation properties of the Dedekind eta function.

Atkin [2] needed the following facts which are less obvious: $h \in M(\Gamma_0(11))$, h has no poles in $\mathbb{H}$, a double pole at 0, and an expansion at ∞ given by,

$$h(\tau) = q + 5q^2 + 19q^3 + 63q^4 + 185q^5 + 502q^6 + 1270q^7 + 3046q^8 + O(q^9).$$

We note explicitly that for this congruence subgroup the genus $g_{\Gamma_0(11)}$ is 1; see, for instance, [6, Ex. 3.1.4(e)].

Express g locally as $g = y(h)$. To determine a holonomic differential equation for $y(h)$, we proceed as described by input step `In[3]` in Example 2.1. This means, we use sufficiently many coefficients $c(n)$ in the local expansion,

$$g(\tau) = y\,(h(\tau)) \text{ where } y(z) := \sum_{n=1}^{\infty} c(n) z^n, \tag{50}$$

to do a computer-assisted guessing of a holonomic differential equation for $y(x)$, resp. $y(h)$. As expected by formula (26), the computations deliver an equation of the form

$$Q_0(h)y(h) + Q_1(h)Dy(h) + Q_2(h)D^2y(h) + \cdots + Q_6(h)D^6y(h) = 0, \tag{51}$$

recalling that $D_h f(h) := hf'(h)$, and where the $Q_j(x)$ are polynomials in x of degree 19.

Remark 5.3 In contrast to using the first 12 coefficients for calling `GuessDE` in input step`In[3]`, to guess (51) we used the first 201 values $c(0)$ to $c(200)$; the CPU time used by Mathematica was about 20 min on a standard laptop.

To prove the correctness of this differential equation, we proceed with the steps of the ModFormDE algorithm. The differential equation (51) is already put into the form required by Step 0. To carry out Step 1, we set $\Gamma := \Gamma_0(11)$,

$$Y_k := g\frac{G_1^k}{G_2^k}, u_j := \frac{d^j p_1}{dh^j}, \text{ and } v_j := \frac{d^j p_2}{dh^j},$$

and, noticing $k = 2$, use the formulas (32), (33), (34), and analogous ones for $D_h^4 y$, $D_h^5 y$, and $D_h^6 y$, to convert (51) into the form,

$$\begin{aligned} 0 &= \alpha_2(h, u_0, \ldots, u_4, v_0, \ldots, v_4)Y_2 + \alpha_1(h, u_0, \ldots, u_4, v_0, \ldots, v_4)Y_1 \\ &+ \alpha_0(h, u_0, \ldots, u_4, v_0, \ldots, v_4)Y_0, \end{aligned}$$

where the α_i are polynomials in the modular functions u_i, v_j, and h in $M(\Gamma)$. Finally, in the remaining Step 2 we have to show that the α_i are all zero. To this end, we will use various degree estimates together with the following useful fact, which is easy to prove.

Lemma 5.4 *Let* $f_1, \ldots, f_n$ *be modular functions in* $M(\Gamma)$ *with* NofPoles$(f_i) \leq m_i$. *Suppose* $p(X_1, X_2, \ldots, X_n)$ *is a polynomial in* $\mathbb{C}[X_1, \ldots, X_n]$ *with degrees* $\deg_{X_i}(p) = d_i$. *Then*

$$\text{NofPoles}(p(f_1, \ldots, f_n)) \leq m_1 d_1 + \cdots + m_n d_n.$$

Proof Let

$$S := \{P \in X(\Gamma) : \text{ there exists a } j \text{ such that } f_j \text{ has a pole at } P\}$$

be the set of points where $f_1, \dots, f_n$ have poles. Suppose $S = \{P_1, \dots, P_r\}$ and let $\operatorname{pord}_P(f_j) := \max(-\operatorname{ord}_P(f_j), 0)$ be the pole order of f_j at $P \in X(\Gamma)$. Then,

$$\sum_{j=1}^{r} \operatorname{pord}_{P_j}(p(f_1, f_2, \dots, f_n)) \leq \sum_{j=1}^{r}\sum_{i=1}^{n} d_i \operatorname{pord}_{P_j}(f_i) = \sum_{i=1}^{n} d_i \sum_{j=1}^{r} \operatorname{pord}_{P_j}(f_i)$$
$$= \sum_{i=1}^{n} d_i \operatorname{NofPoles}(f_i) \leq \sum_{i=1}^{n} d_i m_i.$$

□

To apply the lemma, we need bounds on the number of poles of u_i and v_j. This information is supplied by Theorem 6.2. To apply it, besides the number of cusps of $X(\Gamma_0(11))$ which equals 2, we also need to know the number of elliptic points of $X(\Gamma)$. But $\operatorname{NofElliptic}(\Gamma_0(11)) = 0$; see, for instance, [6, Ex. 3.1.4(c,d)]. Consequently, by Theorem 6.2, for $j \in \{1, \dots, 4\}$,

$$\operatorname{NofPoles}(u_0) \leq 8 =: m_{1,0}, \qquad \operatorname{NofPoles}(v_0) \leq 8 =: m_{2,0}$$
$$\operatorname{NofPoles}(u_j) \leq 16j + 4 =: m_{1,j} \qquad \operatorname{NofPoles}(v_j) \leq 16j + 4 =: m_{2,j}.$$

We collect this information in a tuple,

$$\mu := (2, m_{1,0}, \dots, m_{1,4}, m_{2,0}, \dots, m_{2,4}) = (2, 8, 20, 36, 52, 68, 8, 20, 36, 52, 68).$$

Finally, we need the degrees of the α_i with respect to each variable. Set for $i = 0, 1, 2$,

$$B_i := (\deg_x(\alpha_i), \deg_{u_0}(\alpha_i), \dots, \deg_{u_4}(\alpha_i), \deg_{v_0}(\alpha_i), \dots, \deg_{v_4}(\alpha_i)).$$

By inspection of the computed polynomials α_i we have,

$$B_0 = (23, 4, 2, 1, 1, 0, 3, 2, 1, 1, 1), \quad B_1 = (23, 5, 2, 1, 1, 1, 2, 1, 1, 1, 0), \text{ and}$$
$$B_2 = (22, 4, 2, 1, 1, 0, 2, 1, 1, 0, 0).$$

Now we are ready to apply Lemma 5.4 to obtain

$$\operatorname{NofPoles}(\alpha_i) \leq m_i, \quad i = 0, 1, 2,$$

where the m_i are given by

$$m_0 = B_0 \cdot \mu^t = 426, \quad m_1 = B_1 \cdot \mu^t = 406, \quad m_2 = B_2 \cdot \mu^t = 276.$$

Consequently, to prove $\alpha_i = 0$, it suffices to compute the q-expansions of α_i up to q^{m_i} and to see whether all these coefficients are zero. The actual computations show that this is indeed the case for $i = 0, 1, 2$.

Despite not really small, the bounds m_i are still of reasonable size. This indicates that our method will be feasible also for applications of even higher complexity.

5.3 New Applications of the ModFormDE Algorithm

Obviously the ModFormDE algorithm can be used as a tool to produce new proofs of known identities, for instance, of relations stemming from local expansions of modular forms in terms of modular functions. But ModFormDE can also be used as a tool for discovery. In this regard we are planning to do a more systematic investigation in future work. To illustrate the aspect of discovery we restrict to present a family of three identities, provable with ModFormDE, which were discovered using holonomic methods as described in Sect. 2, and which in this form we were not able to spot in the literature. Still we believe that the identities (52), (54), and (55) can be derived from the rich web of theta series identities beautifully presented in Shaun Cooper's monograph [4].

For $q = e^{2\pi i \tau}$, $\tau \in \mathbb{H}$, consider the normalized Eisenstein series

$$E_2(\tau) = 1 - 24 \sum_{n=1}^{\infty} \sum_{d|n} d\, q^n = 1 - 24q - 72q^2 - 96q^3 - 168q^4 - 144q^5 - 288q^6 + O(q^7);$$

in Ramanujan's notation, $P(q) := E_2(\tau)$. Define

$$Z_N(\tau) := \frac{N E_2(N\tau) - E_2(\tau)}{N - 1}, \quad N \in \mathbb{Z}_{\geq 2}.$$

These functions are modular forms in $M_2(\Gamma_0(N))$; see, e.g., [4].

For our first example, expand

$$g := \sqrt{Z_2} = 1 + 12q - 60q^2 + 768q^3 - 11004q^4 + 178200q^5 + O(q^6)$$

locally in terms of

$$t_2(\tau) := q \prod_{j=1}^{\infty} \frac{(1 - q^{2j})^{24}}{(1 - q^j)^{24}},$$

which is a Hauptmodul for $\Gamma_0(2)$. Using the ModFormDE algorithm one can prove that[13]

$$\sqrt{Z_2} = (1 + 64t_2)^{1/4} \cdot {}_2F_1\left(\begin{matrix}1/4, 1/4\\ 1\end{matrix}; -64t_2\right). \tag{52}$$

Remark 5.5 This formula is a relative to (16); moreover, Clausen's formula [1, Ex. 2.9.13] implies,

$$Z_2 = \sqrt{1 + 64t_2} \cdot {}_3F_2\left(\begin{matrix}1/2, 1/2, 1/2\\ 1, 1\end{matrix}; -64t_2\right). \tag{53}$$

As a second example, expand

$$g := \sqrt{Z_3} = 1 + 6q + 6q^3 + 6q^4 + 12q^7 + 6q^9 + 6q^{12} + 12q^{13} + O(q^{16})$$

locally in terms of

$$t_3(\tau) := q \prod_{j=1}^{\infty} \frac{(1 - q^{3j})^{12}}{(1 - q^j)^{12}},$$

which is a Hauptmodul for $\Gamma_0(3)$. Using the ModFormDE algorithm one can prove that[14]

$$\sqrt{Z_3} = (1 + 27t_3)^{1/3} \cdot {}_2F_1\left(\begin{matrix}1/3, 1/3\\ 1\end{matrix}; -27t_3\right). \tag{54}$$

As a third example, expand

$$g := \sqrt{Z_4} = 1 + 4q + 4q^2 + 4q^4 + 8q^5 + 4q^8 + 4q^9 + 8q^{10} + 8q^{13} + O(q^{16})$$

locally in terms of

$$t_4(\tau) := q \prod_{j=1}^{\infty} \frac{(1 - q^{4j})^{8}}{(1 - q^j)^{8}},$$

which is a Hauptmodul for $\Gamma_0(4)$. Using the ModFormDE algorithm one can prove that[15]

$$\sqrt{Z_4} = (1 + 16t_4)^{1/2} \cdot {}_2F_1\left(\begin{matrix}1/2, 1/2\\ 1\end{matrix}; -16t_4\right). \tag{55}$$

Remark 5.6 This formula is a relative to (17); moreover,

$$Z_4 = (1 + 16t_4) \cdot \sum_{n=0}^{\infty} a(n)(-16t_4)^n, \tag{56}$$

---

[13] $(1 + 64t_2)^{1/4} = 1 + 16q - 192q^2 + 6560q^3 - 230976q^4 + O(q^5)$.
[14] $(1 + 27t_3)^{1/3} = 1 + 9q + 27q^2 + 81q^3 + 198q^4 + O(q^5)$.
[15] $(1 + 16t_2)^{1/2} = 1 + 8q + 32q^2 + 96q^3 + 256q^4 + O(q^5)$.

where

$$a(n) = \sum_{k=0}^{n} \binom{2k}{k}^2 \binom{2n-2k}{n-k}^2.$$

This prominent combinatorial sequence is entry A036917 in Sloane's online database [15]; for a recent study, in the context of new congruences for Apéry-like sequences, see Zhi-Hong Sun [17].

6 Main Theorems for Even and Odd Weight

Our algorithm ModFormDE consists of two steps: after Step 1, which transforms the conjectured differential equation into Yang form (38), in Step 2 one has to prove (41) for the coefficients $\alpha_j \in R \subseteq M(\Gamma)$. To this end, one invokes bounds for the number of poles of the elements in R. In this section we state the two main theorems, one for k even and one for k odd, which provide such bounds by determining bounds for the number of poles of the generators of R,

$$h, p_1, p_2, \frac{dp_1}{dh}, \frac{dp_2}{dh}, \ldots, \frac{d^m p_1}{dh^m}, \frac{d^m p_2}{dh^m}, \text{ a.s.o.}$$

To state the main results of this section, Theorem 6.2 and Theorem 6.3, we need to make some preparations.

In Sect. 3.1 we defined cusps. We need to recall another standard notion from modular group actions.

Definition 6.1 *Let $P = [p] \in X(\Gamma)$ with $p \in \mathbb{H}$. Then P is an elliptic point of $X(\Gamma)$ (resp. for Γ), if*

$$\{\gamma \in \{\pm I\}\Gamma : \gamma p = p\} \supsetneq \{\left(\begin{smallmatrix}1 & 0\\ 0 & 1\end{smallmatrix}\right), \left(\begin{smallmatrix}-1 & 0\\ 0 & -1\end{smallmatrix}\right)\}. \tag{57}$$

Define

$$\text{NofCusps}(\Gamma) := \text{ no. of cusps of } X(\Gamma), \tag{58}$$
$$\text{NofElliptic}(\Gamma) := \text{ no. of elliptic points of } X(\Gamma). \tag{59}$$

In addition, for a modular function $f \in M(\Gamma)$ define,

$$\text{NofPoles}(f) := \text{ number of poles } P \in X(\Gamma) \text{ of } \hat{f}, \tag{60}$$

multiplicities of poles are counted.

For the whole section we assume that p_1 and p_2 are defined as in (31).

To state the following theorem, we need to extend definition (60) to modular forms with even weight; this is done by the relation (86) in Definition 7.12. Using these definitions, (60) and (86), the bounds we obtained for even weight k are as follows.

Theorem 6.2 *For a congruence subgroup* Γ *let* $h \in M(\Gamma)$ *and* $g \in M_k(\Gamma)$. *Let* g_Γ *be the genus of* $X(\Gamma)$. *Then, if* k *is even, one has,*

$$\mathrm{NofPoles}(p_1) \leq 4\,\mathrm{NofPoles}(h) + 2\,\mathrm{NofPoles}(g) + (k+4)(g_\Gamma - 1), \tag{61}$$

$$\begin{aligned}\mathrm{NofPoles}(p_2) \leq\ & 2\,\mathrm{NofElliptic}(\Gamma) + 2\,\mathrm{NofCusps}(\Gamma)\\ & + 2\,\mathrm{NofPoles}(h) + 7\,\mathrm{NofPoles}(g) + (3k+4)(g_\Gamma - 1),\end{aligned} \tag{62}$$

and for the derivatives where $j \geq 1$,

$$\begin{aligned}\mathrm{NofPoles}\left(\frac{d^j p_1}{dh^j}\right) \leq\ & (8j+2)\,\mathrm{NofPoles}(h) + 2(j+1)\,\mathrm{NofPoles}(g)\\ & + (jk+8j+k+2)(g_\Gamma - 1),\end{aligned} \tag{63}$$

$$\begin{aligned}\mathrm{NofPoles}\left(\frac{d^j p_2}{dh^j}\right) \leq\ & 2(j+1)\,\mathrm{NofElliptic}(\Gamma) + 2(j+1)\,\mathrm{NofCusps}(\Gamma)\\ & + 6j\,\mathrm{NofPoles}(h) + 7(j+1)\,\mathrm{NofPoles}(g)\\ & + (3jk+8j+3k+2)(g_\Gamma - 1).\end{aligned} \tag{64}$$

The bounds we obtained for odd weight k are as follows.

Theorem 6.3 *For a congruence subgroup* Γ *let* $h \in M(\Gamma)$ *and* $g \in M_k(\Gamma)$. *Let* g_Γ *be the genus of* $X(\Gamma)$. *Then, if* k *is odd, one has,*

$$\mathrm{NofPoles}(p_1) \leq 4\,\mathrm{NofPoles}(h) + 2\,\mathrm{NofPoles}(g^2) + (2k+4)(g_\Gamma - 1), \tag{65}$$

$$\begin{aligned}\mathrm{NofPoles}(p_2) \leq\ & 2\,\mathrm{NofElliptic}(\Gamma) + 2\,\mathrm{NofCusps}(\Gamma)\\ & + 2\,\mathrm{NofPoles}(h) + 7\,\mathrm{NofPoles}(g^2) + (6k+4)(g_\Gamma - 1),\end{aligned} \tag{66}$$

and for the derivatives where $j \geq 1$,

$$\begin{aligned}\mathrm{NofPoles}\left(\frac{d^j p_1}{dh^j}\right) \leq\ & (8j+2)\,\mathrm{NofPoles}(h) + 2(j+1)\,\mathrm{NofPoles}(g^2)\\ & + (2jk+8j+2k+2)(g_\Gamma - 1),\end{aligned} \tag{67}$$

$$\begin{aligned}\mathrm{NofPoles}\left(\frac{d^j p_2}{dh^j}\right) \leq\ & 2(j+1)\,\mathrm{NofElliptic}(\Gamma) + 2(j+1)\,\mathrm{NofCusps}(\Gamma)\\ & + 6j\,\mathrm{NofPoles}(h) + 7(j+1)\,\mathrm{NofPoles}(g^2)\\ & + (6jk+8j+6k+2)(g_\Gamma - 1).\end{aligned} \tag{68}$$

The rest of our paper is devoted to the proofs of these statements. Section 7 introduces the mathematical requirements needed. Section 8 proves the bound for p_1, Sect. 9 for p_2. The bounds for the derivatives of p_1 and p_2 are given in Sect. 10. These considerations are completed by a proof summary in Sect. 11.

7 Locals Expansions and Orders

The rest of this section is devoted to proving these bounds. To this end, we first consider local expansions which are used in a crucial way, also for defining NofPoles(g) for modular forms g with even weight; see Definition 7.12.

7.1 Local Expansions and Orders

To estimate bounds for the possible number of poles of modular functions we use local series expansions in terms of charts. These charts z_P, as defined below, are homeomorphisms between open subsets of $X(\Gamma)$ and of $\mathbb{C}$. More details on the topology used, in particular, why these charts make $X(\Gamma)$ into a Riemann surface, can be found, for instance, in [6, sec. 2.2, 2.3 and 2.4].

Given $P = [p] \in X(\Gamma)$ for some $p \in \hat{\mathbb{H}} = \mathbb{H} \cup \mathbb{Q} \cup \{\infty\}$, we consider charts $z_p : U_P \to \mathbb{C}$ with $z_p([\tau]) := z_p(\tau)$ defined as usual by

$$z_p(\tau) := \tau - p, \text{ if } p \in \mathbb{H} \text{ is no elliptic point,} \tag{69}$$

or by

$$z_p(\tau) := \left(\frac{\tau - p}{\tau - \overline{p}}\right)^{h(p)}, \text{ if } p \in \mathbb{H} \text{ is an elliptic point (cf. Def. 6.1),} \tag{70}$$

or, by

$$z_p(\tau) := e^{2\pi i \gamma_0^{-1}\tau/w}, \text{ if } p = \frac{a_0}{c_0} = \gamma_0\infty \in \mathbb{Q} \cup \{\infty\}, \tag{71}$$

where $\gamma_0 = \left(\begin{smallmatrix} a_0 & b_0 \\ c_0 & d_0 \end{smallmatrix}\right) \in \mathrm{SL}_2(\mathbb{Z})$ and $w = w_{\gamma_0}(\Gamma)$; see (20). Here $U_P \subseteq X(N)$ is a neighborhood of $P = [p]$ such that $U_P = \{[\tau] : \tau \in V_0\}$, where $V_0 \subseteq \hat{\mathbb{H}}$ is suitable open neighborhood of p in the given topology of $\hat{\mathbb{H}}$. Notice that defining $z_p([\tau]) := z_p(\tau)$, we overloaded the meaning: besides being a map on open subsets of $X(\Gamma)$, z_p is also an analytic function on open subsets of $\hat{\mathbb{H}}$.

Furthermore, the periods $h(p)$ equal either 2 or 3; we also note explicitly that all these charts are centered at 0; i.e.,

$$z_p(P) = z_p(p) = 0. \tag{72}$$

We need to describe the behavior of charts under the change of orbit representatives.

Lemma 7.1 *Given $p \in \hat{\mathbb{H}}$, let $r := \gamma p$ for $\gamma = \left(\begin{smallmatrix} a & b \\ c & d \end{smallmatrix}\right) \in \Gamma$. Then the charts $z_p : V_0 \to \mathbb{C}$ relate to the charts $z_r : \gamma(V_0) \to \mathbb{C}$ as follows:*

(i) if $p \in \mathbb{H}$ is no elliptic point, then r is no elliptic point, and for $\tau \in V_0$,

$$z_r(\gamma\tau) = \gamma\tau - r = \frac{1}{cp+d} \cdot \frac{z_p(\tau)}{c\tau+d}; \tag{73}$$

(ii) if $p \in \mathbb{H}$ is an elliptic point, then r is an elliptic point,[16] *and for $\tau \in V_0$,*

$$z_r(\gamma\tau) = \left(\frac{\gamma\tau - r}{\gamma\tau - \overline{r}}\right)^{h(r)} = \left(\frac{c\overline{p}+d}{cp+d}\right)^{h(p)} \cdot z_p(\tau); \tag{74}$$

(iii) if $p = \gamma_0\infty \in \hat{\mathbb{Q}}$, $\gamma_0 \in \mathrm{SL}_2(\mathbb{Z})$, is a cusp, then r is a cusp, and for $\tau \in V_0$,

$$z_r(\gamma\tau) = e^{2\pi i(\gamma\gamma_0)^{-1}(\gamma\tau)/w} = z_p(\tau). \tag{75}$$

Proof By straight-forward verifications. □

Before turning to local series representations, we recall the notion of order ("of vanishing") of Laurent series.

Definition 7.2 (order of a Laurent series) *Let $\varphi(z) = \sum_{n=M}^{\infty} c(n)z^n$ be a Laurent series with $M \in \mathbb{Z}$ and $c(M) \neq 0$:*

$$\operatorname{ord} \varphi(z) := M.$$

Remark 7.3 In this article we use several notions of order. When referring to the order of a Laurent series in powers of z, we always include the argument z explicitly; i.e., we write $\operatorname{ord} \varphi(z)$ instead of $\operatorname{ord} \varphi$.

The first part of the following lemma is implied by the fact that the z_p are local charts. The second part, the order invariance, follows from Riemann surface point of view, namely, by connecting $\operatorname{Ord}_x f$ with the notion of multiplicity; see the Sect. 3.2 for further details. For an alternative proof of this invariance one can use the argument from the proof of Lemma 7.6 when $k = 0$.

Lemma 7.4 *Given a non-zero $t \in M(\Gamma)$, and $P = [p] \in X(\Gamma)$, $p \in \hat{\mathbb{H}}$. Then there exists a Laurent series*

$$\tilde{t}_p(z) := \sum_{n=M_p}^{\infty} a_p(n)z^n \tag{76}$$

[16] Notice that for the periods, $h(p) = h(r)$.

such that

$$t(\tau) = \tilde{t}_p(z_p(\tau)), \quad \tau \in V_0,$$

where $V_0 \subseteq \hat{\mathbb{H}}$ *is a suitable neighborhood of* p*. Moreover, for any* $\gamma \in \Gamma$,

$$M_p = \operatorname{ord} \tilde{t}_p(z) = \operatorname{ord} \tilde{t}_{\gamma p}(z) = M_{\gamma p};$$

i.e., the order, $\operatorname{ord} \tilde{t}_p(z)$*, is independent from the choice of* p *as a representative of* $P = [p]$.

The following definition generalizes (21) from cusps $P = [a/c]$, $a/c \in \hat{\mathbb{Q}}$, to general points (Γ-orbits) $P = [p] \in X(\Gamma)$:

Definition 7.5 (order of a modular function at at point) *For non-zero* $t \in M(\Gamma)$ *and* $P = [p] \in X(\Gamma)$, $p \in \hat{\mathbb{H}}$*:*

$$\operatorname{Ord}_P t := \operatorname{ord} \tilde{t}_p(z), \tag{77}$$

where $\tilde{t}_p(z)$ *is as in Lemma 7.4.*

The next two lemmas generalize Lemma 7.4 to modular forms.

Lemma 7.6 *Given a non-zero* $F \in M_k(\Gamma)$ *with* k *even, and* $P = [p] \in X(\Gamma)$, $p \in \hat{\mathbb{H}}$*. Then there exists a Laurent series*

$$\tilde{F}_p(z) := \sum_{n=N_p}^{\infty} b_p(n) z^n \tag{78}$$

such that

$$F(\tau) = z_p'(\tau)^{k/2}\, \tilde{F}_p(z_p(\tau)), \quad \tau \in V_0, \tag{79}$$

where $V_0 \subseteq \hat{\mathbb{H}}$ *is a suitable neighborhood of* p*. Moreover, for any* $\gamma \in \Gamma$,

$$N_p = \operatorname{ord} \tilde{F}_p(z) = \operatorname{ord} \tilde{F}_{\gamma p}(z) = N_{\gamma p};$$

i.e., the order, $\operatorname{ord} \tilde{F}_p(z)$*, is independent from the choice of* p *as a representative of* $P = [p]$.

Proof Take $t \in M(\Gamma)$ with $\tilde{t}_p(z)$ as in (76). Consequently,

$$t'(\tau) = z_p'(\tau) \sum_{n=M_p}^{\infty} n\, a_p(n) z_p(\tau)^{n-1} \in M_2(\Gamma), \tag{80}$$

and applying Lemma 7.4 to the modular function $F/(t')^{k/2} \in M(\Gamma)$ proves the first part of the statement. To prove the invariance of the order when choosing different orbit representatives, assume that $r = \gamma p$ for $\gamma = \left(\begin{smallmatrix} a & b \\ c & d \end{smallmatrix}\right) \in \Gamma$. If $p \in \mathbb{H}$ is not an elliptic point, for the expansion with respect to r we have by (73),

$$F(\gamma\tau) = z_r'(\gamma\tau)^{k/2}\, \tilde{F}_r(z_r(\gamma\tau)) = \tilde{F}_r\Big(\frac{1}{cp+d}\cdot\frac{z_p(\tau)}{c\tau+d}\Big).$$

This, together with the modular transformation property, implies

$$\begin{aligned} F(\tau) &= \frac{1}{(c\tau+d)^k}F(\gamma\tau) = \frac{1}{(c\tau+d)^k}\tilde{F}_r\Big(\frac{1}{cp+d}\cdot\frac{z_p(\tau)}{c\tau+d}\Big) \\ &= \tilde{F}_p(z_p(\tau)), \end{aligned}$$

where the last line is by (79) with $z_p'(\tau) = 1$. Hence by $-d/c \notin \mathbb{H}$, and observing that

$$\frac{1}{c\tau+d} = \frac{1}{cp+d} - \frac{c(\tau-p)}{(cp+d)^2} + O\left((\tau-p)^2\right) = \frac{1}{cp+d} + O(z_p(\tau)),$$

we have $\operatorname{ord}\tilde{F}_r = \operatorname{ord}\tilde{F}_p$. Second, suppose $p \in \mathbb{H}$ is an elliptic point. Then by (74),

$$z_r(\gamma\tau) = \xi(p)z_p(\tau)\text{ for }\xi(p) := \Big(\frac{c\overline{p}+d}{cp+d}\Big)^{h(p)}.$$

Hence,

$$\begin{aligned} F(\gamma\tau) &= z_r'(\gamma\tau)^{k/2}\, \tilde{F}_r(z_r(\gamma\tau)) = \Big(\frac{\xi(p)}{(\gamma\tau)'}\Big)^{k/2} z_p'(\tau)^{k/2}\tilde{F}_r(\xi(p)z_p(\tau)) \\ &= \xi(p)^{k/2}(c\tau+d)^k z_p'(\tau)^{k/2}\tilde{F}_r(\xi(p)z_p(\tau)). \end{aligned}$$

This implies, similarly to above,

$$\begin{aligned} F(\tau) &= \frac{1}{(c\tau+d)^k}F(\gamma\tau) = \xi(p)^{k/2}z_p'(\tau)^{k/2}\tilde{F}_r(\xi(p)z_p(\tau)) \\ &= z_p'(\tau)^{k/2}\tilde{F}_p(z_p(\tau)); \end{aligned}$$

which gives $\operatorname{ord}\tilde{F}_r(z) = \operatorname{ord}\tilde{F}_p(z)$. Finally, suppose $p = \gamma_0\infty \in \hat{\mathbb{Q}}$, $\gamma_0 \in \mathrm{SL}_2(\mathbb{Z})$. Then applying (75) gives,

$$\begin{aligned} F(\gamma\tau) &= z_r'(\gamma\tau)^{k/2}\, \tilde{F}_r(z_r(\gamma\tau)) = \Big(\frac{z_p'(\tau)}{(\gamma\tau)'}\Big)^{k/2} \tilde{F}_r(z_p(\tau)) \\ &= (c\tau+d)^k\, z_p'(\tau)^{k/2}\tilde{F}_r(z_p(\tau)). \end{aligned}$$

Invoking the modular transformation property as above, we obtain

$$F(\tau) = \frac{1}{(c\tau + d)^k} F(\gamma\tau) = z'_p(\tau)^{k/2} \tilde{F}_r(z_p(\tau))$$
$$= z'_p(\tau)^{k/2} \tilde{F}_p(z_p(\tau));$$

which implies ord $\tilde{F}_r(z) =$ ord $\tilde{F}_p(z)$, and which completes the proof of the lemma. □

To state the analogue of Lemma 7.6 for odd k, we need the square root of a Laurent series, which is a Puiseux series defined as follows.

Definition 7.7 *Let* $G(z) = \sum_{n=N}^{\infty} c(n) z^n$ *with* ord $G(z) = N$. *Then*

$$G(z)^{1/2} := \sqrt{c(N)}\, z^{N/2} (1 + \psi(z))^{1/2} = \sqrt{c(N)}\, z^{N/2} \sum_{\ell=0}^{\infty} \binom{1/2}{\ell} \psi(z)^{\ell},$$

where $\psi(z) = \frac{1}{c(N)} \sum_{n=1}^{\infty} c(N+n) z^n$, *and where for* $\sqrt{c(N)}$ *we choose the principal branch.*

Lemma 7.8 *Given a non-zero* $F \in M_k(\Gamma)$ *with k odd, and* $P = [p] \in X(\Gamma)$, $p \in \hat{\mathbb{H}}$. *Then*

$$F(\tau) = z'_p(\tau)^{k/2}\, \tilde{F}_p(z_p(\tau)), \quad \tau \in V_0, \tag{81}$$

with

$$\tilde{F}_p(z) := \tilde{G}_p(z)^{1/2};$$

where $\tilde{G}_p(z)$ *is the Laurent series such that, according to Lemma 7.6,*

$$G(\tau) := F(\tau)^2 = z'_p(\tau)^k\, \tilde{G}_p(z_p(\tau)) \in M_{2k}(\Gamma), \quad \tau \in V_0, \tag{82}$$

where $V_0 \subseteq \hat{\mathbb{H}}$ *is a suitable neighborhood of p, and where the root* $z'_p(z)^{1/2}$ *is chosen as the appropriate branch.*

Proof The existence of the Laurent series $\tilde{G}_(z)$ such that (82) is owing to Lemma 7.6 since $G(\tau) := F(\tau)^2 \in M_{2k}(\Gamma)$. The rest follows from Definition 7.7. □

For our analysis of poles it will be convenient to extend Definition 7.5 to modular forms.

Definition 7.9 *(order of a modular form at a point) For non-zero* $F \in M_k(\Gamma)$ *with k even, and* $P = [p] \in X(\Gamma)$, $p \in \hat{\mathbb{H}}$:

$$\mathrm{Ord}_P\, F := \mathrm{ord}\, \tilde{F}_p(z), \tag{83}$$

where $\tilde{F}_p$ *is as in Lemma 7.6.*

Remark 7.10 This definition can be extended to k odd, but we do not need it here.

Remark 7.11 A more important remark concerns the fact that if $F \in M(\Gamma)$ (i.e., if $k = 0$), one has,

$$\mathrm{Ord}_P\, F = \mathrm{Ord}_P\, \hat{F}, \tag{84}$$

where $P = [p] \in X(\Gamma)$. Here the order on the right side is the order at P of the induced meromorphic function $\hat{F} : X(\Gamma) \to \hat{\mathbb{C}}$ defined on the Riemann surface $X = X(\Gamma)$; see Sect. 3.2. In this case, one has, according to Lemma 3.1,

$$\sum_{P \in X(\Gamma)} \mathrm{Ord}_P\, F = 0. \tag{85}$$

We will need to generalize this fact to the relation (90) for modular forms $F \in M_k(\Gamma)$; this is the main motivation to introduce the order as in Definition 7.9

Before stating and proving (90), we extend the notion NofPoles(F) from modular functions to modular forms.

Definition 7.12 *Let $F \in M_k(\Gamma)$ with k even.*

$$\mathrm{NofPoles}(F) := - \sum_{\substack{[p] \in X(\Gamma) \\ \mathrm{ord}\, \tilde{F}_p(z) < 0}} \mathrm{Ord}_{[p]}\, F, \tag{86}$$

where $\tilde{F}_p(z)$ is the Laurent series defined in Lemma 7.6.

Remark 7.13 When $k = 0$; i.e., if F is a modular function in $M(\Gamma)$, then this coincides with the definition (60). This means, then NofPoles(F) is nothing but the number of poles of the induced function $\hat{F}$ on $X(\Gamma)$.

We conclude this section by a fact which is useful in applying the algorithm ModFormDE; see, for example, Sect. 5. Given a modular form of even weight, it relates our two different notions of orders taken at cusps.

Lemma 7.14 *Let $F \in M_{2k}(\Gamma)$ and $P = [a/c] \in X(\Gamma)$, $p = a/c \in \hat{\mathbb{Q}}$. Then,*

$$\mathrm{ord}_{a/c}\, F = \mathrm{Ord}_{[a/c]}\, F + k. \tag{87}$$

Proof By Lemma 7.6,

$$F(\tau) = z_{a/c}'(\tau)^k \sum_{n=N_{a/c}}^{\infty} b(n) z_{a/c}(\tau)^n, \tag{88}$$

where $N_{a/c} = \mathrm{Ord}_{[a/c]}\, F$. If $\gamma_0 = \left(\begin{smallmatrix} a & b \\ c & d \end{smallmatrix}\right) \in \mathrm{SL}_2(\mathbb{Z})$ then $\gamma_0 \infty = a/c$, and by (71),

$$z_p(\tau) := e^{2\pi i \gamma_0^{-1}\tau / w_0} \text{ where } w_0 = w_{\gamma_0}(\Gamma).$$

Define,

$$q_{w_0} := z_{a/c}(\gamma_0\tau) = e^{2\pi i\tau/w_0}, \text{ and thus, } (c_0\tau + d_0)^{-2} z'_{a/c}(\gamma_0\tau) = \frac{2\pi i}{w_0} q_{w_0}. \quad (89)$$

The weight-$2k$ action applied to F, and using (88) gives,

$$\begin{aligned}(F|_{2k}\gamma_0)(\tau) &= (c_0\tau + d_0)^{-2k} z'_{a/c}(\gamma_0\tau)^k \sum_{n=N_{a/c}}^{\infty} b(n) q_{w_0}^n \\ &= \left(\frac{2\pi i}{w_0}\right)^k q_{w_0}^k \sum_{n=N_{a/c}}^{\infty} b(n) q_{w_0}^n,\end{aligned}$$

where the last equality is by (89). Comparing this to the definition (21), proves the statement. □

Notice that for $k = 0$, Lemma 7.14 turns into (84) where P is a cusp.

7.2 A Bound from the Riemann-Hurwitz Formula

A last major ingredient for our analysis of poles is the following proposition which is an application of the Riemann-Hurwitz formula.

Proposition 7.15 *Let $F \in M_k(\Gamma)$ be non-zero with k even. Then*

$$\sum_{P \in X(\Gamma)} \mathrm{Ord}_P\, F = k(g_\Gamma - 1), \quad (90)$$

where g_Γ is the genus of the compact Riemann surface $X(\Gamma)$.

Remark 7.16 Notice that the sum on the left side is well-defined. Namely, suppose that $\mathrm{ord}_P\, F \neq 0$ for infinitely finitely many $P \in X(\Gamma)$. Choosing a bounded fundamental domain $\mathcal{F} \subset \hat{\mathbb{H}}$ for $X(\Gamma)$, this would imply that F would have infinitely many poles or zeros in $\mathcal{F}$. Let us assume the latter.[17] Then this set has a limit point in $\hat{\mathbb{H}}$. Since F is non-zero, this limit point must be in $\hat{\mathbb{Q}}$. But property (19) in the definition of modular forms implies that the poles of F cannot cluster at any $a/c \in \hat{\mathbb{Q}}$; see, e.g., the remark in [6, Def. 3.2.1].

We will apply Proposition 7.15 in the following form.

Corollary 7.17 *Let $F \in M_k(\Gamma)$, k even, such that for $P = [p] \in X(\Gamma)$,*

$$F(\tau) = z'_p(\tau)^{k/2} \tilde{F}_p(z_p(\tau)),$$

[17] Otherwise, consider $1/F$.

where $\tilde{F}_p(z)$ is a Laurent series as in Lemma 7.6. Then

$$\sum_{[p]\in X(\Gamma)} \operatorname{ord} \tilde{F}_p(z) = k(g_\Gamma - 1), \tag{91}$$

where g_Γ is the genus of the compact Riemann surface $X(\Gamma)$.

Convention. When here and in the following the domain of a sum is specified as "$[p] \in X(\Gamma)$", then this is understood as follows: For each $P = [p] \in X(\Gamma)$ take exactly one representative $p \in \hat{\mathbb{H}}$; the sum then runs over all such p.

Before we prove Proposition 7.15, we prepare with a few facts.

Lemma 7.18 *Let F and G be non-zero modular forms in $M_k(\Gamma)$ with k even. Then for $P \in X(\Gamma)$,*

$$\operatorname{Ord}_P F - \operatorname{Ord}_P G = \operatorname{Ord}_P(F/G). \tag{92}$$

Proof Apply Lemma 7.6 together with Lemma 7.4. □

Lemma 7.19 *Let F and G be non-zero modular forms in $M_k(\Gamma)$ with k even. Then*

$$\sum_{P\in X(\Gamma)} \operatorname{Ord}_P F = \sum_{P\in X(\Gamma)} \operatorname{Ord}_P G. \tag{93}$$

Proof In view of $F/G \in M(\Gamma)$, apply Lemma 3.1 together with Lemma 7.18 □

As already mentioned, for the proof of Proposition 7.15 we utilize the Riemann-Hurwitz formula for a non-constant meromorphic function $\varphi : X(\Gamma) \to \hat{\mathbb{C}}$ which has exactly n poles in $X(\Gamma)$, counting multiplicities; see, e.g., [10, Ch. 2, Thm. 4.16][18]:

$$\sum_{P\in X(\Gamma)} (\operatorname{mult}_P \varphi - 1) = 2\, g_\Gamma + 2n - 2. \tag{94}$$

Here g_Γ denotes the genus of $X(\Gamma)$, and $\operatorname{mult}_P \varphi$ is the usual multiplicity of φ at P; we recall its exact definition in Sect. 3.2.

Now we are ready to prove the sum estimate (90).

Proof of Proposition 7.15 For $k = 0$ this is (24). To prove the statement for weight $k = 2$, take some $t \in M(\Gamma)$ and suppose $\hat{t}$ has exactly n poles in $X(\Gamma)$. For $P \in X(\Gamma)$, notice that

$$\operatorname{mult}_P \hat{t} = \begin{cases} -\operatorname{Ord}_P t, & \text{if } P \text{ is a pole of } \hat{t} \\ \operatorname{Ord}_P t' + 1, & \text{otherwise} \end{cases}. \tag{95}$$

[18] Actually, the special case of [10, Ch. 2, Thm. 4.16] we need, $Y = \hat{\mathbb{C}}$, was given by Riemann; e.g., [5].

The second equality is a consequence of using Definition 7.9 on $t' \in M_2(\Gamma)$ together with (80). In addition, we need that at a pole $P \in X(\Gamma)$ of $\hat{t}$,

$$\mathrm{Ord}_P\, t' - \mathrm{Ord}_P\, t = -1.$$

Using these properties and defining,

$$\mathrm{Poles}(t) := \{P \in X(\Gamma) : P \text{ is a pole of } \hat{t}\}, \tag{96}$$

we obtain by (95),

$$\begin{aligned}\sum_{P\in \mathrm{Poles}(t)} (\mathrm{mult}_P\, \hat{t} - 1) &= \sum_{P\in \mathrm{Poles}(t)} (-\,\mathrm{Ord}_P\, t - 1)\\ = -2 \sum_{P\in \mathrm{Poles}(t)} \mathrm{Ord}_P\, t &+ \sum_{P\in \mathrm{Poles}(t)} \mathrm{Ord}_P\, t' = 2n + \sum_{P\in \mathrm{Poles}(t)} \mathrm{Ord}_P\, t',\end{aligned}$$

and,

$$\sum_{P\in X(\Gamma)\setminus \mathrm{Poles}(t)} (\mathrm{mult}_P\, \hat{t} - 1) = \sum_{P\in X(\Gamma)\setminus \mathrm{Poles}(t)} \mathrm{Ord}_P\, t'.$$

Combining these two sums we obtain, as a consequence of Lemma 3.1 and Lemma 7.18,

$$\begin{aligned}0 &= \sum_{P\in X(\Gamma)} \mathrm{Ord}_P(F/t') = \sum_{P\in X(\Gamma)} \mathrm{Ord}_P\, F - \sum_{P\in X(\Gamma)} \mathrm{Ord}_P\, t'\\ &= \sum_{P\in X(\Gamma)} \mathrm{Ord}_P\, F + 2n - \sum_{P\in X(\Gamma)} (\mathrm{mult}_P\, \hat{t} - 1)\\ &= \sum_{P\in X(\Gamma)} \mathrm{Ord}_P\, F - 2g_\Gamma + 2.\end{aligned}$$

This proves (7.15) for $F \in M_2(\Gamma)$; for the last equality we invoked the Riemann-Hurwitz formula (94).

To prove (7.15) for $F \in M_k(\Gamma)$ where $k \geq 2$ is even, take some $g \in M_2(\Gamma)$. Then $f := F/g^{k/2} \in M(\Gamma)$, and one has,

$$\begin{aligned}0 &\overset{(24)}{=} \sum_{P\in X(\Gamma)} \mathrm{Ord}_P\, f \overset{(92)}{=} \sum_{P\in X(\Gamma)} \mathrm{Ord}_P\, F - \frac{k}{2} \sum_{P\in X(\Gamma)} \mathrm{Ord}_P\, g\\ &= \sum_{P\in X(\Gamma)} \mathrm{Ord}_P\, F - \frac{k}{2}(2g_\Gamma - 2)),\end{aligned}$$

where the last line is by the $k = 2$ part already proven. This completes the proof of Proposition 7.15. □

8 Bounds for p_1

For the whole section we assume that p_1 and p_2 are defined as in (31), but now with h replaced by $t \in M(\Gamma)$ and g replaced by $F \in M_k(\Gamma)$. Throughout, $x = q^{1/w_0}$ with $q = e^{2\pi i\tau}$.

8.1 Rewriting p_1

We need to rewrite $p_1 \in M(\Gamma)$ in a form which is convenient for the analysis of poles.

Let $P = [p] \in X(\Gamma)$ be fixed. For $t \in M(\Gamma)$ let

$$t(\tau) = T_p(z_p(\tau)), \quad \text{where } T_p(z) := \tilde{t}_p(z) \tag{97}$$

is the Laurent series as defined in (76).

Derivatives of a Laurent series $T(z) = \sum_{n=M}^{\infty} a(n)z^n$ are defined as usual by

$$T^{(j)}(z) := \sum_{n=M}^{\infty} (n)_j\, a(n)z^{n-j}, \quad j = 0, 1, \ldots$$

In particular, $T(z) = T^{(0)}(z)$, $T'(z) = T^{(1)}(z)$, and $T''(z) = T^{(2)}(z)$. Hence, relation (80) turns into,

$$D_x t(\tau) = \frac{w_0}{2\pi i} t'(\tau) = \frac{w_0}{2\pi i} z_p'(\tau) T_p^{(1)}(z_p(\tau)). \tag{98}$$

For $D_x^2 t$ one has,

$$D_x^2 t = \frac{w_0^2}{(2\pi i)^2}\Big(z_p'' T_p^{(1)}(z_p) + (z_p')^2 T_p^{(2)}(z_p)\Big), \tag{99}$$

where we omitted the argument τ.

Given the same fixed $p \in \hat{\mathbb{H}}$ as above in the orbit $P = [p]$:

Case A: For $F \in M_k(\Gamma)$, k even, let

$$F(\tau) = z_p'(\tau)^{k/2} f_{0,p}(z_p(\tau)), \quad \text{where } f_{0,p}(z) := \tilde{F}_p(z) \tag{100}$$

is the Laurent series defined as in (78).

Case B: For $F \in M_k(\Gamma)$, k odd, let

$$F(\tau) = z_p'(\tau)^{k/2} f_{1,p}(z_p(\tau)), \quad \text{where } f_{1,p}(z) := \tilde{G}_p(z)^{1/2} \tag{101}$$

is the square root of the Laurent series $\tilde{G}_p(z)$ defined as in (82) such that

$$G(\tau) := F(\tau)^2 = z_p'(\tau)^k \, \tilde{G}_p(z_p(\tau)) \in M_{2k}(\Gamma). \tag{102}$$

Combining both Cases A and B, one has[19]

$$\frac{D_x F}{F} = \frac{w_0}{2\pi i}\Big(\frac{z_p' \, f_{\delta,p}^{(1)}(z_p)}{f_{\delta,p}(z_p)} + \frac{k}{2} \cdot \frac{z_p''}{z_p'}\Big), \tag{103}$$

where $\delta = 0$ if k is even, and $\delta = 1$ if k is odd.

Lemma 8.1 *Given $t \in M(\Gamma)$, $F \in M_k(\Gamma)$, and $P = [p] \in X(\Gamma)$, $p \in \hat{\mathbb{H}}$. Then there exists a Laurent series,*

$$P_{1,p}(z) := -1 + \frac{T_p(z)}{T_p^{(1)}(z)}\Big(\frac{T_p^{(2)}(z)}{T_p^{(1)}(z)} - \frac{2}{k}\frac{f_{\delta,p}^{(1)}(z)}{f_{\delta,p}(z)}\Big), \tag{104}$$

where $\delta = 0$ if k is even, and $\delta = 1$ if k is odd, such that

$$p_1(\tau) = P_{1,p}(z_p(\tau)), \quad \tau \in V_0, \tag{105}$$

V_0 being a suitable open neighborhood of p.

Proof By (31), with t instead of h and with F instead of g, and using

$$D_x G_1 = \frac{D_x^2 t}{t} - \frac{(D_x t)^2}{t^2},$$

one has, applying (98), (98), and (103),

$$\begin{aligned}
p_1 &= \Big(\frac{t}{D_x t}\Big)^2\Big(\frac{D_x^2 t}{t} - \frac{(D_x t)^2}{t^2}\Big) - \frac{2}{k}\frac{t}{D_x t}\frac{D_x F}{F} \\
&= -1 + \frac{t}{D_x t}\Big(\frac{D_x^2 t}{D_x t} - \frac{2}{k}\frac{D_x F}{F}\Big) = -1 + \frac{2\pi i}{w_0}\frac{T_p(z_p)}{z_p' T_p^{(1)}(z_p)} \\
&\times \Bigg(\frac{w_0}{2\pi i}\frac{z_p'' T_p^{(1)}(z_p) + (z_p')^2 T_p^{(2)}(z_p)}{z_p' T_p^{(1)}(z_p)} - \frac{2}{k}\frac{w_0}{2\pi i}\Big(\frac{z_p' \, f_{\delta,p}^{(1)}(z_p)}{f_{\delta,p}(z_p)} + \frac{k}{2} \cdot \frac{z_p''}{z_p'}\Big)\Bigg) \\
&= -1 + \frac{T_p(z_p)}{z_p' T_p^{(1)}(z_p)}\Big(\frac{z_p' T_p^{(2)}(z_p)}{T_p^{(1)}(z_p)} - \frac{2}{k}\frac{z_p' \, f_{\delta,p}^{(1)}(z_p)}{f_{\delta,p}(z_p)}\Big).
\end{aligned}$$

□

[19] We again omit the argument τ.

8.2 Bounds for the Poles of p_1

To estimate the number of poles of $p_1 \in M(\Gamma)$, Lemma 8.1 implies,

$$\begin{aligned}\mathrm{NofPoles}(p_1) &= -\sum_{P\in\mathrm{Poles}(p_1)} \mathrm{Ord}_P\, p_1 = -\sum_{[p]\in\mathrm{Poles}(p_1)} \mathrm{ord}\, P_{1,p}(z)\\ &\leq -\sum_{[p]\in\mathrm{Poles}(p_1)} \mathrm{ord}\,\frac{T_p(z)}{T_p^{(1)}(z)} - \sum_{[p]\in\mathrm{Poles}(p_1)} \mathrm{ord}\,\frac{T_p^{(2)}(z)}{T_p^{(1)}(z)}\\ &\quad -\sum_{[p]\in\mathrm{Poles}(p_1)} \mathrm{ord}\,\frac{f_{\delta,p}^{(1)}(z)}{f_{\delta,p}(z)},\end{aligned} \tag{106}$$

with $\delta = 0$ if k is even, and $\delta = 1$ if k is odd.

To proceed with our pole estimation, we treat each of the sums in (106) separately. To this end, it will be convenient to define

$$\pi(x) := \begin{cases} x, & \text{if } x < 0\\ 0, & \text{if } x \geq 0\end{cases}, \quad \text{and } \zeta(x) := \begin{cases} x, & \text{if } x > 0\\ 0, & \text{if } x \leq 0\end{cases}.$$

In addition, the following three facts will be useful.

Lemma 8.2

$$\sum_{[p]\in X(\Gamma)} \mathrm{ord}\, T_p^{(1)}(z) = \sum_{[p]\in X(\Gamma)} \pi(\mathrm{ord}\, T_p^{(1)}(z)) + \sum_{[p]\in X(\Gamma)} \zeta(\mathrm{ord}\, T_p^{(1)}(z)) = 2(g_\Gamma - 1).$$

Proof In view of (98), the statement is proved by applying Corollary 7.17 to $t' \in M_2(\Gamma)$. □

The second lemma we want to list explicitly is trivial, but useful.

Lemma 8.3 *Let $f(z) = \sum_{n=M}^{\infty} c(n)z^n$ be a Laurent series, then:*

$$\mathrm{ord}\, f(z) < 0 \Rightarrow \mathrm{ord}\, f^{(j)}(z) = \mathrm{ord}\, f(z) - j \;\; \textit{for } j \in \mathbb{Z}_{\geq 0}.$$

Finally, we need the following lemma:

Lemma 8.4 *Let $f : X \to \mathbb{C} \cup \{\infty\}$ be a meromorphic function on a compact Riemann surface X. Let $f_P(z)$ be local Laurent series respresentations of f at each $P \in X$. For $k \in \mathbb{Z}_{\geq 0}$ define*

$$\begin{aligned} S_1 &:= \{P \in X : f_P(0) = \infty\},\\ S_2(k) &:= \{P \in X : f_P(0) \neq \infty,\, f_P^{(k)}(0) = 0\},\\ S_3(k) &:= \{P \in X : f_P(0) \neq \infty,\, f_P^{(k)}(0) \neq 0\}.\end{aligned}$$

Then for $j, k \in \mathbb{Z}_{\geq 0}$,

$$-\sum_{P \in X} \frac{f_P^{(j)}(z)}{f_P^{(k)}(z)} \leq |S_1|(j-k) + \sum_{P \in S_2(k)} \operatorname{ord} f_P^{(k)}(z).$$

Proof First we note that as a disjoint union, $S_1 \cup S_2(k) \cup S_3(k) = X$. The following facts are simple verifications:

$$-\sum_{P \in S_1} \operatorname{ord} \frac{f^{(j)}(z)}{f^{(k)}(z)} = |S_1|(j-k), \ -\sum_{P \in S_2(k)} \operatorname{ord} \frac{f^{(j)}(z)}{f^{(k)}(z)} \leq \sum_{P \in S_2(k)} \operatorname{ord} f^{(k)}(z),$$

$$\begin{aligned} -\sum_{P \in S_3(k)} \operatorname{ord} \frac{f^{(j)}(z)}{f^{(k)}(z)} &= -\sum_{P \in S_3(k)} \operatorname{ord} f_P^{(j)}(z) + \sum_{P \in S_3(k)} \operatorname{ord} f_P^{(k)}(z) \\ &= -\sum_{P \in S_3(k)} \operatorname{ord} f_P^{(j)}(z) \leq 0. \end{aligned}$$

Combining these facts gives the statement. □

We begin with bounds for the first two sums on the right side of (106).

Lemma 8.5 *For* $j \in \mathbb{Z}_{\geq 0}$, $j \neq 1$,

$$-\sum_{[p] \in \mathrm{Poles}(p_1)} \operatorname{ord} \frac{T_p^{(j)}(z)}{T_p^{(1)}(z)} \leq 2g_\Gamma - 2 - \sum_{[p] \in X(\Gamma)} \pi(\operatorname{ord} T_p(z)) + j \sum_{\substack{[p] \in X(\Gamma) \\ \operatorname{ord} T_p(z) < 0}} 1.$$

Proof By Lemma 8.4 we have

$$-\sum_{[p] \in X} \frac{T_p^{(j)}(z)}{T_p^{(1)}(z)} \leq |S_1|(j-1) + \sum_{[p] \in S_2(k)} \operatorname{ord} T_p^{(1)}(z).$$

Note that

$$\sum_{[p] \in S_2(k)} \operatorname{ord} T_p^{(1)}(z) = \sum_{[p] \in X} \zeta(\operatorname{ord} T_p^{(1)}(z))$$

Also note that by Lemma 8.2,

$$\sum_{[p] \in X} \zeta(\operatorname{ord} T_p^{(1)}(z)) = 2g_\Gamma - 2 - \sum_{[p] \in X(\Gamma)} \pi(\operatorname{ord} T_p^{(1)}(z)).$$

Now the statement follows by Lemma 8.3 in the version,

$$-\sum_{[p] \in X(\Gamma)} \pi(\operatorname{ord} T_p^{(1)}(z)) = -\sum_{[p] \in X(\Gamma)} \pi(\operatorname{ord} T_p(z)) + \sum_{\substack{[p] \in X(\Gamma) \\ \operatorname{ord} T_p(z) < 0}} 1. \tag{107}$$

together with

$$|S_1| = \sum_{\substack{[p]\in X(\Gamma)\\ \operatorname{ord} T_p(z)<0}} 1.$$

□

The following simplification of the upper bound is straightforward.

Corollary 8.6 *For $j \in \mathbb{Z}_{\geq 0}$, $j \neq 1$,*

$$-\sum_{[p]\in \mathrm{Poles}(p_1)} \operatorname{ord} \frac{T_p^{(j)}(z)}{T_p^{(1)}(z)} \leq 2g_\Gamma - 2 + (j+1)\ \mathrm{NofPoles}(t).$$

Proof Obviously,

$$\sum_{\substack{[p]\in X(\Gamma)\\ \operatorname{ord} T_p(z)<0}} 1 \leq -\sum_{[p]\in X(\Gamma)} \pi(\operatorname{ord} T_p(z))$$

and, by definition (97),

$$-\sum_{[p]\in X(\Gamma)} \pi(\operatorname{ord} T_p(z)) = \mathrm{NofPoles}(t).$$

This implies the corollary. □

Next we treat the even case of the third sum in (106).

Lemma 8.7 *For $j \in \mathbb{Z}_{\geq 1}$ and $i \in \{1, 2\}$:*

$$-\sum_{[p]\in \mathrm{Poles}(p_i)} \operatorname{ord} \frac{f_{0,p}^{(j)}(z)}{f_{0,p}(z)} \leq k(g_\Gamma - 1) - \sum_{[p]\in X(\Gamma)} \pi(\operatorname{ord} f_{0,p}(z)) + j \sum_{\substack{[p]\in X(\Gamma)\\ \operatorname{ord} f_{0,p}(z)<0}} 1.$$

Proof The proof works analogously to that of Lemma 8.5; the only difference is that one uses Lemma 8.2 for general even k in the version,

$$\sum_{[p]\in X(\Gamma)} \operatorname{ord} f_{0,p}(z) = \sum_{[p]\in X(\Gamma)} \pi(\operatorname{ord} f_{0,p}(z)) + \sum_{[p]\in X(\Gamma)} \zeta(\operatorname{ord} f_{0,p}(z)) = k(g_\Gamma - 1).$$

□

Corollary 8.8 *For $j \in \mathbb{Z}_{\geq 1}$ and $i \in \{1, 2\}$:*

$$-\sum_{[p]\in \mathrm{Poles}(p_i)} \operatorname{ord} \frac{f_{0,p}^{(j)}(z)}{f_{0,p}(z)} \leq k(g_\Gamma - 1) + (j+1)\,\mathrm{NofPoles}(F).$$

Proof Obviously,

$$\sum_{\substack{[p]\in X(\Gamma)\\ \operatorname{ord} f_{0,p}(z)<0}} 1 \leq -\sum_{[p]\in X(\Gamma)} \pi(\operatorname{ord} f_{0,p}(z)).$$

By Definition 7.12,

$$-\sum_{[p]\in X(\Gamma)} \pi(\operatorname{ord} f_{0,p}(z)) = \operatorname{NofPoles}(F).$$

This implies the corollary. □

Case (1a), k even (i.e., $\delta = 0$): Applying Corollary 8.6 and 8.8 to (106) gives,

$$\begin{aligned}\operatorname{NofPoles}(p_1) &\leq 2g_\Gamma - 2 + \operatorname{NofPoles}(t) + 2g_\Gamma - 2 + 3\operatorname{NofPoles}(t)\\ &\quad + k(g_\Gamma - 1) + 2\operatorname{NofPoles}(F)\\ &= (k+4)(g_\Gamma - 1) + 4\operatorname{NofPoles}(t) + 2\operatorname{NofPoles}(F). \qquad (108)\end{aligned}$$

For Case (1b), i.e., k odd and $\delta = 1$, we need one more observation. Recalling the definition (101), one has,

$$\frac{f_{1,p}^{(1)}(z)}{f_{1,p}(z)} = \frac{1}{2}\frac{\tilde{G}_p^{(1)}(z)}{\tilde{G}_p(z)}, \qquad (109)$$

with $f_{1,p}(z) = \tilde{G}_p(z)^{1/2}$, where $\tilde{G}_p(z)$ is the Laurent series such that for $G := F^2 \in M_{2k}(\Gamma)$:

$$G(\tau) = z_p'(\tau)^k\, \tilde{G}_p(z_p(\tau)).$$

Consequently, Corollary 8.8 with $j = 1$ carries over to this situation with $2k$ instead of k and $G = F^2$ instead of F; namely:

Corollary 8.9 *For $j \in \mathbb{Z}_{\geq 1}$ and $i \in \{1, 2\}$,*

$$-\sum_{[p]\in \operatorname{Poles}(p_i)} \operatorname{ord} \frac{f_{1,p}^{(1)}(z)}{f_{1,p}(z)} \leq 2k(g_\Gamma - 1) + 2\operatorname{NofPoles}(F^2).$$

Now we establish the bound estimate for odd k.

Case (1b), k odd (i.e., $\delta = 1$): Applying Corollary 8.6 and 8.9 to (106) gives,

$$\begin{aligned}\operatorname{NofPoles}(p_1) &\leq 2g_\Gamma - 2 + \operatorname{NofPoles}(t) + 2g_\Gamma - 2 + 3\operatorname{NofPoles}(t)\\ &\quad + 2k(g_\Gamma - 1) + 2\operatorname{NofPoles}(F^2)\\ &= (2k+4)(g_\Gamma - 1) + 4\operatorname{NofPoles}(t) + 2\operatorname{NofPoles}(F^2). \qquad (110)\end{aligned}$$

9 Bounds for p_2

As in Sect. 8 we assume that p_1 and p_2 are defined as in (31), but now again with h replaced by t, and g replaced by F. Again, $x = q^{1/w_0}$ with $q = e^{2\pi i\tau}$.

9.1 Rewriting p_2

As with p_1, we first need to rewrite $p_2 \in M(\Gamma)$ in a form which is convenient for the analysis of poles.

Let $P = [p] \in X(\Gamma)$ be fixed. Again as in Sect. 8 we assume that for $t \in M(\Gamma)$ the Laurent series $T_p(z)$ is defined as in (97) such that

$$t(\tau) = T_p(z_p(\tau)), \quad \text{where } T_p(z) := \tilde{t}_p(z); \tag{111}$$

for $F \in M_k(\Gamma)$, the Laurent series $f_{\delta,p}(z_p(\tau))$, $\delta \in \{0, 1\}$,

$$F(\tau) = z_p'(\tau)^{k/2} f_{\delta,p}(z_p(\tau)) \text{ where } f_{\delta,p}(z) := \tilde{F}_p(z),$$

are defined as in (100) if k is even, and as in (101) if k is odd. In the even case, we take $\delta = 0$, in the odd case $\delta = 1$.

For p_1 we used the formula (103) for $D_q F/F$. For p_2 we also need,

$$\frac{D_x^2 F}{F} = \Big(\frac{w_0}{2\pi i}\Big)^2 \Big(\frac{(z_p')^2 f_{\delta,p}^{(2)}(z_p)}{f_{\delta,p}(z_p)} + (k+1)\cdot\frac{z_p'' f_{\delta,p}^{(1)}(z_p)}{f_{\delta,p}(z_p)} + \frac{k}{2}\cdot\frac{z_p^{(3)}}{z_p'} + \frac{k(k-2)}{4}\cdot\frac{(z_p'')^2}{(z_p')^2}\Big),$$

which, as (103), can be derived by straightforward computation.

This relation together with (103) gives,

$$\begin{aligned} p_2 &= \frac{1}{G_1^2}\Big(D_x G_2 - \frac{1}{k}G_2^2\Big) = \Big(\frac{t}{D_x t}\Big)^2 \Big(D_x \frac{D_x F}{F} - \frac{1}{k}\Big(\frac{D_x F}{F}\Big)^2\Big) \\ &= \Big(\frac{t}{D_x t}\Big)^2 \Big(\frac{D_x^2 F}{F} - (1+\frac{1}{k})\Big(\frac{D_x F}{F}\Big)^2\Big) = \frac{T_p(z_p)^2}{T_p^{(1)}(z_p)^2} \\ &\times \Big(\frac{f_{\delta,p}^{(2)}(z_p)}{f_{\delta,p}(z_p)} - (1+\frac{1}{k})\frac{f_{\delta,p}^{(1)}(z_p)^2}{f_{\delta,p}(z_p)^2} + \frac{k z_p^{(3)}}{2(z_p')^3} - \frac{3k(z_p'')^2}{4(z_p')^4}\Big). \end{aligned} \tag{112}$$

We need to consider in more detail the expression,

$$\ell(z_p^{(1)}(\tau), z_p^{(2)}(\tau), z_p^{(3)}(\tau)) := \frac{z_p^{(3)}(\tau)}{z_p'(\tau)^3} - \frac{3z_p''(\tau)^2}{2z_p'(\tau)^4}.$$

If $[p] \in X(\Gamma)$ is an ordinary point, $z_p(\tau) = \tau - p$ and

$$\ell(z_p^{(1)}(\tau), z_p^{(2)}(\tau), z_p^{(3)}(\tau)) = 0 \text{ for all } \tau \in V_0.$$

If $[p] \in X(\Gamma)$ is an elliptic point of order 2, $z_p(\tau) = (\frac{\tau-p}{\tau-\bar{p}})^2$, and one has,

$$\ell(z_p^{(1)}(\tau), z_p^{(2)}(\tau), z_p^{(3)}(\tau)) = -\frac{3}{8}\frac{1}{z_p(\tau)^2}.$$

If $[p] \in X(\Gamma)$ is an elliptic point of order 3, $z_p(\tau) = (\frac{\tau-p}{\tau-\bar{p}})^3$, and one has,

$$\ell(z_p^{(1)}(\tau), z_p^{(2)}(\tau), z_p^{(3)}(\tau)) = -\frac{4}{9}\frac{1}{z_p(\tau)^2}.$$

If $[p] \in X(\Gamma)$ is a cusp; then $p = a/c = \gamma\infty$, $\gamma = \left(\begin{smallmatrix} a & b \\ c & d \end{smallmatrix}\right) \in \mathrm{SL}_2(\mathbb{Z})$, and $z_p(\tau) := e^{2\pi i \gamma^{-1}\tau/w}$ with $w = w_\gamma(\Gamma)$ as in (20). In this case,

$$\ell(z_p^{(1)}(\tau), z_p^{(2)}(\tau), z_p^{(3)}(\tau)) = -\frac{1}{2}\frac{1}{z_p(\tau)^2}.$$

This leads us to summarize in a definition.

Definition 9.1 *Define $c : X(\Gamma) \to \mathbb{C}$ for $P = [p] \in X(\Gamma)$ as follows:*

$$c(P) := c(p) := \begin{cases} 0, & \text{if } P \text{ is an ordinary point} \\ -3/8, & \text{if } P \text{ is elliptic of order } 2 \\ -4/9, & \text{if } P \text{ is elliptic of order } 3 \\ -1/2, & \text{if } P \text{ is a cusp, } P = [a/c]. \end{cases}.$$

Summarizing, we obtained a Laurent series representation of p_2:

Lemma 9.2 *Given $t \in M(\Gamma)$, $F \in M_k(\Gamma)$, and $P = [p] \in X(\Gamma)$, $p \in \hat{\mathbb{H}}$. Then there exists a Laurent series,*

$$P_{2,p}(z) := \frac{T_p(z)^2}{T_p^{(1)}(z)^2}\Big(\frac{f_{\delta,p}^{(2)}(z)}{f_{\delta,p}(z)} - (1+\frac{1}{k}) \cdot \frac{f_{\delta,p}^{(1)}(z)^2}{f_{\delta,p}(z)^2} + \frac{k}{2} \cdot \frac{c(p)}{z^2}\Big) \tag{113}$$

where c is as in Definition 9.1, and where $\delta = 0$ if k is even, and $\delta = 1$ if k is odd, such that

$$p_2(\tau) = P_{2,p}(z_p(\tau)), \quad \tau \in V_0, \tag{114}$$

V_0 being a suitable open neighborhood of p.

9.2 Bounds for the Poles of p_2

To estimate the number of poles of $p_2 \in M(\Gamma)$, Lemma 9.2 implies,

$$\begin{aligned}
\mathrm{NofPoles}(p_2) &= -\sum_{P\in\mathrm{Poles}(p_2)} \mathrm{Ord}_P\, p_2 = -\sum_{[p]\in\mathrm{Poles}(p_2)} \mathrm{ord}\, P_{2,p}(z) \\
&\le -\sum_{[p]\in\mathrm{Poles}(p_2)} \mathrm{ord}\,\frac{T_p(z)^2}{T_p^{(1)}(z)^2} - \sum_{[p]\in\mathrm{Poles}(p_2)} \mathrm{ord}\,\frac{f_{\delta,p}^{(2)}(z)}{f_{\delta,p}(z)} \\
&\quad - \sum_{[p]\in\mathrm{Poles}(p_2)} \mathrm{ord}\,\frac{f_{\delta,p}^{(1)}(z)^2}{f_{\delta,p}(z)^2} + \sum_{\substack{[p]\in\mathrm{Poles}(p_2)\\ c(p)\neq 0}} 2,
\end{aligned} \tag{115}$$

with $\delta = 0$ if k is even, and $\delta = 1$ if k is odd.

As a consequence, similarly to the Cases (1a) and (1b), we obtain bounds for p_2: Case (2a), k even (i.e., $\delta = 0$): Applying Corollary 8.6 and 8.8 to (115) gives,

$$\begin{aligned}
\mathrm{NofPoles}(p_2) &\le 4g_\Gamma - 4 + 2\,\mathrm{NofPoles}(t) + k(g_\Gamma - 1) + 3\,\mathrm{NofPoles}(F) \\
&\quad + 2k(g_\Gamma - 1) + 4\,\mathrm{NofPoles}(F) \\
&\quad + 2\ \mathrm{NofCusps}(\Gamma) + 2\ \mathrm{NofElliptic}(\Gamma). \\
&= (3k+4)(g_\Gamma - 1) + 2\,\mathrm{NofPoles}(t) + 7\,\mathrm{NofPoles}(F) \\
&\quad + 2\ \mathrm{NofCusps}(\Gamma) + 2\ \mathrm{NofElliptic}(\Gamma).
\end{aligned} \tag{116}$$

For Case (2b), i.e., k odd and $\delta = 1$, we need one more observation. As with (109), we recall the definition (101) and obtain,

$$\frac{f_{1,p}^{(2)}(z)}{f_{1,p}(z)} = -\frac{1}{4}\frac{\tilde{G}_p^{(1)}(z)^2}{\tilde{G}_p(z)^2} + \frac{1}{2}\frac{\tilde{G}_p^{(2)}(z)}{\tilde{G}_p(z)}, \tag{117}$$

with $f_{1,p}(z) = \tilde{G}_p(z)^{1/2}$, where $\tilde{G}_p(z)$ is the Laurent series such that for $G := F^2 \in M_{2k}(\Gamma)$:

$$G(\tau) = z_p'(\tau)^k\, \tilde{G}_p(z_p(\tau)).$$

Owing to Corollary 8.8, relation (117) implies,

$$\begin{aligned}
-\sum_{[p]\in\mathrm{Poles}(p_2)} \mathrm{ord}\,\frac{f_{1,p}^{(2)}(z)}{f_{1,p}(z)} &\le 4k(g_\Gamma - 1) + 2\cdot 2\,\mathrm{NofPoles}(F^2) \\
&\quad + 2k(g_\Gamma - 1) + (2+1)\,\mathrm{NofPoles}(F^2) \\
&= 6k(g_\Gamma - 1) + 7\,\mathrm{NofPoles}(F^2).
\end{aligned} \tag{118}$$

Now we are ready to establish the bound estimate for p_2 for odd k.

Case (2b), k odd (i.e., $\delta = 1$): Applying (118), Corollary 8.6 and 8.9 to (115) gives,

$$\begin{aligned}\mathrm{NofPoles}(p_2) &\leq 4g_\Gamma - 4 + 2\,\mathrm{NofPoles}(t) + 6k(g_\Gamma - 1) + 7\,\mathrm{NofPoles}(F^2)\\ &\quad + 4k(g_\Gamma - 1) + 4\,\mathrm{NofPoles}(F^2).\\ &\quad + 2\ \mathrm{NofCusps}(\Gamma) + 2\ \mathrm{NofElliptic}(\Gamma).\\ &= (10k+4)(g_\Gamma - 1) + 2\,\mathrm{NofPoles}(t) + 11\,\mathrm{NofPoles}(F^2) \quad (119)\\ &\quad + 2\ \mathrm{NofCusps}(\Gamma) + 2\ \mathrm{NofElliptic}(\Gamma).\end{aligned}$$

We want to point out that, when setting up such bound estimates, a proper organization of the terms involved can be important. For example, we can improve upon (119) as follows.

To bound the right-hand side of (9.2), instead of treating $\phi_p(z) := \frac{f_{1,p}^{(2)}(z)}{f_{1,p}(z)}$ and $\psi_p(z) := \frac{f_{1,p}^{(1)}(z)^2}{f_{1,p}(z)^2}$ separately to obtain,

$$-\sum_{[p]\in\mathrm{Poles}(p_2)} \mathrm{ord}\,\phi_p(z) - \sum_{[p]\in\mathrm{Poles}(p_2)} \mathrm{ord}\,\psi_p(z) \leq 10k(g_\Gamma - 1) + 11\,\mathrm{NofPoles}(F^2),$$

one can combine them,

$$\phi_p(z) - \Big(1 + \frac{1}{k}\Big)\psi_p(z) \overset{(109),(117)}{=} -\frac{2k+1}{4k}\frac{\tilde{G}_p^{(1)}(z)^2}{\tilde{G}_p(z)^2} + \frac{1}{2}\frac{\tilde{G}_p^{(2)}(z)}{\tilde{G}_p(z)},$$

to obtain, using Corollary 8.8 in its version for F^2 instead of F,

$$\begin{aligned}-\sum_{[p]\in\mathrm{Poles}(p_2)} \mathrm{ord}\Big(\phi_p(z) - (1+\frac{1}{k})\psi_p(z)\Big) &\leq 4k(g_\Gamma - 1) + 2(1+1)\,\mathrm{NofPoles}(F^2)\\ &\quad + 2k(g_\Gamma - 1) + (2+1)\,\mathrm{NofPoles}(F^2)\\ &= 6k(g_\Gamma - 1) + 7\,\mathrm{NofPoles}(F^2).\end{aligned}$$

This simple reorganization leads to an improvement of the bound estimate (119):

Case (2b), k odd, improved version:

$$\begin{aligned}\mathrm{NofPoles}(p_2) &\leq (6k+4)(g_\Gamma - 1) + 2\,\mathrm{NofPoles}(t) + 7\,\mathrm{NofPoles}(F^2) \quad (120)\\ &\quad + 2\ \mathrm{NofCusps}(\Gamma) + 2\ \mathrm{NofElliptic}(\Gamma).\end{aligned}$$

10 Bounds for the Derivatives of p_1 and p_2

As in Sect. 8 and Sect. 9 we assume that p_1 and p_2 are defined as in (31), but again with h replaced by $t \in M(\Gamma)$ and $g \in M_k(\Gamma)$ replaced by F. When presenting (31) we noted that the $p_j \in M(\Gamma)$ are also algebraic functions in h, resp. t.

To complete the proof of Theorem 6.2 we need to bound the number of poles of the derivatives $\frac{d^j p_i}{dt^j} \in M(\Gamma)$, $i \in \{1, 2\}$ and $j \in \mathbb{Z}_{\geq 1}$. To this end, given H and t in $M(\Gamma)$ where $H = r(t)$ is a function in t, we will show that such bounds for $\frac{d^j H}{dt^j}$ can be expressed in terms of pole bounds for H and t.

We begin with the case $j = 1$, noting that

$$\frac{dH}{dt} = r'(t) = \frac{H'}{t'} \in M(\Gamma) \text{ owing to } H'(\tau) = r'(t(\tau))t'(\tau). \tag{121}$$

To state our first lemma, we need define a number of "different" poles of a modular form F with even weight. Via the relation (79), this number is related to the number of pairwise different poles of F contained in a fundamental domain.

Definition 10.1 *Let $F \in M_k(\Gamma)$ with k even. Then*

$$\text{NofDiffPoles}(F) := \sum_{\substack{P \in X(\Gamma) \\ \text{Ord}_P F < 0}} 1 = \sum_{\substack{[p] \in X(\Gamma) \\ \text{ord}\, \tilde{F}_p(z) < 0}} 1, \tag{122}$$

where $\tilde{F}_p(z)$ is the Laurent series defined in Lemma 7.6.

Lemma 10.2

$$\text{NofPoles}\left(\frac{dH}{dt}\right) \leq 2g_\Gamma - 2 + \text{NofPoles}(H) + \text{NofPoles}(t) + \text{NofDiffPoles}(H) + \text{NofDiffPoles}(t).$$

Proof

$$\begin{aligned} -\sum_{P \in \text{Poles}(\frac{dH}{dt})} \text{Ord}_P\left(\frac{dH}{dt}\right) &\leq -\sum_{P \in X(\Gamma)} \pi\left(\text{Ord}_P\left(\frac{dH}{dt}\right)\right) \\ &\overset{(121)}{=} -\sum_{P \in X(\Gamma)} \pi(\text{Ord}_P H') + \sum_{P \in X(\Gamma)} \zeta(\text{Ord}_P t') \\ &\overset{(90)}{=} -\sum_{P \in X(\Gamma)} \pi(\text{Ord}_P H') + 2(g_\Gamma - 1) - \sum_{P \in X(\Gamma)} \pi(\text{Ord}_P t'). \end{aligned}$$

Observing that, by Lemma 8.3,

$$-\sum_{P \in X(\Gamma)} \pi(\text{Ord}_P H') = -\sum_{P \in X(\Gamma)} \pi(\text{Ord}_P H) + \sum_{\substack{P \in X(\Gamma) \\ \text{Ord}_P H < 0}} 1,$$

together with the analogous relation for t instead of H, completes the proof of the lemma. □

Lemma 10.3 *For* $j \in \mathbb{Z}_{\geq 1}$,

$$\mathrm{NofPoles}\left(\frac{d^j H}{dt^j}\right) \leq 2j(g_\Gamma - 1) + \mathrm{NofPoles}(H) + j\ \mathrm{NofPoles}(t) + \sum_{m=0}^{j-1} \mathrm{NofDiffPoles}\left(\frac{d^m H}{dt^m}\right) + j\ \mathrm{NofDiffPoles}(t).$$

Proof The case $j = 1$ is Lemma 10.2. Assuming the statement holds for $j - 1$, we show it holds for j. By Lemma 10.2,

$$\begin{aligned}\mathrm{NofPoles}\left(\frac{d^j H}{dt^j}\right) &\leq 2g_\Gamma - 2 + \mathrm{NofPoles}\left(\frac{d^{j-1} H}{dt^{j-1}}\right) + \mathrm{NofPoles}(t) \\ &\quad + \mathrm{NofDiffPoles}\left(\frac{d^{j-1} H}{dt^{j-1}}\right) + \mathrm{NofDiffPoles}(t) \\ &= 2g_\Gamma - 2 + \Big(2(j-1)(g_\Gamma - 1) + \mathrm{NofPoles}(H) \\ &\quad + (j-1)\,\mathrm{NofPoles}(t) + \sum_{m=0}^{j-2} \mathrm{NofDiffPoles}\left(\frac{d^m H}{dt^m}\right) \\ &\quad + (j-1)\,\mathrm{NofDiffPoles}(t)\Big) + \mathrm{NofPoles}(t) \\ &\quad + \mathrm{NofDiffPoles}\left(\frac{d^{j-1} H}{dt^{j-1}}\right) + \mathrm{NofDiffPoles}(t);\end{aligned}$$

for the equality we applied the induction hypothesis. □

For the next lemma we need the counterpart to Definition 10.1.

Definition 10.4 *Let* $F \in M_k(\Gamma)$ *with* k *even. Then*

$$\mathrm{NofDiffZeros}(F) := \sum_{\substack{P \in X(\Gamma) \\ \mathrm{Ord}_P\, F > 0}} 1 = \sum_{\substack{[p] \in X(\Gamma) \\ \mathrm{ord}_p\, \tilde{F}_p(z) > 0}} 1, \tag{123}$$

where $\tilde{F}_p(z)$ *is the Laurent series defined in Lemma 7.6.*

Lemma 10.5 *For* $j \in \mathbb{Z}_{\geq 1}$,

$$\mathrm{NofDiffPoles}\left(\frac{d^j H}{dt^j}\right) \leq \mathrm{NofDiffPoles}(H) + \mathrm{NofDiffZeros}(t'). \tag{124}$$

Proof Given $P = [p] \in X(\Gamma)$, let $H(\tau) = \tilde{H}_p(z_p(\tau))$ and $t(\tau) = \tilde{t}_p(z_p(\tau))$ with Laurent series on the right hand sides as defined in Lemma 7.4. Then $H'(\tau) = z_p'(\tau)(\tilde{H}_p)'(z_p(\tau))$, and owing to $\frac{dH}{dt} = H'/t'$,

$$\text{NofDiffPoles}\left(\frac{dH}{dt}\right)=\sum_{\substack{P\in X(\Gamma)\\ \text{Ord}_P\, H'/t'<0}}1\leq\sum_{\substack{[p]\in X(\Gamma)\\ \text{ord}(\tilde{H}_p)'(z)<0}}1+\sum_{\substack{[p]\in X(\Gamma)\\ \text{ord}(\tilde{t}_p)'(z)>0}}1$$

$$=\sum_{\substack{[p]\in X(\Gamma)\\ \text{ord}(\tilde{H}_p)(z)<0}}1+\text{NofDiffZeros}(t')=\text{NofDiffPoles}(H)+\text{NofDiffZeros}(t').$$

The general case $j \geq 1$ follows by mathematical induction. We omit its (technical) details; instead we sketch the essential structure underlying the induction step. Because of $\frac{d^jH}{dt^j} \in M(\Gamma)$, for each fixed $j \in \mathbb{Z}_{\geq 1}$ and $[p] \in X(\Gamma)$ there is a representation $\frac{d^jr}{dt^j}(t(\tau)) = \frac{d^jH}{dt^j}(t(\tau)) = L_p(z_p(\tau))$, where $L_p(z)$ is a Laurent series. To obtain further insight into this representation, we use Faá Di Bruno's Formula,

$$H^{(j)}(\tau)=\frac{d^j}{d\tau^j}r(t(\tau))=\sum_{i=1}^{j}\frac{d^iH(t(\tau))}{dt^i}\cdot B_{j,i}(t'(\tau),\dots,t^{(j-i+1)}(\tau)),$$

with $B_{j,i}(x_1,\dots,x_{j-i+1})$ being the (partial) Bell polynomials. With this formula one finds that

$$L_p(z)=\sum_{i=1}^{j}c_{i,p}(z)\cdot(\tilde{H}_p)^{(i)}(z)\text{ with }c_{i,p}(z)=\frac{C_{i,p}}{(\tilde{t}_p)'(z)^{2j-1}},$$

where the $C_{i,p}$ are polynomials in $(\tilde{t}_p)'(z), (\tilde{t}_p)''(z), \dots$, such that for each monomial, *constant* $\cdot\, (\tilde{t}_p)'(z)^{\alpha_1}(\tilde{t}_p)''(z)^{\alpha_2}\dots$, occurring as a summand in $C_{i,p}$, one has $1\cdot\alpha_1 + 2\cdot\alpha_2+\cdots \leq 2j-2$. This property guarantees that no further poles are introduced. We give the L_p for $j = 1, 2, 3$ explicitly:

$$L_p(z)=\frac{1}{(\tilde{t}_p)'(z)}\cdot(\tilde{H}_p)'(z),\quad\text{if } j=1;$$

$$L_p(z)=-\frac{(\tilde{t}_p)''(z)}{(\tilde{t}_p)'(z)^3}\cdot(\tilde{H}_p)'(z)+\frac{(\tilde{t}_p)'(z)}{(\tilde{t}_p)'(z)^3}\cdot(\tilde{H}_p)''(z),\quad\text{if } j=2;$$

$$\begin{aligned}L_p(z)=&\frac{3(\tilde{t}_p)''(z)^2+(\tilde{t}_p)'(z)(\tilde{t}_p)^{(3)}(z)}{(\tilde{t}_p)'(z)^5}\cdot(\tilde{H}_p)'(z)\\&-\frac{3(\tilde{t}_p)'(z)(\tilde{t}_p)''(z)}{(\tilde{t}_p)'(z)^5}\cdot(\tilde{H}_p)''(z)+\frac{(\tilde{t}_p)'(z)^2}{(\tilde{t}_p)'(z)^5}\cdot(\tilde{H}_p)^{(3)}(z),\quad\text{if } j=3.\end{aligned}$$

Further details of the induction proof are left to the reader. □

Lemma 10.5 implies for $j \in \mathbb{Z}_{\geq 1}$,

$$\sum_{m=1}^{j-1} \text{NofDiffPoles}\left(\frac{d^m H}{dt^m}\right) \overset{(124)}{\leq} (j-1)(\text{NofDiffPoles}(H) + \text{NofDiffZeros}(t')).$$

We use this to simplify the right side of Lemma 10.3,

$$\begin{aligned} \text{NofPoles}\left(\frac{d^j H}{dt^j}\right) &\leq 2j(g_\Gamma - 1) + \text{NofPoles}(H) + j\ \text{NofPoles}(t) \\ &+ j\ \text{NofDiffPoles}(H) + (j-1)\,\text{NofDiffZeros}(t') \\ &+ j\,\text{NofDiffPoles}(t). \end{aligned} \tag{125}$$

For further simplification, observe that

$$\begin{aligned} \text{NofDiffZeros}(t') &\leq \sum_{P \in X(\Gamma)} \zeta(\text{Ord}_P\, t') \overset{(90)}{=} 2(g_\Gamma - 1) - \sum_{P \in X(\Gamma)} \pi(\text{Ord}_P\, t') \\ &= 2(g_\Gamma - 1) + \text{NofPoles}(t') = 2(g_\Gamma - 1) + \text{NofDiffPoles}(t) + \text{NofPoles}(t), \end{aligned}$$

where the last equality is by Lemma 8.3. Using this on (125) one obtains,

$$\begin{aligned} \text{NofPoles}\left(\frac{d^j H}{dt^j}\right) &\leq 2(2j-1)(g_\Gamma - 1) + \text{NofPoles}(H) + (2j-1)\,\text{NofPoles}(t) \\ &+ j\ \text{NofDiffPoles}(H) + (2j-1)\,\text{NofDiffPoles}(t). \end{aligned} \tag{126}$$

Finally, as another step of simplification, we apply

$$\text{NofDiffPoles}(H) \leq \text{NofPoles}(H) \text{ and } \text{NofDiffPoles}(t) \leq \text{NofPoles}(t),$$

which reduces (126) to

Lemma 10.6 *For $H, t \in M(\Gamma)$ and $j \in \mathbb{Z}_{\geq 1}$,*

$$\begin{aligned} \text{NofPoles}\left(\frac{d^j H}{dt^j}\right) &\leq (2j-1)(2g_\Gamma - 2) \\ &+ (j+1)\,\text{NofPoles}(H) + (4j-2)\,\text{NofPoles}(t). \end{aligned} \tag{127}$$

11 The Proofs of Theorem 6.2 and Theorem 6.3 Summarized

First we collect all the ingredients to prove Theorem 6.2 whose statements are for the case of even weight k. Here, and in the summary for odd weight k, the connection to Theorem 6.2 and Theorem 6.3 is by back-substitution $t \to h$ and $F \to g$.

The derivations resulting in (108), resp. (116), prove the bounds (61) for p_1, resp. (62) for p_2, of Theorem 6.2. Applying Lemma 10.6 for even k to $H := p_1$ gives,

$$\begin{aligned}
\mathrm{NofPoles}\left(\frac{d^j p_1}{dt^j}\right) &\leq (2j-1)(2g_\Gamma - 2) \\
&\quad + (j+1)\,\mathrm{NofPoles}(p_1) + (4j-2)\,\mathrm{NofPoles}(t) \\
&\overset{(108)}{\leq} (2j-1)(2g_\Gamma - 2) + (4j-2)\,\mathrm{NofPoles}(t) \\
&\quad + (j+1)\Big((k+4)(g_\Gamma - 1) + 4\,\mathrm{NofPoles}(t) + 2\,\mathrm{NofPoles}(F)\Big) \\
&= (jk + 8j + k + 2)(g_\Gamma - 1) \\
&\quad + 2(4j+1)\,\mathrm{NofPoles}(t) + 2(j+1)\,\mathrm{NofPoles}(F);
\end{aligned}$$

which is (63) of Theorem 6.2 For even k, applying Lemma 10.6 to $H := p_2$, using (116), gives (64) of Theorem 6.2.

Finally, we collect all the ingredients to prove Theorem 6.3 whose statements are for the case of odd weight k.

The derivations resulting in (110), resp. (120), prove the bounds (65) for p_1, resp. (66) for p_2, of Theorem 6.3. Applying Lemma 10.6 for odd k to $H := p_1$ gives,

$$\begin{aligned}
\mathrm{NofPoles}\left(\frac{d^j p_1}{dt^j}\right) &\leq (2j-1)(2g_\Gamma - 2) \\
&\quad + (j+1)\,\mathrm{NofPoles}(p_1) + (4j-2)\,\mathrm{NofPoles}(t) \\
&\overset{(110)}{\leq} (2j-1)(2g_\Gamma - 2) + (4j-2)\,\mathrm{NofPoles}(t) \\
&\quad + (j+1)\Big(4\,\mathrm{NofPoles}(t) + 2\,\mathrm{NofPoles}(F^2) + (2k+4)(g_\Gamma - 1)\Big) \\
&= 2(1 + 4j + k + jk)(g_\Gamma - 1) \\
&\quad + 2(4j+1)\,\mathrm{NofPoles}(t) + 2(j+1)\,\mathrm{NofPoles}(F^2);
\end{aligned}$$

which is (67) of Theorem 6.2 For odd k, applying Lemma 10.6 to $H := p_2$, using (120), gives (68) of Theorem 6.3.

This completes the proofs of Theorem 6.2 and Theorem 6.3.

12 Conclusion

In this paper we focused on the mathematics underlying our algorithm ModFormDE. With regard to possible applications, we feel there is quite some potential waiting for further exploration; see also Sect. 5.3. There will be also the need to supplement such investigations by algorithmic developments, in particular, by supporting software. For instance, in our illustrating example in Sect. 5.1 we have seen that we still need software to determine $\mathrm{NofPoles}(g^2)$ automatically.

Acknowledgment In October 2019, while working on parts of this paper, the first named author enjoyed the overwhelming hospitality of William Y.C. Chen and his team at the Center for Applied

Mathematics, Tianjin University.—We want to thank the anonymous referees for their careful reading of the manuscript and manifold suggestions for improvement.

13 Appendix

13.1 Linear Independence of the Yang Functions

Proposition 13.1 *Let Γ be a congruence subgroup. Let $g \in M_k(\Gamma)$ with $k \geq 1$, and $h \in M(\Gamma)$. Then the Yang functions,*

$$\left\{1, \frac{G_2}{G_1}, \frac{G_2^2}{G_1^2}, \ldots, \frac{G_2^m}{G_1^m}\right\}, \quad m \geq 0,$$

as functions on $\mathbb{H}$ are linearly independent over the field $M(\Gamma)$ of modular functions for Γ.

Before proving Proposition 13.1 we prove a lemma.

Lemma 13.2 *Let $a_0(\tau), \ldots, a_m(\tau)$ be functions on $\mathbb{H}$ with period $r \in \mathbb{Z}_{\geq 1}$; i.e.,*

$$a_j(\tau + r) = a(\tau), \quad \tau \in \mathbb{H}, \quad j = 0, \ldots, m.$$

If for fixed integers $c, d \in \mathbb{Z}$, $c \neq 0$,

$$a_0(\tau) + \frac{a_1(\tau)}{c\tau + d} + \frac{a_2(\tau)}{(c\tau + d)^2} + \cdots + \frac{a_m(\tau)}{(c\tau + d)^m} = 0, \quad \tau \in \mathbb{H}, \tag{128}$$

then

$$a_0(\tau) = a_1(\tau) = a_2(\tau) = \cdots = a_m(\tau) = 0, \quad \tau \in \mathbb{H}.$$

Proof Consider the Vandermonde matrix

$$A = \begin{pmatrix} 1 & \frac{1}{c\tau+d} & \frac{1}{(c\tau+d)^2} & \cdots & \frac{1}{(c\tau+d)^m} \\ 1 & \frac{1}{c(\tau+r)+d} & \frac{1}{(c(\tau+r)+d)^2} & \cdots & \frac{1}{(c(\tau+r)+d)^m} \\ 1 & \frac{1}{c(\tau+2r)+d} & \frac{1}{(c(\tau+2r)+d)^2} & \cdots & \frac{1}{(c(\tau+2r)+d)^m} \\ \vdots & \vdots & \vdots & \ddots & \vdots \\ 1 & \frac{1}{c(\tau+mr)+d} & \frac{1}{(c(\tau+mr)+d)^2} & \cdots & \frac{1}{(c(\tau+mr)+d)^m} \end{pmatrix},$$

which is invertible since it has a non-zero determinant:

$$\det(A) = \prod_{1 \leq i < j \leq m+1} \left(\frac{1}{c(\tau + jr) + d} - \frac{1}{c(\tau + ir) + d} \right) \neq 0.$$

Hence a relation like (128) with $a_j(\tau)$ not all zero cannot exist. □

Proof of Proposition 13.1 Let $a_0(\tau), \ldots, a_m(\tau)$ be modular functions in $M(\Gamma)$. By induction on $m \geq 0$, we will prove the statement in the following form: Suppose that

$$a_0(\tau) + a_1(\tau)\frac{G_2(\tau)}{G_1(\tau)} + \cdots + a_m(\tau)\frac{G_2(\tau)^m}{G_1(\tau)^m} = 0, \quad \tau \in \mathbb{H}, \tag{129}$$

then $a_j = 0$ for $0 \leq j \leq m$.

The statement is true for $m = 0$. Assuming its truth up to $m = N - 1$, we prove it is true also for $m = N$.

In our argument we use as a crucial fact that there exists an common period $r \in \mathbb{Z}_{\geq 1}$ such that $a_j(\tau + r) = a_j(\tau)$ and $G_j(\tau + r) = G_j(\tau)$. For $\gamma = \left(\begin{smallmatrix} a & b \\ c & d \end{smallmatrix}\right) \in \Gamma$,

$$\frac{G_2(\gamma\tau)}{G_1(\gamma\tau)} = \frac{G_2(\tau)}{G_1(\tau)} + \frac{c'k(c\tau + d)^{-1}}{G_1(\tau)}$$

with $c' := c/(2\pi i)$. Now suppose a relation of type (129) holds for $m = N$; i.e.,

$$a_0(\tau) + a_1(\tau)\frac{G_2(\tau)}{G_1(\tau)} + \cdots + a_N(\tau)\frac{G_2(\tau)^N}{G_1(\tau)^N} = 0, \quad \tau \in \mathbb{H}. \tag{130}$$

Applying $\gamma = \left(\begin{smallmatrix} a & b \\ c & d \end{smallmatrix}\right) \in \Gamma$ to this equation yields,

$$a_0(\tau) + a_1(\tau)\Big(\frac{G_2(\tau)}{G_1(\tau)} + \frac{c'k(c\tau + d)^{-1}}{G_1(\tau)}\Big) + \cdots + a_N(\tau)\Big(\frac{G_2(\tau)}{G_1(\tau)} + \frac{c'k(c\tau + d)^{-1}}{G_1(\tau)}\Big)^N = 0.$$

This can be rewritten into the form,

$$b_0(\tau) + \frac{b_1(\tau)}{c\tau + d} + \cdots + \frac{b_{N-1}(\tau)}{(c\tau + d)^{N-1}} + \frac{(c'k)^N a_N(\tau)}{G_1(\tau)(c\tau + d)^N} = 0, \quad \tau \in \mathbb{H}.$$

Here we use the common periodicity r, $a_j(\tau + r) = a_j(\tau)$ and $G_j(\tau + r) = G_j(\tau)$, which produces periodic coefficients; i.e., $b_j(\tau + r) = b_j(\tau)$. Hence, by Lemma 13.2, the $b_0, \ldots, b_{N-1}$ are all 0, which implies that also $a_N = 0$. This reduces (130) to

$$a_0(\tau) + a_1(\tau)\frac{G_2(\tau)}{G_1(\tau)} + \cdots + a_{N-1}(\tau)\frac{G_2(\tau)^{N-1}}{G_1(\tau)^{N-1}} = 0, \quad \tau \in \mathbb{H},$$

which by the induction hypothesis implies $a_0 = \cdots = a_{N-1} = 0$. This completes the proof of Proposition 13.1. □

13.2 Computational Details for the ModFormDE Example in Sect. 5.1

This section presents computational parts of our exemplification of algorithm ModFormDE in Sect. 5.1.

13.2.1 Cusps of the Congruence Group $\Gamma(2, 4, 2)$

With the Magma system, the cusps [0], [1], and [∞] of $\Gamma = \Gamma(2, 4, 2)$, together with each width, can be computed as follows:

```
> G:=CongruenceSubgroup([2,4,2]);
> Cusps(G);
[
    oo,
    0,
    1
]
> Widths(G);
[ 2, 2, 2 ]
```

13.2.2 Expansion of g^2 at the cusp [1] of $X(\Gamma(2, 4, 2))$

By Lemma 1.13 in [13] and because of $(c\tau + d)^{-1/2}\eta\left(\frac{a\tau+b}{c\tau+d}\right) = \epsilon(a, b, c, d)\eta(\tau)$, where $\epsilon(a, b, c, d)$ is a 24-th root of unity, we have for all $A, B, C, D \in \mathbb{Z}$ such that $AD - BC \neq 0$:

$$\left(\frac{\gcd(A, C)}{AD - BC}(C\tau + D)\right)^{-1/2} \eta\left(\frac{A\tau + B}{C\tau + D}\right)$$
$$= \epsilon(A/\gcd(A, C), -y, C/\gcd(A, C), x)\eta\left(\frac{\gcd(A, C)\tau + Bx + Dy}{(AD - BC)\gcd(A, C)^{-1}}\right).$$

Here x, y are any integers such that $Ax + Cy = \gcd(A, C)$. This formula together with $\eta(\tau) = q^{1/24}\prod_{n=1}^{\infty}(1 - q^n)$ implies that

$$(C\tau + D)^{-1/2}\eta\left(\frac{A\tau + B}{C\tau + D}\right) = q^{\frac{\gcd(A,C)^2}{24(AD-BC)}}\left(u + O\left(q^{\frac{\gcd(A,C)^2}{AD-BC}}\right)\right), \tag{131}$$

where $u \in \mathbb{C}$ is a non-zero constant.

Now, owing to (43),

$$\tau^{-2}g\left(\frac{\tau - 1}{\tau}\right)^2 = 2^{-4}\frac{\left(\tau^{-1/2}\eta\left(\frac{\tau-1}{\tau}\right)\right)^{20}}{\left((2\tau)^{-1/2}\eta\left(\frac{\tau-1}{2\tau}\right)\right)^8\left(\tau^{-1/2}\eta\left(\frac{2\tau-2}{\tau}\right)\right)^8}$$

we have by (131):

$$\tau^{-2} g\Big(\frac{\tau-1}{\tau}\Big)^2 = \frac{(q^{1/24}(u_1+O(q)))^{20}}{(q^{1/(24\cdot 2)}(u_2+O(q^{1/2})))^8 (q^{1/(24\cdot 2)}(u_3+O(q^{1/2})))^8}$$
$$= q^{20/24-8/48-8/48}(u+O(q^{1/2})) = q^{1/2}(u+O(q^{1/2})),$$

where u and the u_j are non-zero complex numbers.

References

1. George E. Andrews, Richard Askey, and Ranjan Roy. *Special Functions*. Cambridge University Press, 1999.
2. Arthur O.L. Atkin. Proof of a conjecture of Ramanujan. *Glasgow Math. J.*, 8:14–32, 1967.
3. Jonathan M. Borwein and Peter B. Borwein. *Pi and the AGM*. John Wiley & Sons, 1987.
4. Shaun Cooper. *Ramanujan's Theta Functions*. Springer, 2017.
5. Richard Dedekind and Heinrich Weber. *Theory of Algebraic Functions in One Variable. Translated and introduced by John Stillwell*. History of Mathematics, Vol. 39. American Mathematical Society, 2012.
6. Fred Diamond and Jerry Shurman. *A First Course in Modular Forms*. Springer, 2005.
7. Hershel M. Farkas and Irwin Kra. *Riemann Surfaces*, volume 71 of *Graduate Texts in Mathematics*. Springer, 1980.
8. Otto Forster. *Lectures on Riemann Surfaces*. Springer, 1981.
9. Manuel Kauers and Peter Paule. *The Concrete Tetrahedron*. Springer, 2011.
10. Rick Miranda. *Algebraic Curves and Riemann Surfaces*, volume 5 of *Grad. Stud. Math.* AMS, 1995.
11. Ken Ono. *The Web of Modularity: Arithmetic of the Coefficients of Modular Forms and q-series*. CBMS Regional Conference Series in Mathematics, Number 102, 2004.
12. Peter Paule and Cristian-Silviu Radu. Holonomic relations for modular functions and forms: First guess, then prove. *Intl. J. Number Theory*, 2020. Open access (47 pages): https://doi.org/10.1142/S1793042120400278.
13. Cristian-Silviu Radu. An algorithmic approach to Ramanujan's congruences. *The Ramanujan J.*, 20:215–251, 2009.
14. Srinivasa Ramanujan. Modular equations and approximations to π. *Quart. J. Math.*, 45:350–372, 1914.
15. Neil J. A. Sloane, editor. *The On-Line Encyclopedia of Integer Sequences*. https://oeis.org
16. Richard P. Stanley. *Enumerative Combinatorics, Volume 2*, Cambridge University Press, 1999.
17. Zhi-Hong Sun. New congruences involving Apéry-like numbers. Preprint (24 pages), 2020. https://arxiv.org/abs/2004.07172v2
18. Yifan Yang. On differential equations satisfied by modular forms. *Math. Z*, 246:1–19, 2004.
19. Don Zagier. Elliptic modular forms and their applications. In Ranestad K. et al. (eds.), *The 1-2-3 of Modular Forms*, pages 1–103. Universitext Springer, 2008.
20. Doron Zeilberger. A holonomic systems approach to special functions identities. *J. Comput. Appl. Math.*, 32:321–368, 1990.

A Case Study for $\zeta(4)$

Carsten Schneider and Wadim Zudilin

Abstract Using symbolic summation tools in the setting of difference rings, we prove a two-parametric identity that relates rational approximations to $\zeta(4)$.

Kingdom:	Mathematical constants
Class:	Periods
Family:	Multiple zeta values
Genus:	Single zeta values
Species:	Even zeta values

1 Introduction

The quantity

$$\zeta(4) = \sum_{k=1}^{\infty} \frac{1}{k^4} = \frac{\pi^4}{90}$$

is a somewhat typical representative of even zeta values—the values of Riemann's zeta function at positive even integers. It is shadowed by the far more famous $\zeta(2) = \pi^2/6$, which was a main subject of Euler's resolution of Basel's problem, and $\zeta(3)$—an *objet de l'étude* of Apéry's iconic proof of the irrationality of the latter (and also of $\zeta(2)$) [4, 30]. Though known to be irrational (and transcendental!), $\zeta(4)$ serves as

Date: 17 April 2020. *Revised*: 22 September 2020
Research partially supported by the Austrian Science Fund (FWF) grants SFB F5006-N15 and F5009-N15 in the framework of the Special Research Program "Algorithmic and Enumerative Combinatorics".

C. Schneider
Research Institute for Symbolic Computation, Johannes Kepler University Linz, Altenberger Str. 69, 4040 Linz, Austria
e-mail: carsten.schneider@risc.jku.at

W. Zudilin (✉)
Department of Mathematics, IMAPP, Radboud University, PO Box 9010, 6500 GL Nijmegen, Netherlands
e-mail: w.zudilin@math.ru.nl

A. Bostan and K. Raschel (eds.), *Transcendence in Algebra, Combinatorics, Geometry and Number Theory*, Springer Proceedings in Mathematics & Statistics 373,
https://doi.org/10.1007/978-3-030-84304-5_17

a natural guinea pig for extending Apéry's machinery to other zeta values. Apéry-type approximations to the number were discovered and rediscovered on several occasions [8, 29, 33], however they were not good enough to draw conclusions about its irrationality. An unexpected difficulty to control the 'true' arithmetic of those rational approximations to $\zeta(4)$ generated further research [14, 34], which eventually led to producing sufficient approximations and establishing a new world record for the irrationality measure of π^4 [15].

In this note we turn our attention to a rational side of the coin and prove the following two-parametric identity.

Theorem 1 *For integers $n \ge m \ge 0$, define two rational functions*

$$R(t) = R_{n,m}(t) = (-1)^m \Big(t + \frac{n}{2}\Big) \frac{(t-n)_m}{m!} \frac{(t-2n+m)_{2n-m}}{(2n-m)!} \times \frac{(t+n+1)_n}{(t)_{n+1}} \frac{(t+n+1)_{2n-m}}{(t)_{2n-m+1}} \left(\frac{n!}{(t)_{n+1}}\right)^2$$

and

$$\tilde{R}(t) = \tilde{R}_{n,m}(t) = \frac{n!\,(t-n)_{2n-m}}{(t)_{n+1}(t)_{2n-m+1}} \sum_{j=0}^{n} \binom{n}{j}^2 \binom{2n-m+j}{n} \frac{(t-j)_n}{n!}. \tag{1}$$

Then

$$-\frac{1}{3} \sum_{\nu=n-m+1}^{\infty} \frac{\mathrm{d}R(t)}{\mathrm{d}t}\bigg|_{t=\nu} = \frac{1}{6} \sum_{\nu=1}^{\infty} \frac{\mathrm{d}^2\tilde{R}(t)}{\mathrm{d}t^2}\bigg|_{t=\nu}. \tag{2}$$

The $m = n$ instance of (2) was stated as Problem 1 in [34].

The fact that both sides of (2) are linear forms in 1 and $\zeta(4)$ with rational coefficients is verifiable by standard techniques [14, 33, 34] which employ the partial-fraction decomposition of the rational functions. A remarkable outcome of this identity is the *coincidence* of two different-looking rational approximations to the zeta value. Such coincidences are often a source of deep algorithmic and analytical developments—check [9] for another exploration of this theme (see also [6]).

The main difficulty in establishing equality (2) (in contrast to tackling, for example, Apéry's sums in [21] for $\zeta(3)$) is that its both sides are not hypergeometric functions but rather *derivatives* of hypergeometric functions. Another issue is that the summation range on the left-hand side is somewhat unnatural.

2 Symbolic Summation

Denote by $Z_l(n, m)$ and $Z_r(n, m)$ the left- and right-hand sides of (2), respectively. In order to prove the identity (2) we proceed as follows.

(A) We compute the linear recurrence

$$a_0(n,m)Z(n,m) + a_1(n,m)Z(n,m+1) + a_2(n,m)Z(n,m+2) = 0 \quad (3)$$

with

$$\begin{aligned} a_0(n,m) &= (2n-m)^5,\\ a_1(n,m) &= -(4n-2m-1)(6n^4-24n^3m+22n^2m^2-8nm^3+m^4-24n^3\\ &\quad +30n^2m-14nm^2+2m^3+8n^2-10nm+2m^2-4n+m),\\ a_2(n,m) &= -(2n-m-1)^3(4n-m)(m+2), \end{aligned} \quad (4)$$

which holds simultaneously for $Z(n,m) = Z_l(n,m)$ and $Z(n,m) = Z_r(n,m)$ for all $n, m \in \mathbb{Z}_{\geq 0}$ with $n-2 \geq m \geq 0$. In addition, we observe that $a_2(n,m) \neq 0$ for all $n, m \in \mathbb{Z}_{\geq 0}$ with $0 \leq m < n$.

(B) We show that the following initial values hold:

$$Z_l(n,0) = Z_r(n,0) \quad \text{for all } n \geq 0, \quad (5)$$
$$Z_l(n,1) = Z_r(n,1) \quad \text{for all } n \geq 1. \quad (6)$$

Combined with **(A)** this proves that $Z_l(n,m) = Z_r(n,m)$ holds true fo all $n \geq m \geq 0$.

In order to carry out the steps **(A)** and **(B)**, advanced symbolic summation techniques in the setting of difference rings are utilized. Among them the following three summation paradigms play a decisive role, that are available within the summation package `Sigma` [22].

(i) Creative telescoping. Given a sum $F(m) = \sum_{\nu=a}^{b} f(m,\nu)$ and $\delta \in \mathbb{Z}_{\geq 0}$, one searches for polynomials $c_0(m), \ldots, c_\delta(m)$, free of ν, and $g(m,\nu)$ such that

$$g(m,\nu+1) - g(m,\nu) = c_0(m)f(m,\nu) + c_1(m)f(m+1,\nu) + \cdots + c_\delta(m)f(m+\delta,\nu) \quad (7)$$

holds for all $a \leq \nu \leq b$. Thus summing (7) over ν one obtains the recurrence

$$g(m,b+1) - g(m,a) = c_0(m)F(m) + c_1(m)F(m+1) + \cdots + c_\delta(m)F(m+\delta). \quad (8)$$

By specializing a, b further—e.g., to $a = 0$ and $b = m$, or sending b to ∞ if the limit exists—one obtains recurrence relations for more specific sums. The computed creative telescoping solution $(c_0(m), \ldots, c_\delta(m), g(m,\nu))$ is also called a proof certificate for the recurrence (8) found: usually it allows one to verify that $F(m)$ is a solution of (8) by simple polynomial arithmetic, without analyzing the usually complicated computation steps of the underlying summation algorithm. The algorithmic version of creative telescoping has been introduced in [18, 32] for hypergeometric sums. In order to prove (2), we will employ a generalized machinery for creative telescoping [26] where the summand can be composed not only in terms of hypergeometric products, but of indefinite nested sums defined over hypergeometric products. We emphasize that all recurrences produced below (using the `Sigma`-command `GenerateRecurrence`) are accompanied by such proof certificates which guarantee the correctness of all the calculations. Since the output is rather large and can be easily reproduced with `Sigma`, any explicit printout of the proof certificates is skipped.

(ii) Recurrence solving. Given a linear recurrence of the form (8), one can search for solutions that are expressible within certain classes function spaces. Using the `Sigma`-command `SolveRecurrence` one can search for hypergeometric solutions [17, 18] and, more generally, for all solutions that are expressible in terms of indefinite nested sums defined over hypergeometric products. Such solutions are also called d'Alembertian solutions [2, 19], a subclass of Liouvillian solutions [11].

(iii) Simplification of expressions. Within `Sigma` the expressions in terms of indefinite nested sums defined over hypergeometric products are represented in the setting of difference rings and fields [12, 23, 27]. Utilizing this difference ring machinery [24, 28] (compare also [10]) one can apply, e.g., the `Sigma`-command `SigmaReduce` to an expression in terms of indefinite nested sums. Then the output is a simplified expression where the arising sums and products (except products such as $(-1)^m$) are independent among each other as functions of their external parameter. In particular, the input expression evaluates to zero (from a certain point on) if and only if `Sigma` reduces the expression to the zero-expression.

These summation paradigms can be used to transform a definite (multi-)sum to an expression in terms of indefinite nested sums by deriving a linear recurrence, solving the recurrence found in terms of indefinite nested sums, and, in case that sufficiently many solutions are found, combining them to an expression that evaluates to the same sequence as the input sum. Recently this machinery has been used for large scale problems coming from particle physics (see, e.g., [1] and references therein). In this regard, also the package `EvaluateMultiSum` [25], which automatizes this summation mechanism, has been utilized non-trivially in the sections below.

In the following sections we present the main steps of our proof for Theorem 1 that is based on the above summation algorithms. All the necessary calculation steps are collected in a Mathematica notebook that can be accessed via[1]

https://www.risc.jku.at/people/cschneid/data/SchneiderZudilinMMA.nb.

3 A Linear Recurrence in *m* for the Left-Hand Side

In order to activate the summation package `Sigma`, the sums arising in (2) have to be tailored to an appropriate input format. As it turns out below, one can carry out the differentiation by introducing additionally the harmonic numbers

[1] In case that the reader does not have access to Mathematica, we supplement the pdf file SchneiderZudilinMMA.pdf (same www-path!) that contains all the calculations in printed form.

$$S_a(n) = \sum_{k=1}^{n} \frac{1}{k^a}$$

of order $a \in \mathbb{Z}_{\geq 0}$. Though we see no natural way to obtain such a representation for the full summation range ν with $n - m + 1 \leq \nu$, splitting it into the ranges over ν with $n - m + 1 \leq \nu \leq 2n - m - 1$ and $2n - m \leq \nu$ makes the job well. More precisely, we split the left-hand side of (2) into the two subsums

$$W_1(n, m) = \sum_{\nu=2n-m+1}^{\infty} \frac{\mathrm{d}R_{n,m}(t)}{\mathrm{d}t}\bigg|_{t=\nu} = \sum_{\nu=1}^{\infty} \frac{\mathrm{d}R_{n,m}(t + 2n - m)}{\mathrm{d}t}\bigg|_{t=\nu}$$

and

$$W_2(n, m) = \sum_{\nu=n-m+1}^{2n-m} \frac{\mathrm{d}R_{n,m}(t)}{\mathrm{d}t}\bigg|_{t=\nu} = \sum_{\nu=1}^{n} \frac{\mathrm{d}R_{n,m}(t + n - m)}{\mathrm{d}t}\bigg|_{t=\nu},$$

so that

$$Z_l(n, m) = -\frac{1}{3}\big(W_1(n, m) + W_2(n, m)\big). \tag{9}$$

Observe that

$$\begin{aligned} R_{n,m}(t + 2n - m) =&(-1)^m \left(t + 2n - m + \frac{n}{2}\right) \frac{(t + n - m)_m}{m!} \frac{(t)_{2n-m}}{(2n - m)!} \\ &\times \frac{(t + 3n - m + 1)_n}{(t + 2n - m)_{n+1}} \frac{(t + 3n - m + 1)_{2n-m}}{(t)_{2n-m+1}} \left(\frac{n!}{(t + 2n - m)_{n+1}}\right)^2 \end{aligned}$$

and

$$\begin{aligned} R_{n,m}(t + n - m) = &(-1)^m \left(t + n - m + \frac{n}{2}\right) \frac{(t - m)_m}{m!} \frac{(t - n)_{2n-m}}{(2n - m)!} \\ &\times \frac{(t + 2n - m + 1)_n}{(t + n - m)_{n+1}} \frac{(t + 2n - m + 1)_{2n-m}}{(t + n - m)_{2n-m+1}} \left(\frac{n!}{(t + n - m)_{n+1}}\right)^2. \end{aligned}$$

By definition all Pochhammer symbols in the former expression are of the form $(t + x)_k$ for some $x \in \mathbb{Z}_{>0}$ and $k \geq 0$. Thus, we can apply the formula

$$\frac{\mathrm{d}}{\mathrm{d}t}(x + t)_k\big|_{t=\nu} = (x + \nu)_k\big(S_1(\nu + x + k - 1) - S_1(\nu + x - 1)\big) \tag{10}$$

for $\nu \in \mathbb{Z}$ with $x + \nu \in \mathbb{Z}_{>0}$ which follows from the product-rule of differentiation. Employing this formula we get for all $\nu = 1, 2, \ldots$ the following representation:

$$F_1(n,m,\nu) = \frac{\mathrm{d}}{\mathrm{d}t} R_{n,m}(t+2n-m)\Big|_{t=\nu}$$
$$= \frac{(-1)^m n!^2 (1+\nu)_{-1-m+2n}(-m+n+\nu)_m(1-m+3n+\nu)_n(1-m+3n+\nu)_{-m+2n}}{2\,m!(-m+2n)!(-m+2n+\nu)_{1+n}^3(-m+2n+\nu)_{1-m+2n}}$$
$$\times\Big(-6\nu+\nu(-2m+5n+2\nu)\big(-S_1(\nu)-S_1(-m+n+\nu)+5S_1(-m+2n+\nu)$$
$$-5S_1(-m+3n+\nu)-S_1(-2m+4n+\nu)+S_1(n+\nu)+S_1(-m+4n+\nu)$$
$$+S_1(-2m+5n+\nu)\big)+\frac{5n(m-2n)}{m-2n-\nu}+\frac{n(-2m+3n)}{n+\nu}+\frac{3n(m-n)}{-m+n+\nu}\Big).$$

Further, we prepare the summand of $W_2(n,m)$. Notice that the rule (10) cannot be applied to the arising factor $(t-n)_{2n-m}$. However we can easily overcome this issue by using the following elementary identity: For $\nu \in \mathbb{Z}_{>0}$ with $1 \le \nu \le n$ and any differentiable function $f(t)$, we have

$$\frac{\mathrm{d}}{\mathrm{d}t}\big((t-n)_{2n-m} f(t)\big)\Big|_{t=\nu} = (-1)^{n-\nu} f(\nu)(\nu+n-m-1)!(n-\nu)!\,. \qquad (11)$$

Therefore, for all $\nu \in \mathbb{Z}_{>0}$ with $1 \le \nu \le n$ we get

$$F_2(n,m,\nu) = \frac{\mathrm{d}R_{n,m}(t+n-m)}{\mathrm{d}t}$$
$$= (-1)^m\Big(\nu+n-m+\frac{n}{2}\Big)\frac{(\nu-m)_m}{m!}\,\frac{(-1)^{n-\nu}(\nu+n-m-1)!(n-\nu)!}{(2n-m)!}$$
$$\times\frac{(\nu+2n-m+1)_n}{(\nu+n-m)_{n+1}}\,\frac{(\nu+2n-m+1)_{2n-m}}{(\nu+n-m)_{2n-m+1}}\Big(\frac{n!}{(\nu+n-m)_{n+1}}\Big)^2.$$

Because of the factor $(\nu-m)_m$, we have $F_2(\nu)=0$ for all $\nu \in \mathbb{Z}_{>0}$ with $1 \le \nu \le m$. Consequently, $W_1(n,m)$ and $W_2(n,m)$ can be written as

$$W_1(n,m) = \sum_{\nu=1}^{\infty} F_1(\nu) \quad \text{and} \quad W_2(n,m) = \sum_{\nu=m+1}^{n} F_2(\nu) = \sum_{\nu=1}^{n-m} F_2(\nu+m),$$

where the summands $F_1(\nu)$ and $F_2(\nu)$ are given in terms of hypergeometric products and linear combinations of harmonic numbers. Since these sums fit the input class of `Sigma`, we can apply the command `GenerateRecurrence` to both sums and compute for $0 \le m \le n$ the recurrences

$$a_0(n,m)W_s(n,m)+a_1(n,m)W_s(n,m+1)+a_2(n,m)W_s(n,m+2)=r_s(n,m) \quad \text{for } s=1,2,$$

where the coefficients are given in (4) and where $r_1(n,m) = -r_2(n,m)$ is too large to be reproduced here (verification of the latter equality required an extra simplification step with `Sigma`). To compute the recurrence for the *hypergeometric* sum $W_2(n,m)$ one can alternatively use the Mathematica package `fastZeil` [16] based on [32].

Thus, $Z_l(n,m)$ given in (9) is a solution of the recurrence (3). For this part we needed 15 min to compute both recurrences and to combine them to (3).

4 A Linear Recurrence in m for the Right-Hand Side

In order to calculate a linear recurrence for $Z_r(n,m)$ we follow the same strategy as for $Z_l(n,m)$ in Sect. 3 by utilizing more advanced summation tools of `Sigma`. Collecting all products in (1) to

$$G_{n,m,j}(t) = \frac{n!\,(t-n)_{2n-m}}{(t)_{n+1}(t)_{2n-m+1}}\binom{n}{j}^2\binom{2n-m+j}{n}\frac{(t-j)_n}{n!},$$

the right-hand side of (2) can be rewritten as

$$Z_r(n,m) := \frac{1}{6}\sum_{\nu=1}^{\infty}\sum_{j=0}^{n}\frac{\mathrm{d}^2}{\mathrm{d}t^2}G_{n,m,j}(t)\Big|_{t=\nu}.$$

Similarly to the previous section, we split the sum further into subsums (see (13) for the final split) such that the differential operator acting on the summands can be replaced by modified summands in terms of harmonic numbers. On the first step, we write

$$Z_r(n,m) = \frac{1}{6}\big(C_1(n,m) + C_2(n,m)\big)$$

with

$$C_1(n,m) = \sum_{\nu=1}^{\infty}\sum_{j=0}^{n}\frac{\mathrm{d}^2}{\mathrm{d}t^2}G_{n,m,j}(t+n)\Big|_{t=\nu} \quad \text{and} \quad C_2(n,m) = \sum_{\nu=1}^{n}\sum_{j=0}^{n}\frac{\mathrm{d}^2}{\mathrm{d}t^2}G_{n,m,j}(t)\Big|_{t=\nu}$$

and apply, as before, formula (10) and its relatives to get a monster summand of $C_1(n,m)$ (that fills two pages) in terms of the harmonic numbers of order 1 and 2. For illustration we print out only a few lines:

$$\begin{aligned} G_1(n,m,j,\nu) &= \frac{\mathrm{d}^2}{\mathrm{d}t^2}G_{n,m,j}(t+n)\Big|_{t=\nu} \\ &= \frac{\binom{n}{j}^2\binom{j-m+2n}{n}(\nu)_{-m+2n}(-j+n+\nu)_n}{(n+\nu)_{1+n}(n+\nu)_{1-m+2n}}\Big(\cdots \\ &\quad + S_1(-j+n+\nu)^2 + S_1(-j+2n+\nu)^2 + S_1(-m+2n+\nu)^2 \\ &\quad + S_1(n+\nu)\frac{4(-j^2mn+2j^2n^2+\cdots+m\nu^3-7n\nu^3-2\nu^4)}{\nu(n+\nu)(-j+n+\nu)(-j+2n+\nu)(-m+2n+\nu)} \\ &\quad +\cdots\Big). \end{aligned}$$

In order to tackle the summand of $C_2(n,m)$, we have to differentiate $G_{n,m,j}(t)$ twice. With $p(t) = (t-n)_{2n-m}$ and

$$q(t)=\frac{G_{n,m}(t)}{p(t)}=\frac{n!}{(t)_{n+1}(t)_{2n-m+1}}\binom{n}{j}^2\binom{2n-m+j}{n}\frac{(t-j)_n}{n!} \tag{12}$$

we conclude that for all $1\le\nu\le n$ we have

$$\begin{aligned}\tilde{G}(\nu)=\frac{\mathrm{d}^2}{\mathrm{d}t^2}G_{n,m,j}(t)\Big|_{t=\nu}&=q(t)\frac{\mathrm{d}^2p(t)}{\mathrm{d}t^2}+2\frac{\mathrm{d}p(t)}{\mathrm{d}t}\frac{\mathrm{d}q(t)}{\mathrm{d}t}+p(t)\frac{\mathrm{d}^2q(t)}{\mathrm{d}t^2}\Big|_{t=\nu}\\&=q(t)\frac{\mathrm{d}^2p(t)}{\mathrm{d}t^2}+2\frac{\mathrm{d}p(t)}{\mathrm{d}t}\frac{\mathrm{d}q(t)}{\mathrm{d}t}\Big|_{t=\nu};\end{aligned}$$

the last equality follows since $p(t)|_{t=\nu}=0$ for all $1\le\nu\le n$. Similarly to (11), we can use in addition the following calculation: For $\nu\in\mathbb{Z}_{>0}$ and $1\le\nu\le n$, we have

$$\frac{\mathrm{d}}{\mathrm{d}t}(t-n)_{2n-m}\Big|_{t=\nu}=h(t)\Big|_{t=\nu}\quad\text{and}\quad\frac{1}{2}\frac{\mathrm{d}^2}{\mathrm{d}t^2}(t-n)_{2n-m}\Big|_{t=\nu}=\frac{\mathrm{d}}{\mathrm{d}t}h(t)\Big|_{t=\nu}$$

with

$$h(t)=\frac{(-1)^{n-\nu}\Gamma(t+n-m)(\nu-t+1)_{n-\nu}}{\Gamma(t-\nu+1)}.$$

In particular, if $\nu>j$, we can apply the rule (10) to all Pochhammer symbols in (12):

$$\begin{aligned}G_2(n,m,j,\nu)&=\tilde{G}(\nu)\\&=2q(t)\frac{\mathrm{d}}{\mathrm{d}t}h(t)+2h(t)\frac{\mathrm{d}}{\mathrm{d}t}q(t)\Big|_{t=\nu}\\&=\frac{2(-1)^{n+\nu}\binom{n}{j}^2\binom{j-m+2n}{n}(1)_{n-\nu}(2)_{-1-m+n+\nu}(1-j+\nu)_{-1+n}}{\nu^3(-m+n+\nu)^2(1+\nu)_n(1+\nu)_{-m+2n}}\\&\quad\times\Big(\nu(-j+\nu)(-m+n+\nu)\Big(\frac{1}{j-n-\nu}-S_1(-j+\nu)+S_1(-j+n+\nu)\Big)\\&\quad+\nu(-j+\nu)(-m+n+\nu)\big(-S_1(-m+2n+\nu)+S_1(\nu)\big)\\&\quad+\nu(-j+\nu)(-m+n+\nu)\big(S_1(\nu)-S_1(n+\nu)\big)\\&\quad-\nu(-j+\nu)+\nu(-m+n+\nu)\\&\quad+2(j-\nu)(-m+n+\nu)+\nu(-j+\nu)(-m+n+\nu)\\&\quad+\nu(-j+\nu)(-m+n+\nu)\big(-1+S_1(-m+n+\nu)\big)\\&\quad-\nu(-j+\nu)(-m+n+\nu)S_1(n-\nu)\Big).\end{aligned}$$

For $1\le\nu\le j$, we use $q(\nu)=0$ and apply the rule

$$\frac{\mathrm{d}}{\mathrm{d}t}\big((t-j)_nf(t)\big)\Big|_{t=\nu}=f(\nu)(\nu-j)_{j-\nu}(n+\nu-j-1)!$$

(compare with (11)) valid for any differentiable function $f(t)$, in place of (10), to (12). It follows that

$$G_3(n, m, j, \nu) = \tilde{G}(\nu) = 2\binom{n}{j}^2\binom{2n-m+j}{n} \times \frac{(-1)^{n+\nu}(n+\nu-m-1)!(n-\nu)!(n+\nu-j-1)!(\nu-j)_{j-\nu}}{(\nu)_{1+n}(\nu)_{2n-m+1}}.$$

Therefore,

$$C_2(n, m) = \sum_{\nu=1}^{n}\sum_{j=0}^{n} \frac{d^2}{dt^2} G_{n,m}(t)\Big|_{t=\nu} = \sum_{j=0}^{n-1}\sum_{\nu=j+1}^{n} G_2(n, m, j, \nu) + \sum_{j=1}^{n}\sum_{\nu=1}^{j} G_3(n, m, j, \nu),$$

hence

$$\begin{aligned} Z_r(n, m) &= \frac{1}{6}\big(C_1(n, m) + C_2(n, m)\big) \\ &= \frac{1}{6}\Bigg(\sum_{j=0}^{n}\sum_{\nu=1}^{\infty} G_1(n, m, j, \nu) + \sum_{j=0}^{n-1}\sum_{\nu=j+1}^{n} G_2(n, m, j, \nu) \\ &\quad + \sum_{j=1}^{n}\sum_{\nu=1}^{j} G_3(n, m, j, \nu)\Bigg). \end{aligned} \tag{13}$$

Denote by $A_1(n, m)$, $A_2(n, m)$ and $A_3(n, m)$ the three resulting sums in (13) and use `Sigma` to compute three linear recurrences of $A_s(n, m)$ with $s = 1, 2, 3$. A routine calculation demonstrates that each of the recurrences found can be brought to the form

$$a_0(n, m)A_s(n, m) + a_1(n, m)A_s(n, m+1) + a_2(n, m)A_s(n, m+2) = u_s(n, m), \tag{14}$$

where the coefficients are given in (4) and where only the right-hand sides $u_s(n, m)$ for $s = 1, 2, 3$ differ. As an illustration, we provide with details about how we treat

$$A_1(n, m) = C_1(n, m) = \sum_{j=0}^{n}\sum_{\nu=1}^{\infty} G_1(n, m, j, \nu).$$

In the first step, `Sigma` is used to compute a linear recurrence of the inner sum

$$c(n,m,j)=\sum_{\nu=1}^{\infty}G_1(n,m,j,\nu) \tag{15}$$

in j,

$$\begin{aligned}
&(j-n)^2(j-n+1)^2(j-m+2n+1)(j-m+2n+2)c(n,m,j)\\
&-(j-n+1)^2(j-m+2n+2)\big(2j^3-2j^2m+2jmn-3jn^2+mn^2-2n^3\\
&\qquad+8j^2-5jm-2jn+4mn-7n^2+11j-3m-4n+5\big)c(n,m,j+1)\\
&+(j+2)^3(j-2n+1)(j-m+n+2)^2c(n,m,j+2)=r(n,m,j),
\end{aligned} \tag{16}$$

and one additional recurrence with one shift in m and one shift in j,

$$\begin{aligned}
&(j-n)^2(j-m+2n+1)\big(j^3+jm^2-j^2m-m^3-2jmn+4m^2n-4mn^2\\
&\qquad+2j^2-jm-2jn+2mn-4n^2+j-2n\big)c(n,m,j)\\
&-(j+1)^3(j-2n)(j-m+n+1)^2c(n,m,j+1)\\
&-(j-n)^2(j-m+2n)(j-m+2n+1)(m+1)(m-2n)c(n,m+1,j)=s(n,m,j);
\end{aligned} \tag{17}$$

here the right-hand sides $r(n,m,j)$ and $s(n,m,j)$ are large expressions in terms of hypergeometric products and the harmonic numbers $S_1(n)$, $S_1(2n)$, $S_1(n-j)$, $S_1(2n-j)$, $S_1(2n-m)$, $S_1(3n-m)$. Finally, we use new algorithms that are described in [5] and that are built on ideas from [3, 20]. Activating these new features of `Sigma` we can compute the linear recurrence (14) with $s=1$ where the right-hand side $u_0(n,m)$ is an expression in terms of the harmonic numbers $S_1(n)$, $S_1(2n)$, $S_1(2n-m)$, $S_1(3n-m)$, the infinite sums

$$c(n,m,0),\ c(n,m,1),\ c(n,m,n+1) \tag{18}$$

and the definite sums

$$\begin{aligned}
&\sum_{i=0}^{n}\binom{n}{i}^2\binom{2n-m+i}{n}\frac{(n-i+1)_n}{(2n-i)^k} && \text{for } k=0,1,2,\\
&\sum_{i=0}^{n}\binom{n}{i}^2\binom{2n-m+i}{n}\frac{(n-i+1)_n}{(2n-i)^k}S_1(n-i) && \text{for } k=0,1,\\
&\sum_{i=0}^{n}\binom{n}{i}^2\binom{2n-m+i}{n}\frac{(n-i+1)_n}{(2n-i)^k}S_1(2n-i) && \text{for } k=0,1.
\end{aligned} \tag{19}$$

Note that all these definite sums in (19) are *not* expressible in terms of hypergeometric products and indefinite nested sums defined over such products. For example, the linear recurrence for the last sum in (19) with $k=0$ computed with `Sigma` has order 5 and has not even a hypergeometric product solution. We further remark that the above approach is connected to the classical holonomic summation approach [31] and their improvements given in [7, 13]. In all these traditional versions one needs systems composed by homogeneous recurrences. However, the transformations of (16) and (17) to such a form would lead to gigantic recurrence systems and the computation of the desired linear recurrence (3) would be out of scope.

Using this refined holonomic summation approach with `Sigma`, we needed in total 10 min to derive the recurrence for $A_1(n, m)$ which holds for all $0 \leq m \leq n$. Similarly, one can compute for the other two double sums $A_2(n, m)$ and $A_3(n, m)$ the recurrence (14) in 15 and 2 min, respectively, which hold for all $0 \leq m \leq n-2$. Here the right-hand sides $u_2(n, m), u_3(n, m)$ consist of similar definite sums as given in (19). Adding up (14) corresponding to $s = 1, 2, 3$, results in a linear recurrence for $Z_r(n, m)$ with (3) on the left-hand side and

$$u(n, m) = \frac{1}{6}\big(u_1(n, m) + u_2(n, m) + u_3(n, m)\big)$$

on the right-hand side which holds for all $0 \leq m \leq n-2$. It remains to show that the inhomogeneous part evaluates to zero, $u(n, m) = 0$ for $0 \leq m \leq n-2$. As indicated earlier, the expression $u(n, m)$ is composed by

- the infinite sums (18) with (15);
- finite definite sums like those given in (19).

A verification for all $n-2 \geq m \geq 0$ looks rather challenging. However, using the toolbox of `Sigma`, this task can be accomplished automatically in 16 min of calculation time. First, we treat the infinite sums by merging them to one big infinite sum and then compute a linear recurrence for it, which happens to be completely solvable in terms of indefinite nested sums. This reduces all the infinite sums to indefinite nested sums. The finite definite sums are a tougher nut to crack. Internally, all sums (including (19)) are first considered as indefinite nested versions (with a common upper bound, say a). Then a finite subset of the sums arising is calculated with the command `SigmaReduce` such that there are no dependences among them and such that all the remaining sums can be represented in terms of these independent sums. It turns out that all sums (with a now replaced by the 'synchronized' upper bound $n-3$) cancel and only one definite sum remains. Activating the package `EvaluateMultiSums` [25] (that combines automatically the available summation tools of `Sigma`) this remaining sum simplifies to

$$\begin{aligned}
&\sum_{i=1}^{n-3} \frac{(-1)^i}{i}\binom{n}{i}\binom{2n-m+i}{n} \\
&= \frac{(-1)^n\binom{3n-m-2}{n-2}}{2(n-2)(n-1)^2n^2}\big(-4m - 4m^2 + 12n + 30mn + 6m^2n - 54n^2 - 43mn^2 - 7m^2n^2 \\
&\quad + 70n^3 + 40mn^3 + 4m^2n^3 - 54n^4 - 19mn^4 - m^2n^4 + 22n^5 + 4mn^5 - 4n^6\big) \\
&\quad - \binom{2n-m}{n}\big(S_1(2n-m) + S_1(n) - S_1(n-m)\big).
\end{aligned}$$

In a nutshell, $u(n, m)$ can be reduced to an expression given purely in terms of indefinite nested sums, which after further simplifications collapses to zero. This shows that not only the left-hand side but also the right-hand side of (2) satisfies the same recurrence (3). The verification of this fact took in total 43 min.

5 Dealing with the Initial Values

In order to verify (2), it remains to show (5) and (6). For (5) we proceed as follows. First, we compute for $Z_l(n, 0)$ the recurrence

$$\begin{aligned}&-16(2n+1)^4 Z_l(n,0) - (n+1)^4 Z_l(n+1,0)\\&= -\frac{(-1)^n n!^8 (1+2n)_{2n}(1+4n)\big(831+5265n+12601n^2+13499n^3+5460n^4\big)}{48(2n+1)!^5}.\end{aligned} \tag{20}$$

Internally, we follow the strategy in Sect. 3: we use the representation from (9) to get

$$Z_l(n,0) = -\frac{1}{3}\big(W_1(n,0) + W_2(n,0)\big)$$

and, for $W_1(n, 0)$ and $W_2(n, 0)$, compute two recurrences, where both have the *same* homogeneous part. Thus adding the inhomogeneous parts and simplifying the result further leads to (20). Solving this recurrence leads, for any $n \geq 0$, to the closed form

$$\begin{aligned}Z_l(n,0) = {}&\frac{(-1)^n}{30720}\big(105U_9(n) + 955U_8(n) + 3095U_7(n) + 2045U_6(n)\\&-12140U_5(n) - 27300U_4(n) + 12288\zeta(2)^2\big)\binom{2n}{n}^4\\&+\frac{(-1)^n(4n+1)(5460n^4+13499n^3+12601n^2+5265n+831)\binom{4n}{2n}}{768(2n+1)^9\binom{2n}{n}^4}\end{aligned} \tag{21}$$

in terms of indefinite nested sums

$$U_k(n) = \sum_{i=0}^{n} \frac{\binom{4i}{2i}}{(2i+1)^k\binom{2i}{i}^8} \quad \text{with } k = 1, 2, \ldots. \tag{22}$$

Similarly to Sect. 4, we use the sum representation in (13) with $m = 0$ encoded by $A_0(n, 0) + \cdots + A_3(n, 0)$ to compute the recurrence

$$
\begin{aligned}
&16(n+1)^3(2n+1)^4(4n+3)(4n+5)(5460n^4+35339n^3+85858n^2+92804n+37656)Z_r(n,0)\\
&+(357913920n^{13}+5716680688n^{12}+41762423804n^{11}+184637211081n^{10}\\
&\quad+550778114541n^9+1169740743051n^8+1818232366245n^7+2092705983417n^6\\
&\quad+1782121652067n^5+1108272850929n^4+488951050619n^3\\
&\quad+144869028586n^2+25833166356n+2094206184)Z_r(n+1,0)\\
&+8(n+2)^4(2n+3)^5(5460n^4+13499n^3+12601n^2+5265n+831)Z_r(n+2,0)=0
\end{aligned}
\tag{23}
$$

which holds true for all $n \geq 0$. Furthermore, we verify that $Z_l(n, 0)$ is also a solution of this recurrence by plugging its representation (21) into the recurrence and checking that the expression simplifies to zero. Finally, we verify that the first two initial values of $Z_l(n, 0)$ and $Z_r(n, 0)$ agree:

$$
Z_l(0,0)=Z_r(0,0)=\frac{2}{5}\zeta(2)^2,\quad Z_l(1,0)=Z_r(1,0)=\frac{277}{16}-\frac{32}{5}\zeta(2)^2;
$$

to determine these evaluations again `Sigma` has been utilized. Together with the fact that the leading coefficient in (23) is nonzero for all $n \geq 0$, this implies that (5) holds.

To verify (6), we repeat the same game for $Z_l(n, 1)$ and $Z_r(n, 1)$: namely, we find the closed form representation

$$
\begin{aligned}
Z_l(n,1)=&\frac{3n(-1)^n}{40960}\big(105U_9(n)+955U_8(n)+3095U_7(n)+2045U_6(n)\\
&-12140U_5(n)-27300U_4(n)+12288\zeta(2)^2\big)\binom{2n}{n}^4\\
&-\frac{(-1)^n\binom{4n}{2n}}{1024n^3(2n+1)^9\binom{2n}{n}^4}(16n^9+116544n^8+398115n^7+587145n^6\\
&+490329n^5+255555n^4+86016n^3+18432n^2+2304n+128)
\end{aligned}
\tag{24}
$$

valid for all $n \geq 1$. In addition, we compute a recurrence of order 2 for $Z_r(n, 1)$ and, as above, verify that $Z_l(n, 1)$ is also its solution (by plugging in the representation (24)). Together with the initial values

$$
Z_l(1,1)=Z_r(1,1)=-13+\frac{24}{5}\zeta(2)^2,\quad Z_l(2,1)=Z_r(2,1)=\frac{4090247}{1944}-\frac{3888}{5}\zeta(2)^2
$$

this implies that (6) holds as well and completes the proof of (2). We note that the verification of each initial value problem, (5) and (6), took about 25 min.

6 Summary

Summarizing, the full proof of (2) took in total around 2 hours (excluding all the human trials and errors to find the tailored paths described above, and days to physically write this paper).

The initial values (21) and (24) are given through $2\zeta(2)^2/5 = \zeta(4)$, hypergeometric products and the indefinite nested sums (22) with $k = 4, 5, 6, 7, 8, 9$. Thus, feeding the recurrence (3) with all this stuff we get the following corollary.

Theorem 2 *For any* $n \geq m \geq 0$, *both sides of* $Z_l(n, m) = Z_r(n, m)$ *can be expressed* (*and computed in linear time*) *in terms of* $\zeta(4)$ *and* $U_4(n), \ldots, U_9(n)$ *in* (22).

The project [15] implicitly suggests that there can be further—more general(!)—forms of (2), with more than two independent parameters. We have tried (unsuccessfully) to find some but cannot even figure out how to adopt (2) to the case $m > n$.

Acknowledgements This project commenced during the joint visit of the authors in the Max Planck Institute for Mathematics (Bonn) in 2007 and went on during the second author's visit in the Research Institute for Symbolic Computation (Linz) in February 2020. We thank the staff of these institutes for providing such excellent conditions for research.

References

1. Ablinger, J., Behring, A., Blümlein, J., De Freitas, A., von Manteuffel, A., Schneider, C.: Calculating three loop ladder and V-topologies for massive operator matrix elements by Computer Algebra. Comput. Phys. Comm. **202**, 33–112 (2016)
2. Abramov, S.A., Petkovšek, M.: D'Alembertian solutions of linear differential and difference equations. In: von zur Gathen, J. (ed.) Proceedings of ISSAC'94, pp. 169–174. ACM Press (1994)
3. Andrews, G.E., Paule, P., Schneider, C.: Plane partitions VI: Stembridge's TSPP Theorem. Adv. Appl. Math. **34**:4, 709–739 (2005)
4. Apéry, R.: Irrationalité de $\zeta(2)$ et $\zeta(3)$. Astérisque **61**, 11–13 (1979)
5. Blümlein, J., Round, M., Schneider, C.: Refined holonomic summation algorithms in Particle Physics. In: Zima, E., Schneider, C. (eds.) Advances in Computer Algebra (WWCA 2016). Springer Proceedings in Mathematics and Statistics, vol. 226, pp. 51–91. Springer (2018)
6. Bostan, A., Chamizo, F., Sundqvist, M.P.: On an integral identity (2020). arXiv:2002.10682 [math.CA]
7. Chyzak, F.: An extension of Zeilberger's fast algorithm to general holonomic functions. Discrete Math. **217**, 115–134 (2000)
8. Cohen, H.: Accélération de la convergence de certaines récurrences linéaires. Sém. Théorie Nombres Bordeaux, exp. 16 (1980–81)
9. Ekhad, S.B., Zeilberger, D., Zudilin, W.: Two definite integrals that are definitely (and surprisingly!) equal. Math. Intelligencer **42**, 10–11 (2020)
10. Hardouin, C., Singer, M.F.: Differential Galois theory of linear difference equations. Math. Ann. **342**:2, 333–377 (2008)
11. Hendriks, P.A., Singer, M.F.: Solving difference equations in finite terms. J. Symbolic Comput. **27**:3, 239–259 (1999)
12. Karr, M.: Summation in finite terms. J. Assoc. Comput. Machinery **28**, 305–350 (1981)

13. Koutschan, C.: Creative telescoping for holonomic functions. In: Schneider, C., Blümlein, J. (eds.) Computer Algebra in Quantum Field Theory: Integration, Summation and Special Functions. Texts and Monographs in Symbolic Computation, pp. 171–194. Springer (2013)
14. Krattenthaler, C., Rivoal, T.: Hypergéométrie et fonction zêta de Riemann. Mem. Amer. Math. Soc. **186**:875 (2007)
15. Marcovecchio, R., Zudilin, W.: Hypergeometric rational approximations to $\zeta(4)$. Proc. Edinburgh Math. Soc. **63**:2, 374–397 (2020)
16. Paule, P., Schorn, M.: A Mathematica version of Zeilberger's algorithm for proving binomial coefficient identities. J. Symbolic Comput. **20**:5-6, 673–698 (1995)
17. Petkovšek, M.: Hypergeometric solutions of linear recurrences with polynomial coefficients. J. Symbolic Comput. **14**:2–3, 243–264 (1992)
18. Petkovšek, M., Wilf, H.S., Zeilberger, D.: $A = B$. A K Peters, Wellesley, MA (1996)
19. Schneider, C.: Symbolic summation in difference fields. PhD Thesis. Technical Report 01-17, RISC-Linz, J. Kepler University (2001)
20. Schneider, C.: A new Sigma approach to multi-summation. Adv. Appl. Math. **34**:4, 740–767 (2005)
21. Schneider, C.: Apéry's double sum is plain sailing indeed. Electron. J. Combin. **14**, N5 (2007)
22. Schneider, C.: Symbolic summation assists combinatorics. Sém. Lothar. Combin. **56**:B56b, 1–36 (2007)
23. Schneider, C.: A refined difference field theory for symbolic summation. J. Symbolic Comput. **43**:9, 611–644 (2008)
24. Schneider, C.: Parameterized telescoping proves algebraic independence of sums. Ann. Comb. **14**, 533–552 (2010)
25. Schneider, C.: Simplifying multiple sums in difference fields. In: Schneider, C., Blümlein, J. (eds.) Computer Algebra in Quantum Field Theory: Integration, Summation and Special Functions. Texts and Monographs in Symbolic Computation, pp. 325–360. Springer (2013)
26. Schneider, C.: Fast algorithms for refined parameterized telescoping in difference fields. In: Weimann, M., Guitierrez, J., Schicho, J. (eds.) Computer Algebra and Polynomials. Lecture Notes in Computer Science, vol. 8942, pp. 157–191. Springer (2015)
27. Schneider, C.: A difference ring theory for symbolic summation. J. Symbolic Comput. **72**, 82–127 (2016)
28. Schneider, C.: Summation theory II: Characterizations of $R\Pi\Sigma$-extensions and algorithmic aspects. J. Symbolic Comput. **80**:3, 616–664 (2017)
29. Sorokin, V.N.: One algorithm for fast calculation of π^4. Russian Academy of Sciences, M.V. Keldysh Institute for Applied Mathematics, Moscow (2002)
30. van der Poorten, A.: A proof that Euler missed... Apéry's proof of the irrationality of $\zeta(3)$. Math. Intelligencer **1**:4, 195–203 (1978/79)
31. Zeilberger, D.: A holonomic systems approach to special functions identities. J. Comput. Appl. Math. **32**, 321–368 (1990)
32. Zeilberger, D.: The method of creative telescoping. J. Symbolic Comput. **11**, 195–204 (1991)
33. Zudilin, W.: Well-poised hypergeometric service for diophantine problems of zeta values. J. Théorie Nombres Bordeaux **15**:2, 593–626 (2003)
34. Zudilin, W.: A hypergeometric problem. J. Comput. Appl. Math. **233**, 856–857 (2009)

Support of an Algebraic Series as the Range of a Recursive Sequence

Jason P. Bell

Abstract Let p be prime and let K be a field of characteristic p. We characterize when the support of an algebraic power series with coefficients in K is equal to the range of a sequence $f : \mathbb{N} \to \mathbb{N}$, where f either satisfies a linear recurrence with constant coefficients or is strictly increasing and is P-recursive. In addition, we prove related results about automatic subsets of the natural numbers.

Keywords Christol's theorem · Algebraic functions · P-recursive sequences · D-finite power series · Holonomic sequences · Automatic sets · Automata

2010 Mathematics Subject Classification: 11B85 · 68Q45

1 Introduction

The Skolem-Mahler-Lech theorem (see Lech [Lec53] and also Hansel [Han86] and [Sch03, Corollary 7.2]) is a fundamental result on linear recurrences. It states that if K is a field of characteristic zero and $f(n)$ is a K-valued sequence satisfying a linear recurrence with constant coefficients then the set of n such that $f(n) = 0$ is a finite union of infinite arithmetic progressions of the form $a\mathbb{N} + b$ along with a finite set.

Since the generating functions of sequences satisfying linear recurrences with constant coefficients are precisely the power series expansions of rational functions that do not have a pole at the origin, this result completely characterizes the sets that can arise as the set of zero coefficients in a rational power series: they are precisely the sets of natural numbers whose characteristic function is eventually periodic. Since the complement of an eventually periodic set is again eventually periodic, one can recast

The author was supported by NSERC Grant RGPIN-2016-03632.

J. P. Bell (✉)
Department of Pure Mathematics, University of Waterloo,
Waterloo, ON N2L 3G1, Canada
e-mail: jpbell@uwaterloo.ca

A. Bostan and K. Raschel (eds.), *Transcendence in Algebra, Combinatorics, Geometry and Number Theory*, Springer Proceedings in Mathematics & Statistics 373,
https://doi.org/10.1007/978-3-030-84304-5_18

this result as one that characterizes the possible support sets of rational functions over a field of characteristic zero (the support of a series is the set of indices for which the coefficient is nonzero): they are again the sets of natural numbers whose characteristic function is eventually periodic.

When one tries to extend the result of Skolem-Mahler-Lech to power series expansions of algebraic functions, the problem becomes significantly more difficult. Indeed, there is currently no characterization of which subsets of the natural numbers can arise as the support set of an algebraic power series, although Rubel [Rub83, Problem 16] has a more general conjecture that implies that the Skolem-Mahler-Lech result should extend to algebraic power series as well, and so an algebraic power series over a field of characteristic zero should have support whose characteristic function is eventually periodic. This conjecture of Rubel remains open, but there is a special case due to Bézivin [Béz89] and Methfessel [Met00], which shows that the set of zero coefficients of a power series with coefficients in a characteristic zero field that satisfies a homogeneous linear differential equation with rational function coefficients is a finite union of infinite arithmetic progressions along with a set of Banach density zero.

In positive characteristic, the situation is much better understood. If K is a field of characteristic $p > 0$ and $F(x) = \sum f(n)x^n \in K[[x]]$ is an algebraic power series, then the support of $F(x)$ is known to be a p-automatic set; that is, there is a finite-state machine that takes as input the base-p expansion of n, reading the digits from right to left, and accepts precisely those n for which $f(n) \neq 0$. In positive characteristic, however, work of Derksen [Der07] shows that the correct analogue of the Skolem-Mahler-Lech theorem is quite subtle and with algebraic power series, the support sets can be more complicated still. For example,

$$F(x) = \sum x^{p^j} \in \mathbb{F}_p[[x]]$$

is algebraic, satisfying the equation $F(x)^p = F(x) - x$ and it has support set $\{1, p, p^2, \ldots\}$. We observe that this set has the additional special property that it is the set of values of the geometric sequence $f(n) = p^n$.

In this paper, we investigate which sequences of natural numbers $f(n)$ that satisfy a linear recurrence (with constant coefficients) have the property that $f(\mathbb{N})$ can occur as the support set of an algebraic power series over a field in positive characteristic. In fact, we are able to classify such support sets when we instead work with increasing P-recursive sequences as well, which includes a large class of interesting combinatorial sequences that are not covered under the umbrella of sequences satisfying linear recurrences with constant coefficients. We recall that an integer-valued sequence $f(n)$ is P-recursive, if there exist $d \geq 0$ and integer polynomials $P_0(x), \ldots, P_d(x)$, not all zero, such that

$$\sum_{i=0}^{d} P_i(n) f(n-i) = 0.$$

This is equivalent to the generating function $F(x) = \sum f(n)x^n$ being D-finite (i.e., satisfying a homogeneous linear differential equation with rational function coefficients). We refer the reader to Stanley [Sta99, Chapter 6] for further background on and interesting examples of P-recursive sequences.

Our main result is the following theorem concerning support sets of algebraic series over positive characteristic fields.

Theorem 1.1. *Let p be a prime, let K be a field of characteristic p, and let $G(x) = \sum g(n)x^n \in K[[x]]$ be an algebraic power series. Suppose that the support of $g(n)$ is the range of a sequence $f(n)$ such that either:*

(a) $f(n)$ satisfies a linear recurrence with constant coefficients;
(b) $f(n)$ is strictly increasing and P-recursive.

Then there exist some $a \geq 1$ and $N \geq 0$ such that for each $b \in \{0, \ldots, a-1\}$ the sequence $n \mapsto f(an+b)$ writes for $n \geq N$ either as:

(1) a constant sequence;
(2) an arithmetic sequence $n \mapsto Cn + C'$ with C, C' integers and $C > 0$;
(3) a geometric sequence $n \mapsto Ck^{an} + C'$ for some rational constants C, C' with $C > 0$ and some positive rational number a with k^a equal to an integer greater than 1.

In particular, Theorem 1.1 shows that—up to basic changes of variables and sums—every algebraic power series over a finite field of characteristic $p > 0$ whose support satisfies a linear recurrence with constant coefficients can be built up from the power series $x + x^p + x^{p^2} + \cdots$ and the rational function $1 + x + x^2 + \cdots$. Although this is perhaps what one who is familiar with algebraic series over positive characteristic fields would expect, we nevertheless provide a proof as the proof itself is not simple and the result showcases the underlying structure of the ring of algebraic power series over positive characteristic fields.

We note that, although we need the hypothesis that the sequence $f(n)$ is strictly increasing in the P-recursive case, this hypothesis is somewhat natural as it captures the idea that the set $f(\mathbb{N})$ is "parametrized" by the sequence f. We note, however, that positive-valued P-recursive sequences need not be weakly increasing. This can be seen from the fact that if $\sum f(n)x^n$ and $\sum g(n)x^n$ are both D-finite then so is $\sum h(n)x^n$, where $h(2n) = f(n)$ and $h(2n+1) = g(n)$ for $n \geq 0$, and so, for example, if we take $h(2n) = 2^n$ and $h(2n+1) = n$ for $n \geq 0$, then we obtain a non-increasing example.

In light of work on algebraic power series [Chr79, CKMR80, AB12], asking when the support set of an algebraic power series over a positive characteristic field is parametrized by a sequence satisfying a linear recurrence with constant coefficients is the same as asking which p-automatic subsets of $\mathbb{N}$ have the property that they are precisely the range of a sequence of natural numbers satisfying a linear recurrence with constant coefficients. As a result much of the paper will involve the study of arithmetical properties of automatic sets. Thus in Sects. 2 and 3, we first give background on automata and sparse automatic sets. In Sect. 4, we give background on a powerful result of Laurent [Lau87] concerning sequences satisfying linear recurrences with constant coefficients, which will be instrumental in proving Theorem 1.1. Finally, in Sect. 5, we prove general results on automatic sets (see Theorem 5.1), which we will then apply to obtain the proof of Theorem 1.1.

2 Background on Automata and Automatic Sets

In this section we give some background on finite-state automata and k-automatic sequences and sets. Further information and proofs of basic facts that we state without proof can be found in the comprehensive book of Allouche and Shallit [AS03].

For us, an *alphabet* will simply be a non-empty set Σ and a *word* over the alphabet Σ will be an element of the free monoid Σ^* generated by Σ. A *deterministic finite automaton with output* (DFAO) is a 6-tuple

$$\Gamma = (Q, \Sigma, \delta, q_0, \Delta, \tau),$$

where Q is a finite set of states, Σ is a finite input alphabet, δ is the transition function from $\Sigma \times Q$ to Q, $q_0 \in Q$ is the initial state, Δ is an output alphabet, and τ is the output function from Q to Δ. More intuitively, one can think of a DFAO as a finite directed graph in which the vertices are the elements of Q, and for each vertex $q \in Q$ and each $x \in \Sigma$ we have a directed arrow with label x from the state q to the state $\delta(x, q)$. This framework allows us to go associate a function from Σ^* to Δ as follows. Given $w \in \Sigma^*$, we begin at the initial state q_0 and then, reading w from right to left, we obtain a path in this directed graph by moving vertex to vertex as we read the letters of w and follow the directed edges with the appropriate labels. After we finish reading w, we end up at some state $q \in Q$ and we then apply τ to obtain an output in Δ.

Since we are interested in linear recurrences, we give an example of a DFAO in Fig. 1 that generates the map f from $\{0, 1\}^*$ to $\Delta = \{0, 1\}$, where f is 1 precisely when the string w is the binary expansion (where we allow no leading zeros in the binary expansion) of a number of the form $3 \cdot 2^n + 1$ with $n \geq 1$. In particular, this is the set of numbers whose binary expansions are of the form 110^a1 with $a \geq 0$.

Let $k \geq 2$ be a natural number and let Σ_k be the alphabet $\{0, 1, \ldots, k-1\}$. For every natural number n, there is a word $w = (n)_k \in \Sigma_k^*$, which is the base-k expansion of n, where we define $(0)_k$ to be the empty word; conversely, given a non-empty word $w \in \Sigma_k^*$ with no leading zeros there is a natural number $n = [w]_k$, which is the natural number whose base-k expansion is w. In the case when w is the empty word, we take $[w]_k = 0$. A sequence $a : \mathbb{N} \to \Delta$ is called *k-automatic* if there exists a DFAO $\Gamma = (Q, \Sigma_k, \delta, q_0, \Delta, \tau)$ such that for each $n \in \mathbb{N}$, $a(n)$ can be computed from Γ using by feeding the word $(n)_k$ into Γ, reading the digits from right to left. We then say that a subset $S \subseteq \mathbb{N}$ is a *k-automatic set* if the characteristic function of S, $\chi_S : \mathbb{N} \to \{0, 1\}$ defines a k-automatic sequence.

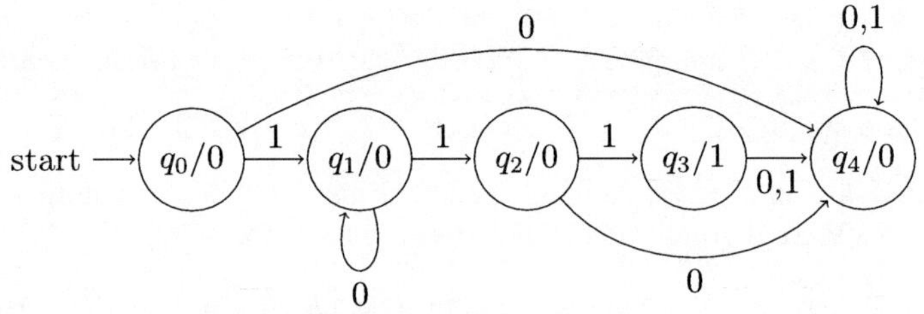

Fig. 1 The DFAO generating the characteristic function of the set of numbers of the form $3 \cdot 2^n + 1$ with $n \geq 1$

There is a nice dichotomy for the behaviour of automatic sets, which is proved in Eilenberg [Eil74, Theorem 5.4, p. 112]. It shows that if S is an infinite k-automatic subset of $\mathbb{N}$ then either it is *syndetic* (i.e., it has uniformly bounded gaps between successive elements) or it has reasonably large gaps occurring infinitely often. We give a precise statement below.

Proposition 2.1. *(Eilenberg) Let $S = \{x_1, x_2, \ldots\}$ be an infinite k-automatic subset of $\mathbb{N}$ with $x_1 < x_2 < \cdots$. Then either $x_{n+1} - x_n = O(1)$ or $\limsup x_{n+1}/x_n > 1$.*

In addition to Eilenberg's result, we need a basic result about closure properties of automatic sets.

Lemma 2.2. *Let $k \geq 2$ be a natural number, let a and b be rational numbers, and let S and T be k-automatic subsets of $\mathbb{N}$. Then the following sets are also k-automatic:*

(1) $\{n : an + b \in S\}$;
(2) $\{an + b : n \in S\} \cap \mathbb{N}$;
(3) $S \cap T$, $S \cup T$, S^c.

Since the sets $\mathbb{N} = \{n : n \geq 0\}$ and $\{k^{an} : n \geq 0\}$ are automatic when a is a rational number such that k^a is an integer greater than 1, we can use the closure properties above to immediately obtain the "if" direction in Theorem 5.1.

3 Sparse Automatic Sets

In this section, we give a brief summary of basic facts about sparse languages and sparse sets, which will play a key role in the proof of Theorem 5.1. Sparse automatic sets and related concepts have been studied by many authors (see [GKRS10] and references therein). In order to define a sparse automatic set, we first state a well-known dichotomy concerning the "growth" of such sets S.

Proposition 3.1. *Let $S \subseteq \mathbb{N}$ be a k-automatic set and let $\pi_S(x) = \#\{n \in S : n \leq x\}$ for $x \geq 0$. Then one of the following alternatives must hold:*

(1) there exists $d \geq 1$ such that $\pi_S(n) = O((\log n)^d)$ as $n \to \infty$; or
(2) there exists a real number $\alpha > 0$ such that $\pi_S(n) > n^\alpha$ for all sufficiently large n.

(See, for example, [GKRS10, §2.3] or [BM17, Proposition 7.1]).

Definition 3.2. Let $k \geq 2$ be a natural number and let S be a k-automatic set. We say that S is *sparse* if condition (1) in Proposition 3.1 holds.

Lemma 3.3. *A sparse k-automatic set is expressible as a finite union of sets of the form $[v_1 w_1^* v_2 w_2^* \dots v_s w_s^* v_{s+1}]_k$, where $s \geq 0$, the v_i are possibly trivial words, and the w_i are non-trivial words over the alphabet $\{0, 1, \dots, k-1\}$.*

The components of sparse sets described in Lemma 3.3 sets are related to *p-normal sets* in [Der07] and when $k = p$, they coincide with the sets

$$U_p(v_{s+1}, v_s, \dots, v_1; w_s, \dots, w_1)$$

that are defined in [Der07, Definition 7.8].

We need a result on sparse subsets of the natural numbers.

Lemma 3.4. *Let $k \geq 2$ be a natural number and let S be a non-empty sparse k-automatic set of natural numbers. Then S is a finite union of k-automatic subset $S_1 \cup S_2 \cup \dots \cup S_m$ such that for each $i \in \{1, 2, \dots, m\}$ there exist $s \geq 0$, $c_0, \dots, c_s \in \mathbb{Q}$ such that $(k^\ell - 1)c_i \in \mathbb{Z}$ for some $\ell \geq 0$, $c_0 + c_1 + \dots + c_s \in \mathbb{Z}_{\geq 0}$ and positive integers $\delta_1, \dots, \delta_s$ such that*

$$S_i = \left\{c_0 + c_1 k^{\delta_s n_s} + c_2 k^{\delta_s n_s + \delta_{s-1} n_{s-1}} \dots + c_s k^{\delta_s n_s + \dots + \delta_1 n_1} : n_1, \dots, n_s \geq 0\right\}. \quad (3.1)$$

Moreover $n \geq c_0$ for all $n \in S$ and $c_0 \in S$ if and only if $s = 0$.

Proof. This result is due to Ginsburg and Spanier [GS66] (see also [AB19, Remark 3.4]). □

We note that the pumping lemma (see Allouche and Shallit [AS03, Lemma 4.2.1]) shows that every infinite k-automatic set contains a subset of the form $\{[v_1 w_1^n v_2]_k : n \geq 0\}$ and so Lemma 3.4 gives the useful well-known result.

Lemma 3.5. *Let S be an infinite k-automatic subset of $\mathbb{N}$. Then there exist rational numbers c_0, c_1 with $c_1 > 0$ and a positive integer δ such that $c_0 + c_1 k^{\delta n} \in S$ for every n. Moreover, $c_1 + c_0 \in \mathbb{N}$ and $c_1(k^\delta - 1) \in \mathbb{N}$.*

Proof. This follows immediately from the pumping lemma (see Allouche and Shallit [AS03, Lemma 4.2.1]) and Lemma 3.4. The fact that $c_1 + c_0$ is in $\mathbb{N}$ and $(k^\delta - 1)c_1 \in \mathbb{N}$ both follow from noting that $c_0 + c_1 k^{\delta n}$ is a nonnegative integer for every n. □

4 Linear Recurrences and Laurent's Theorem

Given a field K, a K-valued sequence $f : \mathbb{N} \to K$ is said to satisfy a *linear recurrence* (with constant coefficients) over K if there is some $d \geq 1$ and $c_1, \ldots, c_d \in K$ such that $f(n) = \sum_{i=1}^{d} c_i f(n-i)$ for all sufficiently large n. This is equivalent to the existence of some $\lambda_0, \ldots, \lambda_q \in \bar{K} \setminus \{0\}$ and polynomials $P_0(x), \ldots, P_q(x) \in \bar{K}[x]$ such that for n sufficiently large

$$f(n) = \sum_{i=0}^{q} P_i(n)\lambda_i^n. \tag{4.2}$$

If the λ_i are pairwise distinct and nonzero and λ_i/λ_j is not a root of unity for $i \neq j$, then we say that the sequence satisfies a *non-degenerate* linear recurrence (with constant coefficients). It is well-known that by passing to subsequences along arithmetic progressions one may work with non-degenerate linear recurrences. We provide a proof of this result for the reader's convenience.

Lemma 4.1. *Let $f(n)$ be a sequence satisfying a linear recurrence with constant coefficients. Then there exists a positive integer a such that for each $b \in \{0, 1, \ldots, a-1\}$ the sequence $f_b(n) := f(an+b)$ satisfies a linear recurrence with constant coefficients and is non-degenerate.*

Proof. By Eq. (4.2), there exist $\lambda_0, \ldots, \lambda_q \in \bar{K} \setminus \{0\}$ and polynomials $P_0(x), \ldots, P_q(x) \in \bar{K}[x]$ such that $f(n) = \sum_{i=0}^{q} P_i(n)\lambda_i^n$ for all sufficiently large n. We let a denote the least common multiple of the orders of the quotients λ_i/λ_j which are roots of unity. Then we may partition $\{0, 1, \ldots, q\}$ into non-empty sets $S_1, \ldots, S_k$ such that for each $\ell \in \{1, \ldots, k\}$ there exists $\alpha_\ell \in K \setminus \{0\}$ such that $\lambda_i^a = \alpha_\ell$ for all $i \in S_\ell$. Then by construction α_i/α_j is not a root of unity for $i \neq j$. We let $b \in \{0, \ldots, a-1\}$. Then for n sufficiently large we have

$$f_b(n) = \sum_{i=0}^{q} P_i(n)\lambda_i^{an+b} = \sum_{\ell=1}^{k} \left(\sum_{i \in S_\ell} P_i(n)\lambda_i^b \right) \alpha_\ell^n.$$

Observe that for $\ell \in \{1, \ldots, k\}$, $T_\ell(x) := \sum_{i \in S_\ell} P_i(x)\lambda_i^b$ is a polynomial in $K[x]$ and so $f_b(n)$ is non-degenerate. The result now follows. □

In the case that $f(n)$ is as in Eq. (4.2) and is non-degenerate, we may assume without loss of generality that $\lambda_0 = 1$ and that $\lambda_1, \ldots, \lambda_q$ are not roots of unity and that P_0 is allowed to be zero, whereas $P_1, \ldots, P_q$ are nonzero. Then if $f(n)$ is not a polynomial we necessarily have $q \geq 1$. If

$$g(n) = \sum_{i=0}^{q'} Q_i(n)\gamma_i^n. \tag{4.3}$$

satisfies a non-degenerate linear recurrence with $1 = \gamma_0, \gamma_1, \ldots, \gamma_{q'}$ nonzero, Q_0 allowed to be zero but $Q_1, \ldots, Q_{q'}$ nonzero, and if $f(n)$ is non-degenerate and as given in Eq. (4.2), we say that the sequences $f(n)$ and $g(n)$ are *related* if $q = q'$ and there exist nonzero integers a and b such that after reordering $\gamma_1, \ldots, \gamma_q$ we have $\lambda_i^a = \gamma_i^b$ for $i = 1, \ldots, q$. We say they are *doubly related* if there are at least two distinct orderings of $\gamma_1, \ldots, \gamma_q$ with respect to which they are related; otherwise they are *simply related* (see [Sch03, Sec. 11]).

The next result is a famous result about polynomial-exponential equations, known as Laurent's theorem and due to Laurent [Lau87, Théorème 3]. The statement below is a reformulation that can be found in a paper of Schmidt [Sch03, Theorem 11.2].

Proposition 4.2. *(Laurent's theorem.) Let $f(n)$, $g(n)$ be sequences satisfying non-degenerate linear recurrences given by Eqs. (4.2) and (4.3) with $q, q' \geq 1$ and consider the equation*

$$f(n) = g(m)$$

in the integers n, m. This equation has only finitely many solutions unless $f(n)$ and $g(n)$ are related. When f and g are simply related, then after a suitable reordering of the γ_i all but finitely many solutions satisfy the system of equations $P_i(n)\lambda_i^n = Q_i(m)\gamma_i^m$ for $i = 0, \ldots, q$.

We recall that a set of natural numbers S has zero density if $\limsup_{x\to\infty} \pi_S(x)/x = 0$, where $\pi_S(x)$ is the number of elements in S that are at most x. Observe that if k is a positive integer and $S_1, \ldots, S_k$ are subsets of the natural numbers of zero density then for each $\varepsilon > 0$ we have $\pi_{S_j}(x) \leq \epsilon x$ for all x sufficiently large and so

$$\pi_S(x) \leq \sum_{j=1}^{k} \pi_{S_i}(x) \leq k\epsilon x$$

for x sufficiently large. Since $\varepsilon > 0$ is arbitrary, we then see that a finite union of zero density sets is again of zero density. We require the following result, whose only subtlety lies in the fact that we do not assume that the sequence satisfying a linear recurrence is increasing in the lemma below.

Lemma 4.3. *If $f : \mathbb{N} \to \mathbb{N}$ satisfies a non-degenerate linear recurrence with constant coefficients then $f(\mathbb{N})$ has zero density unless $f(n) = an + b$ for some non-negative integers a and b with $a > 0$.*

Proof. If $f(\mathbb{N})$ has positive density then there is some integer $B > 0$ such that the density of $f(\mathbb{N})$ is strictly greater than $1/B$. Hence there are infinitely many integers j such that $f(\mathbb{N}) \cap \{Bj, Bj + 1, \ldots, Bj + B - 1\}$ has at least two elements. In particular, there is some $C \in \{1, \ldots, B\}$ for which there are infinitely many integer pairs (m, n) with $f(m) + C = f(n)$. Now if $f(n)$ is not a polynomial then $f(n)$ has the form given in Eq. (4.2) with $q \geq 1$. If we let $g(n) = f(n) + C$, then by Laurent's theorem (Proposition 4.2), we have $P_0(n) = P_0(n) + C$, which is impossible. It follows that $f(n)$ is a polynomial and since $f(\mathbb{N})$ is infinite, it is eventually increasing.

Then if $f(n)$ has degree ≥ 2 then $f(n+1) - f(n) \to \infty$, a contradiction. Hence $f(n)$ is linear, as required. □

As an immediate consequence we get the following lemma.

Lemma 4.4. *Let $f_1(n), \ldots, f_s(n)$ be $\mathbb{N}$-valued sequences satisfying non-degenerate linear recurrences with constant coefficients and suppose that $f_1(n)$ is a polynomial in n of degree at least 2 and that no $f_i(n)$ is affine. Then $S = \{f_i(n)\colon n \geq 0, i = 1, \ldots, s\}$ is not k-automatic.*

Proof. For each i, since the sequence $f_i(n)$ is not affine, its range has zero density and thus S has zero density as a finite union of such sets. In particular, if we see S as the of an increasing sequence x_i, then

$$\limsup_{i\to\infty}(x_{i+1} - x_i) = \infty.$$

Moreover, as $f_1(n)$ is eventually increasing, for each i there is some largest m_i satisfying $f_1(m_i) \leq x_i$ and m_i is a non-decreasing sequence which tends to infinity. But for i large enough, and hence m_i large enough, $f_1(m_i) < f_1(m_i + 1)$. From this inequality we derive $x_{i+1} \leq f_1(m_i + 1)$ since $f_1(m_i + 1)$ is in S. As a consequence, we obtain

$$\limsup_{i\to\infty} x_{i+1}/x_i \leq \limsup_{i\to\infty} f(m_i + 1)/f(m_i) = 1.$$

Hence by Proposition 2.1 S cannot be a k-automatic set. □

5 Arithmetical Properties of Automatic Sets

In this section, we prove the following general result about automatic sets.

Theorem 5.1. *Let $k \geq 2$ and let $f : \mathbb{N} \to \mathbb{N}$ be a sequence that either satisfies a linear recurrence with constant coefficients or is strictly increasing and P-recursive. Then $f(\mathbb{N})$ is a k-automatic subset of $\mathbb{N}$ if and only if there exist some $a \geq 1$ and $N \geq 0$ such that for each $b \in \{0, \ldots, a-1\}$ the sequence $n \mapsto f(an + b)$ writes for $n \geq N$ either as:*

(1) a constant sequence;
(2) an arithmetic sequence $n \mapsto Cn + C'$ with C, C' integers and $C > 0$;
(3) a geometric sequence $n \mapsto Ck^{an} + C'$ for some rational constants C, C' with $C > 0$ and some positive rational number a with k^a equal to an integer greater than 1.

We point out that many special cases of Theorem 5.1 have occurred in the literature, as arithmetical properties of k-automatic sets are natural objects of study. Ritchie [Rit63] showed that the set of squares is not an automatic set (see, also,

Minsky-Pappert [MP66]). This was extended by Cobham [Cob72] (see also Eilenberg [Eil74]), who showed (although not formally recorded) that if $f(n)$ is a polynomial in n of degree at least 2 then the set $f(\mathbb{N})$ is not an automatic set. Cobham [Cob69] proved a famous theorem on sets that are automatic with respect to two independent bases, which in particular shows things such as $\{3^n : n \geq 0\}$ is not k-automatic unless k is a power of 3. Beyond this, however, there does not seem to be much known in the direction of Theorem 5.1.

We divide the proof of Theorem 5.1 into two cases: the first case being when $f(n)$ satisfies a linear recurrence with constant coefficients and the second case being when $f(n)$ is strictly increasing and P-recursive.

Proof of Theorem 5.1 *for* $f(n)$ *a linear recurrence with constant coefficients.* It suffices to prove the "only if" direction of this result, with the "if" direction described after Lemma 2.2. Suppose that $f(n)$ satisfies a linear recurrence with constant coefficients. Then by Lemma 4.1, there exists a natural number $a \geq 1$ such that for each $b \in \{0, 1 \ldots, a-1\}$ we have $f_b(n) := f(an+b)$ satisfies a non-degenerate linear recurrence with constant coefficients. We claim that each $f_b(n)$ is either constant of the form $f_b(n) = cn + d$ or $f_b(n) = Ck^{dn} + C'$. To see this, let $Y \subseteq \{0, 1, \ldots, a-1\}$ be the set of b for which $f_b(n)$ is constant or of one of these forms. Then if $Y = \{0, 1, \ldots, a-1\}$ then we are done. Otherwise, since the set $S_b := \{f_b(n) : n \geq 0\}$ is k-automatic for every $b \in Y$, we see that $S \setminus_{b \in Y} S_b$ is k-automatic by Lemma 2.2 and it is equal to

$$S' := \bigcup_{b \notin Y} S_b.$$

Now by assumption $Y \neq \{0, 1, \ldots, a-1\}$ and since the sequences $f_b(n)$ are non-degenerate and $f(n)$ takes values in $\mathbb{N}$, we see that S' is infinite. By construction, S' is also automatic. For $b \in \{0, 1, \ldots, a-1\} \setminus Y$, we have $f_b(n) = \sum_{i=0}^{q} P_{i,b}(n)\lambda_i^n$ with λ_i/λ_j not a root of unity for $i \neq j$. Thus by Lemma 3.5 S' contains a subset of the form

$$U := \{ck^{am} + d : m \geq 0\}$$

with $c > 0$. If $f_b(n)$ is a polynomial for some $b \notin Y$ then S' is not k-automatic by Lemma 4.4. Thus we may assume that $f_b(n)$ is not a polynomial for $b \notin Y$.

Because the complement of Y is finite and U is an infinite set, there is some $b \notin Y$ such that $U \cap \{f_b(n) : n \geq 0\}$ is infinite. Then since $f_b(n)$ is not a polynomial, there is some $q \geq 1$ such that for n sufficiently large $f_b(n) = \sum_{i=0}^{q} P_{i,b}(n)\lambda_i^n$, with $1 = \lambda_0, \ldots, \lambda_q$ in $\bar{\mathbb{Q}}$ nonzero and where $P_{0,b}$ may be zero but $P_{1,b}, \ldots, P_{q,b}$ are not. In particular, if we let $g(n)$ be the sequence $g(n) = ck^{an} + d$, we see that $f_b(n) = g(m)$ has infinitely many integer solutions. Then by Proposition 4.2, $f_b(n)$ and $g(n)$ are related and so $q = 1$ and $\lambda_1 = k^\beta$ for some positive rational number β. Moreover, since $f_b(n)$ and $g(n)$ are necessarily simply related, $P_{0,b}(x) = d$ and $P_{1,b}(x) = c$. Hence $f_b(n) = d + ck^{\beta n}$. As $f_b(n)$ takes integral values, k^β is necessarily a rational number and even an integer. Moreover, as $U \cap f_b(\mathbb{N})$ is an infinite subset of $\mathbb{N}$, k^β is an integer ≥ 2. But this contradicts the fact that $f_b(n)$ is not of the form $Ck^{dn} + C'$. The result now follows. □

We next consider the P-recursive case of Theorem 5.1. To handle this case, we make use of the results from Sect. 3 along with a result of Bézivin [Béz86, Théorème 4], which we now state.

Theorem 5.2. *Let K be a field of characteristic zero, let G be a finitely generated subgroup of K^*, and let $F(x) = \sum f(n)x^n$ be a power series with coefficients in K. Suppose that:*

(1) F is D-finite;
(2) there exists a positive integer m and sequences $c_j(n)$, $j = 1, \ldots, m$, taking values in $G \cup \{0\}$ such that $a(n) = \sum_{j=1}^{m} c_j(n)$.

Then F is the power series expansion of a rational function and if $F(x)$ is not a polynomial then each of its poles is a simple pole.

Proof of Theorem 5.1 when $f(n)$ is P-recursive and increasing. Let $S = f(\mathbb{N})$. It suffices to show the "only if" direction of the theorem holds. To do this, we divide the proof into two cases. The first case is when the set S is sparse. In this case, by Lemma 3.4 there is some $m \geq 1$ such that S is a finite union of subsets $S_1, \ldots, S_m$ of natural numbers with

$$S_j = \left\{ \sum_{i=1}^{d_j} c_{i,j} k^{n_i \delta_{i,j}} : n_1, \ldots, n_{d_j} \geq 0 \right\},$$

where the $c_{i,j}$ are rational numbers and $\delta_{i,j}$ are nonnegative integers. In particular if we take G to be the subgroup of $\mathbb{Q}^*$ that is generated by the nonzero $c_{i,j}$ and k and if we let $D = \max(d_1, \ldots, d_m)$, then we see that each $f(n)$ is the sum of at most D elements of G. Theorem 5.2 then implies that $F(x)$ is a rational function with simple poles and so $f(n)$ satisfies a linear recurrence with constant coefficients.

Next suppose that S is not sparse. In this case, by Proposition 3.1, there is some $\alpha > 0$ such that $\pi_S(x) \geq x^\alpha$ for x large. If particular, if $f(n) = m$ then for n large, $n = \pi_S(m) > m^\alpha$ and so $f(n) = m < n^{1/\alpha}$ for all n sufficiently large. Hence

$$F(x) := \sum f(n)x^n \in \mathbb{Z}[[x]]$$

has the property that $f(n) = O(n^d)$ for some $d \geq 1$ and is D-finite. In particular, since a D-finite power series has finitely many singularities (see Flajolet and Sedgewick [FS09, VII.9.1]) and since $F(x)$ converges inside the unit circle, by the Pólya-Carlson theorem [Car21] (see also [Seg08, §6.5]), $F(x)$ is a rational function and so $f(n)$ satisfies a linear recurrence with constant coefficients. □

We can now use the main results of the paper to prove Theorem 1.1.

Proof of Theorem 1.1. Suppose that $G(x) \in K[[x]]$ is algebraic. Then by [AB12] the support set S of G is p-automatic. Then by Theorem 5.1 we obtain the desired result. □

Acknowledgments I thank Jeffrey Shallit for asking the question that eventually led to Theorem 5.1. In addition, I am indebted to the anonymous referees who gave numerous suggestions that greatly improved this paper.

References

[AB12] B. Adamczewski and J. Bell, On vanishing coefficients of algebraic power series over fields of positive characteristic. *Invent. Math.* **187** (2012), no. 2, 343–393.

[AB19] S. Albayrak and J. Bell, *A refinement of Christol's theorem for algebraic power series.* Available online at arXiv:1909.02942.

[AS03] J.-P. Allouche and J. Shallit, *Automatic Sequences. Theory, Applications, Generalizations.* Cambridge University Press, Cambridge, 2003.

[BM17] J. P. Bell and R. Moosa, F-sets and finite automata. *J. Théor. Nombres Bordeaux* **31** (2019), no. 1, 101–130.

[Béz86] J.-P. Bézivin, Sur un théorème de G. Pólya. *J. Reine Angew. Math.* **364** (1986), 60–68.

[Béz89] J.-P. Bézivin, Une généralisation du théorème de Skolem-Mahler-Lech. *Quart. J. Math. Oxford Ser.* (2) **40** (1989), no. 158, 133–138.

[Car21] F. Carlson, Über ganzwertige Funktionen. *Math. Z.* **11** (1–2) (1921) 1–23.

[Chr79] G. Christol, Ensembles presque périodiques k-reconnaissables. *Theoret. Comput. Sci.* **9** (1979), 141–145.

[CKMR80] G. Christol, T. Kamae, M. Mendès France, G. Rauzy, Suites algébriques, automates et substitutions. *Bull. Soc. Math. France* **108** (1980), 401–419.

[Cob69] A. Cobham, On the base-dependence of sets of numbers recognizable by finite automata. *Math. Systems Theory* **3** (1969), 186–192.

[Cob72] A. Cobham, Uniform tag sequences. *Math. Systems Theory* **6** (1972), 164–192.

[Der07] H. Derksen, A Skolem-Mahler-Lech theorem in positive characteristic and finite automata. *Invent. Math.* **168** (2007), no. 1. 175–224.

[Eil74] S. Eilenberg, *Automata, Languages and Machines, vol. A, B.* Academic Press, New York, 1974.

[FS09] P. Flajolet and R. Sedgewick, *Analytic Combinatorics.* Cambridge University Press, Cambridge, 2009.

[GKRS10] P. Gawrychowski, D. Krieger, N. Rampersad, and J. Shallit, Finding the growth rate of a regular or context-free language in polynomial time. *Int. J. Found. Comput. Sci.* **21** (2010), 597–618.

[GS66] S. Ginsburg and E. Spanier, Bounded regular sets. *Proc. Amer. Math. Soc.* **17** (1966), 1043–1049.

[Han86] G. Hansel, Une démonstration simple du théorème de Skolem-Mahler-Lech. *Theoret. Comput. Sci.* **43** (1986), no. 1, 91–98.

[Lau87] M. Laurent, Équations exponentielles polynômes et suites récurrentes linéaires. *Astérisque* 147–148 (1987), 121–139.

[Lec53] C. Lech, A note on recurring series. *Ark. Mat.* **2** (1953), 417–421.

[Met00] C. Methfessel, On the zeros of recurrence sequences with non-constant coefficients. *Arch. Math. (Basel)* **74** (2000), no. 3, 201–206.

[MP66] M. Minsky and S. Papert, Unrecognizable sets of numbers. *J. Assoc. Comput. Mach.* **13** (1966), 281–286.

[Rit63] R. W. Ritchie, Finite automata and the set of squares. *J. Assoc. Comput. Mach.* **10** (1963), 528–531.

[Rub83] L. A. Rubel, Some research problems about algebraic differential equations. *Trans. Amer. Math. Soc.* **280** (1983), no. 1, 43–52.

[Sch03] W. Schmidt, *Linear Recurrence Sequences, Diophantine Approximation* (Cetraro, Italy, 2000). Lecture Notes in Math. 1819, Springer-Verlag Berlin Heidelberg, 2003, pp. 171–247.

[Seg08] S. L. Segal, *Nine Introductions in Complex Analysis.* Revised edition. North-Holland Mathematics Studies, 208. Elsevier Science B.V., Amsterdam, 2008.

[Sta99] R. Stanley, *Enumerative combinatorics. Vol. 2*. Cambridge Studies in Advanced Mathematics, vol. 62. Cambridge University Press, Cambridge, 1999.

X-coordinates of Pell Equations in Various Sequences

Florian Luca

Abstract In this paper, we survey some results concerning the occurrence of interesting arithmetic numbers, like Fibonacci numbers and repdigits in X-coordinates of Pell equations.

1 Intersections of Linear Recurrences

Let $\mathbf{u} := \{u_n\}_{n\ge 0}$ and $\mathbf{v} := \{v_n\}_{n\ge 0}$ be linearly recurrent sequences of integers. That is, there exist positive integers k, ℓ and integers $a_1,\ldots,a_k,b_1,\ldots,b_\ell$ such that both recurrences

$$u_{n+k} = a_1 u_{n+k-1} + \cdots + a_k u_n \tag{1}$$
$$v_{n+\ell} = b_1 v_{n+\ell-1} + \cdots + b_\ell v_n \tag{2}$$

hold for all $n \ge 0$. It is then well-known that there exist distinct algebraic numbers $\alpha_1,\ldots,\alpha_r$ and $\beta_1,\ldots,\beta_s$ and polynomials $P_1,\ldots,P_r,\ Q_1,\ldots,Q_s$ such that both formulas

$$u_n = P_1(n)\alpha_1^n + \cdots + P_r(n)\alpha_r^n, \tag{3}$$
$$v_n = Q_1(n)\beta_1^n + \cdots + Q_s(n)\beta_s^n \tag{4}$$

hold for all $n \ge 0$. If the recurrence relation (1) is *minimal*, that is $\mathbf{u}$ does not satisfy a linear recurrence of order less than k, then in the formula (3) the numbers $\alpha_1,\ldots,\alpha_r$ are all the roots of *the characteristic polynomial*

F. Luca (✉)
School of Maths, Wits University, Johannesburg, South Africa
e-mail: florian.luca@wits.ac.za

Research Group in Algebraic Structures and Applications, King Abdulaziz University, Jeddah, Saudi Arabia

Max Planck Institute for Software Systems, Saarbrücken, Germany

Centro de Ciencias Matemáticas, UNAM, Morelia, Mexico

A. Bostan and K. Raschel (eds.), *Transcendence in Algebra, Combinatorics, Geometry and Number Theory*, Springer Proceedings in Mathematics & Statistics 373,
https://doi.org/10.1007/978-3-030-84304-5_19

$$P_{\mathbf{u}}(X) = X^k - a_1 X^{k-1} - \cdots - a_k = \prod_{i=1}^{k} (X - \alpha_i)^{\sigma_i}$$

and $P_i(X)$ are polynomials with complex coefficients of degree $\sigma_i - 1$, for $i = 1, \ldots, k$, where σ_i is the multiplicity of α_i as a root of $P_{\mathbf{u}}(X)$. In practice, the coefficients of $P_i(X)$ for $i = 1, \ldots, r$ can be found by using the initial values $u_0, \ldots, u_{k-1}$ of $\mathbf{u}$. Similar considerations apply to $\mathbf{v}$. The linearly recurrent sequence $\mathbf{u}$ is said to have a *dominant root* if, up to relabelling of the roots $\alpha_1, \ldots, \alpha_r$, the inequality $|\alpha_1| > \max\{1, |\alpha_2|, \ldots, |\alpha_r|\}$ holds. In this case, α_1 is called "the dominant root" of $\mathbf{u}$. A question which has received considerable interest is to decide whether two linearly recurrent sequences have only finitely or infinitely many common values. That is, whether the equation $u_n = v_m$ has only finitely many positive integer solutions (m, n). Below is a qualitative example due to Mignotte [21].

Theorem 1.1. *Assume that* $\mathbf{u}$ *and* $\mathbf{v}$ *are linearly recurrent sequences whose general terms have representations as in* (3) *and* (4) *and which have dominant roots* α_1 *and* β_1*, respectively. There exists an effectively computable constant* n_0 *such that if* $u_n = v_m$ *holds with* $n \geq n_0$*, then* $P_1(n)\alpha_1^n = Q_1(m)\beta_1^m$*. If in addition, if this last equation has infinitely many positive integer solutions* (n, m)*, then* α_1 *and* β_1 *are multiplicatively dependent; that is, there exists integers* x, y *not both zero such that* $\alpha_1^x = \beta_1^y$*.*

In the above statement and in what follows an effectively computable value n_0 means that one can write down a concrete bound for n_0 once the coefficients of the recurrences and the initial terms are given.

In light of Theorem 1.1 above, it follows that if $\mathbf{u}$ and $\mathbf{v}$ have dominant roots which are multiplicatively independent, then the equation $u_n = v_m$ has only finitely many positive integer solutions (n, m) and they are all effectively computable. To compute them, one uses Baker's method of linear forms in logarithms (see [1], for example). Given $\mathbf{u}$ and $\mathbf{v}$ with the above conditions (that they have dominant roots which are multiplicatively independent), Baker's method produces a numerical bound on $\max\{n, m\}$ over all the positive integer solutions (n, m) of the Diophantine equation $u_n = v_m$. In practice these bounds are huge and they need to be reduced using techniques from continued fractions or the LLL-algorithm (see Chapters 2.3.3 and 2.3.4 of Cohen's book [6]) in order to bring them to small enough values where one can find the solutions by simply enumerating them in the remaining small range. It still remains to bound the number of solutions $u_n = v_m$, or to at least show that it has very few "large" solutions. This has been done recently by Bennett and Pintér [3]. To state their result, we need one definition. For an algebraic number α of minimal polynomial

$$a_0 X^d + \cdots + a_d = a_0 \prod_{i=1}^{d} (X - \alpha^{(i)}) \in \mathbb{Z}[X]$$

where $a_0 > 0$, define the *height* of α as

$$h(\alpha) := \frac{1}{d}\left(\log a_0 + \sum_{i=1}^{d} \log \max\{|\alpha^{(i)}|, 1\}\right).$$

The following theorem is the result of Bennet and Pintér's paper [3].

Theorem 1.2. *Assume that* **u** *and* **v** *are given linearly recurrent sequences with dominant roots* α_1 *and* β_1 *which are multiplicatively independent, whose general terms are given by* (3) *and* (4) *with nonzero algebraic numbers* $P_1, \ldots, P_r, Q_1, \ldots, Q_s$. *Assume further that* $P_1 \neq Q_1$. *Put*

$$M := \max\{h(P_i), h(Q_j) : 1 \leq i \leq r, \ 1 \leq j \leq s\}, \quad N := \max\{r, s, \log|\beta_1|, 3\}. \tag{5}$$

Then there exists an effectively computable constant C *such that if*

$$\log|\alpha_1| > CM \log|\beta_1| \log^3 N, \tag{6}$$

then there is at most one pair of positive integers (n, m) *such that* $u_n = v_m$ *and* $P_1\alpha_1^n = Q_1\beta_1^m$.

The above theorem has the advantage that it guarantees that the Diophantine equation $u_n = v_m$ has at most one solution when only one of the sequences say **v** is fully known while we only have some clues about the second one **u**. That is, assume that **v** is given. Assume also that **u** is not given, but that we know r and we know bounds on the heights of the coefficients $P_1, \ldots, P_r$. Then M and N in (10) are determined. What we are missing to fully know **u** are the roots $\alpha_1, \ldots, \alpha_r$. Then Theorem 1.2 says that if $|\alpha_1|$ is large enough such that inequality (6) is satisfied, then the equation $u_n = v_m$ has at most one positive integer solution (n, m).

We will indicate some practical examples of this theorem in the next section.

2 X-coordinates of Pell Equations in Various Sequences

Let $d > 1$ be an integer which is not a square. The Diophantine equation

$$X^2 - dY^2 = \pm 1 \tag{7}$$

in positive integer unknowns (X, Y) is called the *Pell equation*. What is important is finding all its positive integer solutions (X, Y) once d is given. It is an important equation which has been intensively studied in many papers in the literature. There are even books about it like [2]. It is known that Eq. (7) always has positive integer solutions (X, Y). Letting (X_1, Y_1) be the minimal positive integer solution of the above equation (by minimal we mean with X_1 minimal, or Y_1), all others are obtained via the formula

$$X_n + \sqrt{d}Y_n = (X_1 + \sqrt{d}Y_1)^n \quad \text{for some integer} \quad n \geq 1. \tag{8}$$

In particular,

$$X_n = \frac{\alpha_1^n + \alpha_2^n}{2}, \quad \text{where} \quad \alpha_1 = X_1 + \sqrt{d}Y_1 \quad \text{and} \quad \alpha_2 = X_1 - \sqrt{d}Y_1. \tag{9}$$

Thus, $\mathbf{X} := \{X_n\}_{n \geq 1}$ satisfies formula (3) with $P_1 := 1/2$, $Q_1 := 1/2$. The two roots α_1, α_2 are roots of the quadratic $X^2 - (2X_1)X + \varepsilon$, where $\varepsilon = X_1^2 - dY_1^2 \in \{\pm 1\}$ and clearly $\alpha_1 > \max\{1, |\alpha_2|\}$. We are in position to apply Theorem 1.2. The following is an immediate corollary to Theorem 1.2 which we have not explicitly seen in the literature.

Corollary 2.1. *Assume that* $\mathbf{v}$ *is a given linearly recurrent sequence with dominant root* β_1 *whose general term is given by formula* (4) *with nonzero algebraic numbers* $Q_1, \ldots, Q_s$. *Assume that* $Q_1 \neq 1/2$. *Put*

$$M := \max\{2, h(Q_j) : 1 \leq j \leq s\}, \quad N := \max\{s, \log|\beta_1|, 3\}. \tag{10}$$

Then there exists an absolute constant C *which is effectively computable such that if* d *is a squarefree number with*

$$\log d > 2CM \log|\beta_1| \log^3 N, \tag{11}$$

and the sequence $\mathbf{X} := \{X_n\}_{n \geq 1}$ *of* X*-coordinates of the Pell equation* (7) *has the property that* $\alpha_1 := X_1 + \sqrt{d}Y_1$ *is multiplicatively independent over* β_1, *then there is at most one pair of positive integers* (n, m) *such that* $X_n = v_m$ *and* $\alpha_1^n/2 = Q_1\beta_1^m$.

In practice, the condition $Q_1 \neq 1/2$ can easily be omitted. The condition that α_1 and β_1 are mutliplicatively independent can also be committed. Indeed, if $\alpha_1^x = \beta_1^y$ for some integers x, y not both zero, then $x \neq 0$ (because $|\beta_1| > 1$). Furthermore, for a nonzero integer x, α_1^x is irrational and spans the quadratic field $\mathbb{Q}(\sqrt{d})$. Thus, if β_1 is rational, then the above equation is impossible, while if β_1 is irrational, then d is a squarefree number such that the quadratic field $\mathbb{Q}(\sqrt{d})$ is contained in the field $\mathbb{Q}(\beta_1)$. Thus, d can take only finitely many values. In particular, $d < C_1$ for some computable constant C_1. Enlarging the constant C in (6), we may assume that any d satisfying inequality (6) exceeds C. Thus, the assumption that α_1 and β_1 are multiplicatively independent is automatically satisfied once d is large enough. The same goes for the condition $\alpha_1^n = (2Q_1)\beta_1^m$ since it leads to $\mathbb{Q}(\sqrt{d}) \subseteq \mathbb{Q}(\beta_1, Q_1)$, which again gives $d < C_1$ for some appropriate C_1.

3 Concrete Examples with Fibonacci Numbers and Variations

In light of the results and remarks from Sect. 2, the following problem is effectively solvable. Let $\mathbf{v}$ be a linearly recurrent sequence of integers. Then there exist only finitely many squarefree integers d such that by putting $\mathbf{X} := \{X_n\}_{n\geq 1}$ for the sequence of X-coordinates of the Pell equation (7), then the equation $X_n = v_m$ has at least two positive integer solutions (m, n). Note that any integer $v \geq 3$ is the X–coordinate of a Pell equation corresponding to some squarefree positive integer d. Indeed, just write $v^2 - 1 = dw^2$ for some squarefree integer $d > 1$ and positive integer w and then note that the pair of positive integers $(X, Y) := (v, w)$ satisfies $X^2 - dY^2 = 1$, which is why asking of $X_n = v_m$ to have at least two positive integer solutions (n, m) is a natural question. Having bounded the largest such possible d, say by $d < C_2$, one can loop through all squarefree integers $d < C_2$ and find all solutions of the Diophantine equations $X_n = v_m$ for each such d and select the ones that have at least two such solutions (for the same d).

The following choices of $\mathbf{v}$ have been treated in the literature mostly by the present author and his co-authors. Let $\mathbf{v} := \mathbf{F}$ be the sequence of Fibonacci numbers given by $F_0 = 0,\ F_1 = 1$ and $F_{n+2} = F_{n+1} + F_n$ for all $n \geq 0$. The following result was obtained in [20].

Theorem 3.1. *The equation $X_n = F_m$ has at most one positive integer solution (n, m) except when $d = 2$ for which $X_1 = F_1 = F_2 = 1,\ X_2 = 3 = F_4$.*

The above theorem implies that the only nontrivial solutions of the equation

$$(F_n^2 \pm 1)(F_m^2 \pm 1) = x^2$$

are for $(n, m) = (1, 4),\ (2, 4)$, where by *nontrivial* we mean that $F_n \neq F_m$ and $x \neq 0$ (note that $F_n \neq F_m$ entails $n \neq m$ but also $(n, m) \neq (1, 2),\ (2, 1)$).

There are various generalisations of the Fibonacci sequence. One of them is the Tribonacci sequence $\mathbf{T} := \{T_n\}_{n\geq 0}$ which is given by $T_0 = 0,\ T_1 = 1,\ T_2 = 1$ and $T_{n+3} = F_{n+2} + T_{n+1} + T_n$ for all $n \geq 0$. In [19], the authors have treated the case $\mathbf{v} := \mathbf{T}$ obtaining the following result.

Theorem 3.2. *The equation $X_n = T_m$ has at most one positive integer solution (n, m) except when $d = 2$ for which $X_1 = T_1 = T_2 = 1,\ X_3 = 7 = T_5$ and $d = 3$ for which $X_1 = 2 = T_3$ and $X_3 = 7 = T_5$.*

The k-generalized Fibonacci sequence $\mathbf{F}^{(k)} := \{F_n^{(k)}\}_n$ is a common generalization of the Fibonacci and Tribonacci sequences. It satisfies the recurrence $F_{n+k}^{(k)} = F_{n+k-1}^{(k)} + \cdots + F_n^{(k)}$ and it has initial values $0, 0, \ldots, 0, 1$ (there are $k - 1$ zeros in the previous string). It is convenient to shift the string of zeros into the past in such a way that the last one is $F_0^{(k)} = 0$. Hence, $F_j^{(k)} = 0$ for $j = -(k-2), -(k-1), \ldots, -1, 0$ and $F_1^{(k)} = 1$. When $k = 2$ and 3 it coincides with the Fibonacci and Tribonacci sequences, respectively. Although the formalism involving Theorem 1.2 no longer applies, the number of solutions of the equation $X_n = F_m^{(k)}$ has been studied and the following result has been obtained in [5]:

Theorem 3.3. *Let $k \geq 4$ be a fixed integer. Let $d \geq 2$ be a square-free integer. Assume that*

$$X_{n_1} = F_{m_1}^{(k)}, \quad \text{and} \quad X_{n_2} = F_{m_2}^{(k)} \tag{12}$$

for positive integers $m_2 > m_1 \geq 2$ and $n_2 > n_1 \geq 1$, where X_n is the X–coordinate of the nth solution of the Pell equation (7). Put $\varepsilon := X_1^2 - dY_1^2 \in \{\pm 1\}$. Then, either:

(i) $n_1 = 1,\ n_2 = 2,\ m_1 = (k+3)/2,\ m_2 = k+2$ and $\epsilon = 1$; or
(ii) $n_1 = 1,\ n_2 = 3,\ k = 3 \times 2^{a+1} + 3a - 5,\ m_1 = 3 \times 2^a + a - 1,\ m_2 = 9 \times 2^a + 3a - 5$ for some positive integer a and $\epsilon = 1$.

The equation $X_n = v_m$ has been studied even when **v** is not a linearly recurrent sequence but rather the set of values of products or sums of two members from a linearly recurrent sequence. For example, the case $\mathbf{v} := \mathbf{F} \cdot \mathbf{F} = \{F_\ell F_m : 1 \leq \ell, m\}$ which is the set of numbers which are products of two Fibonacci numbers has been studied in [13].

Theorem 3.4. *For each squarefree integer $d \geq 2$ there is at most one n such that*

$$X_n = F_\ell F_m,$$

except for $d = 2,\ 3,\ 5$ for which $X_1 = 1,\ X_2 = 3$ (for $d = 2$), $X_1 = 2,\ X_2 = 9$ ($d = 5$), $X_1 = 2,\ X_3 = 26$ ($d = 3$), respectively.

The case $\mathbf{v} := \mathbf{F} + \mathbf{F} = \{F_\ell + F_m : 1 \leq \ell, m\}$ of the numbers which are sums of two Fibonacci numbers has been studied in [11]:

Theorem 3.5. *For each squarefree integer $d \geq 2$ there is at most one n such that*

$$X_n = F_\ell + F_m$$

except for $d \in \{2, 3, 5, 11, 30\}$.

4 Concrete Examples with Repdigits

A repdigit is a positive integer whose base 10-representation has one repeated digit. Thus, a repdigit is of the form

$$N = a\left(\frac{10^m - 1}{9}\right) \quad \text{for some} \quad a \in \{1, \ldots, 9\} \quad \text{and some} \quad m \geq 1.$$

The repdigit 99 is an X-coordinate of the Pell equation for $d = 2$ since

$$99^2 - 1 = (99 - 1)(99 + 1) = 98 \cdot 100 = 2 \cdot (70)^2.$$

It turns out that $99 = X_3$ for the Pell equation $X^2 - dY^2 = 1$, namely for which we only consider the solutions with $+1$ in the right–hand side. Clearly, $X_1 = 3$ is also a repdigit. The sequence $\mathbf{v}$ of repdigits is not a linearly recurrent sequence but it is the union of 9 linearly recurrent sequence by fixing the value of $a \in \{1, \ldots, 9\}$. They all satisfy formula (4) with $s = 2$, $(\beta_1, \beta_2) = (10, 1)$ and $(Q_1, Q_2) = (a/9, -a/9)$. The following result concerning the equation $X_n = v_m$ was proved in [7].

Theorem 4.1. *(X_n, Y_n) be the nth solution of the Diophantine equation*

$$X^2 - dY^2 = 1$$

and let $\mathbf{v}$ be the increasing sequence of all base 10-repdigits. The equation $X_n = v_m$ has at most one solution n except for:

(i) $d = 2$ for which $n \in \{1, 3\}$;
(ii) $d = 3$ for which $n \in \{1, 2\}$.

One can generalize redigits to arbitrary bases. Namely, given an integer $b \geq 2$, a repdigit in base b is an integer of the form

$$N = a\left(\frac{b^m - 1}{b - 1}\right) \quad \text{for some} \quad m \geq 1 \quad \text{and} \quad a \in \{1, \ldots, b - 1\}.$$

One can ask whether one can generalise Theorem 4.1 to the case when $\mathbf{v}$ is the sequence of all base b-repdigits. This was indeed done in effective form in [8].

Theorem 4.2. *Let (X_n, Y_n) be the nth solution of the Diophantine equation*

$$X^2 - dY^2 = 1.$$

Let $\mathbf{v}$ be the increasing sequence of base b-repdigits for some integer $b \geq 2$. If $X_n = v_m$ has two positive integer solutions (n, m), then

$$d < \exp\left((10b)^{10^5}\right).$$

The above bound on d is not satisfactory since it is too large to allow to perform calculations for any particular value of b. It has been recently improved to

$$d < \exp(6 \times 10^{27} (\log(2b))^3)$$

in [12]. Using it, all solutions from the equations covered by Theorem 4.2 for all bases $b \in [2, 100]$ have been found in [12].

5 The Analogous Problem with Y-coordinates

Given a linearly recurrent sequence $\mathbf{v}$, one may ask the same question as in Sect. 2 for coordinates Y of the Pell equation (7) corresponding to d but the answer is different. Indeed, it is easy to construct infinitely many d such that the equation $Y_n = v_m$ has two positive integer solutions (n, m). Namely, assume that $1 \in \mathbf{v}$. Take $d = u^2 - 1$, where u will be determined later. Then $(X_1, Y_1) = (u, 1)$ and $(X_2, Y_2) = (2X_1^2 - 1, 2X_1Y_1) = (2u^2 - 1, 2u)$. Hence, if also $2u \in \mathbf{v}$, then for this d, we have that the containment $Y_n \in \mathbf{v}$ holds for both $n = 1, 2$. Thus, if $\mathbf{v}$ contains 1 and infinitely many even numbers, then there are infinitely many d such that $Y_n \in \mathbf{v}$ for both $n = 1, 2$. Note that the positive integers d resulting in this way are not necessarily squarefree. One may ask if this is best possible, meaning whether for particular interesting sets of positive integers $\mathbf{v}$, the containment $Y_n \in \mathbf{v}$ holds for three or more values of n only for a finite set of d. In case $\mathbf{v}$ is a binary recurrent sequence satisfying certain technical conditions, an affirmative answer has been given recently in [9]:

Theorem 5.1. *Let* $\mathbf{v} := \{v_n\}_{n \geq 1}$ *be a binary recurrent sequence whose characteristic equation has real roots. Let* $d > 1$ *be an integer which not a square and let* (X_n, Y_n) *be the sequence of positive integer solutions to* $X^2 - dY^2 = 1$. *Then the equation* $Y_n = u_m$ *has at most two positive integer solutions* (n, m) *provided* $d > d_0$, *where* $d_0 := d_0(\mathbf{v})$ *is some effectively computable constant depending on* $\mathbf{v}$.

One may ask how "effectively computable" is the above statement and whether one can find, in concrete cases of sequences $\mathbf{v}$, all d's such that the equation $Y_n = v_m$ has three or more positive integer solutions (n, m). The only concrete result we are aware of is the following from [10].

Theorem 5.2. *Let* $\mathbf{v} := \{2^n - 1\}_{n \geq 1}$. *There is no squarefree integer* $d > 1$ *such that if* $\{Y_n\}_{n \geq 1}$ *is the sequence of* Y*-coordinates for the Pell equation* $X^2 - dY^2 = 1$, *then the equation* $Y_n = 2^m - 1$ *has at least three positive integer solutions* (n, m). *There are infinitely many* d *such that the above equation has two positive integer solutions* (n, m).

Infinitely many d such that $Y_n = v_m$ has two solutions are given by $d = 2^k - 1$ for which $Y_1 = 1 = 2^1 - 1$ and $Y_3 = 2^{2k+2} - 1$ for any $k \geq 1$.

One may ask how does the above Theorem 5.1 reconcile with the Bennett and Pintér result Theorem 1.2. Well, take $\mathbf{u} := \mathbf{Y}$. Formula (8) implies

$$Y_n = \frac{\alpha_1^n - \alpha_2^n}{2\sqrt{d}},$$

so $r := 2$, $P_1 := 1/2\sqrt{d}$, $P_2 := -1/2\sqrt{d}$. Let us look at the parameter

$$M := \max\{h(P_i), h(Q_j) : 1 \leq i, j \leq 2\}.$$

For us $h(Q_j)$ are bounded for $j = 1, 2$ but $h(P_i) > 0.5 \log d$ are not bounded for $i = 1, 2$. So, for us we have $M = (0.5 + o(1)) \log d$ as $d \to \infty$. We also have the parameter

$$N = \max\{h, k, 3, \log |\beta_1|, M\},$$

which for us for large d equals M because all the other quantities inside the max are bounded. The condition (11) translates to

$$\log \alpha_1 \geq C_1 \log d (\log \log d)^3;$$

that is

$$\alpha_1 > d^{C_1 (\log \log d)^3}$$

where C_1 is some absolute constant. This is an unusual condition on the smallest solution of the Pell equation. It is known that the inequality $\alpha_1 \leq \exp(C_2 d^{1/2} \log d)$ holds with some absolute constant C_2 (see, for example, [14] where the above inequality is stated with $C_2 := 3$) and it is believed that for a lot of d's the fundamental solution of the Pell equation is indeed this large (see the bottom of page 185 in [17] for a precise conjecture). So, for such d, inequality (11) will hold. But there are many d's for which it fails like the ones given by quadratic surds of the type $d = a^2 \pm 1$ with some integer $a \geq 3$, etc. and for these ones we have $\alpha_1 < d^2$, so the above lower bound does not hold for such large d. It turns out (from our examples), that it is exactly these d's which give two solutions instead of just one and these are infinitely many.

6 Some Ideas Concerning the Proofs

We give some ideas about how to find all d's such that $X_n = v_m$ has two solutions (n, m). Recall that X_n is given by formula (9). With the notations from (3) we have $X_n = P_1 \alpha_1^n + O(\alpha_1^{-n})$, where $P_1 = 1/2$ and $\alpha_1 = X_1 + \sqrt{d} Y_1$. In all our examples, $v_m = Q_1 \beta_1^m + O(\beta_1^{(1-\delta)m})$, where $Q_1 > 0$, $\beta_1 > 1$ and $\delta > 0$ are given. For example, when $v_m = F_m$ is the mth Fibonacci number we have $Q_1 = 1/\sqrt{5}$, $\beta_1 = (1 + \sqrt{5})/2$, $\delta = 2$, while when $v_m = a(b^m - 1)/(b - 1)$ with $a \in \{1, \ldots, b - 1\}$ is the repdigit of length m in base b with repeated digit a, we have $Q_1 = a/(b - 1)$, $\beta_1 = b$ and $\delta = 1$. Equation $X_n = v_m$ leads to the approximation

$$|(P_1 Q_1^{-1}) \alpha_1^n \beta_1^{-m} - 1| = O(\beta_1^{-\delta m}).$$

The right–hand side tends to zero. The left–hand side is of the form $e^{\Gamma} - 1$, where $\Gamma = n \log \alpha_1 - m \log \beta_1 + \log(P_1 Q_1^{-1})$. Hence, the above estimate leads to

$$|n \log \alpha_1 - m \log \beta_1 + \log(P_1 Q_1^{-1})| = O(\beta_1^{-\delta m}). \quad (13)$$

If the left–hand side is non-zero, Baker's method gives us a lower bound on it of the type

$$\exp(-C_1 \log \alpha \log \beta_1 \log h(P_1 Q_1^{-1})) \log \max\{m, n\})$$

with some explicit positive constant C_1 which is absolute. We assume that $m > n$, for otherwise $n \le m$ and size arguments then show that $\log \alpha_1 \le \log \beta_1 + O(1)$. Since β_1 is fixed, we get that α_1 is bounded, therefore so is d. If α_1 is bounded, Baker's bound tells us that the right–hand side of exceeds $\exp(-C_2 \log m)$, where C_2 is a constant which now incorporates $\log \alpha_1$ which is of order $O(\log \alpha_1)$. Thus, we get $\delta m \log \beta_1 \le C_2 \log m + O(1)$, which bounds m. So, assume that α_1 is not known. In this case it seems that we are in trouble. However, we are assuming that there are two solutions (n_i, m_i) to the equation $X_{n_i} = v_{m_i}$ for $i = 1, 2$. Writing (13) for $(n, m) = (n_1, m_1)$, (n_2, m_2) we get two small linear forms. Assuming $n_2 > n_1$ (so also $m_2 \ge m_1$) and taking a linear combination of them to eliminate the term containing $\log \alpha_1$, we get

$$|(m_1 n_2 - m_2 n_1) \log \beta_1 - (n_2 - n_1) \log(P_1 Q_1^{-1})| = O(n_2 \beta_1^{-\delta m_1}).$$

In the left–hand side, everything is known except the coefficients. Applying Baker's method, the left–hand side is at least as large as $\exp(-C_3 \log m_2)$, where now C_3 is absolute and explicit. We thus get that $\delta m_1 \log \beta_1 = O(\log m_2 + \log \log m_2)$. Since in fact $\log \alpha_1 = O(m_1 \log \beta_1)$, we get $\log \alpha_1 = O(\log m_2)$. Returning to (13) with $(n, m) = (n_2, m_2)$, we get $m_2 = O(\log \alpha_1 \log m_2) = O((\log m_2)^2)$, which bounds m_2. Since m_2 is the "largest" unknown, this bounds everything. In a nut-shell this is the main idea. One needs to pay attention to details, for example, to justify that the linear forms one is applying Baker's inequality to are not zero to begin with and then one needs to lower the upper bounds of the unknowns obtains by Baker's method which are quite large (larger than 10^{20}) to ranges where one can carry on a search for the solutions and each one of these tasks might pose its own challenges.

7 Extensions, Generalisations and Related Results

One may ask whether one can apply the same methods to study the equation $u_n = v_m$ for other parametric families of sequences resembling the sequences $\mathbf{X} = \{X_n\}_{n \ge 1}$ of X-coordinates to solutions of the Pell equation (7). Well, first of all there are some variations of the Pell equation (7) such that the equation

$$X^2 - dY^2 = \pm 4 \tag{14}$$

to be solved in positive integers (X, Y) once a nonsquare positive integer $d \ge 2$ is given. If (X, Y) is a solution of (7), then $(2X, 2Y)$ is a solution of (14) but for various d's there are positive integer solutions (X, Y) of (14) that do not come from solutions of the Eq. (7). Such is the case when $d = 5$, for which all solutions to the

Eq. (14) are of the form (L_n, F_n) for some positive integer n, where F_n is the nth Fibonacci number and L_n is it's Lucas companion given by $L_0 = 2,\ L_1 = 1$ and $L_{n+2} = L_{n+1} + L_n$ for all $n \geq 0$. When F_n is even (or, equivalently, when n is a multiple of 3), $(X, Y) = (L_n/2, F_n/2)$ give all the solutions of the Pell equation (7) with $d = 5$. The general solution (X_n, Y_n) of Eq. (14) has a similar formula as (8), so most of the results mentioned in this paper can (and some have already been) reworked to find all instances in which $X_n = v_m$ has two positive integer solutions (n, m) in case $\mathbf{v}$ is the sequence of Fibonacci, or Tribonacci numbers, or repdigits, etc. A different direction is be to replace $\mathbf{v}$ by some other sequence not necessarily binary recurrent. When $\mathbf{X} := \{X_n\}_{n\geq 1}$ is the sequence of X-coordinates of the Pell equation $X^2 - dY^2 = 1$, Ljunggren [18] showed that there are at most two positive integers n such that X_n is a square. This was improved later in [22] where it was shown that in fact there is at most one such n except for $d = 1785$, for which both X_1 and X_2 are squares. Laishram, Luca and Sias have a recent paper [16] in which they show that the equation $X_n = m!$ implies that $n = 1$.

Finally, another interesting parametric family of linearly recurrent sequences is given by the number of points of a fixed elliptic curve E defined over a finite field $\mathbb{F}_q$ with q-elements upon field extensions from $\mathbb{F}_q$ to $\mathbb{F}_{q^n}$. To fix ideas, let q be a prime power and let E be given by $y^2 = x^3 + Ax + B$. The condition that the above equation represents an elliptic curve is that $4A^3 + 27B^2$ is not zero in $\mathbb{F}_q$. All elliptic curves over $\mathbb{F}_p$ when $p > 3$ is prime have such a representation (for $p = 2, 3$, the so called *Weierstrass equation* can be more complicated). The set of points $(x, y) \in \mathbb{F}_q^2$ together with the point at infinity form a group denoted $E(\mathbb{F}_q)$ whose cardinality is of the form $\#E(\mathbb{F}_q) = q + 1 - a$, where $a \in [-2\sqrt{q}, 2\sqrt{q}]$. Keeping the equation of the curve E but allowing for $(x, y) \in \mathbb{F}_{q^n}$ together with the point at infinity, we get a group of points whose cardinality is $\#E(\mathbb{F}_{q^n}) = q^n + 1 - (\alpha^n + \overline{\alpha}^n)$, where $\alpha, \overline{\alpha}$ are the two roots of the quadratic $x^2 - ax + q = 0$. Since the sequence $\#E(\mathbb{F}_{q^n})$ depends only on q and a, we can denote it as $E_n(q, a) := \#E(\mathbb{F}_{q^n})$. Thus, $\{E_n(q, a)\}_{n\geq 0}$ is a linearly recurrent sequence of order $k = r = 4$ which clearly has q as a dominant root. Given another interesting sequence $\mathbf{v}$, one may ask what can one say about pairs (q, a) where q is a prime powers and a is an integer in the interval $[-2\sqrt{q}, 2\sqrt{q}]$ such that the equation $E_n(q, a) = v_m$ has at least two positive integer solutions (n, m). The following is the main result of [4]:

Theorem 7.1. *The only solutions (q, a) with q a prime power and $a \in [-2\sqrt{q}, 2\sqrt{q}]$ of the system of Diophantine equations*

$$E_{m_1}(q, a) = F_{n_1}, \qquad E_{m_2}(q, a) = F_{n_2},$$

with $1 \leq m_1 < m_2$ are

$$\begin{aligned}
&E_1(2,1) = F_3, \ E_2(2,1) = F_6;\\
&E_1(2,2) = F_2, \ E_2(2,2) = F_5, \ E_3(2,2) = F_7;\\
&E_1(4,2) = F_4, \ E_2(4,2) = F_8;\\
&E_1(5,3) = F_4, \ E_3(5,3) = F_{12};\\
&E_1(7,3) = F_5, \ E_2(7,3) = F_{10}.
\end{aligned} \tag{15}$$

Examples of actual curves with the above number of points are, respectively:

$$\begin{aligned}
C_1 &:= \{(x,y) \in \mathbb{F}_2^2 : y^2 + xy = x^3 + x^2 + 1\} = \{\infty, (0,1)\};\\
C_2 &:= \{(x,y) \in \mathbb{F}_2^2 : y^2 + y = x^3 + x + 1\} = \{\infty\};\\
C_3 &:= \{(x,y) \in (\mathbb{F}_2[\theta]/(\theta^2 + \theta + 1))^2 : y^2 + y = x^3 + \theta x\}\\
&= \{\infty, (0,0), (0,1)\};\\
C_4 &:= \{(x,y) \in \mathbb{F}_5^2 : y^2 = x^3 + 4x + 2\} = \{\infty, (3,1), (3,4)\};\\
C_5 &:= \{(x,y) \in \mathbb{F}_7^2 : y^2 = x^3 + x + 1\} = \{\infty, (0,1), (0,6), (2,3), (2,4)\}.
\end{aligned}$$

Finally, we conclude by mentioning that the perfect squares in the sequence $\{E_n(q,a)\}_{n\geq 0}$ have also been recently investigated in [15]. There it is shown that either the sequence $\mathbf{u} := \{E_n(q,a)\}_{n\geq 0}$ contains at most 10^{200} perfect squares, or it contains infinitely many including all values of n which are multiples of 24. In conclusion, there is no shortage of interesting problems arising from imposing that two "generic" sequences have several members in common although solving some of the problems resulting in this way might involve quite a lot of heavy theoretical and computational machinery.

Acknowledgements I thank the anonymous referees for a careful reading of the manuscript and for comments which improved the quality of this paper.

References

1. A. Baker and G. Wüstholz, *Logarithmic forms and group varieties*, J. Reine Angew. Math. **442** (1993), 19–62.
2. E. J. Barbeau, *Pell's equation*, Springer-Verlag, New York, 2003.
3. M. A. Bennet and Á. Pintér, *Intersections of recurrent sequences*, Proc. Amer. Math. Soc. **143** (2015), 2347–2353.
4. Yu. F. Bilu, C. A Gómez, J. C. Gómez and F. Luca, *Elliptic curves over finite fields with Fibonacci numbers of points*, New York J. Math. **26** (2020), 711–734.
5. M. Ddamulira and F. Luca, *On the x-coordinates of Pell equations which are k-generalized Fibonacci numbers*, J. Number Theory **207** (2020), 156–195.
6. H. Cohen, Number Theory, Volume I: Tools and Diophantine Equations, Graduate Texts in Mathematics 239, Springer, 2007.
7. A. Dossavi-Yovo, F. Luca and A. Togbé, *On the x-coordinate of Pell equations which are rep-digits*, Publ. Math. Debrecen **88** (2016), 381–399.
8. B. Faye and F. Luca, *On x-coordinates of Pell equations which are repdigits*, Fibonacci Quart. **56** (2018), 52–62.

9. B. Faye and F. Luca, *On Y-coordinates of Pell equations which are members of a fixed binary recurrence*, New York J. Math. **26** (2020), 184–206.
10. B. Faye and F. Luca, *On Y-coordinates of Pell equations which are rep-digits in base* 2, Glas. Mat. Ser. III **55**(75) (2020), 1–12.
11. C.-A. Gómez and F. Luca, *Zeckendorf representations with at most two terms to x-coordinates of Pell equations*, Sci. China Math. **63** (2020), 627–642.
12. C. A. Gómez, F. Luca and F. S. Zottor, *On X-coordinates of Pell equations which are repdigits* Res. Number Theor. **6**:41 (2020)
13. B. Kafle, F. Luca, A. Montejano, L. Szalay and A. Togbé, *On the x-coordinates of Pell equations which are products of two Fibonacci numbers*, J. Number Theory **203** (2019), 310–333.
14. M. Křížek and F. Luca, *On the congruence* $n^2 \equiv 1 \pmod{\phi(n)^2}$, Proc. Amer. Math. Soc. **129** (2001), 2191–2196.
15. K. C. Chim and F. Luca, *Perfect squares representing the number of rational points on elliptic curves over finite field extensions*, Finite Fields Appl. **67** (2020), 101725.
16. S. Laishram, F. Luca and M. Sias, *On members of Lucas sequences which are products of factorials*, Monatsh. Math. **193** (2020), 329–359.
17. H. W. Lenstra, *Solving the Pell equation*, Notices Amer. Math. Soc. **49** (2002), 182–192.
18. W. Ljunggren, *Über die Gleichung* $x^4 - Dy^2 = 1$, Arch. Math. Naturvid. **45** (1942), 61–70.
19. F. Luca, A. Montejano, L. Szalay and A. Togbé, *On the X-coordinates of Pell equations which are Tribonacci numbers*, Acta Arith. **179** (2017), 25–35.
20. F. Luca and A. Togbé, *On the x-coordinates of Pell equations which are Fibonacci numbers*, Math. Scand. **122** (2018), 18–30.
21. M. Mignotte, *Intersection des images de certaines suites récurrentes linéaires*, Theoret. Comput. Sci. **7** (1978), 117–122.
22. Q. Sun and P. Z. Yuan, *A note on the Diophantine equation* $x^4 - Dy^2 = 1$, Sichuan Daxue Xuebao **34** (1997), 265–268.

A Conditional Proof of the Leopoldt Conjecture for CM Fields

Preda Mihăilescu

Par les gosses battus, par l'ivrogne qui rentre
Par l'âne qui reçoit des coups de pied au ventre
Par l'âne qui reçoit des coups de pied au ventre
Et par l'humiliation de l'innocent châtié
Par la vierge vendue qu'on a déshabillée
Par le fils dont la mère a été insultée
Je vous salue, Marie[a]

To Theres and Seraina, to the memory of Marta

[a] Francis Jammes: Prière. Music by Georges Brassens

Abstract In this short paper, we reduce the proof of the Leopoldt Conjecture to the proof of the fact that Iwasawa's μ-constant vanishes for all CM $\mathbb{Z}_p$-extensions, an assumption that will be proved in a separate paper.

Keywords 11R23 Iwasawa theory · 11R27 units

1 Introduction

Leopoldt suggested in his seminal paper [9] that the p-adic regulator of abelian extensions of $\mathbb{Q}$ never vanishes. This fact was proved by Brumer [3] in 1967, using a plan of Ax [1], as soon as Baker had proved his Archimedean version of the approximation Theorem for linear forms in logarithms [2]: it remained to adapt Baker's proof to the p-adic topology. The fact that Leopoldt's observation should hold in arbitrary number fields became soon a largely accepted conjecture. It should

P. Mihăilescu (✉)
Mathematisches Institut der Universität Göttingen, Göttingen, Germany
e-mail: preda@uni-math.gwdg.de

A. Bostan and K. Raschel (eds.), *Transcendence in Algebra, Combinatorics, Geometry and Number Theory*, Springer Proceedings in Mathematics & Statistics 373,
https://doi.org/10.1007/978-3-030-84304-5_20

however be noted, that Leopoldt's definition of the p-adic regulator as a formal adaptation of the global regulator, in which logarithms are taken p-addically, only makes sense in totally real or in CM fields. For non-CM fields, there is no simple and canonical definition of the regulator known: therefore, the Leopoldt conjecture for this case should be related to the equality of $\mathbb{Z}$- and $\mathbb{Z}_p$-ranks of the units, i.e., to the vanishing of the Leopoldt defect $\mathcal{D}(\mathbb{K})$, as defined below.

In this paper we shall assume the following

Assumption 1 *Let $\mathbb{L}/\mathbb{K}$ be a CM $\mathbb{Z}_p$-extension of the CM number field $\mathbb{K}$. Then Iwasawa's constant μ vanishes for the galois group of the maximal p-abelian unramified extension of $\mathbb{L}$.*

From this fact which we prove elsewhere, we derive the

Theorem 1 *For odd primes p, the Leopoldt Conjecture holds in CM extensions $\mathbb{K}/\mathbb{Q}$.*

1.1 Historical Notes

Since 1967 various attempts have been undertaken for extending the results of [3] to non abelian extensions, using class field theory, Diophantine approximation or both. The following very succinct list is intended to give an overview of various approaches, rather than being an extensive list of results on Leopoldt's conjecture. In [5], Greenberg notes a relation between the Leopoldt Conjecture and a special case of the Greenberg Conjecture: he shows that Leopoldt's Conjecture implies that the T-part $A_\infty(T)$ is finite for totally real fields, i.e. the Greenberg Conjecture holds for the T—part.

Emsalem, Kisilevsky and Wales [4] use group representations and Baker theory for proving the Conjecture for some small non abelian groups; this direction of research has been continued in some further papers by Emsalem or Emsalem and coauthors. Jaulent proves in [8] the Conjecture for some fields of small discriminants. Currently the strongest result on the Leopoldt Conjecture is based on Diophantine approximation; it was achieved by Waldschmidt [11], who proved that if r is the $\mathbb{Z}$—rank of the units in the field $\mathbb{K}$, then the Leopoldt defect satisfies $\mathcal{D}(\mathbb{K}) \leq r/2$.

1.2 Notations and Fundamental Facts

Let p be an odd rational prime. In this paper we denote number fields with black board bold letters—$\mathbb{K}$, $\mathbb{F}$, $\mathbb{L}$, $\mathbb{Q}$, etc. The cyclotomic $\mathbb{Z}_p$-extension of a number field $\mathbb{K}$ will be denoted by $\mathbb{K}_\infty$. We let $\Gamma = \mathrm{Gal}(\mathbb{K}_\infty/\mathbb{K})$ be generated topologically by $\tau \in \Gamma$ and write $T = \tau - 1$; $\tau_n = (T+1)^{p^n}$ for the power of τ that generates the fixing group of $\mathbb{K}_n$, and $\omega_n = \tau_n - 1$; for $j > 0$, we let $\nu_{n+j,n} = \omega_{n+j}/\omega_n$. Let $\mathbb{H}_n = \mathbb{H}(\mathbb{K}_n)$ be the maximal p-abelian unramified extensions of $\mathbb{K}_n$, for all $n \in \mathbb{N}$ and $\mathbb{H}(\mathbb{K}_\infty) = \cup_n \mathbb{H}_n$.

Dirichlet's unit theorem states that, up to torsion made up by the roots of unity $W(\mathbb{K}) \subset \mathbb{K}^\times$, the units $E = \mathcal{O}(\mathbb{K})^\times$ are a free $\mathbb{Z}$—module of $\mathbb{Z}$—rank $r = r_1 + r_2 - 1$. As usual, r_1 and r_2 are the numbers of real, resp. pairs of complex conjugate embeddings $\mathbb{K} \hookrightarrow \mathbb{C}$. In the case of a galois CM extension $\mathbb{K}$, which is an imaginary quadratic extension of a real galois field, we obviously have $r = r_2 - 1 = [\mathbb{K} : \mathbb{Q}]/2 - 1$.

We consider the set $P = \{\mathfrak{p} \subset \mathcal{O}(\mathbb{K}) : (p) \subset \mathfrak{p}\}$ of prime ideals above p and let

$$\mathfrak{K}_p = \mathfrak{K}_p(\mathbb{K}) = \prod_{\mathfrak{p} \in P} \mathbb{K}_\mathfrak{p} = \mathbb{K} \otimes_\mathbb{Q} \mathbb{Q}_p$$

be the product of all completions of $\mathbb{K}$ at primes above p, an algebra that we may also denote by p-idéles, for obvious reasons. When $\mathbb{K}$ is galois with group Δ, the primes in P are conjugate and if $D(\mathfrak{P}) \subset \Delta = \mathrm{Gal}(\mathbb{K}/\mathbb{Q})$ is the decomposition group of one of them, then a set $C \subset \Delta$ of coset representatives for G/Δ will permute the primes above p, so that $P = C\mathfrak{P}$. Let $\iota : \mathbb{K} \hookrightarrow \mathfrak{K}_p$ be the diagonal embedding and $\iota_\mathfrak{p}(x)$ be the projection of $\iota(x)$ in the completion at $\mathfrak{p} \in P$. If $y \in \mathfrak{K}_p$, then $y_\mathfrak{p} := \iota_\mathfrak{p}(y)$ is simply the component of y in $\mathbb{K}_\mathfrak{p}$.

We write $U \subset \mathfrak{K}_p^\times$ for the group of units, thus the product of local units at the same completions; then E embeds diagonally via $\iota : E \hookrightarrow U$. Let $\overline{E} = \overline{\iota(E)} = \bigcap_{n>0} \iota(E) \cdot U^{p^n} \subset U$ be the p-adic closure of $\iota(E)$; this is a $\mathbb{Z}_p$—module with $\mathbb{Z}_p\text{-rk}(\overline{E}) \leq \mathbb{Z}\text{-rk}(E) = r = r_1 + r_2 - 1$. The Leopoldt defect is herewith defined by

$$\mathcal{D}(\mathbb{K}) := \mathbb{Z}\text{-rk}(E(\mathbb{K})) - \mathbb{Z}_p\text{-rk}(\overline{E}(\mathbb{K})), \tag{1}$$

and it is always non-negative. Indeed, any rational linear combination of units translates into a linear combination in the completion, while the converse may not be true. In the case of totally real fields, one can define p-adic regulators—see e.g. [12], §4—and shows that their vanishing is equivalent to the defect being non null. Since this property can be verified also in arbitrary fields, where there may be no canonical definition of a regulator, it is convenient to declare the general

Conjecture 1 (Leopoldt) *For arbitrary number fields* $\mathbb{K}$*, the Leopoldt defect* $\mathcal{D}(\mathbb{K}) = 0$.

The connection of Leopoldt's conjecture to class field theory was noted by Iwasawa in [6]. He shows that if $\Omega(\mathbb{K}) \supset \mathbb{K}_\infty$ is the maximal p-abelian p-ramified extension of $\mathbb{K}$, then

$$\mathrm{Gal}(\Omega(\mathbb{K})/\mathbb{K}) \sim \mathbb{Z}_p^n, \quad \text{where } n = r_2 + 1 + \mathcal{D}(\mathbf{K}); \tag{2}$$

the proof of this fact is in any text book on cyclotomy and Iwasawa theory—e.g. [12], §13.

The definition allows one to verify the following

Fact 1 *Let* $\mathbb{Q} \subset \mathbb{F} \subset \mathbb{K}$ *be a tower of number fields. If* $\mathcal{D}(\mathbb{F}) > 0$*, then* $\mathcal{D}(\mathbb{K}) > 0$.

Proof The proof follows by linear algebra. Let $\mathcal{E}_{\mathbb{F}} = \{e_1; e_2; \ldots; e_{r(\mathbb{F})}\} \subset E(\mathbb{F})$ be a set of fundamental units for $\mathbb{F}$; it can be extended to a set of units of $\mathbb{K}$ of maximal $\mathbb{Z}$-rank $r(\mathbb{K})$, the Dirichlet rank of $\mathbb{K}$. Let this be $\mathcal{E}_{\mathbb{K}} = \mathcal{E}_{\mathbb{F}} \cup \{e'_1, \ldots, e'_{r(\mathbb{K})-r(\mathbb{F})}\}$. We note that $\mathcal{E}(\mathbb{F})$, $\mathcal{E}(\mathbb{K})$ also span $\bar{E}(\mathbb{F})$ and $\bar{E}(\mathbb{K})$, as $\mathbb{Z}_p$-modules. The Leopoldt defect is positive for a field, if the respective spanning set of units displays a p-adic linear dependence. If $\mathcal{D}(\mathbb{F}) > 0$, there is thus a relation $\prod_{j=1}^{r(\mathbb{F})} e_j^{c_j} = 1$; $c_j \in \mathbb{Z}_p$ and not all $c_j = 0$. Sice $\mathcal{E}(\mathbb{F}) \subset \mathcal{E}(\mathbb{K})$, this induces also a non trivial linear dependence in $\mathcal{E}(\mathbb{K})$, and thus $\mathcal{D}(\mathbb{K}) > 0$. □

If the claim of Theorem 1 is false—for instance for some CM field $\mathbb{F}$ –, t+his fact allows us to assume that there also exists some field $\mathbb{K} \supset \mathbb{F}$, which is CM, galois and contains the p-th roots of unity, and in which the Leopoldt conjecture fails. The units of a CM field $\mathbb{K}$ are, up to torsion, the same as the units $E(\mathbb{K}^+)$ of its maximal real subfield; therefore the Leopoldt defect vanishes in $\mathbb{K}$ iff it vanishes in $\mathbb{K}^+$. For $\mathbb{K}^+$, Iwasawa's result (2) means that

$$\mathbb{Z}_p\text{-rk}\left(\mathrm{Gal}(\Omega(\mathbb{K}^+)/\mathbb{K}^+_\infty)\right) = \mathcal{D}(\mathbb{K}^+) = \mathcal{D}(\mathbb{K}).$$

Since $\mathbb{K}^+$ is totally real, the extensions in $\Omega(\mathbb{K}^+)$ must also be real. For each $\mathbb{Z}_p$-subextension $\mathbb{L}^+ \subset \Omega(\mathbb{K}^+)$, it follows that the compositum $\mathbb{L} = \mathbb{K} \cdot \mathbb{L}^+$ is a CM $\mathbb{Z}_p$-extension. Conversely, if a CM field has other CM $\mathbb{Z}_p$-extension than the cyclotomic one, then it follows that $\mathbb{K}^+$ has totally real $\mathbb{Z}_p$-extensions which are not the cyclotomic one, and thus $\mathcal{D}(\mathbb{K}^+) = \mathcal{D}(\mathbb{K}) > 0$. We have thus proven

Lemma 1 *A CM field has more than one CM $\mathbb{Z}_p$-extension (the cyclotomic one), iff $\mathcal{D}(\mathbb{K}) > 0$.*

2 Proof of Theorem 1

We assume that $\mathbb{K}$ is a CM field containing the p-th roots of unity and having a positive Leopoldt defect. In view of Lemma 1, it lays at hand to assume that $\mathbb{K}$ has a CM $\mathbb{Z}_p$-extension different from the cyclotomic one, and deduce the existence of a CM $\mathbb{Z}_p$-extension in which $\mu > 0$. This will be made by using a variation of the idea used by Iwasawa in [7] for displaying examples of $\mathbb{Z}_p$-extensions with positive μ-constant; a complete account of Iwasawa's example can also be found in [10], §13.5—and one can see there also why we need to adapt the argument to our given context. Like for Iwasawa, our construction requires a $\mathbb{Z}_p$-extension in which some prime is totally split—but unlike the case of his example, one such prime will actually suffice. We start by showing that extensions with totally split primes must exist if $\mathcal{D}(\mathbb{K}) > 0$:

Lemma 2 *Let $\mathbb{K}$ be a CM field with $\mathcal{D}(\mathbb{K}) > 0$ and let $\mathfrak{q} \subset \mathbb{K}_j$ for some $j \geq 0$ be a totally split prime above the rational prime $q \neq p$. Then there exists a CM $\mathbb{Z}_p$-extension $\mathbb{L} \supset \mathbb{K}_j$ in which $\mathfrak{q}$ is totally split.*

Proof Under the given premise, all $\mathbb{Z}_p$-subextensions $\mathbb{L} \subset \Omega(\mathbb{K}^+)$ are CM over $\mathbb{K}$. Let $\mathbb{M} = \Omega(\mathbb{K}^+) \cdot \mathbb{K}$, a field with galois group $X_M = \mathrm{Gal}(\mathbb{M}/\mathbb{K})$ of rank $\mathbb{Z}_p$-rk$(X_M) = 1 + \mathcal{D}(\mathbb{K}) > 1$. Let $D = D(\mathfrak{q}) \subset X_M$ be the decomposition group of $\mathfrak{q}$; since $p \neq q$, it follows that $\mathbb{Q}_q$ has only one, unramified $\mathbb{Z}_p$-extension, and thus $\mathbb{Z}_p$-rk$(D) \leq 1$. We have in fact equality, since all primes different from p are inert in the cyclotomic $\mathbb{Z}_p$-extension, beyond some level $\mathbb{K}_{n(q)}$. Let $\mathbb{M}_q = \mathbb{M}^{D(\mathfrak{q})}$; this is a field in which $\mathfrak{q}$ is totally split. In particular $\mathbb{K}_j \subset \mathbb{M}_q$ and, for rank reasons, there is at least one CM $\mathbb{Z}_p$-extension $\mathbb{L} \subset \mathbb{M}_q$, which confirms the claim of the Lemma. □

With this we can proceed with the

Proof of Theorem 1 Let $\mathbb{K}$ be a CM field with positive Leopoldt defect, which contains a p-th root of unity ζ and such that[1] $\mathbb{K} \supsetneq \mathbb{Q}[\zeta]$. By Lemma 2, we may choose some prime $q \neq p$ which is totally split in $\mathbb{K}$, let $\mathfrak{q} \subset \mathbb{K}$ be a prime above it, and find a CM $\mathbb{Z}_p$-extension $\mathbb{L} \supset \mathbb{K}$ in which $\mathfrak{q}$ is totally split. Since $\mathbb{K}$ contains the p-th roots of unity and q is totally split in $\mathbb{K}$, it follows that $q \equiv 1 \mod p$, and the q-th cyclotomic extension $\mathbb{Q}[\zeta_q]$ contains a subfield of degree p: let this be $\mathbb{F} \subset \mathbb{Q}[\zeta_q]$, so $[\mathbb{F} : \mathbb{Q}] = p$, and let the galois group be $\Phi = \langle \nu \rangle = \mathrm{Gal}(\mathbb{F}/\mathbb{Q})$. Since q is totally split in $\mathbb{K}$ and ramified in $\mathbb{F}$, we have $\mathbb{F} \cap \mathbb{K} = \mathbb{Q}$ and thus $\mathbb{K}' := \mathbb{K} \cdot \mathbb{F}$ is an extension of degree p of $\mathbb{K}$; likewise, $\mathbb{L}' := \mathbb{L} \cdot \mathbb{F} \supset \mathbb{L}$ has degree p over $\mathbb{L}$. Since $\mathbb{F}$ is a real cyclic field, disjoint from $\mathbb{K}$, it follows that $\mathbb{K}'$ is CM too, and so is $\mathbb{L}'$. The Theorem will follow from the Fact 1, if we show that $\mu(A_\infty(\mathbb{L}')) > 0$.

We note first some elementary facts on $\mathbb{K}'$. Let $\mathbf{C} = \mathbb{Q}[\zeta]$ and $\mathbb{F}_1 = \mathbb{F} \cdot \mathbf{C}$. The classical theory of Gauss sums implies that there is a Gauss sum $\tau(\chi) \in \mathbb{F}_1$ to a character of conductor q and order p, such that $\mathbb{F}_1 = \mathbf{C}[\tau(\chi)]$. Moreover, $\kappa := \tau^p(\chi) \in \mathbf{C}$ and there is a prime $Q \subset \mathbf{C}$ above q such that

$$(\kappa) = Q^{\theta_s}, \quad \theta_s = \sum_{c=1}^{p-1} c\sigma_c^{-1}, \tag{3}$$

with $\sigma_c : \zeta \mapsto \zeta^c$ being an additive indexing of the automorphisms of $\mathbf{C}$.

Let $\mathbb{L}_n \subset \mathbb{L}$ denote the intermediate extensions with $[\mathbb{L}_n : \mathbb{K}] = p^n$, let $\mathbb{L}'_n = \mathbb{L}_n \cdot \mathbb{F} \subset \mathbb{L}'$ and $A(\mathbb{L}_n)$, $A(\mathbb{L}'_n)$ denote, like usual, the p-parts of the respective class groups. With the previous definition, we have

$$\mathbb{L}'_n = \mathbb{L}_n[\kappa^{1/p}] = \mathbb{F} \cdot \mathbb{L}_n, \quad \forall n \in \mathbb{N}. \tag{4}$$

Since $\mathfrak{q}$ is totally split in $\mathbb{L}/\mathbb{K}$, there are at least p^n pairs of complex conjugate primes of $\mathbb{L}_n$ that ramify in $\mathbb{L}'_n$. Let $(\mathfrak{q}_n, \bar{\mathfrak{q}}_n)$ be such a pair and let $\Gamma = \mathrm{Gal}(\mathbb{L}/\mathbb{K}) = \mathrm{Gal}(\mathbb{L}'/\mathbb{K}')$ have topological generator τ, acting by restriction on $\mathbb{L}'_n$. Let $(\mathfrak{Q}_n, \bar{\mathfrak{Q}}_n) \subset \mathbb{L}'_n$ be the pair of ramified, conjugate primes of $\mathbb{L}'_n$ above the primes of $\mathbb{L}_n$ that we have chosen. For each $\mathfrak{q}_n$, since the primes above q do not ramify in $\mathbb{L}_n/\mathbf{C}$, we have

[1] This fact is of course a consequence of the result of Baker and Brumer, which confirms the Leopoldt Conjecture for abelian fields, but we do not need this here.

$$v_{\mathfrak{q}_n}(\kappa) \in \{1, 2, \ldots, p-1\}. \tag{5}$$

Let $a_n = [\mathfrak{q}_n/\bar{\mathfrak{q}}_n] \in A^-(\mathbb{L}_n)$ and $b_n = [\mathfrak{Q}_n/\bar{\mathfrak{Q}}_n] \in A^-(\mathbb{L}'_n)$ be the classes generated by these pairs of primes in the class groups of $\mathbb{L}_n, \mathbb{L}'_n$, respectively. The following fact is a consequence of the general phenomenon, that capitulation kernels on minus part groups are trivial, with the exception of classes related to Kummer radicals:

Fact 2 *The orders* $\mathrm{ord}(b_n) = p \cdot \mathrm{ord}(\mathfrak{q}_n)$ *and in general,* $\mathrm{ord}(b_n^v) = p\mathrm{ord}(a_n^v)$ *for* $v \in \mathbb{Z}_p[T]\backslash(p, \omega_n)\mathbb{Z}_p[T]$.

Proof Indeed, $\mathrm{ord}(b_n) \geq \mathrm{ord}(a_n) = \mathbf{N}_{\mathbb{L}'_n/\mathbb{L}_n}(b_n)$. If the orders were equal, say $\mathrm{ord}(\mathfrak{q}_n^{1-\jmath}) = \mathrm{ord}\left(\mathfrak{Q}^{1-\jmath}\right) = p^l$, then let $\alpha \in \mathfrak{q}^{p^l(1-\jmath)}$ and $\beta \in \mathfrak{Q}^{p^l(1-\jmath)}$ be generators of these principal ideals, which verify $\beta^{1+\jmath} = \alpha^{1+\jmath} = 1$. From $\mathfrak{Q}_n^p = \mathfrak{q}_n$, we deduce that $(\beta^p) = (\alpha)$, so $\beta^p = \eta\alpha$, with η a root of unity, as follows from Kronecker's Unit Theorem. After eventually modifying α by a root of unity, we obtain $\beta^p = \alpha$, so $\mathbb{L}'_n = \mathbb{L}_n[\alpha^{1/p}] = \mathbb{L}_n[\kappa^{1/p}]$, as a cyclic Kummer extension. By Kummer theory, it follows that there is some $w \in \mathbb{L}_n^\times$ and a c coprime to p, such that

$$\alpha = \kappa^c \cdot w^p. \tag{6}$$

If $l > 0$, it would follow that κ is the p-th power of an ideal, which contradicts (5). So $l = 0$ and $\mathfrak{q}_n$ is principal. But then we obtain a new contradiction, since α is only divisible by pair of the primes above q, while κ is divisible by all of them. If $\mathfrak{r}|\kappa$ but $v_{\mathfrak{r}}(\alpha) = 0$, then the identity implies that $v_{\mathfrak{r}}(\kappa) \equiv 0 \bmod p$, which contradicts again (5). This confirms the first statement.

The same argument holds when replacing b_n by b_n^v: here we note that $p \nmid v$, otherwise $b_n^v \in A_n$ and there is nothing to prove. Like before, assuming that $\mathrm{ord}(b_n^v) = \mathrm{ord}(a_n^v)$, we let $p^l = \mathrm{ord}(\mathfrak{q}_n^v)$ and choose generators

$$\beta \in \mathfrak{Q}_n^{(1-\jmath)p^l v}, \quad \alpha \in \mathfrak{q}_n^{(1-\jmath)p^l v},$$

such that, after a possible modification by a root of unity, we have $\beta^p = \alpha$ and thus a new variant of (6), with other values of α, w and c, will hold. If $l > 0$, we deduce again that κ is the p-th power of an ideal. Otherwise, we obtain a contradiction by showing that all the primes above q which do not divide α—and this set is non empty, since $\mathbb{K} \supsetneq \mathbf{C}$—must occur at a power divisible by p in κ, which contradicts (5). This completes the proof. □

This fact shows that $\mathrm{ord}(b_n^v) \geq p$ for all $v \in \mathbb{Z}/(p \cdot \mathbb{Z})[T]/(\omega_n)$, hence p-rk$(\Lambda b_n) \geq p^n$ and herewith

$$|A^-(\mathbb{L}'_n)| \geq |\Lambda b_n| \geq p^{p^n}.$$

Iwasawa's formula [12], Proposition 13.13 states that for sufficiently large n, the last valuation only depends on the constants μ, λ, ν of the extension $\mathbb{L}'/\mathbb{L}$, namely

$$v_p(|A^-(\mathbb{L}'_n)|) = \nu^- + \lambda^- n + \mu^- p^n \geq p^n.$$

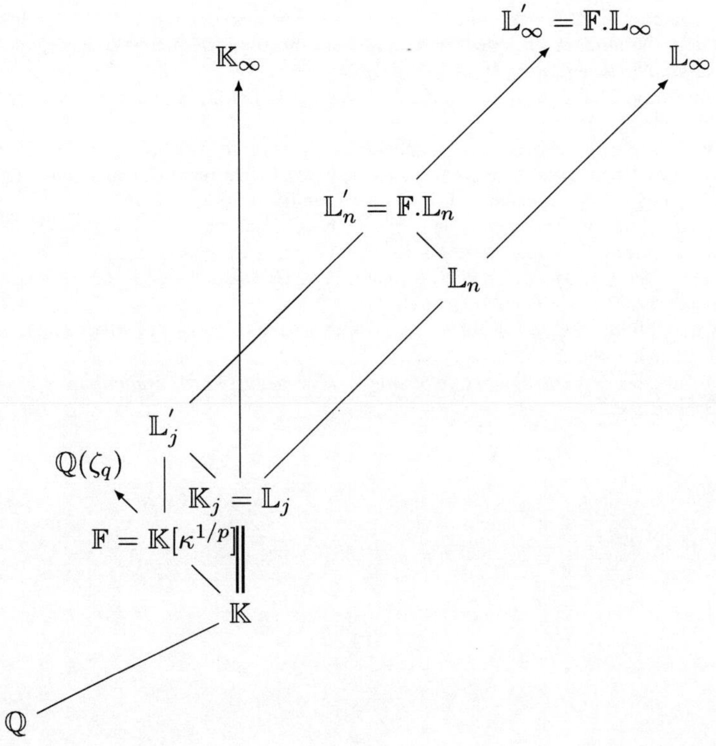

Fig. 1 The Thaine split shift of $\mathbb{L}$

Since $\lim_{n\to\infty} \frac{\nu^- + \lambda^- n}{p^n} = 0$ for any constant λ^-, ν^-, it follows that

$$\mu^-(\mathbb{L}') = \lim_{n\to\infty} \frac{v_p(|A^-(\mathbb{L}'_n)|)}{p^n} \geq \lim_n \frac{p^n}{p^n} = 1,$$

so the μ^--invariant of $\mathbb{L}'$ is non-vanishing. Since $\mathbb{L}'$ is a CM $\mathbb{Z}_p$-extension of $\mathbb{K}'$, this is a contradiction to Fact 1—which confirms the claim of the Theorem 1. □

References

1. J. Ax. On the units of an algebraic number field. *Illinois Journal of Mathematics*, 9:584–589, 1965
2. A. Baker. Linear forms in the logarithms of algebraic numbers I, II, III. *Mathematika*, 13, 14:204–216; 102–107, 220–228, 1966, 67
3. A. Brumer. On the units of algebraic number fields. *Mathematika*, 14:121–124, 1967

4. M. Emsalem, H. Kisilevsky, and D. Wales. Indépendance linéaire sur $\overline{\mathbb{Q}}$ de logarithmes p - adiques de nombres algébriques et rang p - adique du groupe des unités d'un corps de nombres. *Journal of Number Theory*, 19:384–391, 1984
5. R. Greenberg. On the Iwasawa invariants of totally real fields. *American Journal of Mathematics*, 98:263–284, 1976
6. K. Iwasawa. On $\mathbb{Z}_\ell$ - extensions of number fields. *Ann. Math. Second Series*, 98:247–326, 1973
7. K. Iwasawa. On the μ-invariants of cyclotomic fields. In *Algebraic Geometry and Commutative Algebra*, pages 1–11, Kinokuniya, Tokio, 1973. In honor of Y. Akizuki
8. J. Jaulent. Note sur la conjecture de Leopoldt. http://front.math.ucdavis.edu/0712.2995, 2007
9. H. Leopoldt. Zur Artihmetik in Abelschen Zahlkörper. *J. Reine Angew. Math*, 209:54–71, 1962
10. S. Lang. *Cyclotomic fields I and II*, volume 121 of *Graduate Texts in Mathematics*. Springer, combined Second Edition edition, 1990
11. M. Waldschmidt. Transcendence et éxponentielles en plusieures variables. *Inventiones Mathematicae*, 63, 1981
12. L. Washington. *Introduction to Cyclotomic Fields*, volume 83 of *Graduate Texts in Mathematics*. Springer, 1996

Siegel's Problem for E-Functions of Order 2

T. Rivoal and J. Roques

Abstract E-functions are entire functions with algebraic Taylor coefficients at the origin satisfying certain arithmetic conditions, and solutions of linear differential equations with coefficients in $\overline{\mathbb{Q}}(z)$; they naturally generalize the exponential function. Siegel and Shidlovsky proved a deep transcendence result for their values at algebraic points. Since then, a lot of work has been devoted to apply their theorem to special E-functions, in particular the hypergeometric ones. In fact, Siegel asked whether any E-function can be expressed as a polynomial in z and generalized confluent hypergeometric series. As a first positive step, Shidlovsky proved that E-functions with order of the differential equation equal to 1 are in $\overline{\mathbb{Q}}[z]e^{\overline{\mathbb{Q}}z}$. In this paper, we give a new proof of a result of Gorelov that any E-function (in the strict sense) with order ≤ 2 can be written in the form predicted by Siegel with confluent hypergeometric functions ${}_1F_1[\alpha; \beta; \lambda z]$ for suitable $\alpha, \beta \in \mathbb{Q}$ and $\lambda \in \overline{\mathbb{Q}}$. Gorelov's result is in fact more general as it holds for E-functions in the large sense. Our proof makes use of André's results on the singularities of the minimal differential equations satisfied by E-functions, together with a rigidity criterion for (irregular) differential systems recently obtained by Bloch-Esnault and Arinkin. An *ad-hoc* version of this criterion had already been used by Katz in his study of confluent hypergeometric series. Siegel's question remains unanswered for orders ≥ 3.

1 Introduction

We fix an embedding of $\overline{\mathbb{Q}}$ into $\mathbb{C}$. An E-function (in the strict sense) is a power series

$$f(z) = \sum_{n=0}^{\infty} \frac{a_n}{n!} z^n \in \overline{\mathbb{Q}}[[z]]$$

T. Rivoal (✉)
Institut Fourier, CNRS et Université Grenoble Alpes, CS 40700, 38058 Grenoble cedex 9, France
e-mail: Tanguy.Rivoal@univ-grenoble-alpes.fr

J. Roques
Université de Lyon, Université Claude Bernard Lyon 1, CNRS UMR 5208,
Institut Camille Jordan, 69622 Villeurbanne, France
e-mail: Julien.Roques@univ-lyon1.fr

A. Bostan and K. Raschel (eds.), *Transcendence in Algebra, Combinatorics, Geometry and Number Theory*, Springer Proceedings in Mathematics & Statistics 373,
https://doi.org/10.1007/978-3-030-84304-5_21

such that:

(1) $f(z)$ satisfies a non-zero linear differential equation with coefficients in $\overline{\mathbb{Q}}(z)$;
(2) there exists $C > 0$ such that for any $\sigma \in \mathrm{Gal}(\overline{\mathbb{Q}}/\mathbb{Q})$, we have $|\sigma(a_n)| \leq C^{n+1}$; and there exists a sequence of positive integers d_n such that $d_n \leq C^{n+1}$ and $d_n a_m$ is an algebraic integer for all $m \leq n$.

If $a_n \in \mathbb{Z}$, the conditions in (2) reduce to $|a_n| \leq C^{n+1}$. This class of arithmetic power series was defined by Siegel [18] (in a slightly more general way) to mimic the Diophantine properties of the exponential function, and his program was later completed by Shidlovsky [20]. Throughout the paper, we set $\theta = z\frac{d}{dz}$ and by "solution of a differential operator $\mathcal{L} \in \overline{\mathbb{Q}}(z)[\frac{d}{dz}]$", it must be understood "solution of the differential equation $\mathcal{L}y(z) = 0$". More recently, André [2] and Beukers [4] gave a new impulse to the Diophantine theory of E-functions, whose prototypical example is the generalized confluent hypergeometric function

$$ {}_pF_p\left[\begin{matrix}\alpha_1, \ldots, \alpha_p \\ \beta_1, \ldots, \beta_p\end{matrix}; z\right] = \sum_{n=0}^{\infty} \frac{(\alpha_1)_n \cdots (\alpha_p)_n}{n!(\beta_1)_n \cdots (\beta_p)_n} z^n $$

with $\alpha_j, \beta_j \in \mathbb{Q}$, and none of the β's is a negative integer. If $p = 1$, it is a solution of the differential operator $\theta(\theta + \beta - 1) - z(\theta + \alpha)$ and when $\alpha = \beta$, this is simply $\exp(z)$.

The present paper is concerned with the following classical questions: What are E-functions? Are they related to confluent hypergeometric functions? In fact, Siegel [19, p. 58] asked the following question: can any E-function be represented as a multivariate polynomial, with coefficients in $\overline{\mathbb{Q}}[z]$, in finitely many confluent hypergeometric series of the form ${}_pF_p[\underline{a}; \underline{b}; \lambda z]$, for various $p \geq 1$, $\underline{a}, \underline{b} \in \overline{\mathbb{Q}}^p$ and $\lambda \in \overline{\mathbb{Q}}$? See also [20, p. 84]. ([1])

In the recent papers [16, 17], we studied the structural properties of differential equations satisfied by strict E-functions, in the light of [2]. In particular, as a consequence of the main result of [16], we proved that any strict E-function $f(z)$ solution of an inhomogeneous linear equation

$$ f'(z) = u(z)f(z) + v(z) \tag{1.1} $$

of order 1 is essentially hypergeometric, where $u(z) \in \overline{\mathbb{Q}}(z)^{\times}$ and $v(z) \in \overline{\mathbb{Q}}(z)$. More precisely, there exist some $a(z), b(z) \in \overline{\mathbb{Q}}[z, z^{-1}]$, $\beta \in \{1\} \cup \mathbb{Q} \setminus \mathbb{Z}$ and $\lambda \in \overline{\mathbb{Q}}$ such that

$$ f(z) = a(z)\,{}_1F_1(1; \beta; \lambda z) + b(z). \tag{1.2} $$

This solved a problem, raised by Shidlovsky [20, p. 184], and already partially solved by André in [2, p. 724]. If $v(z) = 0$, then in fact $b(z) = 0$, $\beta = 1$ and $a(z) \in \overline{\mathbb{Q}}[z]$,

[1] In this problem, Siegel referred to his original definition of E-functions, which are slightly more general than the strict ones used in this paper, which are themselves called E^*-functions in [20]. It is widely believed that both class are identical, but our proof holds only in the strict sense.

so that $f(z) = a(z)e^{\lambda z}$, a result due to Shidlovsky [20, p. 184]. When $v(z) \neq 0$, any solution of (1.1) is also solution of the differential operator of order 2 given by $\frac{d}{dz}(\frac{1}{v(z)}\frac{d}{dz} - \frac{u(z)}{v(z)})$. We were not aware at that time that Gorelov had already solved Shidlovsky's problem in [11, p. 139, Theorem 2] for E-functions in Siegel's original sense.

It is thus natural to wonder if something similar to (1.2) can be said of E-functions solutions of differential equations of order 2. Actually, a first case was considered in [16]. Indeed, the following result is a direct consequence of [16, Theorem 4]. Again, this result has been proved by Gorelov in [12, p. 515, Theorem 2] for E-functions in Siegel's original sense.

Theorem 1. *Let $f(z)$ be a strict E-function solution of a non-zero linear differential operator of order 2 with coefficients in $\overline{\mathbb{Q}}(z)$ and reducible over $\overline{\mathbb{Q}}(z)$. Then, there exist $a(z), b(z) \in \overline{\mathbb{Q}}[z, z^{-1}]$, $\beta \in \{1\} \cup \mathbb{Q} \setminus \mathbb{Z}$ and $\lambda, \mu \in \overline{\mathbb{Q}}$ such that*

$$f(z) = a(z)e^{\mu z}{}_1F_1(1; \beta; \lambda z) + b(z)e^{\mu z}. \tag{1.3}$$

We now state the result if we remove the reducibility hypothesis in the previous result. We emphasize that Theorem 2 below is a particular case of a result of Gorelov [12, p. 514, Theorem 1], who in fact did not make the distinction between the reducible and irreducible case, which is in accordance with the remarks made after the theorem.

Theorem 2. *Let $f(z)$ be a strict E-function solution of a non-zero linear differential operator of order 2 with coefficients in $\overline{\mathbb{Q}}(z)$ and irreducible over $\overline{\mathbb{Q}}(z)$. Then, we have*

$$f(z) = a(z)e^{\mu z}{}_1F_1(\alpha; \beta; \lambda z) + b(z)e^{\mu z}{}_1F_1'(\alpha; \beta; \lambda z) \tag{1.4}$$

where $a(z), b(z) \in \overline{\mathbb{Q}}(z)$, $\lambda \in \overline{\mathbb{Q}}^\times$, $\mu \in \overline{\mathbb{Q}}$, and $\alpha \in \mathbb{Q}$, $\beta \in \mathbb{Q} \setminus \mathbb{Z}_{\leq 0}$ are such that $\alpha - \beta \notin \mathbb{Z}$.

Note that (1.3) is of the form (1.4) because we have the relation $z\,{}_1F_1'(1; \beta; z) = (z - \beta + 1)\,{}_1F_1(1; \beta; z) + \beta - 1$. Moreover, ${}_1F_1'(\alpha; \beta; z) = \frac{\alpha}{\beta}\,{}_1F_1(\alpha + 1; \beta + 1; z)$. This explains why such results give a positive solution to Siegel's problem for orders ≤ 2. Gorelov proved a stronger version of Theorem 2: he showed that $f(z)$ can be assumed to be an E-function in Siegel's original sense and moreover that the conclusion holds with some $a(z), b(z) \in \overline{\mathbb{Q}}[z]$. On the other hand, he did not state that $\alpha - \beta \notin \mathbb{Z}$. We observe that in fact a strict E-function of order 2 may have a representation of the form (1.4) with $a(z)$ and $b(z)$ not necessarily polynomials, for instance $\varphi(z) := \frac{2}{z}\,{}_1F_1(1; 1/2; z) - \frac{1}{z}\,{}_1F_1'(1; 1/2; z) = \frac{4}{3} + \frac{16z}{15} + \frac{16z^2}{35} + \ldots$ is such an E-function. This does not contradict Gorelov's stronger "polynomial coefficients" version, because a strict E-function does not necessarily admit a unique representation of the form (1.4); indeed, it is readily proved that $\varphi(z) = 2\,{}_1F_1'(1; 3/2; z)$.

The main contribution of this paper is our proof of Theorem 2, which is quite different from that of Gorelov, even though he also used André's theory at some point. We did not try to reprove his version in full. We hope our point of view will be useful for further studies in this field.

We now illustrate Theorem 2 with the non-hypergeometric E-function

$$a(z) = \sum_{n=0}^{\infty} \frac{1}{n!}\Big(\sum_{k=0}^{n} \binom{n}{k}\binom{n+k}{n}\Big) z^n. \tag{1.5}$$

It was brought to our attention by F. Beukers during a lecture he gave in June 2016 at the conference *Automates and Number Theory* held at Porquerolles, where he asked if $a(z)$ was related in some way to hypergeometric series. Since it is solution of the irreducible differential operator $z(\frac{d}{dz})^2 - (6z-1)\frac{d}{dz} + (z-3)$, Theorem 2 applies to it and the answer is yes. Let us give (1.4) in this case. Since $\sum_{n=0}^{\infty}\big(\sum_{k=0}^{n}\binom{n}{k}\binom{n+k}{n}\big)z^n = \frac{1}{\sqrt{1-6z+z^2}}$ (see [15, §3]), we have

$$a(z) = \frac{1}{2i\pi}\int_L \frac{e^{zx}}{\sqrt{1-6x+x^2}}dx$$

where L is a "vertical" straight line leaving the roots of $1-6x+x^2$ to its left; see [10, §§4.2–4.3]. With the change of variable $x = t+3$ and with L' a "vertical" straight line leaving the roots of t^2-8 to its left, we obtain

$$a(z) = \frac{1}{2i\pi}\int_{L'} \frac{e^{z(t+3)}}{\sqrt{t^2-8}}dt = e^{3z}\sum_{n=0}^{\infty}\frac{(2z^2)^n}{n!^2} = e^{3z}\cdot {}_0F_1(\cdot\,;1;2z^2) \tag{1.6}$$

$$= e^{(3-2\sqrt{2})z}\cdot {}_1F_1(1/2\,;1;4\sqrt{2}z). \tag{1.7}$$

The hypergeometric function on the right of (1.6) is not formally of the form suggested by Theorem 2, but this is the case of (1.7). The equality of both expressions is a consequence of an hypergeometric identity between Kummer M function and Bessel I_0 function [1, p. 509, 13.6.1]. In passing, we obtain the binomial identity

$$\sum_{k=0}^{n}\binom{n}{k}\binom{n+k}{n} = \sum_{k=0}^{n}\binom{n}{k}\binom{2k}{k}\sqrt{2}^k(3-2\sqrt{2})^{n-k},\ n \geq 0, \tag{1.8}$$

after multiplication of the two (implicit) power series in (1.7). Conversely, (1.8) could be proved first (by means of Zeilberger's algorithm) and then (1.7) would follow again.

We don't know if results similar to Theorems 1 and 2 could be obtained for E-functions of higher order, even for the order 3; the methods of this paper do not give enough informations to conclude. In fact, Siegel's question might have a negative answer in general and as the order increases, one needs to add more and more special functions to the hypergeometric ones. For instance, $\sum_{n=0}^{\infty}\frac{1}{n!}\big(\sum_{k=0}^{n}\binom{n}{k}^2\binom{n+k}{n}\big)z^n$ is solution of $z^2y'''(z) + (3z-11z^2)y''(z) + (1-22z-z^2)y'(z) - (3+z)y(z) = 0$.

Although not hypergeometric, can it be expressed using confluent hypergeometric series as requested in Siegel's problem, like $a(z)$ above, or is it of a different nature? We refer to [13, 14] for further results on this problem.

The paper is organized as follows. In Sections 2 (hypergeometric operators), 3 (Fuchs relation) and 4 (rigidity), we collect some results from the literature that we need for the proof of Theorem 2, which is given in Sect. 5. Some aspects of the proof are similar with certain results of Katz in [8], though the methods are not written in the same language; we include these computations for the sake of completeness. Finally, in Sect. 6, we make some remarks on E-operators and G-operators. From now on, any E-function is understood to be in the strict sense.

2 The Differential Operators $H_{\alpha;\beta,\gamma}$ and $L_{\alpha,\beta,\gamma,\lambda,\mu}$

In this section, we gather some results on two explicit hypergeometric operators. They will be used in the proof of Theorem 2 later on.

2.1 The Confluent Hypergeometric Operator $H_{\alpha;\beta,\gamma}$

We recall that $\theta = z\frac{d}{dz}$. The confluent hypergeometric operator with parameters $\alpha, \beta, \gamma \in \mathbb{C}$ is the linear differential operator given by

$$H_{\alpha;\beta,\gamma} = (\theta + \beta - 1)(\theta + \gamma - 1) - z(\theta + \alpha).$$

It has at most two singularities on $\mathbb{P}^1(\mathbb{C})$, namely 0 and ∞. The former is a regular singular point, the latter is an irregular singular point. More precisely, the slopes of the Newton polygon of $H_{\alpha;\beta,\gamma}$ at ∞ are 0 and 1, both with multiplicity 1. We denote by

$$Y' = A_{\alpha;\beta,\gamma} Y \tag{2.1}$$

the differential system associated to $H_{\alpha;\beta,\gamma}$.

For later use, we shall now describe the general form of the formal solutions of the differential system (2.1) as predicted by Turittin's theorem [21, Theorem 3.54].

We first assume that $\beta - \gamma \notin \mathbb{Z}$. Then, at 0, the differential system (2.1) admits a fundamental matrix of formal solutions of the form $F_0(z)z^{\Gamma_0}$ where

$$F_0(z) \in \mathrm{GL}_2(\mathbb{C}((z))) \quad \text{and} \quad \Gamma_0 = \begin{pmatrix} 1-\beta & 0 \\ 0 & 1-\gamma \end{pmatrix}.$$

At ∞, the differential system (2.1) admits a fundamental matrix of formal solutions of the form $F_\infty(z)z^{\Gamma_\infty}e^{\Delta z}$ where

$$F_\infty(z) \in \mathrm{GL}_2(\mathbb{C}((z^{-1}))), \quad \Gamma_\infty = \begin{pmatrix} \alpha-\beta-\gamma+1 & 0 \\ 0 & -\alpha \end{pmatrix} \quad \text{and} \quad \Delta = \begin{pmatrix} 1 & 0 \\ 0 & 0 \end{pmatrix}.$$

We now assume that $\beta - \gamma \in \mathbb{Z}$. If $\alpha - \beta \notin \mathbb{Z}$, then, at 0, the differential system (2.1) admits a fundamental matrix of formal solutions of the form $F_0(z)z^{\Gamma_0}$ where

$$F_0(z) \in \mathrm{GL}_2(\mathbb{C}((z))) \quad \text{and} \quad \Gamma_0 = \begin{pmatrix} 1-\beta & 1 \\ 0 & 1-\beta \end{pmatrix}.$$

Remark. If the condition $\alpha - \beta \notin \mathbb{Z}$ is not satisfied, then Γ_0 may be diagonalizable. Consider for instance the case $\alpha = \gamma - 1$ and $\beta - 1 = \alpha - 1$.

We shall now consider a special case with $\alpha - \beta \in \mathbb{Z}$, namely $\alpha = \beta - 1 = \gamma - 1$. In this case, at 0, the differential system (2.1) admits a fundamental matrix of formal solutions of the form $F_0(z)z^{\Gamma_0}$ where

$$F_0(z) \in \mathrm{GL}_2(\mathbb{C}((z))) \quad \text{and} \quad \Gamma_0 = \begin{pmatrix} 1-\beta & 1 \\ 0 & 1-\beta \end{pmatrix}.$$

This can be seen by direct calculation using that, in the present case, we have $H_{\alpha;\beta,\gamma} = (\theta + \alpha - z)(\theta + \alpha)$. At ∞, the differential system (2.1) admits a fundamental matrix of formal solutions of the form $F_\infty(z)z^{\Gamma_\infty}e^{\Delta z}$ where

$$F_\infty(z) \in \mathrm{GL}_2(\mathbb{C}((z^{-1}))), \quad \Gamma_\infty = \begin{pmatrix} 1-\beta & 0 \\ 0 & 1-\beta \end{pmatrix} \quad \text{and} \quad \Delta = \begin{pmatrix} 1 & 0 \\ 0 & 0 \end{pmatrix}.$$

2.2 The Operator $L_{\alpha,\beta,\gamma,\lambda,\mu}$

For any $\alpha, \beta, \gamma, \lambda, \mu \in \mathbb{C}$, with $\lambda \neq 0$, we consider the linear differential operator given by

$$L_{\alpha;\beta,\gamma;\lambda,\mu} = (\theta + \beta - 1 - \mu z)(\theta + \gamma - 1 - \mu z) - \lambda z(\theta + \alpha - \mu z).$$

There is a simple relation between the operators $L_{\alpha;\beta,\gamma;\lambda,\mu}$ and $H_{\alpha;\beta,\gamma}$: we have

$$H_{\alpha;\beta,\gamma}\big(y(z)\big) = 0 \Longleftrightarrow L_{\alpha;\beta,\gamma;\lambda,\mu}\big(y(\lambda z)e^{\mu z}\big) = 0.$$

We denote by

$$Y' = A_{\alpha;\beta,\gamma;\lambda,\mu}Y \tag{2.2}$$

the differential system associated to $L_{\alpha;\beta,\gamma;\lambda,\mu}$.

The results of Sect. 2.1 imply the following facts.

We first assume that $\beta - \gamma \notin \mathbb{Z}$. Then, at 0, the differential system (2.2) admits a fundamental matrix of formal solutions of the form $F_0(z)z^{\Gamma_0}$ where

$$F_0(z) \in \mathrm{GL}_2(\mathbb{C}((z))) \quad \text{and} \quad \Gamma_0 = \begin{pmatrix} 1-\beta & 0 \\ 0 & 1-\gamma \end{pmatrix}.$$

At ∞, the differential system (2.2) admits a fundamental matrix of formal solutions of the form $F_\infty(z)z^{\Gamma_\infty}e^{\Delta z}$ where

$$F_\infty(z) \in \mathrm{GL}_2(\mathbb{C}((z^{-1}))), \quad \Gamma_\infty = \begin{pmatrix} \alpha-\beta-\gamma+1 & 0 \\ 0 & -\alpha \end{pmatrix} \quad \text{and} \quad \Delta = \begin{pmatrix} \lambda+\mu & 0 \\ 0 & \mu \end{pmatrix}.$$

We now assume that $\beta - \gamma \in \mathbb{Z}$. If $\alpha - \beta \notin \mathbb{Z}$, then, at 0, the differential system (2.2) admits a fundamental matrix of formal solutions of the form $F_0(z)z^{\Gamma_0}$ where

$$F_0(z) \in \mathrm{GL}_2(\mathbb{C}((z))) \quad \text{and} \quad \Gamma_0 = \begin{pmatrix} 1-\beta & 1 \\ 0 & 1-\beta \end{pmatrix}.$$

If $\alpha = \beta - 1 = \gamma - 1$, then, at 0, the differential system (2.2) admits a fundamental matrix of formal solutions of the form $F_0(z)z^{\Gamma_0}$ where

$$F_0(z) \in \mathrm{GL}_2(\mathbb{C}((z))) \quad \text{and} \quad \Gamma_0 = \begin{pmatrix} 1-\beta & 1 \\ 0 & 1-\beta \end{pmatrix}.$$

At ∞, the differential system (2.2) admits a fundamental matrix of formal solutions of the form $F_\infty(z)z^{\Gamma_\infty}e^{\Delta z}$ where

$$F_\infty(z) \in \mathrm{GL}_2(\mathbb{C}((z^{-1}))), \quad \Gamma_\infty = \begin{pmatrix} \alpha-\beta-\gamma+1 & 0 \\ 0 & -\alpha \end{pmatrix} \quad \text{and} \quad \Delta = \begin{pmatrix} \lambda+\mu & 0 \\ 0 & \mu \end{pmatrix}.$$

3 Fuchs' Relation

In this section, we state a "Fuchs relation" between the exponents of certain (possibly irregular) differential systems.

Proposition 1. *Let us consider a differential system $Y' = AY$ with $A \in M_2(\mathbb{C}(z))$. We assume that the following properties are satisfied:*

(1) *$Y' = AY$ has only apparent singularities on $\mathbb{C}^\times$;*
(2) *$Y' = AY$ has a basis of solutions at 0 of the form $F_0(z)z^{\Gamma_0}$ where $F_0(z) \in \mathrm{GL}_2(\mathbb{C}((z)))$ and $\Gamma_0 \in M_2(\mathbb{C})$ is upper-triangular.*
(3) *$Y' = AY$ admits a basis of formal solutions at ∞ of the form $F_\infty(z)z^{\Gamma_\infty}e^{\Delta z}$ where $F_\infty(z) \in \mathrm{GL}_2(\mathbb{C}((z^{-1})))$, $\Gamma_\infty \in M_2(\mathbb{C})$ is upper-triangular, $\Delta = \mathrm{diag}(\theta_1, \theta_2) \in M_2(\mathbb{C})$ is diagonal, and Γ_∞ and Δ commute.*

Then, the trace of $\Gamma_0 - \Gamma_\infty$ *belongs to* $\mathbb{Z}$.

Proof. The monodromy of $Y' = AY$ at 0 with respect to the fundamental matrix of solutions $F_0(z)z^{\Gamma_0}$ is given by $M_0 = e^{2\pi i \Gamma_0}$. Its monodromy at ∞ with respect to the same fundamental matrix of solutions is of the form $M_\infty = Pe^{-2\pi i \Gamma_\infty} S_\infty P^{-1}$ where $P \in \mathrm{GL}_2(\mathbb{C})$ and where $S_\infty \in \mathrm{GL}_2(\mathbb{C})$ is a product of Stokes matrices (see [21, Proposition 8.12]). In particular, S_∞ is unipotent and, hence, $\det(S_\infty) = 1$. But, we have $M_0 M_\infty = I_2$ because $Y' = AY$ has only apparent singularities on $\mathbb{C}^\times$ and, hence, trivial monodromies around each point of $\mathbb{C}^\times$. It follows that $\det(M_0 M_\infty) = \det(I_2) = 1$ *i.e.* $e^{2\pi i \, \mathrm{tr}(\Gamma_0 - \Gamma_\infty)} = 1$. Whence the result. □

4 A Reminder on Rigidity

This section is a reminder about the notion of rigidity of (possibly irregular) differential systems. We start by recalling the notions of formal and rational equivalences.

4.1 Formal Equivalence

Recall that two differential systems

$$Y' = AY \text{ and } Y' = BY \text{ with } A, B \in M_n(\mathbb{C}(z)) \tag{4.1}$$

are formally equivalent at 0 if there exists $R \in \mathrm{GL}_n(\mathbb{C}((z)))$ such that

$$B = R^{-1}AR - R^{-1}R'. \tag{4.2}$$

This means that one gets $Y' = BY$ from $Y' = AY$ by replacing Y by RY.

More generally, there is a notion of formal equivalence at any $s \in \mathbb{P}^1(\mathbb{C})$. More precisely, we denote by $\widehat{K_s}$ the field of formal Laurent series at $s \in \mathbb{P}^1(\mathbb{C})$. We say that the differential systems (4.1) are formally equivalent at s if there exists $R \in \mathrm{GL}_n(\widehat{K_s})$ such that Eq. (4.2) holds.

4.2 Rational Equivalence

The differential systems (4.1) are called rationally equivalent if there exists $R \in \mathrm{GL}_n(\mathbb{C}(z))$ such that Eq. (4.2) holds.

Of course, "rationally equivalent" implies "formally equivalent at any $s \in \mathbb{P}^1(\mathbb{C})$", but the converse is not true in general. This is where rigidity comes into play.

4.3 Rigidity

We say that a given differential system

$$Y' = AY \text{ with } A \in M_n(\mathbb{C}(z)) \tag{4.3}$$

is rigid if, for any differential system $Y' = BY$ with $B \in M_n(\mathbb{C}(z))$, the fact that $Y' = AY$ is *formally equivalent* to $Y' = BY$ at each $s \in \mathbb{P}^1(\mathbb{C})$ implies that $Y' = AY$ and $Y' = BY$ are *rationally equivalent*.

If $Y' = AY$ is irreducible over $\mathbb{C}(z)$, there is a numerical rigidity criterion, which can be stated as follows. We denote by $\otimes$ the Kronecker tensor product on $M_n(\mathbb{C})$, *i.e.*, for $C, D \in M_n(\mathbb{C}(z))$, the tensor product $C \otimes D \in M_{n^2}(\mathbb{C})$ is defined by

$$C \otimes D = \begin{pmatrix} c_{1,1}D & \cdots & c_{1,n}D \\ \vdots & \ddots & \vdots \\ c_{n,1}D & \cdots & c_{n,n}D \end{pmatrix}.$$

We define the "internal End" of $Y' = AY$ as the differential system $Y' = \mathcal{E}(A)Y$ with

$$\mathcal{E}(A) = A \otimes I_n - I_n \otimes A^t \in M_{n^2}(\mathbb{C}).$$

We set

$$\mathrm{rig}(A) = 2n^2 - \sum_{s\in\mathbb{P}^1(\mathbb{C})} \big(\mathrm{irr}(\mathcal{E}(A), s) + n^2 - \dim_{\mathbb{C}} \mathrm{sol}(\mathcal{E}(A), s)\big)$$

where $\mathrm{irr}(\mathcal{E}(A), s)$ is Malgrange irregularity of $Y' = \mathcal{E}(A)Y$ at s, and where $\mathrm{sol}(\mathcal{E}(A), s)$ is the $\mathbb{C}$-vector space of the solutions in $M_{n,1}(\widehat{K_s})$ of $Y' = \mathcal{E}(A)Y$. We recall that the Malgrange irregularity $\mathrm{irr}(B, s)$ at s of a given differential system $Y' = BY$, with $B \in M_n(\mathbb{C}(z))$, is the height of its Newton polygon. Equivalently, it is equal to the sum of the slopes (counted with multiplicities) of the Newton polygon at s of $Y' = BY$.

Theorem 3 ([3, Proposition 3.4], [5, Theorems 4.7 and 4.10]). *Assume that $Y' = AY$ is irreducible. Then, it is rigid if and only if* $\mathrm{rig}(A) = 2$.

We will also use the following inequality.

Theorem 4 ([3, Remark following Proposition 3.4]). *Assume that $Y' = AY$ is irreducible. Then, we have* $\mathrm{rig}(A) \leq 2$.

5 Proof of Theorem 2

We are now in position to prove our main result. For simplicity, we split the proof into two steps.

5.1 First Step

We first prove the following result.

Theorem 5. *Let us consider a differential system $Y' = AY$ with $A \in M_2(\mathbb{C}(z))$. We assume that this differential system is irreducible and that the following properties are satisfied:*

(1) *$Y' = AY$ has at most apparent singularities on $\mathbb{C}^\times$;*
(2) *$Y' = AY$ has at most a regular singularity at 0;*
(3) *the slopes at ∞ of $Y' = AY$ are included in $\{0, 1\}$.*

Then, the differential system $Y' = AY$ is rationally equivalent to $Y' = A_{\alpha;\beta,\gamma;\lambda,\mu}Y$ for some $\alpha, \beta, \gamma, \lambda, \mu \in \mathbb{C}$ with $\lambda \neq 0$.

Proof. Turittin's theorem yields the following facts:

(1) $Y' = AY$ has a basis of solutions at 0 of the form $F_0(z)z^{\Gamma_0}$ where $F_0(z) \in \mathrm{GL}_2(\mathbb{C}((z)))$ and $\Gamma_0 \in M_2(\mathbb{C})$ is upper-triangular;
(2) $Y' = AY$ admits a basis of formal solutions at ∞ of the form $F_\infty(z)z^{\Gamma_\infty}e^{\Delta z}$ where $F_\infty(z) \in \mathrm{GL}_2(\mathbb{C}((z^{-1})))$, $\Gamma_\infty \in M_2(\mathbb{C})$ is upper-triangular, $\Delta = \mathrm{diag}(\theta_1, \theta_2) \in M_2(\mathbb{C})$ is diagonal, and Γ_∞ and Δ commute.

Hence, we have:

- at 0, the differential system $Y' = AY$ is formally equivalent to $Y' = \frac{\Gamma_0}{z}Y$;
- at ∞, the differential system $Y' = AY$ is formally equivalent to $Y' = \left(\frac{\Gamma_\infty}{z} + \Delta\right)Y$.

Therefore, setting $B = \mathcal{E}(A)$, we see that

- at 0, the differential system $Y' = BY$ is formally equivalent to $Y' = B_0Y$ with

$$B_0 = \frac{\Gamma_0 \otimes I_2 - I_2 \otimes \Gamma_0^t}{z}.$$

- at ∞, the differential system $Y' = BY$ is formally equivalent to $Y' = B_\infty Y$ with

$$B_\infty = \left(\frac{\Gamma_\infty}{z} + \Delta\right) \otimes I_2 - I_2 \otimes \left(\frac{\Gamma_\infty}{z} + \Delta\right)^t.$$

Note that $\theta_1 \neq \theta_2$. Indeed, assume at the contrary that $\theta_1 = \theta_2$. Then, the differential system $Y' = AY$ is rationally equivalent to $Y' = \left(\frac{\Gamma_0}{z} + \theta_1 I_2\right) Y$ (because the differential system satisfied by $F_0(z)z^{\Gamma_0}e^{-\theta_1 z}$ is regular singular on $\mathbb{P}^1(\mathbb{C})$, with at most apparent singularities on $\mathbb{C}^\times$ and, hence, is of the form $R(z)z^{\Gamma_0}$ for some $R(z) \in \mathrm{GL}_2(\mathbb{C}(z))$). Therefore, the differential system $Y' = AY$ is reducible. This is a contradiction.

Since Δ and Γ_∞ commute, the fact that $\theta_1 \neq \theta_2$ implies that Γ_∞ is diagonal:

$$\Gamma_\infty = \begin{pmatrix} \gamma_{\infty,1} & 0 \\ 0 & \gamma_{\infty,2} \end{pmatrix}.$$

We now split our study in several cases, but the idea of the proof will be the same in any cases: we will prove that the differential system $Y' = AY$ is rigid and formally equivalent to $Y' = A_{\alpha;\beta,\gamma;\lambda,\mu} Y$ for some $\alpha, \beta, \gamma, \lambda, \mu \in \mathbb{C}$, with $\lambda \neq 0$, at any $s \in \mathbb{P}^1(\mathbb{C})$. We will conclude that these systems are actually rationally equivalent by rigidity.

The Case Γ_0 Diagonal Non resonant. We assume that Γ_0 is a diagonal matrix

$$\Gamma_0 = \begin{pmatrix} \gamma_{0,1} & 0 \\ 0 & \gamma_{0,2} \end{pmatrix}$$

such that $\gamma_{0,2} - \gamma_{0,1} \notin \mathbb{Z}$. It follows that

$$B_0 = \begin{pmatrix} 0 & 0 & 0 & 0 \\ 0 & \frac{\gamma_{0,1}-\gamma_{0,2}}{z} & 0 & 0 \\ 0 & 0 & \frac{\gamma_{0,2}-\gamma_{0,1}}{z} & 0 \\ 0 & 0 & 0 & 0 \end{pmatrix}$$

and

$$B_\infty = \begin{pmatrix} 0 & 0 & 0 & 0 \\ 0 & \frac{\gamma_{\infty,1}-\gamma_{\infty,2}}{z} + \theta_1 - \theta_2 & 0 & 0 \\ 0 & 0 & -\left(\frac{\gamma_{\infty,1}-\gamma_{\infty,2}}{z} + \theta_1 - \theta_2\right) & 0 \\ 0 & 0 & 0 & 0 \end{pmatrix}.$$

Then, we have

$$\mathrm{irr}(B, 0) = \mathrm{irr}(B_0, 0) = 0 \text{ and } \dim_{\mathbb{C}} \mathrm{sol}(B, 0) = \dim_{\mathbb{C}} \mathrm{sol}(B_0, 0) = 2.$$

Moreover, we have

$$\mathrm{irr}(B, \infty) = \mathrm{irr}(B_\infty, \infty) = 2 \text{ and } \dim_{\mathbb{C}} \mathrm{sol}(B, \infty) = \dim_{\mathbb{C}} \mathrm{sol}(B_\infty, \infty) = 2.$$

So, we get

$$\mathrm{rig}(A) = 2 \cdot 4 - (0 + 4 - 2) - (2 + 4 - 2) = 2.$$

Therefore, $Y' = AY$ is rigid in virtue of Theorem 3.

We now consider $\alpha, \beta, \gamma, \lambda, \mu \in \mathbb{C}$ such that

$$\begin{cases} 1-\beta & \equiv \gamma_{0,1} \mod \mathbb{Z} \\ 1-\gamma & \equiv \gamma_{0,2} \mod \mathbb{Z} \\ \alpha-\beta-\gamma+1 & \equiv \gamma_{\infty,1} \mod \mathbb{Z} \\ -\alpha & \equiv \gamma_{\infty,2} \mod \mathbb{Z} \\ \lambda+\mu & = \theta_1 \\ \mu & = \theta_2 \end{cases}.$$

We can indeed solve this system of equations because, in virtue of Proposition 1, we have $\gamma_{0,1} + \gamma_{0,2} - \gamma_{\infty,1} - \gamma_{\infty,2} \equiv 0 \mod \mathbb{Z}$. Note that $\lambda \neq 0$ because $\theta_1 \neq \theta_2$. Note also that $\beta - \gamma \equiv \gamma_{0,2} - \gamma_{0,1} \not\equiv 0 \mod \mathbb{Z}$. Using Sect. 2, we see that $Y' = AY$ is formally equivalent at 0 and ∞ to $Y' = A_{\alpha;\beta,\gamma;\lambda,\mu}Y$. Therefore, by rigidity, these two systems are rationally equivalent.

The Case Γ_0 ***Diagonal and Resonant is Impossible.*** We assume that Γ_0 is a diagonal matrix

$$\Gamma_0 = \begin{pmatrix} \gamma_{0,1} & 0 \\ 0 & \gamma_{0,2} \end{pmatrix}$$

such that $\gamma_{0,2} - \gamma_{0,2} \in \mathbb{Z}$. Then, the equality $\mathrm{rig}(A) \leq 2$ (see Theorem 4 above) reads as follows:

$$8 - (0+4-4) - (2+4-\dim_{\mathbb{C}} \mathrm{sol}(B,\infty)) \leq 2$$

i.e. $\dim_{\mathbb{C}} \mathrm{sol}(B,\infty) \leq 0$. This is a contradiction because $\dim_{\mathbb{C}} \mathrm{sol}(B,\infty) \geq 2$.

The Case Γ_0 ***Non diagonal.*** Up to conjugation, we can assume that

$$\Gamma_0 = \begin{pmatrix} \gamma_{0,1} & 1 \\ 0 & \gamma_{0,1} \end{pmatrix}.$$

Then, we have

$$B_0 = \begin{pmatrix} 0 & 0 & 1/z & 0 \\ -1/z & 0 & 0 & 1/z \\ 0 & 0 & 0 & 0 \\ 0 & 0 & -1/z & 0 \end{pmatrix},$$

whose Jordan normal form is given by

$$\begin{pmatrix} 0 & 1/z & 0 & 0 \\ 0 & 0 & 1/z & 0 \\ 0 & 0 & 0 & 0 \\ 0 & 0 & 0 & 0 \end{pmatrix}.$$

Moreover, we have

$$B_\infty = \begin{pmatrix} 0 & 0 & 0 & 0 \\ 0 & \frac{\gamma_{\infty,1}-\gamma_{\infty,2}}{z} + \theta_1 - \theta_2 & 0 & 0 \\ 0 & 0 & -(\frac{\gamma_{\infty,1}-\gamma_{\infty,2}}{z} + \theta_1 - \theta_2) & 0 \\ 0 & 0 & 0 & 0 \end{pmatrix}.$$

Then, we have

$$\mathrm{irr}(B, 0) = \mathrm{irr}(B_0, 0) = 0 \text{ and } \dim_{\mathbb{C}} \mathrm{sol}(B, 0) = \dim_{\mathbb{C}} \mathrm{sol}(B_0, 0) = 2.$$

Moreover, we have

$$\mathrm{irr}(B, \infty) = \mathrm{irr}(B_\infty, \infty) = 2 \text{ and } \dim_{\mathbb{C}} \mathrm{sol}(B, \infty) = \dim_{\mathbb{C}} \mathrm{sol}(B_\infty, \infty) = 2.$$

So, we have

$$\mathrm{rig}(A) = 2 \cdot 4 - (0 + 4 - 2) - (2 + 4 - 2) = 2.$$

Therefore, $Y' = AY$ is rigid in virtue of Theorem 3.

We consider $\alpha, \beta, \gamma, \lambda, \mu \in \mathbb{C}$ such that

$$\begin{cases} 1 - \beta & \equiv \gamma_{0,1} \mod \mathbb{Z} \\ 1 - \gamma & \equiv \gamma_{0,1} \mod \mathbb{Z} \\ \alpha - \beta - \gamma + 1 & \equiv \gamma_{\infty,1} \mod \mathbb{Z} \\ -\alpha & \equiv \gamma_{\infty,2} \mod \mathbb{Z} \\ \lambda + \mu & = \theta_1 \\ \mu & = \theta_2 \end{cases}.$$

We can indeed solve this system of equations because, in virtue of Proposition 1, we have $\gamma_{0,1} + \gamma_{0,1} - \gamma_{\infty,1} - \gamma_{\infty,2} \equiv 0 \mod \mathbb{Z}$. Note that $\lambda \neq 0$ because $\theta_1 \neq \theta_2$. Note also that $\beta - \gamma \in \mathbb{Z}$. If $\alpha - \beta \in \mathbb{Z}$, then the congruence $\alpha - \beta - \gamma + 1 \equiv \gamma_{\infty,1} \mod \mathbb{Z}$ implies that $\beta \equiv \gamma \equiv -\gamma_{\infty,1} \mod \mathbb{Z}$ but $\beta \equiv \gamma \equiv -\gamma_{0,1} \mod \mathbb{Z}$ so $\gamma_{0,1} \equiv \gamma_{\infty,1} \mod \mathbb{Z}$. Now, the congruence $\gamma_{0,1} + \gamma_{0,1} - \gamma_{\infty,1} - \gamma_{\infty,2} \equiv 0 \mod \mathbb{Z}$ implies that $\gamma_{\infty,2} \equiv \gamma_{\infty,1} \mod \mathbb{Z}$ hence $\alpha \equiv \beta \equiv \gamma \mod \mathbb{Z}$. Therefore, we can and will assume that $\alpha = \beta - 1 = \gamma - 1$. Using Sect. 2, we see that $Y' = AY$ is formally equivalent at 0 and ∞ to $Y' = A_{\alpha;\beta,\gamma;\lambda,\mu} Y$. By rigidity, these two systems are rationally equivalent. □

5.2 Second Step

Let $f(z)$ be an E-function as in Theorem 2. By hypothesis, $f(z)$ satisfies a non-zero linear differential equation of order ≤ 2 with coefficients in $\overline{\mathbb{Q}}(z)$.

If $f(z)$ satisfies a non-zero differential equation of order 1 with coefficients in $\overline{\mathbb{Q}}(z)$, then it is well-known that $f(z) = a(z)e^{\alpha z}$ with $a(z) \in \overline{\mathbb{Q}}(z)$ and $\alpha \in \overline{\mathbb{Q}}$. Whence the result.

We shall now assume that $f(z)$ does not satisfy a differential equation of order 1 with coefficients in $\overline{\mathbb{Q}}(z)$. Then, according to [2, Theorem 4.3], the vector $F(z) = (f(z), f'(z))^t$ satisfies some linear differential system $F'(z) = A(z)F(z)$ for some $A(z) \in M_2(\overline{\mathbb{Q}}(z))$, with the following properties:

(1) $Y' = AY$ has only apparent singularities on $\mathbb{C}^\times$;
(2) $Y' = AY$ has a basis of solutions at 0 of the form $F_0(z)z^{\Gamma_0}$ where $F_0(z) \in \mathrm{GL}_2(\overline{\mathbb{Q}}((z)))$ and $\Gamma_0 \in M_2(\mathbb{Q})$ is upper-triangular;
(3) $Y' = AY$ admits a basis of formal solutions at ∞ of the form $F_\infty(z)z^{\Gamma_\infty}e^{\Delta z}$ where $F_\infty(z) \in \mathrm{GL}_2(\overline{\mathbb{Q}}((z)))$, $\Gamma_\infty \in M_2(\mathbb{Q})$ is upper-triangular, $\Delta \in M_2(\overline{\mathbb{Q}})$ is diagonal and such that $\Gamma_\infty\Delta = \Delta\Gamma_\infty$.

If $Y' = AY$ is reducible over $\mathbb{C}(z)$ or, equivalently, over $\overline{\mathbb{Q}}(z)$, then we are in the situation of Theorem 1, which is a consequence of [16, Theorem 4]. We observe here that the series ${}_1F_1(1; \gamma; \lambda z)$ is a solution of $L_{1;1,\gamma;\lambda,0}$.

We shall now assume that $Y' = AY$ is irreducible over $\mathbb{C}(z)$ or, equivalently, over $\overline{\mathbb{Q}}(z)$. Then, according to Theorem 5, there exists $R(z) \in \mathrm{GL}_2(\mathbb{C}(z))$ such that

$$A = R^{-1}A_{\alpha;\beta,\gamma;\lambda,\mu}R - R^{-1}R'.$$

But both β and γ are congruent to one of the eigenvalues of Γ_0 modulo $\mathbb{Z}$. Therefore, β and γ belong to $\mathbb{Q}$. A similar argument yields that $\alpha \in \mathbb{Q}$, $\lambda \in \overline{\mathbb{Q}}^\times$ and $\mu \in \overline{\mathbb{Q}}$. In particular, $A_{\alpha;\beta,\gamma;\lambda,\mu}$ has coefficients in $\overline{\mathbb{Q}}(z)$ and, hence, one can assume that $R(z)$ has coefficients in $\overline{\mathbb{Q}}(z)$.

We observe that RF is a vector solution of $Y' = A_{\alpha;\beta,\gamma;\lambda,\mu}Y$, and any such solutions are of the form $(h(z), h'(z))^t$ where $h(z)$ is a solution of $L_{\alpha;\beta,\gamma;\lambda,\mu}$. Hence, there exist $a(z), b(z) \in \overline{\mathbb{Q}}(z)$ such that

$$f(z) = a(z)h(z) + b(z)h'(z).$$

Since the entries of RF belong to $\overline{\mathbb{Q}}((z))$, we see that $h(z)$ also belongs to $\overline{\mathbb{Q}}((z))$. As noted at the beginning of Sect. 2.2, there exist a solution $g(z) \in \overline{\mathbb{Q}}((z))$ of $H_{\alpha;\beta,\gamma}$ such that

$$h(z) = g(\lambda z)e^{\mu z}.$$

By assumption, the differential system $Y' = AY$ is irreducible over $\overline{\mathbb{Q}}(z)$, so that the differential operators $L_{\alpha;\beta,\gamma;\lambda,\mu}$ and, hence, $H_{\alpha;\beta,\gamma}$ are irreducible over $\overline{\mathbb{Q}}(z)$. This implies that $\alpha - \beta \notin \mathbb{Z}$ and $\alpha - \gamma \notin \mathbb{Z}$. Using Sect. 2.1, we see that $H_{\alpha;\beta,\gamma}$ has a nonzero solution in $\overline{\mathbb{Q}}((z))$ if and only if $\beta \in \mathbb{Z}$ or $\gamma \in \mathbb{Z}$. Assume for instance that $\beta \in \mathbb{Z}$. Moreover, if $\beta - \gamma \in \mathbb{Z}$, we can assume that $\gamma \geq \beta$. Then, the $\overline{\mathbb{Q}}$-vector space of solutions of $H_{\alpha;\beta,\gamma}$ in $\overline{\mathbb{Q}}((z))$ is generated by

$$z^{1-\beta}\sum_{n=0}^{\infty}\frac{(\alpha-\beta+1)_n}{n!(\gamma-\beta+1)_n}z^n = z^{1-\beta}{}_1F_1(\alpha-\beta+1;\gamma-\beta+1;z).$$

Hence, up to a multiplicative constant in $\overline{\mathbb{Q}}^{\times}$, we have $g(z) = z^{1-\beta}{}_1F_1(\alpha-\beta+1;\gamma-\beta+1;z)$. This completes the proof of Theorem 2.

6 Some Remarks on E-Operators and G-Operators

Finally, though this is not the subject of this paper, we mention that similar problems have been formulated for G-functions.

A G-function at $z = 0$ is a power series $f(z) = \sum_{n=0}^{\infty} a_n z^n \in \overline{\mathbb{Q}}[[z]]$ such that $\sum_{n=0}^{\infty} \frac{a_n}{n!} z^n$ is an E-function. Both classes of functions have been first introduced by Siegel, and recently André [2] showed the deep relations that exist between E and G-functions. A non-zero minimal equation in $\overline{\mathbb{Q}}[z, \frac{d}{dz}]$ satisfied by a G-function is called a G-operator. From results of André, Chudnovky and Katz (see [2, 6, 7]), we know that a G-operator is Fuchsian with rational exponents at its singularities, and that all its solutions at any point $\alpha \in \overline{\mathbb{Q}} \cup \{\infty\}$ are essentially G-functions of the variable $z - \alpha$ or $1/z$ if $\alpha = \infty$. André [2] defined an E-operator as a differential operator in $\overline{\mathbb{Q}}[z, \frac{d}{dz}]$ such that its Fourier-Laplace transform is a G-operator. We recall that the Fourier-Laplace transform $\widehat{\mathcal{L}} \in \overline{\mathbb{Q}}[z, \frac{d}{dz}]$ of an operator $\mathcal{L} \in \overline{\mathbb{Q}}[z, \frac{d}{dz}]$ is the image of $\mathcal{L}$ by the automorphism of the Weyl algebra $\overline{\mathbb{Q}}[z, \frac{d}{dz}]$ defined by $z \mapsto -\frac{d}{dz}$ and $\frac{d}{dz} \mapsto z$. Any E-function is solution of an E-operator, which is not necessarily minimal for the degree in $\frac{d}{dz}$ but is minimal for the degree in z. André proved that the leading polynomial of an E-operator is z^m for some integer $m \geq 0$, i.e., that 0 is its only possible finite singularity. It follows that the minimal non-zero differential equation of a given E-function has only apparent finite non-zero singularities. That property was crucial for the results proved here.

Using the André-Chudnovky-Katz Theorem, it is easy to prove that G-functions of differential order 1 are algebraic functions over $\overline{\mathbb{Q}}(z)$ of the form $z^m \prod_{j=1}^{d}(z-\alpha_j)^{e_j}$, $m \in \mathbb{N}$, $\alpha_j \in \overline{\mathbb{Q}}^{\times}$, $e_j \in \mathbb{Z}$. Dwork conjectured that a globally nilpotent differential operator in $\overline{\mathbb{Q}}(z)[\frac{d}{dz}]$ of order 2 either has algebraic solutions or there exists an algebraic pullback to Gauss's hypergeometric equation with rational parameters. This conjecture was disproved by Krammer [9]. We shall not define the notion of "global nilpotence" here. Let us simply say that G-operators are conjectured to be exactly the globally nilpotent operators in $\overline{\mathbb{Q}}(z)[\frac{d}{dz}]$, and that G-operators coming from geometry are known to be globally nilpotent (Katz [7]). Hence, Krammer's result rules out the possibility to describe all G-functions of order 2 with algebraic functions and algebraic pullbacks of Gauss's hypergeometric series only. This is in clear contrast with Gorelov's results.

Acknowledgments We thanks the referees for their comments that helped us to remove some inaccuracies.

References

1. M. Abramowitz, I. Stegun, *Handbook of Mathematical Functions with Formulas, Graphs, and Mathematical Tables*, 9th edition, 1970.
2. Y. André, *Séries Gevrey de type arithmétique I. Théorèmes de pureté et de dualité*, Ann. of Math. **151** (2000), 705–740.
3. D. Arinkin, *Rigid irregular connections on* $\mathbb{P}^1$ Compositio Math. **146** (2010), 1323–1338.
4. F. Beukers, *A refined version of the Siegel-Shidlovskii theorem*, Annals of Math. **163**.1 (2006), 369–379.
5. S. Bloch, H. Esnault, *Local Fourier Transforms and Rigidity*, Asian J. Math., Vol. 8, no. 4 (2004), 587–606.
6. G. V. Chudnovsky, *On applications of diophantine approximations*, Proc. Natl. Acad. Sci. USA **81** (1984), 7261–7265.
7. N. Katz, *Nilpotent connections and the monodromy theorem: applications of a result of Turritin*, Publ. Math. IHES, **32** (1970), 232–355.
8. N. Katz, *Exponential sums and differential equations*, Annals of Mathematical Studies, Princeton 1990.
9. D. Krammer, *An example of an arithmetic Fuchsian group*, J. reine angew. Math. **473** (1996), 69–85.
10. S. Fischler, T. Rivoal, *Arithmetic theory of E-operators*, Journal de l'École polytechnique - Mathématiques **3** (2016), 31–65.
11. V. A. Gorelov, *On the algebraic independence of values of E-functions at singular points and the Siegel conjecture*, Mat. Notes **67**.2 (2000), 174–190.
12. V. A. Gorelov, *On the Siegel conjecture for second-order homogeneous linear differential equations*, Math. Notes **75**.4 (2004), 513–529.
13. V. A. Gorelov, *On the structure of the set of E-functions satisfying linear differential equations of second order*, Math. Notes **78**.3 (2005), 304–319.
14. V. A. Gorelov, *On the algebraic independence of values of E-functions at singular points and the Siegel conjecture*, Mat. Notes **67**.2 (2000), 174–190.
15. É. Reyssat, *Irrationalité de* $\zeta(3)$ *selon Apéry*, Séminaire Delange-Pisot-Poitou, Théorie des nombres, **20** (1978–1979), Exposé No. 6, 6 p. Available at http://archive.numdam.org
16. T. Rivoal, J. Roques, *On the algebraic dependence of E-functions*, Bull. London Math. Soc. **48**.2 (2016), 271–279.
17. T. Rivoal, J. Roques, *Holomorphic solutions of E-operators*, Israel J. Math. **220**.1 (2017), 275–282
18. C. L. Siegel, *Über einige Anwendungen Diophantischer Approximationen*, Abh. Preuss. Akad.Wiss., Phys.-Math. Kl. (1929–30), no. 1, 1–70.
19. C. L. Siegel, *Transcendental Numbers*, Annals of Mathematical Studies, Princeton 1949.
20. A. B. Shidlovsky, *Transcendental Numbers*, W. de Gruyter Studies in Mathematics **12**, 1989.
21. M. van der Put, M. F. Singer, *Galois Theory of Linear Differential Equations*, Grundlehren der mathematischen Wissenschaften, Vol. 328, Springer, 2003.

Irrationality and Transcendence of Alternating Series via Continued Fractions

Jonathan Sondow

Abstract Euler gave recipes for converting alternating series of two types, I and II, into *equivalent* continued fractions, i.e., ones whose convergents equal the partial sums. A condition we prove for irrationality of a continued fraction then allows easy proofs that e, sin 1, and the primorial constant are irrational. Our main result is that, if a series of type II is equivalent to a *simple* continued fraction, then the sum is transcendental and its irrationality measure exceeds 2. We construct all $\aleph_0^{\aleph_0} = \mathfrak{c}$ such series and recover the transcendence of the Davison–Shallit and Cahen constants. Along the way, we mention π, the golden ratio, Fermat, Fibonacci, and Liouville numbers, Sylvester's sequence, Pierce expansions, Mahler's method, Engel series, and theorems of Lambert, Sierpiński, and Thue-Siegel-Roth. We also make three conjectures.

1 *Introduction*

In a 1979 lecture on the Life and Work of Leonhard Euler, André Weil suggested "that our students of mathematics would profit much more from a study of Euler's *Introductio in Analysin Infinitorum*, rather than of the available modern textbooks" [17, p. xii]. The last chapter of the *Introductio* is "On Continued Fractions." In it, after giving their form, Euler "next look[s] for an equivalent expression in the usual way of expressing fractions" and derives formulas for the convergents. He then converts a continued fraction into an *equivalent* alternating series, i.e., one whose partial sums equal the convergents. He "can now consider the converse problem. Given an

This manuscript was submitted posthumously.
The author passed away on January 16, 2020.

J. Sondow (✉)
New York, USA

A. Bostan and K. Raschel (eds.), *Transcendence in Algebra, Combinatorics, Geometry and Number Theory*, Springer Proceedings in Mathematics & Statistics 373,
https://doi.org/10.1007/978-3-030-84304-5_22

alternating series, find a continued fraction such that the series representing the value of the continued fraction is the given series."

In Proposition 1 and Theorem 1, we recall Euler's solutions for alternating series of two types, I and II. Lemma 1, a simplification of Nathan's theorem on irrationality of a continued fraction, then yields conditions for irrationality of the sum of a type I or II series. They easily imply the irrationality of e, sin 1, and the shifted-Fermat-number and primorial constants, and give a simple proof of Sierpiński's theorem.

Our main result is that, if a type II series is equivalent to a *simple* continued fraction, then the sum has irrationality measure greater than 2, and so must be transcendental, by the Thue-Siegel-Roth theorem on rational approximations to algebraic numbers.

Corollary 1 constructs all such series and shows that their sums form a continuum of distinct transcendental numbers, including the Davison-Shallit constant.

Corollary 2 gives explicitly the simple continued fractions for "naturally-occurring" transcendental numbers in a doubly-infinite family which contains Cahen's constant.

Finally, Proposition 2 provides irrationality and transcendence conditions for families of *non*-alternating series, including the Kellogg-Curtiss constant. Here the proofs involve partial sums instead of continued fractions.

Along the way, we encounter π, Fibonacci and golden rectangle numbers, an alternating Liouville constant, Sylvester's sequence, Pierce expansions, Mahler's method, and Engel series. We also make three conjectures; one on e^{-1} is an analog of Sondow's conjecture on e, recently proven by Berndt, Kim, and Zaharescu.

The rest of the paper is organized as follows. Lemma 1 and Proposition 1 are in Sect. 2; Theorem 1, Corollary 1, and Conjectures 1 and 2 are in Sect. 3; Corollary 2 is in Sect. 4; and Proposition 2 and Conjecture 3 are in Sect. 5.

2 Continued Fractions and Irrationality

In 1761 Lambert [26] derived a continued fraction for $\tan x$ and showed that its value is irrational for rational $x \neq 0$. Since $\tan \frac{\pi}{4} = 1$ is rational, Lambert had established that π *is irrational.* For modern treatments of his proof, see [15, §3.6] and [25].

Let us denote the positive integers by $\mathbb{N}$ and the rational numbers by $\mathbb{Q}$. Lemma 1 provides a sufficient condition for irrationality of the value of a continued fraction with all elements in $\mathbb{N}$. (Lambert's has both positive and negative elements.) The statement and quick proof are simplifications of Nathan's theorem in [28].

Lemma 1 (Irrationality Lemma) *Let α be the value of a continued fraction*

$$\alpha = \cfrac{b_1}{a_1 + \cfrac{b_2}{a_2 + \cdots}},$$

where $a_n \in \mathbb{N}$ and $b_n \in \mathbb{N}$ and $a_n \geq b_n$ for $n = 1, 2, 3, \ldots$. Then $\alpha \notin \mathbb{Q}$.

Proof If $\alpha \in \mathbb{Q}$, define the nth "tail" of α to be the value of the continued fraction

$$\alpha_n := \cfrac{b_{n+1}}{a_{n+1} + \cfrac{b_{n+2}}{a_{n+2} + \ddots}}, \quad \text{so} \quad \alpha_n = \frac{b_{n+1}}{a_{n+1} + \alpha_{n+1}}, \tag{1}$$

for all $n \geq 0$. The hypotheses ensure that $0 < \alpha_n < 1$ for all $n \geq 0$. As $\alpha_0 = \alpha$, and $\alpha_n \in \mathbb{Q}$ implies $\alpha_{n+1} \in \mathbb{Q}$, we can write $\alpha_n = u_n/v_n$, where u_n and v_n are coprime positive integers with $u_n < v_n$. Thus from (1) we get

$$\frac{u_{n+1}}{v_{n+1}} = \alpha_{n+1} = \frac{v_n b_{n+1} - u_n a_{n+1}}{u_n},$$

so $u_{n+1} < v_{n+1} \leq u_n$. But then $(u_n)_{n \geq 0}$ is a strictly decreasing, infinite sequence of positive integers, which is impossible. Therefore, $\alpha \notin \mathbb{Q}$. ■

For instance, if $a_n = 1$ and $b_n = 1$ for all n, then by Lemma 1

$$\alpha = \cfrac{1}{1 + \cfrac{1}{1 + \ddots}} = \frac{1}{1+\alpha} > 0 \quad \Longrightarrow \quad \alpha = \frac{\sqrt{5}-1}{2} \notin \mathbb{Q}.$$

Thus *the golden ratio* $\varphi := \alpha^{-1}$ *is irrational.* For more on φ, see Examples 2 and 6.

Lemma 1 generalizes the irrationality of an infinite *simple* continued fraction, i.e., one with all *partial numerators* $b_n = 1$ and all *partial quotients* (or *partial denominators*) $a_n \in \mathbb{N}$.

Our hypothesis $a_n \geq b_n$ is weaker than Nathan's $a_n > b_n$. Ours is also *sharp*: with the even weaker hypothesis $a_n \geq b_n - 1$, the lemma would be false, e.g.,

$$\alpha = \cfrac{2}{1 + \cfrac{2}{1 + \ddots}} = \frac{2}{1+\alpha} > 0 \quad \Longrightarrow \quad \alpha = 1 \in \mathbb{Q}.$$

Lemma 1 holds more generally when $a_n \geq b_n$ for all *sufficiently large* n. There is also a condition for irrationality of a continued fraction with both positive and negative integers a_n and b_n, namely, that $|a_n| \geq |b_n| + 1$; see, e.g., [15, §3.6]. We have chosen simplicity over generality here and elsewhere in the paper.

We now apply the Irrationality Lemma to our first kind of alternating series, type I.

Proposition 1 *Let* $B_0 < B_1 < B_2 < \cdots$ *be positive integers.*

(i). Then there is an equivalence

$$\alpha := \frac{1}{B_0} - \frac{1}{B_1} + \frac{1}{B_2} - \cdots \cong \cfrac{1}{B_0 + \cfrac{B_0^2}{B_1 - B_0 + \cfrac{B_1^2}{B_2 - B_1 + \cdots}}}. \quad (2)$$

(ii). Suppose that

$$B_{n+1} \geq B_n(B_n + 1) \text{ for all } n \geq 0. \quad (3)$$

Then the sum α *is irrational.*

Proof (i). Euler establishes the equivalence in [17, §369]; for example,

$$\frac{1}{B_0} - \frac{1}{B_1} = \cfrac{1}{B_0 + \cfrac{B_0^2}{B_1 - B_0}}.$$

(ii). Set $a_1 = B_0, b_1 = 1, a_{n+1} = B_n - B_{n-1}$, and $b_{n+1} = B_{n-1}^2$ for $n \geq 1$. Then (3) guarantees that $a_n \geq b_n$ for all n, so by Lemma 1 the value of the continued fraction in (2) is irrational. By (i), that value equals the sum α, so $\alpha \notin \mathbb{Q}$. ■

Proposition 1 provides an easy proof of *Sierpiński's theorem*, which states that, *if* (3) *holds with all* $B_n \in \mathbb{N}$, *then* $\alpha := \sum_{n=0}^{\infty}(-1)^n B_n^{-1} \notin \mathbb{Q}$. Sierpiński [34] (see also Cahen [9]) showed moreover that *such a representation of any irrational number* α *in* $(0, 1)$ *exists and is unique*. For extensions of his theorem, see Badea [2], Duverney [13], and Nyblom [29].

Note that *part (ii) and Sierpiński's theorem are sharp*: if $B_{n+1} + 1 = B_n(B_n + 1)$ for all $n \geq 0$, then $(B_{n+1} + 1)^{-1} = B_n^{-1} - (B_n + 1)^{-1}$, so by telescoping

$$\sum_{n=0}^{\infty} \frac{(-1)^n}{B_n} = \sum_{n=0}^{\infty} \left(\frac{(-1)^n}{B_n + 1} + \frac{(-1)^n}{B_{n+1} + 1} \right) = \frac{1}{B_0 + 1} \in \mathbb{Q}.$$

Example 1 The *Fermat numbers* $F_n = 2^{2^n} + 1$ form the sequence [35, A000215]

$$(F_n)_{n \geq 0} = 3, 5, 17, 257, 65537, 4294967297, 18446744073709551617, \ldots.$$

Let us define the *shifted-Fermat-number constant* F to be the alternating sum of reciprocals of the numbers $F_n - 2$ (for them, see [35, A051179])

$$F := \sum_{n=0}^{\infty} \frac{(-1)^n}{F_n - 2} = \sum_{n=0}^{\infty} \frac{(-1)^n}{2^{2^n} - 1} = 1 - \frac{1}{3} + \frac{1}{15} - \frac{1}{255} + \cdots = 0.7294270\ldots.$$

The numbers $B_n := 2^{2^n} - 1$ satisfy (3), so *the shifted-Fermat-number constant F is irrational.* For a generalization with a different proof, take $\epsilon = -1$ in [13, Corollary 3.3]. We return to F in Example 5.

The next section studies irrationality and transcendence of our second kind of alternating series, type II, which is a special case of type I.

3 Simple Continued Fractions and Transcendence

Our main results are Theorem 1 and Corollaries 1 and 2. We denote the algebraic numbers by $\mathbb{A}$ (others denote them by $\overline{\mathbb{Q}}$, the algebraic closure of $\mathbb{Q}$).

Theorem 1 *Fix positive integers $A_0, A_1, A_2, \ldots$, with $A_n \geq 2$ for all $n \geq 1$.*

(i). For any positive real numbers $x_0, x_1, x_2, \ldots$, we have the equivalence between an alternating series and a continued fraction

$$\alpha := \sum_{n=0}^{\infty} \frac{(-1)^n}{A_0 A_1 \cdots A_n} \cong \cfrac{x_0}{A_0 x_0 + \cfrac{A_0 x_0 x_1}{(A_1 - 1)x_1 + \cfrac{A_1 x_1 x_2}{(A_2 - 1)x_2 + \cdots}}}. \tag{4}$$

(ii). If $A_{n+1} > A_n$ for all $n \geq 0$, then α is irrational.

(iii). If the continued fraction is simple for some $x_0, x_1, x_2, \ldots$, then α is a transcendental number, with irrationality measure $\mu(\alpha) \geq 2.5$.

The *irrationality measure* (or *irrationality exponent*) $\mu(\rho)$ of a real number ρ is defined as (see [3, 4, 7], [8, §1.4], [15, Chapter 9], [18, §2.22], [36])

$$\mu(\rho) := \sup\left\{\mu > 0 : 0 < \left|\rho - \frac{p}{q}\right| < \frac{1}{q^{\mu}} \text{ for infinitely many } \frac{p}{q} \in \mathbb{Q}\right\}. \tag{5}$$

By the famous *Thue-Siegel-Roth theorem* [1], [8, p. 22], [15, p. 147], [18, p. 172], [22, p. 176]

$$\mu(\rho) \begin{cases} = 1 & \text{if } \rho \text{ is rational,} \\ = 2 & \text{if } \rho \text{ is irrational, but algebraic,} \\ \geq 2 & \text{if } \rho \text{ is transcendental.} \end{cases}$$

Proof (of Theorem 1)

(i). Apply Proposition 1, part (i), with $B_n := A_0 A_1 \cdots A_n$ for $n \geq 0$. Since $B_n - B_{n-1} = (A_n - 1)A_0 A_1 \cdots A_{n-1}$, cancelling the common factors A_0, $A_0 A_1$, $A_0 A_1 A_2, \ldots$ in the resulting continued fraction gives

$$\frac{1}{A_0} - \frac{1}{A_0A_1} + \frac{1}{A_0A_1A_2} - \cdots \cong \cfrac{1}{A_0 + \cfrac{A_0^{\cancel{2}}}{(A_1-1)\cancel{A_0} + \cfrac{A_0^{\cancel{2}}A_1^2}{(A_2-1)A_0A_1 + \ddots}}}$$

$$\cong \cfrac{1}{A_0 + \cfrac{A_0}{A_1 - 1 + \cfrac{\cancel{A_0}A_1^{\cancel{2}}}{(A_2-1)\cancel{A_0A_1} + \ddots}}} \cong \cfrac{1}{A_0 + \cfrac{A_0}{A_1 - 1 + \cfrac{A_1}{A_2 - 1 + \ddots}}},$$

where "$\cong$" between two continued fractions means they are *equivalent*, i.e., they have the same convergents (see [15, p. 25]; for two numerical continued fractions which are equivalent but not equal, see Example 4 below). This proves the special case of (i) in which all $x_n = 1$ (compare to [17, §370]). The general case follows by cancelling the common factors $x_0, x_1, x_2, \ldots$ in (4).

(ii). In Lemma 1, we take $a_1 := A_0$, $b_1 := 1$, $a_n := A_{n-1} - 1$, and $b_n := A_{n-2}$ for $n \geq 2$. Then $A_{n+1} > A_n$ implies $a_n \geq b_n$ for all $n \geq 1$, so $\alpha \notin \mathbb{Q}$.

(iii). (Compare to the proof of [12, Theorem 3].) Redefining $a_1, a_2, \ldots$, we write the simple continued fraction for α, and its nth convergent, as usual as

$$\alpha = 0 + \cfrac{1}{a_1 + \cfrac{1}{a_2 + \ddots}} = [0,\ a_1,\ a_2,\ \ldots] \quad \text{and} \quad \frac{p_n}{q_n} = [0,\ a_1,\ a_2,\ \ldots,\ a_n].$$

The hypothesis in (iii) means that

$$\sum_{i=0}^{n} \frac{(-1)^i}{A_0A_1 \cdots A_i} = \frac{p_{n+1}}{q_{n+1}} \quad \text{for } n \geq 0. \tag{6}$$

A classical theorem [22, Theorem 150] and relation (6) imply, respectively, that

$$\frac{(-1)^n}{q_nq_{n+1}} = \frac{p_{n+1}}{q_{n+1}} - \frac{p_n}{q_n} = \frac{(-1)^n}{A_0A_1 \cdots A_n} \tag{7}$$

for $n \geq 1$. Hence $q_nq_{n+1} = A_0A_1 \cdots A_n$; since $q_0 = 1$ and $q_1 = A_0$, this also holds for $n = 0$. It follows that the divisibility $q_nq_{n+1} \mid q_{n+1}q_{n+2}$ holds; hence $q_n \mid q_{n+2}$. A standard identity [22, Theorem 149] is

$$q_{n+2} = a_{n+2}q_{n+1} + q_n, \tag{8}$$

so $q_n \mid a_{n+2}q_{n+1}$. Multiplying (7) by q_nq_{n+1}, we deduce that $\gcd(q_n, q_{n+1}) = 1$, so $q_n \mid a_{n+2}$. Define $w_0, w_1, \ldots$ in $\mathbb{N}$ by $w_0 = a_1$ and $w_{n+1}q_n = a_{n+2}$ for $n \geq 0$. By a "simple lemma" [12, Lemma 2],

$$w_n q_{n-1} \geq \sqrt{q_n} \quad \text{for infinitely many } n. \tag{9}$$

Now, from (6), a classical inequality [8, p. 24], the equality $a_{n+1} = w_n q_{n-1}$, and (9), respectively, we see that

$$0 < \left| \alpha - \frac{p_n}{q_n} \right| < \frac{1}{a_{n+1} q_n^2} = \frac{1}{w_n q_{n-1} q_n^2} \leq \frac{1}{q_n^{5/2}}$$

infinitely often. This and definition (5) imply $\mu(\alpha) \geq 2.5$. By the Thue-Siegel-Roth theorem, $\mu(\rho) \leq 2$ if $\rho \in \mathbb{A}$, so $\alpha \notin \mathbb{A}$. This completes the proof of Theorem 1. ■

Note that *the hypothesis in (ii) is sharp*: if $A_n = A_0 > 1$ for all $n > 0$, then the series in (4) is geometric, with sum $\alpha = (A_0 + 1)^{-1} \in \mathbb{Q}$. Also, in (ii) the inequality $A_{n+1} > A_n$ is much weaker than that in (3) with $B_n = A_0 A_1 \cdots A_n$, which amounts to $A_{n+1} > A_0 A_1 \cdots A_n$. Compare Examples 1 and 5.

For any strictly increasing sequence of positive integers $A_0 < A_1 < A_2 < \cdots$, finite or infinite, the alternating sum

$$\alpha := \frac{1}{A_0} - \frac{1}{A_0 A_1} + \frac{1}{A_0 A_1 A_2} - \cdots$$

is called the *Pierce expansion* of α. *Any number* $\alpha \in (0, 1)$ *has a unique Pierce expansion, which is infinite if, and only if,* α *is irrational* [30–32, 34]. The "only if" part follows immediately from (ii).

Example 2 The Pierce expansion of φ^{-1} begins [35, A118242, A006276]

$$\frac{1}{\varphi} = \frac{1}{1} - \frac{1}{1 \cdot 2} + \frac{1}{1 \cdot 2 \cdot 4} - \frac{1}{1 \cdot 2 \cdot 4 \cdot 17} + \frac{1}{1 \cdot 2 \cdot 4 \cdot 17 \cdot 19} - \cdots.$$

As $\varphi^{-1} \in \mathbb{A}$, we see that *the hypothesis in (iii) cannot be omitted.* Combined with the next example, this shows that, *if the Pierce expansion of* $\alpha \notin \mathbb{Q}$ *is not equivalent to a simple continued fraction, then* $\alpha \in \mathbb{A}$ *is possible, but so is* $\alpha \notin \mathbb{A}$.

Example 3 Euler [17, p. 325] says, "Something especially deserving of our attention is the number e ...." The Taylor series $e^t = \sum_{n=0}^{\infty} t^n n!^{-1}$ and (i) lead to the Pierce expansion of e^{-1} and the equivalence

$$e^{-1} = \sum_{n=2}^{\infty} \frac{(-1)^n}{2 \cdot 3 \cdot 4 \cdots n} \cong \cfrac{x_0}{2x_0 + \cfrac{2x_0 x_1}{2x_1 + \cfrac{3x_1 x_2}{3x_2 + \cfrac{4x_2 x_3}{4x_3 + \ddots}}}}. \tag{10}$$

Part (ii) now gives an easy proof that *e is irrational.* The Taylor series for $\sin t$ and $\cos t$ lead to similar proofs that $\sin \frac{1}{k}$ *and* $\cos \frac{1}{k}$ *are irrational for all* $k \in \mathbb{N}$.

From (10) we also see that a strong converse to (iii) is not true. Namely, *although* $e^{-1} \notin \mathbb{A}$ (because $e \notin \mathbb{A}$ by Hermite [15, §12.14]), *the type II series for* e^{-1} *in* (10) *is not equivalent to a simple continued fraction.* Indeed, when $x_0, x_1, \ldots$ are chosen so that all partial numerators in the continued fraction for e^{-1} in (10) equal 1

$$\frac{1}{e} = \cfrac{1}{2 + \cfrac{2(1/2)}{2(1/2) + \cfrac{3(1/2)(2/3)}{3(2/3) + \cfrac{4(2/3)(3/8)}{4(3/8) + \ddots}}}} = \cfrac{1}{2 + \cfrac{1}{1 + \cfrac{1}{2 + \cfrac{1}{(3/2) + \ddots}}}} \tag{11}$$

the partial quotients do not all lie in $\mathbb{N}$. For a weaker converse to (iii), which is also not true, see Example 8.

By (11), the simple continued fraction for e^{-1} begins $e^{-1} = [0, 2, 1, 2, \ldots]$. From (10) (or by inspection), the first four convergents are also partial sums of the Taylor series $e^{-1} = \sum_{n=0}^{\infty}(-1)^n n!^{-1}$.

Conjecture 1 Only four partial sums of the Taylor series for e^{-1} are convergents to e^{-1}, namely, 0, 1/2, 1/3, and 3/8.

Conjecture 1 is an analog for e^{-1} of the fact that *only two partial sums of the Taylor series for e are convergents to e, namely,* 2 *and* 8/3. This property of e was conjectured by Sondow [36], partially proven by him and Schalm [37], and recently proven in full by Berndt, Kim, and Zaharescu [6].

Example 4 An analog of series (10) for e^{-1}, with the factorial $n!$ replaced by the primorial $p_n\#$, is "the constant obtained through Pierce retro-expansion of the prime sequence" [35, A132120], which we dub the *primorial constant*

$$P := \sum_{n=1}^{\infty} \frac{(-1)^{n-1}}{p_n\#} = \frac{1}{2} - \frac{1}{2\cdot 3} + \frac{1}{2\cdot 3\cdot 5} - \frac{1}{2\cdot 3\cdot 5\cdot 7} + \frac{1}{2\cdot 3\cdot 5\cdot 7\cdot 11} - \cdots$$
$$= \frac{1}{2} - \frac{1}{6} + \frac{1}{30} - \frac{1}{210} + \frac{1}{2310} - \cdots = 0.3623062223\ldots.$$

Proposition 1, part (i), and Theorem 1, parts (i) and (ii), imply that

$$P = \cfrac{1}{2 + \cfrac{2^2}{4 + \cfrac{6^2}{24 + \cfrac{30^2}{180 + \cfrac{210^2}{2100 + \ddots}}}}} \cong \cfrac{1}{2 + \cfrac{2}{2 + \cfrac{3}{4 + \cfrac{5}{6 + \cfrac{7}{10 + \ddots}}}}} \notin \mathbb{Q}.$$

Conjecture 2 The primorial constant P is transcendental.

Example 5 By induction, for $n \geq 0$ the shifted Fermat number $F_n - 2$ can be factored as the product of all smaller Fermat numbers

$$F_n - 2 = 2^{2^n} - 1 = \prod_{k=0}^{n-1} \left(2^{2^k} + 1\right) = F_0 F_1 \cdots F_{n-1}, \tag{12}$$

where the empty product equals 1 when $n = 0$. (From (12) Pólya deduced that $F_0, F_1, F_2, \ldots$ are pairwise coprime, thereby giving an alternate proof to Euclid's theorem on the infinitude of the primes [22, §2.4].) The constant F in Example 1 thus has Pierce expansion

$$F = \sum_{n=0}^{\infty} \frac{(-1)^n}{F_0 F_1 \cdots F_{n-1}} = \frac{1}{1} - \frac{1}{1 \cdot 3} + \frac{1}{1 \cdot 3 \cdot 5} - \frac{1}{1 \cdot 3 \cdot 5 \cdot 17} + \cdots.$$

Part (ii) of Theorem 1 now gives a second proof that $F \notin \mathbb{Q}$. Moreover, parts (i) of Proposition 1 and Theorem 1 yield the equivalent continued fractions

$$F = \cfrac{1}{1 + \cfrac{1^2}{2 + \cfrac{3^2}{12 + \cfrac{15^2}{240 + \ddots}}}} \cong \cfrac{1}{1 + \cfrac{1}{2 + \cfrac{3}{4 + \cfrac{5}{16 + \ddots}}}}.$$

Theorem 1 does not yield $F \notin \mathbb{A}$, but Duverney [16] has proven it by other methods.

Remark 1 *Non*-alternating series involving F_n have also been studied. In 1963, Golomb [21] proved that *the sum* $G := \sum_{n=0}^{\infty} F_n^{-1}$ *is irrational*. Two years later, Mahler [27] remarked that G *is in fact transcendental*, as a consequence of a general theorem he proved in 1929—see [14, pp. 194–195]. (*Mahler's method* [15, §12.3] proves the transcendence of values, at certain algebraic points, of functions that satisfy a type of functional equation.) Recently, Coons [10] showed that G *has irrationality measure* $\mu(G) = 2$. In the pre-Mahler year 1916, Kempner [24] proved that *the number* $\kappa := \sum_{n=0}^{\infty} (F_n - 1)^{-1} = \sum_{n=0}^{\infty} 2^{-2^n}$ *is transcendental*; see Adamczewski [1] for five proofs with interesting comments. (The second proof applies Mahler's method to the function $f(x) := \sum_{n=0}^{\infty} x^{2^n}$, which is defined when $|x| < 1$, satisfies the functional equation $f(x^2) = f(x) - x$, and has the value $f(1/2) = \kappa$.)

The next example shows that *the sufficient condition for transcendence of the sum of a type II series in Theorem* 1 *does not extend to the more general type I series in Proposition* 1.

Example 6 Let $(f_n)_{n \geq 0} = 1, 1, 2, 3, 5, 8, 13, \ldots$ be the positive *Fibonacci numbers* [35, A000045], defined by $f_0 = 1$, $f_1 = 1$, and $f_{n+1} = f_n + f_{n-1}$ for $n \geq 1$. The product $B_n := f_n f_{n+1}$ is a *golden rectangle number* [35, A001654]. The difference between successive golden rectangle numbers is a square:

$$B_n - B_{n-1} = f_n f_{n+1} - f_{n-1} f_n = f_n(f_{n+1} - f_{n-1}) = f_n^2. \tag{13}$$

Therefore, using Proposition 1, part (i), and cancelling common factors $f_1^2, f_2^2, \ldots,$ we obtain the equivalence

$$\alpha := \sum_{n=0}^{\infty} \frac{(-1)^n}{f_n f_{n+1}} \cong \cfrac{1}{f_0 f_1 + \cfrac{f_0^2 \cancel{f_1^2}}{\cancel{f_1^2} + \cfrac{\cancel{f_1^2}\cancel{f_2^2}}{\cancel{f_2^2} + \ddots}}} = [0, 1, 1, 1, \ldots].$$

The latter is the simple continued fraction expansion of $\alpha = \varphi^{-1} \in \mathbb{A}$. This shows that, *given* $B_0 < B_1 < B_2 < \cdots$ *in* $\mathbb{N}$, *the sum of the series* $\alpha := \sum_{n=0}^{\infty}(-1)^n B_n^{-1}$ *might not be transcendental, even if the series is equivalent to a simple continued fraction.* (However, *if in addition* B_{n-1} *divides* B_n *for all* $n \geq 1$, *then* $\alpha \notin \mathbb{A}$, by Theorem 1 with $A_0 := B_0$ and $A_n := B_n/B_{n-1}$ for $n \geq 1$.)

Remark 2 Example 6 is a special case of the following well-known fact. *For any irrational number* ρ *with simple continued fraction expansion* $\rho = [a_0, a_1, a_2, \ldots]$ *and nth convergent* p_n/q_n, *there is an equivalence*

$$\rho = a_0 + \sum_{n=0}^{\infty} \frac{(-1)^n}{q_n q_{n+1}} \cong [a_0, a_1, a_2, \ldots].$$

(*Proof* Replacing ρ with $\rho - a_0$, we may assume that $a_0 = 0$. Note that $q_0 = 1$. Setting $B_n = q_n q_{n+1}$, we use (8) to get $B_n - B_{n-1} = a_{n+1} q_n^2$, generalizing relation (13). The rest of the proof is like the argument in Example 6, and is omitted.)

By part (i) of Theorem 1, *if the continued fraction in* (4) *is simple for some* $x_0, x_1, \ldots,$ *then the series in* (4) *is equivalent to a simple continued fraction*, i.e., (6) holds. Conversely, it is not hard to show by induction that, *if* (6) *holds, then the continued fraction in* (4) *is simple for some* $x_0, x_1, \ldots$. For instance, if the partial sums A_0^{-1} and $A_0^{-1} - (A_0 A_1)^{-1}$ equal the convergents a_1^{-1} and $(a_1 + a_2^{-1})^{-1}$, respectively, then $A_0 = a_1$ and $(A_1 - 1)A_0^{-1} = a_2 \in \mathbb{N}$, so the choices $x_0 = 1$ and $x_1 = A_0^{-1}$ give the finite simple continued fraction

$$\cfrac{x_0}{A_0 x_0 + \cfrac{A_0 x_0 x_1}{(A_1 - 1)x_1}} = \cfrac{1}{A_0 + \cfrac{1}{(A_1 - 1)A_0^{-1}}} = [0, a_1, a_2].$$

We now give a method for constructing all examples of Theorem 1, part (iii).

Corollary 1 *(i). Construct a sequence of positive integers* $(A_n)_{n \geq 0}$ *in three steps.*

Step 1. Choose a sequence $(M_n)_{n \geq 0}$ *with all* $M_n \in \mathbb{N}$.

Step 2. Let $(N_n)_{n\geq 1}$ *satisfy the recursion*

$$N_1 = 1, N_2 = M_0, \quad \textit{and} \quad N_{n+2} = (M_n N_{n+1} + 1)N_n \quad \textit{for} \quad n \geq 1. \tag{14}$$

Step 3. Define $(A_n)_{n\geq 0}$ *by*

$$A_0 = M_0 \quad \textit{and} \quad A_n = M_n N_{n+1} + 1 \quad \textit{for} \quad n \geq 1. \tag{15}$$

Then there exists $(x_n)_{n\geq 0}$ *such that* (4) *is an equivalence between an alternating series and a simple continued fraction, namely,*

$$\alpha := \sum_{n=0}^{\infty} \frac{(-1)^n}{A_0 A_1 \cdots A_n} \cong [0, M_0, M_1 N_1, M_2 N_2, M_3 N_3, \ldots]. \tag{16}$$

(ii). Conversely, if the continued fraction in (4) *is simple for some* $(x_n)_{n\geq 0}$*, then the sequence* $(A_n)_{n\geq 0}$ *in* (4) *can be constructed by Steps 1, 2, 3.*

(iii). The series in (16) *is the Pierce expansion of* α*, that is,* $A_{n+1} > A_n$ *for* $n \geq 0$*.*

(iv). Distinct sequences $(M_n)_{n\geq 0} \neq (M'_n)_{n\geq 0}$ *in Step 1 lead to distinct transcendental numbers* $\alpha \neq \alpha'$ *in* (16)*. In particular, if* $\mathbb{S}$ *denotes the set of real numbers* α *whose Pierce expansion is equivalent to a simple continued fraction, then* $\#\mathbb{S} = \aleph_0^{\aleph_0} = \mathfrak{c}$.

Proof By definition, the continued fraction in (4) is simple if, and only if,

$$\begin{aligned} &(a),\ x_0 = 1,\\ &(b),\ A_n x_n x_{n+1} = 1 \quad \text{for } n \geq 0,\\ &(c),\ A_0 x_0 \in \mathbb{N}, \quad \text{and}\\ &(d),\ (A_n - 1)x_n \in \mathbb{N} \quad \text{for } n \geq 1. \end{aligned}$$

(i), Set $x_0 = 1$ and $x_n = N_n/N_{n+1}$ for $n \geq 1$. From formulas (15) and (14) we get $A_n = N_{n+2}/N_n$ for $n \geq 1$. It is now easy to verify (a), (b), (c), and (d). Observing that $(A_n - 1)x_n = M_n N_n$ for $n \geq 1$, the equivalence (4) gives (16). This proves (i).

(ii), Assume (a), (b), (c), and (d). Then $A_n \in \mathbb{N}$ implies that $x_n \in \mathbb{Q}$ for $n \geq 1$, so $x_n = N_n/D_n$, where $N_n \in \mathbb{N}$ and $D_n \in \mathbb{N}$, with $\gcd(N_n, D_n) = 1$. From (a) and (b), we get $N_1 = 1$ and $D_1 = A_0$. From (d), we see that $D_n \mid (A_n - 1)$ for $n \geq 1$, so there exists $M_n \in \mathbb{N}$ such that $A_n = M_n D_n + 1$. Since (b) implies $A_n N_n N_{n+1} = D_n D_{n+1}$, we get

$$(M_n D_n + 1)N_n N_{n+1} = D_n D_{n+1} \quad \text{for } n \geq 1. \tag{17}$$

Consequently, $N_{n+1} \mid D_n D_{n+1}$, so $N_{n+1} \mid D_n$. Also, $D_n \mid (M_n D_n + 1)N_n N_{n+1}$, so $D_n \mid N_{n+1}$. Thus $D_n = N_{n+1}$ for all $n \geq 1$; in particular, $N_2 = D_1 = A_0$.

Making replacements in (17) and in $A_n = M_n D_n + 1$, we obtain (14) and (15), respectively. This proves (ii).

(iii), Note that (14) and (15) give $A_{n+1} = M_{n+1} A_n N_n + 1 > A_n$ for $n \geq 0$.

(iv), By Theorem 1, the sum α is transcendental. It now suffices to show that, given $\alpha = [0, M_0, M_1 N_1, M_2 N_2, \dots]$ and $\alpha' = [0, M'_0, M'_1 N'_1, M'_2 N'_2, \dots]$, if $\alpha = \alpha'$, then $M_n = M'_n$ for all $n \geq 0$. By the uniqueness of simple continued fraction expansion, $M_0 = M'_0$ and $M_k N_k = M'_k N'_k$ for $k \geq 1$. Using (14), the rest of the proof is an easy induction, which we omit. This completes the proof of the corollary. ■

Example 7 Choosing the constant sequence $M_n = 1$ yields $N_1 = 1, N_2 = 1$, and $N_{n+2} = (N_{n+1} + 1) N_n$ for $n \geq 1$. Then $A_0 = 1$ and $A_n = N_{n+1} + 1$ for $n \geq 1$, so

$$A_n = 1, 2, 3, 4, 9, 28, 225, 6076, 1361025, \dots$$

(see [35, A007704]). By (iv), we recover the transcendence of the *Davison-Shallit constant* [12, Example A] (see also [18, pp. 436, 445], [35, A242724])

$$D := \sum_{n=0}^{\infty} \frac{(-1)^n}{A_0 A_1 \cdots A_n} = 1 - \frac{1}{2} + \frac{1}{6} - \frac{1}{24} + \frac{1}{216} - \cdots = 0.62946502045\dots$$

and, by (i), the expansion [12, p. 122], [35, A006277]

$$D = [0, 1, N_1, N_2, N_3, \dots] = [0, 1, 1, 1, 2, 3, 8, 27, 224, 6075, 1361024, \dots].$$

Example 8 Let us define an *alternating Liouville constant* by the series

$$\begin{aligned} \lambda := \sum_{n=2}^{\infty} \frac{(-1)^n}{10^{n!}} &= \frac{1}{10^2} - \frac{1}{10^6} + \frac{1}{10^{24}} - \frac{1}{10^{120}} + \cdots \\ &= 0.009999000000000000000000000\underbrace{99\dots9}_{96}00\dots. \end{aligned}$$

For $n = 1, 2, 3, \dots$, the nth partial sum of the series satisfies

$$\frac{P_n}{Q_n} := \sum_{k=2}^{n+1} \frac{(-1)^k}{10^{k!}} \implies 0 < \left| \lambda - \frac{P_n}{Q_n} \right| < \frac{1}{10^{(n+2)!}} = \frac{1}{Q_n^{n+2}}.$$

From this and (5), we infer that λ has irrationality measure $\mu(\lambda) = \infty$. By definition, λ is therefore a *Liouville number*, so *Liouville's theorem* [8, §1.4], [15, §9.3], [22, §11.7] (or its descendant, the Thue-Siegel-Roth theorem) implies λ is transcendental.

On the other hand, its Pierce expansion

$$\lambda = \sum_{n=0}^{\infty} \frac{(-1)^n}{A_0 A_1 \cdots A_n} = \frac{1}{10^{2!}} - \frac{1}{10^{2!}10^{3!-2!}} + \frac{1}{10^{2!}10^{3!-2!}10^{4!-3!}} - \cdots \qquad (18)$$

cannot be constructed from any sequence $(M_n)_{n\geq 0}$ as in (i). (*Proof.* If it could, then $M_0 = A_0 = 10^{2!}$ would imply $M_1 10^{2!} + 1 = A_1 = 10^{3!-2!}$, contradicting $M_1 \in \mathbb{N}$.)

Hence by (ii) a converse to Theorem 1, part (iii), weaker than the false converse in Example 3, is also not true. Namely, *although* $\lambda \notin \mathbb{A}$ *and* $\mu(\lambda) \geq 2.5$, *the type II series for* λ *in* (18) *is not equivalent to a simple continued fraction.*

More positively, one can show that, *if a sequence* $(M_n)_{n\geq 0}$ *in (i) grows sufficiently rapidly, then the sum* α *in the equivalence* (16) *is a Liouville number.*

The next section gives further applications of Theorem 1.

4 Sylvester's Sequence and Cahen's Constant

There are not many "naturally-occurring" transcendental numbers for which the simple continued fraction is known explicitly. They include the beautiful expansions

$$\begin{aligned} e - 1 &= [1, 1, 2, 1, 1, 4, 1, 1, 6, 1, 1, 8, 1, 1, 10, 1, 1, 12, 1, 1, 14, \ldots], \\ \tan 1 &= [1, 1, 1, 3, 1, 5, 1, 7, 1, 9, 1, 11, 1, 13, 1, 15, 1, 17, 1, 19, \ldots], \\ 1/\tanh 1 &= [1, 3, 5, 7, 9, 11, 13, 15, 17, 19, 21, 23, 25, 27, 29, 31, 33, \ldots], \\ I_0(2)/I_1(2) &= [1, 2, 3, 4, 5, 6, 7, 8, 9, 10, 11, 12, 13, 14, 15, 16, 17, 18, \ldots], \end{aligned}$$

and those of $e^{2/q}$, $\tan \frac{1}{q}$, $\tanh \frac{1}{q}$, and $I_{\frac{p}{q}}(\frac{2}{q})/I_{1+\frac{p}{q}}(\frac{2}{q})$, for p and q in $\mathbb{N}$, where $I_c(x)$ is a modified, or hyperbolic, Bessel function of the first kind [15, Chapter 3]. References to several others are given in [12, §V].

Theorem 1 yields a doubly-infinite family of such numbers. We define them by a natural recursion, independently of Corollary 1.

Corollary 2 *Fix* $k \in \mathbb{N}$ *and* $\ell \in \mathbb{N}$. *For* $n \geq 0$, *define* $s_n = s_n(k, \ell)$ *by the recurrence*

$$s_0 = k \;\; \text{and} \;\; s_n = (s_0 s_1 \cdots s_{n-1})^{\ell} + 1 \;\; \text{for} \;\; n \geq 1. \qquad (19)$$

(i), Then there is an equivalence

$$C_{k,\ell} := \sum_{n=0}^{\infty} \frac{(-1)^n}{s_{n+1} - 1} \cong [a_0, a_1, a_2, \ldots],$$

where the partial quotients of the simple continued fraction are

$$a_0 = 0, \; a_1 = s_0^{\ell}, \; \text{and} \; a_{n+1} = (s_n^{\ell} - 1) \prod_{i=0}^{n-1} (s_i^{\ell})^{(-1)^{n+i}} \in \mathbb{N} \;\; \text{for} \;\; n \geq 1.$$

(ii), *The sum* $C_{k,\ell}$ *is transcendental, and* $C_{k,\ell} = C_{k',\ell'}$ *only when* $(k,\ell) = (k',\ell')$.
(iii), *The double-exponential lower bound* $a_n > (k^\ell + 1)^{(\ell+1)^{n-4}}$ *holds for all* $n \geq 4$.
(iv), *There are the summations*

$$\sum_{n=0}^{\infty} \frac{s_n^\ell - 1}{s_{n+1} - 1} = 1 \quad \textit{and} \quad \sum_{n=0}^{\infty} \frac{s_{2n+1}^\ell - 1}{s_{2n+2} - 1} = C_{k,\ell}.$$

(v), *Taking* $\ell = 1$ *gives*

$$C_{k,1} = \frac{1}{s_0} - \frac{1}{s_0 s_1} + \frac{1}{s_0 s_1 s_2} - \frac{1}{s_0 s_1 s_2 s_3} + \frac{1}{s_0 s_1 s_2 s_3 s_4} - \frac{1}{s_0 s_1 s_2 s_3 s_4 s_5} + \cdots$$
$$\cong [0, s_0, 1, (s_0)^2, (s_1)^2, (s_0 s_2)^2, (s_1 s_3)^2, (s_0 s_2 s_4)^2, (s_1 s_3 s_5)^2, \ldots].$$

(vi), *For odd* $n \geq 1$ *and even* $m \geq 2$, *the partial quotients* a_n *and* a_m *of* $C_{k,1}$ *are coprime.*

Proof (i), Set $A_n := s_n^\ell$ for $n \geq 0$. Then (19) gives $s_{n+1} - 1 = A_0 A_1 \cdots A_n$, so by Theorem 1, for any $x_0, x_1, \ldots$ in $\mathbb{R}^+$ there is an equivalence

$$C_{k,\ell} = \sum_{n=0}^{\infty} \frac{(-1)^n}{s_0^\ell s_1^\ell \cdots s_n^\ell} \cong \cfrac{x_0}{s_0^\ell x_0 + \cfrac{s_0^\ell x_0 x_1}{(s_1^\ell - 1)x_1 + \cfrac{s_1^\ell x_1 x_2}{(s_2^\ell - 1)x_2 + \cdots}}}.$$

The partial numerators equal 1 when $x_0 = 1$ and $x_{n+1} = (s_n^\ell x_n)^{-1}$ for $n \geq 0$. By induction, the solution of this recursion is

$$x_n = \prod_{i=0}^{n-1} (s_i^\ell)^{(-1)^{n+i}} \quad \text{for } n \geq 1.$$

The partial quotients are then $a_0 = 0$, $a_1 = s_0^\ell x_0 = s_0^\ell$, and $a_{n+1} = (s_n^\ell - 1)x_n$ for $n \geq 1$. Substituting $s_n^\ell - 1 = (s_0^\ell s_1^\ell \cdots s_{n-1}^\ell + 1)^\ell - 1$ and expanding the binomial, the 1s cancel, so $s_0^\ell s_1^\ell \cdots s_{n-1}^\ell$ divides $s_n^\ell - 1$ and $a_{n+1} \in \mathbb{N}$. This proves (i).

(ii), Theorem 1 and (i) imply $C_{k,\ell} \notin \mathbb{A}$. From $a_1 = k^\ell$ and $a_2 = ((k^\ell + 1)^\ell - 1)k^{-\ell}$, we deduce that $C_{k,\ell} \neq C_{k',\ell'}$ when $(k,\ell) \neq (k',\ell')$. This proves (ii).

(iii), Let $\alpha_n := s_n - 1$. Then (19) implies $\alpha_{n+1} = \alpha_n(\alpha_n + 1)^\ell > \alpha_n^{\ell+1}$ for $n \geq 1$. As $\alpha_2 = k^\ell(k^\ell + 1)^\ell \geq k^\ell + 1$, induction yields $\alpha_n \geq (k^\ell + 1)^{(\ell+1)^{n-2}}$ for $n \geq 2$. Since (i) implies $a_n \geq s_{n-3}^{2\ell} > \alpha_{n-3}^{2\ell} \geq \alpha_{n-3}^{\ell+1}$, we get (iii).

(iv), For $n > 0$, definition (19) implies $s_{n+1} - 1 = (s_n - 1)s_n^\ell$, so

$$\frac{1}{s_n - 1} - \frac{1}{s_{n+1} - 1} = \frac{s_n^\ell - 1}{s_{n+1} - 1} \quad \text{for } n \geq 1. \tag{20}$$

Hence the first series in (iv) telescopes to $(s_0^\ell - 1)(s_1 - 1)^{-1} + (s_1 - 1)^{-1} = 1$. Replacing n with $2n + 1$ in (20), we sum from $n = 0$ to ∞ and obtain the second equality in (iv).

(v), Set $\ell = 1$ in parts (i) and (ii).

(vi), Recursion (19) yields $\gcd(s_i, s_j) = 1$ for $i \neq j$, so (ii) follows from (i). This completes the proof of the corollary. ∎

Example 9 Take $(k, \ell) = (1, 1)$. *Sylvester's sequence* [39, 40] is defined as

$$(S_n)_{n\geq 0} := (s_{n+1}(1, 1))_{n\geq 0} = 2, 3, 7, 43, 1807, 3263443, 10650056950807, \ldots$$

(see [12, p. 123], [18, pp. 436, 444], [20], [35, A000058]). Sylvester's sequence satisfies the recursion $S_0 = 2$ and $S_{n+1} = (S_n - 1)S_n + 1$ for $n \geq 0$.

Likewise, $C := C_{1,1}$ defines *Cahen's constant* [9], [18, §6.7], [35, A118227]

$$\begin{aligned} C = \sum_{n=0}^{\infty} \frac{(-1)^n}{S_n - 1} &= 1 - \frac{1}{2} + \frac{1}{6} - \frac{1}{42} + \frac{1}{1806} - \cdots = 0.643410546288338\ldots \\ = 1 - \sum_{n=1}^{\infty} \frac{(-1)^{n-1}}{S_0 S_1 \cdots S_{n-1}} &= 1 - \frac{1}{2} + \frac{1}{2\cdot 3} - \frac{1}{2\cdot 3\cdot 7} + \frac{1}{2\cdot 3\cdot 7\cdot 43} - \cdots . \end{aligned}$$

Corollary 2 recovers $C \notin \mathbb{A}$ from [12] and gives the expansion [35, A006279]

$$\begin{aligned} C &= [0, 1, 1, 1, (S_0)^2, (S_1)^2, (S_0 S_2)^2, (S_1 S_3)^2, (S_0 S_2 S_4)^2, (S_1 S_3 S_5)^2, \ldots] \quad (21)\\ &= [0,\ 1,\ 1,\ 1,\ 2^2,\ 3^2,\ 14^2,\ 129^2,\ 25298^2,\ 420984147^2,\ \ldots]. \end{aligned}$$

Since $\alpha_n := S_n - 1$ satisfies $\alpha_{n+1} - \alpha_n = \alpha_n^2$ and $\sum_{n=0}^{\infty}(-1)^n \alpha_n^{-1} = C$, Proposition 1 and Theorem 1 give, respectively, the continued fractions

$$C = \cfrac{1}{1 + \cfrac{1^2}{1^2 + \cfrac{2^2}{2^2 + \cfrac{6^2}{6^2 + \cfrac{42^2}{42^2 + \ddots}}}}} \cong \cfrac{1}{1 + \cfrac{1}{1 + \cfrac{2}{2 + \cfrac{3}{6 + \cfrac{7}{42 + \ddots}}}}}.$$

In his 1891 paper "A remark on an expansion of numbers which has some similarities with continued fractions" [9], Cahen defined C and showed that it is irrational. Exactly 100 years later, as an example of their "self-similar" (or "self-generating" [18, §6.7], [19, §6.7]) simple continued fractions, Davison and Shallit [12] proved that C is transcendental and that $C = [0, 1, q_0^2, q_1^2, q_2^2, \ldots]$. (This expansion agrees with (21), by (8) and induction.) For generalizations of [12], see Becker [5] and Töpfer [41].

Example 10 Corollary 2 shows that the Cahen-type constant $C_{k,1} = [0, k, 1, \dots]$, so

$$1 > C_{1,1} > \frac{1}{2} > C_{2,1} > \frac{1}{3} > C_{3,1} > \frac{1}{4} > C_{4,1} > \cdots .$$

When $k = 2$ we have $s_0(2, 1) = 2 = s_1(1, 1) = S_0$. It follows that in general $s_{n+1}(2, 1) = s_{n+2}(1, 1) = S_{n+1}$, so

$$C_{2,1} = \sum_{n=0}^{\infty} \frac{(-1)^n}{s_{n+1}(2, 1) - 1} = \sum_{n=0}^{\infty} \frac{(-1)^n}{S_{n+1} - 1} = 1 - \sum_{n=0}^{\infty} \frac{(-1)^n}{S_n - 1} = 1 - C.$$

By Corollary 2,

$$C_{2,1} = [0,\ 2,\ 1,\ 2^2,\ 3^2,\ 14^2,\ 129^2,\ 25298^2,\ 420984147^2,\ \dots] \notin \mathbb{A}.$$

Example 11 For an example with $\ell > 1$, we take $(k, \ell) = (1, 2)$ to get

$$(s_{n+1}(1, 2))_{n \ge 0} = 2, 5, 101, 1020101, 1061522231810040101, \dots$$

(see [33] and [35, A231830]). Then $C_{1,2}$ is the transcendental number

$$\begin{aligned} C_{1,2} &= 1 - \frac{1}{2^2} + \frac{1}{2^2 \cdot 5^2} - \frac{1}{2^2 \cdot 5^2 \cdot 101^2} + \cdots \\ &= 1 - \frac{1}{4} + \frac{1}{100} - \frac{1}{1020100} + \cdots = 0.759999019703\dots . \end{aligned}$$

Here $\alpha_n := s_{n+1}(1, 2) - 1$ satisfies $\alpha_{n+1} - \alpha_n = \alpha_n^2(\alpha_n + 2)$, so Proposition 1, Theorem 1, and Corollary 2 give the continued fractions

$$\begin{aligned} C_{1,2} &= \cfrac{1}{1 + \cfrac{1^2}{1^2 \cdot 3 + \cfrac{4^2}{4^2 \cdot 6 + \cfrac{100^2}{100^2 \cdot 102 + \ddots}}}} \cong \cfrac{1}{1 + \cfrac{1}{3 + \cfrac{4}{24 + \cfrac{25}{10200 + \ddots}}}} \\ &\cong [0, 1^2, 2^2 - 1, (5^2 - 1)2^{-2}, (101^2 - 1)2^2 5^{-2}, (1020101^2 - 1)2^{-2} 5^2 101^{-2}, \\ &\qquad (1061522231810040101^2 - 1)2^2 5^{-2} 101^2 1020101^{-2}, \dots] \\ &= [0,\ 1,\ 3,\ 6,\ 1632,\ 637563750,\ 1767398865801083661443214432, \dots]. \end{aligned}$$

Our final section studies series of *positive* terms involving Sylvester-type sequences.

5 Some Non-alternating Series

Another series formed from Sylvester's sequence is the sum of reciprocals. Setting $(k, \ell) = (1, 1)$ in (20), the right-hand side is then S_n^{-1}, so the series telescopes to

$$\sum_{n=0}^{\infty} \frac{1}{S_n} = \sum_{n=0}^{\infty} \left(\frac{1}{S_n - 1} - \frac{1}{S_{n+1} - 1} \right) = \frac{1}{S_0 - 1} = 1, \tag{22}$$

a rational number. By contrast, the corresponding alternating sum

$$\sum_{n=0}^{\infty} \frac{(-1)^n}{S_n} = \sum_{n=0}^{\infty} \frac{(-1)^n}{S_n - 1} - \sum_{n=0}^{\infty} \frac{(-1)^n}{S_{n+1} - 1} = C - (1 - C) = 2C - 1$$

is transcendental, as are the non-alternating sums

$$\sum_{n=0}^{\infty} \frac{1}{S_{2n}} = \sum_{n=0}^{\infty} \left(\frac{1}{S_{2n} - 1} - \frac{1}{S_{2n+1} - 1} \right) = C \tag{23}$$

and $\sum_{n=0}^{\infty} S_{2n+1}^{-1} = 1 - C$.

Finch asked, "What can be said about $\sum_{n=0}^{\infty} (S_n - 1)^{-1} = 1.6910302067\ldots$?" [18, p. 436]. We denote this constant by

$$K := \sum_{n=0}^{\infty} \frac{1}{S_n - 1} = 1 + \sum_{n=1}^{\infty} \frac{1}{S_0 S_1 \cdots S_{n-1}}$$

and we name it the *Kellogg-Curtiss constant*, because Kellogg conjectured [23], and Curtiss proved [11], the following bound on solutions to a unit fraction equation:

$$x_i \in \mathbb{N} \text{ and } \sum_{i=0}^{n} \frac{1}{x_i} = 1 \quad \Longrightarrow \quad \max_{0 \le i \le n} x_i \le S_n - 1.$$

Remark 3 By (22), one solution of the equation $\sum_{i=0}^{\infty} x_i^{-1} = 1$ is $x_i = S_i$. In fact, this is the solution provided by the "greedy Egyptian fraction algorithm"—see Soundararajan [38]. Likewise, the greedy Egyptian fraction expansion of Cahen's constant C is series (23) with $x_i = S_{2i}$.

The following general result shows in particular that K is irrational.

Proposition 2 *For* $k \in \mathbb{N}$ *and* $\ell \in \mathbb{N}$, *define the* Kellogg-Curtiss-type constant

$$K_{k,\ell} := \sum_{n=0}^{\infty} \frac{1}{s_{n+1}(k, \ell) - 1},$$

where the Sylvester-type sequence $(s_n(k,\ell))_{n\ge 0}$ *is defined in Corollary 2.*

(i), *Then* $K_{k,\ell} \notin \mathbb{Q}$. *In particular, the Kellogg-Curtiss constant* $K = K_{1,1} = 1 + K_{2,1}$ *is irrational.*
(ii), *If* $\ell \ge 2$, *then* $K_{k,\ell}$ *is transcendental and* $\mu(K_{k,\ell}) \ge 3$.

We could prove (i) from the fact that, *given a non-decreasing sequence of positive integers* $A_0, A_1, \ldots,$ *the* Engel series $\sum_{n=0}^{\infty}(A_0A_1\cdots A_n)^{-1}$ *converges to an irrational number if (and only if)* A_n *tends to infinity with* n (see, e.g., [15, §2.2]). Instead, we give a mostly self-contained proof. It uses partial sums instead of continued fractions (compare to Example 8).

Proof (of Proposition 2*)* Let us fix integers $k \ge 1$ and $\ell \ge 1$, and write s_n in place of $s_n(k,\ell)$. Then for $n \ge 1$, the nth partial sum of the series for $K_{k,\ell}$ is, in lowest terms,

$$\frac{P_n}{Q_n} := \sum_{i=0}^{n-1} \frac{1}{s_{i+1}-1} = \sum_{i=0}^{n-1} \frac{1}{s_0^\ell s_1^\ell \cdots s_i^\ell} \quad \Longrightarrow \quad Q_n = s_0^\ell s_1^\ell \cdots s_{n-1}^\ell = s_n - 1.$$

With this value of Q_n we see that

$$\begin{aligned} 0 < K_{k,\ell} - \frac{P_n}{Q_n} &= \sum_{i=n}^{\infty} \frac{1}{s_0^\ell s_1^\ell \cdots s_i^\ell} = \frac{1}{Q_n}\sum_{j=0}^{\infty}\frac{1}{s_n^\ell \cdots s_{n+j}^\ell} \\ &< \frac{1}{Q_n}\sum_{j=0}^{\infty}\frac{1}{(s_n^\ell)^{j+1}} = \frac{1}{Q_n}\frac{1}{s_n^\ell - 1} \le \frac{1}{Q_n^{\ell+1}}. \end{aligned} \tag{24}$$

(i). If $K_{k,\ell} \in \mathbb{Q}$, say $K_{k,\ell} = P/Q$, then

$$K_{k,\ell} - \frac{P_n}{Q_n} = \frac{P}{Q} - \frac{P_n}{Q_n} \ge \frac{1}{QQ_n} > \frac{1}{Q_n^2} \tag{25}$$

for n so large that $Q_n > Q$. But $\ell \ge 1$, so (25) contradicts (24). Therefore, $K_{k,\ell} \notin \mathbb{Q}$.

(ii). From (24), we infer that $\mu(K_{k,\ell}) \ge \ell + 1$. If $\ell \ge 2$, then $\mu(K_{k,\ell}) \ge 3$, so by the Thue-Siegel-Roth theorem, $K_{k,\ell} \notin \mathbb{A}$. This completes the proof of the proposition. ■

By a similar argument (also not using Theorem 1 or continued fractions), $C_{k,\ell} \notin \mathbb{A}$ for $\ell \ge 2$. The case $\ell = 1$ though (which includes Cahen's constant C) would seem to require using Theorem 1, as in the proof of Corollary 2. However, Duverney [16] has found a proof that $C \notin \mathbb{A}$ which is similar to that of Proposition 2, part (ii). He uses relation (23) and the fact that $S_{2n+2} > \frac{1}{8}S_{2n}^4$, which follows from $S_{n+1} > \frac{1}{2}S_n^2$.

Duverney has also answered Finch's question by pointing out that, as a special case of a result of Becker [5, p. 186, Remark (ii)], *the Kellogg-Curtiss constant K is transcendental.*

Conjecture 3 For $k \geq 1$, the Kellogg-Curtiss-type constant $K_{k,1}$ is transcendental.

Acknowledgments I thank Daniel Duverney and Michael Nyblom for (independently) pointing out the series which shows that Sierpiński's theorem is sharp. I am also indebted to Duverney for generalizing my earlier special case of Corollary 1. I thank Steven Finch for comments on the manuscript and for discussions on [12]. Finally, I am grateful to Yohei Tachiya for a proof that K is irrational.

References

1. B. Adamczewski, The many faces of the Kempner number, *J. Integer Seq.* **16** (2013) Article 13.2.15.
2. C. Badea, A theorem on irrationality of infinite series and applications, *Acta Arith.* **63** (1993) 313–323.
3. N. D. Baruah, B. C. Berndt, H. H. Chan, Ramanujan's series for $1/\pi$: a survey, *Amer. Math. Monthly* **116** (2009) 567–587.
4. K. Beanland, J. W. Roberts, C. Stevenson, Modifications of Thomae's function and differentiability, *Amer. Math. Monthly* **116** (2009) 531–535.
5. P.-G. Becker, Algebraic independence of the values of certain series by Mahler's method, *Monatsh. Math.* **114** (1992) 183–198.
6. B. C. Berndt, S. Kim, A. Zaharescu, Diophantine approximation of the exponential function and Sondow's conjecture, *Adv. Math.* **248** (2013) 1298–1331.
7. J. M. Borwein, S. T. Chapman, I prefer Pi: a brief history and anthology of articles in the American Mathematical Monthly, *Amer. Math. Monthly* **122** (2015) 195–216.
8. J. Borwein, A. van der Poorten, J. Shallit, W. Zudilin, *Neverending Fractions: An Introduction to Continued Fractions*. Cambridge Univ. Press, Cambridge, 2014.
9. E. Cahen, Note sur un développement des quantités numériques, qui presente quelque analogie avec celui des fractions continues, *Nouv. Ann. Math.* **10** (1891) 508–514; also available at http://archive.numdam.org/ARCHIVE/NAM/NAM_1891_3_10_/NAM_1891_3_10__508_0/NAM_1891_3_10__508_0.pdf.
10. M. Coons, On the rational approximation of the sum of the reciprocals of the Fermat numbers, *Ramanujan J.* **30** (2013) 39–65.
11. D. R. Curtiss, On Kellogg's Diophantine problem, *Amer. Math. Monthly* **29** (1922) 380–387; also available at http://www.jstor.org/stable/2299023.
12. J. L. Davison, J. O. Shallit, Continued fractions for some alternating series, *Monatsh. Math.* **111** (1991) 119–126.
13. D. Duverney, Irrationality of fast converging series of rational numbers, *J. Math. Sci. Univ. Tokyo* **8** (2001) 275–316.
14. ———, Transcendence of a fast converging series of rational numbers, *Math. Proc. Cambridge Philos. Soc.* **130** (2001) 193–207.
15. ———, *Number Theory: An Elementary Introduction Through Diophantine Problems*. Monographs in Number Theory, Vol. 4. World Scientific, Singapore, 2010.
16. ———, Transcendence of Cahen's constant and related numbers, preprint, 2015.
17. L. Euler, *Introduction to Analysis of the Infinite*. English trans. and Introduction by J. D. Blanton. Springer, New York, 1988.
18. S. R. Finch, *Mathematical Constants*. Encyclopedia of Mathematics and its Applications 94. Cambridge Univ. Press, Cambridge, 2003.

19. ———, Errata and addenda to *Mathematical Constants*, e-print (2020), available at http://arxiv.org/abs/2001.00578.
20. S. W. Golomb, On certain nonlinear recurring sequences, *Amer. Math. Monthly* **70** (1963) 403–405.
21. ———, On the sum of the reciprocals of the Fermat numbers and related irrationalities, *anad. J. Math.* **15** (1963) 475–478.
22. G. H. Hardy, E. M. Wright, *An Introduction to the Theory of Numbers*. Fifth edition. Oxford Univ. Press, Oxford, 1979.
23. O. D. Kellogg, On a Diophantine problem, *Amer. Math. Monthly* **28** (1921) 300–303.
24. A. J. Kempner, On transcendental numbers, *Trans. Amer. Math. Soc.* **17** (1916) 476– 482.
25. M. Laczkovich, On Lambert's proof of the irrationality of π, *Amer. Math. Monthly* **104** (1997) 439–443.
26. J. H. Lambert, Mémoire sur quelques propriétés remarquables des quantités transcendantes circulaires et logarithmiques, 1768, in L. Berggren, J. Borwein, P. Borwein, *Pi, a Source Book*, Springer-Verlag, New York, 1997, pp. 129–146.
27. K. Mahler, Remarks on a paper by W. Schwarz, *J. Number Theory* **1** (1969) 512–521 .
28. J. A. Nathan, The irrationality of e^x for nonzero rational x, *Amer. Math. Monthly* **105** (1998) 762–763.
29. M. A. Nyblom, An extension of a result of Sierpiński, *J. Number Theory* **105** (2004) 49–59.
30. J. Paradís, P. Viader, L. Bibiloni, Approximation to quadratic irrationals and their Pierce expansions, *Fib. Quart.* **36** (1998) 146–153.
31. ———, A mathematical excursion: from the three-door problem to a Cantor-type set, *Amer. Math. Monthly* **106** (1999) 241–251.
32. T. A. Pierce, On an algorithm and its use in approximating roots of algebraic equations, *Amer. Math. Monthly* **36** (1929) 523–525.
33. S. A. Shirali, A family portrait of primes–a case study in discrimination, *Math. Mag.* **70** (1997) 263–272.
34. W. Sierpiński, Sur un algorithme pour développer les nombres réels en séries rapidement convergentes, *Bull. Internat. Acad. Sci. Lettres Cracovie Sér. A* (1911) 113–117.
35. N. J. A. Sloane, The On-Line Encyclopedia of Integer Sequences, available at https://oeis.org.
36. J. Sondow, A geometric proof that e is irrational and a new measure of its irrationality, *Amer. Math. Monthly* **113** (2006) 637–641 [article], **114** (2007) 659 [editor's endnote].
37. J. Sondow, K. Schalm, Which partial sums of the Taylor series for e are convergents to e? (and a link to the primes 2, 5, 13, 37, 463). Part II, in *Gems in Experimental Mathematics*, T. Amdeberhan, L.A. Medina, V.H. Moll, eds. Contemp. Math., Vol. 517. American Mathematical Society, Providence, 2010, 349–363.
38. K. Soundararajan, Approximating 1 from below using n Egyptian fractions, e-print (2005), available at http://arxiv.org/abs/math/0502247.
39. J. J. Sylvester, On a point in the theory of vulgar fractions, *Amer. J. Math.* **3** (1880) 332–335; also available at http://www.jstor.org/stable/2369261.
40. ———, Postscript to note on a point in vulgar fractions, *Amer. J. Math.* **3** (1880) 388–389; also available at http://www.jstor.org/stable/2369265.
41. T. Töpfer, On the transcendence and algebraic independence of certain continued fractions, *Monatsh. Math.* **117** (1994) 255–262.

On the Transcendence of Critical Hecke L-Values

Johannes Sprang

Abstract Euler's remarkable formula gives an explicit expression for the values of the Riemann zeta function at positive even integers as non-zero rational multiples of powers of the period $2\pi i$. The transcendence of $2\pi i$ implies immediately the transcendence of all these even zeta values. Hecke L-functions are of great importance in number theory. They can be seen as a common generalization of the Riemann zeta function, Dirichlet L-functions and Dedekind zeta functions. The values of these L-functions for which a generalization of Euler's formula is expected are called critical L-values. In this small survey, we summarize known results on the algebraicity of critical Hecke L-values up to explicit periods. Afterwards, we collect transcendence results of the involved periods and deduce (conditional) transcendence results for all critical Hecke L-values.

1 Introduction

It is a remarkable fact that all special values of the Riemann zeta function at the negative integers are rational. More precisely, we have the explicit formula

$$\zeta(1-n) = -(-1)^n \frac{B_n}{n} \text{ for } n \geq 1, \tag{1}$$

where B_n are the Bernoulli numbers defined by the generating series

$$\frac{z}{\exp(z)-1} = \sum_{n=0}^{\infty} B_n \frac{z^n}{n!}.$$

The author has been supported by the SFB 1085 "Higher Invariants" funded by the DFG.

J. Sprang (✉)
Fakultät für Mathematik, Universität Regensburg, 93040 Regensburg, Germany
e-mail: johannes.sprang@mathematik.uni-regensburg.de

A. Bostan and K. Raschel (eds.), *Transcendence in Algebra, Combinatorics, Geometry and Number Theory*, Springer Proceedings in Mathematics & Statistics 373,
https://doi.org/10.1007/978-3-030-84304-5_23

The functional equation

$$\Gamma_\infty(s)\zeta(s) = \Gamma_\infty(1-s)\zeta(1-s), \quad \Gamma_\infty(s) := \Gamma(s/2)\pi^{-s/2},$$

allows us to deduce a closed formula for the values of the Riemann zeta function at all positive even integers

$$\zeta(2n) = -\frac{(2\pi i)^{2n}}{2(2n)!}B_{2n}. \tag{2}$$

The existence of poles of the Gamma function at non-positive integers is the reason why we cannot deduce a similar closed formula for the positive odd zeta values. We call an integer $n \in \mathbb{Z}$ *critical* for the Riemann zeta function if neither $\Gamma_\infty(s)$ nor $\Gamma_\infty(1-s)$ has a pole at $s = n$, otherwise we call it *non-critical*. The distinction between critical and non-critical L-values goes back to Deligne, who defined these notions for arbitrary motivic L-functions depending on the existence of poles in the Gamma factors in the (mostly conjectural) functional equation of the motivic L-function. But before we turn to more general L-values, let us return for a moment to the values of the Riemann zeta function. The positive critical values of the Riemann zeta function are exactly the values where Euler's formula (2) applies. Euler's formula (2) shows that the positive critical values of the Riemann zeta function are non-zero rational multiples of powers of the period $2\pi i$. Together with Lindemann's theorem this implies the transcendence of all positive critical values of the Riemann zeta function. This raises immediately the following two questions:

(a) What can be said about the algebraicity of more general L-values up to explicit periods?
(b) What can be said about the transcendence of these periods?

For general motivic L-functions, these questions are out of scope and part of deep conjectures in arithmetic geometry. The first question is related to the Deligne conjecture (see [Del79]) for critical motivic L-values, while the second question is related to Grothendieck's period conjecture. Even the existence of the functional equation for motivic L-functions, which enters the definition of the notions 'critical' and 'non-critical', is a very delicate question which is only known in particular cases.

In this note, we concentrate on a certain class of L-functions which are of great importance for number theory, namely Hecke L-functions. They can be seen as a common generalization of the Riemann zeta function, Dirichlet L-functions and Dedekind zeta functions. We will use this small expository note to summarize known results on the algebraicity of critical Hecke L-values up to explicit periods. Afterwards, we will discuss the transcendental properties of these periods and deduce (conditional) transcendence results for all critical L-values.

In this note, we will only speak about transcendence results for critical L-values. The study of the transcendental properties of non-critical L-values is much more challenging even in the case of the Riemann zeta function. For results in this direction, see [Apé79, BR01, Riv00, Zud01, FSZ19, LY20].

2 Hecke L-Functions

An immediate generalization of the Riemann zeta function is the class of Dirichlet L-functions

$$L(\chi, s) := \sum_{n=1}^{\infty} \frac{\chi(n)}{n^s}, \qquad \text{for } \Re(s) > 1.$$

for a Dirichlet character $\chi : (\mathbb{Z}/f\mathbb{Z})^\times \to \mathbb{C}^\times$, where we set $\chi(n) := \chi(n + f\mathbb{Z})$ if $\gcd(n, f) = 1$, and $\chi(n) = 0$ if $\gcd(n, f) \neq 1$. Dirichlet L-functions are obtained by twisting the Riemann zeta function by a character. On the other hand, the Dedekind zeta function of a number field K generalizes the Riemann zeta function to arbitrary number fields

$$\zeta_K(s) := \sum_{0 \neq \mathfrak{a} \subseteq \mathcal{O}_K} \frac{1}{N(\mathfrak{a})^s}, \qquad \text{for } \Re(s) > 1,$$

where the sum runs over all non-zero ideals of $\mathcal{O}_K$ and $N(\mathfrak{a})$ denotes the norm of $\mathfrak{a} \subseteq \mathcal{O}_K$. Hecke L-functions can be seen as a generalization of Dirichlet L-functions to arbitrary number fields, i.e. by twisting Dedekind zeta functions by certain characters. In the following section, we biefly recall the basic definitions and properties of Hecke characters. We refer to [Sch88] for a more detailed discussion.

We start by fixing the following notation: Let K be a number field with ring of integers $\mathcal{O}_K$. We fix the algebraic closure $\overline{\mathbb{Q}} \subseteq \mathbb{C}$ of $\mathbb{Q}$ in $\mathbb{C}$ and we write $J_K := \mathrm{Hom}_{\mathbb{Q}}(K, \overline{\mathbb{Q}})$ for the set of field embeddings. For a given ideal $\mathfrak{f} \subseteq \mathcal{O}_K$ we will write $\mathcal{I}_\mathfrak{f}$ for the group of all fractional ideals prime to $\mathfrak{f}$. We denote by $\mathcal{P}_\mathfrak{f} \subseteq \mathcal{I}_\mathfrak{f}$ the subgroup of principal fractional ideals generated by $\xi \in K^\times$ which satisfy $\xi \equiv 1 \bmod^\times \mathfrak{f}$ and which are totally positive at the real places of K. Here, the notation $\xi \equiv 1 \bmod^\times \mathfrak{f}$ means that $\nu_\mathfrak{p}(\xi - 1) \geq \nu_\mathfrak{p}(\mathfrak{f})$ for all prime ideals $\mathfrak{p}$ dividing $\mathfrak{f}$. An element $\xi \in \mathcal{O}_K$ is called *totally positive* if $\sigma(\xi) > 0$ for all real embeddings $\sigma : K \hookrightarrow \mathbb{R}$.

Definition 2.1. An *algebraic Hecke character* of conductor dividing $\mathfrak{f}$ and infinity type $T = \sum_{\sigma \in J_K} n(\sigma)\sigma \in \mathbb{Z}[J_K]$ is a homomorphism $\chi : \mathcal{I}_\mathfrak{f} \to \overline{\mathbb{Q}}^\times$ such that

$$\chi((\xi)) = \xi^T := \prod_{\sigma \in J_K} \sigma(\xi)^{n(\sigma)}$$

for all $(\xi) \in \mathcal{P}_\mathfrak{f}$. If $\mathfrak{f}'$ divides $\mathfrak{f}$, we may view every Hecke character of conductor dividing $\mathfrak{f}'$ as a Hecke character of conductor dividing $\mathfrak{f}$ by restriction along $\mathcal{I}_\mathfrak{f} \subseteq \mathcal{I}_{\mathfrak{f}'}$. In this case, we say that $\chi : \mathcal{I}_\mathfrak{f} \to \overline{\mathbb{Q}}^\times$ is induced from $\mathfrak{f}'$. The smallest ideal $\mathfrak{f}'$ with respect to divisibility such that χ is induced from $\mathfrak{f}'$ is called *the conductor* of χ and denoted by $\mathfrak{f}_\chi$. The Hecke character is called *finite* if $T = 0$. Note that the finite Hecke characters are exactly the ones of finite order.

It is not difficult to see that the finite Hecke characters of the number field $\mathbb{Q}$ are exactly the Dirichlet characters.

Example 2.2. Let $K = \mathbb{Q}$, $\mathfrak{f} = f\mathbb{Z}$. Every fractional ideal $\mathfrak{a} \in \mathcal{I}_\mathfrak{f}$ can be written as $\mathfrak{a} = (a/b)$ with positive integers a, b satisfying $(a, f) = (b, f) = 1$. The map

$$\mathcal{I}_\mathfrak{f} \to (\mathbb{Z}/f\mathbb{Z})^\times, \quad \mathfrak{a} = (a/b) \mapsto ab^{-1} \mod f$$

is well-defined, surjective and its kernel is given by

$$\{(a/b) \in \mathcal{I}_\mathfrak{f} \mid a \equiv b \mod f\} = \mathcal{P}_\mathfrak{f}.$$

The induced isomorphism

$$\mathcal{I}_\mathfrak{f}/\mathcal{P}_\mathfrak{f} \cong (\mathbb{Z}/f\mathbb{Z})^\times$$

allows us to identify finite Hecke characters of $\mathbb{Q}$ with Dirichlet characters.

The following provides an important example of infinite Hecke characters.

Example 2.3. Let us fix an ideal $\mathfrak{f} \subseteq \mathcal{O}_K$. For a non-zero ideal $\mathfrak{a} \subseteq \mathcal{O}_K$, let us define the norm by $N(\mathfrak{a}) := [\mathcal{O}_K : \mathfrak{a}]$. The norm is multiplicative on ideals and hence it extends uniquely to a multiplicative function

$$N\colon \mathcal{I}_\mathfrak{f} \to \mathbb{Q}^\times$$

on the group $\mathcal{I}_\mathfrak{f}$ of fractional ideals prime to $\mathfrak{f}$. The norm of a principal ideal $(\xi) \in \mathcal{P}_\mathfrak{f}$ coincides with the usual field norm $N_{K/\mathbb{Q}}(\xi) = \prod_{\sigma \in J_K} \sigma(\xi)$. Thus, we have

$$N((\xi)) = N_{K/\mathbb{Q}}(\xi) = \xi^{\underline{1}}$$

for $\underline{1} = \sum_{\sigma \in J_K} \sigma \in \mathbb{Z}[J_K]$ and we see that N is a Hecke character of infinity type $\underline{1}$. The Hecke character N is called the *Norm character*.

Let us observe that not every $T \in \mathbb{Z}[J_K]$ can occur as an infinity type of a Hecke character. If χ is an algebraic Hecke character of infinity type T, then

$$\xi^T = 1 \quad \text{for all totally positive units } \xi \in 1 + \mathfrak{f}\mathcal{O}_L. \tag{3}$$

Since the totally positive units are of finite index in the group of all units, (3) implies that for every embedding $\tau\colon \overline{\mathbb{Q}} \hookrightarrow \mathbb{C}$ the integer

$$w = n(\sigma) + n(\overline{\sigma}) \tag{4}$$

is independent of τ and σ. Here, $\overline{(\cdot)}$ denotes the complex conjugation on $\overline{\mathbb{C}}$ induced by the embedding τ. The integer w is called the *weight* of χ. Let us note that the norm character introduced in Example 2.3 has weight 2.

In the following example, we will have a closer look at Hecke characters of fields admitting a real embedding.

Example 2.4. Let χ be an algebraic Hecke character of infinity type $T = \sum_{\sigma \in J_K} n(\sigma)\sigma$ of a field K with a real embedding $K \hookrightarrow \mathbb{R}$. In Eq. (4), we can choose the embedding τ for a given $\sigma \in J_K$ in such a way that $\tau \circ \sigma : K \hookrightarrow \mathbb{R}$ is a real embedding. In particular, we get $n(\sigma) = n(\overline{\sigma})$ for each $\sigma \in J_K$. This shows $n(\sigma) = \frac{w}{2}$ for all $\sigma \in J_K$. Thus,

$$N(\cdot)^{-\frac{w}{2}}\chi$$

is a Hecke character of weight 0 and hence it is a finite Hecke character. This proves that all algebraic Hecke characters of a field with a real embedding are products of the norm character and a finite Hecke character. In particular, for $K = \mathbb{Q}$ all algebraic Hecke characters are products of Dirichlet characters and integral powers of the norm character. In this case, the norm character admits the following explicit description. As in Example 2.2, every $\mathfrak{a} \in \mathcal{I}_{\mathfrak{f}}$ can be written as $\mathfrak{a} = (a/b)$ with positive integers a, b and $(a, b) = 1$. For such an ideal, the norm character is given by the identity, i.e.

$$N((a/b)) = a/b.$$

To any number field K and any algebraic Hecke character χ we associate a Hecke L-function

$$L(\chi, s) := \sum_{0 \neq \mathfrak{a} \subseteq \mathcal{O}_K} \frac{\chi(\mathfrak{a})}{N(\mathfrak{a})^s}, \quad \text{for } \Re(s) \text{ sufficiently large.}$$

To formulate the functional equation for Hecke L-functions, we have to introduce the Gamma factors at the infinite places, see also [Sch88, Ch.0, §6]:

Definition 2.5. Let χ be a Hecke character of conductor dividing $\mathfrak{f}$ and weight w of a number field K.

(a) Since $\chi((\xi)) = 1$ for any totally positive unit $\xi \in 1 + \mathfrak{f}$, there exists for every real place v of K an $\epsilon_v \in \{0, 1\}$ such that for every unit $\xi \in 1 + \mathfrak{f}$

$$\chi((\xi)) = \prod_{v \text{ real}} \operatorname{sign}(v(\xi))^{\epsilon_v + \frac{w}{2}}.$$

The number $\epsilon_v \in \{0, 1\}$ will be called *the parity of* χ at v. The character is called *totally even* (respectively *totally odd*) if $\epsilon_v = 0$ (respectively $\epsilon_v = 1$) for all real embeddings v of K. For a totally even/odd character χ we will write $\epsilon \in \{0, 1\}$ for its parity (at any of its real places).

(b) For a real place v of K set

$$\Gamma_v(\chi, s) := \pi^{-\frac{1}{2}(s+\epsilon_v-w/2)}\Gamma\left(\frac{s+\epsilon_v-w/2}{2}\right),$$

and for a complex place v of K corresponding to a pair of embeddings $\sigma, \overline{\sigma} : K \to \mathbb{C}$ set

$$\Gamma_\nu(\chi, s) := 2(2\pi)^{-(s-\min(n(\sigma),n(\overline{\sigma})))}\Gamma(s - \min(n(\sigma), n(\overline{\sigma}))).$$

The Gamma factor of χ is defined as the product over the local Gamma factors at all infinite places

$$\Gamma_\infty(\chi, s) := \prod_{\nu \text{ infinite}} \Gamma_\nu(\chi, s).$$

By multiplication with the Gamma factors, we obtain the *completed Hecke L-function* of the character χ

$$\Lambda(\chi, s) := (|d_K|N(\mathfrak{f}_\chi))^{s/2}\Gamma_\infty(\chi, s)L(\chi, s),$$

where d_K denotes the discriminant of the number field K. It can now be shown that the completed Hecke L-function satisfies the functional equation

$$\Lambda(\chi, s) = W(\chi)\Lambda(\chi^{-1}, 1 - s),$$

where $W(\chi)$ is a certain explicit local factor with $|W(\chi)| = 1$.

3 Critical Hecke L-Values and Periods

In this section, we define 'critical' and 'non-critical' Hecke L-values. Afterwards, we determine all cases where critical L-values exist. Finally, we summarize known results on the algebraicity of critical Hecke L-values up to explicit periods.

Twisting by the norm character allows us to reduce the question about values of Hecke L-functions at arbitrary integers to the question about Hecke L-values at zero

$$L(\chi, n) = L(\chi \cdot N(\cdot)^{-n}, 0).$$

Similar to the case of the Riemann zeta function, we define the notions 'critical' and 'non-critical' for Hecke L-values. By the above observation, it is enough to define the notion 'critical' and 'non-critical' for the value $L(\chi, s)$ at $s = 0$.

Definition 3.1. Let χ be a Hecke character of a number field K. We call $L(\chi, s)$ *critical* at $s = 0$ if neither $\Gamma_\infty(\chi, s)$ nor $\Gamma_\infty(\chi^{-1}, 1 - s)$ has a pole at $s = 0$. Otherwise, the value is called *non-critical*. In the following, we will often say that the Hecke character χ is critical (respectively non-critical) if $L(\chi, s)$ is critical (respectively non-critical) at $s = 0$.

Example 3.2. In Example 2.4, we have already seen that every algebraic Hecke character χ of the number field $\mathbb{Q}$ is of the form $\chi = \chi_0 \cdot N(\cdot)^m$ with χ_0 a finite character (i.e. a Dirichlet character) and $m \in \mathbb{Z}$. Let us write ϵ for the parity of χ and observe that χ is even if and only if the Dirichlet character χ_0 is even. According to

Definition 2.5, the Gamma factors appearing in the functional equation are given by the formulas

$$\Gamma_\infty(\chi, s) = \pi^{-\frac{1}{2}(s+\epsilon-m)}\Gamma\left(\frac{s+\epsilon-m}{2}\right),$$
$$\Gamma_\infty(\chi^{-1}, 1-s) = \pi^{-\frac{1}{2}(1-s+\epsilon+m)}\Gamma\left(\frac{1-s+\epsilon+m}{2}\right).$$

Since the poles of the Gamma function are located at the non-positive integers, we see that χ is critical if and only if neither $\frac{\epsilon+m}{2}$ nor $\frac{1+\epsilon-m}{2}$ is a non-positive integer. Thus, χ is critical if and only if one of the following two cases holds:

(a) $m \geq 0$ and $m \not\equiv \epsilon \mod 2$,
(b) $m < 0$ and $m \equiv \epsilon \mod 2$.

The formula

$$L(\chi, 0) = L(\chi_0, -m)$$

allows us to express the Hecke L-value $L(\chi, 0)$ in terms of Dirichlet L-values. Let us write f for the conductor of the Dirichlet character χ_0 and define the generalized Bernoulli numbers B_{n,χ_0} by the generating series

$$\sum_{a=1}^{f} \frac{\chi_0(a)te^{at}}{e^{ft}-1} = \sum_{n=0}^{\infty} B_{n,\chi_0}\frac{t^n}{n!}.$$

The values of Dirichlet L-functions at negative integers can be described explicitly in terms of generalized Bernoulli numbers:

$$L(\chi_0, 1-n) = -\frac{B_{\chi_0,n}}{n}, \quad \text{for } n \geq 1.$$

Together with the functional equation of Hecke L-functions, the above discussion gives an explicit description of all critical Hecke L-values for $K = \mathbb{Q}$.

In a next step, we will give an explicit criterion to decide if a given Hecke character is critical. We have already seen that the infinity type $T = \sum_{\sigma \in J_K} n(\sigma)\sigma$ of an algebraic Hecke character χ satisfies the property that the number

$$n(\sigma) + n(\overline{\sigma})$$

does not depend on the choice of an embedding $\tau : \overline{\mathbb{Q}} \hookrightarrow \mathbb{C}$. This implies that $n(\sigma)$ does only depend on $\sigma|_{K_0}$, where K_0 is the maximal subfield of K with a unique involution $c\colon K_0 \to K_0$ corresponding to complex conjugation under each embedding $K_0 \hookrightarrow \mathbb{C}$. Fields with this property fall naturally into two classes; totally real fields and CM fields. Here, let us recall:

Definition 3.3. A number field K is *totally real* if all embeddings $K \hookrightarrow \mathbb{C}$ factor through $\mathbb{R}$. It is called *totally imaginary* if none of its embeddings $K \hookrightarrow \mathbb{C}$ factors through $\mathbb{R}$. A number field K_0 is called a *CM field* if and only if it is an imaginary quadratic extension of a totally real field. A *CM type* of a CM field K_0 is a subset $\Sigma \subseteq J_{K_0}$ satisfying $\Sigma \sqcup \overline{\Sigma} = J_{K_0}$. Here, we write $\overline{(\cdot)}$ for the complex conjugation given by any embedding $\overline{\mathbb{Q}} \subseteq \mathbb{C}$.

For a subfield $K_0 \subseteq K$, the restriction $(\cdot)|_{K_0}\colon J_K \to J_{K_0}$ of embeddings induces a map

$$\mathrm{Ind}_{K_0}^K\colon \mathbb{Z}[J_{K_0}] \to \mathbb{Z}[J_K], \quad \sum_{\sigma\in J_{K_0}} n_0(\sigma)\sigma \mapsto \sum_{\sigma\in J_K} n_0(\sigma|_{K_0})\sigma|_{K_0}.$$

We say that T is *induced from* K_0 if $T = \mathrm{Ind}_{K_0}^K T_0$ for some $T_0 \in \mathbb{Z}[J_{K_0}]$. It can be deduced from the discussion preceding the Definition 3.3 that the infinity type of an algebraic Hecke character is always induced from a subfield which is either totally real or a CM field.

Before we determine all critical Hecke characters, let us recall that the weight $w = n(\sigma) + n(\overline{\sigma})$ of a Hecke character is always an even integer if the field K is totally real. Let us also recall that we write ϵ for the parity of a totally even/odd character.

Proposition 3.4. *Let K be a number field and χ an algebraic Hecke character of infinity type $T = \sum_{\sigma\in J_K} n(\sigma)\sigma$. The Hecke character χ is critical if and only if one of the following cases holds:*

(a) K is totally real, χ is totally odd or totally even, $\frac{w}{2} < 0$ and $\frac{w}{2} \equiv \epsilon \mod 2$,
(b) K is totally real, χ is totally odd or totally even, $\frac{w}{2} \geq 0$ and $\frac{w}{2} \not\equiv \epsilon \mod 2$,
(c) K is totally imaginary and T induces a disjoint decomposition

$$J_K = \left\{\sigma \in J_K \mid n(\sigma) < \min\left(0, \frac{w}{2}\right)\right\} \sqcup \left\{\sigma \in J_K \mid \max\left(0, \frac{w}{2}\right) < n(\sigma)\right\}.$$

Proof. This can be easily deduced from the explicit formulas for the Gamma factors at the infinite places, see [Sch88, Chapter II, §2.0.2] for the totally real case and [Sch88, Chapter II, §2.0.3] for the totally imaginary case. □

Corollary 3.5. *If χ is a critical Hecke character of a totally imaginary field, then $T = \sum_{\sigma\in J_K} n(\sigma)\sigma$ is induced from the infinity type $T_0 = \sum_{\sigma\in J_{K_0}} n_0(\sigma)\sigma \in \mathbb{Z}[J_{K_0}]$ of a CM subfield and*

$$\Sigma_0 := \left\{\sigma_0 \in J_{K_0} \mid n_0(\sigma_0) < \frac{w}{2}\right\}$$

is a CM type of K_0.

Proof. We have already seen that T is induced from a subfield K_0 which is either totally real or a CM field. If K_0 were totally real, we would have $n(\sigma) = \frac{w}{2}$ for all $\sigma \in J_K$ contradicting Proposition 3.4. So K_0 is a CM field and there exists

$T_0 = \sum_{\sigma \in J_{K_0}} n_0(\sigma)\sigma \in \mathbb{Z}[J_{K_0}]$ such that $n(\sigma) = n_0(\sigma|_{K_0})$ for all $\sigma \in J_K$. Again by Proposition 3.4, we know that for each $\sigma \in J_{K_0}$ either $n_0(\sigma) < \frac{w}{2}$ or $n_0(\sigma) > \frac{w}{2}$. On the other hand, we have

$$n_0(\sigma) + n_0(\overline{\sigma}) = w.$$

This implies that exactly one of σ and $\overline{\sigma}$ is contained in Σ_0. □

In particular, critical Hecke L-values do only exist if the underlying number field is either totally real or totally imaginary. In the following, we summarize all known results on the algebraicity of critical L-values up to explicit periods. Let us start with the case of totally real fields, which is due to Siegel and Klingen:

Theorem 3.6 (Siegel–Klingen, [Sie37, § 23], [Kli62, p. 271]). *Let K be a totally real field and χ a critical Hecke character which is either totally odd or totally even.*

(a) If χ has negative weight, then

$$L(\chi, 0) \in (2\pi i)^{-\frac{w}{2}}\overline{\mathbb{Q}}^{\times}.$$

(b) If χ has positive weight, then

$$L(\chi, 0) \in \overline{\mathbb{Q}}^{\times}.$$

Let us now turn our attention to the totally imaginary case. In this case, the appearing periods are slightly more complicated and related to periods of abelian varieties with complex multiplication. Let us start by introducing these periods. Let K be a totally imaginary field, χ a critical Hecke character of infinity type T of the number field K. The infinity type T is induced from an infinity type $T_0 = \sum_{\sigma \in J_{K_0}} n_0(\sigma)\sigma \in \mathbb{Z}[J_{K_0}]$ with associated CM type $\Sigma_0 := \{\sigma \mid n_0(\sigma) < 0\}$. Let us consider the complex torus

$$\mathbb{C}^{\Sigma_0}/\mathcal{O}_{K_0},$$

where we view $\mathcal{O}_{K_0}$ as a lattice in $\mathbb{C}^{\Sigma_0}$ using the embedding

$$\mathcal{O}_{K_0} \to \mathbb{C}^{\Sigma_0}, \quad \xi \mapsto (\sigma(\xi))_\sigma.$$

By the theory of complex multiplication, we know that $\mathbb{C}^{\Sigma_0}/\mathcal{O}_{K_0}$ are the complex points of an abelian variety A defined over $\overline{\mathbb{Q}}$ with complex multiplication by $\mathcal{O}_{K_0}$

$$\mathcal{O}_{K_0} \to \mathrm{End}_{\overline{\mathbb{Q}}}(A).$$

The Lie algebra $\mathrm{Lie}(A)$ decomposes canonically into 1-dimensional $\overline{\mathbb{Q}}$-vector spaces

$$\mathrm{Lie}(A) = \bigoplus_{\sigma \in \Sigma_0} \mathrm{Lie}(A)(\sigma)$$

such that $\mathcal{O}_{K_0}$ acts on $\mathrm{Lie}(A)(\sigma)$ through the embedding $\sigma : \mathcal{O}_{K_0} \to \overline{\mathbb{Q}}$. By dualizing we get a decomposition of the cotangent space ω_A

$$\omega_A = \bigoplus_{\sigma \in \Sigma_0} \omega_A(\sigma).$$

For each $\sigma \in \Sigma_0$, let us choose a 1-form $\omega_0(\sigma)$ generating the 1-dimensional $\overline{\mathbb{Q}}$-vector space $\omega_A(\sigma)$, i.e.:

$$\omega_A(\sigma) = \omega_0(\sigma)\overline{\mathbb{Q}}.$$

The first homology $H_1(A(\mathbb{C}), \mathbb{Z})$ of $A(\mathbb{C})$ is canonically isomorphic to $\mathcal{O}_{K_0}$. Let us write $\gamma \in H_1(A(\mathbb{C}), \mathbb{Z})$ for the element corresponding to $1 \in \mathcal{O}_{K_0}$. By integration of the global 1-forms $\omega_0(\sigma)$ over γ, we obtain periods of A

$$\Omega_\sigma := \int_\gamma \omega_0(\sigma). \tag{5}$$

Let us observe that the period Ω_σ is well-defined up to multiplication by a non-zero algebraic number.

The following example is a special case of the Chowla–Selberg formula which relates the periods of CM elliptic curves to products of values of the Gamma function, see [Gro78, SC67]:

Example 3.7. The complex points of the elliptic curve E with Weierstraß equation $y^2 = 4x^3 - 4x$ are given by $E(\mathbb{C}) = \mathbb{C}/\mathbb{Z}[i]$. The period associated with the algebraic differential $\omega = \frac{dx}{y}$ is given by the explicit formula

$$\Omega = \int_\gamma \omega = 2\int_1^\infty \frac{dt}{\sqrt{4t^3 - 4t}} = \frac{\Gamma(1/4)^2}{2^{3/2}\pi^{1/2}}.$$

The first result on the algebraicity of critical Hecke L-values is due to Damerell, [Dam70]. He was able to prove the algebraicity of critical Hecke L-values of imaginary quadratic fields up to powers of $2\pi i$ and products of periods of CM elliptic curves. This has been generalized by work of Katz and Shimura to CM fields, see [Kat78, Shi75]. Colmez could prove the algebraicity of critical Hecke L-values for certain non-CM extensions of imaginary quadratic fields up to certain explicit periods in [Col89]. The general totally imaginary case has been announced by Harder in [HS85], but the details never appeared. Using a completely different approach, we have settled the general case of arbitrary totally imaginary fields in a recent joint work with Guido Kings, [KS19]. Independently, Bergeron–Charollois–Garcia have developed new methods to prove the algebraicity of critical Hecke L-values in certain cases (work in progress, private communication, see also [BCG20]).

Theorem 3.8 ([KS19, Corollary 4.12])**.** *Let K be a totally imaginary field and χ a critical Hecke character with infinity type T induced by an infinity type T_0 of a CM* *subfield K_0. Let us keep the notation introduced above. In particular, let us write Ω_σ for the periods of an abelian variety with complex multiplication by the CM field K_0, see (5). Then*

$$L(\chi,0) \in \prod_{\sigma\in\Sigma_0}\left(\frac{\Omega_\sigma^{n_0(\overline{\sigma})-n_0(\sigma)}}{(2\pi i)^{n_0(\overline{\sigma})}}\right)^{[K:K_0]}\overline{\mathbb{Q}}.$$

4 Transcendence of Critical Hecke L-Values

In Sect. 3, we have summarized results on the algebraicity of critical Hecke L-values up to explicit periods. These results naturally raise the question of the transcendence of the periods involved. In this section, we summarize known results in this direction and deduce (conditional) transcendence results for all critical Hecke L-values.

Let us start with the case of totally real fields. An immediate corollary of the theorems of Siegel–Klingen and Lindemann's theorem is the following:

Corollary 4.1. *Let K be a totally real field and χ a critical Hecke character. Then, $L(\chi,0)$ is transcendental if and only if χ has negative weight.*

It remains to discuss the case of totally imaginary fields. The results of Sect. 3 raise the question about the algebraic independence of $2\pi i$ and the periods of CM abelian varieties.

Conjecture 4.2. Let A be an abelian variety defined over $\overline{\mathbb{Q}}$ of dimension d with complex multiplication by a CM field. Let $\omega_1,\dots,\omega_d$ be a basis of $H^0(A,\Omega^1_A)$ and $0\neq\gamma\in H_1(A(\mathbb{C}),\mathbb{Z})$. Define the periods

$$\Omega_i := \int_\gamma \omega_i .$$

Then $\pi,\Omega_1,\dots,\Omega_d$ are algebraically independent.

The transcendence of periods of elliptic curves has been shown by Schneider, [Sch37]. Baker, Coates and Masser have extended the results of Scheider to gain linear independence results for periods of elliptic curves, see [Mas75].

For general abelian varieties, the linear independence over $\overline{\mathbb{Q}}$ of the periods $\Omega_1,\dots,\Omega_d$ is known by the celebrated Analytic Subgroup Theorem of Wüstholz, see [Wüs87, Theorem 2] and [BW07]. Unfortunately, little is known about the algebraic independence of the periods $\pi,\Omega_1,\dots,\Omega_d$. An important result of Chudnovsky proves Conjecture 4.2 for elliptic curves with complex multiplication:

Theorem 4.3 [Chu84, Ch. 7, Theorem 3.1]**.** *Let E be an elliptic curve with complex multiplication and ω a non-zero algebraic differential form with associated period Ω. Then π and Ω are algebraically independent.*

Let us observe that Conjecture 4.2 is implied by Grothendieck's period conjecture, see [Yos03, Appendix III, Thm. 1] and observe that the Mumford-Tate group is known to coincide with the motivic Galois group, see [And96]. The fact that the dimension of the Mumford–Tate group is an upper bound for the transcendental degree of the periods has been shown by Deligne, see [Del82, Corollary 6.4].

Finally, we summarize the results of Sect. 3 to get a full (conditional) statement on the transcendence of critical Hecke L-values for totally imaginary fields. Thanks to Chudnovsky's theorem, it is an unconditional result if K_0 is imaginary quadratic:

Corollary 4.4. *Let K be a totally imaginary field and χ a critical Hecke character, whose infinity type is lifted from the CM subfield $K_0 \subseteq K$. If Conjecture 4.2 holds for abelian varieties with CM by K_0 then $L(\chi, 0)$ is either transcendental or zero.*

Proof. If the L-value is non-zero, we have according to Theorem 3.8

$$L(\chi, 0) \in \prod_{\sigma \in \Sigma_0} \left(\frac{\Omega_\sigma^{n_0(\overline{\sigma}) - n_0(\sigma)}}{(2\pi i)^{n_0(\overline{\sigma})}} \right)^{[K:K_0]} \overline{\mathbb{Q}}^\times.$$

By Conjecture 4.2, the numbers $(2\pi i)$ and $(\Omega_\sigma)_{\sigma \in \Sigma_0}$ are algebraically independent. Finally, let us remark that the fact that χ is critical implies that $n_0(\overline{\sigma}) - n_0(\sigma) \neq 0$. Indeed, $n_0(\sigma) = n_0(\overline{\sigma})$ would imply $n_0(\sigma) = n_0(\overline{\sigma}) = \frac{w}{2}$, in contradiction with Proposition 3.4 (c). Thus, $\prod_{\sigma \in \Sigma_0} \left(\frac{\Omega_\sigma^{n_0(\overline{\sigma}) - n_0(\sigma)}}{(2\pi i)^{n_0(\overline{\sigma})}} \right)$ is a non-trivial product of algebraically independent numbers and hence transcendental. □

Acknowledgements I would like to take this opportunity to thank Alin Bostan and Kilian Raschel for the excellent organization of the conference *Transient Transcendence in Transylvania*. Furthermore, I am grateful to the referees for all valuable comments, remarks and suggestions for improvement.

References

[And96] Y. André, *Pour une théorie inconditionnelle des motifs*, Inst. Hautes Études Sci. Publ. Math. (1996), no. 83, 5–49.

[Apé79] R. Apéry, *Irrationalité de $\zeta(2)$* et $\zeta(3)$, Astérisque (1979), no. 61, 11–13, Luminy Conference on Arithmetic.

[BCG20] N. Bergeron, P. Charollois, and L. E. Garcia, *Transgressions of the Euler class and Eisenstein cohomology of GLN(Z)*, Japanese Journal of Mathematics **15** (2020), no. 2, 311–379.

[BR01] K. Ball and T. Rivoal, *Irrationalité d'une infinité de valeurs de la fonction zêta aux entiers impairs*, Invent. Math. **146** (2001), no. 1, 193–207.

[BW07] A. Baker and G. Wüstholz, *Logarithmic forms and Diophantine geometry*, New Mathematical Monographs, vol. 9, Cambridge University Press, Cambridge, 2007.

[Chu84] G. V. Chudnovsky, *Contributions to the theory of transcendental numbers*, Mathematical Surveys and Monographs, vol. 19, American Mathematical Society, Providence, RI, 1984.

[Col89] P. Colmez, *Algébricité de valeurs spéciales de fonctions L*, Invent. Math. **95** (1989), no. 1, 161–205.

[Dam70] R. M. Damerell, *L-functions of elliptic curves with complex multiplication. I*, Acta Arith. **17** (1970), 287–301.

[Del79] P. Deligne, *Valeurs de fonctions L et périodes d'intégrales*, Automorphic forms, representations and L-functions (Proc. Sympos. Pure Math., Oregon State Univ., Corvallis, Ore., 1977), Part 2, Proc. Sympos. Pure Math., XXXIII, Amer. Math. Soc., Providence, R.I., 1979, With an appendix by N. Koblitz and A. Ogus, pp. 313–346.

[Del82] P. Deligne, *Hodge cycles on abelian varieties*, Hodge Cycles, Motives, and Shimura Varieties, Lecture Notes in Math., vol. 900, Springer, Berlin, 1982, pp. 9–100.

[FSZ19] S. Fischler, J. Sprang, and W. Zudilin, *Many odd zeta values are irrational*, Compositio Mathematica **155** (2019), no. 5, 938–952.

[Gro78] B. H. Gross, *On the periods of abelian integrals and a formula of Chowla and Selberg*, Invent. Math. **45** (1978), no. 2, 193–211, With an appendix by David E. Rohrlich.

[HS85] G. Harder and N. Schappacher, *Special values of Hecke L-functions and abelian integrals*, Workshop Bonn 1984 (Bonn, 1984), Lecture Notes in Math., vol. 1111, Springer, Berlin, 1985, pp. 17–49.

[Kat78] N. M. Katz, *p-adic L-functions for CM fields*, Invent. Math.**49** (1978), no. 3, 199–297.

[Kli62] H. Klingen, *Über die Werte der Dedekindschen Zetafunktion*, Math. Ann. **145** (1961/62), 265–272.

[KS19] G. Kings and J. Sprang, *Eisenstein-Kronecker classes, integrality of critical values of Hecke L-functions and p-adic interpolation*, 2019.

[LY20] L. Lai and P. Yu, *A note on the number of irrational odd zeta values*, Compositio Mathematica **156** (2020), no. 8, 1699–1717.

[Mas75] D. Masser, *Elliptic functions and transcendence*, Lecture Notes in Mathematics, Vol. 437, Springer-Verlag, Berlin-New York, 1975.

[Riv00] T. Rivoal, *La fonction zêta de Riemann prend une infinité de valeurs irrationnelles aux entiers impairs*, C. R. Acad. Sci. Paris Sér. I Math. **331** (2000), no. 4, 267–270.

[SC67] Atle Selberg and S. Chowla, *On Epstein's zeta-function*, J. Reine Angew. Math. **227** (1967), 86–110.

[Sch37] T. Schneider, *Arithmetische Untersuchungen elliptischer Integrale*, Math. Ann. **113** (1937), no. 1, 1–13.

[Sch88] N. Schappacher, *Periods of Hecke characters*, Lecture Notes in Mathematics, vol. 1301, Springer-Verlag, Berlin, 1988.

[Shi75] G. Shimura, *On some arithmetic properties of modular forms of one and several variables*, Ann. of Math. (2) **102** (1975), no. 3, 491–515.

[Sie37] C.-L. Siegel, *Über die analytische Theorie der quadratischen Formen. III*, Ann. of Math. (2) **38** (1937), no. 1, 212–291.

[Wüs87] G. Wüstholz, *Algebraic groups, Hodge theory, and transcendence*, Proceedings of the International Congress of Mathematicians, Vol. 1, 2 (Berkeley, Calif., 1986), Amer. Math. Soc., Providence, RI, 1987, pp. 476–483.

[Yos03] H. Yoshida, *Absolute CM-periods*, Mathematical Surveys and Monographs, vol. 106, American Mathematical Society, Providence, RI, 2003, notes by J. Milne.

[Zud01] W. Zudilin, *One of the numbers* $\zeta(5)$, $\zeta(7)$, $\zeta(9)$, $\zeta(11)$ *is irrational*, Uspekhi Mat. Nauk **56** (2001), no. 4(340), 149–150.